STRESS: NEUROENDOCRINOLOGY AND NEUROBIOLOGY

TOOLS FOR ALL YOUR TEACHING NEEDS
textbooks.elsevier.com

ACADEMIC
PRESS

STRESS: NEUROENDOCRINOLOGY AND NEUROBIOLOGY

Handbook of Stress, Volume 2

Edited by

GEORGE FINK

Florey Institute of Neuroscience and Mental Health,
University of Melbourne,
Parkville, Victoria, Australia

AMSTERDAM • BOSTON • HEIDELBERG • LONDON
NEW YORK • OXFORD • PARIS • SAN DIEGO
SAN FRANCISCO • SINGAPORE • SYDNEY • TOKYO
Academic Press is an imprint of Elsevier

Academic Press is an imprint of Elsevier
125 London Wall, London EC2Y 5AS, United Kingdom
525 B Street, Suite 1800, San Diego, CA 92101-4495, United States
50 Hampshire Street, 5th Floor, Cambridge, MA 02139, United States
The Boulevard, Langford Lane, Kidlington, Oxford OX5 1GB, United Kingdom

Library of Congress Cataloging-in-Publication Data
A catalog record for this book is available from the Library of Congress

British Library Cataloguing-in-Publication Data
A catalogue record for this book is available from the British Library

ISBN: 978-0-12-802175-0

For information on all Academic Press publications
visit our website at https://www.elsevier.com/

Working together
to grow libraries in
developing countries

www.elsevier.com • www.bookaid.org

Publisher: Mara Conner
Acquisition Editor: Natalie Farra
Editorial Project Manager: Kathy Padilla
Production Project Manager: Edward Taylor
Designer: Inês Maria Cruz

Typeset by TNQ Books and Journals

Contents

I

NEUROENDOCRINE CONTROL OF THE STRESS RESPONSE

1. Stress Neuroendocrinology: Highlights and Controversies

G. FINK

2. Limbic Forebrain Modulation of Neuroendocrine Responses to Emotional Stress

J.J. RADLEY, S.B. JOHNSON AND P.E. SAWCHENKO

3. Adrenergic Neurons in the CNS

C.P. SEVIGNY, C. MENUET, A.Y. FONG, J.K. BASSI, A.A. CONNELLY AND A.M. ALLEN

4. Noradrenergic Control of Arousal and Stress

C.W. BERRIDGE AND R.C. SPENCER

5. Evolution and Phylogeny of the Corticotropin-Releasing Factor Family of Peptides

D.A. LOVEJOY AND O.M. MICHALEC

6. Corticotropin-Releasing Factor and Urocortin Receptors

D.E. GRIGORIADIS

II
ENDOCRINE SYSTEMS AND MECHANISMS IN STRESS CONTROL

19. Stress, Glucocorticoids, and Brain Development in Rodent Models
C.M. MCCORMICK AND T.E. HODGES

III

DIURNAL, SEASONAL, AND ULTRADIAN SYSTEMS

List of Contributors

G. Aguilera National Institutes of Health, Bethesda, MD, United States

A.M. Allen The University of Melbourne, Parkville, VIC, Australia

C. Anacker Columbia University, New York, NY, United States

F.A. Antoni Egis Pharmaceuticals PLC, Budapest, Hungary

M. Bader Max-Delbruck-Center for Molecular Medicine, Berlin-Buch, Germany

O.C. Baltatu Anhembi Morumbi University – Laureate International Universities, Sao Paulo, Brazil

M.S. Bartlang University of Würzburg, Würzburg, Germany

J.K. Bassi The University of Melbourne, Parkville, VIC, Australia

C.M. Bauer Tufts University, Medford, MA, United States

K. Beck VA NJ Health Care System, East Orange, NJ, United States; Rutgers Medical School, Newark, NJ, United States

C.W. Berridge University of Wisconsin–Madison, Madison, WI, United States

B. Boari University of Ferrara, Ferrara, Italy

J.C. Borniger The Ohio State University – Wexner Medical Center, Columbus, OH, United States

M.E. Bowers Icahn School of Medicine at Mount Sinai, New York, NY, United States

R. Bowman Sacred Heart University, Fairfield, CT, United States

J.C. Buckingham Brunel University London, Uxbridge, Middlesex, United Kingdom

L.A. Campos Anhembi Morumbi University – Laureate International Universities, Sao Paulo, Brazil

L.A. Carvalho University College London, United Kingdom

A. Chen Max Planck Institute of Psychiatry, Munich, Germany; Weizmann Institute of Science, Rehovot, Israel

Y.M. Cisse The Ohio State University – Wexner Medical Center Columbus, OH, United States

A.A. Connelly The University of Melbourne, Parkville, VIC, Australia

R. de Bruijn Tufts University, Medford, MA, United States

F.H. de Jong Erasmus MC, Rotterdam, The Netherlands

E.R. de Kloet Leiden University Medical Center, Leiden, The Netherlands

F.S. den Boon University Medical Center Utrecht, Utrecht, The Netherlands

G. Fink University of Melbourne, Parkville, VIC, Australia

J.D. Flory Icahn School of Medicine at Mount Sinai, New York, NY, United States

R.J. Flower Queen Mary University of London, London, United Kingdom

A.Y. Fong The University of Melbourne, Parkville, VIC, Australia

J.W. Funder Hudson Institute of Medical Research (Formerly Prince Henry's Institute of Medical Research), Clayton, VIC, Australia; Monash University, Clayton, VIC, Australia

J. Gomez National Institute on Drug Abuse, Baltimore, MD, United States

H. Gong Second Military Medical University, Shanghai, P.R. China

K. Goonan Butler Hospital, Providence, RI, United States

D.E. Grigoriadis Neurocrine Biosciences Inc., San Diego, CA, United States

R.J. Handa Colorado State University, Fort Collins, CO, United States

J.E. Hassell Jr. University of Colorado Boulder, Boulder, CO, United States

T.E. Hodges Brock University, St. Catharines, ON, Canada

J. Hofland Erasmus MC, Rotterdam, The Netherlands

M.A. Holschbach Colorado State University, Fort Collins, CO, United States

O. Issler Weizmann Institute of Science, Rehovot, Israel; Max-Planck Institute of Psychiatry, Munich, Germany; Icahn School of Medicine at Mount Sinai, New York, NY, United States

C.-L. Jiang Second Military Medical University, Shanghai, P.R. China

M. Joëls University Medical Center Utrecht, Utrecht, The Netherlands

P.L. Johnson Indiana University School of Medicine, Indianapolis, IN, United States

S.B. Johnson University of Iowa, Iowa City, IA, United States

H. Karst University Medical Center Utrecht, Utrecht, The Netherlands

A.M. Khan UTEP Systems Neuroscience Laboratory, El Paso, TX, United States; University of Southern California, Los Angeles, CA, United States

A. Korosi SILS-CNS, University of Amsterdam, Amsterdam, The Netherlands

H.J. Krugers SILS-CNS, University of Amsterdam, Amsterdam, The Netherlands

I. Kyrou   Aston University, Birmingham, United Kingdom; University Hospital Coventry and Warwickshire NHS Trust WISDEM, Coventry, United Kingdom; University of Warwick, Coventry, United Kingdom; Harokopio University, Athens, Greece

C.R. Lattin   Tufts University, Medford, MA, United States

S.L. Lightman   University of Bristol, Bristol, United Kingdom

L. Liu   Second Military Medical University, Shanghai, P.R. China

D.A. Lovejoy   University of Toronto, Toronto, ON, Canada

C.A. Lowry   University of Colorado Boulder, Boulder, CO, United States

P.J. Lucassen   SILS-CNS, University of Amsterdam, Amsterdam, The Netherlands

V. Luine   Hunter College of CUNY, New York, NY, United States

G.B. Lundkvist   Max Planck Institute for Biology of Ageing, Cologne, Germany

F. Manfredini   University of Ferrara, Ferrara, Italy

R. Manfredini   University of Ferrara, Ferrara, Italy

L.B. Martin   University of South Florida, Tampa, FL, United States

C.M. McCormick   Brock University, St. Catharines, ON, Canada

O.C. Meijer   Leiden University Medical Center, Leiden, The Netherlands

C. Menuet   The University of Melbourne, Parkville, VIC, Australia

O.M. Michalec   University of Toronto, Toronto, ON, Canada

N. Mishra   The Hebrew University of Jerusalem, Jerusalem, Israel

R.J. Nelson   The Ohio State University – Wexner Medical Center, Columbus, OH, United States

N. Nikkheslat   King's College London, United Kingdom

C.A. Oomen   SILS-CNS, University of Amsterdam, Amsterdam, The Netherlands

N.V. Ortiz Zacarias   Leiden Academic Center for Drug Research, Leiden, The Netherlands

C.M. Pariante   King's College London, United Kingdom

E.D. Paul   Max Planck Institute of Psychiatry, Munich, Germany

J. Pooley   University of Bristol, Bristol, United Kingdom

L.H. Price   Butler Hospital, Providence, RI, United States; Alpert Medical School of Brown University, Providence, RI, United States

J.C. Pruessner   McGill University, Montreal, QC, Canada

J.J. Radley   University of Iowa, Iowa City, IA, United States

H.S. Randeva   Aston University, Birmingham, United Kingdom; University Hospital Coventry and Warwickshire NHS Trust WISDEM, Coventry, United Kingdom; University of Warwick, Coventry, United Kingdom

M.E. Rhodes   Saint Vincent College, Latrobe, PA, United States

K.K. Ridout   Butler Hospital, Providence, RI, United States; Alpert Medical School of Brown University, Providence, RI, United States

S.J. Ridout   Butler Hospital, Providence, RI, United States; Alpert Medical School of Brown University, Providence, RI, United States

L.M. Romero   Tufts University, Medford, MA, United States

A. Roy   University of Calgary, Calgary, AB, Canada

R.N. Roy   Saba University School of Medicine, Saba, Netherlands-Antilles

G. Russell   University of Bristol, Bristol, United Kingdom

R. Salmi   General Hospital of Ferrara, Ferrara, Italy

R.A. Sarabdjitsingh   University Medical Center Utrecht, Utrecht, The Netherlands

D.K. Sarkar   Rutgers, The State University of New Jersey, New Brunswick, NJ, United States

P.E. Sawchenko   The Salk Institute for Biological Studies, San Diego, CA, United States

M.J.M. Schaaf   Leiden University, Leiden, The Netherlands

J.R. Seckl   University of Edinburgh, Edinburgh, United Kingdom

C.P. Sevigny   The University of Melbourne, Parkville, VIC, Australia

A. Shekhar   Indiana University School of Medicine, Indianapolis, IN, United States

H. Soreq   The Hebrew University of Jerusalem, Jerusalem, Israel

R.C. Spencer   University of Wisconsin–Madison, Madison, WI, United States

F. Spiga   University of Bristol, Bristol, United Kingdom

C.M. Stoney   National Institutes of Health, Bethesda, MD, United States

R. Tiseo   University of Ferrara, Ferrara, Italy

C. Tsigos   Harokopio University, Athens, Greece

A.R. Tyrka   Butler Hospital, Providence, RI, United States; Alpert Medical School of Brown University, Providence, RI, United States

E.M. Walker   UTEP Systems Neuroscience Laboratory, El Paso, TX, United States

A.G. Watts   University of Southern California, Los Angeles, CA, United States

O.T. Wolf   Ruhr University Bochum, Bochum, Germany

P.S.M. Yamashita   University of Colorado Boulder, Boulder, CO, United States

R. Yehuda   Icahn School of Medicine at Mount Sinai, New York, NY, United States; James J. Peters Veterans Affairs Medical Center, Bronx, NY, United States

H. Zangrossi Jr.   University of São Paulo, Ribeirão Preto, Brazil

E.P. Zorrilla   The Scripps Research Institute, La Jolla, CA, United States

P.A. Zunszain   King's College London, United Kingdom

Preface

Stress is not something to be avoided. Indeed, it cannot be avoided, since just staying alive creates some demand for life-maintaining energy. Complete freedom from stress can be expected only after death. *Hans Selye*

Complementary to Volume One of the Handbook of Stress Series this second volume is focused on the neuroendocrinology and neurobiology of stress. The stress response comprises a set of integrated cascades in the nervous, endocrine, and immune defense systems. The neuroendocrine stress response system in vertebrates is comprised of two arms: the hypothalamic-pituitary-adrenal axis (HPA) and the autonomic, predominantly sympathetic, nervous system. Both arms transmit the output of central nervous activity, starting with arousal, alarm, or anxiety, and resulting in the release into the systemic circulation of the monoamines, adrenaline and noradrenaline, and the adrenocortical glucocorticoids, cortisol or corticosterone. The glucocorticoids and monoamines act synergistically to increase the concentrations of blood glucose, and adrenaline and noradrenaline increase cardiac output and shunt blood toward skeletal muscles thus facilitating the "fight or flight" response to a perceived or real threat. Adrenal glucocorticoid release is stimulated by pituitary adrenocorticotropin (ACTH) secreted in response to synergistic stimulation of corticotropes by corticotrophin-releasing factor-41 (CRF) and arginine vasopressin (AVP). The two neuropeptides ("neurohormones") are synthesized in the hypothalamus and transported to the anterior pituitary gland by the hypophyseal portal vessels. ACTH release is moderated or restored to normal basal values after termination of a stressful stimulus by the negative feedback action of glucocorticoids on the hypothalamus and pituitary corticotropes. CRF through its CRF-RI receptor also affects the activity of central, especially autonomic (monoaminergic), neurons. Perhaps relevant to memory and possibly cell damage, the concentration of glucocorticoids to which individual brain cells are exposed is "micromanaged" by the 11beta-hydroxysteroid dehydrogenase enzymes that convert inert to active glucocorticoids and vice versa. In addition to glucocorticoid negative feedback, the circadian rhythm of ACTH and glucocorticoid secretion is also a canonical function of the HPA system.

The central stress command and control systems for the HPA and autonomic nervous system receive input from the sensory systems and comprise the primary components of the limbic system (hypothalamus, amygdala, bed nucleus of the stria terminalis, hippocampus, basal ganglia, thalamus, and cingulate gyrus), the prefrontal cortex, locus coeruleus, and other regions of the forebrain and hindbrain. The neuroanatomical and functional interactions between the HPA and the autonomic nervous control systems are covered in detail in this volume as are the ways in which stress neuropeptides, CRF, the urocortins, and AVP, and central monoamines are involved in mounting stress and related behavioral responses.

We were fortunate to be able to enlist as authors a galaxy of scientific stars to provide cutting edge information on the above as well as on themes such as microRNAs and other epigenetic mechanisms; the parasympathetic system; age–stress interactions; telomeres; aldosterone and mineralocorticoid receptors; androgens and estrogen; angiotensin; annexin; stress and alcohol; anxiety; depression; panic; learning; and memory. Also reviewed are posttraumatic stress disorder; synthetic glucocorticoids; corticosteroid receptors including nongenomic receptors; CRF antagonists; seasonal and ultradian rhythms, and the effect of circadian rhythms on cardiovascular and other stress-related events.

Molecular genetics (optogenetics in particular) together with the use of sophisticated behavioral phenotypes has made a significant impact on the precision with which we now understand the physiological and pathological processes of the stress response including fear, fight and flight, and freezing. Step advances have been made in our understanding, for example, of the central and peripheral actions of CRF and urocortin and their receptors; the role of central CRF in the overall control of the mental, behavioral, as well as neuroendocrine responses to stress; the cardiac peptides; the cytokines and other inflammatory and immune signaling molecules; and the complex neuropeptide and monoamine regulation of hunger and satiety. Modern brain imaging has enabled us to begin to understand the neural basis of stress and its consequences in the human. The genomics of stress vulnerability and its clinically adverse effects such as anxiety and major depression is uncertain and therefore remains an irresistible and necessary challenge, as does the precise role of epigenetics.

Notwithstanding the large gaps in our knowledge that still remain, the adage formerly applied to diabetes mellitus might now apply to stress: *"Understand stress and you will understand medicine."*

I am pleased to take this opportunity to acknowledge with gratitude the enthusiasm, patience, and invaluable support of Elsevier staff, Natalie Farra, Kathy Padilla, and Kristi Anderson in the preparation of this as well as the first volume of the *Handbook of Stress*. My best thanks go to all the authors who in times of stringent research budgets have generously given of their time to produce excellent chapters without which this work could not have been published. Finally, I thank Ann Elizabeth, my wife, for seemingly eternal patience, support, and forbearance.

George Fink
Florey Institute of Neuroscience and Mental Health,
University of Melbourne,
Parkville, Victoria, Australia
June 2016.

NEUROENDOCRINE CONTROL OF THE STRESS RESPONSE

CHAPTER

1

Stress Neuroendocrinology: Highlights and Controversies

G. Fink

University of Melbourne, Parkville, VIC, Australia

OUTLINE

Abstract

Stress in mammals triggers a neuroendocrine response mediated by the hypothalamic–pituitary–adrenal (HPA) axis and the autonomic nervous system (ANS). Increased activity of these two systems induces behavioral, cardiovascular, endocrine, and metabolic cascades that enable the individual to fight or flee and cope with stress. Our understanding of stress and stress response mechanisms is generally robust. However, several themes remain uncertain/controversial and perhaps deserve further scrutiny before they achieve canonical status. These themes include perinatal exposure effects on the health of the adult, genetic susceptibility to stress, the neurochemistry of posttraumatic stress disorder (ANS versus the HPA), significance of hippocampal volume for major depressive disorder and other mental disorders, and the role of stress in the etiology of gastroduodenal ulcers. All five themes are of clinical and therapeutic significance, pose fundamental questions about stress mechanisms and offer important areas for future research.

- Stress in mammals triggers a neuroendocrine response mediated by the hypothalamic–pituitary–adrenal (HPA) axis and the autonomic nervous system (ANS).

- Increased activity of these two systems induces behavioral, cardiovascular, endocrine, and metabolic cascades that enable the individual to fight or flee and cope with stress.

- Our understanding of stress and stress response mechanisms is generally robust. However, several themes remain uncertain and/or controversial and deserve further scrutiny before they achieve canonical status.

- These themes include:
  - The "thrifty phenotype hypothesis" and other perinatal factors such as maternal obesity and diabetes that may adversely affect health in adult life;
  - Genetic susceptibility (gene × environment interactions) to stress, which is at an early stage of development;
  - The neurochemistry of posttraumatic stress disorder (PTSD)…nearly three decades of research suggest that the ANS plays a dominant and more consistent role in PTSD than the HPA axis;
  - Significance of hippocampal volume for major depressive and other mental disorders…the weight of evidence suggests that changes in hippocampal volume are reversible and a function of state rather than trait;
  - Notwithstanding the importance of *Helicobacter pylori*, stress (especially trauma and burns) plays an important role in the etiology of gastroduodenal ulcers.

- All five themes are of clinical and therapeutic significance, pose fundamental questions about stress mechanisms, and offer important areas for future research.

INTRODUCTION

Stress concepts, definitions, and history have been covered in Volume 1 of this Handbook Series and in numerous reviews.[1–3] Briefly, the two founders of stress concepts, mechanisms, and impact were Walter Bradford Cannon at Harvard and Hans Selye at McGill. Cannon[4] introduced the phrase "fight or flight" and showed how these responses to a stressful challenge were dependent on activation of the sympatho-neuronal and sympatho-adrenomedullary systems (SAS). Cannon also introduced the concept of homeostasis. Based on observations in humans and experimental animals, Hans Selye advanced the first, generic definition of stress: "Stress is the non-specific response of the body to any demand."[5,6] Although criticized by some because of the claim of non-specificity, Selye's definition applies to all known stress responses in the three major phylogenetic domains.[1]

Selye was also the first to recognize that homeostasis, "stability through constancy," could not by itself ensure stability of body systems under stress. He coined the term heterostasis (from the Greek heteros or other) as the process by which a new steady state was achieved by adaptive mechanisms. Heterostasis, could be regarded as the precursor for the concept of allostasis, "stability through change" brought about by central neural regulation of the set points that adjust physiological parameters to meet the stressful challenge. Allostasis[7–9] has partly overtaken homeostasis, "stability through constancy" that dominated physiological and medical thinking since the 19th century influenced, for example, by Claude Bernard's precept of the "fixity of the milieu interieur."[10]

This second volume of the Handbook of Stress Series is focused on the neuroendocrine, autonomic nervous, and other neurobiological systems/mechanisms involved in the stress response. Because the details are covered in numerous previous papers and reviews[2,3,11–14] and in several chapters in this volume, the present chapter is confined to a brief outline of the neural mechanisms involved in the stress response followed by a review of some highlights and controversies.

NEUROENDOCRINE AND AUTONOMIC NERVOUS CONTROL OF STRESS RESPONSE

The ANS and the HPA system subserve the afferent and efferent limbs of the stress response in vertebrates and are also central to maintaining homeostasis and effecting allostasis. Controlled by the brain, and utilizing as neurotransmitters epinephrine and norepinephrine (sympathetic nervous component) or acetylcholine (parasympathetic component), the role of the ANS in fight-or-flight and homeostasis (especially cardiovascular) was clearly explained by Walter Cannon. Recent advances and especially the role of the oft-neglected parasympathetic system, especially in brain–heart interactions, are reviewed by McEwen[3] and Silvani et al.[15]

The CNS control of the anterior pituitary gland is mediated by neurohormones synthesized and released from hypothalamic neurons and transported to the anterior pituitary gland by the hypophysial portal vessels. Proof of the neurohumoral hypothesis of anterior pituitary control came, first, from the elegant pituitary stalk section and pituitary grafting experiments of Geoffrey Harris and Dora Jacobsohn, secondly, the characterization of some of the neurohormones by Andrew Schally and Roger Guillemin, for which they were awarded the 1977 Nobel Prize for Physiology and Medicine, and, thirdly, the demonstration, first by my laboratory, that these neurohormones were indeed released into hypophysial portal blood.[12,16–20] Corticotropin releasing factor (CRF), a 41 amino acid peptide that mediates neural control of adrenocorticotropic hormone (ACTH) release from pituitary corticotropes, was isolated and sequenced by Wylie Vale and associates in 1981.[21] The brain→hypothalamus→anterior pituitary ACTH release system is under the negative feedback control by adrenal glucocorticoids, cortisol in the human, corticosterone in rodents, released from the adrenal cortex in response to ACTH.[11,12,16,22]

A series of physiological studies, including measurements of neurohormone release into hypophysial portal blood in vivo, confirmed earlier views that arginine vasopressin (AVP) acts synergistically with CRF-41 to control ACTH release.[12,23–26] In fact acute (within 2.5 h) glucocorticoid feedback inhibition of ACTH release is mediated by inhibition of AVP, not CRF-41, release into hypophysial portal blood together with the blockade of pituitary responsiveness to CRF-41.[25] The amount of glucocorticoid available for cells is "micromanaged" by 11-beta-hydroxysteroid dehydrogenase (HSD-11β) enzymes of which there are two isoforms. First, HSD11B1 which reduces cortisone to the active hormone cortisol that activates glucocorticoid receptors. Second, HSD11B2 which oxidizes cortisol to cortisone and thereby prevents illicit activation of mineralocorticoid receptors (for details see Chapter 34 by Jonathan Seckl in this volume).

THRIFTY PHENOTYPE HYPOTHESIS: VARIANCE BETWEEN STUDIES AND BROADENED TO INCLUDE MATERNAL OBESITY AND DIABETES

Of the epidemiological studies that have generated new stress concepts, the most important perhaps remains that of David Barker,[27] which led to the hypothesis that fetal undernutrition in middle to late gestation programs later coronary heart disease. This concept was soon extended by Hales and Barker[28] in their "thrifty phenotype hypothesis." The latter proposes an association between poor fetal and infant growth and the subsequent development of type 2 diabetes and the metabolic syndrome, which afflict communities in epidemic proportions. Poor nutrition in early life, it is postulated, produces permanent changes in glucose–insulin metabolism. These changes include

insulin resistance (and possibly defective insulin secretion) which, combined with effects of obesity, ageing, and physical inactivity, are the most important factors in generating type 2 diabetes. Many studies have confirmed Barker's initial epidemiological evidence, although the strength of the relationship has varied between studies. Furthermore, as well as intrauterine growth restriction, maternal obesity, and diabetes can also be associated with disease of the individual later in life.[29,30] Paternal exposures too can result in the later development of metabolic disorders in the offspring. Human and animal studies suggest intergenerational transmission of the maternal or paternal phenotype.[30] Several changes in epigenetic marks identified in offspring provide a likely mechanism by which early molecular changes result in persistent phenotypic changes,[31] although additional mechanisms yet to be uncovered cannot be excluded.

Early life exposures, fetal and perinatal, could result in biological embedding by way of at least three mechanisms: brain plasticity, epigenetics (especially DNA methylation), and hormonal action. There are numerous examples of brain plasticity, perhaps one of the most dramatic being the recruitment of new neurons and the increase in hippocampal grey matter in scatter-hoarding animals which need to remember the location of food caches[32,33] and London cab drivers[34,35] for whom spatial memory is of paramount importance. Hormonal effects are exemplified by the possible effect of excess androgen in causing the polycystic ovary syndrome phenotype.[36]

GENETIC SUSCEPTIBILITY TO STRESS: GENE×ENVIRONMENT INTERACTION

With recent advances in genomics, the concept of susceptibility genes that increase the vulnerability of individuals to stressful life events has attracted considerable research interest. Thus, for example, the work of Avshalom Caspi and associates[37] suggested that a polymorphism in the monoamine oxidase A (MAOA) gene promoter, which reduces MAOA expression, influences vulnerability to environmental stress and that this can be initiated by childhood abuse. Furthermore, in what seemed to be a landmark paper, Caspi et al.[38] reported that a polymorphism in the promoter of the serotonin transporter (SERT or 5-HTT) gene can render individuals more susceptible to stressful life events. Because SERT is also the therapeutic target for antidepressant drugs, the selective serotonin reuptake inhibitors, the publication by Caspi et al.[38] led to a flurry of research publications and subsequent meta-analyses to assess the replicability of the findings. However, the overall conclusions from primary and meta-analyses on genetic × environmental interactions suggest that it is unlikely that there is a stable gene × environment interaction involving the SERT, life stress, and mental disorders.[39] Whether the uncertainty regarding the suggested link between the SERT gene promoter polymorphism and susceptibility to stress reflects an artifact or problems in post hoc analysis is discussed by Rutter et al.[40] who conclude that "the totality of the evidence on GXE is supportive of its reality but more work is needed to understand properly how 5HTT allelic variations affect response to stressors and to maltreatment." In fact, recent data suggest that there is a stable gene × environment interaction involving MAOA, abuse exposure, and antisocial behavior across the life course[41] as proposed in the earlier paper by Caspi et al.[37]

In contrast to the uncertainty of SERT gene promoter polymorphisms and susceptibility to stress (above), seemingly robust genomic links have been reported between enhanced activity in human SERT gene variants and autism.[42] Indeed, strikingly high-expressing SERT genotypes were associated with increased tactile hyperresponsiveness in autistic spectrum disorder (ASD) suggesting that genetic variation that increases SERT function may specifically impact somatosensory processing in ASD.[43]

Single nucleotide polymorphisms in the gene for the CRF receptor-1 have been tentatively linked to suicidal thoughts and behavior, depression, panic disorder, and fear.[44–49] Recent genetic and functional magnetic resonance imaging (fMRI) studies have shown that a functional deletion variant of ADRA2B, the gene encoding the alpha-2B adrenergic receptor, is related to enhanced emotional memory in healthy humans and enhanced traumatic memory in war victims.[50] The ADRA2B deletion variant, which acts primarily as a loss-of-function polymorphism, is related to increased responsiveness and connectivity of brain regions implicated in emotional memory.[50,51] In normal subjects, only carriers of ADRA2B deletion variant showed increased phasic amygdala responses to acute psychological stress, illustrating that genetic effects on brain function can be context (state) dependent.[52] In twin studies, Tambs and associates[53] have investigated the genetic interactions and the five types of anxiety disorder. A recent report by Zannas et al.[54] suggests that cumulative lifetime stress may accelerate epigenetic aging, possibly driven by glucocorticoid-induced DNA methylation.

Taken together, the above examples illustrate that our understanding of gene × environment interactions in stress and stress-related disorders is still at an early phase. Nonetheless, these findings, and especially the apparent positive linkage between human SERT gene variants and ASD, have heuristic importance for further research on genetic–environmental interactions that determine the response to stress and the etiology of mental disorders.

POSTTRAUMATIC STRESS DISORDER: AUTONOMIC NERVOUS SYSTEM APPARENTLY DOMINANT

PTSD is a condition in which a traumatic event is persistently re-experienced in the form of intrusive recollections, dreams, or dissociative flashback episodes. Cues to the event lead to distress and are avoided, and there are symptoms of increased arousal. To meet the diagnostic criteria of the DSM-5, the full symptom picture must be present for more than one month, and the disturbance must cause clinically significant distress or impairment in social, occupational, or other areas of functioning.

PTSD, although recorded since the 3rd century BCE, was only accepted officially as a mental disorder in 1980, consequent on the psychological trauma experienced by US Vietnam veterans and their demand for compensation. PTSD is a common condition that occurs frequently in civilian (e.g., rape trauma syndrome, battered woman syndrome, and abused child syndrome) as well as in military trauma response syndromes. The importance and prevalence of PTSD are such that the disorder is covered in several chapters in Volume 1 and Volume 2 of this Handbook Series. Here I review briefly the evidence that the central ANS rather than the HPA plays a dominant role in PTSD. Neurobiologically, this distinguishes PTSD from major depressive disorder (MDD).

Neuroendocrine Correlates: No Consistent Evidence for Hypocortisolemia in Posttraumatic Stress Disorder

Mason et al.[55] first reported that mean urinary free cortisol concentrations during hospitalization of Vietnam combat veterans were significantly lower in PTSD than in MDD, bipolar I manic disorder and undifferentiated schizophrenia. The low, stable cortisol levels in these PTSD patients were remarkable because the overt signs in PTSD of anxiety and depression would on first principles be expected to be associated with cortisol elevations. Mason et al.[55] concluded that "the findings suggest a possible role of defensive organization as a basis for the low, constricted cortisol levels in PTSD and paranoid schizophrenia patients." The findings of Mason et al.[55] were broadly confirmed by subsequent independent studies. Low dose dexamethasone suppression test in some studies suggested that hypocortisolemia associated with PTSD reflected increased central sensitivity to glucocorticoids with consequent greater negative feedback inhibition.[56–58] By 2002, the concept that PTSD was associated with hypocortisolemia had almost achieved canonical status.[57,59]

The observation of low cortisol in PTSD, precipitated by extreme stress, appeared to contradict the popular "glucocorticoid cascade hypothesis" which posits that stress-induced increased plasma cortisol concentrations (hypercortisolemia) result in damage to brain regions such as the hippocampus that are involved in memory and cognition.[3,59] However, the early findings of Mason et al.[55] have not been universally replicated. The details of meta-analyses of cortisol levels in the various PTSD groups have been reviewed previously.[60] In summary, the data show that when compared with levels in normal controls, ambient cortisol levels in PTSD over a 24-h period have been reported as significantly lower, significantly higher or not significantly different.[56,61–64]

In reflecting on the between-study variance in cortisol levels in PTSD, Yehuda[56] pointed out that most cortisol levels, high or low, in individuals with PTSD are within the normal endocrine range, and not suggestive of endocrine pathology. That is, in endocrine disorders, Yehuda[56] cogently argues pathological hormonal changes are generally strikingly obvious and robust, whereas in psychiatric disorders, neuroendocrine alterations may be subtle, and therefore difficult to determine with any degree of certainty or reproducibility.

Autonomic Nervous System Correlates of PTSD: Robust Evidence for Increased Sympathetic Activity and Arousal in PTSD

In contrast to the significant between-study differences in HPA function tests in patients with PTSD, hyperarousal, and increased SAS activity (e.g., increased skin conductance, heart rate, blood pressure, and catecholamine secretion) appear to be robust and consistent features of PTSD.[60] Hyperactivity of the peripheral sympathetic and central adrenergic system in PTSD is supported by the clinical improvement that occurs in patients with PTSD in response to agents that reduce the centrally hyperactive noradrenergic state. This is exemplified by the beneficial effects of clonidine, the alpha 2 adrenergic receptor agonist that reduces noradrenaline release, and prazosin, the alpha 1 postsynaptic adrenergic receptor blocker.[65–68] Furthermore, the noradrenergic alpha 2-antagonist, yohimbine, which increases peripheral and central noradrenergic activity in man, increases flashbacks, intrusive thoughts, emotional numbness, difficulty in concentrating, and panic symptoms in combat

veterans with PTSD.[65,66,69] In a recent review, Arnsten et al.[70] advanced a model which might explain why blocking alpha 1 receptors with prazosin, or stimulating alpha 2A receptors with guanfacine or clonidine can be useful in reducing the symptoms of PTSD. Placebo-controlled trials have shown that prazosin is helpful in veterans, active duty soldiers, and civilians with PTSD, including improvement of prefrontal cortical symptoms such as impaired concentration and impulse control. Recent genetic and fMRI studies have shown that a functional deletion variant of ADRA2B, the gene encoding the alpha-2B adrenergic receptor, is related to enhanced emotional memory in healthy humans and enhanced traumatic memory in war victims (see Genetic Susceptibility to Stress: Gene × Environment Interaction section).

Noradrenergic neurons interact with several neuronal types, including GABAergic, dopaminergic, and serotonergic, and serotonin reuptake inhibitors are sometimes helpful in moderating PTSD. Indeed, notwithstanding the importance of noradrenergic mechanisms, it is important to emphasize that the neurobiology of PTSD is complex and involves several central neurotransmitter systems.[69] The mechanism of noradrenergic action remains uncertain, as does the precise neural circuitry involved in PTSD. Adrenergic receptors in the hypothalamus, glucocorticoid, and CRF receptors in the locus ceruleus and adrenergic innervation of the adrenal gland suggest that both central and peripheral catecholamine systems and the HPA axis may be linked in a way that could subserve PTSD-related changes in the two arms of the stress response system.[71,72] Liberzon et al.[71] posit that PTSD might represent a syndrome, or "final common pathway," which reflects abnormalities in a number of different neurobiological modalities. However, in their study of Vietnam veterans with and without PTSD, Liberzon et al.[71] found that although challenges activated multiple stress response systems in the PTSD patients, the systems were not activated in an integrated fashion. This may be consonant with the pharmacological and cognitive findings which suggest that under normal circumstances noradrenergic systems can influence the magnitude of the HPA axis response to stress but that in major depression HPA axis activation appears to be autonomous of noradrenergic influence.[72]

Experimental studies in animals and fMRI in man are making some headway in elucidating the central neural circuits involved in arousal and PTSD. Thus, for example, neuroimaging studies in PTSD have led to a neurobiological model in which reductions in the inhibitory activity of the ventromedial prefrontal cortex (PFC) are considered to lead to heightened amygdala activity in response to threat.[71] Recent fMRI studies suggest that PTSD is associated with a reduction in activity of the ventral anterior cingulate gyrus that is specifically linked to engagement of arousal networks as assessed by increases in skin conduction in response to neuropsychological challenge.[73]

Conclusions

In summary, although often associated with depression, PTSD is characterized by intense arousal with concomitant activation of the ANS, especially the SAS. In contrast, the role of the HPA and cortisol in PTSD remains uncertain as a result of major differences between studies and patient groups. Given this fact, it is perhaps surprising that clinicians and laboratory scientists have, in their research on PTSD, apparently focused on the HPA at the expense of the ANS and especially central noradrenergic and serotonergic mechanisms that are involved in PTSD, arousal, and anxiety.[74] Indicators of altered ANS activity (e.g., electrodermal, cardiac, and blood gas measures) may prove to be useful supplementary diagnostic markers and possible endophenotypes in genetically focused studies of PTSD.[75]

HUMAN HIPPOCAMPUS AND MAJOR DEPRESSIVE DISORDER: DOES HIPPOCAMPAL SIZE MATTER?

The short answer to this question is "possibly, although the present data are inconsistent and the mechanism is uncertain." Much is made of hippocampal volume and MDD. Indeed a current Medline search draws down nearly 300 papers on the subject. Here I review what appears to be a rather confused status of whether hippocampal volume is a consistent and functionally significant feature of MDD.

Background to Hippocampal Volume Reduction in Major Depressive Disorder

Three sets of interrelated findings led to the proposition that MDD is associated with reduced hippocampal volume. First, hypercortisolemia and resistance to glucocorticoid suppression of the HPA (i.e., disruption of normal glucocorticoid negative feedback) was considered a key biological correlate of MDD.[76–79] Second, work by Sapolsky, McEwen, and associates confirmed earlier findings that high doses of glucocorticoids appeared to damage or destroy hippocampal neurons and/or render the neurons vulnerable to metabolic insult.[80–83] Third, Sheline et al.[78] using MRI showed that hippocampal volumes in subjects with a history of MDD were significantly reduced bilaterally compared to those in nondepressed control subjects. The degree of hippocampal volume reduction correlated with total duration of major depression. Sheline et al.[78] concluded "These results suggest that depression is associated with hippocampal atrophy, perhaps due to a progressive process mediated by glucocorticoid neurotoxicity."

The glucocorticoid neurotoxicity hypothesis would, in principle, offer a coherent biological basis for our understanding of MDD and its relationship to stress. Indeed, structural/neuroplastic changes in key components of the limbic system, such as the PFC, amygdala, and hippocampus, offer an alternative or additional mechanism to the chemical (i.e., monoamine) hypothesis of MDD.[84,85] However, although several volumetric MRI studies[86–89] have confirmed the initial findings of Sheline et al.,[78] a substantial number have failed to show significant hippocampal atrophy in subjects with major depression compared to healthy controls.[87,90–93]

Magnetic Resonance Imaging Findings in Conflict

Vythilingam et al.,[87] for example, report that of 14 studies of hippocampal volume in MDD, 8 were unable to find significant between-group differences, whereas 6 found significantly smaller left or left and right hippocampus. Reduced hippocampal volume was correlated with total duration of depression in some but not all studies. Differences in hippocampal volume in bipolar disorder (BPD) have also been mixed, with reports of increased, decreased, and unchanged hippocampal volumes.[87] Indeed, Ladouceur et al.[94] reported increased hippocampal and parahippocampal volume in individuals with high risk of BPD. Sheline,[95] on the basis of a literature review, concluded that "volumetric brain studies exhibit inconsistency in measurements from study to study." These inconsistencies were attributed to clinical and methodological sources of variability.[60]

Recent reviews suggest that reduction in hippocampal volume, considered by some to reflect the neurotoxic effect of prolonged depression, PTSD, or chronic stress,[59] is not a consistent finding in major depression.[79,96] In a meta-analysis of 2418 patients with MDD and 1974 healthy individuals, Koolschijn et al.[97] reported that MDD patients showed large volume reductions in frontal regions, especially in the anterior cingulate and orbitofrontal cortex with smaller reductions in PFC. Similarly, based on their monumental review of the literature, Savitz and Drevets[91] conclude that although hippocampal volume reduction in MDD has been widely reported, a significant number of studies have failed to find evidence of hippocampal atrophy in depressed patients and that "the majority of studies reporting evidence of hippocampal atrophy have made use of elderly, middle-aged or chronically ill populations." In patients with BPD, most studies found no reduction of hippocampal volume.[91] Hypometabolism of the dorsal PFC is one of the most robust findings in both MDD and BPD.[91] A reduction in hippocampal volume cannot automatically be attributed to high levels of cortisol: indeed, reduced hippocampal volume has also been reported in PTSD patients[98] in which cortisol levels are assumed to be normal or lower than normal (see Posttraumatic Stress Disorder: Autonomic Nervous System Apparently Dominant section), and in patients with cardiovascular disease.[79] Twin studies of male Vietnam veterans with and without PTSD and their combat-unexposed identical (monozygotic) co-twins showed that subtle neurologic dysfunction in PTSD does not reflect brain damage acquired along with the PTSD but, instead, represents a familial vulnerability factor, which likely antedates the traumatic exposure.[99]

To date, studies examining the hippocampus in MDD have been mostly cross-sectional.[100] Longitudinal prospective studies that track patients over disease onset are required. In this respect, Chen et al.[101] appear to be the first to report smaller hippocampal volume in healthy girls at high familial risk of depression. A recent longitudinal study of adolescents found that smaller hippocampal volumes preceded clinical symptoms of depression in at-risk adolescents, particularly in those who experienced high levels of adversity during childhood.[98]

On the basis of their human postmortem brain studies, Swaab et al.[92] conclude "In depressed subjects or in patients treated with synthetic corticosteroids the hippocampus is intact and does not show any indication of neuropathological alterations or major structural damage. We conclude that in the human brain there is no conclusive evidence that corticosteroids would be neurotoxic for the Hippocampus." Other studies have confirmed the absence of hippocampal cell loss in MDD or steroid treated subjects.[102]

Mechanisms of Hippocampal Volume Changes and Reversibility

The mechanisms responsible for hippocampal volume loss in some subjects with MDD remain unclear.[84] Genetic factors have been associated with reduced hippocampal volume.[101] Stockmeier et al.[102] found that MDD is associated with a significant increase in the packing density of glia, pyramidal neurons, and granule cell neurons in all hippocampal subfields and the dentate gyrus. Pyramidal neuron soma size was also significantly decreased. Stockmeier et al.[102] suggest that a significant reduction in neurophil as well as hippocampal water loss (i.e., shift in fluid balance between the ventricles and brain tissue) may account

for decreased hippocampal volume detected in some MRI studies in MDD.

Adult neurogenesis has been regarded as another mechanism that could conceivably affect hippocampal volume. However, neurogenesis adds relatively few neurons per day, and its suppression if such inhibition exists in depressed patients is likely to be too modest to provide a significant contribution to the hippocampal volume reduction reported in some depressed patients. Recent findings from behavioral studies in rodents argue against a role for neurogenesis in MDD. That is, inhibition of cell proliferation in the dentate gyrus is not associated with development of anhedonic-like symptoms in rats or sensitivity to unpredictable chronic mild stress in mice.[103,104]

Neurotropic and other growth factors have been implicated in hippocampal neuroplasticity and in the actions of antidepressants.[105] Changes in hippocampal structure when present in MDD would appear to be state-dependent rather than trait-dependent. Thus, Zhao et al.,[106] using MRI shape analysis, reported significant differences in the mid-body of the left hippocampus between depressed and control subjects, although this difference was not apparent in subjects who had recovered from depression. The reversibility (and therefore state-dependence) of hippocampal volume reduction receives support from findings in patients with Cushing's disease in which effective treatment resulting in reduction of plasma cortisol concentrations was associated with an increase of up to 10% in hippocampal volume.[92,107]

Impact of Normal Variability of Hippocampal Volume

Assessment of the possible relationships between stress, hippocampal volume, and depression is confounded by the normal variability in hippocampal volume. Thus, based on their seminal MRI study, Lupien et al.[98] reported that hippocampal volume is as variable in young as in older adults, suggesting that smaller hippocampal volume attributed to the ageing process could in fact be determined early in life. Furthermore, within similar age groups, the percentage difference in hippocampal volume between the individuals with the smallest hippocampal volume and the group average was greater than the percentage difference reported between psychiatric populations and normal controls.[98] These observations led Lupien et al.[98] to conclude, "Taken together, these results confront the notion of hippocampal atrophy in humans and raise the possibility that predetermined interindividual differences in hippocampal volume in humans may determine the vulnerability for age-related cognitive impairments or psychopathology throughout the lifetime."

Functional Changes and Relevance for Cognition and Dementia

Major depression is frequently associated with cognitive and especially memory deficits that are attributable to hippocampal dysfunction. Thus, for example, in their study of 38 medication-free nonelderly depressed outpatients without alcohol dependence or adverse experiences in childhood, Vythilingam et al.[87] found that MDD patients had normal hippocampal volumes but also had focal declarative memory deficits that reflect hippocampal dysfunction. Treatment with antidepressants significantly improved memory and depression but did not alter hippocampal volume.[87] Similarly, hippocampal and parahippocampal gyrus volumes do not differentiate between elderly patients with mild to moderate depressive disorders and those without depression.[108] This suggests that mechanisms other than hippocampal volume reductions are associated with cognitive and especially memory deficits in elderly depressed patients. Avila et al.[108] speculate "Future MRI studies involving larger samples, and taking into consideration genetic markers, clinical comorbidities and the age of the depressive symptoms onset, will likely shed further light on the pathophysiology of depression and its associated cognitive deficits in the elderly, as well as on the relationship between this clinical condition and Alzheimer's Disease."

Lack of Illness Specificity

Structural hippocampal changes are not specific for MDD. Several groups have reported smaller hippocampal volumes in PTSD as well as in adults with early childhood trauma, depressed women with childhood sexual or physical abuse and in subjects with alcohol dependence.[87] Structural brain changes in the hippocampus have also been reported in patients with schizophrenia. Meisenzahl et al.[109] carried out a head-to-head comparison of hippocampal volume in patients with schizophrenia and those with MDD, with the same duration of illness. Bilateral hippocampal volume reductions were detected in both schizophrenic and depressed patients compared with healthy control subjects. Hippocampal volumes were significantly lower in schizophrenic patients compared to MDD patients. As in MDD, hippocampal volume changes in schizophrenia are reversible on illness remission,[110] emphasizing the fact that changes in hippocampal volume when present are state not trait-dependent.

Conclusions

The early seminal studies by Bruce McEwen and Robert Sapolsky, which suggested a biological substrate for the effects of glucocorticoids on mood and cognition,[3,82,83,111] attained clinical relevance when Sheline et al.[78] and other groups provided MRI evidence for volume reduction in the hippocampus of patients with MDD. However, the reduction in hippocampal volume is not a consistent finding: a substantial number of studies of MDD and BPD report no significant reduction. Several meta-analyses point to heterogeneity of subjects as a possible cause for these inconsistencies. The duration of and the number of depressive illness episodes are also significant factors in determining hippocampal volume reduction.[100,112]

The crucial importance of patient homogeneity is illustrated by Sheline et al.[78] who confined their study to right-handed females with a mean and median age of 68 years recruited from the Memory and Aging Project of the Alzheimer's Disease Research Center and the outpatient psychiatry service at the Washington University School of Medicine. Each depressed subject was matched using a case–control design for age and educational level, and the groups were matched overall for height (a predictor of overall brain size). "The choice was made to select all women because it eliminated brain differences due to gender, decreased the possibility of hypertension and occult cardiovascular disease, and increased the ability to obtain subjects, although at the cost of generalizability."[78] The study by Sheline et al.[78] has important heuristic value for future clinical studies on the possible relationship between hippocampal volume, mood disorders and cognition.

Additional concerns regarding the possible causal association between MDD, reduced hippocampal volume, and hypercortisolemia are, first, the reversibility of reduced hippocampal volume suggesting that there is no permanent structural change and that any change in hippocampal volume is state-dependent rather than trait-dependent. Second, hippocampal volume reduction lacks illness specificity. Further incisive research appears to be required to establish the clinical fidelity and significance of reduced hippocampal volume in mental illness, and, indeed, whether hippocampal atrophy may be a risk factor for or a consequence of depressive disorders.[89]

GASTRODUODENAL (PEPTIC) ULCERS: STRESS-*HELICOBACTER PYLORI* INTERACTIONS

Gastric ulceration is one of the three key elements of Hans Selye's general adaptation syndrome.[5,113] Stress was regarded as the main cause of gastroduodenal ulceration until Barry Marshall and Robin Warren made their 2005 Nobel Prize winning discovery that a substantial proportion of gastric and duodenal ulcers appeared to be caused by *H. pylori*[114,115] Nonetheless, several studies show that stress may play a significant role in peptic ulceration, either alone or as a factor that predisposes to ulcer induction by *H. pylori*.[113,116,117] This is exemplified by the fact that the Hanshin-Awaji earthquake, which occurred on January 17, 1995, was followed by a significant increase in the number of people with peptic ulceration.[116,118] In most people, ulceration developed in conjunction with *H. pylori*. However, among the physically injured, stomach ulcers developed independently of this infectious agent. Extensive burns also lead to stress ulcers that are likely to be independent of *H. pylori* status.[119] Creed[116] argues that there is a clear relationship with stress, which must be considered alongside other risk factors for peptic ulceration: *H. pylori* infection, nonsteroidal antiinflammatory drugs, smoking, and an inherited predisposition.

Most people living with *H. pylori* never develop ulcers, about 30% of patients with ulcer do not harbor *H. pylori* infection and some patients in whom *H. pylori* colonization has been eliminated by antibiotics subsequently develop new ulcers.[120] These facts suggest that other risk factors, including gastric acid hypersecretion, smoking, nonsteroidal antiinflammatory drugs, psychological stress, and genetic predisposition, play a part in ulcer formation. This issue has been addressed by Levenstein et al.[121] in a prospective study of a population-based Danish cohort ($n = 2410$) which showed that psychological stress increased the incidence of peptic ulcer, in part by influencing health risk behaviors. Stress had similar effects on ulcers associated with *H. pylori* infection and those unrelated to either *H. pylori* or use of nonsteroidal antiinflammatory drugs.

The precise mechanism by which stress can cause gastroduodenal ulcers has yet to be determined. However, in addition to the possible direct or indirect (e.g., involving cytokines) effects of HPA-induced elevation of glucocorticoid levels on the gastroduodenal mucosa (e.g., Levenstein et al.[121]), central and peripheral dopaminergic mechanisms seem to remain contenders.[122–125] Perhaps incredulously, the clue to a central role of dopamine came from clinical observations and hypotheses regarding the etiology of Parkinson's disease and schizophrenia. Thus, in 1965, Strang[126] noted an association between Parkinson's disease, characterized by central dopamine deficiency, and gastroduodenal ulceration. This was especially the case in young Parkinsonian patients compared with control subjects. Pari passu, duodenal ulcers are rare in patients with schizophrenia a condition in which dopaminergic activity is thought to be increased.[123] Experimental evidence for a central role of dopamine deficiency as a cause of duodenal ulceration comes from the effects of the dopaminergic neurotoxin, l-methyl-4-phenyl-1,2,3,6-tetrahydropyridine (MPTP). The latter induces nigrostriatal dopamine depletion

the magnitude of which correlates with the severity of the duodenal ulcers.[123] In addition to MPTP, dopamine agonists and antagonists have been used extensively to investigate the role of dopamine deficiency or excess in causing or protecting against gastroduodenal ulceration.[123–125]

In a critical review of the interaction between stress and *H. pylori* in the causation of peptic ulcers Susan Levenstein[127] concludes "Peptic ulcer is a valuable model for understanding the interactions among psychosocial, socioeconomic, behavioral, and infectious factors in causing disease. The discovery of *H. pylori* may serve, paradoxically, as a stimulus to researchers for whom the concepts of psychology and infection are not necessarily a contradiction in terms." Just as high cholesterol is not "the cause" of myocardial infarction but one of many risk factors, *H. pylori* might be one among several risk factors for ulcer formation.[120] The etiology of gastric and duodenal ulceration and the causal switch in dogma from all stress to all bacteria followed by a sober realization that both factors may play a role have heuristic value for our understanding of disease pathogenesis.

GENERAL CONCLUSIONS AND OBSERVATIONS

Stress research has made huge advances since Hans Selye's note to Nature in 1936.[5] This is especially the case with respect to the fundamental neuroendocrinology of the stress response. However, there are still many unknowns. Thus, for example, our understanding of the apparently straightforward HPA/glucocorticoid response to stress has been confounded by PTSD in which patients may have high, low, or normal cortisol levels. The central and peripheral autonomic system seems to be dominant. Our understanding of gene × environment interactions is at an early stage: infinitely more research is required using more precise phenotyping and sophisticated genetic techniques, such as, for example, gene editing. Despite the power of modern functional brain imaging, our understanding of the relationships between changes in the nervous system and signals in the endocrine system remain poor. That is, the significance of changes in hippocampal volume for stress neuroendocrinology (and HPA activity in particular), mood, and cognition remains uncertain. The glucocorticoid vulnerability and neurotoxicity hypothesis has almost achieved canonical status. However, although there is no doubt about the veracity of the experimental data, is the hypothesis correct for mood and cognitive disorders in the human? What precisely does a change in the volume of the hippocampus or other brain region signify functionally? Are we missing some major conceptual and factual points? The tension between concept, fact, and dogma is illustrated by the relative importance of stress compared with *H. pylori* in the causation of gastro-duodenal (peptic) ulcers. Whether there are any biological interactions between these two pathogenic factors or whether they act separately remains to be determined.

So, in terms of understanding stress and behavior, have we progressed significantly beyond the century-old Yerkes–Dodson Law which, simply put, states that the relationship between arousal and behavioral performance can be linear or curvilinear depending upon the difficulty of the task?[128] This skeptical view has a brighter side, in that the stress field, far from settled, offers numerous clinically relevant opportunities and challenges for researchers especially in areas of genetics, genomics, epigenetics, neural plasticity, the perinatal factors that affect health and disease in later life, and the translation of information gained from molecular synaptology to the medical bedside, society, culture, economics, and politics.

Acknowledgments

I am grateful for facilities provided by the Florey Institute of Neuroscience and Mental Health, University of Melbourne, Victoria, Australia and the Weizmann Institute, Rehovot, Israel. Some sections of this chapter are based on updated versions of a recent review (Fink, 2011[60]).

References

1. Fink G. Stress: definition and history. In: Squire LR, ed. *Encyclopedia of Neuroscience*. vol. 9. Oxford: Academic Press/Elsevier; 2009:549–555.
2. Fink G. Stress, definitions, mechanisms, and effects outline: lessons from anxiety. In: Fink G, ed. *Stress: Concepts, Cognition, Emotion, and Behavior*. Stress: Concepts, Cognition, Emotion, and Behavior (Volume 1 of the Handbook of Stress Series); San Diego: Elsevier, Inc.; 2016:3–11.
3. McEwen BS. Physiology and neurobiology of stress and adaptation: central role of the brain. *Physiol Rev*. 2007;87:873–904.
4. Cannon WB. *The Wisdom of the Body*. New York: Norton; 1932.
5. Selye H. A syndrome produced by diverse nocuous agents. *Nature*. 1936;138:32.
6. Selye H. *Stress in Health and Disease*. Stoneham, MA: Butterworth; 1976.
7. Schulkin J, ed. *Allostasis, Homeostasis, and the Costs of Physiological Adaptation*. Cambridge, UK: Cambridge University Press; 2004.
8. McEwen BS. Stress, adaptation, and disease. Allostasis and allostatic load. *Ann N Y Acad Sci*. 1998;840:33–44.
9. Sterling P, Eyer J. Allostasis: a new paradigm to explain arousal pathology. In: Fisher S, Reason J, eds. *Handbook of Life Stress, Cognition, and Health*. New York: John Wiley and Sons; 1988:629–649.
10. Bernard C. *Leçons sur les propriétés physiologiques et les altérations pathologiques des liquides de l'organisme*. Paris: Baillière; 1859.
11. Fink G. Neuroendocrine feedback control systems: an introduction. In: Fink G, Pfaff DW, Levine JE, eds. *Handbook of Neuroendocrinology*. London, Waltham, San Diego: Academic Press, Elsevier; 2012:55–72.
12. Fink G. Neural control of the anterior lobe of the pituitary gland (pars distalis). In: Fink G, Pfaff DW, Levine JE, eds. *Handbook of Neuroendocrinology*. London, Waltham, San Diego: Academic Press, Elsevier; 2012:97–138.
13. Fortress AM, Hamlett ED, Vazey EM, et al. Designer receptors enhance memory in a mouse model of Down syndrome. *J Neurosci*. 2015;35:1343–1353. http://dx.doi.org/10.1523/JNEUROSCI.2658-14.2015.
14. Mah L, Szabuniewicz C, Fiocco AJ. Can anxiety damage the brain? *Curr Opin Psychiatry*. 2016;29:56–63.
15. Silvani A, Calandra-Buonaura G, Dampney RAL, Cortelli P. Brain–heart interactions: physiology and clinical implications. *Phil Trans R Soc A Math Phys Eng Sci*. 2016;374(2067). http://dx.doi.org/10.1098/rsta.2015.0181. pii:20150181.
16. Harris GW. *Neural Control of the Pituitary Gland*. London: Edward Arnold; 1955.
17. Fink G. 60 Years of neuroendocrinology: memoir: Harris' neuroendocrine revolution: of portal vessels and self-priming. *J Endocrinol*. 2015;226:T13–T24. http://dx.doi.org/10.1530/JOE-15-0130. Epub 2015 May 12.
18. Fink G. Inadvertent collaboration. *Nature*. 1977;269:747–748.
19. Fink G, Sheward WJ. Neuropeptide release in vivo: measurement in hypophysial portal blood. In: Fink G, Harmar AJ, eds. *Neuropeptides: A Methodology*. Chichester: John Wiley & Sons Ltd.; 1989:157–188.
20. Fink G, Smith JR, Tibballs J. Corticotrophin releasing factor in hypophysial portal blood of rats. *Nature*. 1971;203:467–468.
21. Vale W, Spiess J, Rivier C, Rivier J. Characterization of a 41 residue ovine hypothalamic peptide that stimulates the secretion of corticotropin and beta-endorphin. *Science*. 1981;213:1394–1397.
22. Fink G. Feedback actions of target hormones on hypothalamus and pituitary with special reference to gonadal steroids. *Annu Rev Physiol*. 1979;41:571–585.
23. Antoni FA. Vasopressinergic control of pituitary adrenocorticotropin secretion comes of age. *Front Neuroendocrinol*. 1993;14:76–122.
24. Antoni FA, Fink G, Sheward WJ. Corticotrophin-releasing peptides in rat hypophysial portal blood after paraventricular lesions: a marked reduction in the concentration of corticotrophin-releasing factor-41, but no change in vasopressin. *J Endocrinol*. 1990;125:175–183.
25. Fink G, Robinson IC, Tannahill LA. Effects of adrenalectomy and glucocorticoids on the peptides CRF-41, AVP and oxytocin in rat hypophysial portal blood. *J Physiol*. 1988;401:329–345.
26. Sheward WJ, Fink G. Effects of corticosterone on the secretion of corticotrophin-releasing factor, arginine vasopressin and oxytocin into hypophysial portal blood in long-term hypophysectomized rats. *J Endocrinol*. 1991;129:91–98.
27. Barker DJ. Fetal origins of coronary heart disease. *Br Med J*. 1995;311:171–174.
28. Hales NC, Barker DJ. The thrifty phenotype hypothesis. *Br Med Bull*. 2001;60:5–20.
29. Simmons R. Developmental origins of adult metabolic disease: concepts and controversies. *Trends Endocrinol Metab*. 2005;16(8):390–394.
30. Rando OJ, Simmons RA. I'm eating for two: parental dietary effects on offspring metabolism. *Cell*. 2015;161:93–105. http://dx.doi.org/10.1016/j.cell.2015.02.021.
31. Keating ST, El-Osta A. Epigenetics and metabolism. *Circ Res*. 2015;116:715–736. http://dx.doi.org/10.1161/CIRCRESAHA.116.303936.
32. Barnea A, Nottebohm F. Seasonal recruitment of hippocampal neurons in adult free-ranging black-capped chickadees. *Proc Natl Acad Sci USA*. 1994;91:11217–11221.
33. Barnea A, Mishal A, Nottebohm F. Social and spatial changes induce multiple survival regimes for new neurons in two regions of the adult brain: an anatomical representation of time? *Behav Brain Res*. 2006;167:63–74.
34. Maguire EA, Woollett K, Spiers HJ. London taxi drivers and bus drivers: a structural MRI and neuropsychological analysis. *Hippocampus*. 2006;16:1091–1101.

35. Woollett K, Spiers HJ, Maguire EA. Talent in the taxi: a model system for exploring expertise. *Philos Trans R Soc Lond B Biol Sci.* 2009;364:1407–1416.

36. Gur EB, Karadeniz M, Turan GA. Fetal programming of polycystic ovary syndrome. *World J Diabetes.* 2015;6:936–942.

37. Caspi A, McClay J, Moffitt TE, et al. Role of genotype in the cycle of violence in maltreated children. *Science.* 2002;297:851–854.

38. Caspi A, Sudgen K, Moffit TE, et al. Influence of life stress on depression: moderation by a polymorphism in the 5-HTT gene. *Science.* 2003;301:386–389.

39. Fergusson DM, Horwood LJ, Miller AL, Kennedy MA. Life stress, 5-HTTLPR and mental disorder: findings from a 30-year longitudinal study. *Br J Psychiatry.* 2011;198:129–135. http://dx.doi.org/10.1192/bjp.bp.110.085993.

40. Rutter M, Thapar A, Pickles A. Gene-environment interactions: biologically valid pathway or artifact? *Arch Gen Psychiatry.* 2009;66:1287–1289.

41. Fergusson DM, Boden JM, Horwood LJ, Miller A, Kennedy MA. Moderating role of the MAOA genotype in antisocial behavior. *Br J Psychiatry.* 2012;200:116–123.

42. Prasad HC, Steiner JA, Sutcliffe JS, Blakely RD. Enhanced activity of human serotonin transporter variants associated with autism. *Philos Trans R Soc Lond B Biol Sci.* 2009;364:163–173.

43. Schauder KB, Muller CL, Veenstra-VanderWeele J, Cascio CJ. Genetic variation in serotonin transporter modulates tactile hyperresponsiveness in ASD. *Res Autism Spectr Disord.* 2015;10:93–100.

44. Ben-Efraim YJ, Wasserman D, Wasserman J, Sokolowski M. Gene-environment interactions between CRHR1 variants and physical assault in suicide attempts. *Genes Brain Behav.* 2011;10:663–672. http://dx.doi.org/10.1111/j.1601-183X.2011.00703.x. Epub 2011 Jun 9.

45. Ishitobi Y, Nakayama S, Yamaguchi K, et al. Association of CRHR1 and CRHR2 with major depressive disorder and panic disorder in a Japanese population. *Am J Med Genet B Neuropsychiatr Genet.* 2012;159B:429–436. http://dx.doi.org/10.1002/ajmg.b.32046. Epub 2012 Mar 29.

46. Wasserman D, Sokolowski M, Rozanov V, Wasserman J. The CRHR1 gene: a marker for suicidality in depressed males exposed to low stress. *Genes Brain Behav.* 2008;7:14–19. Epub 2007 Mar 21.

47. Wasserman D, Wasserman J, Rozanov V, Sokolowski M. Depression in suicidal males: genetic risk variants in the CRHR1 gene. *Genes Brain Behav.* 2009;8:72–79. http://dx.doi.org/10.1111/j.1601-183X.2008.00446.x.

48. Wasserman D, Wasserman J, Sokolowski M. Genetics of HPA-axis, depression and suicidality. *Eur Psychiatry.* 2010;25:278–280. http://dx.doi.org/10.1016/j.eurpsy.2009.12.016. Epub 2010 May 4. Review.

49. Weber H, Richter J, Straube B, et al. Allelic variation in CRHR1 predisposes to panic disorder: evidence for biased fear processing. *Mol Psychiatry.* September 1, 2015. http://dx.doi.org/10.1038/mp.2015.125. Epub ahead of print.

50. Rasch B, Spalek K, Buholzer S, et al. A genetic variation of the noradrenergic system is related to differential amygdala activation during encoding of emotional memories. *Proc Natl Acad Sci USA.* 2009;106:19191–19196.

51. de Quervain DJ, Kolassa IT, Ertl V, et al. A deletion variant of the alpha2b-adrenoceptor is related to emotional memory in Europeans and Africans. *Nat Neurosci.* 2007;10:1137–1139.

52. Cousijn H, Rijpkema M, Qin S, et al. Acute stress modulates genotype effects on amygdala processing in humans. *Proc Natl Acad Sci USA.* 2010;107:9867–9872.

53. Tambs K, Czajkowsky N, Røysamb E, et al. Structure of genetic and environmental risk factors for dimensional representations of DSM-IV anxiety disorders. *Br J Psychiatry.* 2009;195:301–307.

54. Zannas AS, Arloth J, Carrillo-Roa T, et al. Lifetime stress accelerates epigenetic aging in an urban, African American cohort: relevance of glucocorticoid signaling. *Genome Biol.* 2015;16:266. http://dx.doi.org/10.1186/s13059-015-0828-5.

55. Mason JW, Giller EL, Kosten TR, Ostroff RB, Podd L. Urinary free-cortisol levels in posttraumatic stress disorder patients. *J Nerv Ment Dis.* 1986;174:145–149.

56. Yehuda R. Advances in understanding neuroendocrine alterations in PTSD and their therapeutic implications. *Ann N Y Acad Sci.* 2006;1071:137–166.

57. Yehuda R. Current status of cortisol findings in posttraumatic stress disorder. *Psychiatr Clin North Am.* 2002;25:341–368.

58. Metzger LJ, Carson MA, Lasko NB, et al. Basal and suppressed salivary cortisol in female Vietnam nurse veterans with and without PTSD. *Psychiatry Res.* 2008;161:330–335.

59. Lupien SJ, McEwen BS, Gunnar MR, Heim C. Effects of stress through-out the lifespan on the brain, behavior and cognition. *Nat Rev Neurosci.* 2009;10:434–445.

60. Fink G. Stress controversies: post-traumatic stress disorder, hippocampal volume, gastroduodenal ulceration. *J Neuroendocrinol.* 2011;23:107–117.

61. Young EA, Breslau N. Cortisol and catecholamines in posttraumatic stress disorder: an epidemiologic community study. *Arch Gen Psychiatry.* 2004;61:394–401.

62. Young EA, Tolman R, Witkowski K, Kaplan G. Salivary cortisol and post-traumatic stress disorder in a low-income community sample of women. *Biol Psychiatry.* 2004;55:621–626.

63. Eckart C, Engler H, Riether C, Kolassa S, Elbert T, Kolassa I-T. No PTSD-related differences in diurnal cortisol profiles of genocide survivors. *Psychoneuroendocrinology.* 2009;34:523–531.

64. Inslichta SS, Charles R, Marmarc CR, et al. Increased cortisol in women with intimate partner violence-related posttraumatic stress disorder. *Psychoneuroendocrinology.* 2006;31:825–838.

65. Rasmusson AM, Hauger RL, Morgan III CA, et al. Low baseline and yohimbine-stimulated plasma neuropeptide Y (NPY) levels in combat-related PTSD. *Biol Psychiatry.* 2000;47:526–539.

66. Southwick SM, Bremner JD, Rasmusson A, Morgan III CA, Arnsten A, Charney DS. Role of norepinephrine in the pathophysiology and treatment of posttraumatic stress disorder. *Biol Psychiatry.* 1999;46:1192–1204.

67. Strawn JR, Geracioti Jr TD. Noradrenergic dysfunction and the psycho-pharmacology of posttraumatic stress disorder. *Depress Anxiety.* 2008;25:260–271.

68. Ziegenhorn AA, Roepke S, Schommer NC, et al. Clonidine improves hyperarousal in borderline personality disorder with or without comorbid posttraumatic stress disorder: a randomized, double-blind, placebo-controlled trial. *J Clin Psychopharmacol.* 2009;29:170–173.

69. Zoladz PR, Conrad CD, Fleshner M, Diamond DM. Acute episodes of predator exposure in conjunction with chronic social instability as an animal model of post-traumatic stress disorder. *Stress.* 2008;11:259–281.

70. Arnsten AF, Raskind MA, Taylor FB, Connor DF. The effects of stress exposure on prefrontal cortex: translating basic research into successful treatments for post-traumatic stress disorder. *Neurobiol Stress.* 2015;1:89–99.

71. Liberzon I, Abelson JL, Flagel SB, Raz J, Young EA. Neuroendocrine and psychophysiologic responses in PTSD: a symptom provocation study. *Neuropsychopharmacology.* 1999;21:40–50.

72. Young EA, Abelson JL, Cameron OG. Interaction of brain noradrenergic system and the hypothalamic–pituitary–adrenal (HPA) axis in man. *Psychoneuroendocrinology.* 2005;30:807–814.

73. Felmingham KL, Williams LM, Kemp AH, Rennie C, Gordon E, Bryant RA. Anterior cingulate activity to salient stimuli is modulated by autonomic arousal in posttraumatic stress disorder. *Psychiatry Res.* 2009;173:59–62.

74. Krystal JH, Neumeister A. Noradrenergic and serotonergic mechanisms in the neurobiology of posttraumatic stress disorder and resilience. *Brain Res*. 2009;1293:13–23.

75. Blechert J, Michael T, Grossman P, Lajtman M, Wilhelm FH. Autonomic and respiratory characteristics of posttraumatic stress disorder and panic disorder. *Psychosom Med*. 2007;69:935–943.

76. Carroll BJ, Feinberg M, Greden JF, et al. A specific laboratory test for the diagnosis of melancholia. Standardization, validation, and clinical utility. *Arch Gen Psychiatry*. 1981;38:15–22.

77. Young EA, Haskett RF, Murphy-Weinberg V, Watson SJ, Akil H. Loss of glucocorticoid fast feedback in depression. *Arch Gen Psychiatry*. 1991;48:693–699.

78. Sheline YI, Po W, Wang PW, Gadots MH, Csernansky JG. Hippocampal atrophy in recurrent major depression. *Proc Natl Acad Sci USA*. 1996;93:3908–3913.

79. Rubin RT, Carroll BJ. Depression and manic-depressive illness. In: Fink G, ed. *Encyclopedia of Stress*. vol. 1. Oxford: Academic Press/Elsevier; 2007:744–754.

80. Sapolsky RM. A mechanism for glucocorticoid toxicity in the hippocampus: increased neuronal vulnerability to metabolic insults. *J Neurosci*. 1985;5:1228–1232.

81. Sapolsky RM, Krey LC, McEwen BS. Prolonged glucocorticoid exposure reduces hippocampal neuron number: implications for aging. *J Neurosci*. 1985;5:1222–1227.

82. Sapolsky RM. Glucocorticoids and hippocampal atrophy in neuropsychiatric disorders. *Arch Gen Psychiatry*. 2000;57:925–935.

83. Sapolsky RM. The possibility of neurotoxicity in the hippocampus in major depression: a primer on neuron death. *Biol Psychiatry*. 2000;48:755–765.

84. Czeh B, Lucassen PJ. What causes the hippocampal volume decrease in depression? Are neurogenesis, glial changes and apoptosis implicated? *Eur Arch Psychiatry Clin Neurosci*. 2007;257:250–260.

85. Castren E. Is mood chemistry? *Nat Rev Neurosci*. 2005;6:241–246.

86. Abe O, Yamasue H, Kasai K, et al. Voxel-based analyses of gray/white matter volume and diffusion tensor data in major depression. *Psychiatry Res*. 2010;181:64–70.

87. Vythilingam M, Vermetten E, Anderson GM, et al. Hippocampal volume, memory, and cortisol status in major depressive disorder: effects of treatment. *Biol Psychiatry*. 2004;56:101–112.

88. Vasic N, Walter H, Höse A, Wolf RC. Gray matter reduction associated with psychopathology and cognitive dysfunction in unipolar depression: a voxel-based morphometry study. *J Affect Disord*. 2008;109:107–116.

89. Hickie I, Naismith S, Ward PB, et al. Reduced hippocampal volumes and memory loss in patients with early- and late-onset depression. *Br J Psychiatry*. 2005;186:197–202.

90. Conrad CD. Chronic stress-induced hippocampal vulnerability: the glucocorticoid vulnerability hypothesis. *Rev Neurosci*. 2008;19:395–411.

91. Savitz J, Drevets WC. Bipolar and major depressive disorder: neuroimaging the developmental degenerative divide. *Neurosci Biobehav Rev*. 2009;33:699–771.

92. Swaab DF, Bao AM, Lucassen PJ. The stress system in the human brain in depression and neurodegeneration. *Ageing Res Rev*. 2005;4:141–194.

93. Van Petten C. Relationship between hippocampal volume and memory ability in healthy individuals across the lifespan: review and meta-analysis. *Neuropsychologia*. 2004;42:1394–1413.

94. Ladouceur CD, Almeida JR, Birmaher B, et al. Subcortical gray matter volume abnormalities in healthy bipolar offspring: potential neuroanatomical risk marker for bipolar disorder? *J Am Acad Child Adolesc Psychiatry*. 2008;47:532–539.

95. Sheline YI. Neuroimaging studies of mood disorder effects on the brain. *Biol Psychiatry*. 2003;54:338–352.

96. Harrison NA, Critchley HD. Neuroimaging and emotion. In: Fink G, ed. *Encyclopedia of Stress*. vol. 2. Oxford: Academic Press/Elsevier; 2007:870–878.

97. Koolschijn PC, van Haren NE, Lensvelt-Mulders GJ, Hulshoff Pol HE, Kahn RS. Brain volume abnormalities in major depressive disorder: a meta-analysis of magnetic resonance imaging studies. *Hum Brain Mapp*. 2009;30:3719–3735.

98. Lupien SJ, Evans A, Lord C, et al. Hippocampal volume is as variable in young as in older adults: implications for the notion of hippocampal atrophy in humans. *Neuroimage*. 2007;34:479–485.

99. Gurvits TV, Metzger LJ, Lasko NB, et al. Subtle neurologic compromise as a vulnerability factor for combat-related posttraumatic stress disorder: results of a twin study. *Arch Gen Psychiatry*. 2006;63:571–576.

100. McKinnon MC, Yucel K, Nazarov A, MacQueen GM. A meta-analysis examining clinical predictors of hippocampal volume in patients with major depressive disorder. *J Psychiatry Neurosci*. 2009;34:41–54.

101. Chen MC, Hamilton JP, Gotlib IH. Decreased hippocampal volume in healthy girls at risk of depression. *Arch Gen Psychiatry*. 2010;67:270–276.

102. Stockmeier CA, Mahajan GJ, Konick LC, et al. Cellular changes in the post-mortem hippocampus in major depression. *Biol Psychiatry*. 2004;56:640–650.

103. Jayatissa MN, Henningsen K, West MJ, Wiborg O. Decreased cell proliferation in the dentate gyrus does not associate with development of anhedonic-like symptoms in rats. *Brain Res*. 2009;1290:133–141.

104. Surget A, Saxe M, Leman S, et al. Drug-dependent requirement of hippocampal neurogenesis in a model of depression and of antidepressant reversal. *Biol Psychiatry*. 2008;64:293–301.

105. Pittenger C, Duman RS. Stress, depression, and neuroplasticity: a convergence of mechanisms. *Neuropsychopharmacology*. 2008;33:88–109.

106. Zhao Z, Taylor WD, Styner M, Steffens DC, Krishnan KR, MacFall JR. Hippocampus shape analysis and late-life depression. *PLoS One*. 2008;3:e1837.

107. Starkman MN, Giordani B, Gebarski SS, Berent S, Schork MA, Schteingart DE. Decrease in cortisol reverses human hippocampal atrophy following treatment of Cushing's disease. *Biol Psychiatry*. 1999;46:1595–1602.

108. Avila R, Ribeiz S, Duran FL, et al. Effect of temporal lobe structure volume on memory in elderly depressed patients. *Neurobiol Aging*. 2009. http://dx.doi.org/10.1016/j.neurobiolaging.2009.11.004. Epub 2009 Dec 23.

109. Meisenzahl EM, Seifert D, Bottlender R, et al. Differences in hippocampal volume between major depression and schizophrenia: a comparative neuroimaging study. *Eur Arch Psychiatry Clin Neurosci*. 2010;260:127–137.

110. Velakoulis D, Wood SJ, Wong MT, et al. Hippocampal and amygdala volumes according to psychosis stage and diagnosis: a magnetic resonance imaging study of chronic schizophrenia, first-episode psychosis, and ultra-high-risk individuals. *Arch Gen Psychiatry*. 2006;63:139–149.

111. McEwen BS. Central effects of stress hormones in health and disease: understanding the protective and damaging effects of stress and stress mediators. *Eur J Pharmacol*. 2008;583:174–185.

112. Videbech P, Ravnkilde B. Hippocampal volume and depression: a meta-analysis of MRI studies. *Am J Psychiatry*. 2004;161:1957–1966.

113. Murison R, Milde AM. Ulceration, gastric. In: 2nd ed. Fink G, ed. *Encyclopedia of Stress*. vol. 3. Oxford: Academic Press/Elsevier; 2007:787–791.

114. Marshall BJ, Warren JR. Unidentified curved bacilli in the stomach of patients with gastritis and peptic ulceration. *Lancet*. 1984;1:1311–1315.

115. Cover TL, Blaser MJ. *Helicobacter pylori* in health and disease. *Gastroenterology*. 2009;136:1863–1873.

116. Creed F. Somatic disorders. In: Fink G, ed. *Encyclopedia of Stress*. vol. 3. 2nd ed. Oxford: Academic Press/Elsevier; 2007:545–547.

117. Wachirawat W, Hanucharurnkul S, Suriyawongpaisal P, et al. Stress, but not *Helicobacter pylori*, is associated with peptic ulcer disease in a Thai population. *J Med Assoc Thai*. 2003;86:672–685.

118. Matsushima Y, Aoyama N, Fukuda H, et al. Gastric ulcer formation after the Hanshin-Awaji earthquake: a case study of *Helicobacter pylori* infection and stress-induced gastric ulcers. *Helicobacter*. 1999;4:94–99.

119. McColl KE. *Helicobacter pylori*-negative nonsteroidal anti-inflammatory drug-negative ulcer. *Gastroenterol Clin North Am*. 2009;38:353–361.

120. Levenstein S. *Helicobacter pylori* and ulcers. Against reductionism. *BMJ*. 2009;339:b3855. http://dx.doi.org/10.1136/bmj.b3855.

121. Levenstein S, Rosenstock S, Jacobsen RK, Jorgensen T. Psychological stress increases risk for peptic ulcer, regardless of *Helicobacter pylori* infection or use of nonsteroidal anti-inflammatory drugs. *Clin Gastroenterol Hepatol*. 2015;13. http://dx.doi.org/10.1016/j.cgh.2014.07.052. 498–506.e1. Epub 2014 Aug 9.

122. Szabo S. Dopamine disorder in duodenal ulceration. *Lancet*. October 27, 1979;2(8148):880–882.

123. Glavin GB, Szabo S. Dopamine in gastrointestinal disease. *Dig Dis Sci*. 1990;35(9):1153–1161.

124. Ozdemir V, Jamal MM, Osapay K, et al. Cosegregation of gastrointestinal ulcers and schizophrenia in a large national inpatient discharge database: revisiting the "brain-gut axis" hypothesis in ulcer pathogenesis. *J Investig Med*. 2007;55(6):315–320.

125. Rasheed N, Alghasham A. Central dopaminergic system and its implications in stress-mediated neurological disorders and gastric ulcers: short review. *Adv Pharmacol Sci*. 2012;2012:182671. Epub 2012 Sep 13.

126. Strang R. The association of gastro-duodenal ulceration and Parkinson's disease. *Med J Aust*. June 5, 1965;1(23):842–843.

127. Levenstein S. The very model of a modern etiology: a biopsychosocial view of peptic ulcer. *Psychosom Med*. 2000;62:176–185.

128. Diamond DM, Campbell AM, Park CR, Halonen J, Zoladz PR. The temporal dynamics model of emotional memory processing: a synthesis on the neurobiological basis of stress-induced amnesia, flashbulb and traumatic memories, and the Yerkes–Dodson law. *Neural Plast*. 2007;2007:60803. http://dx.doi.org/10.1155/2007/60803. Epub 2007 Mar 28.

CHAPTER

2

Limbic Forebrain Modulation of Neuroendocrine Responses to Emotional Stress

J.J. Radley[1], S.B. Johnson[1], P.E. Sawchenko[2]

[1]University of Iowa, Iowa City, IA, United States; [2]The Salk Institute for Biological Studies, San Diego, CA, United States

Abstract

One major class of stress paradigms termed *emotional* stresses possess distinct cognitive/affective components and model the human experience of fear and anxiety. Their capacity to engage adaptive response systems, notably the hypothalamo–pituitary–adrenal (HPA) axis, requires comparison with past experience, a function fulfilled by a network of interconnected structures in the limbic forebrain, including aspects of the prefrontal cortex, hippocampus, extended amygdala, and midline thalamus. This network may exert both positive and negative modulatory influences on HPA axis responses to acute emotional stresses. These adjustments are mediated indirectly, with proximate relays having been recently identified in the hypothalamus and bed nucleus of the stria terminalis, respectively, with the latter representing a point of convergence of influences from multiple stress-inhibitory limbic forebrain structures. This same network participates in adaptations (habituation, facilitation) to repeated emotional stress, and its activity is thus relevant to the many adverse health consequences of chronic stress exposure.

INTRODUCTION

Stresses of many diverse kinds almost invariably elicit generalized neuroendocrine (hypothalamo–pituitary–adrenal or HPA) and autonomic (sympatho-adrenal) responses that mobilize and redistribute bodily resources to facilitate and complement situation-specific adaptations in combating the particular challenge at hand. Glucocorticoid end products of the HPA axis (cortisol in humans, corticosterone in rodents) are particularly potent hormonal mediators that act broadly, and over extended time domains, to prepare for emergency conditions. Recruitment of this response system facilitates coping with acute insults, but because of their potent catabolic, immunosuppressive, and even neurotoxic, effects, adrenal corticosteroids can present significant health risks if stress exposure is sustained or repeated. This is true of one major class of experimental stress paradigms, termed emotional stresses, which are generally viewed as modeling the day-to-day fear, anxiety, and social stress encountered in modern human existence. The identity of, and degree to which, particular response avenues may be engaged by emotionally stressful events, and hence the pro-pathogenic potential of those events, requires an evaluative function to compare the current threat with past experience and/or to engage appropriate hard-wired, species-specific response programs. This function of comparison, and the capacity for modulation of stress responses that emerges from it, are the province of an interconnected set of structures in the limbic region of the telencephalon. Here we summarize recent work on the functional organization of limbic forebrain circuitry that provides for the modulation of neuroendocrine responses to acute and chronic emotional stress. Other excellent surveys of this topic are available.[42,49,108]

KEY POINTS

- So-called *emotional* stress paradigms model everyday human fear, anxiety
  - Physical restraint as prototype animal model
- Limbic forebrain cell groups engaged in stereotyped manner by emotional stressors
  - Attach emotional "valence" to stressful event
  - Provide comparison with past experience
- Limbic forebrain modulates hypothalamo–pituitary–adrenal (HPA) axis responses to acute emotional stress
  - Mainly implicated in HPA axis inhibition (prefrontal cortex, hippocampus, septum)
  - Effects are indirect, mediated via GABAergic relays
  - Evidence for convergent mediation via bed nucleus of the stria terminalis
- Same network implicated in adaptations (habituation, facilitation) to repeated stress
- Paraventricular thalamic nucleus implicated in "stress memory"

KINDS OF STRESS AND THEIR UNDERPINNINGS

Attempts to categorize animal models of stress, and relate them to human experience, commonly distinguish between what may be termed *physiological* (also referred to as physical, homeostatic or systemic)

and *emotional* (psychological, processive, neurogenic) paradigms.[31,42,94,95] Physiological stressors are generally conceived as targeted disruptions in homeostatic parameters that are transduced by internal (peripheral or central) receptors, whose output is processed mainly at subcortical levels of the nervous system. Commonly employed physiological stress models include cardiovascular (e.g., hemorrhage), metabolic (hypoglycemia), and immune (endotoxin) challenges. Emotional stress models, by contrast, involve stimuli that target one or more exteroceptive modalities and involve distinct cognitive (comparison with past experience) and affective components. Restraint, electrical shock, loud noise, and open field exposure are among the more common of a diverse group of emotional stress models.

Independent of category, stress-related sensory information is conveyed ultimately to the paraventricular nucleus of the hypothalamus (PVH), a crucial structure in stress adaptation (Fig. 2.1). The PVH houses parvocellular neurosecretory neurons characterized by the expression of corticotropin-releasing factor (CRF), which is required for initiation of the endocrine stress cascade by stimulating the release of pituitary corticotropin (ACTH), which, in turn, activates adrenal glucocorticoid secretion.[5,96] Although we focus here on circuitry controlling this central limb of the HPA axis, it may be noted that a separate PVH population projects to CNS cell groups involved in autonomic control, including

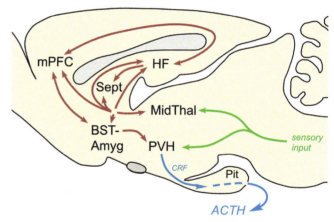

FIGURE 2.1   Limbic forebrain interconnections in relation to hypothalamo–pituitary–adrenal (HPA) axis control circuitry. For the most part, this limbic network is not integrally involved in sensory information processing and does not issue substantial direct connections to the neurosecretory neurons of the paraventricular nucleus of the hypothalamus (PVH) that govern HPA output. Instead, the extensively interconnected network of cell groups in the limbic forebrain concerned with such functions as memory, affect, and anticipatory planning provides a basis for modulating HPA axis output through of comparison with past experience. Other abbreviations: *ACTH*, adrenocorticotropic hormone; *Amyg*, amygdala; *BST*, bed nucleus of the stria terminalis; *CRF*, corticotropin-releasing factor; *HF*, hippocampal formation; *midThal*, midline thalamic nuclei; *mPFC*, medial prefrontal cortex; *Pit*, pituitary gland; *Sept*, septal nuclei.

sympatho-adrenal preganglionic neurons.[92,104] A third major cell type, the magnocellular neurosecretory neurons, produces the hormones oxytocin and vasopressin, both of which are implicated as participating in generalized stress, as well as more situation-specific, responses.[25] The major stimulatory drive to the CRF contingent of the PVH is provided by neural inputs in a position to carry information from most major sensory systems, though dominated by pathways from the nucleus of the solitary tract and forebrain circumventricular structures that convey interoceptive neural and blood-borne information, respectively.[94,105] Glucocorticoids themselves provide the major negative regulation of axis activity, acting at least in part via circuitry originating in the limbic forebrain.[33,52,93]

Many general surveys of immediate-early gene (IEG) induction patterns seen in response to physiological and emotional stressors are available.[17,22,29,34,59,97] In general, physiological challenges are characterized by a dominant activation of subcortical cell groups recognized as nodal points in central autonomic control, including stressor-specific complements of PVH effector neurons and its major sources of interoceptive input, medullary catecholamine neurons, and/or circumventricular structures of the lamina terminalis. Evidence supporting a dependence of PVH responses to immune challenge[34] or hypotensive hemorrhage on the integrity of the former[17] and of the responses to chronic hyperosmotic stimulation on the latter[53] endorses the view that PVH adaptations to physiological stresses are mediated by relatively simple visceral reflex pathways.

What we term here as emotional stressors are distinguished by their recruitment to activation of substantial expanses of non-autonomic portions of the limbic region of the telencephalon, including associated cortical fields, along with cell groups involved in the processing of sensory information specific to the nature of the challenge.[14,22,31,59] They share with physiological stressors a capacity to activate major effector neuron pools of the PVH, and while they do not provoke generalized central autonomic arousal, medullary catecholamine-containing cell groups are quite invariantly activated.[29,38,59,75] This raised the possibility that aminergic neurons might comprise a substrate for a general alarm reaction to *any* novel challenge, analogous to the manner in which catecholamine cells of the sympatho-adrenal system are called into play under emergency conditions.[26] But disruption of ascending aminergic pathways has been found to be ineffective in mitigating PVH responses to a range of emotional stress paradigms[29,60,90] and, indeed, recruitment of medullary catecholamine cell groups by emotional stress appears to be a consequence, rather than a cause, of hypothalamic engagement.[30,60]

Deriving in good measure from seminal studies aimed at dissecting the neural substrates underlying adaptive responses in a conditioned fear paradigm,[57,58] it is generally believed that emotional stress-related information must be processed through the limbic system for evaluation before distribution to appropriate physiological and behavioral response avenues, including the HPA axis. In line with this, IEG profiles elicited by emotional stressors characteristically include an interconnected loop of limbic structures, including aspects of the medial prefrontal cortex, hippocampal formation, amygdala, bed nucleus of the stria terminalis, lateral septum, and midline thalamus (see Fig. 2.1). Lesions or other manipulations of each of these have been shown capable of substantially modifying PVH and/or HPA responses to emotional stressors.[32,33,35,43,70] We are just beginning to understand how these structures may function cooperatively or hierarchically in the regulation of PVH output, and how the diversity of sensory modalities that may be engaged by emotional stressors access a common limbic circuitry.

LIMBIC MODULATION OF THE HPA AXIS

Emotional stressors require interpretation by exteroceptive sensory processing systems and integration with distinct cognitive (comparison with past experience) and affective information processing systems in the brain.[29,42,95] Whereas emotional stressors enlist brainstem and hypothalamic effectors for activation of the sympatho-adrenal and HPA axis output, they also manifest widespread activation in the limbic forebrain, corresponding to a broad array of behavioral changes (e.g., vigilance, fear) that help to facilitate adaptive coping as required by the particular environmental demand.[15,22,29,59] Functional and lesion studies implicate a network of limbic forebrain circuitry in the inhibitory control of HPA activation during emotional stress.[1,19,22,42,48] While the amygdaloid complex, hippocampal formation (HF), and medial prefrontal cortex (mPFC) are most prominently implicated in HPA axis modulation in the rodent brain,[2,33,93] other structures such as septum and paraventricular thalamic nucleus are also regarded as active in this regard.[4,8] Notably, these cell groups generally do not provide substantial direct innervation of the PVH, suggesting that a basic organizing principle of HPA control during emotional stress is that one or multiple intermediaries provide the interface for communication between higher-order modulatory centers and pools of effector neurons.

There are a number of candidate regions within the basal forebrain and hypothalamus that have been identified in anatomical tract-tracing studies as providing nodal points of relay between the limbic forebrain and PVH,[23,40,91,110] though attempts to unravel which of these may be functionally relevant in a particular context has proven a formidable challenge. Moreover, several stress-inhibitory regions in the limbic forebrain, such as mPFC and HF, give rise to predominantly excitatory (glutamatergic) projections,[24,113] suggesting that at least some of the proximate relays to PVH are likely to be inhibitory (GABAergic) in nature. Another complicating factor concerns the extent to which these intermediaries may serve as simple relays, as opposed to playing a more dynamic role in integrating convergent information from multiple sources to the PVH. While such lingering issues may convey a sense that the field is far from achieving a coherent systems level understanding of how HPA axis modulation is accomplished during emotional stress, below we highlight recent studies that may illuminate some organizing principles.

EXCITATORY MODULATION: AMYGDALA

The amygdaloid complex is implicated in a wide array of behavioral, autonomic, and neuroendocrine responses to stress, such as those associated with aversive learning, associative fear conditioning, and anxiety.[2,28,56,62,67,89,99] Information flow through the amygdala is generally considered to begin with sensory input to the lateral amygdala, and proceed through several stations in the basolateral complex (BLA) before reaching the central nucleus (CeA), which provides the principal source of extrinsic amygdaloid connections with behavioral, autonomic, and endocrine effector systems.[76–78] The amygdala exhibits a capacity to facilitate most of these functions in response to emotionally stressful experiences, and in some instances may even promote augmentation or sensitization following repeated stress exposure.

Stress-induced HPA axis activation is generally facilitated or attenuated by CeA stimulation or lesions, respectively[2,7,81,111]; cf. Ref. 82, 16; while others have highlighted the importance of medial amygdala as a contributor to HPA activation.[32] Regardless, the circuitry that provides for stress-stimulatory influences from the amygdala to reach HPA effector neurons in PVH remains to be identified definitively. Because direct amygdaloid projections to PVH appear sparse, this likely involves disynaptic relays via components of the bed nuclei of the stria terminalis (BST)[103,106] or posterior aspects of the hypothalamus.[71,73] The capacity of audiogenic stimuli to evoke HPA activation has provided an opportunity to interrogate how stress-related information from a single, well-defined sensory modality may access HPA effector neurons. Previous work has shown that the amygdala provides the entry point through which HPA axis-evoking auditory stimuli from the acoustic thalamus reach the PVH.[13,56,100] More recently, Campeau and colleagues have identified a

proximate relay in the posterior hypothalamus as being necessary for driving HPA activation during acute exposure to stressful noise.[73] Evidence from other stress paradigms also supports a more general role of the posterior hypothalamus in facilitating HPA output.[6,73] Anatomical evidence supports the possibility that CeA outputs may reach PH, although this scenario is complicated somewhat by the fact that extrinsic projections of CeA are largely GABAergic.[102,106] Nevertheless, the idea has been proposed that intrinsic GABAergic neurons within PH may underlie a mechanism whereby GABAergic inputs from CeA disinhibit output neurons from PH via local interneurons.[71] Another scenario may involve GABAergic outputs from CeA-to-BST that inhibits a GABAergic projection to PVH, proper,[106] thereby leading to HPA axis disinhibition, or possibly via a BST→PH pathway, leading to disinhibition of excitatory glutamatergic neurons in PH.

INHIBITORY MODULATION: HIPPOCAMPAL FORMATION AND MEDIAL PREFRONTAL CORTEX

For almost 50 years, the hippocampus has been the quintessential limbic structure implicated in stress regulation. Attention was first directed to this when autoradiographic evidence revealed prominent adrenal corticosteroid binding in principal hippocampal cell layers, relative to other limbic regions.[65,66] Since then, numerous studies have shown that the hippocampal formation has the capacity to inhibit the HPA axis under both basal and stress conditions, is a prominent site for glucocorticoid receptor-mediated feedback, and is essential for restoring glucocorticoids to baseline levels following the cessation of stress.[39,41,44,47,93,107] A significant advance in conceptualizing stress-inhibitory hippocampal circuitry derived from a series of studies conducted in the 1990s[23,42,91,94] that identified a number of candidate cell groups between the ventral subiculum (i.e., the portion of the hippocampal formation that provides the bulk of subcortical outputs from hippocampus proper) and PVH, thus defining possible routes to account for HPA axis modulation to be tested under a range of conditions. Candidate regions include distinct subdivisions of BST (anterior and posterior), the dorsomedial nucleus of the hypothalamus, and several clusters of GABAergic neurons that reside immediately adjacent to the PVH.[23,91]

The rodent mPFC is implicated in a wide array of higher cognitive functions subserved by the primate prefrontal cortex, including decision-making, working memory, and attentional set shifting. Nevertheless, the mPFC in rodents also plays a role in the modulation of stress-related neuroendocrine and autonomic function.[20,33,72,85,87,88,101,112] mPFC is anatomically positioned to modulate both amygdala and hippocampal activity, providing a means for top-down regulation of limbic information processing. Nevertheless, mPFC is also poised to modulate adaptive responses to stress via a number of anatomical relays to the basal forebrain, hypothalamus, and midbrain.[46,87,98] In keeping with the theme of indirect limbic modulation of HPA activity, none of the mPFC subfields projects substantially to HPA effectors in PVH, proper (Fig. 2.2). Whereas mPFC effects on these visceromotor response systems are generally inhibitory, evidence suggests some regional differentiation by cortical subfield that may account for excitatory modulation of some such responses.[83,85,87] Several recent studies suggest that HPA-inhibitory influences from mPFC are mediated by the prelimbic cortex and reach neuroendocrine effector cells in the hypothalamus via a disynaptic pathway that relays through the anteroventral BST.[87,88] Importantly, BST appears to figure more prominently than by merely providing a waystation between mPFC and PVH, as this structure also receives convergent information from hippocampus and is positioned to integrate the influences of multiple limbic forebrain cell groups on stress-related PVH output (Fig. 2.3).

MECHANISMS OF HPA MODULATION FOLLOWING REPEATED STRESS EXPOSURE

Secretory responses of the HPA axis tend to decline upon repeated exposure to the same (homotypic) stressor but overrespond to a novel (heterotypic) challenge.[8,26,27] These phenomena of habituation and facilitation, respectively, represent adaptations to chronic stress that are highly relevant to understanding and managing the negative health consequences of chronic stress exposure. Whereas plasticity in limbic forebrain circuits normally charged with modulating the HPA axis under acute conditions can result in long-term modifications of HPA axis functioning, recent work highlights the notion that similar substrates may also underlie neuroendocrine adjustments following repeated stress exposure.

The tendency for HPA responses to diminish with repeated exposure to the same (homotypic) stressor has been widely observed.[50,63,80] Habituation may be viewed as an adaptive phenomenon, which serves to protect against adverse catabolic and immunosuppressive effects of sustained elevations in circulating glucocorticoids. When evident, habituation is not generally attributable to

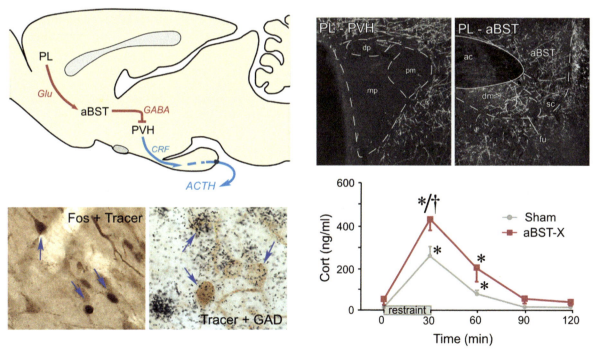

FIGURE 2.2 Indirect mediation of medial prefrontal (prelimbic) modulation of hypothalamo–pituitary–adrenal (HPA) axis response to acute emotional stress. Schematic (upper left) illustrating circuitry for indirect mediation of prelimbic (PL) cortical influences on HPA axis responses to an acute emotional stress (restraint) by GABAergic neurons in the anterior bed nucleus of the stria terminalis (aBST). Supporting evidence includes the facts that PL projects weakly to the paraventricular nucleus of the hypothalamus (PVH), and more robustly to the aBST (upper right), retrograde tracer injections in the PVH label neurons in aBST that are both stress-sensitive (display nuclear Fos induction) and GABAergic (express glutamate decarboxylase, or GAD, mRNA; lower left), and immunotoxin lesions of GABAergic elements in aBST result in exaggerated HPA secretory (corticosterone) responses to acute restraint. *Images and data are modified after Radley JJ, Gosselink, KL, Sawchenko PE. A Discrete GABAergic relay mediates medial prefrontal cortical inhibition of the neuroendocrine stress response.* J Neurosci. 2009;29(22):7330–7340; Radley JJ, Sawchenko PE. *A common substrate for prefrontal and hippocampal inhibition of the neuroendocrine stress response.* J Neurosci. 2011;31(26):9683–9695.

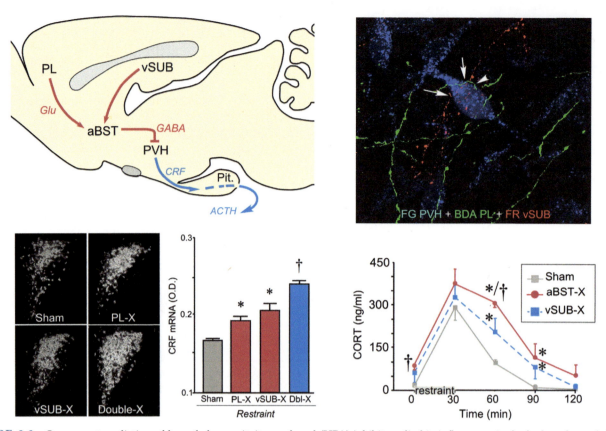

FIGURE 2.3 Convergent mediation of hypothalamo–pituitary–adrenal (HPA)-inhibitory limbic influences via the bed nucleus of the stria terminalis. The anterior bed nucleus of the stria terminalis (aBST) has been identified as a site through which stress-inhibitory influences of at least two major limbic forebrain cell groups, the prefrontal cortex (prelimbic or PL) and hippocampus (vSUB), are integrated and mediated. Supporting evidence includes anatomical demonstration of projections from PL and vSUB contacting identified (retrogradely labeled) paraventricular nucleus of the hypothalamus (PVH)-projecting neurons in aBST (upper right), and the facts that lesions of PL and vSUB exert additive effects on cellular (corticotropin-releasing factor or CRF mRNA; lower left) and secretory (corticosterone; lower right) indices of HPA responses to acute emotional (restraint) stress. *Images and data are modified after Radley JJ, Sawchenko PE. A common substrate for prefrontal and hippocampal inhibition of the neuroendocrine stress response.* J Neurosci. 2011;31(26):9683–9695.

an exhaustion of response capacity, since repeated exposure may sensitize, or facilitate, HPA secretory responses to a novel (heterotypic) insult.[26,27] Decrements in IEG induction with repeated exposure to restraint or immobilization have been described in the PVH, as well as in several extrahypothalamic cell groups, leading to the general belief that habituation is likely to be mediated by decreased neuronal activity in facilitatory afferents (cf. Ref. 13,18,55,68,109). One of the most compelling illustrations of this idea comes from a recent study in which transient inactivation of the PVH-projecting posterior hypothalamus during acute audiogenic stress was found to diminish HPA output and block the development of HPA habituation following repeated stress exposure.[73] While the neurochemical identity of these PVH afferents is yet to be evaluated in this context, these data appear to identify an excitatory pathway that drives HPA activation acutely and exhibit response decrements that underlie habituation under repeated stress conditions.

The most thorough and systematic analysis of the CNS substrates that might account for adaptations to repeated stress derives from the work of Bhatnagar and colleagues.[8,10,11,36,48] These studies identified the posterior part of the paraventricular nucleus of the *thalamus* (pPVT) at the core of an interconnected series of cell groups, including the lateral parabrachial nucleus, several amygdaloid nuclei, and the PVH itself that displayed increased IEG activation profile during HPA axis facilitation associated with heterotypic stress exposure (Fig. 2.4). Lesions of pPVT were also observed to enhance HPA axis activation under a facilitation paradigm,[9] indicating an inhibitory role in HPA regulation. PVT is interconnected with other upstream components of the limbic-PVH inhibitory network, including vSUB, amygdala, and most prominently, mPFC[37,61,69] as well as downstream regions, notably via direct projections to PVH-projecting GABAergic neurons in the anteroventral BST.[86] Thus, PVT may serve as an important interface between pathways conveying stress-related information to the limbic forebrain[12,74] and may play a prominent role in modulating adaptive responses to prolonged challenges by interceding for higher-order cognitive processing systems such as mPFC. Nevertheless, mPFC and hippocampal

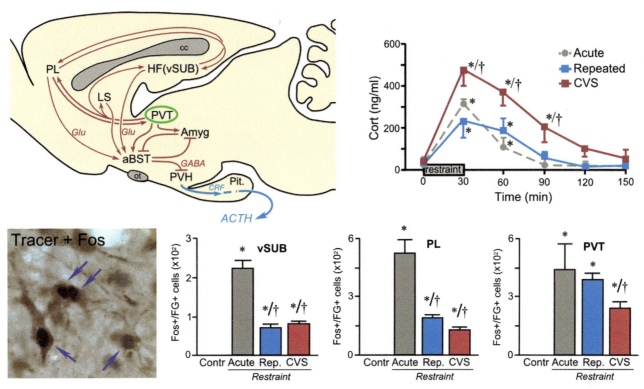

FIGURE 2.4 Involvement of limbic forebrain circuitry in hypothalamo–pituitary–adrenal (HPA) axis adaptations to repeated stress: Central role for the paraventricular thalamic nucleus? Summary of substrates for limbic forebrain modulation of HPA axis responses to emotional stress (upper left). The paraventricular thalamic nucleus (PVT) is highlighted because of its implication as being pivotally involved in HPA axis adaptations to repeated stress. This includes the tendency for axis responses to habituate upon repeated exposure to the same (homotypic) stressor and to be exaggerated upon exposure to novel (heterotypic) insults, as illustrated in modified corticosterone responses under repeated restraint and chronic variable stress (CVS) paradigms (upper right). Recent evidence supporting involvement of the PVT and the anterior bed nucleus of the stria terminalis (aBST) in such adjustments to repeated stress derives from the observation that aBST-projecting neurons in limbic sites implicated in HPA axis inhibition showed diminished activational responses in both repeated restraint and CVS paradigms, except for the PVT whose responsiveness was maintained in repeatedly stressed animals (bottom row). *Image and data modified after Radley JJ, Sawchenko PE. Evidence for involvement of a limbic paraventricular hypothalamic inhibitory network in hypothalamic-pituitary-adrenal axis adaptations to repeated stress.* J Comp Neurol. 2015;523(18):2769–2787.

formation, which are each implicated in the inhibition of HPA activation acutely, display a diminished capacity to provide inhibitory control over the HPA axis under sensitization or facilitation of HPA output.[84,86] The available evidence supports a network model in which structures interact to provide inhibitory control over the HPA axis that diminishes under chronic conditions, whereby HPA activity is increased. However, more work is needed to address whether certain network perturbations have some differential involvement in long-term changes in HPA function.

FUTURE PERSPECTIVE

We have highlighted recent studies that have advanced understanding of the organization of neural circuits capable of modulating HPA response to emotional stressors under acute conditions and that have provided insight into substrates underlying chronic stress-related alterations. In these examples, structures such as the BST and amygdala exhibit the capacity to integrate respective inhibitory

and excitatory influences exerted by the limbic forebrain upon the HPA axis. While beyond the scope of this review, these intermediaries are also poised to integrate glucocorticoid receptor-mediated negative feedback signals from upstream regions. Work in recent years increasingly implicates endocannabinoid signaling in tuning glucocorticoid-dependent feedback in both limbic forebrain regions and within the PVH itself.[21,45] Continuing efforts to understand steroid-dependent feedback should assess (1) whether glucocorticoid effects may be exerted on upstream regions, proximate relays, and/or the PVH, itself, (2) the generality of endocannabinoid signaling involvement in these processes, and (3) the extent to which disruption of these mechanisms may contribute to HPA hyperreactivity following chronic stress exposure. In a related vein, repeated stress and glucocorticoid exposure are well known to induce structural and functional synaptic reorganization throughout the limbic forebrain, thereby altering its modulatory influences on subsequent HPA responses. In general, regressive and progressive synaptic alterations are evident in limbic regions that exert inhibitory (mPFC, hippocampal formation) and excitatory (amygdala) influences, respectively, on stress-induced HPA activity (reviewed in Refs. 64,79,83). This suggests that synaptic alterations in these circuits may contribute to long-term modifications in HPA activity, in addition to altering behavioral and cognitive functions subserved by these regions. This provides a basis for the vicious cycle of feed-forward interactions between stress, limbic structural alterations, and disordered corticosteroid feedback regulation, which has been implicated in a range of neuropsychiatric disorders.

Finally, also worthy of consideration is the fact that there exists a wide degree of individual variation in HPA axis reactivity among humans. For example, patterns of cortisol release vary widely in humans exposed to repeated challenges in standard experimental settings.[51,54] Similar to humans, and depending on the characteristics of stimuli employed in animal models, stressors do not invariably lead to HPA axis overreactivity.[3] While it is clear that such variability of response depends on factors such as genetics, early-life experiences, and history of previous stress exposure, a more coherent understanding of the neural circuitry that provides for HPA axis modulation should inform the mechanisms underlying differential vulnerability to stress-related disorders.

Acknowledgments

Work from our laboratories that is summarized here was supported by NIH grants MH-095972 (JJR), NS-21182, NS-49641, and DK-26741 (PES) and a grant from the Clayton Medical Research Foundation. PES is a Senior Investigator of the Clayton Medical Research Foundation.

References

1. Akana SF, Chu A, Soriano L, Dallman MF. Corticosterone exerts site-specific and state-dependent effects in prefrontal cortex and amygdala on regulation of adrenocorticotropic hormone, insulin and fat depots. *J Neuroendocrinol.* 2001;13(7):625–637.
2. Allen JP, Allen CF. Role of the amygdaloid complexes in the stress-induced release of ACTH in the rat. *Neuroendocrinology.* 1974;15(3–4):220–230.
3. Anisman H, Matheson K. Stress, depression, and anhedonia: caveats concerning animal models. *Neurosci Biobehav Rev.* 2005;29(4–5):525–546.
4. Anthony TE, Dee N, Bernard A, Lerchner W, Heintz N, Anderson DJ. Control of stress-induced persistent anxiety by an extra-amygdala septohypothalamic circuit. *Cell.* 2014;156(3):522–536.
5. Antoni FA. Hypothalamic control of adrenocorticotropin secretion: advances since the discovery of 41-residue corticotropin-releasing factor. *Endocr Rev.* 1986;7(4):351–378.
6. Bailey TW, Dimicco JA. Chemical stimulation of the dorsomedial hypothalamus elevates plasma ACTH in conscious rats. *Am J Physiol – Regul Integr Comp Physiol.* 2001;280(1):R8–R15.
7. Beaulieu S, Di Paolo T, Barden N. Control of ACTH secretion by the central nucleus of the amygdala: implication of the serotoninergic system and its relevance to the glucocorticoid delayed negative feedback mechanism. *Neuroendocrinology.* 1986;44(2):247–254.
8. Bhatnagar S, Dallman M. Neuroanatomical basis for facilitation of hypothalamic-pituitary-adrenal responses to a novel stressor after chronic stress. *Neuroscience.* 1998;84(4):1025–1039.
9. Bhatnagar S, Dallman MF, Roderick RE, Basbaum AI, Taylor BK. The effects of prior chronic stress on cardiovascular responses to acute restraint and formalin injection. *Brain Res.* 1998;797(2):313–320.
10. Bhatnagar S, Huber R, Nowak N, Trotter P. Lesions of the posterior paraventricular thalamus block habituation of hypothalamic-pituitary-adrenal responses to repeated restraint. *J Neuroendocrinol.* 2002;14(5):403–410.
11. Bhatnagar S, Viau V, Chu A, Soriano L, Meijer OC, Dallman MF. A cholecystokinin-mediated pathway to the paraventricular thalamus is recruited in chronically stressed rats and regulates hypothalamic-pituitary-adrenal function. *J Neurosci.* 2000;20(14):5564–5573.
12. Bubser M, Deutch AY. Stress induces Fos expression in neurons of the thalamic paraventricular nucleus that innervate limbic forebrain sites. *Synapse.* 1999;32(1):13–22.
13. Campeau S, Dolan D, Akil H, Watson SJ. c-fos mRNA induction in acute and chronic audiogenic stress: possible role of the orbito-frontal cortex in habituation. *Stress.* 2002;5(2):121–130.
14. Campeau S, Falls WA, Cullinan WE, Helmreich DL, Davis M, Watson SJ. Elicitation and reduction of fear: behavioural and neuroendocrine indices and brain induction of the immediate-early gene c-fos. *Neuroscience.* 1997;78(4):1087–1104.
15. Campeau S, Watson SJ. Neuroendocrine and behavioral responses and brain pattern of c-fos induction associated with audiogenic stress. *J Neuroendocrinol.* 1997;9(8):577–588.
16. Carter RN, Pinnock SB, Herbert J. Does the amygdala modulate adaptation to repeated stress? *Neuroscience.* 2004;126(1):9–19.
17. Chan RKW, Sawchenko PE. Spatially and temporally differentiated patterns of *c-fos* expression in brainstem catecholaminergic cell groups induced by cardiovascular challenges in the rat. *J Comp Neurol.* 1994;348(3):433–460.
18. Chen X, Herbert J. Regional changes in c-fos expression in the basal forebrain and brainstem during adaptation to repeated stress: correlations with cardiovascular, hypothermic and endocrine responses. *Neuroscience.* 1995;64(3):675–685.
19. Christoffel DJ, Golden SA, Dumitriu D, et al. IkappaB kinase regulates social defeat stress-induced synaptic and behavioral plasticity. *J Neurosci.* 2011;31(1):314–321.
20. Crane JW, Ebner K, Day TA. Medial prefrontal cortex suppression of the hypothalamic-pituitary-adrenal axis response to a physical stressor, systemic delivery of interleukin-1. *Eur J Neurosci.* 2003;17(7):1473–1481.
21. Crosby KM, Bains JS. The intricate link between glucocorticoids and endocannabinoids at stress-relevant synapses in the hypothalamus. *Neuroscience.* 2012;204:31–37.
22. Cullinan WE, Herman JP, Battaglia DF, Akil H, Watson SJ. Pattern and time course of immediate early gene expression in rat brain following acute stress. *Neuroscience.* 1995;64(2):477–505.
23. Cullinan WE, Herman JP, Watson SJ. Ventral subicular interaction with the hypothalamic paraventricular nucleus: evidence for a relay in the bed nucleus of the stria terminalis. *J Comp Neurol.* 1993;332(1):1–20.
24. Cullinan WE, Ziegler DR, Herman JP. Functional role of local GABAergic influences on the HPA axis. *Brain Struct Funct.* 2008;213(1–2):63–72.
25. Cunningham Jr ET, Sawchenko PE. Reflex control of magnocellular vasopressin and oxytocin secretion. *Trends Neurosci.* 1991;14(9):406–411.
26. Dallman MF. Stress update Adaptation of the hypothalamic-pituitary-adrenal axis to chronic stress. *Trends Endocrinol Metab.* 1993;4(2):62–69.

27. Dallman MF, Akana SF, Scribner KA, et al. Stress, feedback and facilitation in the hypothalamo-pituitary-adrenal axis. *J Neuroendocrinol*. 1992;4(5):517–526.

28. Davis M. The role of the amygdala in fear and anxiety. *Annu Rev Neurosci*. 1992;15:353–375.

29. Dayas CV, Buller KM, Day TA. Medullary neurones regulate hypothalamic corticotropin-releasing factor cell responses to an emotional stressor. *Neuroscience*. 2001;105(3):707–719.

30. Dayas CV, Buller KM, Day TA. Hypothalamic paraventricular nucleus neurons regulate medullary catecholamine cell responses to restraint stress. *J Comp Neurol*. 2004;478(1):22–34.

31. Dayas CV, Buller KM, Crane JW, Xu Y, Day TA. Stressor categorization: acute physical and psychological stressors elicit distinctive recruitment patterns in the amygdala and in medullary noradrenergic cell groups. *Eur J Neurosci*. 2001;14(7):1143–1152.

32. Dayas CV, Buller KM, Day TA. Neuroendocrine responses to an emotional stressor: evidence for involvement of the medial but not the central amygdala. *Eur J Neurosci*. 1999;11(7):2312–2322.

33. Diorio D, Viau V, Meaney MJ. The role of the medial prefrontal cortex (cingulate gyrus) in the regulation of hypothalamic-pituitary-adrenal responses to stress. *J Neurosci*. 1993;13(9):3839–3847.

34. Ericsson A, Kovács KJ, Sawchenko PE. A functional anatomical analysis of central pathways subserving the effects of interleukin-1 on stress-related neuroendocrine neurons. *J Neurosci*. 1994;14(2):897–913.

35. Gray TS, Piechowski RA, Yracheta JM, Rittenhouse PA, Bethea CL, Van de Kar LD. Ibotenic acid lesions in the bed nucleus of the stria terminalis attenuate conditioned stress-induced increases in prolactin, ACTH and corticosterone. *Neuroendocrinology*. 1993;57(3):517–524.

36. Grissom N, Bhatnagar S. Habituation to repeated stress: get used to it. *Neurobiol Learn Mem*. 2009;92(2):215–224.

37. Heidbreder CA, Groenewegen HJ. The medial prefrontal cortex in the rat: evidence for a dorso-ventral distinction based upon functional and anatomical characteristics. *Neurosci Biobehav Rev*. 2003;27(6):555–579.

38. Helfferich F, Palkovits M. Acute audiogenic stress-induced activation of CRH neurons in the hypothalamic paraventricular nucleus and catecholaminergic neurons in the medulla oblongata. *Brain Res*. 2003;975(1–2):1–9.

39. Herman JP, Cullinan WE, Morano MI, Akil H, Watson SJ. Contribution of the ventral subiculum to inhibitory regulation of the hypothalamo-pituitary-adrenocortical axis. *J Neuroendocrinol*. 1995;7(6):475–482.

40. Herman JP, Figueiredo H, Mueller NK, et al. Central mechanisms of stress integration: hierarchical circuitry controlling hypothalamo-pituitary-adrenocortical responsiveness. *Front Neuroendocrinol*. 2003;24(3):151–180.

41. Herman JP, Schafer MK-H, Young EA, Thompson R, Douglass J, Akil H, et al. Evidence for hippocampal regulation of neuroendocrine neurons of the hypothalamo-pituitary-adrenocortical axis. *J Neurosci*. 1989;9(9):3072–3082.

42. Herman JP, Cullinan WE. Neurocircuitry of stress: central control of the hypothalamo-pituitary-adrenal axis. *Trends Neurosci*. 1997;20(2):78–84.

43. Herman JP, Dolgas CM, Carlson SL. Ventral subiculum regulates hypothalamo-pituitary-adrenocortical and behavioural responses to cognitive stressors. *Neuroscience*. 1998;86(2):449–459.

44. Herman JP, Cullinan WE, Young EA, Akil H, Watson SJ. Selective forebrain fibertract lesions implicate ventral hippocampal structures in tonic regulation of paraventricular nucleus CRH and AVP mRNA expression. *Brain Res*. 1992;592(1–2):228–238.

45. Hill MN, Tasker JG. Endocannabinoid signaling, glucocorticoid-mediated negative feedback, and regulation of the hypothalamic-pituitary-adrenal axis. *Neuroscience*. 2012;204:5–16.

46. Hurley KM, Herbert H, Moga MM, Saper CB. Efferent projections of the infralimbic cortex of the rat. *J Comp Neurol*. 1991;308(2):249–276.

47. Jacobson L, Sapolsky RM. The role of the hippocampus in feedback regulation of the hypothalamo-pituitary-adrenocortical axis. *Endocr Rev*. 1991;12(2):118–134.

48. Jaferi A, Bhatnagar S. Corticosterone can act at the posterior paraventricular thalamus to inhibit hypothalamic-pituitary-adrenal activity in animals that habituate to repeated stress. *Endocrinology*. 2006;147(10):4917–4930.

49. Jankord R, Herman JP. Limbic regulation of hypothalamo-pituitary-adrenocortical function during acute and chronic stress. *Ann NY Acad Sci*. 2008;1148:64–73.

50. Keim KL, Sigg EB. Physiological and biochemical concomitants of restraint stress in rats. *Pharmacol Biochem Behav*. 1976;4(3):289–297.

51. Kirschbaum C, Prussner JC, Stone AA, et al. Persistent high cortisol responses to repeated psychological stress in a subpopulation of healthy men. *Psychosom Med*. 1995;57(5):468–474.

52. Kovács K, Makara GB. Corticosterone and dexamethasone act at different brain sites to inhibit adrenalectomy-induced adrenocorticotropin secretion. *Brain Res*. 1988;474(2):205–210.

53. Kovacs KJ, Sawchenko PE. Mediation of osmoregulatory influences on neuroendocrine corticotropin-releasing factor expression by the ventral lamina terminalis. *Proc Natl Acad Sci USA*. 1993;90(16):7681–7685.

54. Kudielka BM, Wust S. Human models in acute and chronic stress: assessing determinants of individual hypothalamus-pituitary-adrenal axis activity and reactivity. *Stress*. 2010;13(1):1–14.

55. Lachuer J, Delton I, Buda M, Tappaz M. The habituation of brainstem catecholaminergic groups to chronic daily restraint stress is stress specific like that of the hypothalamo-pituitary-adrenal axis. *Brain Res*. 1994;638(1–2):196–202.

56. LeDoux JE. Emotion circuits in the brain. *Annu Rev Neurosci*. 2000;23:155–184.

57. LeDoux JE, Iwata J, Cicchetti P, Reis DJ. Different projections of the central amygdaloid nucleus mediate autonomic and behavioral correlates of conditioned fear. *J Neurosci*. 1988;8(7):2517–2529.

58. LeDoux JE, Sakaguchi A, Reis DJ. Subcortical efferent projections of the medial geniculate nucleus mediate emotional responses conditioned to acoustic stimuli. *J Neurosci*. 1984;4(3):683–698.

59. Li HY, Sawchenko PE. Hypothalamic effector neurons and extended circuitries activated in "neurogenic" stress: a comparison of footshock effects exerted acutely, chronically, and in animals with controlled glucocorticoid levels. *J Comp Neurol*. 1998;393(2):244–266.

60. Li H-Y, Ericsson A, Sawchenko PE. Distinct mechanisms underlie activation of hypothalamic neurosecretory neurons and their medullary catecholaminergic afferents in categorically different stress paradigms. *Proc Natl Acad Sci USA*. 1996;93(6):2359–2364.

61. Li S, Kirouac GJ. Sources of inputs to the anterior and posterior aspects of the paraventricular nucleus of the thalamus. *Brain Struct Funct*. 2012;217(2):257–273.

62. Loewy AD. Forebrain nuclei involved in autonomic control. *Prog Brain Res*. 1991;87:253–268.

63. Mason JW. Corticosteroid response to chair restraint in the monkey. *Am J Physiol*. 1972;222(5):1291–1294.

64. McEwen BS, Morrison JH. The brain on stress: vulnerability and plasticity of the prefrontal cortex over the life course. *Neuron*. 2013;79(1):16–29.

65. McEwen BS, Weiss JM, Schwartz LS. Uptake of corticosterone by rat brain and its concentration by certain limbic structures. *Brain Res*. 1969;16(1):227–241.

66. McEwen BS. Corticosteroids and hippocampal plasticity. *Ann NY Acad Sci*. 1994;746:134–142.

67. McGaugh JL. Memory consolidation and the amygdala: a systems perspective. *Trends Neurosci*. 2002;25(9):456.

68. Melia KR, Ryabinin AE, Schroeder R, Bloom FE, Wilson MC. Induction and habituation of immediate early gene expression in rat brain by acute and repeated restraint stress. *J Neurosci*. 1994;14(10):5929–5938.

69. Moga MM, Weis RP, Moore RY. Efferent projections of the paraventricular thalamic nucleus in the rat. *J Comp Neurol*. 1995;359(2):221–238.

70. Morin SM, Stotz-Potter EH, DiMicco JA. Injection of muscimol in dorsomedial hypothalamus and stress-induced Fos expression in paraventricular nucleus. *Am J Physiol – Regul Integr Comp Physiol*. 2001;280:R1276–R1284.

71. Myers B, Carvalho-Netto E, Wick-Carlson D, et al. GABAergic signaling within a limbic-hypothalamic circuit integrates social and anxiety-like behavior with stress reactivity. *Neuropsychopharmacology*. 2015;41(6):1530–1539.

72. Neafsey EJ. Prefrontal cortical control of the autonomic nervous system: anatomical and physiological observations. *Prog Brain Res*. 1990;85:147–166.

73. Nyhuis TJ, Masini CV, Day HE, Campeau S. Evidence for the integration of stress-related signals by the rostral posterior hypothalamic nucleus in the regulation of acute and repeated stress-evoked hypothalamo-pituitary-adrenal response in rat. *J Neurosci*. 2016;36(3):795–805.

74. Otake K, Kin K, Nakamura Y. Fos expression in afferents to the rat midline thalamus following immobilization stress. *Neurosci Res*. 2002;43(3):269–282.

75. Pezzone MA, Lee W-S, Hoffman GE, Pezzone KM, Rabin BS. Activation of brainstem catecholaminergic neurons by conditioned and unconditioned aversive stimuli as revealed by c-fos immunoreactivity. *Brain Res*. 1993;608(2):310–318.

76. Pitkänen A, Amaral DG. Organization of the intrinsic connections of the monkey amygdaloid complex: projections originating in the lateral nucleus. *J Comp Neurol*. 1998;398(3):431–458.

77. Pitkanen A, Savander V, LeDoux JE. Organization of intra-amygdaloid circuitries in the rat: an emerging framework for understanding functions of the amygdala. *Trends Neurosci*. 1997;20(11):517–523.

78. Pitkanen A, Stefanacci L, Farb CR, Go GG, LeDoux JE, Amaral DG. Intrinsic connections of the rat amygdaloid complex: projections originating in the lateral nucleus. *J Comp Neurol*. 1995;356(2):288–310.

79. Pittenger C, Duman RS. Stress, depression, and neuroplasticity: a convergence of mechanisms. *Neuropsychopharmacology*. 2008;33(1):88–109.

80. Pollard I, Bassett JR, Cairncross KD. Plasma glucocorticoid elevation and ultrastructural changes in the adenohypophysis of the male rat following prolonged exposure to stress. *Neuroendocrinology*. 1976;21(4):312–330.

81. Prewitt CM, Herman JP. Lesion of the central nucleus of the amygdala decreases basal CRH mRNA expression and stress-induced ACTH release. *Ann NY Acad Sci*. 1994;746:438–440.

82. Prewitt CM, Herman JP. Hypothalamo-pituitary-adrenocortical regulation following lesions of the central nucleus of the amygdala. *Stress*. 1997;1(4):263–280.

83. Radley JJ. Toward a limbic cortical inhibitory network: implications for hypothalamic-pituitary-adrenal responses following chronic stress. *Front Behav Neurosci*. 2012;6:7.

84. Radley JJ, Anderson RM, Hamilton BA, Alcock JA, Romig-Martin SA. Chronic stress-induced alterations of dendritic spine subtypes predict functional decrements in an hypothalamo-pituitary-adrenal-inhibitory prefrontal circuit. *J Neurosci*. 2013;33(36):14379–14391.

85. Radley JJ, Arias CM, Sawchenko PE. Regional differentiation of the medial prefrontal cortex in regulating adaptive responses to acute emotional stress. *J Neurosci*. 2006;26(50):12967–12976.

86. Radley JJ, Sawchenko PE. Evidence for involvement of a limbic paraventricular hypothalamic inhibitory network in hypothalamic-pituitary-adrenal axis adaptations to repeated stress. *J Comp Neurol*. 2015;523(18):2769–2787.

87. Radley JJ, Gosselink KL, Sawchenko PE. A Discrete GABAergic relay mediates medial prefrontal cortical inhibition of the neuroendocrine stress response. *J Neurosci*. 2009;29(22):7330–7340.

88. Radley JJ, Sawchenko PE. A common substrate for prefrontal and hippocampal inhibition of the neuroendocrine stress response. *J Neurosci*. 2011;31(26):9683–9695.

89. Ressler KJ. Amygdala activity, fear, and anxiety: modulation by stress. *Biol Psychiatry*. 2010;67(12):1117–1119.

90. Ritter S, Watts AG, Dinh TT, Sanchez-Watts G, Pedrow C. Immunotoxin lesion of hypothalamically projecting norepinephrine and epinephrine neurons differentially affects circadian and stressor-stimulated corticosterone secretion. *Endocrinology*. 2003;144(4):1357–1367.

91. Roland BL, Sawchenko PE. Local origins of some GABAergic projections to the paraventricular and supraoptic nuclei of the hypothalamus in the rat. *J Comp Neurol*. 1993;332(1):123–143.

92. Saper CB, Loewy AD, Swanson LW, Cowan WM. Direct hypothalamo-autonomic connections. *Brain Res*. 1976;117(2):305–312.

93. Sapolsky RM, Krey LC, McEwen BS. Glucocorticoid sensitive hippocampal neurons are involved in terminating the adrenocortical stress response. *Proc Natl Acad Sci USA*. 1984;81(19): 6174–6177.

94. Sawchenko PE, Brown ER, Chan RK, et al. The paraventricular nucleus of the hypothalamus and the functional neuroanatomy of visceromotor responses to stress. *Prog Brain Res*. 1996;107: 201–222.

95. Sawchenko PE, Li HY, Ericsson A. Circuits and mechanisms governing hypothalamic responses to stress: a tale of two paradigms. *Prog Brain Res*. 2000;122:61–78.

96. Sawchenko PE, Swanson LW. Localization, colocalization, and plasticity of corticotropin-releasing factor immunoreactivity in rat brain. *Fed Proc*. 1985;44(1 Pt 2):221–227.

97. Senba E, Matsunaga K, Tohyama M, Noguchi K. Stress-induced c-fos expression in the rat brain: activation mechanism of sympathetic pathway. *Brain Res Bull*. 1993;31(3–4):329–344.

98. Sesack SR, Deutch AY, Roth RH, Bunney BS. Topographical organization of the efferent projections of the medial prefrontal cortex in the rat: an anterograde tract-tracing study with Phaseolus vulgaris leucoagglutinin. *J Comp Neurol*. 1989;290(2):213–242.

99. Shekhar A, Sajdyk TJ, Gehlert DR, Rainnie DG. The amygdala, panic disorder, and cardiovascular responses. *Ann NY Acad Sci*. 2003;985:308–325.

100. Sullivan GM, Apergis J, Bush DE, Johnson LR, Hou M, Ledoux JE. Lesions in the bed nucleus of the stria terminalis disrupt corticosterone and freezing responses elicited by a contextual but not by a specific cue-conditioned fear stimulus. *Neuroscience*. 2004;128(1):7–14.

101. Sullivan RM, Gratton A. Lateralized effects of medial prefrontal cortex lesions on neuroendocrine and autonomic stress responses in rats. *J Neurosci*. 1999;19(7):2834–2840.

102. Sun N, Cassell MD. Intrinsic GABAergic neurons in the rat central extended amygdala. *J Comp Neurol*. 1993;330(3):381–404.

103. Sun N, Roberts L, Cassell MD. Rat central amygdaloid nucleus projections to the bed nucleus of the stria terminalis. *Brain Res Bull*. 1991;27(5):651–662.

104. Swanson LW, Kuypers HGJM. The paraventricular nucleus of the hypothalamus: cytoarchitectonic subdivisions and organization of projections to the pituitary, dorsal vagal complex, and spinal cord as demonstrated by retrograde fluorescence double-labeling methods. *J Comp Neurol*. 1980;194(3):555–570.

105. Swanson LW, Sawchenko PE. Hypothalamic integration: organization of the paraventricular and supraoptic nuclei. *Annu Rev Neurosci*. 1983;6:269–324.

106. Tsubouchi K, Tsumori T, Yokota S, Okunishi H, Yasui Y. A disynaptic pathway from the central amygdaloid nucleus to the paraventricular hypothalamic nucleus via the parastrial nucleus in the rat. *Neurosci Res*. 2007;59(4):390–398.

107. Tuvnes FA, Steffenach HA, Murison R, Moser MB, Moser EI. Selective hippocampal lesions do not increase adrenocortical activity. *J Neurosci*. 2003;23(10):4345–4354.

108. Ulrich-Lai YM, Herman JP. Neural regulation of endocrine and autonomic stress responses. *Nat Rev Neurosci*. 2009;10(6):397–409.

109. Umemoto S, Kawai Y, Senba E. Differential regulation of IEGs in the rat PVH in single and repeated stress models. *Neuroreport*. 1994;6(1):201–204.

110. Van de Kar LD, Blair ML. Forebrain pathways mediating stress-induced hormone secretion. *Front Neuroendocrinol*. 1999;20(1):1–48.

111. Van de Kar LD, Piechowski RA, Rittenhouse PA, Gray TS. Amygdaloid lesions: differential effect on conditioned stress and immobilization-induced increases in corticosterone and renin secretion. *Neuroendocrinology*. 1991;54(2):89–95.

112. Van Eden CG, Buijs RM. Functional neuroanatomy of the prefrontal cortex: autonomic interactions. *Prog Brain Res*. 2000;126: 49–62.

113. Walaas I, Fonnum F. Biochemical evidence for glutamate as a transmitter in hippocampal efferents to the basal forebrain and hypothalamus in the rat brain. *Neuroscience*. 1980;5(10):1691–1698.

3

Adrenergic Neurons in the CNS

C.P. Sevigny, C. Menuet, A.Y. Fong, J.K. Bassi, A.A. Connelly,
A.M. Allen

The University of Melbourne, Parkville, VIC, Australia

Abstract

Within the central nervous system there are three groups of neurons that express all of the enzymes required for generation of the catecholamine, adrenaline. The cell groups, called the C1, C2, and C3 cell groups, are located in the medulla oblongata. Their axons project widely throughout the CNS, from the sacral spinal cord to the olfactory nucleus. While they are predominantly studied within the context of a spinal projection to sympathetic preganglionic neurons, and a role in cardiovascular regulation, this review outlines data showing a wider range of projections and potential functions that suggest a much broader role for these neurons.

Catecholamines are a class of small neurotransmitter molecules based upon hydroxylation and methylation of the amino acid, tyrosine. The conversion of tyrosine to dopamine, the first molecule in the catecholamine biosynthetic pathway, occurs via the actions of tyrosine hydroxylase (TH) and L-aromatic acid decarboxylase. Dopaminergic neurons, designated by the nomenclature of A8–A16 cell groups, occur in distinct regions of the midbrain, hypothalamus, and olfactory bulb and have established roles in regulation of movement, reward pathways, and neuroendocrine function, amongst other functions. Hydroxylation of dopamine, by dopamine β-hydroxylase (DBH), forms noradrenaline. Noradrenergic neurons in the CNS are designated A1–A7 cell groups and occur in the medulla oblongata, pons, and midbrain. Many of these neuronal groups, exemplified by the well-studied locus coeruleus (A4 and A6), project widely throughout the brain and contribute to the diffuse modulatory systems. Finally, methylation of noradrenaline, by phenylethanolamine-N-methyl transferase (PNMT), produces adrenaline. Adrenaline is known more as a systemic hormone released by the adrenal medulla; however, adrenergic neurons occur in three cell groups in the medulla, designated C1–C3 cell groups. The C1–C3 cell groups are defined by their expression of TH, DBH, and PNMT. Some neurons exist in different brain regions that express PNMT in the absence of TH and DBH[24] and while these may have the ability to produce adrenaline,[45] they are not included as adrenergic cell groups. This chapter will concentrate only on the C1–C3 cell groups (Fig. 3.1).

KEY POINTS

- The central nervous system contains three groups of neurons possessing the capability to synthesize adrenaline—these are the C1, C2, and C3 groups. All groups are likely to use glutamate as their principal transmitter.

- The adrenergic cells project widely throughout the central nervous system, from the sacral spinal cord to the olfactory nucleus. Like the noradrenergic cells of the locus coeruleus, these adrenergic neurons are broadcasting information throughout the brain.

- The best characterized adrenergic cell group is the C1 group in the rostral ventrolateral medulla. These neurons play an important role in the regulation of blood pressure via effects on the sympathetic nervous system.

- The adrenergic neurons are activated by a wide variety of physiological stimuli, including, but not limited to, decreases in blood pressure, altered blood gas composition, glucoprivation, and immune challenges.

LOCALIZATION

C1 neurons, which comprise approximately 70% of the adrenergic neurons in the brain, are a component of a nucleus called the rostral ventrolateral medulla (RVLM). In humans the C1 neurons are located dorsal to the inferior olives and, due to the size of the inferior olives, are displaced dorsally from the ventral surface.[6,31] In other species, as the inferior olives are smaller and more medially located, the RVLM is separated from the ventral surface of the brainstem only by the thin ventrospinocerebellar tract.[46,47] The medial boundary is the lateral paragigantocellular nucleus which separates the C1 neurons from the inferior olivary nucleus. Laterally the C1 neurons are separated from the spinal trigeminal nucleus by the rubrospinal tract. At the dorsal border, the C1 neurons are intermixed with the neurons of the Bötzinger complex, which sits between the RVLM and the compact formation of the nucleus ambiguus. Caudally, the C1 neurons are intermixed with the noradrenergic neurons of the A1 cell group—part of a nucleus called the caudal ventrolateral medulla (CVLM). The rostral border is the facial motor nucleus, where the C1 neurons become intermixed with the CO_2 responsive neurons of the retrotrapezoid nucleus (RTN). Whilst these anatomical boundaries are defined, they are loose as other neuronal groups with different functions and neurochemistry are located within this region called the RVLM.[59]

The C2 cell group, which makes up approximately 15% of the adrenergic neurons in the brain, is part of the nucleus of the solitary tract (NTS). At caudal levels, approximately starting at the level of the mid- to rostral area postrema, the C2 neurons are intermingled with noradrenergic neurons of the A2 cell group. Rostrally the neurons continue for the extent of the NTS. The C2 neurons are located in the medial subnucleus of the NTS and may constitute at least two different cell groups, with the caudal C2 neurons having quite small somata (~10 μm in diameter) compared to the larger (~25 μm in diameter) rostral neurons.

The remainder of the adrenergic neurons, approximately 15%, forms a loose column in the dorsal part of the rostral medulla. The neurons form a column that in the rat is approximately 1.7 mm in length, running from the caudal end of the fourth ventricle to the caudal edge of the dorsal cochlear nucleus. The cells line the floor of the fourth ventricle at their dorsal extent and are bounded by the raphe obscurus ventrally. The C3 neurons can be divided into four distinct columns: a dorsal group at the surface of the fourth ventricle; a median column and medial and lateral columns bilaterally. A detailed description of the distribution of these cell groups has been published.[44]

PROJECTIONS

Axonal terminals expressing PNMT are widely distributed throughout the central nervous system[33] with extensive contributions from all adrenergic cell groups. Due to technical limitations these initial extensive descriptions did not identify which of the C1–C3 groups was projecting to each site. In fact, such descriptions are difficult as the adrenergic neurons are intermingled with neurons of different neurochemistry and so traditional approaches would require anterograde tracing in combination with terminal labeling of an appropriate phenotypic marker. The advent of recombinant viral methods that enable cell- and region-selective transgenic expression of fluorescent proteins, such as green fluorescent protein (GFP), has enabled the specific projections of the adrenergic neurons to be mapped in detail. The *cis*-regulatory region of the DBH promoter contains binding sites, called PRS, for the transcription factor phox2. The demonstration that a small multimer of PRS provided strong and selective expression in phox2 neurons has enabled relatively selective transgene expression in noradrenergic and adrenergic neurons.[35] Using replication-deficient lentiviruses, with fluorophore expression under the control of the PRS multimer, called PRSx8, the projections of C1 and C3 neurons have been mapped in detail.[12,53] The only caveat to these studies is that while catecholaminergic neurons express phox2, some phox2-expressing neurons are not catecholaminergic.[11,56] Examples of these phox2-expressing,

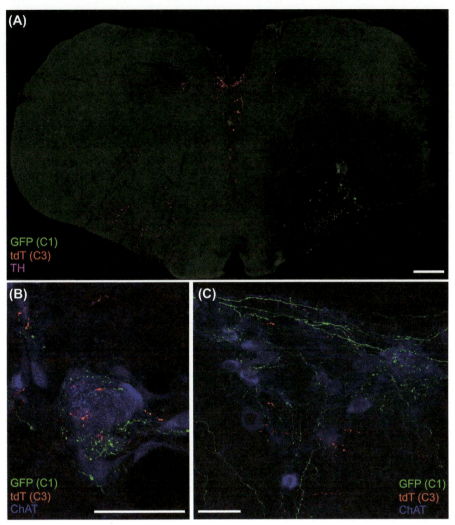

FIGURE 3.1 **Distribution of C1–C3 neuronal cell bodies in the medulla oblongata and their terminals in the intermediolateral cell column of the spinal cord.** Fluorescence photomicrographs of (A) a coronal section through the rostral medulla oblongata, (B) a coronal section through the thoracic T3–4 spinal cord, and (C) a horizontal section through the thoracic T3–4 spinal cord. (A) shows the distribution of tyrosine hydroxylase (TH) immunoreactivity in purple. A subgroup of the C1 neurons on the right of the section are also labeled with green fluorescent protein (GFP) following an injection of a lentivirus inducing expression of GFP under the control of a phox2-selective promoter (adrenergic neurons express phox2). Most of the dorsal midline C3 neurons have been labeled with a red fluorescent protein (tandem dimer tomato (tdT)) following an injection of a lentivirus-inducing expression of tdT under the control of a phox2-selective promoter. The rostral extent of C2 neurons are shown as TH-expressing only (purple) lateral to the red C3 neurons. (B) and (C) show immunoreactivity for choline acetyltransferase (blue) in the sympathetic preganglionic neurons of the intermediolateral cell column. Axon terminals from C3 (red) and C1 (green) neurons are shown in close proximity to these sympathetic preganglionic neurons. In each case the scale bar represents 50 μm.

noncatecholaminergic neurons occur in the RTN, at the rostral extent of the C1 region, and many neurons within the boundary of the NTS, intermingled with the C2 cell group. Technically it is possible to obtain minimal RTN expression when transducing C1 neurons; however, such selectivity is not possible for the C2 neurons and as a consequence, their projections have not been systematically mapped.

Initial studies indicated that the column of C1 neurons could be roughly divided into two compartments on the basis of axonal projections. The majority of spinally projecting C1 neurons were localized to the rostral part of the C1 column, while the hypothalamic-projecting C1 neurons were located more caudally.[58] Subsequent triple-labeling studies indicated that many C1 neurons had axonal collaterals, with a substantial proportion of C1 neurons projecting to the rostral pons and medulla also having collateral projections to the spinal cord.[32] A smaller proportion of hypothalamic-projecting C1 neurons had spinal collaterals. As detailed in Table 3.1, the C1 neurons have extensive projections throughout the CNS.[12] These include the well-characterized projection of the rostral C1 neurons to sympathetic preganglionic neurons of the thoracic and lumbar spinal cord, as well as some other regions in the spinal cord—including the central autonomic area.[12,44,52] As an overall summary, it is clear that most regions throughout the CNS that were

TABLE 3.1   Distribution of PNMT-ir Fibers as Described by Hökfelt and Colleagues[33] (Overall Density Column), C1 Neuron Fibers as Described by Card and Colleagues[12] (C1 Column), and C3 Fibers as Described by Sevigny and Colleagues[53] Throughout the Rat Central Nervous System

			PNMT	C1	C2	C3
Telencephalon		Anterior olfactory N.	+			+
		Septal N.	+			+
		N. tractus diagonalis	+			+
		Bed N. of stria terminalis	+++			+
		Central amygdaloid N.	+		#	×
		Medial amygdaloid N.	++			++
Diencephalon	Hypothalamus	OVLT	+++			++
		Suprachiasmatic N.	+++			++
		Arcuate N.	++	*		++
		Lateral hypothalamus	++	*		+
		Perifornical area	++	*		+
		Medial tuberal N. and terete N.	++++	*		+
		Paraventricular N.	++++	*		+++
		Periventricular N.	+++			+++
		Dorsomedial N.	++	*		++
		Medial preoptic N.	++	*		++
		Anterior hypothalamic N.	+		#	×
		Ventromedial N.	+		#	×
		Retrochiasmatic N.	++			++
		Median eminence	++	*		+
		Supraoptic N.	+	*		++
	Thalamus	Periventricular N.	++++	*		++++
		Rhomboid N.	+++			+++
	Other	Supramammillary region	++		#	×
Mesencephalon and pons		Lateral periaqueductal gray	+++	*		++
		Ventrolateral periaqueductal gray	+++	*		++++
		Substantia nigra (zona compacta)	+		#	×
		Ventral tegmental area	++			++
		Dorsal raphé	++			++
		Central superior N.	++		#	×
		Pontine reticular formation	+		#	×
		Locus coeruleus	+++	*		++
		A7/Kölliker-Fuse N.	+++			++
		Parabrachial N. (dorsal)	+++			+++
		Superior olive	++		#	×
		Corpus trapezoideum	++		#	×
		A5 N.	+++		#	×

TABLE 3.1 Distribution of PNMT-ir Fibers as Described by Hökfelt and Colleagues[33] (Overall Density Column), C1 Neuron Fibers as Described by Card and Colleagues[12] (C1 Column), and C3 Fibers as Described by Sevigny and Colleagues[53] Throughout the Rat Central Nervous System—cont'd

			PNMT	C1	C2	C3
Cerebellum		Purkinje layer	++		#	×
Medulla oblongata		Reticular gigantocellular N.	+		#	×
		Raphé pallidus	+++	*		+
		Raphé magnus	+++			+
		Pyramidal tract	++		#	×
		Raphé obscurus	++	*		+
		Dorsal motor N. of the vagus	++++	*		+++
		Area postrema	++++	*		++++
		N. of the solitary tract (commissural)	++++	*		+++
		N. of the solitary tract (medial)	+++	*		++++
		N. intercalatus	++		#	×
		Caudal ventrolateral medulla (A1)	+++	*		++
		Rostral ventrolateral medulla	++	*		++
		Reticular N.	+	*		×
		Lateral reticular N.	+	*		×
		Decussation of the pyramids	+			×
		N. ambiguus		*		×
Spinal cord	Cervical	Lamina X	++			++
	Thoracic	Lamina X	++++	*		++++
		Lamina IX (IML)	++++	*		++++
	Lumbar	Lamina X	++			++
	Sacral	Lamina X	++			+

Densities of PNMT-immunoreactive fibers and C3 inputs are semiquantified, ranging from lowest (+) to highest (++++) densities. Blank spaces in the C1 column indicate areas not reported as having inputs from C1. # highlight areas that presumably receive an input only from C2, although this remains to be tested. × denotes areas examined and found not to have inputs from C3. *IML*, intermediolateral cell column; *N.*, nucleus; *OVLT*, organum vasculosum of the lamina terminalis; *PNMT*, phenylethanolamine-N-methyl transferase.

reported to receive dense PNMT inputs, receive projections from C1 neurons. In addition to the intermediolateral cell column (IML), the highest densities of C1 input occurred in the hypothalamic para- and periventricular nuclei, the paraventricular nucleus of the thalamus, the dorsal motor nucleus of the vagus, the NTS, and the area postrema.[12] The exceptions, being PNMT-immunoreactive areas without C1 or C3 inputs, are perhaps the more informative and are discussed below in relation to C2 neuron projections.

Like the C1 neurons, the C3 neurons show a topographic organization based upon their axonal projections. Using standard retrograde tracing techniques, neurons in the median and medial C3 columns were found to project to the spinal cord, while rostrally projecting neurons were located in the dorsal and lateral C3 regions.[44] Dual rostral- and spinal-projecting neurons were not observed. Despite

consisting of only a relatively small number of cells, viral tracing studies indicate that the projections of C3 neurons cover most of the CNS, from the anterior olfactory nucleus rostrally, to the sacral spinal cord caudally.[53] A comprehensive list, with a semiquantitative index of terminal density, is provided in Table 3.1, with comparison to the C1 and overall PNMT distributions. Most regions receiving a C1 input also receive a C3 input. Thus all diencephalic regions receiving C1 inputs receive C3 inputs, including the lateral hypothalamus, perifornical area, and the arcuate, medial tuberal, paraventricular, dorsomedial, medial preoptic, and supraoptic nuclei of the hypothalamus. This codistribution extends caudally to the lateral and ventrolateral parts of the periaqueductal gray, locus coeruleus, raphe pallidus and obscurus, rostral and caudal ventrolateral medulla, dorsal motor nucleus of the vagus, NTS, area postrema, and IML. In fact, individual preganglionic neurons in the IML receive both C1 and C3 inputs.[44] Regions receiving C3 neuron inputs but not C1 inputs are rare and must be considered carefully as the work of Card and colleagues[12] does not definitively state that C1 terminals were not in these regions. However, notable C3 inputs, in the absence of reported C1 inputs, occur in the medial amygdala, suprachiasmatic, retrochiasmatic, and periventricular nuclei of the diencephalon, thalamic rhomboid nucleus, and Kölliker-Fuse and dorsal parabrachial nuclei in the pons and lamina X at all levels of the spinal cord.

While the projections of C2 neurons have not been described, Table 3.1 provides some information suggestive of regions that might receive inputs from these neurons only. Whether C2 neurons also project to regions receiving C1/C3 inputs will require a specific study. Key regions which have PNMT-immunoreactive terminals but do not receive inputs from C1 or C3 neurons include the central nucleus of the amygdala, anterior and ventromedial hypothalamus, the supramammillary region, the superior olive and the A5 nucleus in the pons, the Purkinje layer of the cerebellar cortex, and the nucleus intercalatus in the medulla oblongata. However, these exceptions may just be due to technical limitations. For example, using a viral construct with PRSx8 promoter, Abbott and colleagues expressed fluorescently labeled channel rhodopsin in C1 neurons and showed activation of A5 neurons.[3] However, even this elegant approach is not definitive as the fluorescent terminals surrounding A5 neurons did not appear to contain TH and so may have come from the small proportion on non-C1 neurons labeled by this approach.

NEUROCHEMISTRY

In addition to expressing the enzymes required for synthesis of adrenaline, most C1–C3 neurons express the vesicular monoamine transporter 2 (VMAT2).[52,44] While our understanding of the role of catecholamines as transmitters released from these neurons remains unclear, the presence of VMAT2 supports a transmitter function. Interestingly VMAT2 immunoreactivity is not detectable in the somata of most bulbospinal C1 neurons, yet detectable at the axon terminal in the IML.[52] Rostrally projecting C1 neurons do express detectable levels of VMAT2-immunoreactivity in their somata. A possible implication of this observation might be that expression of VMAT2-immunoreactivity reflects somatic versus synaptic release of monoamines.

The primary transmitter employed by C1–C3 neurons is glutamate and more than 70% of C1–C3 neurons express vesicular glutamate transporter 2 (VGlut2)—the remainder probably just expressing levels below the level of detectability of the hybridization method.[57]

Effectively all C1 and C3 neurons with rostrally projecting axons express mRNA for neuropeptide Y

(NPY).[54] A smaller proportion of spinally projecting C3 (~60%) and C1 (~10%) neurons express NPY mRNA and expression in C2 neurons is very rare. Most C1, C2, and C3 neurons express cocaine- and amphetamine-related transcript[10,20] and very high proportions, greater than 80%, express pituitary adenylate cyclase-activating polypeptide.[23] Approximately 20% of C1–C3 neurons express mRNA for pre-protachykinin, the precursor for substance P,[37] and approximately 30–40% express calbindin.[26,27]

The roles of these cotransmitters are not yet clearly determined. While many of the cotransmitters in bulbospinal C1 neurons, for example, can alter the activity of SPNs, the physiological or pathological circumstances under which they might be released to modulate function have not been identified.

Studies on the expression of receptors for different neurotransmitters and modulators have largely been confined to C1 neurons. However, such information remains relatively scarce as the antibodies for identifying receptors, particularly G protein-coupled receptors, suffer from issues related to nonselectivity and the level of mRNA expression tends to be very low making in situ hybridization histochemistry difficult. Very few studies have examined the responses of neurons in the C1–C3 regions to different neuromodulators and then labeled the responsive neurons, using methods such juxtacellular labeling, to determine their adrenergic phenotype. Angiotensin activates neurons in the C1 region via its type 1 receptor (AT$_1$R).[15,39] Using a transgenic mouse, in which the expression of GFP is driven by the AT$_{1A}$R promoter, it has been possible to show that most C1 neurons express this receptor.[16] In humans, angiotensin receptor binding occurs on the pigmented, presumably catecholaminergic, neurons of the RVLM.[6] Extensive studies have demonstrated that C1 and C3 neurons express α_{2A} receptors[29,55] and that activation of these receptors inhibits the activity of C1 neurons. Microinjection of somatostatin into the RVLM (C1 region) inhibits sympathetic nerve activity via action of somatostatin 2A receptors.[7,9] Approximately 50% of bulbospinal C1 neurons express somatostatin 2A receptors.

FUNCTIONS

Knowledge of the functions of adrenergic neurons is again mostly confined to C1 neurons. As detailed in the next section, there is more information about stimuli that activate C2 and C3 neurons, admittedly mostly because they occur in histological sections used to examine C1 responses, but only one direct study examining selective activation of C3 neurons.[44] Using cell-selective viral transduction and expression of optically activated ion channels, we demonstrated that C3 activation induces increased sympathetic activity, perfusion pressure, and heart rate.

Nonselective activation of the RVLM induces a strong, sympathetically mediated increase in blood pressure,[17] while inactivation or lesion of the region attenuates ongoing sympathetic vasomotor activity and reduces blood pressure to spinal levels.[18,28,50] The predominant physiological response to selective activation of C1 neurons, using optogenetics, is also an increase in sympathetic vasomotor activity,[4,5] with sleep-state-dependent cardiorespiratory arousal in conscious rats.[1,8] In conscious mice, selective stimulation of C1 neurons results in respiratory activation and only modest cardiovascular changes due to concurrent stimulation of sympathetic and parasympathetic efferent activity.[2] Given that C1 neurons comprise a significant proportion of RVLM bulbospinal neurons, it was surprising that selective lesions of C1 neurons, using the antibody-directed toxin anti-DBH conjugated saporin, did not dramatically alter resting arterial pressure.[40,41,51] Subsequent microinjection of the GABA$_A$ receptor agonist, muscimol, into the RVLM still decreased blood pressure to spinal levels indicating that remnant, non-C1 neurons in the RVLM play a key role in the maintenance of sympathetic vasomotor tone in the absence of C1 neurons. Following selective C1 neuron lesion, attenuated chemoreceptor, baroreceptor, and Bezold–Jarisch reflex regulation of sympathetic nerve activity was observed. Using viral transduction and pharmacogenetics to acutely inhibit C1 neurons selectively, a more significant decrease in sympathetic nerve activity and blood pressure is observed,[42] suggesting that lesion studies underestimate the relative contribution of C1 neurons to blood pressure control. Possibly during the slow time course of lesion development, the homeostatic maintenance of blood pressure becomes more heavily reliant upon the activity of non-C1 RVLM neurons.

While the predominant observation associated with excitation of C1 neurons is cardiorespiratory activation, it is unlikely that this represents the only function of these neurons. Optogenetic stimulation of C1 neurons increases parasympathetic activity in conscious mice[2]; directly stimulates parasympathetic preganglionic neurons in the dorsal motor nucleus of the vagus[19]; and activates noradrenergic neurons in the locus coeruleus and A5 region.[3,34] Given the widespread projections of C1 neurons, both to other medullary regions and more rostral sites, these studies are likely to represent the tip of the iceberg in terms of understanding the functional relevance of this cell group.

ACTIVATING STIMULI

The most relevant information about the stimuli that activate C1–C3 neurons selectively, as opposed to neurons of unknown neurochemistry within a particular anatomical region, comes from the use of approaches to measure increased expression of the immediate early gene, *c-fos*. Most often the protein product, Fos, is visualized by immunohistochemistry and used as a surrogate marker of neuronal activation. Using this approach it has been demonstrated that C1 neurons are activated by multiple physiological stressors, including hypotension,[14,38] hypoxia,[21] hypoglycemia,[49] hypovolemia,[48] hypotensive hemorrhage,[14] the diving response,[43] and immune challenges.[22] Interestingly while electric footshock induces activation of C1 neurons,[36] other aversive stressors, such as air jet stress[25] and conditioned fear[13] do not activate C1 neurons. This activation of C1 neurons by a diverse range of stressors, combined with the widespread distribution of excitatory projections throughout the neuraxis that arise from these cells, has led Guyenet and colleagues to label them as emergency medical technicians.[30]

In contrast to studies examining C1 neuronal activation specifically, few studies have examined C2 and C3 neurons, and these only tend to have been in passing as they were within the same sections as C1 neurons. For example, Chan and Sawchenko noted that a small proportion of neurons in the rostral NTS that were activated by hypotensive hemorrhage were catecholaminergic.[14]

Initial studies considered that C3 neurons might be an ectopic group of C1 neurons, but it has since been shown that they have different response characteristics reflective of different inputs. While C1 neurons are activated by both hypotension and glucoprivation, C3 neurons are not activated by hypotension and are by glucoprivation.[44]

CONCLUSION

While generally studied within the context of cardiovascular regulation, and specifically sympathetic vasomotor regulation, it is clear that the C1–C3 group of neurons is more diverse. Their axons project throughout the CNS and have the potential to affect multiple circuits, sending information related to multiple physiological stressors and homeostatic processes. In general our understanding relates to the C1 group, with the C2 and C3 neurons as forgotten appendages; however, these groups do seem to have selectivity in both their projections and in the stimuli that activate them. Teasing apart the differential roles for these groups is now more feasible, with the advent of relatively specific viral technologies and a multitude of molecules that can be expressed to modify neuronal activity. While shaded by their noradrenergic cousins in the locus coeruleus, we predict that the global role of the adrenergic neurons in modulating responses to physiological stressors is on the cusp of being fully appreciated.

References

1. Abbott SB, Coates MB, Stornetta RL, Guyenet PG. Optogenetic stimulation of c1 and retrotrapezoid nucleus neurons causes sleep state-dependent cardiorespiratory stimulation and arousal in rats. *Hypertension*. 2013;61:835–841.

2. Abbott SB, DePuy SD, Nguyen T, Coates MB, Stornetta RL, Guyenet PG. Selective optogenetic activation of rostral ventrolateral medullary catecholaminergic neurons produces cardiorespiratory stimulation in conscious mice. *J Neurosci*. 2013;33:3164–3177.

3. Abbott SB, Kanbar R, Bochorishvili G, Coates MB, Stornetta RL, Guyenet PG. C1 neurons excite locus coeruleus and A5 noradrenergic neurons along with sympathetic outflow in rats. *J Physiol*. 2012;590:2897–2915.

4. Abbott SB, Stornetta RL, Fortuna MG, et al. Photostimulation of retrotrapezoid nucleus phox2b-expressing neurons in vivo produces long-lasting activation of breathing in rats. *J Neurosci*. 2009;29:5806–5819.

5. Abbott SB, Stornetta RL, Socolovsky CS, West GH, Guyenet PG. Photostimulation of channel rhodopsin-2 expressing ventrolateral medullary neurons increases sympathetic nerve activity and blood pressure in rats. *J Physiol*. 2009;587:5613–5631.

6. Allen AM, Chai SY, Clevers J, McKinley MJ, Paxinos G, Mendelsohn FA. Localization and characterization of angiotensin II receptor binding and angiotensin converting enzyme in the human medulla oblongata. *J Comp Neurol*. 1988;269:249–264.

7. Bou Farah L, Bowman BR, Bokiniec P, et al. Somatostatin in the rat rostral ventrolateral medulla: origins and mechanism of action. *J Comp Neurol*. 2016;524:323–342.

8. Burke PG, Abbott SB, Coates MB, Viar KE, Stornetta RL, Guyenet PG. Optogenetic stimulation of adrenergic C1 neurons causes sleep state-dependent cardiorespiratory stimulation and arousal with sighs in rats. *Am J Respir Crit Care Med*. 2014;190:1301–1310.

9. Burke PG, Li Q, Costin ML, McMullan S, Pilowsky PM, Goodchild AK. Somatostatin 2A receptor-expressing presympathetic neurons in the rostral ventrolateral medulla maintain blood pressure. *Hypertension*. 2008;52:1127–1133.

10. Burman KJ, Sartor DM, Verberne AJ, Llewellyn-Smith IJ. Cocaine- and amphetamine-regulated transcript in catecholamine and non-catecholamine presympathetic vasomotor neurons of rat rostral ventrolateral medulla. *J Comp Neurol*. 2004;476:19–31.

11. Card JP, Lois J, Sved AF. Distribution and phenotype of Phox2a-containing neurons in the adult sprague-dawley rat. *J Comp Neurol*. 2010;518:2202–2220.

12. Card JP, Sved JC, Craig B, Raizada M, Vazquez J, Sved AF. Efferent projections of rat rostroventrolateral medulla C1 catecholamine neurons: implications for the central control of cardiovascular regulation. *J Comp Neurol*. 2006;499:840–859.

13. Carrive P, Gorissen M. Premotor sympathetic neurons of conditioned fear in the rat. *Eur J Neurosci*. 2008;28:428–446.

14. Chan RK, Sawchenko PE. Spatially and temporally differentiated patterns of c-fos expression in brainstem catecholaminergic cell groups induced by cardiovascular challenges in the rat. *J Comp Neurol*. 1994;348:433–460.

15. Chen D, Bassi JK, Walther T, Thomas WG, Allen AM. Expression of angiotensin type 1A receptors in C1 neurons restores the sympathoexcitation to angiotensin in the rostral ventrolateral medulla of angiotensin type 1A knockout mice. *Hypertension*. 2010;56:143–150.

16. Chen D, Jancovski N, Bassi JK, et al. Angiotensin type 1A receptors in C1 neurons of the rostral ventrolateral medulla modulate the pressor response to aversive stress. *J Neurosci*. 2012;32:2051–2061.

17. Dampney RA, Goodchild AK, Robertson LG, Montgomery W. Role of ventrolateral medulla in vasomotor regulation: a correlative anatomical and physiological study. *Brain Res*. 1982;249:223–235.

18. Dampney RA, Moon EA. Role of ventrolateral medulla in vasomotor response to cerebral ischemia. *Am J Physiol*. 1980;239:H349–H358.

19. DePuy SD, Stornetta RL, Bochorishvili G, et al. Glutamatergic neurotransmission between the C1 neurons and the parasympathetic preganglionic neurons of the dorsal motor nucleus of the vagus. *J Neurosci*. 2013;33:1486–1497.

20. Dun SL, Ng YK, Brailoiu GC, Ling EA, Dun NJ. Cocaine- and amphetamine-regulated transcript peptide-immunoreactivity in adrenergic C1 neurons projecting to the intermediolateral cell column of the rat. *J Chem Neuroanat*. 2002;23:123–132.

21. Erickson JT, Millhorn DE. Hypoxia and electrical stimulation of the carotid sinus nerve induce Fos-like immunoreactivity within catecholaminergic and serotoninergic neurons of the rat brainstem. *J Comp Neurol*. 1994;348:161–182.

22. Ericsson A, Kovacs KJ, Sawchenko PE. A functional anatomical analysis of central pathways subserving the effects of interleukin-1 on stress-related neuroendocrine neurons. *J Neurosci*. 1994;14:897–913.

23. Farnham MM, Li Q, Goodchild AK, Pilowsky PM. PACAP is expressed in sympathoexcitatory bulbospinal C1 neurons of the brain stem and increases sympathetic nerve activity in vivo. *Am J Physiol Regul Integr Comp Physiol*. 2008;294:R1304–R1311.

24. Foster GA, Hokfelt T, Coyle JT, Goldstein M. Immunohistochemical evidence for phenylethanolamine-*N*-methyltransferase-positive/tyrosine hydroxylase-negative neurones in the retina and the posterior hypothalamus of the rat. *Brain Res*. 1985;330:183–188.

25. Furlong TM, McDowall LM, Horiuchi J, Polson JW, Dampney RA. The effect of air puff stress on c-Fos expression in rat hypothalamus and brainstem: central circuitry mediating sympathoexcitation and baroreflex resetting. *Eur J Neurosci*. 2014;39:1429–1438.

26. Goodchild AK, Llewellyn-Smith IJ, Sun QJ, Chalmers J, Cunningham AM, Pilowsky PM. Calbindin-immunoreactive neurons in the reticular formation of the rat brainstem: catecholamine content and spinal projections. *J Comp Neurol*. 2000;424:547–562.

27. Granata AR, Chang HT. Relationship of calbindin D-28k with afferent neurons to the rostral ventrolateral medulla in the rat. *Brain Res*. 1994;645:265–277.

28. Guertzenstein PG, Silver A. Fall in blood pressure produced from discrete regions of the ventral surface of the medulla by glycine and lesions. *J Physiol*. 1974;242:489–503.

29. Guyenet PG, Lynch KR, Allen AM, Rosin DL, Stornetta RL. Alpha2-adrenergic receptors rather than imidazoline binding sites mediate the sympatholytic effect of clonidine in the rostral ventrolateral medulla. In: Trouth CO, Millis RM, Kiwull-Schone H, Schlafke ME, eds. *Ventral Brainstem Mechanisms and Control of Respiration and Blood Pressure*. New York: Marcel Dekker, Inc.; 1995:281–303.

30. Guyenet PG, Stornetta RL, Bochorishvili G, Depuy SD, Burke PG, Abbott SB. C1 neurons: the body's EMTs. *Am J Physiol Regul Integr Comp Physiol*. 2013;305:R187–R204.

31. Halliday GM, Li YW, Joh TH, Cotton RG, Howe PR, Geffen LB, et al. Distribution of monoamine-synthesizing neurons in the human medulla oblongata. *J Comp Neurol*. 1988;273:301–317.

32. Haselton JR, Guyenet PG. Ascending collaterals of medullary barosensitive neurons and C1 cells in rats. *Am J Physiol*. 1990;258:R1051–R1063.

33. Hökfelt T, Johansson O, Goldstein M. Central catecholamine neurons as revealed by immunohistochemistry with special reference to adrenaline neurons. In: Björklund A, Hokfelt T, eds. *Handbook of Chemical Neuroanatomy*. Classical Transmitters in the CNS, Part 1; vol. 2. Amsterdam: Elsevier; 1984:157–276.

34. Holloway BB, Stornetta RL, Bochorishvili G, Erisir A, Viar KE, Guyenet PG. Monosynaptic glutamatergic activation of locus coeruleus and other lower brainstem noradrenergic neurons by the C1 cells in mice. *J Neurosci*. 2013;33:18792–18805.

35. Hwang DY, Hwang MM, Kim HS, Kim KS. Genetically engineered dopamine beta-hydroxylase gene promoters with better PHOX2-binding sites drive significantly enhanced transgene expression in a noradrenergic cell-specific manner. *Mol Ther*. 2005;11:132–141.

36. Li HY, Ericsson A, Sawchenko PE. Distinct mechanisms underlie activation of hypothalamic neurosecretory neurons and their medullary catecholaminergic afferents in categorically different stress paradigms. *Proc Natl Acad Sci USA*. 1996;93:2359–2364.

37. Li Q, Goodchild AK, Seyedabadi M, Pilowsky PM. Pre-protachykinin A mRNA is colocalized with tyrosine hydroxylase-immunoreactivity in bulbospinal neurons. *Neuroscience*. 2005;136:205–216.

38. Li YW, Dampney RA. Expression of Fos-like protein in brain following sustained hypertension and hypotension in conscious rabbits. *Neuroscience*. 1994;61:613–634.

39. Li YW, Guyenet PG. Angiotensin II decreases a resting K+ conductance in rat bulbospinal neurons of the C1 area. *Circ Res*. 1996;78: 274–282.

40. Madden CJ, Ito S, Rinaman L, Wiley RG, Sved AF. Lesions of the C1 catecholaminergic neurons of the ventrolateral medulla in rats using anti-DbetaH-saporin. *Am J Physiol*. 1999;277:R1063–R1075.

41. Madden CJ, Sved AF. Cardiovascular regulation after destruction of the C1 cell group of the rostral ventrolateral medulla in rats. *Am J Physiol Heart Circ Physiol*. 2003;285:H2734–H2748.

42. Marina N, Abdala AP, Korsak A, et al. Control of sympathetic vasomotor tone by catecholaminergic C1 neurones of the rostral ventrolateral medulla oblongata. *Cardiovasc Res*. 2011;91:703–710.

43. McCulloch PF, Panneton WM. Activation of brainstem catecholaminergic neurons during voluntary diving in rats. *Brain Res*. 2003;984:42–53.

44. Menuet C, Sevigny CP, Connelly AA, et al. Catecholaminergic C3 neurons are sympathoexcitatory and involved in glucose homeostasis. *J Neurosci*. 2014;34:15110–15122.

45. Osborne NN, Nesselhut T. Adrenaline: occurrence in the bovine retina. *Neurosci Lett*. 1983;39:33–36.

46. Paxinos G, Franklin KB. *The Mouse Brain in Stereotaxic Coordinates*. Academic Press; 2012.

47. Paxinos G, Watson C. *The Rat Brain in Stereotaxic Coordinates*. Academic Press; 2006.

48. Potts PD, Ludbrook J, Gillman-Gaspari TA, Horiuchi J, Dampney RA. Activation of brain neurons following central hypervolaemia and hypovolaemia: contribution of baroreceptor and non-baroreceptor inputs. *Neuroscience*. 2000;95:499–511.

49. Ritter S, Llewellyn-Smith I, Dinh TT. Subgroups of hindbrain catecholamine neurons are selectively activated by 2-deoxy-D-glucose induced metabolic challenge. *Brain Res*. 1998;805:41–54.

50. Ross CA, Ruggiero DA, Joh TH, Park DH, Reis DJ. Adrenaline synthesizing neurons in the rostral ventrolateral medulla: a possible role in tonic vasomotor control. *Brain Res*. 1983;273:356–361.

51. Schreihofer AM, Stornetta RL, Guyenet PG. Regulation of sympathetic tone and arterial pressure by rostral ventrolateral medulla after depletion of C1 cells in rat. *J Physiol*. 2000;529(Pt 1):221–236.

52. Sevigny CP, Bassi J, Teschemacher AG, et al. C1 neurons in the rat rostral ventrolateral medulla differentially express vesicular monoamine transporter 2 in soma and axonal compartments. *Eur J Neurosci*. 2008;28:1536–1544.

53. Sevigny CP, Bassi J, Williams DA, Anderson CR, Thomas WG, Allen AM. Efferent projections of C3 adrenergic neurons in the rat central nervous system. *J Comp Neurol*. 2012;520:2352–2368.

54. Stornetta RL, Akey PJ, Guyenet PG. Location and electrophysiological characterization of rostral medullary adrenergic neurons that contain neuropeptide Y mRNA in rat medulla. *J Comp Neurol*. 1999;415:482–500.

55. Stornetta RL, Huangfu D, Rosin DL, Lynch KR, Guyenet PG. Alpha-2 adrenergic receptors. Immunohistochemical localization and role in mediating inhibition of adrenergic RVLM presympathetic neurons by catecholamines and clonidine. *Ann N Y Acad Sci*. 1995;763:541–551.

56. Stornetta RL, Moreira TS, Takakura AC, et al. Expression of Phox2b by brainstem neurons involved in chemosensory integration in the adult rat. *J Neurosci*. 2006;26:10305–10314.

57. Stornetta RL, Sevigny CP, Guyenet PG. Vesicular glutamate transporter DNPI/VGLUT2 mRNA is present in C1 and several other groups of brainstem catecholaminergic neurons. *J Comp Neurol*. 2002;444:191–206.

58. Tucker DC, Saper CB, Ruggiero DA, Reis DJ. Organization of central adrenergic pathways: I. Relationships of ventrolateral medullary projections to the hypothalamus and spinal cord. *J Comp Neurol*. 1987;259:591–603.

59. Weber F, Chung S, Beier KT, Xu M, Luo L, Dan Y. Control of REM sleep by ventral medulla GABAergic neurons. *Nature*. 2015;526:435–438.

I. NEUROENDOCRINE CONTROL OF THE STRESS RESPONSE

4

Noradrenergic Control of Arousal and Stress

C.W. Berridge, R.C. Spencer

University of Wisconsin–Madison, Madison, WI, United States

Abstract

Stress impacts a variety of behavioral processes, including arousal and higher cognitive processes. Central noradrenergic systems are highly sensitive to stress and other high-arousal states. Via extensive axon collaterals, brainstem noradrenergic neurons innervate nearly the entire brain, positioning them to participate broadly in behavioral responding in stress. Consistent with this, activation of central noradrenergic signaling elicits an array of stresslike effects, including elevated arousal, impaired prefrontal-dependent cognition, strengthened aversive memories, and impaired preattentive filtering of sensory information. These actions involve noradrenergic α_1- and β-receptor signaling in a diversity of cortical and subcortical regions. Finally, stress-related psychopathology is associated with alterations in norepinephrine (NE)-sensitive behavioral and cognitive function. Evidence suggests that neuroplasticity within both noradrenergic systems and noradrenergic projection targets likely contribute to the enduring nature of stress-related changes in behavior associated with stress-related disorders. These, and other observations, indicate a prominent role of brain NE in behavioral responding in stress and stress-related psychopathology.

Our current view of stress as a behavioral and physiological state elicited by challenging or threatening events arises from nearly a century of research starting with the seminal work of Cannon[12] and Selye.[34] While the psychological features that comprise the state of stress are unclear, a heightened level of readiness for action is a core aspect. Elevated arousal is a prominent component of this preparatory state, defined for the purposes of this review as a heightened sensitivity to environmental stimuli. Importantly, both aversive and appetitive conditions can elicit an activation of prototypical stress systems, leading to the suggestion of both good (eustress) and bad (distress) forms of stress (for review, see Ref. 23). This suggests the working hypothesis that at least a subset of the physiological indices of stress may be independent of affective valence (pleasant vs. unpleasant) and more closely aligned with arousal level, motivational state, and/or the need for action. While the neural bases of stress-related behavior are not fully understood, significant evidence indicates that noradrenergic systems are uniquely sensitive to stress and that these systems contribute to a broad spectrum of behavioral processes in stress and stress-related psychopathology.

KEY POINTS

- Central noradrenergic systems arise from a network of small brainstem nuclei.

- Noradrenergic systems are robustly activated in stress.

- Stress-related activation of central noradrenergic signaling at α_1 and β receptors contributes to a diversity of behaviors in stress, including elevated arousal, impaired prefrontal cognition, strengthened memory for aversive events, and elevated startle.

- Neural plasticity within norepinephrine (NE) systems as well as NE-dependent plasticity in noradrenergic target regions likely contributes to long-term behavioral dysfunction associated with stress-related psychopathology, including posttraumatic stress disorder.

- These observations indicate that noradrenergic signaling is a potentially useful target for the development of novel pharmacological treatment for stress-related behavioral dysfunction.

AROUSAL-PROMOTING ACTIONS OF CENTRAL NOREPINEPHRINE

Locus Coeruleus-Noradrenergic Modulation of Arousal

There are multiple noradrenergic nuclei, with the locus coeruleus (LC) the largest source of brain norepinephrine (NE)[27] (for review, see Ref. 9). The LC is comprised of a small cluster of NE-synthesizing neurons located in the pontine brainstem adjacent to the fourth ventricle (~1500/hemisphere in the rat and 15,000–30,000 in primates).[9] Despite its restricted size, LC neurons extend immensely ramified axons that project widely throughout the neuraxis.[20] Although early observations suggested the LC was the sole source of the noradrenergic innervation of the hippocampus and neocortex, recent studies indicate other noradrenergic nuclei also provide input to select neocortical sites, including the prefrontal cortex (PFC[32]).

Early electrophysiological recordings demonstrated LC neurons evince state-dependent firing, with higher discharge rates in waking than sleep.[19] Importantly, changes in LC neuronal discharge rate precede sleep–wake state transitions suggesting a potentially causal role in the regulation of arousal.[19] Within waking, LC neurons increase discharge rates during periods of elevated arousal, including those associated with reward or stress[19] (for review, see Ref. 9).

NE binds to three major receptor families, α_1, α_2, and β, each comprised of multiple subtypes, as well as the dopamine D4 receptor.[9,28] α_1 and β receptors are thought to exist primarily postsynaptically, whereas α_2 receptors are present both pre- and postsynaptically. Early observations demonstrated sedative effects of drugs that suppress NE signaling (i.e., α_2 agonists and α_1- and β-receptor antagonists; for review, see Ref. 8). However, these approaches fail to identify which noradrenergic nuclei participate in the regulation of arousal. Subsequent research used electrophysiological recordings to guide small infusions (35–150 nL) of drugs immediately adjacent to the LC to examine the arousal-modulating actions of *selective* activation and suppression of LC neuronal activity in anesthetized rats. Using this approach, selective activation of the LC elicited a robust and waking-like activation of forebrain electroencephalographic (EEG) activity that closely tracked the time course of LC activation (Fig. 4.1A).[4] Conversely, pharmacological suppression of LC activity bilaterally elicited a robust increase in EEG indices of sedation (e.g., increased slow-wave activity) in lightly anesthetized rats which also closely tracked drug-induced suppression of LC activity (Fig. 4.1B).[7] More recently, similar results were obtained with optogenetic activation and suppression of LC in unanesthetized animals.[14] Specifically, brief optogenetic stimulation of the LC (1–10 s) elicited rapid transitions from sleep to waking and, within waking, prolonged time spent awake. Conversely, 1-h optogenetic inhibition of the LC decreased time spent awake.

These and other observations demonstrate that LC activity is both *sufficient* and *necessary* for alert waking.

Circuitry: α_1 and β Receptors Promote Arousal in a Network of Subcortical Regions

Subcortically, the medial septal area (MSA), the substantia innominata (SI), the medial preoptic area (MPOA), and the lateral hypothalamus (LH; including LH proper, the dorsomedial hypothalamus, and the perifornical area) participate in the regulation of arousal (for review, see Ref. 9). Each of these regions receives LC-noradrenergic input.[8] In a series of studies aimed at identifying the neurocircuitry and receptor mechanisms involved in the arousal-promoting actions of NE, the wake-promoting effects of α_1- or β-receptor activation in the MSA, the MPOA, or the LH were examined in unanesthetized, sleeping animals. It was observed

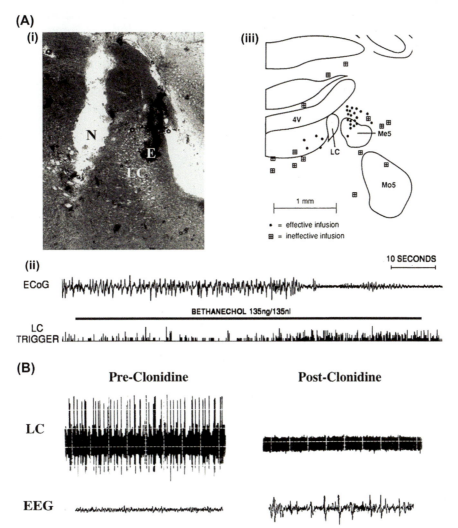

FIGURE 4.1 Effects of locus coeruleus (LC) activation on cortical encephalographic (EEG) (*ECoG*) activity state. Electrophysiological recordings were used to locate the LC and place small (100–150 nL) infusions of drugs at varying distances from the LC in anesthetized rats. Drug effects on both LC neuronal activity and cortical ECoG were simultaneously recorded. (A) Effects of LC activation on EEG indices of arousal. Panels i–iii depict results obtained with peri-LC infusions of the cholinergic agonist, bethanechol, to activate LC neurons. (i) Photomicrograph of a peri-LC infusion site. In this example, the infusion needle was placed lateral of the LC. *N* indicates position within the track created by the infusion needle where the infusate exited the needle. *E* indicates position of the recording electrode within the LC. (ii) Effects of bethanechol infusion on LC activity (*LC Trigger*, bottom trace) and ECoG activity (top trace). Bethanechol infusion (indicated by horizontal bar) increased LC discharge rate approximately two-thirds of the way through the 60-s infusion and several seconds following this EEG activation/desynchronization is observed. EEG recordings were ipsilateral to the manipulated LC. However, identical effects were observed in the contralateral hemisphere. The return of synchronized EEG activity over the subsequent 5–15 min closely followed the return of LC activity rates to preinfusion levels. (iii) Schematic depicting bethanechol infusion sites that were effective or ineffective for EEG activation. Solid circles and shaded boxes indicate sites at which bethanechol infusion either activated or had no effect on EEG, respectively. There is a radius of approximately 500 μm around LC within which infusions, placed either medially or laterally, activated forebrain EEG. Infusions placed immediately anterior to the LC were also ineffective (not shown). These mapping studies strongly argue against drug action at other arousal-related brainstem nuclei (e.g., cholinergic, serotonergic). *Me5*, mesencephalic nucleus of the trigeminal nerve; *Mo5*, motor nucleus of the trigeminal nerve; *Pr5*, principle sensory nucleus of the trigeminal nerve; *4V*, fourth ventricle *(From Berridge CW, Foote SL. Effects of locus coeruleus activation on electroencephalographic activity in neocortex and hippocampus. J Neurosci. 1991;11(10):3135–3145.)* (B) Effects of LC suppression on EEG activation in lightly anesthetized rats. In these studies, peri-LC infusions (35–100 nL) of the α_2 agonist, clonidine, were used to suppress LC neuronal discharge. The top panels (LC) depict oscilloscope traces of a multiunit LC recording before (preclonidine) and after (postclonidine) a peri-LC clonidine infusion. Clonidine completely suppressed LC discharge. LC activity was robustly suppressed and this was followed by the appearance of slow-wave ECoG activity (bottom panels, 25-s of raw ECoG traces). This effect appeared within seconds–minutes of LC suppression and persisted for the entire period in which LC neurons were completely inactive (>60–90 min). Recovery from these EEG effects quickly followed the emergence of minimal LC discharge activity *(From Berridge CW, Page ME, Valentino RJ, Foote SL. Effects of locus coeruleus inactivation on electroencephalographic activity in neocortex and hippocampus. Neuroscience. 1993;55(2):381–393.)*

that α_1- or β-receptor activation in these regions elicits robust increases in waking (Fig. 4.2A).[5,6,33] Furthermore, within all regions the wake-promoting actions of α_1- and β-receptor stimulation are additive (Fig. 4.2B).[5,33]

The SI, situated immediately lateral to both the MSA and MPOA, provides a potent activating influence on cortical EEG, in part through direct cholinergic

projections.[26] Nonetheless, this region is not a site of action for the arousal-promoting effects of NE, α_1- or β agonists, or the indirect noradrenergic agonist, amphetamine.[8] The only exception to this occurs with infusions of high concentrations of NE.[6,13] In this case, the latency to waking is substantially longer and the magnitude of waking reduced, relative to NE infusion into

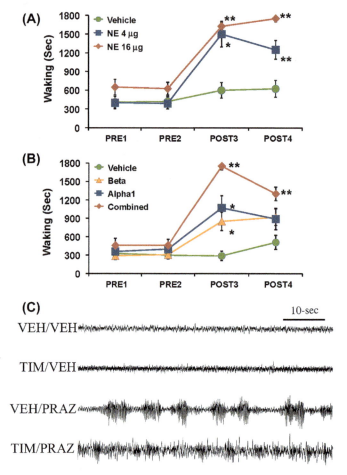

FIGURE 4.2 **NE signaling at α_1 and β receptors within subcortical structures exerts additive arousal-enhancing actions.** (A) Effects of norepinephrine (NE) infusions into the medial preoptic area (MPOA) on time spent awake. Symbols represent mean (±SEM) time (seconds) spent awake per 30-min testing epochs before (PRE1, PRE2) and following (POST1, POST2) unilateral infusions (150 nL) of vehicle, 4 nmol NE or 16 nmol NE. NE elicited dose-dependent increases in time spent awake with the highest dose eliciting near continuous waking. (B) Additive wake-promoting effects of α_1- and β-agonist receptor stimulation. Panels depict the wake-promoting effects of vehicle, 10 nmol of the α_1 agonist, phenylephrine (PHEN), 4 nmol of the β agonist, isoproterenol (ISO), and combined PHEN + ISO (combined) when infused into MPOA. A relatively low dose of each drug was used that elicited only a mild wake-promoting action when administered alone. In the combined treatment group, additive wake-promoting effects of ISO and PHEN were observed. Similar effects are observed in the medial septal area and the lateral hypothalamus. *$p < .05$, **$p < .01$ compared to PRE1; $^+p < .05$ compared to combined *(From Berridge CW, Isaac SO, Espana RA. Additive wake-promoting actions of medial basal forebrain noradrenergic alpha1- and beta-receptor stimulation.* Behav Neurosci. 2003;117(2):350–359.) (C) Synergistic sedative effects of α_1- and β-receptor blockade in stress. Shown are the effects of the β antagonist, timolol (intracerebroventricularly, ICV), the α_1 antagonist, prazosin (intraperitoneally, IP), and combined antagonist treatment on cortical EEG in animals exposed to a novelty stress. Animals were treated 30-min before stress exposure with either: (1) ICV vehicle + IP saline (VEH/VEH); (2) 150 µg ICV timolol + IP saline (TIM/VEH); (3) ICV vehicle + IP 500 µg/kg prazosin (VEH/PRAZ); and (4) combined timolol + prazosin (TIM/PRAZ). In this figure, electroencephalographic (EEG) traces are from the second 5-min epoch of exposure to novelty stress. Vehicle-treated controls displayed behavioral and EEG indices of alert waking throughout the recording session. This was reflected in sustained EEG desynchronization (low-amplitude, high-frequency). β-receptor blockade alone (TIM/VEH) had no effects on EEG activity. α_1-receptor blockade alone (VEH/PRAZ) increased the frequency and duration of sleep spindles (high-voltage spindles), an intermediate form of sedation. Combined α_1-receptor and β-receptor blockade produced profound sedation as indicated by robust increases in large-amplitude, slow-wave activity *(From Berridge CW, Espana RA. Synergistic sedative effects of noradrenergic alpha(1)- and beta-receptor blockade on forebrain electroencephalographic and behavioral indices.* Neuroscience. 2000;99(3):495–505.)

the MPOA,[6] suggesting that at high concentrations, NE diffuses from the SI to the MPOA where it then acts to promote arousal.

Differential Noradrenergic Input Across Arousal-Promoting Regions

The above-described observations provide clear evidence that LC neurons exert a robust excitatory influence on forebrain activity state that involves a network of subcortical sites. However, anatomical tracing studies demonstrate that the LC is not the only source of noradrenergic innervation to these regions and that the proportion of NE innervation arising from the LC vs. other noradrenergic nuclei varies across arousal-related regions.[8] Thus, although the LC provides the majority (~50%) of noradrenergic input to the MSA and MPOA, significant noradrenergic innervation to these areas arises from the A1 and A2 noradrenergic nuclei (~25% each). Conversely, in the case of the LH, the LC provides only a small proportion of the NE innervation (~10%), with significantly larger contributions from A1 and A2.[17,42] The LC appears unique among noradrenergic nuclei in receiving prominent input from forebrain regions associated with higher cognitive and affective function (e.g., PFC, amygdala),[9] allowing for top-down regulation of arousal through this noradrenergic system. These observations suggest that noradrenergic regulation of arousal and arousal-dependent processes involves a complex network of noradrenergic nuclei, receptors, and terminal fields that may be differentially recruited under varying conditions.

Noradrenergic Modulation of Arousal in Stress

The above-described observations suggest that NE signaling likely contributes to stress-related elevations in arousal. In support of this, prevention of LC activation blocks the arousal promoting effects of hypotensive stress in anesthetized rats.[29] Additionally, in unanesthetized animals, pharmacological studies indicate that α_1 and β receptors are necessary for stress-related arousal (Fig. 4.2C).[3] Interestingly, the sedative effects of α_1/β-receptor blockade in stress is not observed in the first few minutes of stress exposure, indicating an activation of additional arousal systems associated with early stressor exposure is sufficient to maintain alert waking.[3] This likely reflects the actions of multiple arousal systems activated in stress, including acetylcholine, serotonin, dopamine, and corticotropin releasing factor. Nonetheless, despite the activation of these additional arousal-promoting systems, blockade of noradrenergic α_1/β-receptor signaling prevents *sustained* elevations in arousal typically observed in stress.[3]

Summary

Noradrenergic systems exert potent arousal-promoting effects that contribute to elevated arousal in stress. This involves additive actions of α_1 and β receptors within a network of subcortical regions. The relative contributions of noradrenergic nuclei to the NE innervation of subcortical arousal-related structures is regionally dependent, with both the MSA and MPOA receiving a majority of input from the LC while the LHA receives a significantly smaller proportion from the LC. Although activation of either α_1 or β receptors in any single region is *sufficient* to elevate behavioral and EEG/EMG indices under low-arousal conditions, blockade of both receptors globally in the brain is required to maximally attenuate arousal in stress.

BROAD CONTRIBUTIONS OF NE TO BEHAVIORAL RESPONDING IN STRESS

A variety of behavioral, motivational, and cognitive processes are highly dependent on arousal state. The above-reviewed information suggests NE systems likely exert *indirect, arousal-dependent* modulatory actions on a range of state-dependent behavioral processes in stress.[9] Additionally, noradrenergic systems project widely in the brain and thus are positioned to *directly* influence a range of behavioral processes in stress via NE receptor signaling within functionally distinct regions. Consistent with these hypotheses, a large body of evidence demonstrates a role for NE in a broad array of stress-related alterations in neuronal signaling, attention, and memory.

Noradrenergic Modulation of Neuronal Information Processing

Under challenging conditions, there is a need for rapid and accurate processing of information arising from multiple sources. Early observations demonstrated that NE exerts strong modulatory actions on evoked neuronal activity ("signals"), relative to spontaneous discharge activity ("noise"), resulting in a net increase in the "signal-to-noise" ratio of neuronal signaling in a variety of brain regions (for review, see Ref. 9). Subsequent work indicated a more complicated array of electrophysiological actions of NE on neuronal signaling, including the facilitation of both inhibitory and excitatory responses to afferent information.[9] These actions are both receptor- and region-dependent. Thus, in sensory regions, α_1 receptors promote excitation and β receptors promote inhibition, while the opposite pattern is observed in the hippocampus (for review, see Ref. 9). More recent work suggests that NE may exert more specific actions on sensory information processing than simply modulating responsiveness to excitatory and inhibitory input, including the modulation of feature extraction properties of sensory neurons.[9,15]

Importantly, the actions of NE on cortical neuronal activity are highly concentration-dependent.[16] For example, in vitro NE exerts an inverted U-shaped facilitation of glutamate-evoked neuronal activity of sensory cortical neurons, with maximal facilitation observed at intermediate concentrations. Recent work in unanesthetized rat, further demonstrates nonmonotonic, dose-dependent actions of NE on sensory information processing within sensory cortex and thalamus.[15]

Combined, these observations indicate that it is inaccurate to classify NE as an "excitatory" or "inhibitory" neurotransmitter. Instead NE acts as a "neuromodulator," exerting a diversity of actions on neuronal signaling throughout the brain that are dependent on cell type, activity state of the neuron, cellular expression of receptors, and rates of noradrenergic signaling. Collectively, these observations suggest that NE plays a critical role in filtering, processing, and responding to salient information in stress.

Modulatory Actions of the LC-Noradrenergic System on Cognitive Processes

Stress impacts a diversity of cognitive and behavioral processes. The previously reviewed actions of the LC efferent system suggest a widespread influence of this noradrenergic pathway on information processing at both the single neuron and neuronal network levels across a variety of LC terminal fields. These actions likely have a critical impact on cognitive and behavioral processes that are of particular importance under demanding (i.e., stressful) conditions requiring rapid and accurate decisions regarding behavioral response selection. Consistent with this, evidence reviewed herein indicates central noradrenergic systems, particularly the LC, play a prominent role in a variety of cognitive processes related to the collection, processing, retention, and utilization of environmental information to guide action, including under stressful conditions.

Working Memory: Opposing Actions of α_2 and α_1 Receptors

Stress impairs working memory in humans and animals (for review, see Ref. 2). An extensive body of work demonstrates a pivotal role for NE signaling in the PFC on working memory.[2] Similar to that observed at the single cell level, the dose-dependency of NE modulation of working memory is nonmonotonic. Thus, *both* inadequate and excessive NE signaling are associated with impaired working memory, while moderate levels of NE signaling are associated with effective working memory (Fig. 4.3).[31] This inverted-U-shaped function reflects the

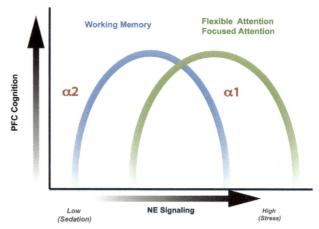

FIGURE 4.3   Norepinephrine (NE) α_2 and α_1 receptors exert differential actions across distinct prefrontal cortex (PFC)-dependent cognitive processes. Schematic depiction of the relationship between increasing rates of NE release in the PFC on distinct PFC-dependent cognitive processes: working memory, focused attention, and flexible attention. At minimal rates of NE signaling, all processes are impaired. At moderates rates of NE release that support moderate levels of arousal associated with alert waking, high-affinity α_2 receptor stimulation promotes working memory. Under higher rates of NE release associated with stress, lower affinity α_1 receptors are engaged, driving an impairment in working memory while simultaneously facilitating focused and flexible attention. Under conditions of high levels arousal and NE signaling elicited with high-dose psychostimulants, all cognitive processes are impaired.

differential activation of NE receptor subtypes within the PFC. Thus, impaired working memory associated with low rates of NE signaling in the PFC (e.g., sedation, NE lesions, aging) is reversed by selective activation of high-affinity post-synaptic α_2 receptors in the PFC.[2] Conversely, under conditions associated with moderate arousal levels and moderate rates of NE release, PFC α_2 receptor blockade degrades working memory.[2] In contrast, at higher rates of PFC NE release, the engagement of lower affinity α_1 receptors elicits a stresslike impairment in working memory, while stress-related working memory dysfunction is reversed by α_1 receptor blockade.[2] These observations indicate that under moderate rates of NE release, high-affinity α_2 receptors in the PFC support working memory, while under higher rates of release lower affinity α_1 receptors degrade working memory (Fig. 4.3). Finally, limited evidence indicates that β_1 and β_2 receptors in the PFC exert opposing actions on working memory; blockade of β_1 or stimulation of β_2 receptors in the PFC improve working memory (for review, see Ref. 38).

Attention: Flexible versus Focused

The ability to switch attention between cues in a context-appropriate manner is essential for the successful attainment of distal goals in a complex and rapidly changing world. The regulation of attentional shifts is highly dependent on the PFC.[10] In humans, the Wisconsin Card

Sort Task is a commonly used test of attentional flexibility. In this task, the cue that predicts success is changed without being signaled to the subject. In rodents, an attentional set shifting task is similarly used to measure the ability to switch attention from a previously rewarded cue to a new reward-predicting cue.[10] Attentional shifts across sensory modality (e.g., olfaction to texture), referred to as extradimensional shifts, are believed to require greater attentional flexibility than intradimensional shifts (e.g., from one odor to a different odor). Depletion of NE *within the PFC* selectively impairs extradimensional attentional set shifting, similar to that seen with working memory.[25] However, in distinct contrast to that seen with working memory, α_1 receptor signaling in the PFC facilitates attentional set shifting, while PFC α_2 and β receptors have minimal effects.[22]

These observations were initially posited to suggest that working memory requires a more focused form of attention relative to attentional set shifting. This hypothesis largely stemmed from earlier observations in monkeys engaged in a focused attention task lacking a need for flexible attentional shifts.[30] In these studies, optimal focused attention was associated with moderate rates of tonic LC activity, while impaired focused attention, associated with visual scanning, was observed under conditions associated with elevated LC activity.[30] Collectively, these two series of studies suggested the hypothesis that under conditions of stress, high rates of NE signaling within the PFC and the subsequent activation of α_1 receptors promote a more flexible mode of scanning attention at the expense of focused attention. Consistent with this, attentional set shifting in rats is improved by activation of LC neurons by the stress-related neuropeptide, corticotropin releasing factor.[35]

While a seemingly plausible hypothesis, more recent observations call into question its validity. Specifically, in rats, clinically relevant doses of the psychostimulant, methylphenidate (Ritalin), elicit moderate increases in NE signaling in the PFC that improve working memory via an α_2-receptor-dependent mechanism (as well as dopamine D1 receptors).[38] However, at doses that moderately exceed clinical relevance and impair working memory, methylphenidate improves both attentional set shifting (flexible) *and* sustained (focused/inflexible) attention via an activation of α_1 receptors.[38,22] Moreover, α_1-receptor activation directly in the PFC improves sustained attention, similar to that observed with attentional set shifting.[38] These observations suggest that in the PFC, α_1 receptors activated at higher rates of NE release improve both focused and flexible attention (Fig. 4.3).

Collectively, the available evidence indicates that NE differentially modulates PFC-dependent cognitive processes. These differential actions are dependent on both rates of release and receptor subtype but not on the degree to which a process requires focused vs. flexible attention (Fig. 4.3). The varying inverted-U dose-dependencies

across cognitive processes likely contribute to the ability of stress to both improve or impair PFC-dependent cognition, depending on conditions (see above).[40,41]

Preattentional Sensory Filtering

Effective goal-directed behavior requires preattentive filtering (gating) of a steady stream of behaviorally irrelevant sensory information. Preattentive sensory filtering is impaired in a variety of psychopathologies, including stress-related disorders [e.g., posttraumatic stress disorder (PTSD)].[21,39] The neurobiology of preattentive sensory filtering has been extensively studied in animals using prepulse inhibition (PPI[21]). In this test, a low-intensity auditory stimulus (prepulse), applied immediately before a high-intensity and startle-eliciting auditory stimulus, reduces the startle response to the high-intensity stimulus. Stress impairs PPI in animals, an action that involves noradrenergic α_1- and β-receptor signaling in a variety of brain regions.[1] α_1-dependent impairment in preattentive filtering contrasts with the attention-promoting actions of these receptors described above.

Arousal-Related Memory

Memory strength is enhanced by stressful, emotionally arousing conditions (for review, see Ref. 24). Circulating glucocorticoids and catecholamines (e.g., epinephrine) participate in this arousal-related enhancement of memory.[18,24] Evidence indicates that peripherally, epinephrine activation of β receptors located on vagal afferents triggers an activation of the noradrenergic nucleus of the solitary tract (NTS, A2), which results in increased NE release in the basolateral nucleus of the amygdala (BLA), in part through the subsequent activation of the LC (for review, see Ref. 18,24). NE signaling in the BLA facilitates arousal-related memory via actions of both α_1 and β receptors.[18,24] For example, posttraining NE infusion into the BLA improves, while β-antagonist infusion impairs, footshock-stress-induced inhibitory avoidance.[18,24] α_1 agonists and antagonists also facilitate or impair, respectively, inhibitory avoidance when infused directly within the basolateral amygdala.[18] The aversive memory enhancing actions of BLA α_1 receptors appears to involve an α_1-dependent enhancement of β-receptor-mediated cAMP production.[18]

These observations indicate a prominent role of NE in the regulation of memory for aversive/arousing events via actions of β and α_1 receptors. Consistent with this, β-receptor blockade in human subjects attenuates the typically observed enhanced memory for emotionally activating images relative to emotionally neutral images.[11]

Stress-Related Neuronal Plasticity

Stress-related psychopathology (e.g., PTSD, depression) is associated with long-term alterations in cognition and behavior, including varying forms of memory.[37] Noradrenergic systems could contribute to this behavioral plasticity in two ways. First, repeated/chronic stress is known to sensitize NE systems to subsequent exposure to novel stressors.[9] Second, actions of NE within functionally distinct terminal fields can initiate long-term changes in gene regulation, neural activity, and behavior. As described above, NE systems elicit long-term alterations in memory and memory-related synaptic efficacy (for review, see Ref. 9). Moreover, stress elicits long-term alterations in rates of gene transcription and protein production, particularly "immediate-early genes," which often serve as transcription factors, regulating gene transcription more broadly (IEGs[9]). Importantly, the activating effects of stressor-induced increases in IEG expression are attenuated with pretreatment of noradrenergic β and α_1 antagonists or LC lesions.[9] These and other observations suggest that NE-dependent alterations in rates of IEG expression likely impact a variety of physiological systems that support higher cognitive processes, including learning and memory.[9]

Summary

Noradrenergic signaling contributes to a range of cognitive actions of stress. This involves α_1- and β-receptor signaling in a network of cortical and subcortical regions associated with specific cognitive processes. NE-dependent long-term alterations in gene transcription likely contribute to sustained cognitive effects of stress.

CLINICAL IMPLICATIONS

The above-described observations indicate that noradrenergic systems impact widespread neural circuits involved in the collection and processing of sensory information. This includes a variety of cognitive, electrophysiological, and genetic actions that are superimposed on NE-induced arousal. Moreover, actions of NE are well-documented to contribute to both acute and long-term behavioral effects of stress. Combined, these observations suggest that NE signaling may contribute to behavioral dysfunction associated with stress-related psychopathology, including depression and PTSD. This hypothesis is consistent with the fact that major depression and PTSD are associated with alterations in arousal and cognitive function. Moreover, in the case of PTSD, this disorder is associated with a hyperreactivity of noradrenergic systems,[36] mirroring stress-related sensitization of noradrenergic systems observed in animals.[9] An excessive reactivity of noradrenergic systems in PTSD suggests a *causal* relationship between NE release and the symptoms of this disorder. In support of this,

drug-induced increases in noradrenergic neurotransmission in PTSD patients elicit both panic and intrusive memories associated with traumatic events.[36] Based on the above reviewed information, a hyperreactivity of NE systems is expected to influence a variety of behavioral processes associated with stress-related psychopathology, including elevated arousal, impaired working memory, and impaired preattentive sensory filtering. Independent of the degree to which noradrenergic systems contribute to the etiology of stress-related disorders, the available evidence suggests that noradrenergic systems represent a highly relevant target for pharmacological treatment of stress-related behavioral dysfunction. Consistent with either hypothesis, noradrenergic selective drugs are commonly used in the treatment of PTSD and depression.[36]

SUMMARY

A defining feature of stress is the need to acquire and process information rapidly and efficiently before making optimal response selection. Central noradrenergic systems are a critical component of information processing and action. This involves two basic categories of action. First, NE initiates behavioral and neuronal activity states appropriate for the collection of sensory information (e.g., waking). Second, within waking the noradrenergic systems modulate a variety of state-dependent processes including sensory information processing, attention, and various forms of memory. At high rates of signaling observed in stress, NE elicits a diversity of cognitive actions that are dependent on the specific noradrenergic terminal field (e.g., suppresses PFC-dependent working memory while improving amygdala-dependent long-term memory for aversive events). Evidence indicates that the sustained behavioral actions of stress likely involve NE-induced changes in gene transcription and long-term alterations in neuronal circuit function. These actions of NE could contribute to a range of symptoms associated with stress-related disorders. Moreover, limited observations suggest that dysregulated NE signaling contributes to the stress-related psychopathology, particularly in the case of PTSD. A better understanding of the behavioral actions of central NE systems may provide novel insight into the development of better pharmacological treatments for stress-related behavioral dysfunction.

Acknowledgments

The author gratefully acknowledges support by the University of Wisconsin–Madison Graduate School and PHS grants MH098631 and MH081843.

References

1. Alsene KM, Rajbhandari AK, Ramaker MJ, Bakshi VP. Discrete forebrain neuronal networks supporting noradrenergic regulation of sensorimotor gating. *Neuropsychopharmacology.* 2011;36(5): 1003–1014.
2. Arnsten AF. The biology of being frazzled. *Science.* 1998; 280(5370):1711–1712.
3. Berridge CW, Espana RA. Synergistic sedative effects of noradrenergic alpha(1)- and beta-receptor blockade on forebrain electroencephalographic and behavioral indices. *Neuroscience.* 2000;99(3):495–505.
4. Berridge CW, Foote SL. Effects of locus coeruleus activation on electroencephalographic activity in neocortex and hippocampus. *J Neurosci.* 1991;11(10):3135–3145.
5. Berridge CW, Isaac SO, Espana RA. Additive wake-promoting actions of medial basal forebrain noradrenergic alpha1- and beta-receptor stimulation. *Behav Neurosci.* 2003;117(2):350–359.
6. Berridge CW, O'Neill J. Differential sensitivity to the wake-promoting actions of norepinephrine within the medial preoptic area and the substantia innominata. *Behav Neurosci.* 2001;115(1):165–174.
7. Berridge CW, Page ME, Valentino RJ, Foote SL. Effects of locus coeruleus inactivation on electroencephalographic activity in neocortex and hippocampus. *Neuroscience.* 1993;55(2):381–393.
8. Berridge CW, Schmeichel BE, Espana RA. Noradrenergic modulation of wakefulness/arousal. *Sleep Med Rev.* 2012;16(2):187–197.
9. Berridge CW, Waterhouse BD. The locus coeruleus-noradrenergic system: modulation of behavioral state and state-dependent cognitive processes. *Brain Res Rev.* 2003;42(1):33–84.
10. Birrell JM, Brown VJ. Medial frontal cortex mediates perceptual attentional set shifting in the rat. *J Neurosci.* 2000;20(11):4320–4324.
11. Cahill L, Prins B, Weber M, McGaugh JL. Beta-adrenergic activation and memory for emotional events. *Nature.* 1994;371(6499):702–704.
12. Cannon WB. The emergency function of the adrenal medulla in pain and the major emotions. *Am J Physiol.* 1914;33:356–372.
13. Cape EG, Jones BE. Differential modulation of high-frequency gamma-electroencephalogram activity and sleep-wake state by noradrenaline and serotonin microinjections into the region of cholinergic basalis neurons. *J Neurosci.* 1998;18(7):2653–2666.
14. Carter ME, Yizhar O, Chikahisa S, et al. Tuning arousal with optogenetic modulation of locus coeruleus neurons. *Nat Neurosci.* 2010;13(12):1526–1533.
15. Devilbiss DM, Page ME, Waterhouse BD. Locus ceruleus regulates sensory encoding by neurons and networks in waking animals. *J Neurosci.* 2006;26(39):9860–9872.
16. Devilbiss DM, Waterhouse BD. Norepinephrine exhibits two distinct profiles of action on sensory cortical neuron responses to excitatory synaptic stimuli. *Synapse.* 2000;37(4):273–282.
17. Espana RA, Reis KM, Valentino RJ, Berridge CW. Organization of hypocretin/orexin efferents to locus coeruleus and basal forebrain arousal-related structures. *J Comp Neurol.* 2005;481(2):160–178.
18. Ferry B, Roozendaal B, McGaugh JL. Role of norepinephrine in mediating stress hormone regulation of long- term memory storage: a critical involvement of the amygdala. *Biol Psychiatry.* 1999;46(9):1140–1152.
19. Foote SL, Aston-Jones G, Bloom FE. Impulse activity of locus coeruleus neurons in awake rats and monkeys is a function of sensory stimulation and arousal. *Proc Natl Acad USA.* 1980;77(5):3033–3037.
20. Foote SL, Bloom FE, Aston-Jones G. Nucleus locus ceruleus: new evidence of anatomical and physiological specificity. *Physiol Rev.* 1983;63(3):844–914.
21. Geyer MA, Swerdlow NR, Mansbach RS, Braff DL. Startle response models of sensorimotor gating and habituation deficits in schizophrenia. *Brain Res Bull.* 1990;25(3):485–498.
22. Lapiz MD, Morilak DA. Noradrenergic modulation of cognitive function in rat medial prefrontal cortex as measured by attentional set shifting capability. *Neuroscience.* 2006;137(3):1039–1049.
23. Marinelli M, Piazza PV. Interaction between glucocorticoid hormones, stress and psychostimulant drugs. *Eur J Neurosci.* 2002;16(3):387–394.
24. McGaugh JL. Memory–a century of consolidation. *Science.* 2000;287(5451):248–251.
25. McGaughy J, Ross RS, Eichenbaum H. Noradrenergic, but not cholinergic, deafferentation of prefrontal cortex impairs attentional set-shifting. *Neuroscience.* 2008;153(1):63–71.
26. Metherate R, Cox CL, Ashe JH. Cellular bases of neocortical activation: modulation of neural oscillations by the nucleus basalis and endogenous acetylcholine. *J Neurosci.* 1992;12(12): 4701–4711.
27. Moore RY, Bloom FE. Central catecholamine neuron systems: anatomy and physiology of the norepinephrine and epinephrine systems. *Annu Rev Neurosci.* 1979;2:113–168.
28. Newman-Tancredi A, Audinot-Bouchez V, Gobert A, Millan MJ. Noradrenaline and adrenaline are high affinity agonists at dopamine D4 receptors. *Eur J Phamacol.* 1997;319(2–3):379–383.
29. Page ME, Berridge CW, Foote SL, Valentino RJ. Corticotropin-releasing factor in the locus coeruleus mediates EEG activation associated with hypotensive stress. *Neurosci Lett.* 1993;164(1–2): 81–84.
30. Rajkowski J, Kubiak P, Aston-Jones G. Locus coeruleus activity in monkey: phasic and tonic changes are associated with altered vigilance. *Brain Res Bull.* 1994;35(5–6):607–616.
31. Robbins TW, Arnsten AF. The neuropsychopharmacology of fronto-executive function: monoaminergic modulation. *Annu Rev Neurosci.* 2009;32:267–287.
32. Robertson SD, Plummer NW, de Marchena J, Jensen P. Developmental origins of central norepinephrine neuron diversity. *Nat Neurosci.* 2013;16(8):1016–1023.
33. Schmeichel BE, Berridge CW. Wake-promoting actions of noradrenergic alpha1-and beta-receptors within the lateral hypothalamic area. *Eur J Neurosci.* 2013;37(6):891–900.
34. Selye H. The general adaptation syndrome and the diseases of adaptation. *J Clin Endocrinol Metab.* 1946;6:117–230.
35. Snyder K, Wang WW, Han R, McFadden K, Valentino RJ. Corticotropin-releasing factor in the norepinephrine nucleus, locus coeruleus, facilitates behavioral flexibility. *Neuropsychopharmacology.* 2012;37(2):520–530.
36. Southwick SM, Bremner JD, Rasmusson A, Morgan III CA, Arnsten A, Charney DS. Role of norepinephrine in the pathophysiology and treatment of posttraumatic stress disorder. *Biol Psychiatry.* 1999;46(9):1192–1204.
37. Southwick SM, Vythilingam M, Charney DS. The psychobiology of depression and resilience to stress: implications for prevention and treatment. *Annu Rev Clin Psychol.* 2005;1:255–291.
38. Spencer RC, Devilbiss DM, Berridge CW. The cognition-enhancing effects of psychostimulants involve direct action in the prefrontal cortex. *Biol Psychiatry.* 2015;77(11):940–950.
39. Stewart LP, White PM. Sensory filtering phenomenology in PTSD. *Depress Anxiety.* 2008;25(1):38–45.
40. Verdejo-Garcia A, Moreno-Padilla M, Garcia-Rios MC, et al. Social stress increases cortisol and hampers attention in adolescents with excess weight. *PLoS One.* 2015;10(4):e0123565.
41. Weerda R, Muehlhan M, Wolf OT, Thiel CM. Effects of acute psychosocial stress on working memory related brain activity in men. *Hum Brain Mapp.* 2010;31(9):1418–1429.
42. Yoshida K, McCormack S, Espana RA, Crocker A, Scammell TE. Afferents to the orexin neurons of the rat brain. *J Comp Neurol.* 2006;494(5):845–861.

5

Evolution and Phylogeny of the Corticotropin-Releasing Factor Family of Peptides

D.A. Lovejoy, O.M. Michalec

University of Toronto, Toronto, ON, Canada

Abstract

The corticotropin-releasing factor (CRF) of peptides in vertebrates consists of four paralogous peptides that evolved along two paralogous lineages. One lineage includes CRF and urotensin-I, which is termed "urocortin" in mammals, "sauvagine" in frogs, and urotensin-I in the fishes. The second lineage consists of urocortin-2 and urocortin-3. CRF and urotensin-I have been evolutionarily selected to coordinate the stress response in the pituitary gland and urophysis, respectively. As a result, they possess a number of derived sequence characteristics that reflect these novel vertebrate functions. Urocortin-2 and urocortin-3, on the other hand, retain a number of the invertebrate CRF-ortholog structural characteristics and regulate energy metabolism separate from the hypothalamus-pituitary-adrenal axis. Current evidence indicates that the CRF and calcitonin ligands shared a common ancestor and may have been derived from an earlier evolving peptide system as a result of lateral gene transfer in a single-celled ancestor of multicellular organisms.

INTRODUCTION

The corticotropin-releasing factor (CRF) family of peptide hormones regulates the chronic energetic requirements of the organismal stress response among vertebrates. However, the presence of CRF-like orthologs in invertebrates indicates that their basic mechanism evolved before the evolution of vertebrates. Current evidence suggests that early in metazoan evolution, likely before the separation of deuterostomes and protostomes, this family of hormones was already ensconced in physiological mechanisms pertaining to homeostasis, particularly osmoregulation and diuresis which are critical elements of the stress response in all organisms.[11,37,38]

The canonical vertebrate hypothalamus–pituitary–adrenal (HPA) axis, the central neuroendocrine mechanism that coordinates central nervous system (CNS) integration of stressful sensory stimuli with the required physiological activity, and ultimately behavior, does not exist in prechordates. Previous studies of CRF-like peptide in invertebrates indicate that only a single CRF-like ancestor was present at the time when basal chordates evolved.[11,35,38] Assuming this to be the situation, then this single form diverged into four separate paralogous CRF-like peptides found in almost all vertebrates studied to date. During subsequence evolution, these peptides have evolved to become associated with the key neuroendocrine components of the stress circuit. Although they took on novel functions, the peptides remained involved in the regulation of energy metabolism, thermoregulation, and reproduction, among others.[32,34,38]

CRF is a 41-amino acid peptide hormone that has been highly conserved throughout time, signifying its importance in the stress regulatory pathway. This hormone plays a critical role in the regulation and integration of the stress pathway among vertebrates.[16,34] In 1955, evidence for a CRF was discovered by two different teams.[22,48] However, it was not until 1981, that Vale and colleagues determined that CRF was largely involved in mediating the stress response system in vertebrates when they isolated the hormone from sheep.[52] A major role of CRF is its involvement in the HPA axis, which is a crucial circuit involved in mediating the vertebrate stress response. The secretion of CRF from the hypothalamus induces the anterior pituitary to release adrenocorticotropic hormone (ACTH), which then stimulates the adrenal cortex to release glucocorticoids such as cortisol, which act to stimulate other brain regions.[16]

EVOLUTION OF THE CRF FAMILY IN CHORDATES AND ROLE WITH THE HPA AXIS

There are four paralogous forms of CRF-like peptides in vertebrates. These include CRF, urotensin-I (termed urocortin in tetrapods, with the exception of frogs, where it is called sauvagine) urocortin-2, and urocortin-3. In nonchordates (i.e., most invertebrates) the CRF-like peptide ortholog is referred to as diuretic hormone (DH) or diuretic peptide (DP).[38]

Although the CRF family of peptides constitutes four paralogous forms in the Chordata, they evolved along two distinct gene expansion pathways. It is postulated that the four CRF vertebrate paralogs arose as a result of two rounds of genome duplication that occurred in basal chordates, termed the 2R hypothesis.[15,24,41] The urochordates, comprising the tunicates and are now considered the sister taxon to the chordates (see Ref. 35,49), have provided new understanding of CRF evolution. The vase tunicate, *Ciona intestinalis*, has emerged as a major model organism to understand chordate genetics and function since the sequencing of its genome has

been completed. There is only a single form of a CRF-like peptide found in *C. intestinalis* and the related species, *Ciona savignyi*, which supports the hypothesis that the first round of genome expansion occurred sometime after the emergence of the tunicates, but before the jawed fishes (Chondrichthyes, Actinopterygians, Sarcopterygians).[32,34,35]

Sometime after the appearance of the urochordates, but before the first true chordates, the Myxini (Agnatha, hagfishes), there was likely at least one major genome duplication. Details are not clear, as there are no CRF-related sequences known from this group. However, current evidence, based on extant sequences, indicates that this duplication event led to the development of two distinct CRF-like peptide lineages that would be retained for the rest of chordate evolution (Fig. 5.1). Although initially both would have been identical, one of these paralogous lineages was selected to regulate the ancestral pro-opiomelanocorticotrophic (POMC) gene and protein from the ancestral pituitary gland along with other regions of the brain including the arcuate nucleus. The initial CRF-like peptide would have evolved to possess characteristics similar to both CRF, on one hand, and urotensin-I, along with its paralogs, urocortin and sauvagine, on the other hand. The second lineage, ultimately retained a number of the DH and DP characteristics found in insects and possessed a combination of residue characteristics that led to the formation of urocortin-2 and urocortin-3.

The significance of this evolution, from a physiological perspective, is that CRF and urotensin-1/urocortin/ sauvagine comprise one paralogous lineage, whereas urocortin-2 and urocortin-3 comprise the second lineage. Because they evolved so early in chordate evolution, it is essential to consider the function of these peptides from an evolutionary perspective. The current nomenclature of these peptides has caused much confusion in recent studies. The initial characterization of urotensin-I in mammals was termed "urocortin" because at that time it was not clear whether the peptide was an ortholog of urotensin-I.[53] However, later studies indicated that it was, in fact, a urotensin-I ortholog.[32,34,38] The further discovery of urocortin-2 and urocortin-3, and the widespread utilization of this nomenclature[25,31,47] compounded this confusion, as from a phylogenetic perspective, these peptides are fundamentally distinct from the originally described urocortin and physiologically they should be treated as such. For this reason, we have continued with the term "urotensin-I" to describe the direct paralog of CRF and have retained the terms urocortin-2 and urocortin-3 to refer to the peptides in the second paralogous lineage (Fig. 5.1).

The time of the second divergence has not been ascertained but presumably occurred before the evolution of the first jawed fishes, the Chondrichthyes (sharks, skates,

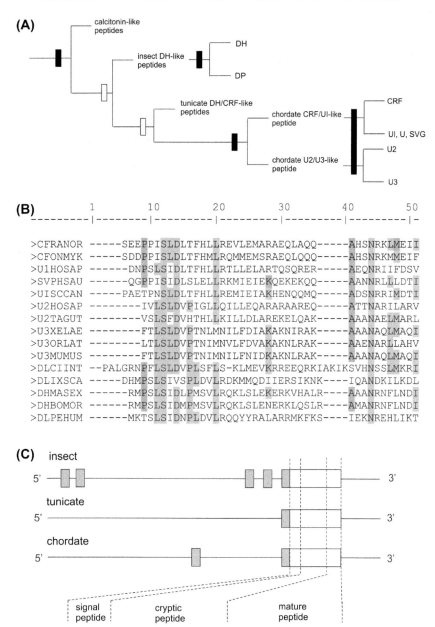

FIGURE 5.1 **Evolution and phylogeny of the corticotropin-releasing factor family of peptides.** (A) Evolutionary scheme: CRF and calcitonin likely shared a common ancestor and through subsequent gene or genomic duplications led to the expansion of the CRF family of peptides. *Black boxes*, genome expansion events; *white boxes*, species expansion events. (B) Alignment of CRF paralogs and orthologs in vertebrates and invertebrates. *CF*, CRF; *DH*, diuretic hormone; *DL*, diuretic hormone-like peptide; *SV*, sauvagine; *UI*, urotensin-I; *U1*, urocortin; *U2*, urocortin 2; *U3*, urocortin 3. (C) Comparison of the gene structure of CRF-like peptide genes. *Gray boxes*: Untranslated regions; *white boxes*, translated regions. *Boxes* indicate exonic regions, whereas the *lines* represent intronic regions.

rays, chimaeras). This reflects our understanding of CRF evolution. Although there are no reports of CRF peptides in the Myxini (hagfishes), an MC4 receptor ortholog (MCa), which binds ACTH, has been reported.[23] A tentative identification of CRF-like peptides in the lamprey has been characterized.[39] Lamprey (*Petromyzon marinus*) possesses a clear CRF peptide that is distinct from urotensin-I (urocortin, sauvagine), urocortin-2, and urocortin-3 sequences in the vertebrates (Fig. 5.1). In addition, the CRF sequence is also found in the holocephalan subclass (*Chimaeras*) of the Chondrichthyes, although two orthologous forms are present.[44] Thus, taken together these findings indicate that the phylogenetically "mature" form of CRF associated with the regulation of the HPA axis evolved sometime between the evolution of the tunicates and lampreys.

These findings are significant. CRF, with respect to urotensin-I and orthologs, urocortin-2 and urocortin-3, is the most conserved of the CRF family of peptides.[32,34,37,38] We have previously posited that once CRF became physiologically entrained with the HPA axis, it was under considerable evolutionary constraint, and therefore, it became the most conserved of the family of peptides in vertebrates. In this respect, CRF may be considered the most derived of the CRF-like peptides in that its function was selected for its role in the regulation of the HPA axis.

However, a complete functional HPA system as found in vertebrates is not found in any invertebrate species.[5] Although, in *C. intestinalis*, a CRF-like ligand[40] and receptor[5] have been found, other elements of the HPA/I (I, interrenal tissue in fishes) axis are missing. Genomic analyses of the *C. intestinalis* or *C. savignyi* genomes have not identified the existence of any POMC-like genes.[5,49] Similar findings are noted in insects. Steroid feedback is a necessary feature of the HPA/I axis. Although few studies have been performed in invertebrates, the initial stages of steroid synthesis are missing in insects.[5] The essential enzymes for the synthesis of vertebrate-like corticoids also not expressed in *C. intestinalis*. However, in some Chondrichthyes, both glucocorticoid- and mineralocorticoid-like receptors are expressed.[6] The hagfish, *Myxine glutinosa*, in contrast, only possesses a single corticoid-like receptor with similarities to both the glucocorticoid and mineralocorticoid receptors.[3,6] Thus, the key components of the HPA/I system were likely been in place in early evolving vertebrates and well before CRF was selected to become the principal neuroendocrine releasing factor to integrate the CNS actions with a peripheral stress response system leading to the evolution of the HPA axis, as we know it, in the vertebrates. Taken together, these studies indicate that the HPA axis, where CRF acts at the neuroendocrine apex, as we understand it, is a derived characteristic of vertebrates.

The urotensin-I peptide lineage, as a direct paralog of CRF, has a different story. Urotensin-I is a 41-amino acid peptide hormone that is secreted from the urophysis. The urophyseal neurosecretory cells release their contents into the fenestrated capillaries of the vascular system similar to the neurohypophysis. This caudal neurosecretory organ is found in all fishes and likely evolved due to osmoregulatory selection pressures due to fishes moving between different salinities.[32] Discovered in 1982, by Lederis and colleagues, urotensin-I was extracted from *Catostomus commersonii*, the white sucker. However, urotensin-I also stimulates the anterior pituitary to release ACTH in the rat (see Refs. 34,29 for discussion). This peptide is involved primarily in osmoregulation; however, it also plays a role in the regulation of the cardiovascular activity and stimulates the interrenal glucocorticoid release in some fishes.[34] Urocortin, the mammalian ortholog of urotensin-I, compared to CRF, has a modified role of osmoregulation as mammals and all tetrapods. Urocortin, however, retains its osmoregulatory function within the brain, also has

lost its urophyseal role.[32,53] Urocortin also functions to inhibit nutrient satiety more strongly than CRF but, it is less effective at promoting acute anxiety-like effects.[47] Taken together, studies indicate that urocortin and CRF work closely together to elicit a combined response to homeostasis in the vertebrates[19] and retain a number of their ancestral attributes (see below).

Sauvagine is a 40-amino acid peptide found among amphibians that was reported by Montecucchi and colleagues in 1979 and 1980 but has been found in other species.[43,46,56] Because of its sequence similarity to urotensin-I and CRF, it is now considered a paralog in the CRF family of peptides.[4,34] The peptide was originally found to have effects on a number of endocrine glands.[42] When administered to rats, this hormone has a number of antidiuretic and hypotensive effects.[17,42] Sauvagine also releases ACTH from the anterior pituitary when administered to rats, and this, therefore, further supported the idea that it was further related to CRF paralogous lineage.[4,32]

Urocortin-2 and urocortin-3, also known as stressocopin and stressocopin-related peptide, respectively, were discovered in 2001[25,31,47] and comprise the second paralogous lineage of the CRF peptide family in vertebrates. Both neuropeptides, which are comprised of 38 amino acids, are structurally related to CRF and urocortin, and bind exclusively to the CRF2 receptor.[47] Urocortin-2 is mainly synthesized in the paraventricular nucleus (PVN) and the locus coeruleus, whereas urocortin-3 is mainly produced in the hypothalamus and the amygdala.[19] Like CRF, urocortin-2 and urocortin-3 regulate glucose levels in the body. Urocortin-2 is expressed mainly in skeletal muscle, and it plays a role in the regulation of insulin sensitivity. Urocortin-3, on the other hand, acts on the B-cells of the pancreas to modulate insulin secretion.[28] These two hormones also work through autocrine and paracrine mechanisms to regulate the energy metabolism in the peripheral tissues.[28]

However, the independent evolution of urocortin-2 and urocortin-3 is not clear and given their strong sequence similarity may have a number of overlapping functions.[37] In the holocephalan, *Callorhinchus milii*, and the chimpanzee, the urocortin-2 gene appears to be a pseudogene as the result of a premature stop codon.[26,44] Moreover, in humans, this gene apparently lacks the amidation signal that is required for full activity, and, thus, it is not clear how functional the endogenous peptide is in humans.[8] Taken together, current evidence suggest that urocortin-2 and urocortin-3 remain under a dynamic selection process and have yet to achieve the evolutionary specialization found in CRF and urotensin-I.

Two CRF receptors exist among vertebrates, and the four paralogous peptide hormones bind to these receptors. The CRF receptors were first identified by Chen and colleagues[9] and have since been determined to be

G protein-coupled receptors (GPCRs). The CRF1 and CRF2 receptor share 70% sequence similarity, with their intracellular and transmembrane domains being most similar.[14] The CRF1 receptor mRNA is highly expressed in the mammalian brain, whereas the CRF2 receptor is more widely expressed in peripheral tissues such as cardiac muscle, GI tract, lung, ovary, and skeletal muscle.[32] In mammals, CRF and urocortin have high affinity to both the CRF1 and two receptors, whereas urocortin-2 and urocortin-3 exert their actions via the CRF2 receptor only. Ligand-receptor studies in nonmammalian vertebrates are few, but generally, the data indicate that the CRF1 receptor is more specialized for the CRF/urotensin-1 (urocortin) paralog line, and the CRF2 receptor is less specialized and retains the ability to bind all CRF family ligands.[28,36]

PRECHORDATE EVOLUTION OF THE CRF FAMILY

Several peptides that share sequence similarity with the CRF family of peptides exist among insects. The first such hormone was discovered in 1989, in the tobacco hornworm, *Manduca sexta*.[27] Despite the observation that this peptide had compelling sequence to sauvagine, there is no evidence to suggest that this particular ortholog peptides has a direct phylogenetic relationship with the vertebrate CRF/urotensin-1 lineage.[38] Subsequent to this report, numerous other similar peptides, dissimilar to sauvagine, yet still retaining key motif elements with the vertebrate CRF family, and involved in osmoregulation and diuresis, were discovered in insects. Generally, these peptides increase the fluid secretion by insect Malpighian tubules and are collectively known as the DHs or DPs. This process is under endocrine control and, therefore, through the stimulation of ion transport, the insects can regulate hemolymph levels and composition, while excreting the harmful wastes.[13]

To date, at least two separate lineages of DHs have been identified among insects.[2,12,38] Blackburn and colleagues[2] isolated a paralog of the DH in *M. sexta*, termed Manse-diuretic peptide II (Manse-DPII). Numerous other discoveries of peptides among insects followed; however, the sequencing of DPs among *Locusta migratoria* revealed that there was direct evidence of this hormone being involved in the endocrine control of diuresis and therefore the whole family was termed the DHs.[30,45]

However, both lineages of peptides appear to be the result of gene or genome expansion events that occurred within insects.[38] The first such hormone was discovered in 1989, in the tobacco hornworm (*M. sexta*), by Kataoka and colleagues, termed Manse-DH. It was noted to have structural similarity to sauvagine, which suggested that

it might be related to the CRF family of peptides found among vertebrates. Although two separate lineages of DHs have been identified among insects where one has a resemblance to the CRF and urocortin/urotensin-1 peptides and the second has features perhaps more similar to the urocortin-2 and urocortin-3 peptides,[2,12,38] this may reflect an insect- and possibly arthropod-specific gene expansion. For example, there is evidence of some DHs being more closely related to the calcitonin family of peptides. The discovery of a 31-amino acid peptide in *Schistocera americana*, revealed sequence similarity to the calcitonin family of peptides with little sequence similarity to the CRF.[13] However, it has also recently been suggested that the DHs, which share sequence similarity with urocortin-2 and urocortin-3, are more closely related to the calcitonin family.[12,21,37,38] The possibility that the CRF family of peptides may have a common origin with the calcitonin family of peptides cannot be discounted.[38]

This paradox was resolved, in part, by the discovery of CRF-like peptides in the tunicate species *C. intestinalis* and *C. savignyi*, which showed a significant primary structural similarity to the both CRF peptides found among the chordates and the insect DHs.[35] There is no evidence of the presence of an HPA/I axis among the tunicates, suggesting that the CRF-like peptides that are found in these organisms played a different role in the stress response, one that was not yet associated with the vertebrate HPA/I circuit[35] but may reflect some of the physiological aspects of the insect DHs. Structurally, the CRF-like peptides in tunicates shows a clear similarity to the insect DHs and are structurally similar to vertebrate urocortin-2 and urocortin-3. Because the tunicates are the sister lineage to the Chordata, this further substantiates that only one CRF-like peptide was inherited by the chordates, and that it then underwent two rounds of genome duplication that occurred sometime before the Actinopterygian-Sarcopterygian split, as is consistent with the 2R hypothesis.[35]

Given the strong evidence that CRF-like peptides were present in both deuterostomes and protostomes, this indicates that a proto-CRF-like peptide was present before this bifurcation. Little is known about the ancestral CRF-like peptides that evolved before this time, although there are some important clues. The CRF receptors, R1 and R2 are most closely related to the calcitonin receptors.[20] The close structural similarity among both the CRF ligands and calcitonin, and their receptors indicate that they have a common ancestry. This suggests that because insects, as protostomes, and chordates, being deuterostomes, a CRF-calcitonin-like peptide was present before the bifurcation of deuterostomes and protostomes.[37,38] Given that both peptides had a clear physiological role at this time in evolution, a simple explanation is that the protofunctional peptide appeared much earlier in metazoan evolution.

Despite this period of evolution, the basic gene structure has not changed that much. The signal peptide, cryptic peptide, and mature peptide are all found in a single exon regardless of the species (Fig. 5.1). Differences lie within the organization of the 5′ untranslated region. In insects, notably *Drosophila*, this may include four exons over two distinct clusters. However, typically in chordates this region occurs across a small upstream exon and part of the major exon. However current, evidence suggests that the tunicates only possess a single exon encompassing both untranslated and translated regions of the gene.[35,38]

Recently, the discovery of the teneurin C-terminal associated peptides (TCAP) family has provided an important clue as to how this came about. TCAP, has a structural similarity to both CRF and calcitonin peptide families.[33,37,38] Moreover TCAP's putative cognate receptor, latrophilin, bears strong structural similarity to the secretin family of GPCRs of which the CRF and calcitonin receptors are members of.[37,54] The latrophilins possess a hormone binding site similar to the CRF receptors and phylogenetic evidence indicates that the secretin family of receptors and likely their ligands, evolved from the adhesion group of GPCRs, of which the latrophilins are a member of.[20] The peptide, TCAP, along with its proprotein precursor, teneurin, and the receptor, latrophilins were likely the result of lateral gene transfer from a prokaryote to a choanoflagellate ancestor of the metazoans.[7,50,51,55] We, therefore, postulate that the ancestral peptide to CRF and calcitonin may have been related to this TCAP peptide which became incorporated into the metazoan genome and later through a number of genome and gene duplications led to the CRF family as we know it.

Acknowledgments

We thank the University of Toronto for graduate funding to Ola M. Michalec and the National Science and Engineering Research Council (NSERC) for funding to Dr. David A. Lovejoy for our research program on the structure and function of CRF-related peptides.

References

1. Deleted in review.
2. Blackburn MB, Kingan TG, Bodnar W, et al. Isolation and identification of a new diuretic peptide from the tobacco horn-worm, *Manduca sexta*. *Biochem Biophys Res Commun*. 1991;181:927–932.
3. Bridgham JT, Carroll SM, Thornton JW. Evolution of hormone receptor complexity by molecular exploitation. *Science*. 2006;312:97–101.
4. Britton D, Hoffman D, Lederis K, Rivier J. A comparison of the behavioural effects of CRF, sauvagine and urotensin 1. *Brain Res*. 1984;304:201–205.
5. Campbell RK, Satoh N, Degnan BM. Piecing together the evolution of the vertebrate endocrine system. *Trends Genet*. 2004;20:359–366.
6. Carroll SM, Bridgham JT, Thornton JW. Evolution of hormone signaling in elasmobranches by exploitation of promiscuous receptors. *Mol Biol Evol*. 2008;25:2643–2652.
7. Chand D, de Lannoy L, Tucker RP, Lovejoy DA. Origin of chordate peptides by horizontal protozoan gene transfer in early metazoans and protists: evolution of the teneurin C-terminal associated peptides. *Gen Comp Endocrin*. 2013;188:144–150.
8. Chand D, Lovejoy DA. Stress and reproduction: controversies and challenges. *Gen Comp Endocrin*. 2011;171:253–257.
9. Chen R, Lewis K, Perrin M, Vale W. Expression cloning of a human corticotropin-releasing factor receptor. *Proc Natl Acad Sci USA*. 1993;90:8967–8971.
10. Deleted in review.
11. Coast GM. Insect diuretic peptides; structures, evolution and actions. *Am Zool*. 1998;38:422–449.
12. Coast GM, Webster SG, Schegg KM, Tobe SS, Schooley DA. The *Drosophila melanogaster* homologue of an insect calcitonin-like diuretic peptide stimulates V-ATPase activity in fruit fly malpighian tubules. *J Exp Biol*. 2001;204:1795–1804.
13. Coast GM, Orchard I, Phillips JE, Schooley DA. Insect diuretic and antidiuretic hormones. *Adv Insect Physiol*. 2002;29:279–409.
14. Dautzenberg FM, Hauger RL. The CRF peptide family and their receptors: yet more partners discovered. *Trends Pharmacol Sci*. 2002;23(2):71–77.
15. Dehal P, Boore J. Two rounds of whole genome duplication in the ancestral vertebrate. *PLoS Biol*. 2005;3(10):e314.
16. De Souza E. Corticotropin-releasing factor receptors: physiology, pharmacology, biochemistry and role in central nervous system and immune disorders. *Psychoneuroendocrinology*. 1995;20(8):789–819.
17. Erspamer V, Falconieri Erspamer G, Improta G, Negri L, de Castiglione R. Sauvagine, a new polypeptide from *Phyllomedusa sauvagei* skin. *Naunyn Schmiedebergs Arch Pharmacol*. 1980;312:265–270.
18. Deleted in review.
19. Fekete E, Zorrilla E. Physiology, pharmacology, and therapeutic relevance of urocortins in mammals: ancient CRF paralogs. *Front Neuroendocrinol*. 2007;28(1):1–27.
20. Fredriksson R, Lagerström MC, Lundin L, Schiöth HB. The G-protein-coupled receptors in the human genome form five main families. Phylogenetic analysis, paralogon groups, ad fingerprints. *Mol Pharmacol*. 2003;63:1256–1272.
21. Furuya K, Mikchak RJ, Schegg KM, et al. Cockroach diuretic hormones: characterization of a calcitonin-like peptide in insects. *Proc Natl Acad Sci USA*. 2000;97:6469–6474.
22. Guillemin R, Rosenberg B. Humoral hypothalamic control of anterior pituitary: a study with combined tissue cultures. *Endocrinology*. 1955;57:599–607.
23. Haitina T, Klovins J, Takahashi A, et al. Functional characterization of two melanocortin (MC) receptors in lamprey showing orthology to the MC1 and MC4 receptor subtypes. *BMC Evol Biol*. 2007;7:101.
24. Holland PW, Garcia-Fernandez J. Hox genes and chordate evolution. *Dev Biol*. 1996;173:282–395.
25. Hsu SY, Hseuh AJW. Human stresscopin and stresscopin-related peptide are selective ligands for the type 2 corticotropin releasing hormone receptor. *Nat Med*. 2001;7:605–611.
26. Ikemoto T, Park MK. A system for receptor functional analysis based on c-fos mRNA expression: analysis of GnRH receptors as a test system. *J Biochem Biophys Methods*. 2007;70(3):349–353.
27. Kataoka H, Troetschler RG, Li JP, Kramer SJ, Carney RL, Schooley DA. Isolation and identification of a diuretic hormone from the tobacco hornworm, *Manduca sexta*. *Proc Natl Acad Sci USA*. 1989;86:2976–2980.
28. Kuperman Y, Chen A. Urocortins: emerging metabolic and energy homeostasis perspectives. *Trends Endocrinol Metab*. 2008;19(4):122–129.
29. Lederis K, Letter A, McMaster D, Moore G, Schlesinger D. Complete amino acid sequence of urotesnin I, a hypotensive and corticotropin-releasing neuropeptide from *Catostomus*. *Science*. 1982;218:162–164.

30. Lehmberg E, Ota RB, Furuya K, et al. Identification of a diuretic hormone of *Locusta migratoria*. *Biochem Biophys Res Commun*. 1991;179(2):1036–1041.

31. Lewis K, Li C, Perrin MH, et al. Identification of urocortin III and additional member of the corticotropin releasing factor family with high affinity for the CRF2 receptor. *Proc Natl Acad Sci USA*. 2001;98:7570–7575.

32. Lovejoy DA. Structural evolution of urotensin-I: retaining ancestral functions before corticotropin-releasing hormone evolution. *Gen Comp Endocrinol*. 2009;164:15–19.

33. Lovejoy DA, Al Chawaf A, Cadinouche A. Teneurin C-terminal associated peptides: an enigmatic family of neuropeptides with structural similarity to the corticotrophin releasing factor and calcitonin family of peptides. *Gen Comp Endocrinol*. 2006;148:299–305.

34. Lovejoy DA, Balment RJ. Evolution and physiology of the corticotropin-releasing factor (CRF) family of neuropeptides in vertebrates. *Gen Comp Endocrinol*. 1999;115:1–22.

35. Lovejoy DA, Barsyte-Lovejoy D. Characterization of a diuretic hormone-like peptide from tunicates: insight into the origins of the vertebrate corticotropin-releasing factor (CRF) family. *Gen Comp Endocrinol*. 2010;165:330–336.

36. Lovejoy DA, Chang B, Lovejoy N, Del Castillo J. Molecular evolution and origin of the corticotrophin-releasing hormone receptors. *J Mol Endocrinol*. 2014;52:43–60.

37. Lovejoy DA, De Lannoy L. Evolution and phylogeny of the corticotropin-releasing factor (CRF) family of peptides: expansion and specialization in the vertebrates. *J Chem Neuroanat*. 2013;54:50–56.

38. Lovejoy DA, Jahan S. Phylogeny and evolution of the corticotropin releasing factor family of peptides. *Gen Comp Endocrinol*. 2006;146:1–8.

39. Lovejoy DA, Manzon R. Evolution of the corticotropin-releasing factor (CRF) family of peptides in vertebrates: novel findings in phylogenetically older fishes. In: *Proceedings of the International Symposium of Fish Endocrinology Buenos Aires Argentina*; 2012.

40. Lovejoy DA, Rotzinger S, Barsyte-Lovejoy D. Evolution of complementary peptide systems: teneurin C-terminal associated peptide (TCAP) and corticotropin-releasing factor (CRF) superfamilies. *Ann N Y Acad Sci*. 2009;1163:215–220.

41. Lundin LG. Evolution of the vertebrate genome as reflected in paralogous chromosome regions in man and house mouse. *Genomics*. 1993;16:1–19.

42. Montecucchi P, Anastasi A, De Castiglione R, Erspamer V. Isolation and amino acid composition of sauvagine. *Int J Pept Protein Res*. 1980;16:191–199.

43. Montecucchi PA, Neschen A, Erspamer V. Structure of sauvagine, a vasoactive peptide from the skin of a frog. *Hoppe-Seyler's Z Physiol Chem*. 1979;360:1178.

44. Nock TG, Chand D, Lovejoy DA. Identification of members of the gonadotropin-releasing hormone (GnRH), corticotropin-releasing factor (CRF) families in the genome of the holocephalan, *Callorhinchus milii* (elephant shark). *Gen Comp Endocrinol*. 2011;171:237–244.

45. Patel M, Hayes TK, Coast GM. Evidence for the hormonal function of a CRF-related diuretic peptide (locusta-DP) in *Locusta migratoria*. *J Exp Biol*. 1995;198:793–804.

46. Perrin MH, Tan LA, Vaughan JM, et al. Characterization of a *Pachymedusa dacnicolor*-Sauvagine analog as a new high-affinity radioligand for corticotropin-releasing factor receptor studies. *J Pharmacol Exp Ther*. 2015;353(2):307–317.

47. Reyes TM, Lewis K, Perrin MH, et al. Urocortin II: a member of the corticotropin releasing factor (CRF) neuropeptide family that is selectively bound by type 2 receptors. *Proc Natl Acad Sci USA*. 2001:2843–2848.

48. Saffran M, Schally A. The release of corticotropin by anterior pituitary tissue in vitro. *Can J Physiol Pharmacol*. 1955;33:408–415.

49. Sherwood NM, Tello JA, Roch GJ. Neuroendocrinology of protochordates: insights from *Ciona* genomics. *Comp Biochem Physiol A Mol Integr Physiol*. 2006;144:254–271.

50. Tucker RP. Horizontal gene transfer in choanoflagellates. *J Exp Zool B Mol Dev Evol*. 2013;320:1–9.

51. Tucker RP, Beckmann J, Leachman NT, Scholer J, Chiquet-Ehrismann R. Phylogenetic analysis of the teneurins: conserved features and premetazoan ancestry. *Mol Biol Evol*. 2012;29:1019–1029.

52. Vale W, Spiess J, Rivier C, Rivier J. Characterization of a 41-residue ovine hypothalamic peptide that stimulates secretion of corticotropin and beta-endorphin. *Science*. 1981;213:1394–1397.

53. Vaughan J, Donaldson C, Bittencourt J, et al. Characterization of urocortin, a novel mammalian neuropeptide related to fish urotensin-I and to CRF. *Nature*. 1995;378:287–292.

54. Woelfle R, D'Aquila AL, Husic M, Pavlovic T, Lovejoy DA. Ancient interaction between the teneurin C-terminal associated peptides (TCAP) and latrophilin ligand-receptor coupling: a role in behavior. *Front Neurosci*. 2015;9:146.

55. Zhang D, de Souza RF, Anantharaman V, Iyer LM, Aravind L. Polymorphic toxin systems: comprehensive characterization of trafficking modes, mechanism of action, immunity and ecology using comparative genomics. *Biol Direct*. 2012;7:18.

56. Zhou Y, Jiang Y, Wang R, et al. PD-sauvagine: a novel sauvagine/corticotropin releasing factor analogue from the skin secretion of the Mexican giant leaf frog, Pachymedusa dacnicolor. *Amino Acids*. 2012;43(3):1147–1156.

6

Corticotropin-Releasing Factor and Urocortin Receptors

D.E. Grigoriadis

Neurocrine Biosciences Inc., San Diego, CA, United States

Abstract

The key mediator of the innate physiological stress response has long been hypothesized to be the hypothalamic peptide corticotropin-releasing factor (CRF). This 41 amino acid peptide modulates the endocrine, autonomic, behavioral, and immune responses to stress through actions at specific receptors both in the central nervous system as well as the periphery. These seven transmembrane proteins belong to the superfamily of G protein–coupled receptors and play a role in central nervous system, cardiovascular, gastrointestinal, and pituitary-adrenal function. Decades of research have implied a direct relationship between stress-related disorders and CRF activity in the brain. As a result, this system has been the focus of many drug discovery programs aimed at identifying selective and specific orally active molecules that can block the pathophysiological increase in activity of this peptide. This review will summarize this system, the evidence supporting the utility of molecules that block the CRF system in human disease and the current state of the drug discovery programs that have yet to produce a viable therapeutic.

INTRODUCTION

More than 2000 years ago, Hippocrates referred to the concept of homeostasis and equilibrium in health and disease as a dysregulation of this equilibrium. He further observed that there were individual differences in the severity of disease symptoms and that some individuals were better able to cope with their disease or illness than others. More than a half century ago, Hans Selye proposed that stress is the nonspecific (autonomic) response of a human body to any demand made upon it, real or perceived and that concept was broadened by Geoffrey Harris who first defined the role of the hypothalamus in the regulation of the pituitary–adrenocortical axis extending the notion that an organism's ability to maintain homeostasis in the face of external stressors was coordinated by a specific portion of the CNS operating in an automatic fashion. During the mid-1950s, Guillemin and Rosenberg, and Saffran and Schally independently recognized the presence of a specific factor in extracts of the hypothalamus that could stimulate the release of corticotropin [adrenocorticotropic hormone (ACTH)] from anterior pituitary cells in vitro. This extract was termed corticotropin-releasing factor (CRF). It was not until 1981, following the development of radioimmunoassays for ACTH and quantitative in vitro methods for assaying hypophysiotropic hormones along with the utility of ion exchange and high performance liquid chromatographic techniques, that enabled a team of Salk Institute scientists led by Wylie Vale to successfully isolate and purify the peptide CRF from sheep hypothalamic extracts.[37] Thus, from the first basic principles in the understanding of an organism's homeostatic response to stress that began centuries ago, this neuropeptide has been studied for its role in mediating the hypothalamic–pituitary–adrenal (HPA) axis and regulating the response to physical, emotional, and environmental stress.

The discovery and structural elucidation of CRF ushered in a unique and focused hypothesis-driven era of scientific study. While the initial understanding of its direct role at the pituitary pointed to the regulation of the endocrine stress response, it was the distribution of the CRF peptide in extrahypothalamic regions of the brain that intensified study into its role as a bona fide neurotransmitter and neuromodulator in the CNS. Multiple reports have detailed the anatomical distribution of both CRF-containing cell bodies and fibers in the CNS and more importantly, in regions of the brain shown to clearly regulate autonomic function and mediate the behavioral responses to stress. For example, in the cortex, high densities of CRF containing neurons that are localized primarily to prefrontal, cingulate, and insular areas, appear to regulate behavioral actions of the peptide. Many studies in rodents have demonstrated that exogenous administration of CRF into the CNS, has demonstrated direct interactions between the CRF system and virtually all monoaminergic neurotransmitter systems including the norepinephrine system in the locus coeruleus, the serotonergic systems in the dorsal and median raphe nuclei, and the dopaminergic system in the ventral tegmental area indicating that this system can impact multiple facets of the stress response of an organism. This critical convergent relationship between the hormone CRF and the monoaminergic systems in the brain further implicates this particular neuropeptide in mediating the central mechanisms through which various stressors can alter behavior.

The cloning of multiple CRF receptor subtypes and the identification of a specific binding protein for the peptide (known as CRF-BP) have also enhanced the ability to dissect the behavioral consequences of activation or inhibition of this system. In addition to the receptors, the identification of endogenous-related family members of the CRF peptide (termed the urocortins) has expanded the notion that this system is under the control of multiple regulatory factors and plays a much more complex role in the CNS than previously thought. This chapter will focus primarily on the molecular biological, pharmacological, and functional characteristics of the CRF system and describe the recent challenges in the discovery and late-stage development of small molecule nonpeptide inhibitors of this system as potential therapeutics.

KEY POINTS

- Corticotropin-releasing factor (CRF) and the peptide family of urocortins activate two subtypes of Class-B G protein–coupled receptors (CRF_1 and CRF_2 receptors) and regulate the endocrine, autonomic, and behavioral response to stress.

- Nonpeptide antagonists for the CRF_1 receptor have been discovered that act allosterically and demonstrate efficacy in a wide range of animal models for stress behaviors.

- The allosteric nature of the nonpeptide interaction requires a long residence time (slow dissociation) of the small molecule on the receptor for optimal efficacy in vivo.

- Multiple small molecules have been examined in Phase 2 human studies for major depressive disorder, social anxiety, generalized anxiety, posttraumatic stress disorder, irritable bowel syndrome, and alcohol dependence.

- Despite the tremendous body of evidence in animal models from different species providing an indication in human psychiatric disease, none of the CRF_1 receptor antagonists to date have progressed beyond Phase 2 compelling the field to rethink the clinical application of these potent and selective compounds.

CORTICOTROPIN-RELEASING FACTOR AND UROCORTIN RECEPTOR FAMILY

As a member of the superfamily of Class-B G-protein coupled receptors, the receptors for CRF and the urocortins have been shown to contain seven putative transmembrane domains and function through the coupling of a stimulatory guanine-nucleotide–binding protein. These receptor subtypes all fall within the now well-described and still-growing, family of "gut-brain" neuropeptide receptors, which includes receptors for calcitonin, vasoactive intestinal peptide, parathyroid hormone, secretin, pituitary adenylate cyclase–activating peptide, glucagon, and growth hormone–releasing factor. While these receptors all share considerable sequence homology, they have been shown to activate a variety of intracellular signaling pathways. Cloning of the CRF receptor subtypes has enabled the pharmacological and biochemical characterization of these proteins and allowed the discovery of potential therapeutics through a variety of chemical screening strategies.

Receptor Family Subtypes

The discovery of the CRF peptide led to the first radiolabeled form of the peptide that was used to probe the localization, distribution, and functional mechanisms of receptors in a variety of tissues from the brain and periphery in many different species. These studies served to define the physiology of the system and enabled a detailed investigation using molecular, biochemical, and behavioral tools to try and understand the role of this system in the normal and pathological state. In addition to the discovery of CRF, Vale and colleagues were the first to clone the CRF_1 receptor in 1993 from a human Cushing anterior pituitary corticotropic adenoma using expression cloning.[10] Several other groups later also identified the CRF_1 form of the receptor from a variety of animal species. Interestingly, all species of CRF_1 receptor mRNAs thus far identified encode proteins of 415 amino acids that are 98% identical to one another. In fact, the family of peptides has been well characterized in species ranging from insects to high-order vertebrates indicating the evolutionary importance of this system.[24] The human CRF_1 receptor gene contains at least two introns and is found in a number of alternative splice forms none of which to date have been found to have any physiological significance. Characteristic of most G protein–coupled receptors there are potential N-linked glycosylation sites on the large N-terminal extracellular domain and these molecularly identified sites confirm the glycosylation profiles determined previously using chemical affinity cross-linking studies. Indeed, the predicted molecular weight of the CRF_1 receptor derived from deglycosylation studies was virtually identical to that obtained from the cloned amino acid sequence. Furthermore, mRNA distribution of the CRF_1 receptor in the CNS of both rat and human tissues correlates extremely well with the previously identified biding sites for radiolabeled CRF using autoradiographic techniques in frozen

brain sections. Alternative splice variants of the CRF_1 receptor have been described localized to discrete tissues. For example, a unique splice variant of the CRF_1 receptor has been reported in human pregnant myometrium and while the pharmacology appears to be similar to the original subtype, the different signaling characteristics suggest that this isoform may be expressed under specific physiological conditions.[14]

Shortly following the cloning and characterization of the CRF_1 receptor, a second subtype of this family was identified. This receptor (termed the CRF_2 receptor subtype) exists as three individual splice variants termed $CRF_{2(a)}$ and $CRF_{2(b)}$ and $CRF_{2(c)}$. These receptor isoforms differ only in the extracellular region where each has a unique N-terminal sequence generating 411, 431, and 397 amino acid proteins for the $CRF_{2(a)}$, $CRF_{2(b)}$, and $CRF_{2(c)}$ receptors, respectively. Thus far, the $CRF_{2(c)}$ receptor has been identified exclusively in human brain and its function still remains to be determined. Comparing the CRF_1 and CRF_2 receptor subtypes, there are very large regions of amino acid identity, particularly between transmembrane domains five and six. This similarity indicates that biochemical action is conserved and these receptors act through G protein coupling and signal transduction.

Detailed distribution and receptor autoradiographic studies have localized CRF-binding sites and CRF receptor mRNA, respectively, in slide-mounted sections of many tissues from a variety of species identifying these proteins in anatomically and physiologically relevant areas. Of primary interest, the CRF_1 receptor is highly expressed in the pituitary gland, specifically in corticotrophs where it regulates the release of ACTH as a key mediator of the stress response. This discrete localization has facilitated the utility of selective receptor antagonists in disorders related to increased HPA activity. This will be detailed later in this chapter.

While the CRF_2 receptor has been documented extensively with respect to its pharmacology, its role in behavior and physiological function has been defined largely through its localization and the effects of the endogenous agonists. Specific chemical tools that block this receptor subtype systemically are few and restricted to peptide analogs (illustrated in the next section). In addition, the distribution of the CRF_2 receptor is isoform-dependent. Specifically, the precise localization the splice variants, $CRF_{2(a)}$ and $CRF_{2(b)}$, indicates discrete anatomical distribution. The predominant form in the CNS is the $CRF_{2(a)}$ receptor where localization studies and those using peptide agonists have implicated a role for this subtype in anxiety and feeding behaviors. The $CRF_{2(b)}$ splice variant is localized almost exclusively in the CNS to nonneuronal elements, such as the choroid plexus of the ventricular system and cerebral arterioles. In the periphery, however, the $CRF_{2(b)}$ receptor is highly expressed in both cardiac and skeletal muscle with lower levels evident in both lung and intestine. Its role peripherally has been described in cardiovascular function and gastrointestinal motility. In fact, for a short period of time, the $CRF_{2(b)}$ receptor was the target of intense study for the treatment of acute decompensated heart failure (ADHF) owing to the very potent vasodilatory effects of the endogenous peptide urocortin 2. Studies performed in rats, sheep, and some initial studies in humans showed early promise as a potential treatment for ADHF.[1,9] The $CRF_{2(c)}$ isoform has yet to be identified in the rodent, however, in the human brain, this isoform was shown to be highly expressed in the septum, amygdala, hippocampus, and frontal cortex.

Endogenous Ligands

Since the elucidation of the amino acid sequence of CRF, the peptide was found to bear strong similarities to two nonmammalian peptides, sauvagine from the frog and urotensin I from the teleost fish. These mammalian and nonmammalian peptides all potently release ACTH from cultured rat pituitary cells directly through the CRF_1 receptor in the anterior pituitary. One striking difference between these peptides was that only the nonmammalian forms retained a high affinity for the CRF_2 receptor. Thus it became highly unlikely that CRF itself was the endogenous ligand for the CRF_2 receptor in mammals. Using antibodies for fish urotensin I, the first mammalian peptide with high affinity for the CRF_2 receptor was discovered and termed urocortin 1. This unique peptide discovered first in the rat was observed to retain its high affinity for the CRF_1 receptor and the CRF-binding protein as well. Furthermore, the peptide was found to be localized in areas corresponding to the distribution of the CRF_2 receptor itself. The human homolog of urocortin 1 was subsequently cloned from a human brain library and defined as the endogenous ligand for the CRF_2 receptor. Since then, two other mammalian peptides have been identified from the mouse and human termed urocortin 2 (UCN2, also referred to as stresscopin-related peptide) and urocortin 3 (UCN3, also referred to as stresscopin). Urocortin two and urocortin 3, unlike urocortin 1, have much higher selectivity for the CRF_2 receptor subtype and have little or no affinity for the CRF-binding protein. A great deal of literature now exists on the discrete and independent physiologic function of the urocortins and their activity on the CRF_2 receptor subtypes (for a comprehensive review see Ref. 34). However, the current paucity of selective small molecule antagonists that could be administered chronically, hinders our ability to fully understand the role of the CRF_2 receptor subtype in normal or pathophysiology. Nonetheless, major advances have been made in elucidating the function of these ligands and the CRF_2 receptor in anxiety, cardiovascular disease, inflammatory disease, and gastrointestinal and feeding disorders.

Interestingly, there are no known endogenous physiological antagonists for any of the CRF receptor subtypes known to date. However, various truncations and modifications of the agonist peptides have resulted in potent peptide antagonists. A large body of work has been dedicated to identifying specific and selective peptide antagonists of these receptor subtypes.[8,35] These ligands were designed to extend duration of action in vivo and demonstrated high affinity and activity for both the CRF_1 and CRF_2 receptor subtypes. Two of these synthetic high-affinity peptides, Antisauvagine-30, an N-terminally truncated form of sauvagine, and the cyclized Astressin-2B have been shown to be selective for the CRF_2 receptor subtype and used to dissect out the specific role of this receptor in a variety of in vitro and in vivo studies. The pharmacological profiles of these peptide agonists and antagonists greatly enhanced the understanding of the physiology of the CRF system and the important role this neurohormone plays in mediating a variety of physiological and pathophysiological responses.

PHARMACOLOGY OF CRF_1 AND CRF_2 RECEPTORS

The pharmacology of the class B GPCRs CRF_1 and CRF_2 have been described, characterized, and studied since the discovery of the peptide itself. Given the importance of the system and the desire to link human pathology to interventions that would have some therapeutic benefit, basic biochemical studies became focused on the distinct interactions of both peptide and nonpeptide ligands. As more and more ligand tools were developed, there was a clear understanding of the relationships between the peptide and small molecule interactions and the functional relationship between the two.

Interaction of Peptide and Nonpeptide Ligands

The peptide interactions of the CRF family of ligands with the receptors has been extensively reviewed and reported to be the same for both receptor subtypes. Studies using the CRF_1 receptor have posited that the C-terminal region of CRF adopts an α-helical conformation, which binds to the extracellular domain of the receptor. This initial low-affinity interaction then allows the binding of the N-terminal portion of CRF to a portion of the transmembrane domain that ultimately translocates the signal to function within the cell. An important concept of this mechanism is that this tethering of the N-terminal portion of CRF greatly increases its local concentration at the receptor. This two-domain binding of the peptide implies that the peptide CRF may remain tethered to the

N-terminus of the receptor when the receptor is in its inactive state.[3,16,29,32] This has direct implications when examining the abilities of non-peptide small molecule ligands to block the function of these receptors for therapeutic benefit. It is interesting to note that truncated forms of CRF such as D-Phe CRF(12-41) or the synthetic peptides Astressin or Astressin-2 bind with very high affinity to the N-terminus of their receptors but since they lack the ability to interact with the transmembrane domain, act as complete antagonists.

Small molecule tools have enabled the understanding of the mechanisms of receptor blockade for this family of receptors. It is important to note from the outset that while there have been thousands of CRF$_1$-specific small molecule receptor antagonists reported in the academic and patent literature, the tremendous effort that was put forth in identifying selective CRF$_2$ receptor antagonists has yet to provide molecules that do not interact with the CRF$_1$ receptor. This is an interesting scientific problem given the high degree of similarity of the two receptor subtypes especially in the transmembrane domain regions. The discovery and characterization of these small molecules will be described below. As will be illustrated in the next sections, these molecules were found to bind well within the transmembrane domain region of the CRF$_1$ receptor with virtually no overlap in the binding sites of the peptides, thus defining these interactions as allosteric to each other. Therefore, any inhibition of CRF$_1$ receptor function by nonpeptide small molecules that are allosteric to the peptide-binding domain must by definition be a result of a physical conformational change in the receptor protein rather than a simple blockade of peptide agonist binding. Furthermore, as postulated above, agonist peptides can be bound to the N-terminus of the receptor protein despite the existence of a small molecule occupying its binding

site in the transmembrane region. This has clear clinical and therapeutic implications. If the peptide is essentially tethered to the N-terminus, then it is positioned to signal through the receptor as soon as the small molecule dissociates from its binding site. Given these dynamics, the longer the small molecule can occupy the receptor and maintain its inactive conformation, the longer the therapeutic benefit. It became clear that in addition to the high affinity and selectivity these small molecules needed as potential therapeutics, they would also have to maintain the property of a long kinetic receptor off-rate to maximize their therapeutic efficacy.

The Argument for Receptor Kinetics

Accounting for the receptor kinetics in the discovery of potential therapeutics is not a novel concept. This has been described in elegant detail for compounds blocking the angiotensin receptor in hypertension.[20] Applying this concept to molecules that block the function of the CRF$_1$ receptor allosterically, adds another dimension to the drug discovery and lead optimization process. Correlating the apparent affinity of the molecules with in vitro and in vivo efficacy became the critical criterion for progressing compounds into advanced stage development. As described above, the receptor occupancy and the pharmacokinetic and pharmacodynamic properties of these molecules all had to be optimized simultaneously. A great deal of work has been published in this area examining the relationships, both physical and experimental, of small molecule interaction that have led to their testing in large clinical trials.[12,33,39] An illustration of this effort is shown in Table 6.1 using molecules that have in fact progressed into Phase 2 human clinical studies. Extracted from previously published reports,

TABLE 6.1 Comparison of the Properties of Select CRF$_1$ Receptor Antagonists That Have Been in Human Clinical Trials for Anxiety or Depression

Compound	Clinical Trials Identifier	[^{125}I]-Sauvagine Binding Affinity (Ki nM)	Dissociation Time (t$_{1/2}$ min)	Kinetic Affinity (Ki nM)	In Vivo ACTH (% Veh AUC @ 120 min)
CP-316,311	NCT00143091	1.9	4.1	12	76
Emicerfont/GW-876008	NCT00397722	20	10	39	87
Pexacerfont/BMS-562086	NCT00135421 NCT00481325	7.4	14	19	96
ONO-2333Ms	NCT00514865	1.2	17	15	78
Verucerfont/GSK-561679	NCT00733980 NCT01018992	1.3	70	3.9	43
NBI 30775/R121919	Ref. 40	2.6	130	0.36	42

Comparison of clinical compounds in their ability to inhibit [^{125}I]-sauvagine, their dissociation times from the receptor and calculated kinetic affinity, and their ability to inhibit ACTH release in adrenalectomized rats. Clinical trials highlighted are those directly testing efficacy of these molecules for anxiety or depression only and listed on www.ClinicalTrials.gov with the exception of NBI-30775 (R121919), which was tested in an open label study in Major Depression. Multiple other exploratory clinical trials exist for each of these compounds.
Data adapted from Fleck BA, Hoare SR, Pick RR, Bradbury MJ, Grigoriadis DE. Binding kinetics redefine the antagonist pharmacology of the corticotropin-releasing factor type 1 receptor. J Pharmacol Exp Ther. *2012;341(2):518–531. doi:10.1124/jpet.111.188714.*

Table 6.1 summarizes the most advanced chemical compounds compared using their binding affinity, dissociation time, the affinity determined via kinetic analysis, and the in vivo efficacy for inhibition of ACTH release. It is clear that all molecules had high affinity for the receptor as defined by [^{125}I]-sauvagine binding. What is also clear is that if the compounds are rank ordered by their receptor off rates or dissociation times, this correlated very well with their ability to block the ACTH release from adrenalectomized rats. Coupled with the notion described above that compounds bind allosterically, and the fact that adrenalectomy causes an uncontrolled release of both CRF from the hypothalamus and ACTH from the pituitary, it is not surprising that compounds would have to have very long residence times on the receptor to effectively block the response of CRF in vivo. The data accumulated in Table 6.1 provides evidence for a link between ligand residence time, defined by receptor off-rate, and offset kinetics and insurmountable/noncompetitive antagonism at the CRF$_1$ receptor.

SMALL MOLECULE NONPEPTIDE CRF$_1$ RECEPTOR ANTAGONISTS

The discovery of small molecule receptor antagonists did not emerge from any advances in the peptide arena. Although peptide antagonists had been synthesized, their poor pharmacokinetics (oral bioavailability, rapid clearance, and minimal brain penetration) and general lack of CRF$_1$ receptor selectivity did not yield any information useful in the development of nonpeptide blockers. Utilizing large screening libraries, many small molecule antagonists were developed and profiled for their pharmacokinetic and pharmacodynamics properties. The first disclosure of a small molecule CRF$_1$ receptor antagonist was in 1991 and since then, numerous low-molecular weight ligands have been developed that potently bind and antagonize the CRF$_1$ receptor (for detailed reviews see[13,15,19]). What is important to note is that the chemical starting points for optimization of these varied structures were as diverse as the chemical libraries amassed following years of small molecule research in the pharmaceutical industry. These structurally distinct, specific and selective CRF$_1$ receptor antagonists were discovered through extensive screening methodologies utilizing these diverse chemical libraries and painstakingly optimized for drug-like characteristics.[21] As eluded to above, these molecules were found to bind well within the transmembrane domain region of the CRF$_1$ receptor with virtually no overlap in the binding sites of the peptides.

First described in 1997, point mutation studies of the CRF$_1$ receptor identified transmembrane domains three and five as critical points of interaction of the

small molecules compared to the binding domains of the peptides.[22] Regardless of the initial chemical starting points identified through large chemical library screening, when these molecules were ultimately modified into compounds that could be tested in the clinic, they all seem to bind to exactly the same pocket deep within the transmembrane domain of this receptor. In fact, this has now been clearly demonstrated in studies that solved the crystal structure of the CRF$_1$ receptor bound to a typical CRF$_1$ receptor antagonist.[17] One inescapable conclusion from all these studies is that so far, this receptor has been shown to have a defined binding pocket for small molecules that can modulate its activity but more importantly, it will not tolerate structures that deviate from this well-defined pocket. As will be hypothesized below, if the current lack of progress of clinical candidates is due to the physical limitations of this particular binding site, then a complete revision of the mechanisms through which we can block this system will have to be understood and exploited.

CLINICAL EXPERIENCE AND RELEVANCE

The first publication of a CRF$_1$ receptor antagonist used in the human population was a small open-label study in a group of major depression patients.[40] This preliminary study energized the field and instigated a tremendous effort in the design and development of subsequent molecules. Multiple compounds were discovered and evaluated clinically for efficacy ranging from major depression, social anxiety generalized anxiety, posttraumatic stress disorder, and irritable bowel syndrome as well as studies in alcohol abuse. In these well-controlled placebo and many times comparator trials, the efficacy of a CRF$_1$ antagonist has not been demonstrated. The clinical indications that have been examined have all been based on the following premise that CRF is the primary regulator of the stress response, and since these diseases have been reported to have a stress component in their etiology or manifestation, a CRF$_1$ receptor blocker would be a viable therapeutic.

Central Nervous System Indications

In the CNS, the stress axis has been implicated in playing a major role in depression and anxiety-related disorders and has been extensively reviewed.[6,7,23] Many patients with major depression have been shown to have a perturbed HPA axis and are hypercortisolemic defined by an abnormal dexamethasone suppression test.[5,18] Since it was demonstrated that CRF is the primary regulator of the stress response through the HPA axis in animals, it was a logical extension that hypersecretion or hyperactivity of CRF in brain might underlie the symptomatology seen in major depression or anxiety. Major clinical observations collectively provided overwhelming evidence that implicated the CRF system in the pathologic development of CNS diseases. For example, the first indication was the observation that the concentration of CRF was found to be significantly higher in the cerebrospinal fluid (CSF) of drug-free individuals with major depression[28] and subsequently, it was found that the levels of CRF in the CSF were elevated while the CRF receptors were downregulated in the brains of suicide victims.[2,27] Furthermore, this elevation could be reversed following successful treatment with electroconvulsive therapy.[26] Since then, a myriad of studies have provided evidence for the role of CRF in depression and anxiety-related disorders.[4,11,30] Beyond all of the animal studies where CRF_1 antagonists have demonstrated efficacy in anxiety and depression behaviors, the now routinely available genetic analyses for disease states have offered a new approach in the selection of candidate populations with which to test the system.[25,31,38] All of these data, evidence generated over the past three decades, created the overwhelming motivation to test this hypothesis in the clinic in a number of diseases of the CNS. To date however, no CRF_1 receptor antagonist has progressed in development beyond Phase 2 proof-of-concept trials. A reevaluation of the hypothesis and specific disease-state or population is therefore warranted.

Endocrine Indications

From a potential therapeutic perspective, while decades of study have focused on CNS and peripheral disorders like irritable bowel syndrome, the potential therapeutic indications at the level of the pituitary have been largely ignored. This is not entirely surprising or unexpected given the large potential patient populations with psychiatric disease, stress disorders, and immune and gastrointestinal disorders. A very small but unmet medical need does however exist for individuals whose HPA axis is chronically activated either through overexpression of hypothalamic CRF or genetic disorders that disrupt the axis. One example of such a disorder is congenital adrenal hyperplasia. This disease was first described in the 1950's and is a complex genetic disorder in which the adrenals produce little or no cortisol. The most predominant form of this disease is a mutation in the 21-hydroxylase enzyme in the adrenal, which is directly responsible for the conversion of steroid precursors ultimately to glucocorticoids. This leads to a significant reduction in the negative feedback loop of cortisol at the level of both the hypothalamus and the pituitary, in effect causing substantial increases in both CRF and

ACTH respectively. The elevated ACTH stimulates the adrenal steroidogenic pathways and without the ability to convert steroid precursors to cortisol, upstream precursors, such as 17α-hydroxyprogesterone, accumulate and generate high levels of androgens that are the root pathology of this disease. Current and most common treatment therapies are direct administration of glucocorticoids and are intended to not only replace the missing cortisol but also decrease the ACTH drive of androgen excess. In order to achieve this level of control, supraphysiological doses of glucocorticoids are required to effectively suppress adrenal androgens leading to issues associated with high steroid use. CRF receptor antagonists, which act at the level of the pituitary and can directly block CRF from stimulating ACTH release, may help in the overall treatment and well-being of these patients. By inhibiting the release of ACTH directly, the pressure on the adrenal to produce androgens should be lessened and administered cortisol may be reduced to physiologic replacement levels. A very recent clinical pharmacology study has been performed and shown some promising results.[36]

SUMMARY AND CONCLUSIONS

Corticotropin-releasing factor has long been described as the primary regulator of the HPA axis and a target for the discovery of molecules that may have therapeutic utility in stress-related disorders. Despite heroic efforts in evaluating multiple CRF_1 receptor antagonists in diseases such as major depressive disorder, generalized anxiety, social anxiety, posttraumatic stress disorder, irritable bowel disease, and stress in alcoholism (see www.Cinicaltrials.gov for details on numerous studies), none of the molecules examined have shown any clinical efficacy. While there may be fundamental issues with selection of the appropriate patient population, the specific design of the clinical trials, or the disease hypotheses themselves, it is evident that the tools we have currently at our disposal cannot demonstrate a therapeutic benefit in psychiatric disease. One common feature of these clinical trials (where actually measured) was that there was an effect on HPA activity. In this patient pool, there has been evidence of inhibition of ACTH release and this at least supports the utility of these agents as potential therapeutics in endocrine disorders where the primary requirement is modulation of ACTH release from the pituitary. The last three decades of intense study of this system should be examined with a critical view and the hypotheses modified to account for what the field has learned from these failures. CRF remains an important biochemical system both in the CNS and in the periphery and deserves a next-generation investigation.

References

1. Adao R, Santos-Ribeiro D, Rademaker MT, Leite-Moreira AF, Bras-Silva C. Urocortin 2 in cardiovascular health and disease. *Drug Discov Today*. 2015;20(7):906–914. http://dx.doi.org/10.1016/j.drudis.2015.02.012.

2. Arato M, Banki CM, Bissette G, Nemeroff CB. Elevated CSF CRF in suicide victims. *Biol Psychiatry*. 1989;25(3):355–359.

3. Assil IQ, Qi LJ, Arai M, Shomali M, Abou-Samra AB. Juxtamembrane region of the amino terminus of the corticotropin releasing factor receptor type 1 is important for ligand interaction. *Biochemistry*. 2001;40(5):1187–1195.

4. Aubry J-M. CRF system and mood disorders. *J Chem Neuroanat*. 2013;54:20–24. http://dx.doi.org/10.1016/j.jchemneu.2013.09.003.

5. Aubry JM, Gervasoni N, Osiek C, et al. The DEX/CRH neuroendocrine test and the prediction of depressive relapse in remitted depressed outpatients. *J Psychiatr Res*. 2006;41(3–4):290–294.

6. Bali A, Singh N, Jaggi AS. Neuropeptides as therapeutic targets to combat stress-associated behavioral and neuroendocrinological effects. *CNS Neurol Disord Drug Targets*. 2014;13(2):347–368. http://dx.doi.org/10.2174/1871527313666140314163920.

7. Binder EB, Nemeroff CB. The CRF system, stress, depression and anxiety-insights from human genetic studies. *Mol Psychiatry*. 2010;15(6):574–588. http://dx.doi.org/10.1038/mp.2009.141.

8. Broadbear JH, Winger G, Rivier JE, Rice KC, Woods JH. Corticotropin-releasing hormone antagonists, astressin B and antalarmin: differing profiles of activity in rhesus monkeys. *Neuropsychopharmacology*. 2004;29(6):1112–1121.

9. Chan WY, Frampton CM, Crozier IG, Troughton RW, Richards AM. Urocortin-2 infusion in acute decompensated heart failure: findings from the UNICORN study (urocortin-2 in the treatment of acute heart failure as an adjunct over conventional therapy). *JACC Heart Fail*. 2013;1(5):433–441. http://dx.doi.org/10.1016/j.jchf.2013.07.003.

10. Chen R, Lewis KA, Perrin MH, Vale WW. Expression cloning of a human corticotropin-releasing-factor receptor. *Proc Natl Acad Sci USA*. 1993;90(19):8967–8971.

11. Devi M, Sharma R. A review on animal models of depression. *Int J Pharm Sci Res*. 2013;4(10):3731–3736. http://dx.doi.org/10.13040/IJPSR.0975-8232.4(10).3731-36.

12. Fleck BA, Hoare SR, Pick RR, Bradbury MJ, Grigoriadis DE. Binding kinetics redefine the antagonist pharmacology of the corticotropin-releasing factor type 1 receptor. *J Pharmacol Exp Ther*. 2012;341(2):518–531. http://dx.doi.org/10.1124/jpet.111.188714.

13. Gilligan PJ, Li YW. Corticotropin-releasing factor antagonists: recent advances and exciting prospects for the treatment of human diseases. *Curr Opin Drug Discov Devel*. 2004;7(4):487–497.

14. Grammatopoulos DK, Dai Y, Randeva HS, et al. A novel spliced variant of the type 1 corticotropin-releasing hormone receptor with a deletion in the seventh transmembrane domain present in the human pregnant term myometrium and fetal membranes. *Mol Endocrinol*. 1999;13(12):2189–2202.

15. Griebel G, Holsboer F. Neuropeptide receptor ligands as drugs for psychiatric diseases: the end of the beginning? *Nat Rev Drug Discov*. 2012;11(6):462–478. http://dx.doi.org/10.1038/nrd3702.

16. Hoare SR, Sullivan SK, Schwarz DA, et al. Ligand affinity for amino-terminal and juxtamembrane domains of the corticotropin releasing factor type I receptor: regulation by G-protein and nonpeptide antagonists. *Biochemistry*. 2004;43(13):3996–4011. http://dx.doi.org/10.1021/bi036110a.

17. Hollenstein K, Kean J, Bortolato A, et al. Structure of class B GPCR corticotropin-releasing factor receptor 1. *Nature*. 2013;499(7459):438–443. http://dx.doi.org/10.1038/nature12357.

18. Ising M, Kunzel HE, Binder EB, Nickel T, Modell S, Holsboer F. The combined dexamethasone/CRH test as a potential surrogate marker in depression. *Prog Neuropsychopharmacol Biol Psychiatry*. 2005;29(6):1085–1093.

19. Kehne J, De LS. Non-peptidic CRF1 receptor antagonists for the treatment of anxiety, depression and stress disorders. *Curr Drug Targets CNS Neurol Disord*. 2002;1:467–493 (Copyright © 2013 U.S. National Library of Medicine.).

20. Lacourciere Y, Asmar R. A comparison of the efficacy and duration of action of candesartan cilexetil and losartan as assessed by clinic and ambulatory blood pressure after a missed dose, in truly hypertensive patients: a placebo-controlled, forced titration study. Candesartan/Losartan study investigators. *Am J Hypertens*. 1999;12(12 Pt 1–2):1181–1187.

21. Lanier M, Williams JP. Small molecule corticotropin-releasing factor antagonists. *Expert Opin Ther Pat*. 2002;12(11):1619–1630.

22. Liaw CW, Grigoriadis DE, Lorang MT, De Souza EB, Maki RA. Localization of agonist- and antagonist-binding domains of human corticotropin-releasing factor receptors. *Mol Endocrinol*. 1997;11(13):2048–2053.

23. Lin E-JD. Neuropeptides as therapeutic targets in anxiety disorders. *Curr Pharm Des*. 2012;18(35):5709–5727.

24. Lovejoy DA, Balment RJ. Evolution and physiology of the corticotropin-releasing factor (CRF) family of neuropeptides in vertebrates. *Gen Comp Endocrinol*. 1999;115(1):1–22. http://dx.doi.org/10.1006/gcen.1999.7298.

25. Menke A, Domschke K, Czamara D, et al. Genome-wide association study of antidepressant treatment-emergent suicidal ideation. *Neuropsychopharmacology*. 2012;37(3):797–807. http://dx.doi.org/10.1038/npp.2011.257.

26. Nemeroff CB, Bissette G, Akil H, Fink M. Neuropeptide concentrations in the cerebrospinal fluid of depressed patients treated with electroconvulsive therapy: corticotropin-releasing factor, β-endorphin and somatostatin. *Br J Psychiatry*. 1991;158:59–63.

27. Nemeroff CB, Owens MJ, Bissett G, Andorn AC, Stanley M. Reduced corticotropin-releasing factor receptor binding sites in the frontal cortex of suicide victims. *Arch Gen Psychiatry*. 1988;45:577–579.

28. Nemeroff CB, Widerlov E, Bissett G, et al. Elevated concentration of CSF corticotropin-releasing factor-like immunoreactivity in depressed patients. *Science*. 1984;226:1342–1344.

29. Nielsen SM, Nielsen LZ, Hjorth SA, Perrin MH, Vale WW. Constitutive activation of tethered-peptide/corticotropin-releasing factor receptor chimeras. *Proc Natl Acad Sci USA*. 2000;97(18):10277–10281.

30. Overstreet DH. Modeling depression in animal models. *Methods Mol Biol*. 2012;829:125–144. http://dx.doi.org/10.1007/978-1-61779-458-2_7 (Psychiatric Disorders).

31. Paez-Pereda M, Hausch F, Holsboer F. Corticotropin releasing factor receptor antagonists for major depressive disorder. *Expert Opin Invest Drugs*. 2011;20(4):519–535. http://dx.doi.org/10.1517/13543784.2011.565330.

32. Perrin MH, Sutton S, Bain DL, Berggren WT, Vale WW. The first extracellular domain of corticotropin releasing factor-R1 contains major binding determinants for urocortin and astressin. *Endocrinology*. 1998;139(2):566–570.

33. Ramsey SJ, Attkins NJ, Fish R, van der Graaf PH. Quantitative pharmacological analysis of antagonist binding kinetics at CRF1 receptors in vitro and in vivo. *Br J Pharmacol*. 2011;164(3):992–1007. http://dx.doi.org/10.1111/j.1476-5381.2011.01390.x.

34. Richards M, Rademaker M. *Urocortins*; 2013.

35. Rivier J, Gulyas J, Kirby D, et al. Potent and long-acting corticotropin releasing factor (CRF) receptor 2 selective peptide competitive antagonists. *J Med Chem*. 2002;45(21):4737–4747.

36. Turcu AF, Spencer-Segal JL, Farber RH, et al. Single-Dose study of a corticotropin-releasing factor receptor-1 antagonist in women with 21-hydroxylase deficiency. *J Clin Endocrinol Metab*. 2016;101(3):1174–1180. http://dx.doi.org/10.1210/jc.2015-3574.

37. Vale W, Spiess J, Rivier C, Rivier J. Characterization of a 41-residue ovine hypothalamic peptide that stimulates secretion of corticotropin and β-endorphin. *Science*. 1981;213:1394–1397.

38. White S, Acierno R, Ruggiero KJ, et al. Association of CRHR1 variants and posttraumatic stress symptoms in hurricane exposed adults. *J Anxiety Disord*. 2013;27(7):678–683. http://dx.doi.org/10.1016/j.janxdis.2013.08.003.

39. Zhou L, Dockens RC, Liu-Kreyche P, Grossman SJ, Iyer RA. In vitro and in vivo metabolism and pharmacokinetics of BMS-562086, a potent and orally bioavailable corticotropin-releasing factor-1 receptor antagonist. *Drug Metab Dispos*. 2012;40(6): 1093–1103. http://dx.doi.org/10.1124/dmd.111.043596.

40. Zobel AW, Nickel T, Kunzel HE, et al. Effects of the high-affinity corticotropin-releasing hormone receptor 1 antagonist R121919 in major depression: the first 20 patients treated. *J Psychiatr Res*. 2000;34(3):171–181.

7

Tracking the Coupling of External Signals to Intracellular Programs Controlling Peptide Synthesis and Release in Hypothalamic Neuroendocrine Neurons

A.M. Khan[1,2], E.M. Walker[1], A.G. Watts[2]

[1]UTEP Systems Neuroscience Laboratory, El Paso, TX, United States;
[2]University of Southern California, Los Angeles, CA, United States

Abstract

Corticotropin-releasing hormone (CRH) neuroendocrine neurons of the paraventricular hypothalamic nucleus constitute the final common pathway for the hypothalamo–pituitary–adrenal axis. These neurons trigger an important cascade of hormone releasing events that allow the organism to respond adaptively to stress, including the release of CRH from the hypothalamus, adrenocorticotropic hormone from the anterior pituitary gland, and cortisol (corticosterone in rats) from the adrenal cortex. Despite the central place of CRH neuroendocrine neurons in this hierarchy, the precise manner by which stimuli engage the brain to drive these neurons remains unclear. In this chapter, we describe experiments establishing functional linkages among a peripheral stressor (glycemic challenge), hindbrain-originating neural circuits, and intracellular programs that control the synthesis and release of CRH neuropeptide. A model is presented based on these studies that offers a testable framework upon which to build, and extensions of these findings to larger brain networks are discussed.

INTRODUCTION

An important goal in neuroendocrinology is to define the functional linkages within brain circuits that allow an organism to respond appropriately to stressful stimuli. In mammals, such responses usually include the activation of corticotropin-releasing hormone (CRH) neurons in the parvicellular paraventricular hypothalamic nucleus (PVH). These neurons constitute the principal neural control system of the hypothalamic–pituitary–adrenal (HPA) axis, and their activation is the primary means by which the neuropeptides CRH and arginine vasopressin (AVP) are released into the portal circulation to trigger adrenocorticotropic hormone (ACTH) secretion from pituitary corticotrophs and, ultimately, glucocorticoid release from the adrenal gland. Understanding the mechanisms by which stressful events trigger neuropeptide secretion and synthesis in CRH neuroendocrine neurons may help explain the remarkable ability of the HPA axis to respond to such events with exquisite fidelity and specificity in the face of unpredictable situations.

KEY POINTS

- The concept of stimulus-response coupling in relation to neuroendocrine responses to stress is introduced.
- The anatomy of the corticotropin-releasing hormone (CRH) neuroendocrine neurons in the paraventricular hypothalamic nucleus is presented.
- The organization of hindbrain-originating catecholamine inputs to the CRH neuroendocrine system is described.
- Experimental studies interrogating catecholamine-CRH neuron coupling in vivo are described.
- Mitogen-activated protein kinase is a rapid intracellular effector in vivo following glycemic challenges.
- Catecholamines are sufficient to recapitulate responses to glycemic challenges.
- Catecholaminergic pathways from hindbrain are necessary for CRH responses.
- A model of signal-effector coupling in CRH neuroendocrine neurons is presented.
- Catecholaminergic pathways from hindbrain are necessary for neuronal responses in the arcuate hypothalamic nucleus during glycemic challenges.
- Future directions for this research are discussed.

In this chapter, we highlight studies that have identified functional linkages within the brain that enable CRH neuroendocrine neurons in the hypothalamus to become activated in response to peripheral signals that trigger stressful events, such as changes in glycemic status. In particular, we highlight a fundamental role of

hindbrain-originating neuronal populations that signal through catecholamine neurotransmitters as being critical in conveying information about peripheral glycemic challenge to CRH neuroendocrine neurons. For additional reviews on related aspects of this topic, please consult Wittmann[84] and Watts and Khan.[81]

THE PVH AND CRH NEUROENDOCRINE NEURONS

The PVH (Fig. 7.1), which contains CRH neuroendocrine neurons, is an important coordinator of neuroendocrine, autonomic, and behavioral responses to a variety of homeostatic challenges.[74] The rat PVH, the focus of this chapter, consists of distinct neuronal subpopulations,[70] each of which projects primarily to unique target regions of the central nervous system.[74] Rat PVH neurons have been classified into two main groups based on their target projections (Fig. 7.1). First, preautonomic neurons that send descending brainstem and spinal projections are found mainly in the dorsal and lateral parvicellular parts, as well as in the ventral zone of the medial parvicellular part, of the PVH (Fig. 7.1D, regions labeled "a"[66]). Second, neuroendocrine neurons include those located in the dorsal zone of the medial parvicellular part of the PVH (Fig. 7.1D, region labeled "b_1") that send projections into the neurohemal zone of the median eminence, where they release a variety of hypophysiotropic factors, including CRH.

CRH neuroendocrine neurons are secretomotor neurons that release CRH as their primary motor output into the portal vasculature to trigger ACTH secretion from pituitary corticotropic cells. ACTH, in turn, triggers cortisol release (corticosterone in rats) from the adrenal cortex (Fig. 7.1A), which can exert negative feedback on CRH synthesis and release.[79] In addition to releasing CRH from their axonal terminals in the median eminence, CRH neuroendocrine neurons must also continually synthesize new CRH neuropeptide and have intracellular programs regulating *Crh* gene expression under basal and stimulated conditions.[83]

HINDBRAIN CATECHOLAMINE INPUTS TO THE CRH NEUROENDOCRINE SYSTEM

Since the pioneering studies by Dahlström and Fuxe[13,14,27–29] and Hökfelt and colleagues,[36] who identified discrete neuronal populations of catecholamine-containing neurons and their axonal terminals in the brain using staining methods established earlier by Falck, Hillarp and colleagues[22,9] (reviewed by Refs. 15,35), much work has been done to establish that the primary recipients of projections from catecholaminergic (CA) neurons include

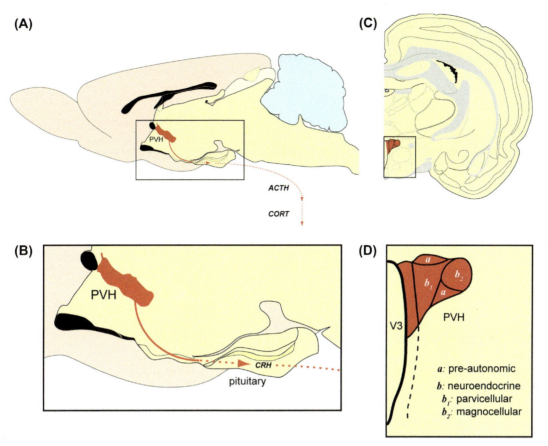

FIGURE 7.1 A view of the paraventricular hypothalamic nucleus (PVH) in the rat. (A) Sagittal view of the rat brain, with the *boxed region* highlighting the PVH (in red). The *boxed region* is enlarged in (B). (C) A coronal section through the brain, at the level of the PVH (in the *boxed region*, in red). The *boxed region* in (C) is enlarged in (D), to highlight subregions that contain preautonomic neurons (*a*) or neuroendocrine neurons (b_1 and b_2). Corticotropin-releasing hormone (CRH) neuroendocrine neurons, the topic of this chapter, are found in region b_1. *All figures adapted from Swanson LW.* Brain Maps: Structure of the Rat Brain. *3rd ed. Amsterdam: Elsevier; 2004; Simmons DM, Swanson LW. Comparison of the spatial distribution of seven types of neuroendocrine neurons in the rat paraventricular nucleus: toward a global 3D model.* J Comp Neurol. *2009;516(5):423–441.*

subpopulations of neurons in the PVH. In particular, beginning with the advent of tract tracing and immunohistochemical methods in the 1980s and development of viral tracing methods since then, the major neural pathways from CA neurons to their targets in the hypothalamus have been elaborated gradually.[74,65,66,75,67,76,12,11,58,8,69]

Fig. 7.2 provides a basic working model of our current conceptualization of how hindbrain-originating afferents to the PVH are organized. Although hindbrain afferents to the rat PVH are far from being mapped systematically, it is well documented that the brain regions of origin for these PVH-projecting CA neuronal populations include the nucleus of the solitary tract (NTS), a principal sensory relay for information from the periphery. In addition to the viscerosensory inputs it receives from cranial nerve afferents, the NTS also receives spinal inputs that likely convey somatosensory information to the NTS (Fig. 7.2A[54]). Recent evidence also suggests that signals encoding slow-onset hypoglycemia, transmitted by glucosensors in the portal mesenteric vein, travel by way of spinal rather than vagal afferents to reach higher

order brain centers, including the NTS[26,2] (reviewed by Ref. 20). Tract tracing experiments combined with immunohistochemical studies have identified several neuropeptides that are colocalized with catecholamines in hindbrain-originating neuronal populations projecting to the PVH, which (in addition to the NTS) also include the locus coeruleus (LC) and ventrolateral medulla (VLM) (Fig. 7.2B; see also Fig. 7.3). In particular, many CA neurons in the hindbrain also express neuropeptide Y,[48] cocaine- and amphetamine-related transcript,[23] and pituitary adenylate cyclase-activating polypeptide.[16]

IN VIVO INTERROGATION OF CATECHOLAMINE-CRH NEURON COUPLING

A major challenge of experimental neuroendocrinology is to test the relationships between signal and effector in vivo, in a manner that allows insight about the identity and nature of the functional linkages that connect

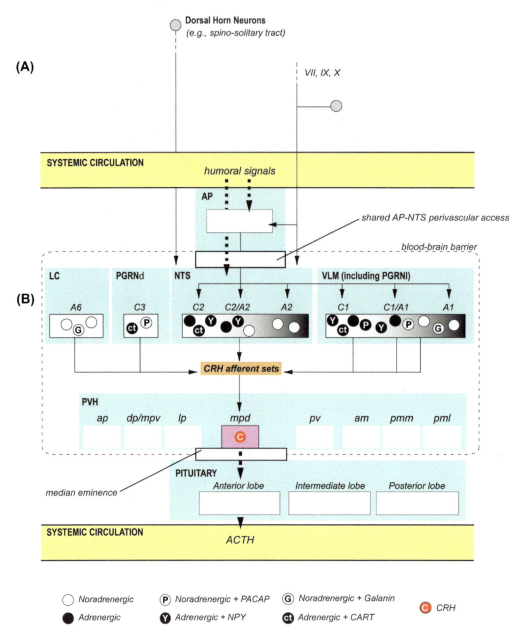

FIGURE 7.2   A model of the connectional organization for hindbrain neural inputs to the paraventricular hypothalamic nucleus in the rat. (A) Portion of the model in which sensory inputs to the hindbrain are shown from peripheral sources, including the spinal cord and cranial nerves VII (facial nerve), IX (glossopharyngeal nerve), and X (vagus nerve). Humoral signals from the circulation pass through a shared perivascular space between the area postrema (AP) to contact the nucleus of the solitary tract (NTS), the hindbrain target of the spinal and cranial nerve inputs. (B) Portion of the model that shows the basic organization of major hindbrain regions described in this chapter In addition to the NTS, these regions include the locus coeruleus (LC), the dorsal portion of the paragigantocellular reticular nucleus (PGRNd), and the ventrolateral medulla (VLM). The latter region also includes portions of the lateral PGRN (PGRNl). Each of these hindbrain regions contains catecholaminergic cell populations designated by the nomenclature (in italics: A6, C3, etc.) used by their original discoverers (see Hindbrain Catecholamine Inputs to the CRH Neuroendocrine System section for details). These catecholaminergic cell populations include specific neuronal phenotypes listed in the legend at the bottom of the figure. Generally, all *open white circles* represent noradrenergic neurons and all *black circles* represent adrenergic neurons, with the letters inside some of these neurons designating various peptides colocalized in the neurons: *CART (ct)*, cocaine- and amphetamine-related transcript; *G*, galanin; *NPY (Y)*, neuropeptide Y; *PACAP (P)*, pituitary adenylate cyclase-activating polypeptide. All of these neuronal populations send axonal projections to CRH (*C*, corticotropin-releasing hormone) neurons in the paraventricular hypothalamic nucleus (PVH); collectively these afferents are designated as "CRH afferent sets." The CRH neuroendocrine neurons receiving these inputs reside in the dorsal zone of the medial parvicellular part of the PVH (mpd). See Swanson[73] for explanations of the abbreviations for the other PVH subregions. The CRH neuroendocrine neurons send their projections to the neurohemal zone of the median eminence to release CRH into the portal vasculature of the anterior lobe of the pituitary gland. This lobe contains pituitary corticotroph cells that, in turn, release adrenocorticotropic hormone (ACTH) into the systemic circulation to ultimately cause the release of corticosterone (cortisol in humans) from the adrenal cortex (not shown). See Hindbrain Catecholamine Inputs to the CRH Neuroendocrine System section for details regarding the primary studies demonstrating catecholaminergic inputs to the CRH neuroendocrine system.

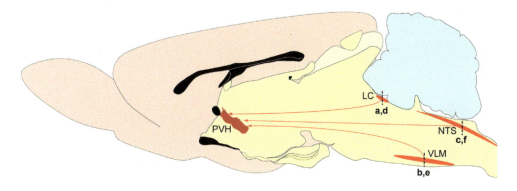

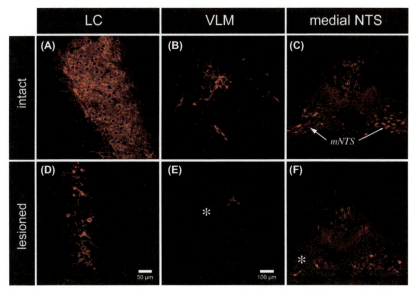

FIGURE 7.3 Overview of hindbrain regions examined in our dopamine-β hydroxylase (DBH)-saporin lesion studies.[41,43] Top: sagittal view of the rat brain adapted from Swanson[73] depicting the PVH[70] as well as three major regions of the hindbrain examined in our studies: the locus coeruleus (LC), ventrolateral medulla (VLM), and nucleus of the solitary tract (NTS). The *lowercase letters* denote the panels shown below the figure. Bottom: Immunostaining for DBH, which in these structures marks noradrenergic neurons. (a, b, c) Images showing DBH-immunoreactive neurons in rats receiving sham lesions of retrogradely transportable immunotoxin into the PVH. (d, e, f) Images of DBH-immunoreactive neurons in rats receiving DBH-saporin immunotoxin injections into the PVH (see text for details). Note the frank losses of neurons in these images (*) relative to the corresponding regions above them in panels a, b, and c.

the two. In terms of the present discussion, it remains unclear how stressful stimuli affecting an organism are conveyed to CRH neuroendocrine neurons in a manner that ensures an adaptive response to the challenge the organism faces. By what route and manner do signals from such stimuli travel within the body to reach these neurons deep within the hypothalamus? What couples the general stimulus with the specific response?

Answers to these questions for CA neuron–CRH neuroendocrine neuron coupling have been addressed for specific organismal challenges that elicit a stress response. These challenges include those categorized as exterosensory stressors—stimuli that originate from outside the organism—or those that are physiological, or interosensory stressors (e.g., see Ref. 18). In particular, hindbrain CA neuronal projections to the PVH have been shown to help orchestrate CRH neuroendocrine

responses to exterosensory stressors such as restraint,[19] as well as interosensory stressors such as glycemic[61,63] and immune challenges.[21,50,5,68,30] Most recently, Eileen Hasser and colleagues have also established a clear role of CA neuronal afferents to the PVH in conveying signals encoding hypoxia to the CRH neuroendocrine system.[44–46]

Although these studies have established linkages among various types of stressor, CA neuronal projections to the PVH, and CRH neuroendocrine responses, a key issue left unaddressed is tracing the linkages further to within the CRH neuroendocrine neuron itself. In other words, once catecholamine signals arrive at CRH neuroendocrine neurons and bind to adrenergic receptors on the cell surface in those neurons, what couples the CA signals to programs that trigger the release of existing CRH neuropeptide from these neurons and programs that initiate new synthesis of CRH neuropeptide from its gene? In order to address these questions, we sought an intracellular marker that would track activation of CRH neurons in the PVH following various challenges. Below, we review the studies in which we identified such a marker and how its activation to glycemic challenge was determined to involve initial recruitment from CA signaling and to require endogenous CA afferents from the hindbrain.

MAP KINASE AS A RAPIDLY ACTIVATED INTRACELLULAR EFFECTOR IN CRH NEUROENDOCRINE NEURONS IN VIVO

To address the stimulus-effector coupling issue with respect to intracellular control of CRH synthesis and release, we sought an experimental system that satisfied two criteria: (1) the stressor we select as an experimental paradigm is already established as being coupled to the CA afferent system driving CRH neuroendocrine responses in vivo and (2) any activation marker we identify is already known to become activated within the time frame of *Crh* gene expression (i.e., within minutes rather than hours). To fulfill the first criterion, we opted to employ a peripheral glycemic challenge as the physiological stimulus. A study performed by Sue Ritter and colleagues served as the basis for this selection, in which they had shown that selective immunotoxin-mediated destruction of hindbrain CA afferents to the PVH abolished CRH neuroendocrine responses to peripheral administration of 2-deoxy-D-glucose (2-DG), a glucose analogue that inhibits glycolysis[61] (reviewed by Ref. 62). In collaboration with her laboratory, we also found that CRH gene expression was also reduced markedly following immunotoxin-mediated CA deafferentation of the PVH.[63]

Although we were interested in identifying an intracellular marker that serves as a reliable tracker of *Crh* expression or CRH peptide release (i.e., the marker as a "sign" of expression or release), we were reminded by Bullock[6] that it would be ideal if the marker itself was also causally linked to the signaling pathway that triggers such expression or release (i.e., a "signal" that exerted influence on these processes). In other words, whereas a sign of neuroendocrine output was important to use as a tool, identifying a sign that also served as a signal that was causally involved in producing that output was an even more important experimental goal. Given that the time course for *Crh* gene expression can occur within as little as 5 min following the onset of a stressful stimulus,[47] we sought a biomarker that had its own activation onset within a similar temporal domain. Detection of the expression of Fos nuclear protein is commonly used as a means to track neuronal activation after the fact in tissue

post mortem.[34,33] However, Fos protein expression usually is not detectable for up to 30–90 min after stimulus onset and is not always driven by neuronal firing rates per se (e.g., see Ref. 53). In contrast, posttranslational modifications that switch signaling intermediates on and off within cells would appear to be better suited as trackers of activation onset in a cell for processes that occur within a few seconds or minutes. Prompted by reports of immunocytochemical detection of phosphorylated mitogen-activated protein (MAP) kinase as a means to track activation,[72] we opted to test whether this enzyme was similarly activated in CRH neuroendocrine neurons following 2-DG administration.[40,82]

We found that the phosphorylated forms of p44/p42 MAPKs (pERK1/2) were rapidly detected in PVHmp cells after intravenous 2-DG administration. Using a combination of immunocytochemistry and in situ hybridization methods that could effectively localize both phospho-ERK immunoreactive protein and CRH mRNA, we found that pERK1/2 immunoreactivity is detectable 10 min within most PVH neurons containing CRH mRNA (79% of mean total CRH cells counted) and also in many non-CRH neurons (46% of mean total sampled cells). Fos protein, in contrast, was not detected within this time period.[40]

CATECHOLAMINES ARE SUFFICIENT TO RECAPITULATE EFFECTS OF GLYCEMIC CHALLENGES ON CRH NEURONS

Given that ERK1/2 are activated in response to peripheral 2-DG administration and that hindbrain CA neuronal projections are required for 2-DG to trigger CRH gene expression or release from PVH neuroendocrine neurons in vivo, we hypothesized that CA transmitters engage phospho-ERK1/2 within CRH neurons. To test this hypothesis, we designed a series of experiments to ask whether acute delivery of the prototypical catecholamine, norepinephrine (NE) to PVH CRH neuroendocrine neurons was sufficient to produce the sequence of outputs and readouts that 2-DG was able to produce in our previous studies (i.e., *Crh* gene expression, ACTH release, and ERK1/2 phosphorylation).

Cole and Sawchenko[10] demonstrated that NE microinjected directly into the PVH of conscious rats was associated with elevated levels of plasma corticosterone at 30 min postinjection as well as increased Fos expression in the PVH at 2 h postinjection. We reasoned that if 2-DG-induced increases in *Crh* expression were mediated by a CA signal, then direct administration of NE would also trigger activation of ERK1/2 in CRH neuroendocrine neurons, in addition to triggering *Crh* expression and CRH neuropeptide release. To test this hypothesis, we surgically implanted indwelling guide cannulas into the brains of rats placed under sedation, targeting the cannulas to reside directly above the PVH using carefully delineated maps of the structure in stereotaxic space. Following surgical recovery, pharmacological trials were performed. Injections of NE into the PVH triggered increases in *Crh* gene expression as well as elevations in plasma ACTH and corticosterone. Importantly, we also found that NE, but not saline injections, also triggered rapid increases in phospho-ERK1/2 expression in PVH neurons immunopositive for CRH.[42]

One of the challenges associated with establishing cause and effect relationships that link stimulus and effector in vivo is to isolate observed effects and attribute them to the appropriate causal factor. In our *post mortem* analysis of brain tissue from NE-injected animals, we observed phospho-ERK activation not only in the PVH but in other brain regions as well. Curiously, we found that the PVH, along with other regions, was activated bilaterally, despite the fact that the NE injections we delivered into the PVH were unilateral. We also found that sites quite distant from the injection site were also activated bilaterally, suggesting that NE injection was associated with sensorimotor feedback along bilaterally distributed circuits that could produce activation of distant sites that might be linked anatomically to the injected region. Interestingly, Cole and Sawchenko[10] also reported this bilateral activation. Together, the observations called into question whether NE delivered to the PVH per se was the reason for the phospho-ERK1/2 activation we observed in that region, or for the concomitant *Crh* expression and ACTH/corticosterone elevations we detected. In sum, using the in vivo experimental paradigm we selected produced results that suggested other factors could be confounding our interpretation of direct effects of NE on CRH neuroendocrine neurons.

To address these confounds, we reasoned that an isolated tissue preparation containing the PVH, maintained ex vivo much the same way as is done for electrophysiological recording experiments, may yield more clarity as to the nature of the actions we observed for NE using the in vivo model. Accordingly, in collaboration with Glenn Hatton and colleagues, we tested the effects of NE using an ex vivo tissue slice preparation containing the PVH first developed in the Hatton laboratory[32]; see Ref. 31. The slices were maintained in an oxygenated solution after being rapidly prepared from freshly dissected brain tissue and, after a requisite equilibrium period in the appropriate temperature- and pH-controlled conditions, the slices were treated with either NE or saline. After treatment, slices were fixed in formaldehyde solution and processed for immunocytochemistry. Rather than freezing and sectioning the slices before staining them with antibodies,[71] we opted to not section the slices further, but instead permeabilize them

with antibodies directly, clear the tissue, and then optically section the immunoreactive neurons in the slice using confocal microscopy. We found that bath-applied NE triggered robust phospho-ERK1/2 immunoreactivity in PVH (including CRH) neurons, which attenuated markedly in the presence of the α_1 adrenergic receptor antagonist, prazosin, or the MAP kinase kinase (MEK) inhibitor, U0126.[42] Thus, in addition to demonstrating ERK activation after NE administration to the slice, the experiments established a causal role for NE in driving this activation through its actions at an adrenergic receptor subtype. In addition, we identified the protein kinase MEK as the upstream mediator of this activation.

Taking the in vivo and ex vivo data together, these studies demonstrated that, at a systems level, PVH-delivered NE is sufficient to account for hindbrain activation of CRH neuroendocrine neurons during glycemic challenge. Moreover, the actions of NE we observed in vivo were unlikely due solely to sensorimotor activation or nonspecific effects at sites distal from the injection. At a cellular level, these data provided the first demonstration that MAP kinase signaling cascades are intracellular transducers of noradrenergic signals in CRH neurons, implicating this signaling pathway as a component of central neuroendocrine responses during glycemic challenge.

CATECHOLAMINERGIC PROJECTIONS FROM HINDBRAIN ARE NECESSARY FOR CRH RESPONSES TO GLYCEMIC CHALLENGES

Although the experiments described in the preceding section established that NE was sufficient to drive MAP kinase activation in CRH neuroendocrine neurons, a few key questions remained. First, it remained unclear whether endogenous CA afferents originating from the hindbrain were required for ERK activation under conditions of glycemic challenge in vivo. Additionally, although an MEK > ERK pathway was identified as being downstream of adrenergic receptor activation in the PVH, it remained unclear how phospho-ERK1/2 could, in turn, be coupled to *Crh* expression or to mechanisms that trigger CRH neuropeptide release. We therefore performed a series of experiments to address these issues.

First, we sought to create targeted lesions of the CA afferent system to establish its necessity in driving CRH neuroendocrine output following glycemic challenge. To achieve this goal, we used the immunotoxin-based lesioning approach used by Ritter and colleagues[61] and first developed by Picklo et al.[59] Unlike early lesion studies, which targeted CA neurons in the central nervous system using reagents such as 6-hydroxydopamine[77] or which took advantage of immune-mediated destruction of antigens via complement cascades,[3] the immunotoxin-based lesioning approach takes advantage of both the targeting specificity of a monoclonal antibody against dopamine β-hydroxylase (DBH), the enzyme that synthesizes norepinephrine, and the ribosome-inactivating properties of certain toxins such as saporin. The anti-DBH-saporin immunotoxin conjugate (DSAP) is readily taken up by noradrenergic nerve terminals and transported retrogradely to the cell bodies of origin, where it ultimately destroys them.

We injected the DSAP conjugate bilaterally into the PVH of adult rats. A control group of animals received injections of saporin that was conjugated to a non-specific monoclonal immunoglobulin, which is not taken up or transported by noradrenergic terminals. Following a recovery period, during which time the DSAP was allowed sufficient time to transport, the animals received either glycemic challenges (intravenous insulin or 2-DG) or saline vehicle. PVH tissue was processed to identify changes in ERK activation and *Crh* expression, and blood plasma was assayed for ACTH. We found that glycemic challenges activated CRH neuroendocrine neurons rapidly, as demonstrated by marked elevations in phospho-ERK1/2 immunoreactivity that was localized within CRH neurons, as well as increased levels of primary transcripts (hnRNA) for CRH. The intravenous challenges also increased plasma levels of ACTH, indicating measurable neuroendocrine activation of the CRH system. In contrast to these increases, rats with complete bilateral lesions of CA afferents in the PVH displayed marked reductions in each of these indices (phospho-ERK1/2, CRH hnRNA, ACTH). These animals also displayed dramatic losses of CA neurons in the hindbrain, including the LC, VLM, and medial subnucleus of the NTS (Fig. 7.3[43]). These results suggest that hindbrain-originating CA afferents are required for the full responses we observed to glycemic challenges.

A key question regarding these observations is whether the deafferentation induced by DSAP injections impaired the ability of the PVH to even mount any cellular responses to glycemic challenge. In other words, the reductions or absence in phospho-ERK1/2 immunoreactivity, CRH hnRNA, or plasma ACTH might be attributable to general impairment of the PVH itself as a result of nonspecific damage to the region during injection. To address this issue, we delivered a mixed (multimodal) stimulus, anesthesia combined with hypertonic saline, to animals receiving either sham or full lesions of the CA system. We found that in lesioned animals, rats exposed to this multimodal stimulus mounted full phospho-ERK1/2 responses that were comparable in levels to intact controls, demonstrating that the PVH is capable of mounting responses to stressors even in the absence of CA afferents, and that ERK activation in this system is stimulus selective.[41]

Although the experiments described above address the question of whether CA afferents are required for ERK activation in CRH neuroendocrine neurons following glycemic challenge, they do not address whether ERK is coupled downstream to intracellular programs controlling CRH peptide synthesis or release. To address this question, we first asked whether ERK was concomitantly activated in cells in conjunction with the transcription factor whose activation is required for *Crh* expression: cyclic AMP response element-binding protein (CREB).[51] Using dual-label immunofluorescence histochemistry, we found that intravenous insulin or 2-DG treatments were associated with increased levels of phospho-CREB immunoreactivity in the nuclei of cells that displayed cytoplasmic phospho-ERK1/2 immunoreactivity.[41] The colocalized signals indicated that under in vivo conditions, ERK activation was occurring in neurons that displayed CREB activation, making it plausible that their mutual involvement could be driving *Crh* expression.

Although concomitant activation of ERK and CREB in the same PVH neurons after glycemic challenges indicated that ERK could be involved upstream of CREB, the correlation itself was not sufficient to establish a causal link between ERK and CREB. To address this issue, we turned again to the ex vivo slice preparation we employed in our experiments using bath-applied NE (described in Catecholamines Are Sufficient to Recapitulate Effects of Glycemic Challenges on CRH Neurons section). We again treated slices maintained ex vivo with NE in the presence or absence of the MEK inhibitor, U0126. In slices receiving only NE, robust increases in the immunoreactivities for phospho-ERK1/2 and phospho-CREB were evident relative to control slices receiving only vehicle treatment. As with the data we obtained in vivo, these signals appeared to be colocalized to the same neurons. Importantly, slices pretreated with U0126 fail not only to display elevated phospho-ERK1/2 immunoreactivity after NE application, but also to display phospho-CREB immunoreactivity.[41] In contrast, U0124, an inactive U0126 analogue that does not inhibit MEK failed to attenuate either phospho-ERK1/2 or phospho-CREB immunoreactivities in the slices. This result suggested that the actions of U0126 were pharmacologically specific to its functions as an MEK inhibitor. These results demonstrated for the first time that NE-induced phospho-CREB signaling was dependent on upstream MEK-mediated signaling in PVH neurons. Thus, for NE signals to drive *Crh* expression, an MEK > ERK signaling component may be required to drive phospho-CREB in vivo.

What about ERK1/2 driving CRH neuropeptide release? Using the ex vivo slice preparation, we next addressed whether ERK1/2 could play a role in this process as well. Since neuropeptide release is believed to require spike activity along axons extending from CRH neuroendocrine neurons to the median eminence, we reasoned that ERK1/2 activity might influence neuronal firing in these neurons. To test this hypothesis, we collaborated with the laboratory of Jaideep Bains to evaluate the role of ERK1/2 in NE-induced spike activity from parvicellular PVH neurons. Using identical concentrations of NE and similar slice incubation conditions as that for our experiments visualizing ERK1/2 and CREB activation, the Bains laboratory bath-applied NE to PVH slices and recorded robust spike activity in response to this treatment. Importantly, bath application of U0126, but not U0124, markedly reduced spike activity compared to slices treated only with NE, demonstrating again a marked dependence upon ERK1/2 for full spiking activity in response to NE.[41]

A MODEL FOR CRH NEUROENDOCRINE NEURON SIGNAL-EFFECTOR COUPLING

Taken together, the studies described in the preceding sections demonstrate that: (1) intravenous glycemic challenges activate ERK1/2 rapidly in CRH neurons of the PVH and trigger increases in CRH expression and plasma ACTH/corticosterone; (2) NE is sufficient to recapitulate these effects when delivered to PVH neurons in vivo; (3) NE can drive ERK1/2 activation in PVH neurons ex vivo, and that this activation is a critical step for neuronal firing in response to NE as well as CREB activation; and (4) hindbrain-originating CA afferents are required for glycemic challenges to engage ERK1/2 in CRH neurons, and augment CRH expression and plasma ACTH levels in vivo.

From these findings, a basic model of the functional linkages that extend from the initial stimulus (glycemic challenge) to the final response (CRH neuroendocrine output) can be constructed (Fig. 7.4). First, various stimulation paradigms, involving either peripheral stimuli delivered in the whole animal intravenously, or central stimuli delivered intracranially, or bath-applied stimuli delivered to the tissue slice, all helped to establish basic stimulus-response relations between glycemic challenge, NE, and intracellular changes in signaling activation, gene expression, or neuronal firing rates (Fig. 7.4; top panel). Second, causal roles between these various elements were established using specific intervention strategies (DSAP lesions, adrenergic receptor antagonist, MEK inhibitor) that helped establish the essential nature for many of these elements (Fig. 7.4; middle panel). Finally, what can be constructed from these stimulation paradigms and intervention strategies is a model (Fig. 7.4; bottom panel) where glycemic challenges recruit hindbrain-originating CA neurons to signal to CRH neuroendocrine neurons. These CA signals engage adrenergic receptors that are, in turn, coupled to the activation of MEK kinase and downstream activation of ERK1/2 and CREB to drive *Crh* expression. Concomitantly, ERK1/2 signaling can influence as yet unidentified molecular substrates mediating neuronal excitability in these neurons to change firing rates and, by extension, the release of CRH neuropeptide from terminal in the median eminence to influence subsequent ACTH release from the anterior pituitary.

A few outstanding aspects of this model remain untested. In particular, whereas we have determined that CA inputs from the hindbrain are required to activate intracellular programs of CRH synthesis and release, the actual mechanistic linkages between synthesis and release programs in CRH neuroendocrine neurons remain obscure. For example, we have not yet established that inhibition of ERK1/2 activation prevents the release of ACTH from pituitary corticotrophs in vivo during glycemic challenges. Further, although it is clear that CREB is a major transcriptional player in the mechanism that drives *Crh* expression, this also has not yet been tested explicitly in vivo in the context of peripheral glycemic challenge. It is likely that intracellular coupling mechanisms are complex[52] within CRH neuroendocrine neurons and are organized in a manner that likely facilitates rapid switching of many signal-effector couplings within the cell to accommodate the diverse afferent signals that arrive to trigger CRH release in vivo under various physiological states.

EXTENDING THE STUDY TO A LARGER NETWORK: FOCUS ON THE ARCUATE HYPOTHALAMIC NUCLEUS

During the course of the experiments described in the preceding sections, we observed phospho-ERK1/2 immunoreactivity in a variety of brain regions in addition to the PVH. In particular, increased phospho-ERK1/2 levels were observed within neurons of the arcuate hypothalamic nucleus (ARH) following intravenous insulin or 2-DG challenge. We were interested in this observation for a few reasons. First, it is well documented that hindbrain CA neurons project widely to many forebrain regions, including the ARH. Therefore, the activation we were observing in the ARH and possibly other regions could be the result of hindbrain CA signals arriving in those locations, much

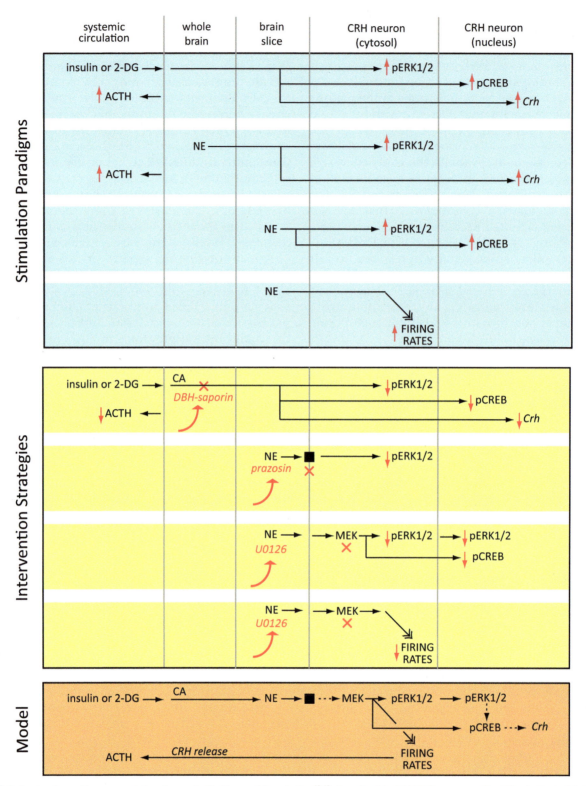

FIGURE 7.4   A schematic summarizing the main findings of the studies[40,42] described in MAP Kinase as a Rapidly Activated Intracellular Effector in CRH Neuroendocrine Neurons In Vivo section, Catecholamines Are Sufficient to Recapitulate Effects of Glycemic Challenges on CRH Neurons section, and Catecholaminergic Projections From Hindbrain Are Necessary for CRH Responses to Glycemic Challenges section of this chapter. The stimulation paradigms in the top panel and intervention strategies in the middle panel lead us to posit the model shown in the lowest panel regarding the functional linkages between glycemic challenges in the systemic circulation, catecholaminergic (CA) afferents in the hindbrain, the release of norepinephrine (NE) from these afferents, and downstream receptor-mediated intracellular signaling cascades involving phosphorylated forms of ERK1/2 (pERK1/2) and CREB (pCREB) to control the expression of the corticotropin-releasing hormone (Crh) gene and firing rates in the neuron. These firing rates, in turn, lead to CRH release and adrenocorticotropin hormone (ACTH) release into the systemic circulation.

as we had shown for the PVH. Indeed, signals through CA afferents are known to influence the expression of multiple neuropeptides in the ARH.[25] Second, CRH neuroendocrine neurons of the PVH receive direct inputs from neuropeptide Y (NPY)-expressing neurons in the ARH,[49] and NPY and dopaminergic neurons in the ARH bear CRH receptors,[7] indicating that the two structures display bidirectional communication links.

Accordingly, we sought to determine if the ERK activation we observed in the ARH also required intact CA afferents.[43] To this end, we examined ARH tissue from the same sham- and DSAP-lesioned animals given insulin, 2-DG, or the multimodal challenges that we studied for the PVH experiments.[41] In these animals, we examined the effects of our PVH lesions on both CA afferents and ERK1/2-activated neurons in the ARH.

The results of these investigations can be summarized as follows. First, extending the findings of our previous study[41] in which we found that the *PVH-targeted* DSAP lesions producing loss of hindbrain-originating CA afferents in the PVH, we also observed losses of these afferents in the ARH. Second, these ARH losses in CA afferents were accompanied with reductions in the levels of insulin- and 2-DG-induced phospho-ERK1/2 in the ARH. Third, similar to our findings for the PVH, we did not observe reductions, in DSAP-lesioned rats, in ARH levels of phospho-ERK1/2 that were induced by multimodal stressors (isoflurane anesthesia and hypertonic saline). Thus, as with the PVH, the dependency of ERK1/2 activation in the ARH upon intact CA afferents from the hindbrain was stimulus-selective.[43]

Alongside our investigation of the dependence of ERK activation upon intact CA afferents in the ARH, we performed a series of control experiments that led to a surprising finding. Specifically, in order to evaluate the specificity of the DSAP lesions in the PVH, we took a spare tissue series from sham- and DSAP-lesioned animals to examine whether the DBH-saporin immunotoxin conjugate affected the integrity of other, non-CA afferent systems innervating the PVH. To this end, we performed triple-label immunocytochemistry to examine fiber systems immunoreactive for DBH, α-melanocyte stimulating hormone (αMSH), and agouti-related peptide (AgRP).[4] We found that whereas DBH-saporin immunotoxin delivery in the PVH resulted in CA afferent destruction, there was no observable reduction in the immunoreactivity for stained axonal fibers and terminals in the PVH for αMSH and AgRP. Indeed, to our surprise, we found an apparent upregulation of these levels that was qualitatively observable in 2-DG-treated animals. To investigate this effect further, we performed high-resolution confocal imaging of triple-labeled material (DBH, αMSH, AgRP)

and quantitated the levels of immunoreactive signal for each antigen. We observed an increase in both the mean gray level and the percentage of area occupied by the signal for αMSH-immunopositive fibers in the PVH in DSAP-lesioned rats receiving 2-DG relative to sham-lesioned controls receiving the same challenge. In contrast, there was no significant difference between these groups for AgRP-immunopositive fibers, although there was a trend towards increased signal in the lesioned cohort.[43]

The increased percentage area occupied by αMSH-immunoreactive signal demonstrated that there was an increase in the coverage of immunoreactive fibers in the PVH following lesion of the CA afferents to this structure. This increase immediately suggested the possibility that there might be greater numbers of αMSH-immunoreactive cell bodies at the level of the ARH, a major source of αMSH inputs to the PVH. However, we did not observe any difference in the abundance of ARH αMSH-immunoreactive cell bodies as a result of PVH-directed CA ablation and systemic 2-DG treatment.[43] It remains possible that such changes in the number of neuronal cell bodies may have occurred in the nucleus of the solitary tract in the hindbrain, another source of αMSH inputs to the PVH that was not examined in our study. Collectively, the increase in axonal fiber coverage and elevated mean gray level in the PVH αMSH signal, along with no observable increase in αMSH cell bodies in the ARH, suggest that the αMSH peptidergic system innervating the PVH is displaying a dramatic form of structural plasticity that may be compensating for the loss of CA afferents to the same structure.

Interestingly, we also found that phospho-ERK1/2 expression in the ARH was not within the αMSH-immunopositive neuronal subpopulation but was within a phenotype that we could not identify precisely for technical reasons. However, our location of the pERK1/2-immunoreactive neurons in the medial ARH is consistent with the location of an NPY/AgRP neuronal phenotype known to be present within medial ARH. This suggests the possibility that 2-DG triggered activation of ERK1/2 within NPY/AgRP neurons, which is consistent with the known effects of 2-DG on upregulating mRNA levels for these peptides in the ARH.[25]

More generally, the results of this study provide a novel view of how the ARH might be functioning in the face of glycemic challenge. Despite a possible role for specific ARH neuropeptide-containing neurons in acutely sensing nutrients and metabolites in circulating cerebrospinal fluid,[55,56] it remains the case that these neurons are also likely part of a much larger and as yet incompletely identified network that also includes the PVH and hindbrain that communicates during various feeding- and metabolism-related contexts.[80] This network is also likely to include hindbrain-originating CA afferents to the PVH, which have been shown to drive feeding responses to peripheral stimuli including ghrelin and 2-DG.[61,17] Our results demonstrate that although certain ARH neuronal subpopulations that have been identified as glucosensing ex vivo[24,64] may function similarly in vivo, they do not appear to respond to 2-DG challenge in the absence of neural (CA afferent) input as determined by their inability to mount a phospho-ERK1/2 response following DSAP-mediated CA deafferentation. If this inability to mount a phospho-ERK1/2 response is any indication of a similar inability of ARH glucosensing neurons to respond electrically to 2-DG under these conditions, this suggests that—in vivo—there is a primacy of neural over humoral routes in conveying signals encoding glycemic challenge to ARH neurons: the cerebrospinal access is not sufficient on its own to do so.

In terms of a larger functional network serving energy balance, a final point is worth noting here. The loss of CA afferents at the level of the ARH following DSAP lesions targeting the PVH suggests strongly the existence of axonal collaterals from CA projections to the PVH that innervate the ARH. Indeed, this collateral organization is likely to involve other forebrain target structures, such as the central nucleus of the amygdala[58] and subregions of the bed nuclei of the terminal stria,[1] although there is also evidence for separate CA projections to other forebrain structures, such as the medial prefrontal cortex.[60] A key task that remains is to map the activation produced by glycemic challenges rigorously to a canonical brain atlas in a manner that takes advantage of computational approaches such as databasing, digital mapping, and neuroinformatics.[39,82,38] Such approaches would allow the seamless registration of mapped activation patterns produced by glycemic challenges with maps of activation—produced in the same reference space—that are generated from experiments using other metabolic interventions, such as those patterns recently mapped for fasting and refeeding.[85]

FUTURE DIRECTIONS

Although much progress has been made in elucidating the functional linkages that tether peripheral stressors to discrete communication lines leading to CRH neuroendocrine output in the living organism, much remains to be investigated about such linkages at the network, regional, and cellular levels. At the network level, it remains unclear if and when the many regions that serve as targets of hindbrain CA afferents are activated during glycemic or other challenges, and whether each target region utilizes a common transduction mechanism

to help mediate specific responses to these challenges. What are the governing parameters that determine the recruitment of one or all of these regions in the mobilization of responses to a particular stressor? Across the CA afferent system, is there a role for colocalized peptides within these afferents in mediating signals to each target structure?

At the regional level, part of what remains undetermined is the influence of the local microenvironment on the long-range signals arriving from the CA system. For the PVH, for example, is there a local network of neural elements that gates the incoming CA signals during glycemic or other challenges? Given that the PVH receives signals from not only the hindbrain but other regions as well (e.g., the ARH) do these signals coordinate in some way to influence the neuroendocrine response?

These questions move further queries inexorably to the cellular level, where it remains unclear how CRH neuroendocrine neurons process and coordinate diverse incoming signals to orchestrate an appropriate response. One of these signals is the glucocorticoid, cortisol (corticosterone in rats), which is well documented to provide complex feedback to control the output of the CRH neuroendocrine system, including *Crh* gene expression[79]; this feedback also requires intact CA afferents to the PVH from the hindbrain.[37] Since evidence has also linked glucocorticoid feedback to the modulation of ERK/12 activation in CRH neuroendocrine neurons,[57] it appears that any receptor-mediated signals upstream of ERK in these neurons will probably also engage in cross-signaling with those as yet unidentified intracellular signaling pathways by which corticosterone exerts its effects on ERK signaling. Although more spadework is required before we can dig deeper into these issues for CRH neuroendocrine neurons without speculation, we commit the small crime here of anticipating that some of these signaling interactions may involve ERK translocation to various subcellular compartments, as has been demonstrated in various cell systems.[78]

Acknowledgments

This work is supported by National Institute of Health (NIH) Grants DK081937 and GM109817 (awarded to AMK), and NS029728 (awarded to AGW); and is also supported by funds awarded to AMK from the Border Biomedical Research Center (5G12RR008124 and 8G12MD007592), and a Keelung Hong Graduate Research Fellowship (awarded to EMW). The authors wish to thank Dr.Kimberly L. Kaminski and Graciela Sanchez-Watts at USC and Dr.Armando Varela-Ramirez at UTEP for technical assistance with the experiments described in this chapter. We also thank Dr. J. Brent Kuzmiski and Dr. Jaideep S. Bains of the Bains laboratory, and Dr. Todd A. Ponzio and Dr. Harold Gainer of the Gainer laboratory, for assistance with our tissue slice experiments. Finally, we dedicate this chapter to the dear memory of our late colleague and mentor, Dr. Glenn I. Hatton, a pioneer in the field of neuroendocrinology who collaborated with us on this project.

References

1. Banihashemi L, Rinaman L. Noradrenergic inputs to the bed nucleus of the stria terminalis and paraventricular nucleus of the hypothalamus underlie hypothalamic-pituitary-adrenal axis but not hypophagic or conditioned avoidance responses to systemic yohimbine. *J Neurosci.* 2006;26(44):11442–11453.

2. Bohland MA, Matveyenko AV, Saberi M, Khan AM, Watts AG, Donovan CM. Activation of hindbrain neurons is mediated by portal-mesenteric vein glucosensors during slow-onset hypoglycemia. *Diabetes.* 2014;63(8):2866–2875.

3. Blessing WW, Costa M, Geffen LB, Fink G. Immune lesions of noradrenergic neurones in rat central nervous system produced by antibodies to dopamine-β-hydroxylase. *Nature.* 1977;267:368–369.

4. Broberger C, Johansen J, Johansson C, Schalling M, Hökfelt T. The neuropeptide Y/agouti gene-related protein (AGRP) brain circuitry in normal, anorectic, and monosodium glutamate-treated mice. *Proc Natl Acad Sci USA.* 1998;95(25):15043–15048.

5. Buller K, Xu Y, Dayas C, Day T. Dorsal and ventral medullary catecholamine cell groups contribute differentially to systemic interleukin-1beta-induced hypothalamic pituitary adrenal axis responses. *Neuroendocrinology.* 2001;73(2):129–138.

6. Bullock TH. Signals and signs in the nervous system: the dynamic anatomy of electrical activity is probably information-rich. *Proc Natl Acad Sci USA.* 1997;94(1):1–6.

7. Campbell RE, Grove KL, Smith MS. Distribution of corticotropin releasing hormone receptor immunoreactivity in the rat hypothalamus: coexpression in neuropeptide Y and dopamine neurons in the arcuate nucleus. *Brain Res.* 2003;973:223–232.

8. Card JP, Sved JC, Craig B, Raizada M, Vazquez J, Sved AF. Efferent projections of rat rostroventrolateral medulla C1 catecholamine neurons: implications for the central control of cardiovascular regulation. *J Comp Neurol.* 2006;499(5):840–859.

9. Carlsson A, Falck B, Hillarp N-A. Cellular localization of brain monoamines. *Acta Physiol Scand.* 1962;56(suppl 196):1–27.

10. Cole RL, Sawchenko PE. Neurotransmitter regulation of cellular activation and neuropeptide gene expression in the paraventricular nucleus of the hypothalamus. *J Neurosci.* 2002;22(3):959–969.

11. Cunningham Jr ET, Bohn MC, Sawchenko PE. Organization of adrenergic inputs to the paraventricular and supraoptic nuclei of the hypothalamus in the rat. *J Comp Neurol.* 1990;292(4):651–657.

12. Cunningham Jr ET, Sawchenko PE. Anatomical specificity of noradrenergic inputs to the paraventricular and supraoptic nuclei of the rat hypothalamus. *J Comp Neurol.* 1988;274(1):60–76.

13. Dahlström A, Fuxe K. Evidence for the existence of monoamine containing neurons in the central nervous system. I: demonstration of monoamines in the cell bodies of brainstem neurons. *Acta Physiol Scand.* 1964;62(suppl 232):1–55.

14. Dahlström A, Fuxe K. Evidence for the existence of monoamine containing neurons in the central nervous system. II: experimentally induced changes in the intraneuronal amine levels of bulbospinal neuron systems. *Acta Physiol Scand.* 1965;62(suppl 247):1–36.

15. Dahlström A, Fuxe K. The autonomic nervous system and the histochemical fluorescence method for the microscopical localization of catecholamines and serotonin. *Brain Res Bull.* 1999;50(5–6):365–367.

16. Das M, Vihlen CS, Légrádi G. Hypothalamic and brainstem sources of pituitary adenylate cyclase-activating polypeptide nerve fibers innervating the hypothalamic paraventricular nucleus in the rat. *J Comp Neurol.* 2007;500(4):761–776.

17. Date Y, Shimbara T, Koda S, et al. Peripheral ghrelin transmits orexigenic signals through the noradrenergic pathway from the hindbrain to the hypothalamus. *Cell Metab.* 2006;4(4):323–331.

18. Dayas CV, Buller KM, Crane JW, Xu Y, Day TA. Stressor categorization: acute physical and psychological stressors elicit distinctive recruitment patterns in the amygdala and in medullary noradrenergic groups. *Eur J Neurosci.* 2001;14(7):1143–1152.

19. Dayas CV, Buller KM, Day TA. Medullary neurones regulate hypothalamic corticotropin-releasing factor cell responses to an emotional stressor. *Neuroscience.* 2001;105(3):707–719.

20. Donovan CM, Watts AG. Peripheral and central glucose sensing in hypoglycemic detection. *Physiology (Bethesda).* 2014;29(5):314–324.

21. Ericsson A, Kovács KJ, Sawchenko PE. A functional anatomical analysis of central pathways subserving the effects of interleukin-1 on stress-related neuroendocrine neurons. *J Neurosci.* 1994;14(2):897–913.

22. Falck B. Observations on the possibilities of the cellular localization of monoamines by a fluorescence method. *Acta Physiol Scand.* 1962;56(suppl 197):1–25.

23. Fekete C, Wittmann G, Liposits Z, Lechan RM. Origin of cocaine- and amphetamine-regulated transcript (CART)-immunoreactive innervation of the hypothalamic paraventricular nucleus. *J Comp Neurol.* 2004;469(3):340–350.

24. Fioramonti X, Contié S, Song Z, Routh VH, Lorsignol A, Pénicaud L. Characterization of glucosensing neuron subpopulations in the arcuate nucleus: integration in neuropeptide Y and pro-opio melanocortin networks? *Diabetes.* 2007;56(5):1219–1227.

25. Fraley GS, Ritter S. Immunolesion of norepinephrine and epinephrine afferents to medial hypothalamus alters basal and 2-deoxy-D-glucose-induced neuropeptide Y and agouti-related protein messenger ribonucleic acid expression in the arcuate nucleus. *Endocrinology.* 2003;144(1):75–83.

26. Fujita S, Donovan CM. Celiac-superior mesenteric ganglionectomy, but not vagotomy, suppresses the sympathoadrenal response to insulin-induced hypoglycemia. *Diabetes.* 2005;54:3258–3264.

27. Fuxe K. Evidence for the existence of monoamine containing neurons in the central nervous system. III: the monoamine nerve terminal. *Z Zellforsch.* 1965;65:573–596.

28. Fuxe K. Evidence for the existence of monoamine neurons in the central nervous system. IV: distribution of monoamine nerve terminals in the central nervous system. *Acta Physiol Scand Suppl.* 1965;247:37.

29. Fuxe K, Hökfelt T, Nilsson O. A fluorescence and electron microscopic study on certain brain regions rich in monoamine terminals. *Am J Anat.* 1965;117:33–45.

30. Gaykema RP, Chen CC, Goehler LE. Organization of immune-responsive medullary projections to the bed nucleus of the stria terminalis, central amygdala, and paraventricular nucleus of the hypothalamus: evidence for parallel viscerosensory pathways in the rat brain. *Brain Res.* 2007;1130(1):130–145.

31. Hatton GI. Hypothalamic neurobiology. In: Dingledine R, ed. *Brain Slices.* New York: Plenum Press; 1984:341–372.

32. Hatton GI, Doran AD, Salm AK, Tweedle CD. Brain slice preparation: hypothalamus. *Brain Res Bull.* 1980;5(4):404–414.

33. Hoffman GE, Lyo D. Anatomical markers of activity in neuroendocrine systems: are we all 'fos-ed out'? *J Neuroendocrinol.* 2002;14(4):259–268.

34. Hoffman GE, Smith MS, Verbalis JG. c-Fos and related immediate early gene products as markers of activity in neuroendocrine systems. *Front Neuroendocrinol.* 1993;14(3):173–213.

35. Hökfelt T. Early attempts to visualize cortical monoamine nerve terminals. *Brain Res*. 2016;1645:8–11. http://dx.doi.org/10.1016/j.brainres.2016.01.024.

36. Hökfelt T, Fuxe K, Goldstein M, Johansson O. Evidence for adrenaline neurons in the rat brain. *Acta Physiol Scand*. 1973;89:286–288.

37. Kaminski KL, Watts AG. Intact catecholamine inputs to the forebrain are required for appropriate regulation of corticotrophin-releasing hormone and vasopressin gene expression by corticosterone in the rat paraventricular nucleus. *J Neuroendocrinol*. 2012;24(12):1517–1526.

38. Khan AM. Controlling feeding behavior by chemical or gene-directed targeting in the brain: what's so spatial about our methods? *Front Neurosci*. 2013;7:182. http://dx.doi.org/10.3389/fnins.2013.00182.

39. Khan AM, Hahn JD, Cheng WC, Watts AG, Burns GA. NeuroScholar's electronic laboratory notebook and its application to neuroendocrinology. *Neuroinformatics*. 2006;4(2):139–162.

40. Khan AM, Watts AG. Intravenous 2-deoxy-D-glucose injection rapidly elevates levels of the phosphorylated forms of p44/p42 mitogen activated protein kinases (extracellularly regulated kinases 1 and 2) in rat hypothalamic parvicellular paraventricular neurons. *Endocrinology*. 2004;145:351–359.

41. Khan AM, Kaminski KL, Sanchez-Watts G, et al. MAP kinases couple hindbrain-derived catecholamine signals to hypothalamic adrenocortical control mechanisms during glycemia-related challenges. *J Neurosci*. 2011;31(50):18479–18491.

42. Khan AM, Ponzio TA, Sanchez-Watts G, Stanley BG, Hatton GI, Watts AG. Catecholaminergic control of mitogen-activated protein kinase signaling in paraventricular neuroendocrine neurons in vivo and in vitro: a proposed role during glycemic challenges. *J Neurosci*. 2007;27(27):7344–7360.

43. Khan AM, Walker EM, Dominguez N, Watts AG. Neural input is critical for arcuate hypothalamic neurons to mount intracellular signaling responses to systemic insulin and deoxyglucose challenges in male rats: implications for communication within feeding and metabolic control networks. *Endocrinology*. 2014;155(2):405–416.

44. King TL, Heesch CM, Clark CG, Kline DD, Hasser EM. Hypoxia activates nucleus tractus solitarii neurons projecting to the paraventricular nucleus of the hypothalamus. *Am J Physiol Regul Integr Comp Physiol*. 2012;302(10):R1219–R1232.

45. King TL, Kline DD, Ruyle BC, Heesch CM, Hasser EM. Acute systemic hypoxia activates hypothalamic paraventricular nucleus-projecting catecholaminergic neurons in the caudal ventrolateral medulla. *Am J Physiol Regul Integr Comp Physiol*. 2013;305:R1112–R1123.

46. King TL, Ruyle BC, Kline DD, Heesch CM, Hasser EM. Catecholaminergic neurons projecting to the paraventricular nucleus of the hypothalamus are essential for cardiorespiratory adjustments to hypoxia. *Am J Physiol Regul Integr Comp Physiol*. 2015;309:R721–R731.

47. Kovács KJ, Sawchenko PE. Sequence of stress-induced alterations in indices of synaptic and transcriptional activation in parvicellular neurosecretory neurons. *J Neurosci*. 1996;16(1):262–273.

48. Levin MC, Sawchenko PE, Howe PRC, Bloom SR, Polak JM. Organization of galanin-immunoreactive inputs to the paraventricular nucleus with special reference to their relationship to catecholaminergic afferents. *J Comp Neurol*. 1987;261:562–582.

49. Li C, Chen P, Smith MS. Corticotropin releasing hormone neurons in the paraventricular nucleus are direct targets for neuropeptide Y neurons in the arcuate nucleus: an anterograde tracing study. *Brain Res*. 2000;854:122–129.

50. Li HY, Ericsson A, Sawchenko PE. Distinct mechanisms underlie activation of hypothalamic neurosecretory neurons and their medullary catecholaminergic afferents in categorically different stress paradigms. *Proc Natl Acad Sci USA*. 1996;93(6):2359–2364.

51. Liu Y, Kamitakahara A, Kim AJ, Aguilera G. Cyclic adenosine 3′,5′-monophosphate responsive element binding protein phosphorylation is required but not sufficient for activation of corticotropin-releasing hormone transcription. *Endocrinology*. 2008;149:3512–3520.

52. Liu Y, Poon V, Sanchez-Watts G, Watts AG, Takemori H, Aguilera G. Salt-inducible kinase is involved in the regulation of corticotropin-releasing hormone transcription in hypothalamic neurons in rats. *Endocrinology*. 2012;153(1):223–233.

53. Luckman SM, Dyball RE, Leng G. Induction of *c-fos* expression in hypothalamic magnocellular neurons requires synaptic activation and not simply increased spike activity. *J Neurosci*. 1994;14:4825–4830.

54. Menétrey D, Basbaum AI. Spinal and trigeminal projections to the nucleus of the solitary tract: a possible substrate for somatovisceral and viscerovisceral reflex activation. *J Comp Neurol*. 1987;255(3):439–450.

55. Mullier A, Bouret SG, Prevot V, Dehouck B. Differential distribution of tight junction proteins suggests a role for tanycytes in blood-hypothalamus barrier regulation in the adult mouse brain. *J Comp Neurol*. 2010;518(7):943–962.

56. Olofsson L, Unger EK, Cheung CC, Xu AW. Modulation of AgRP-neuronal function by SOCS3 as an initiating event in diet-induced hypothalamic leptin resistance. *Proc Natl Acad Sci USA*. 2013;110(8):E697–E706.

57. Osterlund CD, Jarvis E, Chadayammuri A, Unnithan R, Weiser MJ, Spencer RL. Tonic, but not phasic corticosterone, constrains stress activated extracellular-regulated-kinase 1/2 immunoreactivity within the hypothalamic paraventricular nucleus. *J Neuroendocrinol*. 2011;23(12):1241–1251.

58. Petrov T, Krukoff TL, Jhamandas JH. Branching projections of catecholaminergic brainstem neurons to the paraventricular hypothalamic nucleus and the central nucleus of the amygdala in the rat. *Brain Res*. 1993;609(1–2):81–92.

59. Picklo MJ, Wiley RG, Lappi DA, Robertson D. Noradrenergic lesioning with an anti-dopamine β-hydroxylase immunotoxin. *Brain Res*. 1994;666:195–200.

60. Radley JJ, Williams B, Sawchenko PE. Noradrenergic innervation of the dorsal medial prefrontal cortex modulates hypothalamo-pituitary-adrenal responses to acute emotional stress. *J Neurosci*. 2008;28(22):5806–5816.

61. Ritter S, Bugarith K, Dinh TT. Immunotoxic destruction of distinct catecholamine subgroups produces selective impairment of glucoregulatory responses and neuronal activation. *J Comp Neurol*. 2001;432(2):197–216.

62. Ritter S, Dinh TT, Bugarith K, Salter DM. Chemical dissection of brain glucoregulatory circuitry. In: Wiley RG, Lappi DA, eds. *Molecular Neurosurgery with Targeted Toxins*. Totowa, NJ: Humana Press, Inc.; 2005:181–218.

63. Ritter S, Watts AG, Dinh TT, Sanchez-Watts G, Pedrow C. Immunotoxin lesion of hypothalamically projecting norepinephrine and epinephrine neurons differentially affects circadian and stressor-stimulated corticosterone secretion. *Endocrinology*. 2003;144(4):1357–1367.

64. Routh VH, Hao L, Santiago AM, Sheng Z, Zhou C. Hypothalamic glucose sensing: making ends meet. *Front Syst Neurosci*. 2014;8:236.

65. Sawchenko PE, Swanson LW. Central noradrenergic pathways for the integration of hypothalamic neuroendocrine and autonomic responses. *Science*. 1981;214(4521):685–687.

66. Sawchenko PE, Swanson LW. The organization of noradrenergic pathways from the brainstem to the paraventricular and supraoptic nuclei in the rat. *Brain Res*. 1982;257(3):275–325.

67. Sawchenko PE, Swanson LW, Grzanna R, Howe PR, Bloom SR, Polak JM. Colocalization of neuropeptide Y immunoreactivity in brainstem catecholaminergic neurons that project to the paraventricular nucleus of the hypothalamus. *J Comp Neurol*. 1985;241(2):138–153.

68. Schiltz JC, Sawchenko PE. Specificity and generality of the involvement of catecholaminergic afferents in hypothalamic responses to immune insults. *J Comp Neurol*. 2007;502(3):455–467.

69. Sevigny CP, Bassi J, Williams DA, Anderson CR, Thomas WG, Allen AM. Efferent projections of C3 adrenergic neurons in the rat central nervous system. *J Comp Neurol.* 2012;520(11):2352–2368.

70. Simmons DM, Swanson LW. Comparison of the spatial distribution of seven types of neuroendocrine neurons in the rat paraventricular nucleus: toward a global 3D model. *J Comp Neurol.* 2009;516(5):423–441.

71. Smithson KG, Hatton GI. Immunocytochemical identification of electrophysiologically characterized cells. In: Björklund A, Hökfelt T, Wouterlood FG, eds. *Handbook of Chemical Neuroanatomy. Analysis of Neuronal Microcircuits and Synaptic Interactions;* Vol. 8. Amsterdam: Elsevier; 1990:305–350.

72. Swank MW. Phosphorylation of MAP kinase and CREB in mouse cortex and amygdala during taste aversion learning. *Neuroreport.* 2000;11(8):1625–1630.

73. Swanson LW. *Brain Maps: Structure of the Rat Brain.* 3rd ed. Amsterdam: Elsevier; 2004.

74. Swanson LW, Sawchenko PE. Paraventricular nucleus: a site for the integration of neuroendocrine and autonomic mechanisms. *Neuroendocrinology.* 1980;31(6):410–417.

75. Swanson LW, Sawchenko PE, Bérod A, Hartman BK, Helle KB, Vanorden DE. An immunohistochemical study of the organization of catecholaminergic cells and terminal fields in the paraventricular and supraoptic nuclei of the hypothalamus. *J Comp Neurol.* 1981;196(2):271–285.

76. Tucker DC, Saper CB, Ruggiero DA, Reis DJ. Organization of central adrenergic pathways: I. Relationships of ventrolateral medullary projections to the hypothalamus and spinal cord. *J Comp Neurol.* 1987;259(4):591–603.

77. Ungerstedt U. 6-hydroxy-dopamine induced degeneration of central monoamine neurons. *Eur J Pharmacol.* 1968;5:107–110.

78. Wainstein E, Seger R. The dynamic subcellular localization of ERK: mechanisms of translocation and role in various organelles. *Curr Opin Cell Biol.* 2016;39:15–20.

79. Watts AG. Glucocorticoid regulation of peptide genes in neuroendocrine CRH neurons: a complexity beyond negative feedback. *Front Neuroendocrinol.* 2005;26(3–4):109–130.

80. Watts AG, Donovan CM. Sweet talk in the brain: glucosensing, neural networks, and hypoglycemic counterregulation. *Front Neuroendocrinol.* 2010;31(1):32–43.

81. Watts AG, Khan AM. Identifying links in the chain: the dynamic coupling of catecholamines, peptide synthesis, and peptide release in hypothalamic neuroendocrine neurons. *Adv Pharmacol.* 2013;68:421–444.

82. Watts AG, Khan AM, Sanchez-Watts G, Salter D, Neuner CM. Activation in neural networks controlling ingestive behaviors: what does it mean, and how do we map and measure it? *Physiol Behav.* 2006;89(4):501–510.

83. Watts AG, Tanimura S, Sanchez-Watts G. Corticotropin-releasing hormone and arginine vasopressin gene transcription in the hypothalamic paraventricular nucleus of unstressed rats: daily rhythms and their interactions with corticosterone. *Endocrinology.* 2004;145(2):529–540.

84. Wittmann G. Regulation of hypophysiotrophic corticotrophin-releasing hormone- and thyrotropin-releasing hormone-synthesising neurones by brainstem catecholaminergic neurones. *J Neuroendocrinol.* 2008;20(7):952–960.

85. Zséli G, Vida B, Martinez A, Lechan RM, Khan AM, Fekete C. Elucidation of the anatomy of a satiety network: focus on connectivity of the parabrachial nucleus in the adult rat. *J Comp Neurol.* 2016;524(14):2803–2827. http://dx.doi.org/10.1002/cne.23992.

Neural Circuitry of Stress, Fear, and Anxiety: Focus on Extended Amygdala Corticotropin-Releasing Factor Systems

E.D. Paul[1], A. Chen[1,2]

[1]Max Planck Institute of Psychiatry, Munich, Germany; [2]Weizmann Institute of Science, Rehovot, Israel

Abstract

The extended amygdala (EA) contains subpopulations of "extrahypothalamic" corticotropin-releasing factor (CRF) neurons located in the bed nucleus of the stria terminalis and central nucleus of the amygdala. Based on anatomical, cytoarchitectural, neurochemical, electrophysiological, and hodological characteristics, several heterogeneous subpopulations of EA CRF emerge that impact their functional neural circuitry. EA CRF neurons, through widespread connections with serotonergic, dopaminergic, and noradrenergic brainstem nuclei, modulate behavioral responses to emotionally salient stimuli, arousal, and aspects of addiction. Interconnections between EA CRF neurons and central autonomic control regions regulate cardiovascular function, nociception, ingestion, and gastrointestinal function. The bed nucleus of the stria terminalis CRF neurons are critically positioned to filter limbic inputs, including other EA CRF subpopulations, to modulate various aspects of the hypothalamic–pituitary–adrenal axis. Advances in optogenetics, pharmacogenetics, and single-cell profiling in combination with transgenic rodents will allow researchers to unravel the contribution of these heterogeneous EA CRF neurons in stress-related processes.

INTRODUCTION: CORTICOTROPIN-RELEASING FACTOR

Corticotropin-releasing factor (CRF) is a 41-residue polypeptide originally isolated from ovine hypothalamic extracts and characterized as the principal secretagogue of adrenocorticotropin hormone by Vale et al.[1] The biological actions of CRF are mediated through two G protein–coupled receptors, CRF_{R1} and CRF_{R2}, both of which have multiple splice variants; differential expression patterns; and varying affinities towards their cognate ligands, CRF, and the related urocortin (I–III) peptides.[2] The functional role of CRF has predominantly focused on its involvement in activation of the hypothalamic–pituitary–adrenal (HPA) axis; however, soon after the discovery of CRF, numerous investigators have reported widespread distribution of CRF in cortical, limbic, and brainstem structures involved in mediating the affective, autonomic, and neuroendocrine components of the stress response.

Prominent populations of these "extrahypothalamic" CRF neurons reside in the central extended amygdala (EA), a macrostructure of the basal forebrain consisting of the central nucleus of the amygdala (CeA), lateral division of the bed nucleus of the stria terminalis (BST), and intermediary structures.[3,4] In this review, we discuss the EA CRF system with an overarching emphasis on its role as a key hub to integrate polymodal information about real and perceived threats and coordinate the appropriate response through projections that span diverse brain-regulatory systems. We begin by outlining the general organization of the EA CRF system and illustrating various characteristics that reveal heterogeneous subpopulations of EA CRF neurons. Then we provide an overview of the neural circuits of the EA CRF systems that orchestrate affective, neuroendocrine, and autonomic responses to stress.

ORGANIZATION OF THE EA CRF SYSTEM

Subpopulations of EA CRF Neurons Based on Anatomical and Cytoarchitectonic Characteristics

The EA CRF neurons reside in two prominent populations located in the BST and CeA. Within the BST, CRF neurons are present throughout the rostrocaudal extent, with reduced density as the nucleus progresses caudally, and are restricted to two distinct subpopulations in the lateral division: the dorsolateral (dlBST; comparable to the oval nucleus of the BST, ovBST[5]) and ventrolateral (vlBST; comparable to the dorsomedial nucleus of the BST, dmBST, and fusiform nucleus of the BST, fuBST[5]) subgroups.[6–8] These two populations, in addition to physical segregation, are distinguished by morphological and ultrastructural characteristics. For example dlBST, compared to vlBST, neurons contain more primary dendrites and somatic and dendritic spines, suggesting they have greater capability to integrate and process afferents.[6] The dlBST CRF neurons are also smaller in size and contain both dense core and alveolate vesicles, whereas vlBST CRF neurons contain only alveolate vesicles.[6] It is worth mentioning that species-specific differences exist between mice and rats in the distribution of BST CRF neurons, with mice displaying more widespread and less defined CRF populations than rats, but nevertheless still adhering to a dorsal versus ventral distribution.[9]

Inside the CeA, CRF expression is found at all rostrocaudal levels with the densest expression found at midline levels in the lateral nucleus of the CeA (CeL) with only sparse neurons in the medial nucleus of the CeA (CeM).[7,8,10] The CeL CRF neurons are predominantly classified as medium spiny neurons resembling gamma-aminobutyric acid (GABA)-ergic striatal neurons and they share similar cytoarchitectural characteristics with dlBST CRF neurons; in contrast, CeM CRF neurons are large pyramiform-shaped, sparsely spinous, and contain long dendrites.[11,12] Overall, the anatomical locations and cytoarchitechtonic characteristics of EA CRF neurons suggest at least several heterogenous subpopulations exist that likely impact their functional properties.

Subpopulations of EA CRF Neurons Based on Neurochemical Phenotypes

Subpopulations of CRF neurons can also be delineated based on their colocalization with the classical neurotransmitters, GABA and glutamate, although there are conflicting results on the latter. Initial reports using markers of the vesicular glutamate transporter (VGLUT1-3) suggested BST CRF neurons are near unanimously GABAergic (see Ref. 13); however, a comprehensive study using Cre-specific driver lines crossed with reporter lines in combination with various immunohistochemical markers for inhibitory (e.g., GABA, parvalbumin, somatostatin, etc.) and excitatory neurons (e.g., CaMKII and EAAC1) revealed substantial CRF-glutamate colocalized neurons in the dorsal medial and ventral BST, whereas almost all CRF neurons in the dlBST colocalize with GABA.[9] The arrangement in the Ce is straightforward with CRF neurons coexpressing GABA.[14]

Another neurochemical factor distinguishing subpopulations of EA CRF neurons is their colocalization with other neuropeptides in addition to classical neurotransmitters. For example, CRF neurons form subpopulations with neurotensin (dlBST and CeL),[15,16] dynorphin (CeL),[17,18] and somatostatin (CeL).[17] Since the dynamics of neuropeptide signaling vastly differ from the classical neurotransmitters in type and localization of vesicles, stimulation requirements for release, methods for reuptake/clearance, and downstream receptor signaling pathways, CRF (and other coexpressed neuropeptides) likely have slower actions and a more modulatory role on effector targets than their coexpressed amino acid transmitters.[19–21] This arrangement provides for diverse forms of neuromodulation leading to CRF potentially facilitating or attenuating the action of the coreleased amino acid transmitter.[19–21] Because of the dynamics of neuropeptide signaling, CRF release may only be recruited under more salient events (e.g., stress, fear, anxiety) that produce the burst firing of neurons required to promote neuropeptide release and, moreover, CRF release likely produces sustained modulatory effects.

Subpopulations of EA CRF Based on Electrophysiological Properties

More recently, investigators are beginning to classify populations of heterogenous EA CRF neurons based on their electrophysiological characteristics. Hammack et al.[22] have characterized three different types of rat BST neurons (Type I–III) and single-cell RT-qPCR profiling of these neuron types reports that Type III neurons near unanimously express CRF mRNA.[13] Type III neurons also have similar electrophysiological characteristics compared to identified dlBST CRF neurons in a CRF-GFP reporter mouse line,[23,24] further supporting evidence that Type III neurons are putative CRF neurons. Compared to Type I-II neurons in the BST and CRF neurons in the paraventricular nucleus of the hypothalamus (PVN), Type III neurons (putative dlBST CRF neurons),

display more hyperpolarized resting membrane potentials and a higher threshold potential for firing action potentials, suggesting these neurons require strong excitatory input for activation,[13,22] such as that which occurs during exposure to stress.

Using a different CRF-*Tomato* reporter mouse line, Silberman et al. confirmed the presence of Type I-III putative CRF neurons, but found a substantial number of putative CRF neurons that did not fit the Type I-III criteria[25]; therefore, at least in the mouse, additional subpopulations of CRF neurons exist. The same group also investigated Ce CRF neurons and found more homogenous electrophysiological responses when compared to BST CRF neurons.[25] CRF neurons in the CeL, compared to BST, show differences in basal properties such as a greater hyperpolarized resting membrane potential; however, in response to current injection, CeL CRF neurons respond with long latencies to fire action potentials, a property shared by Type III dlBST neurons.[25] A limitation of the above studies is that they only characterize putative CRF neurons located in the dlBST so it is possible that CRF neurons in the vlBST will differ in their electrophysiological properties. A number of factors like neurochemical phenotype, cytoarchitechtonic features, and afferent input may shape the complexity of CRF electrophysiological properties and elucidating these features is an active area of investigation.

So far we outlined the heterogeneity of EA CRF neurons based on anatomical location, cytoarchitectural features, neurochemical phenotypes, and electrophysiological properties. Another factor that distinguishes various EA CRF subpopulations is their different patterns of inputs and projections. Although there are significant areas of overlap in connections with limbic and brainstem structures involved in emotional, neuroendocrine, and autonomic responses to stress, there are important differences that provide insight into functional roles of the various EA CRF subpopulations. We turn to this topic in the following sections by discussing the functional neural circuitry of EA CRF systems and their involvement in emotional responses to stress and other potentially threatening stimuli, addiction, and negative effect during withdrawal, neuroendocrine function, and central autonomic control.

EXTENDING THE AMYGDALA: CRF SYSTEMS AND THE NEURAL CIRCUITRY OF STRESS, FEAR, AND ANXIETY

Much of our current understanding of the EA CRF systems involvement in stress, fear, and anxiety comes from the exhaustive work of the Davis lab on various models of fear-, light-, and CRF-potentiated startle, which they used to dissect the role of the EA in fear and anxiety.

Several excellent reviews are available on the subject so here we provide only a brief overview of their proposed model.[26,27] The hypothetical model (Fig. 8.1), in its most general form, highlights a role for CRF pathways arising from the CeL and projecting to the lateral BST as a critical circuit for producing sustained anxiety responses to distal, long-term threats; whereas the CeM, through its dense and widespread connections to brainstem structures involved in behavioral and autonomic responses to stress, is the key node in a circuit eliciting phasic fear responses to immediate, short-term threats. The basolateral amygdala (BLA) provides information about the nature of the threats to both pathways via glutamatergic inputs to the lateral BST and CeM. Moreover, lateral BST-projecting CeL CRF neurons augment the excitatory input of BLA glutamatergic terminals in the lateral BST by stimulating presynaptic CRF_{R1} receptors. Since the CeL lacks BLA input, CeL CRF neurons receive information about potentially threatening stimuli through cortical and thalamic afferents like the insular cortex and paraventricular nucleus of the thalamus, respectively, which also innervate the BLA. EA CRF systems are also sensitive to corticosterone to varying degrees providing an avenue for the neuroendocrine system to exert influence over EA CRF circuitry.

This neuroanatomical circuitry involved in sustained anxiety and short-term fear responses has been expanded upon by other investigators looking into the role of EA CRF neurons in other models of stress, fear, and anxiety, stress-induced relapse and negative emotional states during withdrawal, central autonomic control, and modulation of the HPA axis. In the following sections, we describe this functional neural circuitry in detail.

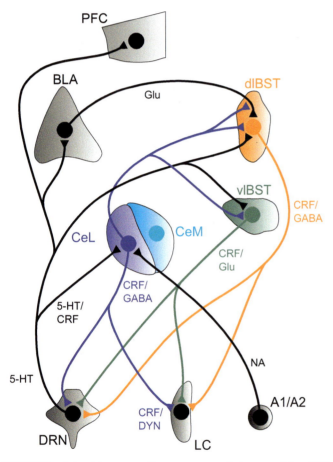

FIGURE 8.1 Schematic illustrates extended amygdala (EA) corticotropin-releasing factor (CRF) connections with brainstem monoaminergic nuclei implicated in facilitating anxietylike behavior and stress-induced arousal (see text for details). *5-HT*, 5-hydroxytryptamine/serotonin; *A1*, A1 noradrenergic nuclei; *A2*, A2 noradrenergic nuclei; *BLA*, basolateral amygdala; *CeL*, central nucleus of the amygdala, lateral division; *CeM*, central nucleus of the amygdala, medial division; *CRF*, corticotropin-releasing factor; *dlBST*, bed nucleus of the stria terminalis, lateral division, dorsal part; *DRN*, dorsal raphe nucleus; *DYN*, dynorphin; *GABA*, gamma-aminobutyric acid; *Glu*, glutamate; *LC*, locus coeruleus; *NA*, noradrenaline; *PFC*, prefrontal cortex; *vlBST*, bed nucleus of the stria terminalis, lateral division, ventral part.

RECIPROCAL CONNECTIONS BETWEEN EA CRF AND SEROTONERGIC SYSTEMS MODULATE ANXIETY

The dorsal raphe nucleus (DRN) contains topographically organized subpopulations of serotonin (5-HT) neurons that provide the majority of serotonergic input to limbic forebrain structures involved in regulating complex emotional responses to stress. The DRN contains a rich plexus of CRF fibers and is one of the few brain structures that contain dense CRF_{R2} expression, even greater than CRF_{R1}[28] The EA CRF neurons are a likely source of these CRF fibers (Fig. 8.1). The CRF neurons of the CeA (CeL and CeM), dlBST, and to a lesser extent the vlBST send projections to the "lateral wings",[10,17,29] an area comprised of the ventrolateral DRN and ventrolateral periaqueductal gray (DRVL/VLPAG), which contains serotonergic neurons involved in eliciting behavioral coping responses to stress such as freezing to conditioned fear stimuli.[30] CeA CRF neurons also innervate the caudal (DRC) and dorsal (DRD) nuclei of the DRN,[17] regions that contain serotonergic neurons that are responsive to a number of anxiogenic stimuli.[31] The DRN also receives input from the dlBST and vLBST with the latter providing more widespread input than the former, and since these regions contain high densities of CRF neurons it is likely that these neurons contribute to this projection.[29]

The functional role of CRF in the DRN depends on a complex interplay between CRF actions on different neuron types (e.g., 5-HT versus GABA), cellular and subcellular localization of CRF receptors, and dynamic changes in CRF receptor distribution driven by environmental factors such as prior stress.[32] Low concentrations of CRF, which preferentially bind the higher affinity CRF_{R1}, tend to inhibit 5-HT activity and release, while high concentrations of CRF bind to the more abundant CRF_{R2} and excite 5-HT activity and release.[32] This arrangement has led to the notion that CRF receptors have opposing functions on DRN 5-HT activity. Similarly, acute stress-induced release of CRF in the DRN leads to reduced 5-HT activity and release via stimulation of CRF_{R1}, whereas chronic stress-induced CRF promotes internalization of CRF_{R1} and trafficking of CRF_{R2} to the plasma membrane, shifting to a CRF_{R2}-mediated neuronal excitation and exaggerated release of 5-HT in the DRN itself and projection regions.[32,33] Several lines of evidence suggest this sensitization of the DRN 5-HT system to CRF via CRF_{R2} stimulation is important for the development of learned helplessness following uncontrollable stress,[32,34,35] immobility during swim stress,[32] and behavioral flexibility following chronic social defeat stress.[32,33] Altogether, excessive or prolonged release of CRF in the DRN, especially in dorsal and caudal regions that contain high CRF_{R2} and are known to be involved in anxiogenesis, may facilitate chronic anxietylike states through elevated activity of 5-HT neurons and 5-HT release in forebrain structures like the BLA and prefrontal cortex (PFC; Fig. 8.1).

The CeA, dlBST, and vlBST CRF neurons receive serotonergic input from the same DRN subregions they innervate[36–39] and these reciprocal connections are thought to be important for stress-induced emotional regulation (Fig. 8.1). The effects of 5-HT in the extended amygdala are complex due to the diversity of 5-HT receptors located on different cell types and in pre- and postsynaptic locations. These interactions have been extensively studied in the dlBST and are briefly summarized.[40] Patch clamp recordings and in vivo microinjection of 5-HT and various 5-HT receptor agonists/antagonists suggest BST neurons respond to 5-HT with both excitation, which is anxiogenic and mediated through $5\text{-}HT_{2A}$, $5\text{-}HT_{2C}$, and/or $5\text{-}HT_7$ receptors, and inhibition, which is anxiolytic and mediated through $5\text{-}HT_{1A}$ receptors. Interestingly, the 5-HT-induced excitation is abolished after application of CRF and shifted towards inhibition after social isolation stress, suggesting that stress-induced increases in BST CRF may impact how BST neurons respond to 5-HT. Unlike acute stress, chronic stress results in anxiety-like behavior that is associated with changes in 5-HT receptor expression patterns that ultimately favor excitatory responses of BST neurons to 5-HT. Taken together with evidence suggesting that BST CRF signaling in the DRC/DRD increases anxiety, a hypothetical model emerges where anxiety provoking stimuli activate dlBST CRF neurons that project to and activate DRC/DRD 5-HT neurons via CRF_{R2}. These 5-HT neurons project back to and inhibit dlBST neurons, creating a negative feedback circuit that may serve to limit stress-induced

activation of dlBST CRF neurons. However, chronic or severe stressors may facilitate the expression of 5-HT receptors that favor excitation, thus increasing dlBST output resulting in chronic anxiety-like states.[40]

It is noteworthy that Type III BST neurons, which are putative CRF neurons, express high levels of the $5\text{-}HT_{2C}$ receptor, relative to the $5\text{-}HT_{1A}$ receptor, and lack expression of $5\text{-}HT_{2A}$ and $5\text{-}HT_7$ receptors.[41] Since $5\text{-}HT_{2C}$ and $5\text{-}HT_{1A}$ receptors mediate depolarizing and hyperpolarizing responses of BST neurons to 5-HT,[41] respectively, putative Type III CRF neurons may be biased towards excitatory responses to 5-HT. An important objective for future research will be to characterize the cellular responses to 5-HT of both dlBST CRF neurons as well as vlBST CRF neurons, which receive serotonergic input, but have not been investigated in detail.

Serotonergic signaling in the CeA is also associated with anxiety/fear states and regulation of HPA axis responses. For example, application of CRF in the "lateral wings" of the DRN (i.e., DRVL/VLPAG) elicits freezing behavior that is associated with a concomitant increase in extracellular 5-HT in the CeA.[42] Exposure to acute restraint stress increases 5-HT release in the CeA and this increase is blocked by intracerebroventricular administration of a nonselective CRF receptor antagonist, presumably acting at CRF receptors located in the DRN.[43] Elevated CeA 5-HT is necessary for the stimulatory effect of the CeA on the HPA axis.[44,45] Together these studies highlight a role for serotonergic inputs to the CeA in mediating anxiety/fear responses as well as activation of the HPA axis, and this increase in 5-HT appears to require CRF stimulation of DRN 5-HT. It is unclear, however, if 5-HT release in the CeA serves as negative feedback to shut down CeA output as proposed for dlBST 5-HT, and whether 5-HT has specific effects on CRF expressing CeA neurons. Notably, the DRN contains a small population of CRF neurons with a portion of them colocalized with 5-HT, and they innervate CeA CRF neurons, suggesting that CRF may be coreleased with 5-HT to influence the activity of CeA CRF (Fig. 8.1).[36]

EA CRF PROJECTIONS TO THE LOCUS COERULEUS FACILITATE STRESS-INDUCED AROUSAL AND ANXIETY STATES

The locus coeruleus (LC) contains noradrenergic neurons that project widely through the brain, especially to forebrain limbic regions, and are essential for mediating arousal to emotionally salient and potentially threatening stimuli.[46,47] CRF neurons in the CeA and BST innervate both noradrenergic and nonnoradrenergic LC dendrites in the rostrolateral peri-LC and CeA

CRF contributes the bulk of CRF input from EA CRF populations (Fig. 8.1).[17,48,49] Many of the EA CRF neurons that project to the LC express the glucocorticoid receptor, implying that circulating corticosteroids regulate these LC-projecting CRF neurons.[50] Stress-induced release of CRF activates LC neurons via CRF_{R1} receptors and consequently increases noradrenaline (NA) in LC targets.[46,47] Stress and CRF are thought to alter LC activity in a manner that promotes high tonic activity, which leads to heightened arousal and scanning of the environment, rather than phasic activity mediated by excitatory amino acids, which results in selective attention to discrete cues.[46,47]

Mounting evidence suggests CeA CRF neurons are central to these stress-induced behavioral states generated by LC high tonic activity. Cardiovascular stress activates CeA and BST CRF neurons and lesions of the CeA, but not BST, eliminate the effects of stress on LC activity, suggesting that CeA CRF drives LC activity in response to this particular stressor.[51] Acute social defeat stress activates LC-projecting CeA CRF neurons and exposure to repeated social defeat stress also activates LC-projecting CeA CRF neurons, but only in mice that exhibit subordinate behavior, suggesting that recruitment of this population of CeA CRF neurons may dictate coping strategies.[52] In a recent elegant study, McCall et al.[53] report that optogenetic stimulation of LC-NA neurons drives high tonic firing, resulting in increased anxietylike behavior in the elevated zero maze as well as conditioned place aversion in the absence of stress. Moreover, selective photostimulation of CeA CRF terminals in the LC mimics the anxiogenic and aversive states produced by high tonic LC activity, and these behavioral effects are attenuated by previous intra-LC blockade of CRF_{R1} receptors.[53] Together, these studies provide definitive evidence that LC-projecting CeA CRF neurons are critical for driving stress-induced arousal and negative affective states through activation of CRF_{R1} receptors on LC noradrenergic neurons.

As noted earlier, over half of the CeA CRF neurons coexpress the opioid precursor, prodynorphin [17,18] and a substantial amount of these CRF-dynorphin neurons project to the LC dendritic zone (Fig. 8.1).[54] CRF (and glutamate) terminals in the LC express κ-opiate receptors (κ-OR), the endogenous cognate receptor for dynorphin, and infusion of a κ-OR agonist into the LC inhibits the cardiovascular stress-evoked activity of LC neurons, which is mediated by CeA CRF.[55] This illustrates that CRF and dynorphin may be coreleased into the LC during stress, with dynorphin opposing the actions of CRF, possibly to fine-tune LC activity to maintain labile arousal states. This arrangement has adaptive value in potentially threatening environments that require vigilance and environmental scanning (mediated by CRF), but require the flexibility to promote selective attention to salient cues that predict impending threats. Noradrenergic terminals arising from the lateral tegmental noradrenergic neurons [i.e., caudal ventrolateral medulla (A1) and nucleus of the solitary tract (A2)] form synapses with CRF-dynorphin neurons, including a population of LC-projecting CRF neurons (Fig. 8.1).[56] Taken together with evidence that NA excites CeA CRF neurons, CRF-NA interactions form a feed-forward system with medullary noradrenergic sources exciting CeA CRF, which in turn activate LC NA neurons, and consequently increasing NA release in target forebrain structures like the BLA.[56] This feed-forward circuit may result in global NA release in order to coordinate adaptive stress response, but dysfunction of this mechanism could lead to maladaptive hypervigilant states such as those observed in stress-related psychopathologies like PTSD.[56]

EA CRF SYSTEM AT THE INTERFACE OF STRESS AND MOTIVATION/REWARD

Decades of research suggest CRF systems impinge on the mesocorticolimbic dopamine (DA) system arising from the ventral tegmental area (VTA) to influence appetitive behavior and reward, including various aspects of addiction like stress-induced relapse. Initial evidence for a role of CRF in addiction came from pharmacological studies showing that administration of CRF_{R1} antagonists systemically, intraventricularly, or directly into the vlBST, but not CeA or PVN, attenuated stress-induced relapse.[57] Administration of CRF directly into the BST mimics the effects of stress on drug-seeking behavior, thus underscoring the BST as an important interface for stress-induced CRF to modify motivational aspects of behavior.[57] Although blockade of CRF receptors in the CeA has no effect on reinstatement, the CeA does provide a source of CRF input to the BST, in addition to intrinsic BST CRF release.[57] The VTA is another key node in this circuit because antagonism of CRF_{R2} receptors in this nucleus attenuates stress-induced relapse and local CRF application mimics the effects of stress on relapse.[57] CRF neurons from the EA (dlBST, vlBST, and CeL) project to the VTA and appear to innervate both dopaminergic and nondopaminergic neurons (e.g., GABAergic interneurons and glutamatergic terminals), suggesting both direct and indirect CRF actions on DA (Fig. 8.2).[58,59] The EA CRF projections to DA neurons utilize both glutamate (vlBST) and GABA (dlBST and CeA) as cotransmitters and the majority (83%) of CRF-DA synapses display asymmetric (excitatory) morphology.[43] CRF directly excites DA neurons via CRF_{R1}/PLC-PKC signaling[60] and increases their neuronal excitability by elevating intracellular calcium concentrations through a CRF_{R2}/PKA-dependent mechanism.[61] CRF actions in

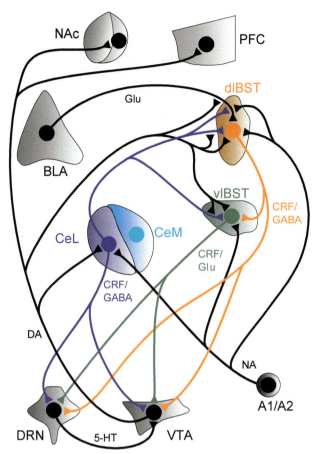

FIGURE 8.2 Schematic outlines extended amygdala (EA) corticotropin-releasing factor (CRF) interactions with dopaminergic and noradrenergic circuits involved in stress-induced relapse and negative emotional states associated with drug withdrawal (see text for details). *DA*, dopamine; *NAc*, nucleus accumbens; *VTA*, ventral tegmental area (see Fig. 8.1 for remaining abbreviations).

the VTA facilitate glutamatergic signaling onto DA neurons and involve CRF_{R2} receptors and the CRF-binding protein.[62] These diverse actions of CRF on VTA DA are likely to have a major impact on dopaminergic output to forebrain structures such as the nucleus accumbens (NAc) and prefrontal cortex (PFC) and consequently approach and avoid behaviors that are central to anxiety and addiction (Fig. 8.2).

An EA CRF-VTA circuit is implicated in other aspects of addiction such as the negative affective states during drug withdrawal as well as escalated drug intake (e.g., binge use).[63] Ethanol consumption and negative affect precipitated by prolonged withdrawal from ethanol are thought to involve elevated levels of CRF in the CeL, presumably from intrinsic sources, as well as CRF_{R1} receptor-dependent modulation of CeA microcircuits.[63,64] Likewise, chronic ethanol exposure modulates dlBST CRF neuron activity to augment the excitatory drive of BST VTA-projecting neurons and this circuit appears to become hyperactive during withdrawal.[25] A recent study using a pharmacogenetic DREADD approach to inhibit

dlBST CRF neurons that project to the VTA revealed a crucial role for this circuit in bingelike ethanol intake.[65]

In addition to these direct projections to the VTA, EA CRF neurons may influence the mesocorticolimbic DA system indirectly through polysynaptic circuits via connections with serotonergic neurons in the DRN and median raphe nucleus, which, in turn modulate DA function in the VTA and NAc (Fig. 8.2).[32,57] This alternative pathway may be especially important in stress-induced reinstatement of drug-seeking behavior.[32,57]

Together these studies highlight the role of the EA and VTA as important neural substrates for the actions of CRF in mediating various aspects of addiction; however, an important question remains in regards to the neural circuitry that drives EA CRF activation in response to stress. The catecholamines, NA and DA, play an important part in this aspect. Tract tracing studies and manipulations of noradrenergic systems in rodent models of stress-induced relapse reveal that medullary sources of noradrenergic neurons (i.e., A1 and A2 nuclei), but not locus coeruleus noradrenergic neurons, project to and activate EA CRF neurons (i.e., vBST and CeL) via stimulation of beta-2 (β2) adrenoceptors (Fig. 8.2).[57,66] VTA DA neurons also send projections to the BST and can directly excite BST CRF neurons via D1 and D2 receptors and may stimulate CRF afferents arising from the CeL.[66] Both NA and DA facilitate glutamatergic input onto VTA-projecting BST neurons through activation of BST CRF interneurons, which in turn bind CRFR1 receptors on glutamatergic afferents.[66] The reciprocal connections between BST CRF and VTA DA neurons may act as a feed-forward system that augments DA signaling in VTA projection regions (e.g., PFC and NAc) during stress and withdrawal states.[66]

EA CRF NEURONS RECEIVES PEPTIDERGIC INPUT FROM THE PARABRACHIAL NUCLEUS

The parabrachial nucleus (PBn) receives viscerosensory afferents from the nucleus of the solitary tract and integrates and relays sensory information primarily related to nociception and gustation to upstream structures, including the BST and CeA. The lateral PBn (LPBn) contains calcitonin gene–related peptide (CGRP) and pituitary adenylate cyclase–activating polypeptide (PACAP) neurons that densely innervate EA CRF neurons[67,68] (Fig. 8.3). Over half of the peptide terminals in the dlBST and CeL display immunoreactivity for both CGRP and PACAP suggesting a high degree of colocalization.[67,68] Additional PACAP terminals innervating dlBST CRF neurons originate in the PVN, suggesting multiple sources of this neuropeptide.[69,70] The dlBST is also innervated by vasoactive intestinal polypeptide (VIP), although the source of this input is unknown.[69,70]

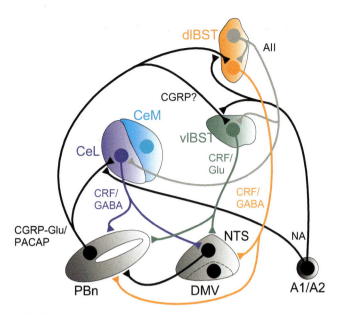

FIGURE 8.3 Hypothetical model shows neuropeptidergic inputs to extended amygdala (EA) corticotropin-releasing factor (CRF) systems involved in facilitating anxiety, fear, and autonomic responses as well as EA CRF projections to autonomic control centers (see text for details). *AII*, angiotensin II; *CGRP*, calcitonin gene related peptide; *DMV*, dorsal motor nucleus of the vagus; *NTS*, nucleus of the solitary tract; *PACAP*, pituitary adenylate cyclase activating polypeptide; *PBn*, parabrachial nucleus (see Fig. 8.1 for remaining abbreviations).

Infusion of CGRP directly into the dlBST is anxiogenic and activates an anxiety-related brain circuit innervated by the BST.[71] Antagonism of BST CGRP receptors blocks contextual fear, but not cued fear, consistent with the idea that EA CRF responds to more diffuse threats.[72] The anxiogenic effects of CGRP are mediated in part by CRF because blockade of CRF_{R1} receptors or siRNA knockdown of CRF in the dlBST attenuated the effects of CGRP on anxietylike behavior.[73] CGRP terminals are also present in the ventral BST; however it is unknown whether they innvervate vlBST CRF neurons.[74] The effects of CGRP on CeL CRF neurons has not been tested, but evidence suggests CGRP in the CeA is involved in pain-related behaviors such as withdrawal reflexes and vocalizations that may involve activation of CRF_{R1}.[75]

Electrophysiological studies report CGRP has inhibitory and excitatory effects in the BST and CeA, respectively, although none of these studies identified CRF neurons so it remains unclear how CGRP alters CRF neurons.[76] Moreover, PBn CGRP afferents to the EA colocalize with glutamate; therefore, in vivo effects of CGRP coreleased with glutamate are likely different than those obtained from local application of CGRP on in vitro slices.[77] Indeed, using an optogenetic strategy to stimulate CGRP afferents in the dlBST the Winder group found excitatory effects, which are modulated by α_2-adrenoreceptors, suggesting that noradrenergic/adrenergic systems may gate CGRP input.[77] Overall,

CGRP facilitates anxietylike responses through inputs to dlBST/CeL CRF neurons as well as nociceptive behaviors through CRF-mediated actions in the CeA.

Converging lines of evidence point to a PACAP-CRF interaction in the BST and CeA that increases anxiety/fear, modulates aspects of nociception, and produces anorexic effects (e.g., weight loss and reduced feeding).[68,78] The interaction between PACAP and EA CRF systems has previously been reviewed[68,78] and is briefly summarized here. PACAP and its cognate receptor, PAC1, are selectively increased in the dlBST, but not CeA, following chronic stress and administration of PACAP in the dlBST or CeA increases anxiety-/fear-related behaviors as well as anorexia suggesting a role in stress-induced suppression of feeding behavior. The behavioral effects of chronic stress are reversed by antagonism of BST PAC1/VPAC2 receptors. PACAP in the CeA also results in hypersensitivity to pain and this effect is attenuated by blockade of PAC1 receptors. PACAP signaling and genetic polymorphisms in the PAC1 gene are linked to PTSD (and other psychiatric disorders) and may involve enhanced BST activity to emotionally salient events, thus highlighting the therapeutic potential of unraveling PACAP function in the EA.[68] Although PACAP directly innervates EA CRF neurons and has similar behavioral effects to CRF, there are a lack of studies on how the two systems interact, making this an important area for future research.

The CRF neurons of the EA send projections to the PBn,[17,79–81] although it is unclear whether these connections are reciprocal with CGRP/PACAP neurons that project to the EA (Fig. 8.3). The projections from the EA are topographically organized in such a manner that the dlBST and CeL CRF systems innervate both the rostral region, which is involved in nociception, cardiovascular, respiratory, and gastric functions and caudal region, which is the taste-responsive area.[17,79–81] The vlBST CRF neurons, on the other hand, innervate primarily the rostral PBn and send only sparse projections to the caudal nucleus.[79,81] Consistent with a role for CRF in taste-responsiveness and the regulation of electrolyte homeostasis, administration of α-helical CRF_{9-41} directly into the LPBn modulates sodium intake (i.e., salt appetite).[82] Other EA neuropeptides such as somatostatin and neurotensin project to the PBn[79–81] and these neuropeptides colocalize with CRF neurons[15–17] so it will be important to determine the relative contribution of these individual neuropeptides and the colocalized subpopulations on central autonomic control.

CRF INTERACTION WITH THE BRAIN RENIN-ANGIOTENSIN SYSTEM

The brain renin-angiotensin system, like its peripheral counterpart, mediates classical functions like cardiovascular regulation and blood volume/electrolyte homeostasis, as well as brain-specific mnemonic functions and affective and neuroendocrine responses to stress. Initial evidence for an interaction between angiotensin II (AII) and CRF systems came from studies reporting that AII directly regulates CRF neurons in the PVN and that blockade of PVN AII receptor type 1 (AT_1), which colocalize with CRF, attenuates the HPA-axis response to stress as well as stress-induced alterations in PVN CRF mRNA and content.[83]

More recent findings suggest the brain angiotensin system also appears to modulate "extrahypothalamic" CRF systems involved in emotional responses to stress. The EA contains AII cell bodies in the dlBST and Me (Fig. 8.3),[84] high levels of AII-immunoreactive fibers,[84] and AII receptor expression, although the receptors appear to be extrinsic in the BST.[85] Administration of the AT_1 antagonist, losartan, directly into the CE blocks stress-induced potentiated fear behavior[86] and enhances fear extinction when given after cued fear conditioning, which is accompanied by reduced AT_1 receptor and c-Fos mRNA levels in the amygdala and BST, respectively.[87] Similar to the PVN, AT_{1a} colocalizes with CE CRF neurons,[88] suggesting that CRF may drive some of the stress-induced changes upon AII activation. Indeed, genetic deletion of AT_{1a} specifically in CRF neurons attenuates consolidation of fear conditioning.[88] An important objective for future work will be to use more targeted optogenetic and chemogenetic techniques to delineate the role of specific populations of CRF neurons (e.g., BST, CE, PVN) in mediating the effects of AII on stress-related emotional, neuroendocrine, and autonomic responses. Clinical studies report beneficial therapeutic effects of drugs that reduce the activity of the brain RAS system in stress-related psychopathologies like PTSD and depression,[89,90] so elucidating the intricacies of AII-CRF interactions in the EA may be fruitful for developing novel and selective treatments.

EA CRF NEURONS AND CENTRAL AUTONOMIC CONTROL

Tracing studies report EA CRF neurons project to the dorsal vagal complex, comprised of the dorsal motor nucleus of the vagus (DMV) and the nucleus of the solitary tract (NTS), and CRF neurons in the dlBST and CeL provide the majority of CRF input, with only sparse innervation arising from vlBST and CeM CRF neurons (Fig. 8.3).[17,91,92] In the case of CeA CRF neurons, a CRF-Cre reporter line suggests these projections are confined to the NTS where they innervate noradrenergic dendrites and perikarya (i.e., A2 noradrenergic neurons).[17] As mentioned above, NTS A2 noradrenergic neurons project to EA CRF neurons and this circuit is involved in stress-induced relapse and anxiety states during withdrawal.

The exact role EA CRF efferents play in the DMV and NTS is unclear as studies selectively manipulating these CRF subpopulations are lacking; however, these structures are well known to be involved in central autonomic control. Central CRF signaling is intimately involved in stress-induced modulation of feeding and gastrointestinal function (i.e., stimulation of colonic motility) and much of this research has emphasized the importance of central CRF systems in the PVN, Barrington's nucleus, and LC in mediating the effects of CRF on the gut–brain axis,[93–95] while EA CRF systems have generally been excluded from this line of research. Perhaps, then, EA CRF systems are upstream of the motor components of stress-induced changes in gastric function and instead integrate viscerosensory information (e.g., nociception) and psychological components of gastrointestinal distress, and through projections to CRF systems in the PVN and LC provide modulatory influences. In support of this hypothesis, recent evidence suggests that CRF acting in the CeA results in hyperalgesia to visceral pain in a rat model of colorectal distension and this effect is blocked by intra-CeA CRF_{R1} antagonism.[96,97] This is consistent with reports of increased CeA CRF following colorectal distension and other manipulations that produce visceral pain.[97] Dysfunction of CeA CRF systems involved in gastrointestinal function and effect may underlie the high comorbidity seen in gastrointestinal disorders such as irritable bowel syndrome and mood/anxiety disorders, accentuating the importance of research into the role of EA CRF circuitry in the gut–brain axis to develop novel therapeutics for medical and psychiatric diseases.[97–99]

EA CRF NEURONS MODULATE HYPOTHALAMIC–PITUITARY–ADRENAL AXIS ACTIVITY

Activation of CRF neurons in the PVN initiates the primary neuroendocrine response to stress, the HPA axis, by triggering ACTH release from the anterior pituitary into the bloodstream.[100] ACTH circulates and stimulates the secretion of corticosteroids (corticosterone in rodents; cortisol in humans) from the adrenals, which regulate homeostasis of diverse physiological systems and ultimately provide negative feedback on the HPA axis at multiple levels.[100] The neurosecretory CRF neurons of the PVN are surrounded by GABAergic neurons that provide tonic inhibition so there is great interest in the neural circuitry that provides stress-excitatory input to the PVN to drive CRF activation and consequent HPA axis responses. Both parvocellular and magnocellular divisions of the PVN contain dense plexuses of CRF fibers that form excitatory synapses on CRF perikarya and dendrites suggesting intrinsic connections between

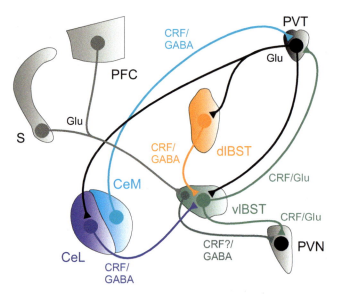

FIGURE 8.4 Summary model depicts EA CRF pathways involved in modulation of HPA axis function (see text for details). *PVN*, paraventricular nucleus of the hypothalamus; *PVT*, paraventricular nucleus of the thalamus; *S*, subiculum (see Fig. 8.1 for remaining abbreviations).

PVN CRF neurons and inputs from extrahypothalamic CRF sources may provide excitatory drive to CRF neurosecretory activity in response to stress.[101]

Tract tracing studies have identified that CRF neurons in the vlBST, which coexpress glutamate, innervate PVN CRF neurosecretory cells placing them in a position to positively regulate the HPA axis (Fig. 8.4).[102,103] This anatomical arrangement is supported by reports that lesions of the anteroventral BST region that contains these excitatory CRF neurons blocks the increase in corticosterone and PVN *c-fos* mRNA following acute stress.[104,105] Likewise, large lesions encompassing the anterior BST, including both the dlBST and vlBST CRF populations produce a 30% decrease in PVN CRF mRNA.[106] Systemic administration of the proinflammatory interleukin-1β activates vlBST neurons, including CRF, that project to the PVN and vlBST lesions attenuate the increase in PVN CRF mRNA and ACTH release.[107] Lesions of the anteroventral BST following chronic stress, however, result in *activation* of the HPA axis suggesting that this region may inhibit the HPA axis following chronic stress.[105] This paradox could be due to different populations of vlBST neurons being recruited during acute versus chronic stress or a chronic stress-induced reorganization of the neural circuitry regulating activity of PVN-projecting vlBST neurons.

In contrast to vlBST CRF, dlBST and CeA CRF neurons are relatively devoid of direct connections to the PVN, but may indirectly influence neuroendocrine responses through projections to vlBST CRF neurons (Fig. 8.4).[17,29,108] The EA CRF neurons display differential sensitivity to circulating corticosterone, with the dlBST/CeL CRF neurons and vlBST CRF neurons responding

to acute and sustained corticosterone elevations, respectively, supporting the idea that EA CRF systems may form a corticosterone-sensitive subsystem and that different populations of EA CRF neurons are recruited during acute versus chronic stress.[109] Finally, it is worth mentioning that the posterior BST integrates inputs from limbic structures like the medial amygdala and ventral subiculum and exerts inhibitory control over the HPA axis,[104] although CRF neurons are likely uninvolved due to minimal expression in the posterior BST.[110] Taken together the vlBST CRF neurons are sensitive to circulating glucocorticoids, acute and chronic stressors, and may integrate inputs from dlBST and CeL CRF populations in order to modulate PVN CRF neurons accordingly.

There is considerable interest in how limbic structures such as the hippocampus and medial prefrontal cortex regulate the activity of the HPA axis in response to stress; however, these structures are largely devoid of direct projections to the PVN CRF neurons so they are thought to influence HPA axis responses through connections with PVN-projecting BST neurons.[100] Several lines of evidence suggest GABAergic neurons in the vlBST may integrate input from limbic structures in order to restrain HPA axis responses to stress. Lesions of the ventral subiculum and medial prefrontal cortex (i.e., prelimbic and anterior cingulate) augment the HPA axis response to stress and diminish the stress-induced activation of GABAergic neurons in the vlBST.[111,112] Tract tracing studies reveal the mPFC and ventral subiculum terminals converge on PVN-projecting vlBST GABAergic neurons and lesions of both limbic structures have an additive effect on HPA axis responses to stress (Fig. 8.4).[111,112] Finally, a comparison of excitotoxic lesions of the ventral subiculum and ablation of vlBST GABAergic neurons on HPA axis responses show that vlBST GABAergic neurons have a more pronounced inhibitory effect on the HPA axis suggesting that the vlBST serves as a final common pathway for the integration of limbic inputs to the HPA axis.[111,112] While the vlBST is unique because of its population of CRF-glutamate containing neurons, the majority of vlBST CRF neurons coexpress GABA[9]; therefore, an important next step will be to characterize the role, if any, of CRF-GABAergic vlBST neurons in providing inhibitory input to the HPA axis.

In addition to limbic structures that modulate HPA axis activity, the paraventricular nucleus of the thalamus (PVT) is implicated in mediating habituation and facilitation of HPA-axis responses to repeated stress and is one of the most "stress-responsive" brain regions.[100] The PVT is strictly innervated by CRF neurons in the vlBST and CeA (primarily CeM)[113] and some of the PVT-projecting neurons are in close apposition to adrenergic terminals, which may provide stress-excitatory input from medullary autonomic centers to this pathway (Fig. 8.4).[114] The PVT in turn sends widespread projections to the EA and these terminals innervate CRF neurons in the dlBST and CeL (Fig. 8.4).[115] Although PVT terminals are found in regions where vlBST CRF neurons are located, the previous study did not assess the proximity of PVT terminals to vlBST CRF neurons.[115] The underlying mechanisms for how PVT projections modulate EA CRF neurons to influence the HPA axis is unclear, but evidence suggests the PVT is involved in habituation of HPA axis responses to repeated, homotypical stressors as well as facilitation of HPA axis responses to novel stressors in the face of repeated stress.[100] In other words, the PVT is sensitive to the chronicity of stressors and may regulate PVN-projecting BST neurons, including CRF neurons, accordingly.

CONCLUSIONS AND FUTURE DIRECTIONS

In this review, we have outlined the various CRF subpopulations located in the EA and their role in coordinating the behavioral, neuroendocrine, and autonomic components of the stress response through widespread interconnections with forebrain limbic structures, brainstem monoaminergic nuclei, and brain regions involved in neuroendocrine and central autonomic control. Based on anatomical location, morphology, neurochemical phenotypes, electrophysiological characteristics, and patterns of connectivity, a concept of heterogeneous EA CRF subpopulations with functional specificity emerges. For example, vlBST CRF neurons, through direct excitatory projections to PVN CRF secretagogue neurons, appear to act as a gateway for limbic and other EA CRF input to modulate HPA axis responses. CeA CRF neurons, on the other hand, are unique in that they appear to be the prime mover for stress-evoked increases in LC activity and arousal/attention processes. The dlBST CRF neurons, compared to vlBST and CeL CRF, are innervated to a greater degree by peptidergic afferents such as AII, CGRP, PACAP, and VIP to name a few. There are also characteristics that suggest a degree of homogeneity between EA CRF populations. CRF neurons in the dlBST and CeL share a number of similarities in their morphology, coexpression with GABA and neurotensin, electrophysiological parameters (e.g., long latency to fire action potentials), and connectivity.

The complexity of these heterogeneous subpopulations, however, offers researchers a multitude of ways to dissect the functional role of these CRF neurons using a combination of approaches (e.g., optogenetics, pharmacogenetics, transgenic mouse models, and viral tools) to selectively manipulate discrete populations of CRF neurons. These approaches have been fruitful in elucidating the role of LC-projecting CeA CRF neurons in anxiety[53] and VTA-projecting BST CRF neurons in binge alcohol drinking.[65] Although these techniques are powerful tools for cell-type specific manipulations, there are important caveats. For example, a recent review of some of the available CRF-Cre transgenic mouse lines revealed ectopic CRF expression in two out of three lines,[116] highlighting a common pitfall of Cre-driver lines having "leaky" expression patterns.

Similarly, while the advantages of gaining genetic control over a specific cell type for experimental manipulation are clear, there is still an important drawback, especially in stress-related research, where only a fraction of a particular neuron type may be activated by a particular stimulus and involved in eliciting a particular response. Optogenetic and pharmacogenetic control over specific cell types capture all the neurons of a given cell type, regardless of their activation state, and thus lack the capability to target functional neuronal ensembles. New techniques that rely on the promoters of activity-related genes, such as the *Fos* gene, to capture neuronal ensembles activated by a particular stimulus provide a solution.[117] Using these new methods in conjunction with optogenetics and pharmacogenetics offers exciting new avenues to elucidate the underlying circuitry of EA CRF subpopulations and their involvement in stress, fear, and anxiety.

References

1. Vale W, Spiess J, Rivier C, Rivier J. Characterization of a 41-residue ovine hypothalamic peptide that stimulates secretion of corticotropin and beta-endorphin. *Science*. 1981;213:1394–1397.
2. Bale TL, Vale WW. Crf and CRF receptors: role in stress responsivity and other behaviors. *Annu Rev Pharmacol Toxicol*. 2004;44:525–557.
3. Alheid GF. Extended amygdala and basal forebrain. *Ann NY Acad Sci*. 2003;985:185–205.
4. De Olmos JS, Heimer L. The concepts of the ventral striatopallidal system and extended amygdala. *Ann NY Acad Sci*. 1999;877:1–32.
5. Dong H-W, Petrovich GD, Swanson LW. Topography of projections from amygdala to bed nuclei of the stria terminalis. *Brain Res Rev*. 2001;38:192–246.
6. Phelix CF, Paull WK. Demonstration of distinct corticotropin releasing factor–containing neuron populations in the bed nucleus of the stria terminalis. A light and electron microscopic immunocytochemical study in the rat. *Histochemistry*. 1990;94:345–364.
7. Cummings S, Elde R, Ells J, Lindall A. Corticotropin-releasing factor immunoreactivity is widely distributed within the central nervous system of the rat: an immunohistochemical study. *J Neurosci*. 1983;3:1355–1368.
8. Swanson LW, Sawchenko PE, Rivier J, Vale WW. Organization of ovine corticotropin-releasing factor immunoreactive cells and fibers in the rat brain: an immunohistochemical study. *Neuroendocrinology*. 1983;36:165–186.
9. Nguyen AQ, Dela Cruz JAD, Sun Y, Holmes TC, Xu X. Genetic cell targeting uncovers specific neuronal types and distinct subregions in the bed nucleus of the stria terminalis. *J Comp Neurol*. 2015. http://dx.doi.org/10.1002/cne.23954.
10. Gray TS, Magnuson DJ. Peptide immunoreactive neurons in the amygdala and the bed nucleus of the stria terminalis project to the midbrain central gray in the rat. *Peptides*. 1992;13:451–460.

11. Cassell MD, Gray TS. Morphology of peptide-immunoreactive neurons in the rat central nucleus of the amygdala. *J Comp Neurol.* 1989;281:320–333.

12. Cassell MD, Gray TS, Kiss JZ. Neuronal architecture in the rat central nucleus of the amygdala: a cytological, hodological, and immunocytochemical study. *J Comp Neurol.* 1986;246:478–499.

13. Dabrowska J, Hazra R, Guo J, DeWitt S, Rainnie D. Central CRF neurons are not created equal: phenotypic differences in CRF-containing neurons of the rat paraventricular hypothalamus and the bed nucleus of the stria terminalis. *Front Neurosci.* 2013;7:1–14.

14. Veinante P, Stoeckel M-E, Freund-Mercier M-J. GABA- and peptide-immunoreactivities co-localize in the rat central extended amygdala. *Neuroreport.* 1997;8:2985–2989.

15. Ju G, Han Z. Coexistence of corticotropin releasing factor and neurotensin within oval nucleus neurons in the bed nuclei of the stria terminalis in the rat. *Neurosci Lett.* 1989;99:246–250.

16. Shimada S, et al. Coexistence of peptides (corticotropin releasing factor/neurotensin and substance P/somatostatin) in the bed nucleus of the stria terminalis and central amygdaloid nucleus of the rat. *Neuroscience.* 1989;30:377–383.

17. Pomrenze MB, et al. A transgenic rat for investigating the anatomy and function of corticotrophin releasing factor circuits. *Front Neurosci.* 2015;9:1–14.

18. Marchant NJ, Densmore VS, Osborne PB. Coexpression of prodynorphin and corticotrophin-releasing hormone in the rat central amygdala: evidence of two distinct endogenous opioid systems in the lateral division. *J Comp Neurol.* 2007;504:702–715.

19. Hökfelt T, et al. Neuropeptides—an overview. *Neuropharmacology.* 2000;39:1337–1356.

20. Mains RE, Eipper BA. *Neuropeptide Functions and Regulation;* 1999.

21. Mains RE, Eipper BA. *The Neuropeptides;* 1999.

22. Hammack SE, Mania I, Rainnie DG. Differential expression of intrinsic membrane currents in defined cell types of the anterolateral bed nucleus of the stria terminalis. *J Neurophysiol.* 2007;98:638–656.

23. Hazra R, et al. A transcriptomic analysis of type I–III neurons in the bed nucleus of the stria terminalis. *Mol Cell Neurosci.* 2011;46:699–709.

24. Martin EI, et al. A novel transgenic mouse for gene-targeting within cells that express corticotropin-releasing factor. *Biol Psychiatry.* 2010;67:1212–1216.

25. Silberman Y, Matthews RT, Winder DGA. Corticotropin releasing factor pathway for ethanol regulation of the ventral tegmental area in the bed nucleus of the stria terminalis. *J Neurosci.* 2013;33:950–960.

26. Walker DL, Toufexis DJ, Davis M. Role of the bed nucleus of the stria terminalis versus the amygdala in fear, stress, and anxiety. *Eur J Pharmacol.* 2003;463:199–216.

27. Walker DL, Miles LA, Davis M. Selective participation of the bed nucleus of the stria terminalis and CRF in sustained anxiety-like versus phasic fear-like responses. *Prog Neuropsychopharmacol Biol Psychiatry.* 2009;33:1291–1308.

28. Chalmers DT, Lovenberg TW, De Souza EB. Localization of novel corticotropin-releasing factor receptor (CRF2) mRNA expression to specific subcortical nuclei in rat brain: comparison with CRF1 receptor mRNA expression. *J Neurosci.* 1995;15:6340–6350.

29. Dong H-W, Petrovich GD, Watts AG, Swanson LW. Basic organization of projections from the oval and fusiform nuclei of the bed nuclei of the stria terminalis in adult rat brain. *J Comp Neurol.* 2001;436:430–455.

30. Paul ED, Johnson PL, Shekhar A, Lowry CA. The Deakin/Graeff hypothesis: focus on serotonergic inhibition of panic. *Neurosci Biobehav Rev.* 2014;46(Pt 3):379–396.

31. Paul ED, Lowry CA. Functional topography of serotonergic systems supports the Deakin/Graeff hypothesis of anxiety and affective disorders. *J Psychopharmacol.* 2013;27:1090–1106.

32. Valentino RJ, Lucki I, Van Bockstaele E. Corticotropin-releasing factor in the dorsal raphe nucleus: linking stress coping and addiction. *Brain Res.* 2010;1314:29–37.

33. Wood SK, et al. Cellular adaptations of dorsal raphe serotonin neurons associated with the development of active coping in response to social stress. *Biol Psychiatry.* 2013;73:1087–1094.

34. Hammack SE, et al. Corticotropin releasing hormone type 2 receptors in the dorsal raphe nucleus mediate the behavioral consequences of uncontrollable stress. *J Neurosci.* 2003;23:1019–1025.

35. Maier SF, Watkins LR. Stressor controllability and learned helplessness: the roles of the dorsal raphe nucleus, serotonin, and corticotropin-releasing factor. *Neurosci Biobehav Rev.* 2005;29:829–841.

36. Commons KG, Connolley RK, Valentino RJA. Neurochemically distinct dorsal raphe-limbic circuit with a potential role in affective disorders. *Neuropsychopharmacology.* 2003;28:206–215.

37. Asan E, et al. The corticotropin-releasing factor (CRF)-system and monoaminergic afferents in the central amygdala: investigations in different mouse strains and comparison with the rat. *Neuroscience.* 2005;131:953–967.

38. Asan E, Steinke M, Lesch K-P. Serotonergic innervation of the amygdala: targets, receptors, and implications for stress and anxiety. *Histochem Cell Biol.* 2013;139:785–813.

39. Phelix CF, Liposits Z, Paull WK. Serotonin-CRF interaction in the bed nucleus of the stria terminalis: a light microscopic double-label immunocytochemical analysis. *Brain Res Bull.* 1992;28:943–948.

40. Hammack SE, et al. The response of neurons in the bed nucleus of the stria terminalis to serotonin: implications for anxiety. *Prog Neuropsychopharmacol Biol Psychiatry.* 2009;33:1309–1320.

41. Guo J-D, Hammack SE, Hazra R, Levita L, Rainnie DG. Bi-directional modulation of bed nucleus of stria terminalis neurons by 5-HT: molecular expression and functional properties of excitatory 5-HT receptor subtypes. *Neuroscience.* 2009;164:1776–1793.

42. Forster GL, et al. Corticotropin-releasing factor in the dorsal raphe elicits temporally distinct serotonergic responses in the limbic system in relation to fear behavior. *Neuroscience.* 2006;141:1047–1055.

43. Mo B, Feng N, Renner K, Forster G. Restraint stress increases serotonin release in the central nucleus of the amygdala via activation of corticotropin-releasing factor receptors. *Brain Res Bull.* 2008;76:493–498.

44. Feldman S, Newman ME, Weidenfeld J. Effects of adrenergic and serotonergic agonists in the amygdala on the hypothalamo-pituitary-adrenocortical axis. *Brain Res Bull.* 2000;52:531–536.

45. Feldman S, Newman ME, Gur E, Weidenfeld J. Role of serotonin in the amygdala in hypothalamo-pituitary-adrenocortical responses. *Neuroreport.* 1998;9:2007–2009.

46. Valentino RJ, Van Bockstaele E. Convergent regulation of locus coeruleus activity as an adaptive response to stress. *Eur J Pharmacol.* 2008;583:194–203.

47. Zitnik GA. Control of arousal through neuropeptide afferents of the locus coeruleus. *Brain Res.* 2016. http://dx.doi.org/10.1016/j.brainres.2015.12.010.

48. Van Bockstaele EJ, Colago EEO, Valentino RJ. Amygdaloid corticotropin-releasing factor targets locus coeruleus dendrites: substrate for the co-ordination of emotional and cognitive limbs of the stress response. *J Neuroendocrinol.* 1998;10:743–758.

49. Van Bockstaele EJ, Peoples J, Valentino RJ. Anatomic basis for differential regulation of the rostrolateral peri–locus coeruleus region by limbic afferents. *Biol Psychiatry.* 1999;46:1352–1363.

50. Lechner SM, Valentino RJ. Glucocorticoid receptor-immunoreactivity in corticotrophin-releasing factor afferents to the locus coeruleus. *Brain Res.* 1999;816:17–28.

51. Curtis AL, Bello NT, Connolly KR, Valentino RJ. Corticotropin-releasing factor neurones of the central nucleus of the amygdala mediate locus coeruleus activation by cardiovascular stress. *J Neuroendocrinol.* 2002;14:667–682.

52. Reyes BAS, Zitnik G, Foster C, Van Bockstaele EJ, Valentino RJ. Social stress engages neurochemically-distinct afferents to the rat locus coeruleus depending on coping strategy. *eNeuro*. 2015;2.

53. McCall JG, et al. Crh engagement of the locus coeruleus noradrenergic system mediates stress-induced anxiety. *Neuron*. 2015;87:605–620.

54. Reyes BAS, Drolet G, Van Bockstaele EJ. Dynorphin and stress-related peptides in rat locus coeruleus: contribution of amygdalar efferents. *J Comp Neurol*. 2008;508:663–675.

55. Kreibich A, et al. Presynaptic inhibition of diverse afferents to the locus ceruleus by κ-opiate receptors: a novel mechanism for regulating the central norepinephrine system. *J Neurosci*. 2008;28:6516–6525.

56. Kravets JL, Reyes BAS, Unterwald EM, Bockstaele EJV. Direct targeting of peptidergic amygdalar neurons by noradrenergic afferents: linking stress-integrative circuitry. *Brain Struct Funct*. 2013;220:541–558.

57. Mantsch JR, Baker DA, Funk D, Lê AD, Shaham Y. Stress-induced reinstatement of drug seeking: 20 years of progress. *Neuropsychopharmacology*. 2016;41:335–356.

58. Rodaros D, Caruana DA, Amir S, Stewart J. Corticotropin-releasing factor projections from limbic forebrain and paraventricular nucleus of the hypothalamus to the region of the ventral tegmental area. *Neuroscience*. 2007;150:8–13.

59. Tagliaferro P, Morales M. Synapses between corticotropin-releasing factor-containing axon terminals and dopaminergic neurons in the ventral tegmental area are predominantly glutamatergic. *J Comp Neurol*. 2008;506:616–626.

60. Wanat MJ, Hopf FW, Stuber GD, Phillips PEM, Bonci A. Corticotropin-releasing factor increases mouse ventral tegmental area dopamine neuron firing through a protein kinase C-dependent enhancement of I_h: CRF increases VTA dopamine neuron firing. *J Physiol*. 2008;586:2157–2170.

61. Riegel AC, Williams JT. Crf facilitates calcium release from intracellular stores in midbrain dopamine neurons. *Neuron*. 2008;57:559–570.

62. Wise RA, Morales M. A ventral tegmental CRF–glutamate–dopamine interaction in addiction. *Brain Res*. 2010;1314:38–43.

63. Zorrilla EP, Logrip ML, Koob GF. Corticotropin releasing factor: a key role in the neurobiology of addiction. *Front Neuroendocrinol*. 2014;35:234–244.

64. Silberman Y, Winder DG. Ethanol and corticotropin releasing factor receptor modulation of central amygdala neurocircuitry: an update and future directions. *Alcohol*. 2015;49:179–184.

65. Rinker JA, et al. Extended amygdala to ventral tegmental area corticotropin-releasing factor circuit controls binge ethanol intake. *Biol Psychiatry*. 2016. http://dx.doi.org/10.1016/j.biopsych.2016.02.029.

66. Silberman Y, Winder DG. Emerging role for corticotropin releasing factor signaling in the bed nucleus of the stria terminalis at the intersection of stress and reward. *Front Psychiatry*. 2013;4:42.

67. Missig G, et al. Parabrachial nucleus (PBn) pituitary adenylate cyclase activating polypeptide (PACAP) signaling in the amygdala: implication for the sensory and behavioral effects of pain. *Neuropharmacology*. 2014;86:38–48.

68. Hammack SE, May V. Pituitary adenylate cyclase activating polypeptide in stress-related disorders: data convergence from animal and human studies. *Biol Psychiatry*. 2015;78:167–177.

69. Kozicz T, Vigh S, Arimura A. Axon terminals containing PACAP- and VIP-immunoreactivity form synapses with CRF-immunoreactive neurons in the dorsolateral division of the bed nucleus of the stria terminalis in the rat. *Brain Res*. 1997;767:109–119.

70. Kozicz T, Vigh S, Arimura A. Immunohistochemical evidence for PACAP and VIP interaction with met-enkephalin and CRF containing neurons in the bed nucleus of the stria terminalis. *Ann NY Acad Sci*. 1998;865:523–528.

71. Sink KS, Walker DL, Yang Y, Davis M. Calcitonin gene-related peptide in the bed nucleus of the stria terminalis produces an anxiety-like pattern of behavior and increases neural activation in anxiety-related structures. *J Neurosci*. 2011;31:1802–1810.

72. Sink KS, Davis M, Walker DL. CGRP antagonist infused into the bed nucleus of the stria terminalis impairs the acquisition and expression of context but not discretely cued fear. *Learn Mem*. 2013;20:730–739.

73. Sink K, Chung A, Ressler K, Davis M, Walker D. Anxiogenic effects of CGRP within the BNST may be mediated by CRF acting at BNST CRFR1 receptors. *Behav Brain Res*. 2013;243:286–293.

74. Dobolyi A, Irwin S, Makara G, Usdin TB, Palkovits M. Calcitonin gene-related peptide-containing pathways in the rat forebrain. *J Comp Neurol*. 2005;489:92–119.

75. Han JS, Adwanikar H, Li Z, Ji G, Neugebauer V. Facilitation of synaptic transmission and pain responses by CGRP in the amygdala of normal rats. *Mol Pain*. 2010;6. http://dx.doi.org/10.1186/1744-8069-6-10.

76. Gungor NZ, Pare D. Cgrp inhibits neurons of the bed nucleus of the stria terminalis: implications for the regulation of fear and anxiety. *J Neurosci*. 2014;34:60–65.

77. Flavin SA, Matthews RT, Wang Q, Muly EC, Winder DG. α2A-Adrenergic receptors filter parabrachial inputs to the bed nucleus of the stria terminalis. *J Neurosci*. 2014;34:9319–9331.

78. Hammack SE, et al. Roles for pituitary adenylate cyclase-activating peptide (PACAP) expression and signaling in the bed nucleus of the stria terminalis (BNST) in mediating the behavioral consequences of chronic stress. *J Mol Neurosci*. 2010;42:327–340.

79. Panguluri S, Saggu S, Lundy R. Comparison of somatostatin and corticotrophin-releasing hormone immunoreactivity in forebrain neurons projecting to taste-responsive and non-responsive regions of the parabrachial nucleus in rat. *Brain Res*. 2009;1298:57–69.

80. Magableh A, Lundy R. Somatostatin and corticotrophin releasing hormone cell types are a major source of descending input from the forebrain to the parabrachial nucleus in mice. *Chem Senses*. 2014;39:673–682.

81. Moga MM, Saper CB, Gray TS. Bed nucleus of the stria terminalis: cytoarchitecture, immunohistochemistry, and projection to the parabrachial nucleus in the rat. *J Comp Neurol*. 1989;283:315–332.

82. Silva EDCE, Fregoneze JB, Johnson AK. Corticotropin-releasing hormone in the lateral parabrachial nucleus inhibits sodium appetite in rats. *Am J Physiol – Regul Integr Comp Physiol*. 2006;290:R1136–R1141.

83. Sommer WH, Saavedra JM. Targeting brain angiotensin and corticotrophin-releasing hormone systems interaction for the treatment of mood and alcohol use disorders. *J Mol Med*. 2008;86:723–728.

84. Lind RW, Swanson LW, Ganten D. Organization of angiotensin II immunoreactive cells and fibers in the rat central nervous system. An immunohistochemical study. *Neuroendocrinology*. 1985;40:2–24.

85. Lenkei Z, Palkovits M, Corvol P, Llorens-Cortès C. Expression of angiotensin Type-1 (AT1) and Type-2 (AT2) receptor mRNAs in the adult rat brain: a functional neuroanatomical review. *Front Neuroendocrinol*. 1997;18:383–439.

86. Marinzalda M, de los A, et al. Fear-potentiated behaviour is modulated by central amygdala angiotensin II AT1 receptors stimulation. *Biomed Res Int*. 2014;2014:e183248.

87. Marvar PJ, et al. Angiotensin type 1 receptor inhibition enhances the extinction of fear memory. *Biol Psychiatry*. 2014;75:864–872.

88. Hurt RC, et al. Angiotensin in type 1a receptors on corticotropin-releasing factor neurons contribute to the expression of conditioned fear. *Genes Brain Behav*. 2015. http://dx.doi.org/10.1111/gbb.12235.

89. Khoury NM, et al. The renin-angiotensin pathway in posttraumatic stress disorder: angiotensin-converting enzyme inhibitors and angiotensin receptor blockers are associated with fewer traumatic stress symptoms. *J Clin Psychiatry*. 2012;73:849–855.

90. Wright JW, Harding JW. Brain renin-angiotensin—a new look at an old system. *Prog Neurobiol*. 2011;95:49–67.

I. NEUROENDOCRINE CONTROL OF THE STRESS RESPONSE

91. Gray TS, Magnuson DJ. Neuropeptide neuronal efferents from the bed nucleus of the stria terminalis and central amygdaloid nucleus to the dorsal vagal complex in the rat. *J Comp Neurol*. 1987;262:365–374.

92. Veening JG, Swanson LW, Sawchenko PE. The organization of projections from the central nucleus of the amygdala to brainstem sites involved in central autonomic regulation: a combined retrograde transport-immunohistochemical study. *Brain Res*. 1984;303:337–357.

93. Taché Y, Brunnhuber S. From Hans Selye's discovery of biological stress to the identification of corticotropin-releasing factor signaling pathways. *Ann NY Acad Sci*. 2008;1148:29–41.

94. Stengel A, Taché Y. Neuroendocrine control of the gut during stress: corticotropin-releasing factor signaling pathways in the spotlight. *Annu Rev Physiol*. 2009;71:219–239.

95. Zorrilla EP, Taché Y, Koob GF. Nibbling at CRF receptor control of feeding and gastrocolonic motility. *Trends Pharmacol Sci*. 2003;24:421–427.

96. Su J, et al. Injection of corticotropin-releasing hormone into the amygdala aggravates visceral nociception and induces noradrenaline release in rats. *Neurogastroenterol Motil*. 2015;27:30–39.

97. Taché Y. Corticotrophin-releasing factor 1 activation in the central amygdale and visceral hyperalgesia. *Neurogastroenterol Motil*. 2015;27:1–6.

98. North CS, Hong BA, Alpers DH. Relationship of functional gastrointestinal disorders and psychiatric disorders: implications for treatment. *World J Gastroenterol*. 2007;13:2020–2027.

99. Roy-Byrne PP, et al. Anxiety disorders and comorbid medical illness. *Gen Hosp Psychiatry*. 2008;30:208–225.

100. Herman JP, et al. In: *Comprehensive Physiology*. John Wiley & Sons, Inc.; 2011.

101. Silverman AJ, Hou-Yu A, Chen WP. Corticotropin-releasing factor synapses within the paraventricular nucleus of the hypothalamus. *Neuroendocrinology*. 1989;49:291–299.

102. Champagne D, Beaulieu J, Drolet G. CRFergic innervation of the paraventricular nucleus of the rat hypothalamus: a tract-tracing study. *J Neuroendocrinol*. 1998;10:119–131.

103. Moga MM, Saper CB. Neuropeptide-immunoreactive neurons projecting to the paraventricular hypothalamic nucleus in the rat. *J Comp Neurol*. 1994;346:137–150.

104. Choi DC, et al. Bed nucleus of the stria terminalis subregions differentially regulate hypothalamic–pituitary–adrenal axis activity: implications for the integration of limbic inputs. *J Neurosci*. 2007;27:2025–2034.

105. Choi DC, et al. The anteroventral bed nucleus of the stria terminalis differentially regulates hypothalamic-pituitary-adrenocortical axis responses to acute and chronic stress. *Endocrinology*. 2008;149:818–826.

106. Herman JP, Cullinan WE, Watson SJ. Involvement of the bed nucleus of the stria terminalis in tonic regulation of paraventricular hypothalamic CRH and AVP mRNA expression. *J Neuroendocrinol*. 1994;6:433–442.

107. Crane JW, Buller KM, Day TA. Evidence that the bed nucleus of the stria terminalis contributes to the modulation of hypophysiotropic corticotropin-releasing factor cell responses to systemic interleukin-1β. *J Comp Neurol*. 2003;467:232–242.

108. Beckerman MA, Van Kempen TA, Justice NJ, Milner TA, Glass MJ. Corticotropin-releasing factor in the mouse central nucleus of the amygdala: ultrastructural distribution in NMDA-NR1 receptor subunit expressing neurons as well as projection neurons to the bed nucleus of the stria terminalis. *Exp Neurol*. 2013;239:120–132.

109. Makino S, Gold PW, Schulkin J. Effects of corticosterone on CRH mRNA and content in the bed nucleus of the stria terminalis; comparison with the effects in the central nucleus of the amygdala and the paraventricular nucleus of the hypothalamus. *Brain Res*. 1994;657:141–149.

110. Bota M, Sporns O, Swanson LW. Neuroinformatics analysis of molecular expression patterns and neuron populations in gray matter regions: the rat BST as a rich exemplar. *Brain Res*. 2012;1450:174–193.

111. Radley JJ, Gosselink KL, Sawchenko PE. A discrete GABAergic relay mediates medial prefrontal cortical inhibition of the neuroendocrine stress response. *J Neurosci*. 2009;29:7330–7340.

112. Radley JJ, Sawchenko PE. A common substrate for prefrontal and hippocampal inhibition of the neuroendocrine stress response. *J Neurosci*. 2011;31:9683–9695.

113. Otake K, Nakamura Y. Sites of origin of corticotropin-releasing factor-like immunoreactive projection fibers to the paraventricular thalamic nucleus in the rat. *Neurosci Lett*. 1995;201:84–86.

114. Otake K, Ruggiero DA, Nakamura Y. Adrenergic innervation of forebrain neurons that project to the paraventricular thalamic nucleus in the rat. *Brain Res*. 1995;697:17–26.

115. Li S, Kirouac GJ. Projections from the paraventricular nucleus of the thalamus to the forebrain, with special emphasis on the extended amygdala. *J Comp Neurol*. 2008;506:263–287.

116. Chen Y, Molet J, Gunn BG, Ressler K, Baram TZ. Diversity of reporter expression patterns in transgenic mouse lines targeting corticotropin-releasing hormone-expressing neurons. *Endocrinology*. 2015;156:4769–4780.

117. Cruz FC, et al. New technologies for examining the role of neuronal ensembles in drug addiction and fear. *Nat Rev Neurosci*. 2013;14:743–754.

9

Vasopressin as a Stress Hormone

F.A. Antoni

Egis Pharmaceuticals PLC, Budapest, Hungary

Abstract

Vasopressin is a small neuropeptide initially identified as the physiologically essential antidiuretic hormone more than 50 years ago. Since then, it has increasingly become apparent that vasopressin is an important hormonal component of the response to stress. In fact, it appears that the antidiuretic effect is only one of several biologically significant actions of vasopressin exerted during the response to stress. This review highlights the main features of vasopressin as a stress hormone produced by relatively simple hypothalamic neurons that release their neurotransmitters into the blood stream and also send axonal projections to key parts of the brain that control the response to stressful environmental challenges. Special focus is on the role of vasopressin in (1) setting the efficacy of adrenal corticosteroid feedback inhibition; (2) the stress of pain; and (3) supporting the response to inflammation.

INTRODUCTION

The hormonal underpinning of the body's reaction to stress, as first formulated by Hans Selye in the 1930s,[92] is one of the most widely studied biological responses. The present treatise will provide an overview of the role of the nonapeptide vasopressin in this process. It will be highlighted that vasopressin is one of three major hormones involved in the preservation of fluid and electrolyte balance in the body, the other two being aldosterone and angiotensin II. A readiness to preempt body fluid loss is a good teleological fit with the stress response, which is geared to facilitate the restoration body homeostasis/allostasis in the face of virtually any challenge that threatens to upset it.[64]

KEY POINTS

- Vasopressin-like peptides serve as hormones that preserve salt–water balance already in invertebrates.

- This ancient trait is preserved in mammals and is part of the neuroendocrine response to stress in which vasopressin plays a fundamental role.

- The set point of feedback inhibition by adrenal corticosteroids in the hypothalamic-pituitary-adrenocortical axis is altered by vasopressin to bring about greater and more prolonged elevations of adrenal corticosteroid concentrations in blood.

- Pain activates hypothalamic neurons producing vasopressin and stress-induced analgesia has a component mediated by V_1 vasopressin receptors.

- The time course and pathological impact of inflammatory conditions is markedly influenced by vasopressin as well as cytokines produced by vasopressinergic neurons of the hypothalamus.

To keep the focus of a concise coverage, references to relevant reviews of the subject will predominate to orient the reader, unless the information is considered to be conceptually novel or unique to the subject matter and thus merit specific citation.

VASOPRESSIN AND ITS DISTRIBUTION IN THE BRAIN AND THE PITUITARY GLAND

Vasopressin and Its Prohormone

Vasopressin was one of the first neuropeptides to be sequenced by Du Vigneaud and colleagues.[75] It is a nonapeptide with an intrapeptide disulfide bond (Fig. 9.1). The porcine pituitaries used to purify the peptide yielded 8-Lys-vasopressin, but it later became clear that 8-Arg-vasopressin (AVP) is the predominant variant in mammals including humans and rodents. The carboxyl-terminal glycine of AVP is amidated. Later work shed light on the structure of the propressorphysin precursor that is biosynthesized and intracellularly processed to give rise to the biologically active hormone secreted by magnocellular neurons of the hypothalamo-hypophyseal tract (HHS).[16] Currently, there is no reason to believe that central AVP-producing neurons projecting to various areas of the brain produce and secrete AVP differently.[72] The notion that an AVP-like peptide is produced in the sympathetic nervous system[43] has not withstood the test of time. However, a paracrine role for vasopressin locally produced in the adrenal medulla seems likely.[40]

Methods of identifying AVP-producing neurons include immunocytochemical localization of AVP, AVP-associated neurophysin (neurophysin II), or the carboxyl-terminal glycopeptide of propressorphysin (copeptin) Box 9.1.[6] Further techniques involve the in situ detection of AVP mRNA. However, these methods often fail to identify nerve terminals and thin fibers with very low amounts of peptide and little if any mRNA. The problem may be overcome by expressing highly fluorescent proteins under the control of a tissue-specific AVP promoter and map AVP projections with higher sensitivity as previously accomplished for OXT neurons.[39] However, as yet, this approach has not been optimized to be sufficiently specific for the detection of AVPergic nerve fibers throughout the brain.[23,108]

Localization of AVP in the Brain

The sites of AVP production in the rat brain are nerve cell bodies in the hypothalamic magnocellular paraventricular nucleus (PVN), supraoptic nucleus (SON), accessory nucleus (AN), the suprachiasmatic nucleus (SCN), the bed nucleus of the stria terminalis (BNST), and the medial amygdaloid nucleus (MA).[26,97,86] With respect to the neuroendocrine stress response, the most significant pathways are the HHS and hypothalamo-infundibular tract (HIS) (Fig. 9.2). Detailed descriptions of these pathways are widely available.[4,9,61,105,110,119,120]

Hypothalamo-Hypophyseal Tract

In brief, axons originating from the hypothalamic magnocellular PVN, SON, and AN terminate in the posterior lobe of the pituitary gland (Fig. 9.2), abutting on fenestrated capillaries that carry the AVP secreted from the nerve endings to the systemic circulation. The discovery of this system and its fundamental physiological role to promote the reabsorption of water in the kidney is reflected by the designation of AVP as an antidiuretic hormone.[113] It is of note here that magnocellular axons coursing through the internal zone of median eminence toward the posterior lobe also release AVP *en passant*, and this AVP finds its way into the pituitary portal blood that irrigates the anterior pituitary gland.[6,27]

Hypothalamo-Infundibular Tract

Parvocellular neurons located in the medial aspect of the hypothalamic paraventricular nucleus coexpress AVP and 41-residue corticotropin-releasing factor (CRF41). The axons from these neurons take the same path toward the median eminence as the magnocellular AVP axons but terminate in the external zone of the median eminence abutting on fenestrated capillaries of the pituitary portal circulation. The two peptides are copackaged and coreleased from the same secretory granules[4,6,105] to regulate the cells of the anterior pituitary gland (Fig. 9.2).

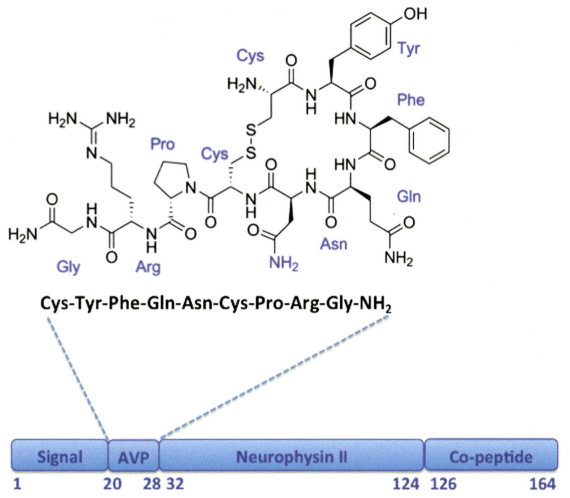

Cys-Tyr-Phe-Gln-Asn-Cys-Pro-Arg-Gly-NH₂

Signal	AVP	Neurophysin II	Co-peptide
1	20 28 32	124 126	164

FIGURE 9.1   The primary structure of 8-Arg-vasopressin and scheme of the propressorphysin precursor. Sites of proteolytic processing occur between the boundaries of the boxes.

BOX 9.1

COPEPTIN

The propressorphysin precursor molecule (Fig. 9.1) is cleaved and packaged in secretory granules in a manner that three peptides are released from neurohypophyseal nerve terminals into the systemic circulation: AVP, neurophysin II, and the carboxyl-terminal glycopeptide (CCP1-39[91]), recently renamed as copeptin.[68] It is of note that measurement of AVP in blood is not commonly used in clinical practice because of the biochemical features of the peptide: its small size, low abundance, and short half-life make testing for AVP labor-intensive. The high-level binding of AVP to platelets adds further complications.[68] By contrast, copeptin (Fig. 9.1) the carboxyl-terminal glycoprotein fragment of the propressorphysin precursor that is coreleased with AVP, is relatively stable in the posterior pituitary[91] as well as blood plasma and is increasingly used as an AVP surrogate in clinical diagnostics.

Other Cell Bodies and Extrahypothalamic Projections

Expression of AVP in the BNST is dependent on androgens; hence males have a higher level of expression than females. Neurons in the BNST and MA project extensively to various sites in the brain.[26,86] Of particular note are the AVPergic pathways to the lateral septum, which are thought to be relevant for social behavior in rats.[19] Another projection of apparent significance originates from the PVN to innervate the ipsilateral CA2 area of the hippocampus.[23] However, due to the scarcity and difficulty of detecting vasopressin in centrally projecting axons, a complete map of extrahypothalamic AVP pathways is yet to be compiled, but see Refs. 97,86 for

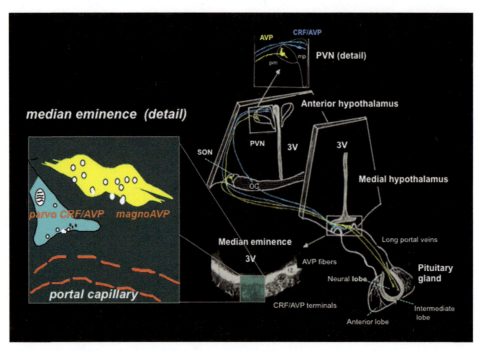

FIGURE 9.2 Schematic representation of the parvocellular hypothalamo-infundibular (blue) and magnocellular hypothalamo-hypophyseal (yellow) vasopressin (AVP) producing neuronal pathways of the brain. The left insert shows a detail of the median eminence where axon terminals of parvocellular neurons (blue) co-release AVP (yellow) and 41-residue corticotropin-releasing factor (CRF41, blue) into the pituitary portal circulation. Magnocellular axons in passage to the posterior lobe of the pituitary gland in the internal zone release AVP (yellow) *en passant* into the pituitary portal circulation. The magnocellular axon terminals in the neural lobe secrete AVP into the systemic circulation. The right insert shows the arrangement of AVP-producing cell bodies in the hypothalamic paraventricular nucleus. *mp*, magnocellular part; *OC*, optic chiasm; *PVN*, paraventricular nucleus; *pm*, parvocellular part; *SON*, supraoptic nucleus; *3V*, third cerebral ventricle.

catalogs using the classical methods. It is of particular relevance in this respect to distinguish between the central projections of hypothalamic AVP neurons and those in the BNST, MA, and SCN and their respective relevance, if any, to the stress response.

In summary the involvement of AVP neurons of the PVN, SON, and AN in the stress response is well established. The contribution of the other AVP-producing cell groups remains to be investigated.

MOLECULAR AND CELLULAR PHYSIOLOGY OF VASOPRESSIN WITH SPECIAL REFERENCE TO ANTERIOR PITUITARY CORTICOTROPE CELLS

Receptors and Ligands

Ligands

Once released, AVP may travel via the extracellular space or the blood stream to engage one of the four types of cell surface receptors for neurohypophyseal hormones. It has been reported that in the brain extracellular enzymatic activities produce bioactive AVP metabolites [pGlu4, Cystine6]-AVP-(4-9) and [pGlu4,Cyt6]vasopressin-(4-8).[83,84] These peptides are biologically active and according to one study, an activator of V_{1a} receptors in the hypothalamic supraoptic nucleus.[36] However, the question of how these AVP-derived peptides are generated and whether or not in quantities relevant for a physiological role remains unanswered.

Properties of Receptors for Vasopressin

Receptors for the neurohypophyseal hormones vasopressin and oxytocin are in the Class A superfamily of heptahelical receptor proteins,[34] previously referred to as G protein-coupled receptors. The neurohypophyseal hormone receptors are produced from four different genes and are designated as vasopressin-1a (V_{1a}), vasopressin-1b (V_{1b}), vasopressin-2 (V_2), and oxytocin (OT),[18,34,53] splice-variant isoforms have also been reported.[111] There is a long list of studies that have contributed to the pharmacologic and functional characterization of these receptors,[62] and it is important to acknowledge the outstanding work of the laboratory of Maurice Manning in the field. Analogs produced and generously distributed by this laboratory have been instrumental in the early pharmacologic differentiation of neurohypophyseal hormone receptors[5] and have been perfected over the years.[62] As yet, a high-resolution structural map of a neurohypophyseal hormone receptor is not available, but given the rapid progress in this area it should not be far away.

Affinity-based analysis of the neurohypophyseal hormone receptors have indicated that V_{1a}, V_{1b}, and V_2 receptors show marked preference (>100×) for AVP over OXT.[62] In contrast, AVP is only 20- to 30-fold less potent than OXT on recombinantly expressed OT receptors.[34,62] The picture becomes more complicated when binding or bioactivity is studied in preparations expressing native receptors. For instance, despite having low affinity in displacing ($K_d \approx 300\,nM$) ^{3}H-AVP from rat pituitary or recombinant V_{1b} receptors,[62] OT stimulated the release of adrenocorticotropic hormone (ACTH) at low nanomolar concentrations from rat anterior pituitary tissue.[5,89] Moreover, it has been argued that this effect is mediated through V_{1b} receptors.[89] When OT is used as the tracer ligand in uterine, pituitary,[5] or hippocampal membrane preparations,[13] AVP is equipotent with OXT as a displacing ligand of ^{3}H-OXT, which is in contrast to results with recombinantly expressed human OT receptors.[62] Yet another example is pain perception: OXT as well as AVP have been reported to be analgesic when administered directly into the brain, the spinal cord, or systemically. Surprisingly, genetic and pharmacologic manipulations both indicate that this effect is mediated by V_{1a} receptors.[90] This kind of atypical pharmacology may be due to heterodimerization between neurohypophyseal receptors[100] or even with other heptahelical receptors such as CRF_1 in corticotrope cells.[117] However, as yet there is no convincing evidence that such heterodimers exist under physiological conditions in vivo.[114]

In summary, the current data indicate that physiological concentrations of AVP can activate all four of the neurohypophyseal hormone receptors.

Tissue Distribution of AVP Receptors

The tissue distribution of the neurohypophyseal hormone receptors is moderately informative: V_{1a}, V_{1b}, as well as OT receptors are highly abundant in the brain.[13,53] Topographically, the enrichment of V_{1b} receptors in the CA2 area of the hippocampus is the most remarkable feature.[19] This area of the brain is associated with social recognition memory and AVP is known to have an important role in this process.[99] The V_{1a} receptor is abundant in peripheral tissues such as vascular smooth muscle and liver.[53] The V_{1b} receptor mRNA is expressed at low levels in the lung and the pancreas.[53] From the point of view of neuroendocrine regulation, it is of interest that the expression of V_{1b} receptors is enhanced by adrenal corticosteroids.[2]

In contrast, to V_1 receptors, V_2 receptors are mainly expressed in the kidney and only found at low abundance in the brain.[53] OT receptors are abundant in various areas of the brain reflecting the numerous CNS actions of OXT—further physiologically relevant sites of expression are the mammary gland and the uterus.[34] The expression of OT receptors in the uterus and the adenohypophysis is prominently controlled by estrogens.[34]

Intracellular Signaling Pathways

General Considerations

Intercellular signals such as hormones and neurotransmitters are detected and processed by target cells through biochemical reactions collectively known as intracellular signaling mechanisms. Since the advent and the wide-scale application of cDNA cloning virtually any protein involved in intracellular signaling may be analyzed in a "cellular laboratory," i.e., heterologous expression system—designated here as "in transfecto."[33] As a result, the information available has expanded exponentially and the major principle distilled after three decades of studies is that the precise mode of operation of a given signaling protein is very strongly dependent on its cellular context. Generalizations, such as "coupling of receptor X to a member of the G protein subfamily G_i (G protein inhibitory to adenylyl cyclase) entails that activation of X will inhibit the biosynthesis of cyclic AMP," are no longer possible.[8] Instead, it is entirely conceivable that by activating the same G_i-coupled receptor in a defined set of neurons, a neurotransmitter will inhibit cAMP formation in the morning and enhance it the evening—the underlying explanation is the rapid circadian change of the cellular context.[8]

To validate observations made "in transfecto" a logical step is to analyze preparations derived from primary tissues in vitro. In the fields of cancer biology or immune regulation, such preparations are accessible with minimally invasive technology even from human patients. Although more laborious, in vitro models derived from animal and human tissues are widely used in neuroendocrine research. However, it is amply clear that even short-term tissue culture can markedly alter the properties of cells,[73] thus the applicability of the results to physiological conditions may be questioned and require further targeted experimental validation in vivo.

It seems well established that the effects of AVP in biological systems are mediated by the four heptahelical receptors listed above. The mechanism of action of such receptors is best understood by examining the intracellular signaling pathways that are activated upon the binding of their ligand(s).[41] Importantly, these reactions are not restricted to primary interactions with G proteins. Moreover, it is also clear that upon activation at the plasma membrane, some heptahelical receptors are rapidly internalized within signaling bodies and continue to generate a signal inside cells.[10,45]

Signaling Reactions of V_1 Receptors
Gq/11

Activation of the V_1-type receptors leads to stimulated phospholipase C through coupling to Gq/11 proteins.[53] The consequent increase in inositol trisphosphate (IP_3) triggers a rise in the concentration of intracellular free Ca^{2+} from intracellular stores via IP_3-receptor Ca^{2+}

channels. In parallel diacylglycerol, the other product of the phospholipase C reaction, enhances the activity of protein kinase C alpha or beta.[53]

The rise in intracellular Ca^{2+} and the activity of protein kinase C may affect a large number of intracellular targets and thus the functional outcome of these signaling events is preeminently dependent on the type of cell targeted by AVP. From the point of view of the stress response, hypothalamic magnocellular AVP neurons, parvocellular CRF41/AVP neurons, and anterior pituitary corticotrope cells appear most relevant. In the case of nerve cells, Ca^{2+} signaling by AVP in the dendritic tree can result in a primed state of excitability[59] without leading to manifest changes in action potential firing patterns.[59,82] The molecular substrates of this effect have not been identified. In the case of anterior pituitary corticotrope cells, relatively high (>1 nM) concentrations of vasopressin can trigger the release of ACTH. When CRF41 is also present, marked synergistic stimulation of ACTH release takes place, and AVP is effective at lower concentrations. The underlying mechanism is the potentiation of the CRF41-induced cAMP signal by AVP.[3,6,10] The amplification of the cAMP signal by AVP beyond levels required for the full activation of ACTH secretion (ca. 3 μM) also renders the corticotrope cells resistant to the inhibitory action of adrenal corticosteroids (Box 9.2).[10,58] Importantly, the effects of AVP to potentiate CRF41-induced ACTH release can be largely mimicked by the pharmacologic activation of protein kinase C. Plausibly, the amplification of the CRF41-induced cAMP response by AVP in corticotrope cells is attributable to an isoform of adenylyl cyclase (AC7) that is stimulated by protein kinase C.[10] Leading on from this, the resistance to inhibition by glucocorticoids is likely to be the consequence of the activation of depolarizing nonselective cation conductances gated directly by high (>10 μM) intracellular levels of cAMP.[10] Similarly, V_{1a} receptor-mediated neuronal depolarization of motoneurons in the facial nucleus has been attributed to the activation of nonselective cation conductances directly gated by cAMP.[82,116]

β-Arrestin

The apparent G protein-independent coupling of V_{1a}[118] as well as V_{1b}[48,53] receptors to β-arrestin has been reported. In the case of the V_{1a} receptor natively expressed in myoblasts, the outcome of coupling to β-arrestin was the activation of extracellular signal-regulated kinase 1 (ERK1) and protection from ischemic stress.[118] In another study cultured rat hippocampal neurons were protected from the neurotoxic effects of glutamate and nutrient deprivation by AVP acting through V_1 receptors.[21] In the case of the V_{1b} receptor recombinantly expressed in CHO cells, β-arrestin-driven internalization was much greater upon exposure to AVP than for V_{1a} receptors.[48] This raises the intriguing possibility that the V_{1b} receptor is also geared to signal away from the plasma membrane as recently reported for a number of heptahelical receptors.[10,45]

Other Pathways

Anecdotal reports of the coupling of V_{1a} receptors to pertussis toxin-sensitive G proteins have been published[82] but not substantiated by biochemical analysis.[82]

Signaling Reactions of V₂ Receptors

The V_2 receptor is well known to stimulate adenylyl cyclase-mediated synthesis of cAMP by coupling to the

---

BOX 9.2

GLUCOCORTICOID FEEDBACK IN THE HYPOTHALAMIC–PITUITARY ADRENOCORTICAL AXIS

The adrenocortical response to stress is governed by a neuroendocrine feedback loop (Fig. 9.3). In brief, neural and hormonal afferent input activated by stress converges on the hypothalamus to evoke the release of CRF41 and AVP from hypothalamic neurons into pituitary portal blood and thus stimulate cells of the adenohypophysis to release adrenocorticotropic hormone (ACTH) into the systemic circulation. ACTH rapidly stimulates the secretion of the adrenocortical steroid hormone cortisol (corticosterone in rodents) into the systemic circulation. In turn, adrenal corticosteroids diminish the secretion of ACTH by exerting inhibitory feedback at the anterior pituitary gland, in the hypothalamus, and further sites in the brain. Inhibitory glucocorticoid feedback is described as early immediate (seconds to minutes), early delayed (20–120 min), and late (>4 h).[50] The parvocellular CRF41/AVP neurons as well as adenohypophyseal corticotrope cells are thought to be direct targets of all three types of corticosteroid inhibition.[6] Importantly, within the CRF41/AVP cells, it is the expression of AVP that is acutely sensitive to the early inhibitory effects of glucocorticoids.[54,60] In contrast, the secretion and expression of AVP by cells of the HHS are not influenced by early feedback inhibition,[54] although late effects may become apparent.[37] The organization of extrahypothalamic neural circuits involved in corticosteroid inhibitory feedback is work in progress.[81]

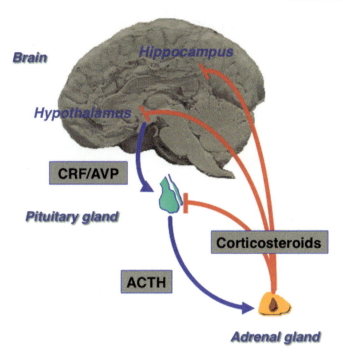

FIGURE 9.3 The neuroendocrine feedback loop of the hypothalamic-pituitary-adrenocortical axis. *Blue arrows* indicate stimulation, red lines with bars inhibition. *ACTH*, adrenocorticotropic hormone; *AVP*, vasopressin; *CRF*, 41-residue corticotropin-releasing factor.

G protein. It is also prominently linked to β-arrestin-dependent signaling.[85] However, there is no significant evidence linking V$_2$ receptors to the stress response, e.g., see Ref. 12.

FUNCTIONS OF VASOPRESSIN

Body Fluid Homeostasis and Connection to the Stress Response

Virtually all forms of stress require a state of cardiovascular readiness and thus protection of blood volume and cardiac output. Activation of the sympathetic nervous system by stress is a well-known mechanism geared to achieve this.[57,76] However, it is also clear that osmoregulation of body fluids via vasopressin-like peptides is a phylogenetically conserved phenomenon that is also prominent in mammals.[1] A quick catalog of the effects of AVP on fluid handling reveals actions promoting fluid conservation in the kidney, the lungs, the brain, the colon, and the salivary glands.[15,74,80]

Neuroendocrinologists have argued for decades whether or not activation of the HHS is an integral part of the stress response.[6] It was broadly agreed that the HHS is activated upon exposure to "physical" stressors such as ether, immobilization, hemorrhage, strenuous exercise, pain, as well as upon the "metabolic" stress of severe insulin-induced hypoglycemia.[6] However, an increase of AVP levels in the systemic circulation,

indicative of an impact of HHS activation out with the brain has not been consistently demonstrated in these paradigms, casting doubt over HHS involvement in the response to stress. Surprisingly, in the last few years clinicians have begun to resolve this problem largely due to the widespread availability of a simple and sensitive immunoassay for the quantification of copeptin in blood plasma[11,68] (Box 9.1). The activation of the HHS AVP system above "normal" in patients with septic shock, lower respiratory tract infection, myocardial infarction, stroke, and metabolic syndrome is beyond reasonable doubt.[28,49,106] The crucial question is what the role of HHS-derived AVP might be in these conditions? With respect to more classical stress paradigms, increases of plasma copeptin were reported in small groups of healthy volunteers exposed to "emotional stressors."[109] Moreover, significant correlation between serum cortisol and copeptin levels were found in male but not female volunteers subjected to social stress.[98] In sum, while intriguing and potentially important, the copeptin data generated so far have posed more questions about the pathophysiological role of HHS-derived AVP than they have answered.[28,49,106]

AVP and the Hypothalamic–Pituitary Adrenocortical Axis

Over the years, the role of AVP in the HPA axis has been examined by a large number of studies deploying various methods including AVP-neutralizing antibodies, surgical/neurochemical lesions of AVP neurons, and AVP antagonists.[6] For the interpretation of such experiments, it is important to bear in mind that current evidence indicates that the neurohypophyseal peptides AVP and OXT reduce the activity of CRF41/AVP neurons of the HIS.[17,79,96] Although the mechanism(s) underlying this effect remains to be clarified, it implies that effective neutralization of the actions of AVP at the hypothalamic level results in an enhancement of the hypothalamic drive to stimulate adrenocortical steroid secretion. By contrast, the selective inhibition of AVP at the pituitary level can effectively suppress HPA activation.[6] Finally, any reduction of adrenocortical steroid output diminishes glucocorticoid feedback inhibition and thus will tend to enhance the hypothalamo-hypophyseal drive during the course of an experiment. This kind of push–pull regulation will lead to a new set point of glucocorticoid feedback that will reflect the blockade of AVP actions indirectly, through a series of complex interactions.[78] Overall, the features listed above make the quantitative interpretation of these studies difficult. Numerous papers and reviews of these approaches to the stress response have been published[6,24,27,35,38,87] providing accounts of the purported role(s) of AVP in different forms of stress. In the following section a summary of the progress in the areas with high clinical translational value and previously not reviewed with a focus on AVP will be covered.

Glucocorticoid Feedback

The neuroendocrine feedback loop that governs the operation of the HPA axis is outlined in Box 9.2. An important feature of this system is that the parvocellular HIS CRF41/AVP neurons are prominent targets of early delayed glucocorticoid feedback inhibition while magnocellular AVP neurons of the HHS are not. Furthermore, within the HIS, the expression of AVP is dynamically regulated by early delayed glucocorticoid feedback inhibition.[60,54] In contrast, this does not appear to be the case in the HHS—only late glucocorticoid inhibition has been demonstrated.[37] However, the dramatic increase of AVP expression seen upon removal of adrenal steroids in the HIS system does not take place in HHS AVP neurons.[4] As the HHS AVP cells are equipped with the same type II glucocorticoid receptor as the HIS CRF41/AVP cells,[52] it is a long-standing conundrum as to exactly how this marked difference in the cellular regulation of AVP gene expression is achieved ?[31,37] The relative insensitivity of HHS AVP cells to glucocorticoid inhibition together with independent measures of HHS AVP activation has led investigators to conclude that forms of HPA activation that are resistant to inhibition by glucocorticoids are dependent on the HHS AVP system.[6,27] A typical example is the stimulation of the HPA axis upon hemorrhage.[79]

The pituitary corticotrope cell is also an important site of glucocorticoid inhibition. Variants of the dexamethasone suppression test, that are used as a diagnostic tool in the clinical setting, essentially probe the efficacy of glucocorticoid feedback at the anterior pituitary level.[115,55] All three phases of glucocorticoid feedback inhibition have been reported in corticotrope cells. However, under physiological conditions, early delayed inhibition appears most relevant.[7] With respect to HPA activation that is resistant to glucocorticoid inhibition, it is worth recalling, that in pituitary corticotrope cells AVP can induce "glucocorticoid escape" from early delayed feedback by switching CRF41 signaling to pathways that generate an exaggerated cAMP response and engage new targets downstream of the cAMP signal.[10] Furthermore, it is important to emphasize that glucocorticoid escape is apparent at concentrations of AVP that are characteristic of magnocellular HHS AVP cells or chronically hyperactive HIS CRF41/AVP cells.

In summary, the expression of AVP is tightly controlled by adrenal corticosteroids in the parvocellular HIS system, but this is not the case for magnocellular AVP cells of the HHS. Concordantly, glucocorticoid escape at the anterior pituitary level requires AVP concentrations normally secreted by magnocellular AVP neurons.

Stress of Pain and AVP

One of the classic stress paradigms is the pain induced by the injection of formalin into the hind paw of rodents. The first evidence for the involvement of AVP in the stimulation of ACTH release in vivo[107] was published with formalin stress. Importantly, synaptic activation of magnocellular AVP neurons in the SON has been demonstrated upon electrical stimulation of somatic afferents or application of noxious stimuli.[25,42,47] Pain increases plasma AVP levels in man.[30,51] More recent work showed[102] that formalin stress as well as adjuvant-induced arthritis increase the expression of AVP in the HIS as well as the HHS. However, the increase of HHS AVP was more pronounced in adjuvant-induced arthritis in keeping with the signs of diminished glucocorticoid suppression of the HPA axis[95] in this condition.

A further aspect of AVP and pain is the finding that both AVP and OT are analgesic when given systemically or intrathecally.[14,90] Is there a connection to stress-induced activation of AVP neurons? A potentially important study reported a role for a V_{1a} receptor haplotype in the perception of pain by human volunteers.[66] However, the haplotype effect was only apparent in stressed males.[66] Similar findings were produced in mice. Exogenous administration of the weak V_1 agonist desmopressin produced analgesia in males with a low level of stress but was ineffective in subjects with higher levels of stress. The interpretation of these findings is that endogenous AVP released during the stress response obliterated the analgesic effect of desmopressin.[66] The results represent a *tour de force* of mouse and human genetics coupled to functional assays. Questions remain: is it AVP or OXT that is the endogenous mediator?[90] If AVP, is it derived from the AVP cells in the BNST and MA that are regulated by testosterone? Which AVP neuronal pathways are involved,[65,86,97] where is the site of action?

AVP and Inflammation

Adrenal corticosteroids are potent endogenous immunomodulators.[63,67,95] In rodents, the active inflammatory phase of experimental autoimmune encephalomyelitis as well as adjuvant-induced arthritis is characterized by extremely high levels of corticosteroids in blood lasting several days.[63,95] The protracted increase of HPA activity is essential for the survival of the affected animals.[63] How are such high levels of corticosteroids, characteristic of severe stress, maintained in the face of glucocorticoid feedback inhibition? Several lines of evidence pointed toward the activation of the magnocellular HHS AVP system to override glucocorticoid feedback inhibition.[6] More recently, this issue was re-examined in rats expressing enhanced green fluorescent protein under the control of the AVP promoter.[88,108] The results clearly showed a marked increase of AVP levels in the HSS as well as the HIS with no sign of a similar enhancement of CRF41 expression.[103] Plasma levels of corticosterone and AVP were also increased in rats with adjuvant-induced arthritis.[103]

These results corroborate the notion that AVP-induced resistance to glucocorticoid feedback is the key to the prolonged increase of the secretion of endogenous adrenal steroids to combat autoimmune inflammation. By contrast, the same authors reported that the acute HPA response to bacterially derived lipopolysaccharide selectively activates the HIS CRF41/AVP cells[101] indicating that the stimulation of HHS AVP expression is induced by different neurohumoral afferentation as that produced by LPS. However, under certain conditions, e.g., depletion of brain macrophages, administration of LPS has been shown to induce c-fos expression in magnocellular neurons of the PVN.[94] Thus the effects of LPS on the HHS system appear conditional. Indeed, interleukin-1β, a major mediator of the effects of LPS in the body, excites AVP neurons in the SON via induction of the production of prostanoids—a classical inflammation-related pathway.[20] A further notable finding is that HHS AVP neurons express and release interleukin-6 into the systemic circulation in response to acute stress.[46] Thus, HHS AVP neurons contribute to the stress response by releasing bioactive products other than AVP.

The picture that emerges from the analysis of HHS and HIS AVP neurons in inflammation is that, on the one hand, heightened activation of these cells allows enhanced and prolonged secretion of adrenal corticosteroids by shifting the normal feedback set point of the HPA axis, thus producing "glucocorticoid escape." On the other hand, in response to stress HHS AVP cells release at least one potent proinflammatory mediator into the blood stream, the production of which is suppressed by adrenal corticosteroids in cells of the immune system.[69,70] A key issue is whether or not the increased levels of corticosteroids balance out the stress-induced proinflammatory processes? If not, do stress-related maladaptational syndromes invariably emerge upon long-term HHS AVP activation? Once more, as in the case of the analgesic actions associated with stress, the genetic make-up of an individual will likely have a major impact on the outcome of these two opposing processes.[63,95] We are entering the era of pharmacogenetics and personalized medicine.

AVP and Behavioral Aspects of Social Stress

Generalized anxiety disorder, melancholic depression, posttraumatic stress disorder, and schizophrenia are relatively common mental illnesses that have been connected to various forms of stress over the years.[29] Swaab and coworkers[104] concluded that in human postmortem material, depression is characterized by the signs of prolonged activation of HIS CRF41/AVP as well as HHS AVP neurons. These signs point to a shift in the set point of the HPA axis, which can be demonstrated in close to 40% of patients with depression.[32] In addition to behavioral changes secondary to the altered dynamics of the HPA axis, what alternative mechanisms may underlie the alterations of behavior mediated by AVP during stress? A significant pointer is the report by Cui and coworkers showing a direct neural pathway between the CA2 hippocampal region and the PVN.[23] The CA2 region is essential for social memory[44] and shows a marked enrichment of V_{1b} receptors when compared to neighboring brain regions.[19] More recently, it was shown that viral expression of the V_{1b} receptor in the CA2 region of male mice with genetic deletion of the V_{1b} receptor restored socially motivated attack behavior without altering anxiety-like behaviors.[77] Facilitation of aggressive behavior is a well-known facet of the stress response[56,112] and it is also conceivable that these effects of AVP require adrenal corticosteroids.[56] It also remains to be clarified which subpopulation(s) of PVN AVP neurons gives rise to axonal projections to the CA2 area?

SUMMARY

The hormonal triad of aldosterone, angiotensin II, and vasopressin has the mission to preserve body fluids and salts, the fundamental requisites of the internal milieu of vertebrates. The neuroendocrine stress response is palpably linked to fluid preservation—ACTH also stimulates aldosterone, vasopressin is released into the general circulation during physical stress, and sympathetic activation increases the formation of angiotensin II. From this rather simple teleologic principle, the plethora of actions of AVP that are connected through stress is truly remarkable. Even more intriguing is the fact that these actions can be traced back to the activity of a few thousand neurons in the hypothalamus. There is still a lot to be learned about the AVP component of stress: The role of epigenetic alterations in the dynamics of the stress response[71]; the mapping in full of the anatomical projections of stress-responsive AVP neurons; the emerging role of AVP in the metabolic syndrome[28]; the reasons behind the lack of efficacy of V_1 receptor antagonists in clinical trials[22,93] are all important topics for the future.

References

1. Acher R. Water homeostasis in the living: molecular organization, osmoregulatory reflexes and evolution. *Ann Endocrinol Paris.* 2002;63:197–218.
2. Aguilera G. The hypothalamic-pituitary-adrenal axis and neuroendocrine responses to stress. *Handb Neuroendocrinol.* 2012:175–196.
3. Aguilera G, Rabadan-Diehl C. Vasopressinergic regulation of the hypothalamic-pituitary-adrenal axis: implications for stress adaptation. *Regul Pept.* 2000;96:23–29.

4. Antoni FA. Hypothalamic regulation of adrenocorticotropin secretion: advances since the discovery of 41-residue corticotropin-releasing factor. *Endocr Rev.* 1986;7:351–378.

5. Antoni FA. Receptors mediating the CRH effects of vasopressin and oxytocin. *Ann NY Acad Sci.* 1987;512:195–204.

6. Antoni FA. Vasopressinergic control of anterior pituitary adrenocorticotropin secretion comes of age. *Front Neuroendocrinol.* 1993;14:76–122.

7. Antoni FA. Calcium checks cyclic AMP — mechanism of corticosteroid feedback in adenohypophysial corticotrophs. *J Neuroendocrinol.* 1996;8:659–672.

8. Antoni FA. Molecular diversity of cyclic AMP signaling. *Front Neuroendocrinol.* 2000;21:103–132.

9. Antoni FA. Role of pituitary regulation. In: Fink G, ed. *Encyclopedia of Stress, Second Edition.* Vol. 3. Oxford: Academic Press; 2007:127–131.

10. Antoni FA. New paradigms in cAMP signalling. *Mol Cell Endocrinol.* 2012;353:3–9.

11. Balanescu S, Kopp P, Gaskill MB, Morgenthaler NG, Schindler C, Rutishauser J. Correlation of plasma copeptin and vasopressin concentrations in hypo-, iso-, and hyperosmolar States. *J Clin Endocrinol Metab.* 2011;96:1046–1052.

12. Balázsfi D, Pintér O, Klausz B, et al. Restoration of peripheral V2 receptor vasopressin signaling fails to correct behavioral changes in Brattleboro rats. *Psychoneuroendocrinology.* 2015;51:11–23.

13. Barberis C, Tribollet E. Vasopressin and oxytocin receptors in the central nervous system. *Crit Rev Neurobiol.* 1996;10:119–154.

14. Berkowitz BA, Sherman S. Characterization of vasopressin analgesia. *J Pharmacol Exp Ther.* 1982;220:329–334.

15. Bridges RJ, Rummel W, Wollenberg P. Effects of vasopressin on electrolyte transport across isolated colon from normal and dexamethasone-treated rats. *J Physiol.* 1984;355:11–23.

16. Brownstein MJ, Russell JT, Gainer H. Synthesis, transport, and release of posterior pituitary hormones. *Science.* 1980;207:373–378.

17. Bulbul M, Babygirija R, Cerjak D, Yoshimoto S, Ludwig K, Takahashi T. Hypothalamic oxytocin attenuates CRF expression via GABA(A) receptors in rats. *Brain Res.* 2011;1387:39–45.

18. Burbach JPH, Adan RAH, Lolait SJ, et al. Molecular neurobiology and pharmacology of the vasopressin oxytocin receptor family. *Cell Mol Neurobiol.* 1995;15:573–595.

19. Caldwell HK, Lee HJ, Macbeth AH, Young 3rd WS. Vasopressin: behavioral roles of an "original" neuropeptide. *Prog Neurobiol.* 2008;84:1–24.

20. Chakfe Y, Zhang Z, Bourque CW. IL-1beta directly excites isolated rat supraoptic neurons via upregulation of the osmosensory cation current. *Am J Physiol Regul Integr Comp Physiol.* 2006;290:R1183–R1190.

21. Chen J, Aguilera G. Vasopressin protects hippocampal neurones in culture against nutrient deprivation or glutamate-induced apoptosis. *J Neuroendocrinol.* 2010;22:1072–1081.

22. Craighead M, Milne R, Campbell-Wan L, et al. Characterization of a novel and selective V1B receptor antagonist. *Prog Brain Res.* 2008;170:527–535.

23. Cui Z, Gerfen CR, Young 3rd WS. Hypothalamic and other connections with dorsal CA2 area of the mouse hippocampus. *J Comp Neurol.* 2013;521:1844–1866.

24. Curley Jr KO, Neuendorff DA, Lewis AW, Rouquette Jr FM, Randel RD, Welsh Jr TH. The effectiveness of vasopressin as an ACTH secretagogue in cattle differs with temperament. *Physiol Behav.* 2010;101:699–704.

25. Day TA, Sibbald JR. Noxious somatic stimuli excite neurosecretory vasopressin cells via A1 cell group. *Am J Physiol.* 1990;258:R1516–R1520.

26. de Vries GJ, Miller MA. Anatomy and function of extrahypothalamic vasopressin systems in the brain. *Prog Brain Res.* 1998;119:3–20.

27. Engelmann M, Landgraf R, Wotjak CT. The hypothalamic-neurohypophysial system regulates the hypothalamic-pituitary-adrenal axis under stress: an old concept revisited. *Front Neuroendocrinol.* 2004;25:132–149.

28. Enhorning S, Struck J, Wirfalt E, Hedblad B, Morgenthaler NG, Melander O. Plasma copeptin, a unifying factor behind the metabolic syndrome. *J Clin Endocrinol Metab.* 2011;96:E1065–E1072.

29. Fink G. In: *Encyclopedia of Stress, Second Edition.* Oxford: Academic Press; 2007.

30. Franceschini R, Leandri M, Cataldi A, et al. Raised plasma arginine vasopressin concentrations during cluster headache attacks. *J Neurol Neurosurg Psychiatry.* 1995;59:381–383.

31. Gainer H. Cell-type specific expression of oxytocin and vasopressin genes: an experimental odyssey. *J Neuroendocrinol.* 2012;24:528–538.

32. Gillespie CF, Nemeroff CB. Hypercortisolemia and depression. *Psychosom Med.* 2005;67(suppl 1):S26–S28.

33. Gilman AG. Nobel Lecture. G proteins and regulation of adenylyl cyclase. *Biosci Rep.* 1995;15:65–97.

34. Gimpl G, Fahrenholz F. The oxytocin receptor system: structure, function, and regulation. *Physiol Rev.* 2001;81:629–683.

35. Goncharova ND. Stress responsiveness of the hypothalamic-pituitary-adrenal axis: age-related features of the vasopressinergic regulation. *Front Endocrinol (Lausanne).* 2013;4:26.

36. Gouzenes L, Dayanithi G, Moos FC. Vasopressin(4-9) fragment activates V1a-type vasopressin receptor in rat supraoptic neurones. *Neuroreport.* 1999;10:1735–1739.

37. Greenwood M, Greenwood MP, Mecawi AS, et al. Transcription factor CREB3L1 mediates cAMP and glucocorticoid regulation of arginine vasopressin gene transcription in the rat hypothalamus. *Mol Brain.* 2015;8:68.

38. Griebel G, Holsboer F. Neuropeptide receptor ligands as drugs for psychiatric diseases: the end of the beginning? *Nat Rev Drug Discov.* 2012;11:462–478.

39. Grinevich V, Knobloch-Bollmann HS, Eliava M, Busnelli M, Chini B. Assembling the puzzle: pathways of oxytocin signaling in the brain. *Biol Psychiatry.* 2016;79,155–164.

40. Guillon G, Grazzini E, Andrez M, et al. Vasopressin: a potent autocrine/paracrine regulator of mammal adrenal functions. *Endocr Res.* 1998;24:703–710.

41. Hall RA, Premont RT, Lefkowitz RJ. Heptahelical receptor signaling: beyond the G protein paradigm. *J Cell Biol.* 1999;145:927–932.

42. Hamamura M, Shibuki K, Yagi K. Noxious inputs to supraoptic neurosecretory cells in the rat. *Neurosci Res.* 1984;2:49–61.

43. Hanley MR, Benton HP, Lightman SL, et al. A vasopressin-like peptide in the mammalian sympathetic nervous system. *Nature.* 1984;309:258–261.

44. Hitti FL, Siegelbaum SA. The hippocampal CA2 region is essential for social memory. *Nature.* 2014;508:88–92.

45. Jalink K, Moolenaar WH. G protein-coupled receptors: the inside story. *Bioessays.* 2010;32:13–16.

46. Jankord R, Zhang R, Flak JN, Solomon MB, Albertz J, Herman JP. Stress activation of IL-6 neurons in the hypothalamus. *Am J Physiol Regul Integr Comp Physiol.* 2010;299:R343–R351.

47. Kannan H, Yamashita H, Koizumi K, Brooks CM. Neuronal activity of the cat supraoptic nucleus is influenced by muscle small diameter afferent (groups III and IV) receptors. *Proc Natl Acad Sci USA.* 1988;85:5744–5748.

48. Kashiwazaki A, Fujiwara Y, Tsuchiya H, Sakai N, Shibata K, Koshimizu T-A. Subcellular localization and internalization of the vasopressin V-1B receptor. *Eur J Pharmacol.* 2015;765:291–299.

49. Katan M, Christ-Crain M. The stress hormone copeptin: a new prognostic biomarker in acute illness. *Swiss Med Wkly.* 2010;140:w13101.

50. Keller-Wood MB, Dallman MF. Corticosteroid inhibition of ACTH secretion. *Endocr Rev.* 1984;5:1–23.

51. Kendler KS, Weitzman RE, Fisher DA. The effect of pain on plasma arginine vasopressin concentrations in man. *Clin Endocrinol (Oxf)*. 1978;8:89–94.

52. Kiss JZ, Van Eekelen JA, Reul JM, Westphal HM, De Kloet ER. Glucocorticoid receptor in magnocellular neurosecretory cells. *Endocrinology*. 1988;122:444–449.

53. Koshimizu T-A, Nakamura K, Egashira N, Hiroyama M, Nonoguchi H, Tanoue A. Vasopressin V1a and V1b receptors: from molecules to physiological systems. *Physiol Rev*. 2012;92:1813–1864.

54. Kovács KJ, Földes A, Sawchenko PE. Glucocorticoid negative feedback selectively targets vasopressin transcription in parvocellular neurosecretory neurons. *J Neurosci*. 2000;20:3843–3852.

55. Krishnan KRR, Rayasam K, Reed D, et al. The corticotropin releasing factor stimulation test in patients with major depression: relationship to dexamethasone suppression test results. *Depression*. 1993;1:133–136.

56. Kruk MR, Halász J, Meelis W, Haller J. Fast positive feedback between the adrenocortical stress response and a brain mechanism involved in aggressive behavior. *Behav Neurosci*. 2004;118:1062–1070.

57. Kvetnansky R, McCarty R. Adrenal medulla. In: Fink G, ed. *Encyclopedia of Stress, Second Edition*. Vol. 3. Oxford: Academic Press; 2007.

58. Lim MC, Shipston MJ, Antoni FA. Posttranslational modulation of glucocorticoid feedback inhibition at the pituitary level. *Endocrinology*. 2002;143:3796–3801.

59. Ludwig M, Leng G. Dendritic peptide release and peptide-dependent behaviours. *Nat Rev Neurosci*. 2006;7:126–136.

60. Ma XM, Aguilera G. Differential regulation of corticotropin-releasing hormone and vasopressin transcription by glucocorticoids. *Endocrinology*. 1999;140:5642–5650.

61. Makara GB, Antoni FA, Stark E, Kárteszi M. Hypothalamic organization of CRF containing structures. *Neuroendocr Perspect*. 1984;3:71–110.

62. Manning M, Misicka A, Olma A, et al. Oxytocin and vasopressin agonists and antagonists as research tools and potential therapeutics. *J Neuroendocrinol*. 2012;24:609–628.

63. Mason D. Genetic variation in the stress response susceptibility to experimental allergic encephalomyelitis and implications for human inflammatory disease. *Immunol Today*. 1991;12:57–60.

64. McEwen BS, Gianaros PJ. Stress- and allostasis-induced brain plasticity. *Annu Rev Med*. 2011;62:431–445.

65. Millan MJ, Schmauss C, Millan MH, Herz A. Vasopressin and oxytocin in the rat spinal cord: analysis of their role in the control of nociception. *Brain Res*. 1984;309:384–388.

66. Mogil JS, Sorge RE, LaCroix-Fralish ML, et al. Pain sensitivity and vasopressin analgesia are mediated by a gene-sex-environment interaction. *Nat Neurosci*. 2011;14:1569–1573.

67. Morand EF, Leech M. Hypothalamic-pituitary-adrenal axis regulation of inflammation in rheumatoid arthritis. *Immunol Cell Biol*. 2001;79:395–399.

68. Morgenthaler NG, Struck J, Jochberger S, Dunser MW. Copeptin: clinical use of a new biomarker. *Trends Endocrinol Metab*. 2008;19:43–49.

69. Munck A, Guyre PM, Holbrook NJ. Physiological functions of glucocorticoids in stress and their relation to pharmacological actions. *Endocr Rev*. 1984;5:25–44.

70. Munck A, Náray-Fejes-Tóth A. Glucocorticoids and stress: permissive and suppressive actions. *Ann NY Acad Sci*. 1994;746:115–130. discussion 131–113.

71. Murgatroyd C, Spengler D. Epigenetic programming of the HPA axis: early life decides. *Stress*. 2011;14:581–589.

72. Murphy D, Konopacka A, Hindmarch C, et al. The hypothalamic-neurohypophyseal system: from genome to physiology. *J Neuroendocrinol*. 2012;24:539–553.

73. Nestor CE, Ottaviano R, Reinhardt D, et al. Rapid reprogramming of epigenetic and transcriptional profiles in mammalian culture systems. *Genome Biol*. 2015;16:11.

74. Niermann H, Amiry-Moghaddam M, Holthoff K, Witte OW, Ottersen OP. A novel role of vasopressin in the brain: modulation of activity-dependent water flux in the neocortex. *J Neurosci*. 2001;21:3045–3051.

75. Ottenhausen M, Bodhinayake I, Banu MA, Stieg PE, Schwartz TH. Vincent du Vigneaud: following the sulfur trail to the discovery of the hormones of the posterior pituitary gland at Cornell Medical College. *J Neurosurg*. 2015:1–5.

76. Pacak K, Palkovits M. Stressor specificity of central neuroendocrine responses: implications for stress-related disorders. *Endocr Rev*. 2001;22:502–548.

77. Pagani JH, Zhao M, Cui Z, et al. Role of the vasopressin 1b receptor in rodent aggressive behavior and synaptic plasticity in hippocampal area CA2. *Mol Psychiatry*. 2015;20:490–499.

78. Peters A, Conrad M, Hubold C, Schweiger U, Fischer B, Fehm HL. The principle of homeostasis in the hypothalamus-pituitary-adrenal system: new insight from positive feedback. *Am J Physiol Regul Integr Comp Physiol*. 2007;293:R83–R98.

79. Plotsky PM. Pathways to the secretion of adrenocorticotropin: a view from the portal. *J Neuroendocrinol*. 1991;3:1–9.

80. Pouzet B, Serradeil-Le Gal C, Bouby N, Maffrand JP, Le Fur G, Bankir L. Selective blockade of vasopressin V-2 receptors reveals significant V-2-mediated water reabsorption in Brattleboro rats with diabetes insipidus. *Nephrol Dial Transplant*. 2001;16:725–734.

81. Radley JJ, Sawchenko PE. Evidence for involvement of a limbic paraventricular hypothalamic inhibitory network in hypothalamic-pituitary-adrenal axis adaptations to repeated stress. *J Comp Neurol*. 2015;523:2769–2787.

82. Raggenbass M. Overview of cellular electrophysiological actions of vasopressin. *Eur J Pharmacol*. 2008;583:243–254.

83. Reijmers LG, Baars AM, Burbach JP, Spruijt BM, van Ree JM. Delayed effect of the vasopressin metabolite VP4-8 on the social memory of sexually naive male rats. *Psychopharmacology (Berl)*. 2001;154:408–414.

84. Reijmers LG, van Ree JM, Spruijt BM, Burbach JP, De Wied D. Vasopressin metabolites: a link between vasopressin and memory? *Prog Brain Res*. 1998;119:523–535.

85. Ren XR, Reiter E, Ahn S, Kim J, Chen W, Lefkowitz RJ. Different G protein-coupled receptor kinases govern G protein and beta-arrestin-mediated signaling of V2 vasopressin receptor. *Proc Natl Acad Sci USA*. 2005;102:1448–1453.

86. Rood BD, Stott RT, You S, Smith CJ, Woodbury ME, De Vries GJ. Site of origin of and sex differences in the vasopressin innervation of the mouse (*Mus musculus*) brain. *J Comp Neurol*. 2013;521:2321–2358.

87. Roper JA, O'Carroll AM, Young III WS, Lolait SJ. The vasopressin Avpr1b receptor: molecular and pharmacological studies. *Stress*. 2011;14:98–115.

88. Satoh K, Oti T, Katoh A, et al. In vivo processing and release into the circulation of GFP fusion protein in arginine vasopressin enhanced GFP transgenic rats: response to osmotic stimulation. *FEBS J*. 2015;282:2488–2499.

89. Schlosser SF, Almeida OFX, Patchev VK, Yassouridis A, Elands J. Oxytocin-stimulated release of adrenocorticotropin from the rat pituitary is mediated by arginine vasopressin receptors of the V1ß type. *Endocrinology*. 1994;135:2058–2063.

90. Schorscher-Petcu A, Sotocinal S, Ciura S, et al. Oxytocin-induced analgesia and scratching are mediated by the Vasopressin-1A receptor in the mouse. *J Neurosci*. 2010;30:8274–8284.

91. Seger MA, Burbach JP. The presence and in vivo biosynthesis of fragments of CPP (the C-terminal glycopeptide of the rat vasopressin precursor) in the hypothalamo-neurohypophyseal system. *Peptides*. 1987;8:757–762.

92. Selye H. *Stress in Health and Disease*. Oxford: Butterworth-Heinemann Ltd; 1976.

93. Serradeil-Le Gal C, Wagnon 3rd J, Tonnerre B, et al. An overview of SSR149415, a selective nonpeptide vasopressin V(1b) receptor antagonist for the treatment of stress-related disorders. *CNS Drug Rev*. 2005;11:53–68.

94. Serrats J, Schiltz JC, Garcia-Bueno B, van Rooijen N, Reyes TM, Sawchenko PE. Dual roles for perivascular macrophages in immune-to-brain signaling. *Neuron*. 2010;65:94–106.

95. Silverman MN, Sternberg EM. Glucocorticoid regulation of inflammation and its behavioral and metabolic correlates: from HPA axis to glucocorticoid receptor dysfunction. *Ann NY Acad Sci*. 2012;1261:55–63.

96. Slattery DA, Neumann ID. No stress please! Mechanisms of stress hyporesponsiveness of the maternal brain. *J Physiol*. 2008;586:377–385.

97. Sofroniew MV. Projections from vasopressin, oxytocin, and neurophysin neurons to neural targets in the rat and human. *J Histochem Cytochem*. 1980;28:475–478.

98. Spanakis EK, Wand GS, Ji N, Golden SH. Association of HPA axis hormones with copeptin after psychological stress differs by sex. *Psychoneuroendocrinology*. 2016;63:254–261.

99. Stevenson EL, Caldwell HK. The vasopressin 1b receptor and the neural regulation of social behavior. *Horm Behav*. 2012;61:277–282.

100. Stoop R, Hegoburu C, van den Burg E. New opportunities in vasopressin and oxytocin research: a perspective from the amygdala. *Annu Rev Neurosci*. 2015;38:369–388.

101. Suzuki H, Kawasaki M, Ohnishi H, Nakamura T, Ueta Y. Regulatory mechanism of the arginine vasopressin-enhanced green fluorescent protein fusion gene expression in acute and chronic stress. *Peptides*. 2009;30:1763–1770.

102. Suzuki H, Kawasaki M, Ohnishi H, et al. Exaggerated response of a vasopressin-enhanced green fluorescent protein transgene to nociceptive stimulation in the rat. *J Neurosci*. 2009;29:13182–13189.

103. Suzuki H, Onaka T, Kasai M, et al. Response of arginine vasopressin-enhanced green fluorescent protein fusion gene in the hypothalamus of adjuvant-induced arthritic rats. *J Neuroendocrinol*. 2009;21:183–190.

104. Swaab DF, Bao AM, Lucassen PJ. The stress system in the human brain in depression and neurodegeneration. *Ageing Res Rev*. 2005;4:141–194.

105. Swanson LW, Sawchenko PE, Lind RW, Rho JH. The CRH motoneuron: differential peptide regulation in neurons with possible synaptic, paracrine, and endocrine outputs. *Ann NY Acad Sci*. 1987;512:12–23.

106. Tasevska I, Enhorning S, Persson M, Nilsson PM, Melander O. Copeptin predicts coronary artery disease cardiovascular and total mortality. *Heart*. 2016,102,127–132.

107. Tilders FJH, Berkenbosch F, Vermes I, Linton EA, Smelik PG. Role of epinephrine and vasopressin in the control of the pituitary-adrenal response to stress. *Fed Proc (FASEB)*. 1985;44:155–160.

108. Ueta Y, Fujihara H, Serino R, et al. Transgenic expression of enhanced green fluorescent protein enables direct visualization for physiological studies of vasopressin neurons and isolated nerve terminals of the rat. *Endocrinology*. 2005;146:406–413.

109. Urwyler SA, Schuetz P, Sailer C, Christ-Crain M. Copeptin as a stress marker prior and after a written examination–the CoEXAM study. *Stress*. 2015;18:134–137.

110. van Leeuwen FW, Verwer RW, Spence H, Evans DA, Burbach JP. The magnocellular neurons of the hypothalamo-neurohypophyseal system display remarkable neuropeptidergic phenotypes leading to novel insights in neuronal cell biology. *Prog Brain Res*. 1998;119:115–126.

111. Vargas KJ, Sarmiento JM, Ehrenfeld P, et al. Postnatal expression of V2 vasopressin receptor splice variants in the rat cerebellum. *Differentiation*. 2009;77:377–385.

112. Veenema AH, Neumann ID. Neurobiological mechanisms of aggression and stress coping: a comparative study in mouse and rat selection lines. *Brain Behav Evol*. 2007;70:274–285.

113. Verney EB. Absorption and excretion of water; the antidiuretic hormone. *Lancet*. 1946;2(739):781.

114. Vischer HF, Castro M, Pin J-P. G protein–coupled receptor multimers: a question still open despite the use of novel approaches. *Mol Pharmacol*. 2015;88:561–571.

115. von Bardeleben U, Holsboer F. Cortisol response to a combined dexamethasone-human corticotropin-releasing hormone challenge in patients with depression. *J Neuroendocrinol*. 1989;1:485–488.

116. Wrobel LJ, Dupre A, Raggenbass M. Excitatory action of vasopressin in the brain of the rat: role of cAMP signaling. *Neuroscience*. 2011;172:177–186.

117. Young SF, Griffante C, Aguilera G. Dimerization between vasopressin V1b and corticotropin releasing hormone type 1 receptors. *Cell Mol Neurobiol*. 2007;27:439–461.

118. Zhu W, Tilley DG, Myers VD, Coleman RC, Feldman AM. Arginine vasopressin enhances cell survival via a G protein-coupled receptor kinase 2/beta-arrestin1/extracellular-regulated kinase 1/2-dependent pathway in H9c2 cells. *Mol Pharmacol*. 2013;84:227–235.

119. Ziegler DR, Herman JP. Neurocircuitry of stress integration: anatomical pathways regulating the hypothalamo-pituitary-adrenocortical axis of the rat. *Integr Comp Biol*. 2002;42:541–551.

120. Zimmerman EA, Nilaver G, Hou-Yu A, Silverman AJ. Vasopressinergic and oxytocinergic pathways in the central nervous system. *Fed Proc*. 1984;43:91–96.

10

Adrenocorticotropic Hormone

M.E. Rhodes

Saint Vincent College, Latrobe, PA, United States

Abstract

Adrenocorticotropic hormone (corticotropin; ACTH) is a 39 amino acid peptide hormone produced by cells of the anterior pituitary gland and carried by the peripheral circulation to its effector organ, the adrenal cortex, where it stimulates the synthesis and secretion of glucocorticoids and, to a more modest extent, mineralocorticoids and adrenal androgens. ACTH is secreted in response to corticotropin-releasing hormone (CRH) from the hypothalamus. The actions of CRH on ACTH release are augmented by another hypothalamic hormone, arginine vasopressin. ACTH production and secretion are regulated by glucocorticoid feedback, with ACTH at the center of a homeostatic network involving the hypothalamus and pituitary. Important brain areas influencing ACTH release include the hypothalamus, amygdala, hippocampus, and prefrontal cortex. As well, ACTH release is modulated by several other stimulatory and inhibitory factors, including neurotransmitters, peptides, and immune factors. ACTH release can become dysregulated by a number of factors including stressors, psychiatric illnesses, endocrine disorders, and various diseases.

INTRODUCTION

The anatomical structures that mediate the stress response are found in both the central nervous system (CNS) and in the peripheral tissues. The principal endocrine effectors of the stress response are released from the hypothalamus, pituitary, and adrenal glands. Corticotropin-releasing hormone (CRH) is released from parvocellular neurons of the paraventricular nucleus (PVN) of the hypothlamus into the hypophyseal pituitary portal system and this hypothalamic hormone is the principal regulator of anterior pituitary adrenocorticotropic hormone (ACTH) synthesis and secretion. The principal target tissue for circulating ACTH is the adrenal cortex, where ACTH stimulates the synthesis and secretion of glucocorticoids (cortisol in humans and corticosterone in rodents such as rats and mice) from the zona fasciculata by binding to melanocortin type 2 receptors (MC2-R). Glucocorticoids are the downstream effectors of the hypothalamic–pituitary–adrenal (HPA) axis and facilitate and regulate the vast array of physiological responses to stress.

KEY POINTS

- Adrenocorticotropic hormone (ACTH) is a 39 amino acid peptide hormone produced following cleavage of proopiomelanocortin (POMC).

- The principal target of circulating ACTH is the adrenal cortex, where ACTH stimulates glucocorticoid synthesis and secretion.

- The regulation of ACTH secretion primarily involves the stimulatory effect of corticotropin-releasing hormone and arginine vasopressin, hypothalamic hormones released directly into the portal blood supplying the anterior pituitary, and the inhibitory effect of glucocorticoids via feedback loops.

- Areas of the brain, including the amygdala, hippocampus, hypothalamus, and prefrontal cortex, regulate ACTH production and secretion.

- Dysregulation of ACTH occurs in endocrinopathies such as Cushing syndrome and Addison disease, as well as other psychiatric illnesses and disease states.

- The emerging role of ACTH and its related POMC-derived peptides in conditions other than those involving the stress response provides important opportunities to better understand their mechanisms of action and to develop new therapeutic uses of these agents.

The focus of this chapter is the melanocortin peptide, ACTH, also known as corticotropin or adrenocorticotropin. ACTH is a 39 amino acid peptide hormone that lies at the center of the HPA axis, and it is the main hormone of interest following cleavage of proopiomelanocortin (POMC). Numerous factors contribute to the stimulation and the inhibition of this important stress hormone.

ADRENOCORTICOTROPIC HORMONE AND PROOPIOMELANOCORTIN

Advances in the measurement of ACTH and its related peptides have elucidated their extensive distribution in areas of the body outside the pituitary gland, with differential processing in different tissues. Tissue-specific differences in the way the peptides are processed after translation give rise to differences in the final secretory products. ACTH and other peptides, including

beta-endorphin, a peptide known for its analgesic and euphoric effects in the brain, are produced in the pituitary from the chemical enzymatic breakdown of a large precursor protein POMC[24,34,45] (Fig. 10.1).

Since the discovery of the sequence and processing of POMC by Shosaku Numa and his colleagues,[38–40] the *POMC* gene has been shown to be expressed in the brain and pituitary of essentially all species that have been studied to date.[36] POMC is a large, 241 amino acid protein (prohormone) produced and processed within anterior pituitary corticotropes that undergoes hydrolytic cleavage at the sites of basic amino acids, such as arginine and lysine, resulting in the production of many hormonally active peptides.[3,21] POMC is also produced by hypothalamic neurons, the intermediate lobe of the pituitary, as well as by the placenta, skin, pancreas, liver, kidney, heart, thymus, lungs, ovaries, testes, spleen, leukocytes, and gastrointestinal tract.[3,36] Within the anterior pituitary, POMC is processed to yield an N-terminal peptide, whose function is unclear, and the peptides ACTH and β-lipotropin (β-LPH). These peptides are generated by the proteolytic cleavage of POMC by prohormone convertase 1 (PC1; also known as prohormone convertase 3), prohormone convertase 2 (PC2), and carboxypeptidase E.[5] The importance of the enzyme PC1 in

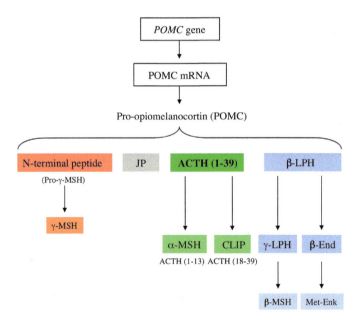

FIGURE 10.1 Proopiomelanocortin (POMC) and its biologically active peptide hormones following processing. Following transcription and translation, POMC is cleaved in a tissue- and species-specific manner by the prohormone convertases 1 and 2 (PC1 and PC2) and by carboxypeptidase E. Products from initial processing include an N-terminal peptide, joining peptide (JP), adrenocorticotropic hormone (ACTH$_{1-39}$), and β-lipotropin (β-LPH). Bioactive products include: ACTH; β-and γ-lipotropins (β-LPH, γ-LPH); α-, β-, and γ-melanocyte-stimulating hormones (α-MSH, β-MSH, γ-MSH); corticotropinlike intermediate lobe peptide (CLIP), β-endorphin (β-End), and met-enkephalin (Met-Enk).

POMC processing is shown by patients with PC1 mutations; these patients exhibit impaired POMC processing, hypocortisolism, childhood obesity, hypogonadism, and diabetes, among other clinical symptoms.[45]

Along with ACTH, β-LPH also is secreted by the anterior pituitary. α-melanocyte-stimulating hormone (α-MSH) and corticotropinlike intermediate lobe peptide are contained within the ACTH molecule. These peptides are especially found in species with developed intermediate lobes (e.g., amphibians, reptiles, and rats); however, they do not appear to be secreted to significant degrees as separate hormones in humans.[21] In humans, MSH can bind to melanocortin-1 receptors (MC1-R) on melanocytes, leading to increased melanin synthesis and darkening of the skin. However, it appears that neither α-MSH nor β-MSH are secreted to significant degrees in humans, suggesting that skin and other tissue-specific processing of ACTH and its peptides provides the source of MSH. In rats, ACTH and α-MSH inhibit food intake through MC4-R, which are highly expressed in the PVN and other hypothalamic nuclei.[46] In humans, ACTH also can activate MC1-R, thus explaining the pigmentary skin changes associated with some endocrine disorders. For example, pale skin is a hallmark of hypopituitarism, resulting from decreased circulating ACTH concentrations. In contrast, hyperpigmentation occurs in patients with primary adrenal insufficiency, resulting from increased ACTH concentrations.[3]

Within the β-LPH molecule exists the amino acid sequence for γ-LPH and β-endorphin, which can be converted to β-MSH and met-enkephalin, respectively. Endorphins and enkephalins are small peptides that bind to opioid receptors mediating analgesia and regulation of the pain response. The physiological roles of the other peptides in neurotransmission, learning and memory, pain responses, and mental disorders continue to be the subject of investigation. For example, studies have suggested key roles for β-MSH and possibly other melanocortins in the regulation of appetite and body weight.[36,45] The physiological role of β-LPH itself remains to be determined.[21]

ADRENOCORTICOTROPIC HORMONE PRODUCTION AND RELEASE FROM PITUITARY CORTICOTROPES

Evidence from light microscopic studies has revealed that within the anterior pituitary ACTH is secreted by basophilic corticotropes that represent 15%–20% of the anterior pituitary cell population.[41] These corticotropes are primarily found near the central mucoid wedge of the anterior pituitary, as well as posteriorly, adjacent to the pars nervosa.[29] When viewed with the electron microscope, ACTH-producing cells appear irregularly

shaped and full of secretory granules. The irregularly shaped corticotropes contain cytoplasmic projections, or cytonemes, that produce contacts between apparently isolated corticotropes within the gland, as well as with the perivascular space capillaries where associated cytoplasmic projections are frequently filled with secretory granules to facilitate release of ACTH into the bloodstream,[11,28] particularly rapid ultradian release of ACTH following stress.[14]

The complete structure of ACTH is known and the 39 amino acid peptide hormone has been synthesized (Fig. 10.2). The entire molecule is not needed for biological activity. The first 16 amino acids beginning with the N-terminal amino acid are all that is required for minimal biological activity. Progressive increases in activity occur as the length of the chain increases until full biological activity is present with a polypeptide over 22 amino acids long. Certain regions within the peptide have been identified as important for biological activity (amino acids 6–10) and binding (amino acids 15–18) to the MC2-R at the adrenal cortex.[18]

The parent ACTH molecule has a circulating half-life of 7–12 min, and basal circulating levels average from 2 to 9 pmol/L in healthy individuals.[34] The principal target of circulating ACTH within the HPA axis is the adrenal cortex, where ACTH stimulates corticosteroid (glucocorticoid) synthesis and secretion from the zona fasciculata. To a more modest extent, ACTH also stimulates the synthesis and secretion of mineralocorticoids and adrenal androgens. Glucocorticoids are the downstream effectors of the HPA axis and regulate physiological responses to stress via ubiquitously distributed intracellular receptors.[42] ACTH produces its effects at the adrenal cortex by activating MC2-R, producing activation of the intracellular cAMP-protein kinase A pathway.[23] Because the specific mechanisms for steroid hormone secretion are not well defined, and because steroids do not accumulate to a large extent in the gland, it is believed that the ability of ACTH to increase glucocorticoid production is mediated

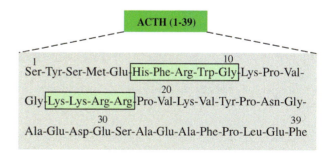

FIGURE 10.2 The complete amino acid sequence of the parent adrenocorticotropic hormone (ACTH$_{1-39}$) molecule. The *green* boxes behind the amino acid sequence highlight regions within the peptide that have been identified as important for biological activity (amino acids 6–10) at the adrenal cortex and binding (amino acids 15–18) to the melanocortin-2 (MC2-R) receptor, also known as the ACTH receptor.

predominantly at the level of de novo synthesis.[45] ACTH activation of MC2-R and the cAMP-protein kinase A pathway induces rapid phosphorylation of the steroid regulatory protein that facilitates transport of cholesterol into the mitochondria for conversion into active steroid hormones, and within the nucleus, induces the transcription of several steroidogenic enzymes necessary for the short- and long-term stimulation of steroidogenesis.[49] Mutations in MC2-R account for approximately 25% of the cases of familial glucocorticoid deficiency, a rare hereditary syndrome of ACTH resistance.[45]

ACTH is secreted into the bloodstream in pulses from pituitary corticotropes with approximately 40 pulses/24 h, correlating with subsequent corticosteroid pulses.[33] The development of immunoradiometric assays specific for intact human and rodent ACTH$_{1-39}$ has improved the reliability of ACTH measurement. Radioimmunoassay and enzyme immunoassay of plasma ACTH and its precursors also are accepted methods for evaluating ACTH secretion, although the antibodies used also recognize some of the shorter ACTH fragments. Therefore, performance differs among the various commercially available assays, and should be considered when interpreting patient or bench work results. Glucocorticoid concentrations in the blood also can be measured as an index of ACTH secretion.

CONTROL OF SECRETION

The regulation of ACTH secretion primarily involves the stimulatory effect of CRH and arginine vasopressin (AVP; also known as antidiuretic hormone), hypothalamic hormones released directly into the portal blood supplying the anterior pituitary, as well as the inhibitory effect of glucocorticoids.[1,22] CRH (corticotropin-releasing factor-41) was first discovered and characterized from ovine hypothalamic extracts by Wylie Vale and his colleagues.[48,51] CRH produces its effects in corticotropes via CRH1 receptors, causing increases in intracellular cAMP and calcium entry via L-type voltage-dependent calcium channels.[30] Increases in these second messengers triggers the release of ACTH secretory vesicles at the plasma membrane. AVP produces its effects in corticotropes via V1b receptors, causing increases in second messengers of the inositol phosphate pathway, as well as calcium.[2,41] Therefore, AVP potentiates CRH-mediated releases of ACTH by a mechanism that involves a molecular crosstalk between protein kinase C activation (caused by AVP) and cAMP-protein kinase A activation (caused by CRH).[2,29,41]

In addition to its effect on ACTH release from vesicles, CRH activation of cAMP-protein kinase A also induces POMC processing via nuclear orphan receptors of the Nur family. In particular, Nur factors Nur77 and Nurr1

are important in the regulation of the entire HPA axis. In the pituitary, Nur 77 and Nurr1 are involved in CRH-dependent induction of POMC mRNA[27,29,41], CRH also stimulates the transcription of c-fos, FosB, and Jun-B, as well as binding to the POMC activator protein-1 (AP-1) site.[29] The pituitary adenylate cyclase-activating protein also increases cAMP production and POMC transcription, perhaps via a common pathway with CRH.[8]

The participation of other regulators in addition to CRH and AVP, that modulate ACTH secretion in the absence of or during stress, cannot be discounted (Fig. 10.3). A number of other factors, such as angiotensin II, catecholamines, ghrelin, growth hormone–releasing hormone (GHRH), and immune factors have been shown to stimulate ACTH secretion and/or augment cleavage of POMC. Pancreatic vasoactive intestinal peptide (VIP) also stimulates ACTH secretion, probably indirectly via CRH secretion, a mechanism which may explain ACTH increases following eating.[33] Indeed, food intake and feeding behaviors appear to be major regulators of HPA axis hormone rhythms.[50]

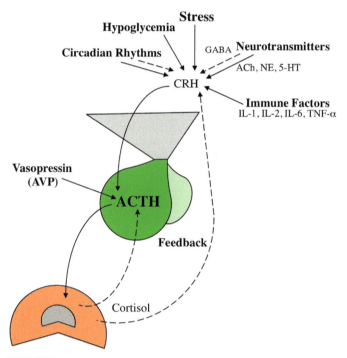

FIGURE 10.3 Factors influencing adrenocorticotropic hormone (ACTH) secretion from the anterior pituitary. Peripheral vasopressin (AVP) and cortisol feedback directly and indirectly (via corticotropic-releasing hormone or CRH) modulate ACTH secretion, while other factors such as stressors, circadian rhythms, immune factors [interleukins 1, 2, and 6 (IL-1, IL-2, and IL-6) and tumor necrosis factor-α], cortisol feedback and neurotransmitters [e.g., acetylcholine (ACh), norepinephrine (NE), serotonin (5-HT), and gamma-aminobutyric acid (GABA)] modulate ACTH release by influencing the portal secretion of CRH and AVP from the paraventricular nucleus (PVN) of the hypothalamus. Dashed lines represent negative feedback or inhibition; solid lines represent stimulation.

In contrast, the endocannabinoid system appears to produce an inhibitory effect on baseline and stimulated ACTH.[50] As evidence, the endocannabinoid anandamide can be detected in human pituitary glands, the cannabinoid receptor type 1 (CB1) receptor is found on human corticotropes, and 14-day administration of the cannabinoid agonist, Δ-9-tetrahydrocannabinol, inhibits the cortisol response to hypoglycemia in humans.[29] As well, antagonism of CB1 causes corticosterone increases in mice.[29] Opioid and opioid peptides also inhibit ACTH release, perhaps by inhibiting CRH release from the hypothalamus.[29] A recent study also suggests that erythropoietin, a cytokine most known for its role in red blood cell production, decreases ACTH concentrations in mice.[16]

Areas of the brain, including the amygdala, hippocampus, and hypothalamus, stimulate and inhibit the HPA axis to different degrees depending on the time of day, season of the year, and physical and environmental stressors.[43] Other brain areas that modulate HPA axis responses include the brainstem, bed nucleus of the stria terminalis, and prefrontal cortex.[31,43] The HPA axis has a normal circadian (24-h) rhythm: In humans, the secretion of CRH, ACTH, and adrenal hormones is greatest between 0700 and 0800h, just before and an hour or so after awakening, and lowest in early morning between 0200 and 0300h. ACTH appears to present the most intense circadian rhythm, with approximately 70% of its secretion taking place between 2400 and 0700h.[4] In nocturnal animals, the secretion of ACTH and adrenal hormones is greatest in the evening prior to their active period. The biological clock responsible for the circadian rhythm of ACTH secretion is located in yet another important brain area, the suprachiasmatic nuclei of the hypothalamus.

HPA axis function also is intimately linked with immune function. Interleukins are chemical messengers secreted by cells of the immune system that affect the status and activity of the rest of the immune system. In addition to their effects on immune function, interleukin-1 (IL-1), IL-2, and IL-6 appear to stimulate HPA axis activity.[33,41,45] IL-1 enhances ACTH release perhaps due to enhancement of CRH release, or by modulating the actions of other ACTH secretagogues. IL-2 augments *POMC* gene expression in the anterior pituitary and enhances ACTH release. The potent ACTH release by IL-6 may derive from stimulation of AVP release.

The HPA axis is very stress-responsive. Both physical and psychological stressors can cause increased activity of this hormone axis. Physical injury, emotional stress, hemorrhage, surgery, trauma, and other physiological challenges activate the axis. In particular, novel stressors (stressors that are new experiences) cause HPA axis activation; and with a person's repeated encountering of the stressor, there is less and less HPA axis response. This is particularly true of demanding tasks, which initially provoke an HPA axis response, but not after the person achieves mastery of the task through training and experience. For example, public speaking in front of a new audience can be a stressful experience to the novice but causes little or no stress, and therefore activation of the HPA axis, in the experienced speaker.

Increased ACTH secretion also is observed in pregnancy and at different periods during the menstrual cycle, because of the effects of estrogen in the circulation. Pathological (abnormal) function of the HPA axis can occur from several causes. These include repeated, uncontrollable environmental stressors; functional psychiatric illnesses such as anorexia; anxiety disorders; major depression and schizophrenia; and disorders of the HPA axis itself, such as hormone-secreting tumors of the pituitary gland and adrenal cortex. Major depression is the best-studied psychiatric illness in this regard: 30%–50% of major depressives have mildly to moderately increased HPA axis activity, consisting of increased blood concentrations of ACTH and cortisol at all times of the day and night, with preservation of the circadian rhythm of these hormones.[43,44] An increasing number of basic and clinical studies also report a dissociation of ACTH and glucocorticoid levels in illness, inflammatory disorders, and psychiatric conditions[7]; however, the mechanisms and physiological and pathological significance of these dissociative findings are unclear and warrant further investigation.

NEGATIVE FEEDBACK MECHANISMS

The mechanisms regulating ACTH during stress are multifactorial, with the stimulatory effect of the hypothalamus, mainly from CRH and AVP, and the inhibitory influence of adrenal glucocorticoids, by both genomic and nongenomic mechanisms, modulating ACTH production and release.[27] This modulation occurs physiologically by two primary processes. The first is "closed-loop" negative feedback of glucocorticoids to hormone receptors in the hippocampus, prefrontal cortex, hypothalamus, and pituitary gland.[20,37,47] This negative feedback acts to suppress the secretion of CRH, ACTH, and cortisol itself, with pituitary ACTH at the center of this homeostatic pathway.[19] The second primary process of ACTH regulation is "open-loop" driving (or braking) of the hypothalamic CRH production and secretion by the CNS, and by metabolic control of CRH secretion independent of glucocorticoids.

The HPA axis also is regulated temporally by three types of feedback: fast, intermediate, and slow feedback. Fast feedback occurs within minutes and involves inhibition of ACTH release by glucocorticoids, probably by nongenomic (second messenger pathway) mechanisms[13,14,26,29,50]; As well, there is some evidence to support the involvement of endocannabinoids in producing this fast feedback.[17,50] Intermediate feedback occurs within 2–10h of initial stimulation, and involves inhibition of CRH synthesis and release, but does not directly influence ACTH synthesis. Slow feedback occurs over longer intervals and results from inhibition of POMC transcription and processing.[26]

CLINICAL PERSPECTIVES

Cushing Syndrome

Cushing syndrome is a clinical condition resulting from chronic elevation of circulating glucocorticoids. Clinical signs and symptoms of Cushing syndrome include obesity of the face and trunk, weakness and atrophy of limb muscles, increased blood pressure, imbalance of glucose metabolism, and psychological changes. Psychiatric symptoms may include depression, mood swings, mania, agitation, and memory disturbances.[21] There are two main types of Cushing syndrome, ACTH-dependent and ACTH-independent. ACTH-dependent Cushing syndrome results from increased pituitary secretion of ACTH, usually from a pituitary tumor (Cushing disease), inappropriate ACTH secretion by nonpituitary tumors, often in the lungs or thyroid gland, and inappropriate CRH secretion by nonhypothalamic tumors, in turn stimulating excessive pituitary ACTH secretion. These conditions, all involving excess ACTH production, cause enlargement of the adrenal glands and excessive cortisol secretion from continuous stimulation.

ACTH-independent Cushing syndrome is caused by primary tumors or abnormalities of the adrenal cortex itself, resulting in excessive cortisol secretion and suppression of ACTH production by the pituitary. Prolonged administration of glucocorticoids for the treatment of certain illnesses also may cause ACTH-independent Cushing syndrome.

The excessive pituitary production of ACTH and adrenal production of cortisol in Cushing disease are only partially suppressible by low-dose dexamethasone, a synthetic and potent glucocorticoid, but more suppressible by higher doses. In contrast, patients with nonpituitary sources of ACTH production will rarely show suppression of ACTH and cortisol by dexamethasone.

Addison Disease

Addison disease is characterized by diminished glucocorticoid production by the adrenal cortex (resulting in adrenal insufficiency) usually resulting from an autoimmune mechanism. Patients with this disease show low levels of circulating cortisol, as well as mineralocorticoids, and high levels of circulating ACTH. Symptoms of Addison disease include muscle weakness, skin hyperpigmentation, loss of appetite, vomiting and diarrhea, fever, fatigue, difficulty concentrating, and lack of initiative. Depression and stress intolerance also may be evident.[21]

Adrenocorticotropic Hormone Resistance Syndromes

Inherited ACTH resistance diseases are rare and include triple A syndrome (Allgrove syndrome) and familial glucocorticoid deficiency (FGD). Clinically, FGD is characterized solely by ACTH resistance while triple A syndrome exhibits a variety of additional characteristics. FGD is caused by mutations of the MC2-R and melanocortin 2 receptor accessory protein (MRAP) genes.[12,32] Therefore, most patients with FGD and triple A syndrome will exhibit adrenal insufficiency and high circulating ACTH levels, while most patients will have no disruption of circulating mineralocorticoids.[9] Other symptoms of triple A syndrome include alacrima (no tear production), achalasia (relaxation of the lower esophageal muscles, preventing food from entering the stomach), and progressive neurological symptoms such as ataxia, spasticity, myopathy, and peripheral neuropathy.[32] Other symptoms of FGD typically present during childhood and include hypoglycemia, hyperpigmentation of the skin, and in some cases, tall stature that may be caused by high-circulating ACTH levels acting at MC1-R in the bone and cartilage.[9,32]

OTHER EFFECTS OF ADRENOCORTICOTROPIC HORMONE

In addition to its endocrine activity, ACTH also influences neuronal activity. In higher animals, decades of research have shown that ACTH and its fragment peptides influence behavior, attention, and learning.[6,15,25] Increased ACTH secretion in Cushing syndromes can be associated with deficits in memory; however, whether the memory impairments result solely from increases in circulating ACTH versus increases in circulating glucocorticoids is unclear. ACTH administered into the brain leads to other, sometimes bizarre, behaviors not observed with peripheral administration. In this case, ACTH increases penile erection and ejaculation, sexual receptivity, yawning, grooming behavior, and stretching.[5] ACTH also has been shown to increase slow-wave sleep and REM sleep in rats.[52] In addition to the adrenal cortex, MC2-R also are expressed in the adipose tissue of rodents, and ACTH affects adipocyte lipolysis and enhances insulin-induced glucose uptake in adipose tissue.[10]

THERAPEUTIC USES

There are currently two ACTH formulations available for use in the US. The first is H.P. Acthar Gel, an injectable formulation used primarily for the treatment of infantile spasms and acute symptoms of multiple sclerosis. However, this injectable formulation also is indicated for rheumatic, allergic, and certain respiratory diseases. The second is cosyntropin (Cortosyn), a synthetic $ACTH_{1-24}$ peptide that is used as a diagnostic agent when screening for adrenal insufficiency. In the UK, $ACTH_{1-24}$ (Synacthen Depot) is available for both diagnostic and therapeutic uses, where it is indicated for short-term conditions where glucocorticoids are indicated in principle, in patients unable to tolerate glucocorticoid therapy or where glucocorticoids have been ineffective.[35]

The discovery of melanocortins within the POMC protein in the 80s and the cloning of melanocortin receptors in the 90s have led to the development and investigation of a number of selective ligands with agonist and antagonist activity at these receptors. Given the ubiquitous distribution of melanocortin receptors within the body, the potential clinical applications of these ligands is interesting and the range of potential therapeutic uses is very broad: erectile dysfunction, obesity, anorexia and bulimia, respiratory syndromes, ischemic events, arthritis, inflammatory bowel disease, neuropathies, and neurodegenerative disease.[5]

The presence of melanocortin receptors on immune cells suggests that new immunomodulatory peptides devoid of steroidogenic actions could be developed that target these receptors.[35] This suggests that ACTH, its related peptides, and new immunomodulatory peptides could find therapeutic use in the treatment of conditions such as Crohn disease, rheumatoid arthritis, systemic lupus erythematosus, and gout. Indeed, many clinical trials are underway to determine the influence of melanocortin drugs on these and other conditions.[35]

CONCLUSION

ACTH and other peptides, including beta-endorphin, MSH peptides, and lipotropins, are produced in the pituitary from the chemical breakdown of the large precursor protein POMC. ACTH is released from the anterior pituitary in response to various stimuli and its release is sustained or inhibited by intricate feedback systems, with ACTH at the heart of this dynamic homeostatic network of "loops." As part of the HPA axis, ACTH is considered one of the major "stress" hormones. It is now apparent that AVP and CRH cooperate as the major factors involved in the regulation of ACTH release, with numerous other peripheral and central factors contributing to this regulation. The emerging role of ACTH and its related POMC-derived peptides in conditions other than those involving the stress response are exciting avenues for exploration into mechanisms and potential therapeutic uses of these agents.

Acknowledgment

This chapter is a revised and expanded adaptation from a previous chapter by Michael E. Rhodes, *Encyclopedia of Stress*, volume 1, pp. 69–72, © 2007, Elsevier, Inc.

References

1. Aguilera G. Regulation of pituitary ACTH secretion during chronic stress. *Front Neuroendocrinol.* 1994;15(4):321–350.
2. Aguilera G, Subburaju S, Young S, Chen J. The parvocellular vasopressinergic system and responsiveness of the hypothalamic pituitary adrenal axis during chronic stress. *Prog Brain Res.* 2008;170:29–39.
3. Amar AP, Weiss MH. Pituitary anatomy and physiology. *Neurosurg Clin N Am.* 2003;14(1):11–23.
4. Batrinos ML. The aging of the endocrine hypothalamus and its dependent endocrine glands. *Hormones.* 2012;11(3):241–253.
5. Bertolini A, Tacchi R, Vergoni AV. Brain effects of melanocortins. *Pharmacol Res.* 2009;59(1):13–47.
6. Born J, Fehm HL, Voigt KH. ACTH and attention in humans: a review. *Neuropsychobiology.* 1986;15(3–4):165–186.
7. Bornstein SR, Engeland WC, Ehrhart-Bornstein M, Herman JP. Dissociation of ACTH and glucocorticoids. *Trends Endocrinol Metab.* 2008;19(5):175–180.
8. Boutillier AL, Monnier D, Koch B, Loeffler JP. Pituitary adenyl cyclase-activating peptide: a hypophysiotropic factor that stimulates proopiomelanocortin gene transcription, and proopiomelanocortin-derived peptide secretion in corticotropic cells. *Neuroendocrinology.* 1994;60(5):493–502.
9. Chan LF, Clark AJ, Metherell LA. Familial glucocorticoid deficiency: advances in the molecular understanding of ACTH action. *Horm Res.* 2008;69(2):75–82.
10. Chan LF, Metherell LA, Clark AJ. Effects of melanocortins on adrenal gland physiology. *Eur J Pharmacol.* 2011;660(1):171–180.
11. Childs GV. Structure-function correlates in the corticotropes of the anterior pituitary. *Front Neuroendocrinol.* 1992;13(3):271–317.
12. Clark AJ, Chan LF, Chung TT, Metherell LA. The genetics of familial glucocorticoid deficiency. *Best Pract Res Clin Endocrinol Metab.* 2009;23(2):159–165.
13. Dallman MF, Akana SF, Cascio CS, Darlington DN, Jacobson L, Levin N. Regulation of ACTH secretion: variations on a theme of B. *Recent Prog Horm Res.* 1987;43:113–173.
14. Dallman MF, Akana SF, Scribner KA, et al. Stress, feedback and facilitation in the hypothalamo-pituitary-adrenal axis. *J Neuroendocrinol.* 1992;4(5):517–526.
15. De Wied D, Jolles J. Neuropeptides derived from pro-opiocortin: behavioral, physiological, and neurochemical effects. *Physiol Rev.* 1982;62(3):976–1059.
16. Dey S, Scullen T, Noguchi CT. Erythropoietin negatively regulates pituitary ACTH secretion. *Brain Res.* 2015;1608:14–20.
17. Di S, Malcher-Lopes R, Halmos KC, Tasker JG. Nongenomic glucocorticoid inhibition via endocannabinoid release in the hypothalamus: a fast feedback mechanism. *J Neurosci.* 2003;23(12):4850–4857.
18. Dores RM, Liang L. Analyzing the activation of the melanocortin-2 receptor of tetrapods. *Gen Comp Endocrinol.* 2014;203:3–9.

19. Fink G. Mechanisms of negative and positive feedback of steroids in the hypothalamic-pituitary system. In: Bittar EE, Bittar A, eds. *Principles of Medical Biology: Molecular and Cellular Endocrinology.* vol. 10A. Greenwich, CT: JAI Press Inc.; 1997:29–100.

20. Fink G, Robinson IC, Tannahill LA. Effects of adrenalectomy and glucocorticoids on the peptides CRF-41, AVP and oxytocin in rat hypophysial portal blood. *J Physiol.* 1988;401:329–345.

21. Geracioti TD, Strawn JR, Ekhator NN, Wortman M. Brain peptides: from laboratory to clinic. In: Rubin RT, Pfaff DW, eds. *Hormone/Behavior Relations of Clinical Importance: Endocrine Systems Interacting With Brain and Behavior.* Amsterdam: Elsevier; 2009:417–463.

22. Herman JP, Flak J, Jankord R. Chronic stress plasticity in the hypothalamic paraventricular nucleus. *Prog Brain Res.* 2008;170:353–364.

23. Hofland J, Delhanty PJ, Steenbergen J, et al. Melanocortin 2 receptor-associated protein (MRAP) and MRAP2 in human adrenocortical tissues: regulation of expression and association with ACTH responsiveness. *J Clin Endocrinol Metab.* 2012;97(5):E747–E754.

24. Javorsky BR, Aron DC, Findling JW, Tyrrell JB. Hypothalamus and pituitary gland. In: Gardner DG, Shoback D, eds. *Greenspan's Basic and Clinical Endocrinology.* 9th ed. New York: McGraw-Hill; 2011:65–114.

25. Jolles J. Neuropeptides and cognitive disorders. *Prog Brain Res.* 1986;65:177–192.

26. Keller-Wood ME, Dallman MF. Corticosteroid inhibition of ACTH secretion. *Endocr Rev.* 1984;5(1):1–24.

27. Laryea G, Muglia L, Arnett M, Muglia LJ. Dissection of glucocorticoid receptor-mediated inhibition of the hypothalamic-pituitary-adrenal axis by gene targeting in mice. *Front Neuroendocrinol.* 2015;36:150–164.

28. Le Tissier PR, Hodson DJ, Lafont C, Fontanaud P, Schaeffer M, Mollard P. Anterior pituitary cell networks. *Front Neuroendocrinol.* 2012;33(3):252–266.

29.. Lim DCT, Grossman A, Khoo B. Normal physiology of ACTH and GH release in the hypothalamus and anterior pituitary in man. In: De Groot LJ, ed. *Endotext. South Dartmouth: MDText.com, Inc.;* 2000.

30. Majzoub JA. Corticotropin-releasing hormone physiology. *Eur J Endocrinol.* 2006;155:S71–S76.

31. McKlveen JM, Myers B, Flak JN, et al. Role of prefrontal cortex glucocorticoid receptors in stress and emotion. *Biol Psychiatry.* 2013;74(9):672–679.

32. Metherell LA, Chan LF, Clark AJ. The genetics of ACTH resistance syndromes. *Best Pract Res Clin Endocrinol Metab.* 2006;20(4):547–560.

33. Mihai R. Physiology of the pituitary, thyroid, parathyroid and adrenal glands. *Surgery.* 2014;32(10):504–512.

34. Molina PE. Anterior pituitary gland. In: Molina PE, ed. *Endocrine Physiology.* 4th ed. New York: McGraw-Hill; 2013:49–72.

35. Montero-Melendez T. ACTH: the forgotten therapy. *Semin Immunol.* 2015;27(3):216–226.

36. Mountjoy KG. Pro-opiomelanocortin (POMC) neurones, POMC-derived peptides, melanocortin receptors and obesity: how understanding of this system has changed over the last decade. *J Neuroendocrinol.* 2015;27(6):406–418.

37. Myers B, McKlveen JM, Herman JP. Neural regulation of the stress response: the many faces of feedback. *Cell Mol Neurobiol.* 2012.

38. Nakanishi S, Inoue A, Kita T, et al. Nucleotide sequence of cloned cDNA for bovine corticotropin-beta-lipotropin precursor. *Nature.* 1979;278(5703):423–427.

39. Nakanishi S, Teranishi Y, Noda M, et al. The protein-coding sequence of the bovine ACTH-beta-LPH precursor gene is split near the signal peptide region. *Nature.* 1980;287(5784):752–755.

40. Numa S, Nakanishi S. Structure and regulation of the messenger RNA coding for the corticotropin/beta-lipotropin precursor. *Biochem Soc Trans.* 1980;8(6):749–751.

41. Perez-Castro C, Renner U, Haedo MR, Stalla GK, Arzt E. Cellular and molecular specificity of pituitary gland physiology. *Physiol Rev.* 2012;92(1):1–38.

42. Rhodes ME, McKlveen JM, Ripepi DR, Gentile NE. Hypothalamic-pituitary-adrenal cortical axis. In: Rubin RT, Pfaff DW, eds. *Hormone/Behavior Relations of Clinical Importance: Endocrine Systems Interacting With Brain and Behavior.* Amsterdam: Elsevier; 2009:47–67.

43. Rhodes ME, Rubin RT. Functional sex differences ('sexual diergism') of central nervous system cholinergic systems, vasopressin, and hypothalamic-pituitary-adrenal axis activity in mammals: a selective review. *Brain Res Brain Res Rev.* 1999;30(2):135–152.

44. Rubin RT. Neuroendocrine aspects of stress in major depression. In: Liberman RP, Yager J, eds. *Stress in Psychiatric Disorders.* New York: Springer; 1994:37–52.

45. Schimmer BP, Funder JW. ACTH, adrenal steroids, and pharmacology of the adrenal cortex. In: Brunton L, Chabner B, Knollman B, eds. *Goodman and Gilman's the Pharmacological Basis of Therapeutics.* 12th ed. New York: McGraw-Hill; 2011:1209–1236.

46. Schulz C, Paulus K, Lobmann R, Dallman M, Lehnert H. Endogenous ACTH, not only alpha-melanocyte-stimulating hormone, reduces food intake mediated by hypothalamic mechanisms. *Am J Physiol Endocrinol Metab.* 2010;298(2):E237–E244.

47. Sheward WJ, Fink G. Effects of corticosterone on the secretion of corticotrophin-releasing factor, arginine vasopressin and oxytocin into hypophysial portal blood in long-term hypophysectomized rats. *J Endocrinol.* 1991;129(1):91–98.

48. Spiess J, Rivier J, Rivier C, Vale W. Primary structure of corticotropin-releasing factor from ovine hypothalamus. *Proc Natl Acad Sci USA.* 1981;78(10):6517–6521.

49. Stocco DM. StAR protein and the regulation of steroid hormone biosynthesis. *Annu Rev Physiol.* 2001;63:193–213.

50. Uchoa ET, Aguilera G, Herman JP, Fiedler JL, Deak T, de Sousa MB. Novel aspects of glucocorticoid actions. *J Neuroendocrinol.* 2014;26(9):557–572.

51. Vale W, Spiess J, Rivier C, Rivier J. Characterization of a 41-residue ovine hypothalamic peptide that stimulates secretion of corticotropin and beta-endorphin. *Science.* 1981;213(4514):1394–1397.

52. Vedder H. Physiology of the hypothalamic-pituitary-adrenocortical axis. In: del Ray A, Chrousos GP, Besedovsky HO, eds. *The Hypothalamic-Pituitary-Adrenal Axis.* vol. 7. Amsterdam: Elsevier; 2008.

11

The Role of MicroRNAs in Stress-Induced Psychopathologies

O. Issler[1,2,3], A. Chen[1,2]

[1]Weizmann Institute of Science, Rehovot, Israel; [2]Max-Planck Institute of Psychiatry, Munich, Germany; [3]Icahn School of Medicine at Mount Sinai, New York, NY, United States

Abstract

Exposure to stress is a known risk factor for disease development, particularly in psychiatric disorders. Such long-lasting effects of stress on an organism's physiology are facilitated by epigenetic processes, and one such mechanism is posttranscriptional regulation by microRNAs (miRNAs). Here, we review the findings of studies in human, animal, and cellular models on miRNAs' involvement in stress-related psychopathologies, focusing on anxiety and depression. A better understanding of the complex genetic and environmental interactions that give rise to the susceptibility to develop psychopathologies, may promote the needed breakthroughs in diagnostics and therapeutics in psychiatry.

INTRODUCTION

Environmental factors, mainly exposure to psychological or physiological stressors, are associated with an increased risk of developing psychiatric disorders. Stress during embryonic development or early life may program brain vulnerability, while stress in adolescence or later in life is associated with triggering the onset of psychiatric disorders. For example, immunological stress caused by a viral infection in utero is linked with increased risk to develop schizophrenia,[80] while exposure to acute stressors in adolescence may trigger the onset of the first psychotic episode of this disease.[13] Most psychiatric disorders display a strong genetic component, but heritability alone only partially explains an individual's risk to develop mental illness. Only a few specific gene mutations have been directly linked to increased susceptibility for a psychopathology, such MeCP2 mutations that are directly linked to Rett syndrome and increased risk of autism.[84] Rather evidence is pointing toward a complex interaction between genetic predisposition and environmental factors being the root of mental illness. For example, the literature reports an interaction between the Sert gene promoter polymorphism and childhood stress in predisposing to medication-resistant depression.[9,66] Environmental factors can lead to changes in gene expression levels via epigenetic mechanisms. These epigenetic changes may mediate the onset of a disease without altering the DNA sequence. Such mechanisms include histone modification, DNA methylation, and posttranscriptional regulation by noncoding RNAs, such as microRNAs (miRNAs), which are the focus of this chapter. Elucidating the role of stress-related processes mediated by miRNAs may promote a better understanding of the pathophysiology and neurobiology of psychiatric disorders. This will potentially promote the highly needed breakthroughs in the development of new drug targets and biomarkers for mental illness.

KEY POINTS

- Dysregulation of epigenetic mechanisms often mediates the adverse effects of stress on anxiety and depression susceptibility.

- miRNAs are modulators of normal stress response and stress-induced anxiety, and are important for fear learning.

- Chronic stress-induced depression alters miRNAs expression profiles, and the effects of antidepressants are induced in part by miRNAs.

- Sperm miRNAs are mediators of transgenerational effects of stress on the offspring behavior and metabolism.

- Circulating miRNAs' fingerprint reflects the anxiety and depression state of the individual and may be used as a diagnostic biomarker.

MICRORNAs

Since their recent discovery, descriptions of novel miRNAs and their widespread role in biological processes are accumulating, and position miRNAs as a prevalent mode of posttranscriptional regulation of gene expression.[1] Each miRNA may regulate hundreds of downstream targets, and collectively miRNAs are predicted to regulate more than half of the protein-coding genes and by that affect many cellular processes in health and disease.[21] Much is already known regarding miRNA biogenesis[38] and function[1,8,41] (Fig. 11.1). Briefly, the mature single-stranded miRNA, of about 22 nucleotides, is incorporated into a miRNA-induced silencing complex and induces translational repression or mRNA destabilization of target mRNAs. MiRNAs can act as an "expression switch" that blocks the expression of their

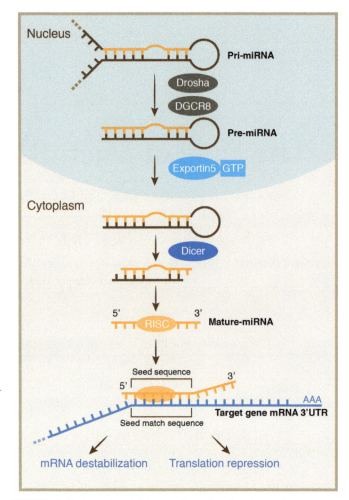

FIGURE 11.1 **miRNA biogenesis and molecular activity.** miRNA genes are transcribed by RNA polymerase II and RNA polymerase III into primary miRNA transcripts (pri-miRNA) and form a distinctive secondary hairpin structure.[5] Pri-miRNAs are transcribed either from dedicated genes or processed from introns of other genes, as individual miRNAs or as miRNAs clusters. This pri-miRNA is then cleaved by a microprocessor complex that contains the RNase III enzyme ribonuclease 3 (called DROSHA) and microprocessor complex subunit DGCR8.[45] Following nuclear processing and cleavage of the RNA molecule, its size is reduced to 70–110 nucleotides. Now referred to as precursor miRNA (pre-miRNA), it is exported to the cytoplasm by Exportin-5 (XPO5) in a complex with Ran-GTP.[81] Here, the pre-miRNA is cleaved by DICER to generate a roughly 22-nucleotide miRNA duplex. Next, DICER and its interaction domain protein Tar RNA-binding protein (TRBP) dissociate from the miRNA duplex to form the active RNA-induced silencing complex (RISC) that performs gene silencing. The double-stranded duplex needs to be separated into the functional guide strand, which is complementary to the target mRNA and the passenger strand, which is subsequently degraded. The functional strand of the mature miRNA binds to form the RISC complex together with Argonaute (AGO2) proteins[11] and guides the complex to target the 3′ untranslated region (3′ UTR) of mRNAs. This inhibits mRNA translation or promotes its degradation. miRNA specificity to their target mRNAs is canonically determined by Watson and Crick base-pairing of a six to eight nucleotide sequence in the 5′ end of the mature miRNA named seed sequence to a complementary seed match sequence in the target gene 3′ UTR.[1,31]

target genes.[1] In such cases, a mutually exclusive expression pattern of the silencing miRNA and its target genes are often observed, as commonly reported in developmental studies.[34] Alternatively, miRNAs can act as "fine tuners" of the expression levels of their target genes, evident by their coexpression with target genes, as often reported in adult tissues.[1,70]

Accumulating evidence suggests that adult brain miRNAs function as endogenous hubs for fine-tuning target gene expression levels, thereby affecting the structure and function of neuronal networks. In the healthy brain, miRNAs play a role in synaptic plasticity,[64,73,78]

and neurodegenerative diseases.[7,17,63] With respect to stress, miRNAs were demonstrated to be involved in the cellular response to stress[18,46,53] and to have a role in an organism's central response to stress. In this chapter we will highlight studies providing evidence for the role of central or circulating miRNAs in stress-related psychopathologies, with a spotlight on depression and anxiety disorders due to the large amount of data on these two disorders that have been accumulated to date. To this end, human samples, animal models, and cellular systems are utilized in a variety of methodical and technical approaches (Fig. 11.2).

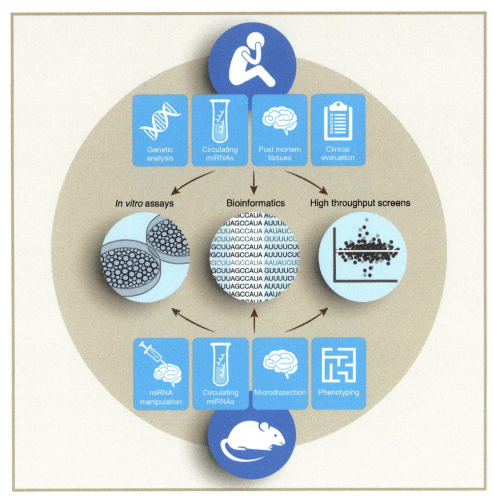

FIGURE 11.2 **Experimental approaches for studying miRNAs in psychopathologies.** In order to reveal the role of miRNAs in stress-related psychopathologies, data is collected from human patients and animal models of mental disorders in a variety of approaches. Several types of studies utilize human samples: genetic association studies, linking chromosomal variation to risk for mental disorder, postmortem studies examining the miRNAs expression profile in specific brain areas of patients, and evaluation of the levels of blood miRNAs for diagnostic purposes. In rodent models for mental disorders, the effects of different paradigms of stress on the miRNome are often tested, along with direct manipulation of groups of specific miRNAs levels. Technically, some approaches are unique to the study of miRNAs and others are modifications of research techniques used to study protein-coding genes. The miRNAs' expression profile is often analyzed using high throughput methods. Bioinformatic analysis is used to identify specific miRNA–target gene interaction. In vitro assays are often carried out to confirm the predicted interactions and for in-depth understanding of miRNA-related molecular mechanisms. Furthermore, mouse models are used to test causal effects of alteration of specific miRNA expression levels on mouse behavior, physiology, and gene expression. By and large, studies combining several approaches and using different models are of the greatest validity.

The Role of miRNAs in Stress-Induced Anxiety

Along with a synchronized neuroendocrine and sympathetic system's activity, the normal response to stress includes increased arousal, fear, and anxiety. After the termination of the stress response, the physiological, endocrine, and behavioral arms of the response to stress normalize the homeostatic balance. However, in some pathological cases, often following chronic stress or robust stressors, the anxiety remains and stabilizes in a form of a psychiatric disorder. There is evidence that miRNAs take part in regulating both the stress-induced anxiety that is part of the physiological response to homeostatic challenge and in psychopathological anxiety conditions.

There are reports from rodent studies on miRNAs' involvement in the central stress response. For example, the expression profile of miRNAs is altered by acute stress in the prefrontal cortex (PFC),[60] the hippocampus,[52] and the amygdala.[30,50,52] Haramati et al. reported an increase in anxiety-like behaviors when ablating the miRNA population by conditionally knocking out the miRNAs' processing gene, Dicer. In addition, a screen of amygdala miRNAs' profile following stress highlighted the interaction between miR-34c and a key gene in the central stress response, corticotropin-releasing factor receptor type 1 (Crfr1).[30] Furthermore, virally mediated overexpression of miR-34 in the adult mouse amygdala protected the mice from stress-induced anxiety as measured behaviorally. Moreover, overexpressing miR-34 in vitro blunted the response of Crfr1 to its endogenous ligand Crf, suggesting that miR-34 functionally regulates the molecular machinery of the response to stress.[30] Taken together, evidence suggests that brain miRNAs take part in the natural, stress-induced anxiety, and potentially also in pathological anxiety.

THE ROLE OF miRNAs IN FEAR CONDITIONING

Based on learning theories, it is hypothesized that the origin of anxiety disorders is from overgeneralization or incorrect associations between neutral signs and fearful stimuli. A common paradigm used to model anxiety-like behaviors in rodents is fear conditioning (FC), where neutral cues and contexts are experimentally associated with instinctive fear responses. Abnormalities in the conditioning itself, cue and content tests, or in extinction can be interpreted as potential models for anxiety disorders. There are several studies that investigate the role of miRNAs in the FC paradigm. The starting point of many of these studies is a high throughput examination of the miRNAs' profile in a certain brain area after the FC paradigm. Some studies followed up the research by an in-depth analysis of the role of a specific miRNA identified in the screen.

Broad evidence for the role of miRNAs in FC comes from a study showing that hippocampal CA1 miRNAs' expression patterns and the expression levels of genes in the miRNAs' biogenesis pathway were changed following contextual FC paradigm.[42] Another study showed, more specifically, that miR-134's interaction with Sirt1 in the hippocampus plays a role in FC.[22] Lentiviral-induced miR-134 knockdown in the hippocampus restored fear learning hampered by the knockout of the transcription regulator of miR-134, Sirt1.[22] Another study showed that miR-128b plays an important role in fear extinction in the infralimbic PFC (ILPFC).[47] MiR-128b is upregulated in mice following fear extinction in the ILPFC and lentiviral-mediated overexpression or knockdown of miR-128b in this brain area facilitates or inhibits fear extinction, respectively. A few potential target genes of miR-128b that can mediate this effect were identified. The authors focus on the Rcs/Arpp21 gene and identify its downregulation after fear extinction. Indeed, knocking down Rcs facilitates fear extinction. Interestingly, miR-128b's precursor is imbedded within an intron of the Rcs gene; therefore, the genes are coregulated and take part in modulating fear extinction.[47] Analyzing the effects of FC on miRNA expression in the rat lateral amygdala demonstrated that downregulation of miR-182 occurs in conjunction with upregulation of its actin-related target genes.[28] These genes are known to be involved in synaptic plasticity, a structural process associated with learning. Blocking miR-182 downregulation by virally overexpressing miR-182 in the lateral amygdala, interfered with the fear memory, suggesting that miR-182 downregulation is essential for this process.[28] An increase in miR-132 levels in mouse hippocampus after FC was also reported, while site specifically reducing miR-132 levels by lentiviruses inhibited the learned freezing behavior.[77] In addition, a study screening for the miRNA's profile in the amygdala shortly after FC-identified upregulation of miR-34a.[15] Next, the authors showed that lentivirally knocking down miR-34a in the amygdala led to decreased freezing in the context test, suggesting that miR-34a is necessary for normal fear consolidation.[15] Subsequently, genes within the Notch signaling pathway were bioinformatically identified to be targets of miR-34a, as confirmed in vitro. Furthermore, Notch1 and its ligand Jag1 are downregulated in the amygdala after FC. Inhibiting Notch1 facilitates and increases Notch signaling, which impairs consolidation. Collectively, decreased

TABLE 11.1 Specific miRNAs Regulating Fear Conditioning

miRNA Number	Effect on Fear Learning	Relevant Target Genes	Associated Brain Area	References
miR-128	Facilitation	Rcs	ILPFC	47
miR-132	Facilitation	?	HC	77
miR-134	Inhibition	Creb	HC	22
miR-182	Inhibition	Cortactin and Rac1	LA	28
miR-19b	Inhibition	Adbr1	BLA	74
miR-33	Inhibition	Gabra4, Kcc2 Gabrb2, Syn2	HC	36
miR-34a	Facilitation	Notch1	BLA	15

BLA, basolateral amygdala; HC, hippocampus; ILPFC, infralimbic prefrontal cortex; LA, lateral amygdala.

Notch signaling enabled partly by increased miR-34a facilitated fear learning consolidation.[15] Recently, Volk et al. identified an upregulation in miR-19b levels in the amygdala of mice exposed to chronic social defeat stress, a model for inducing anxiety and depression-like behaviors. Bioinformatics, in vitro and in vivo studies, showed that miR-19b targets a key gene in the stress-related neuroadrenergic circuit, the adrenergic receptor beta 1 (Adbr1).[74] Lentiviral overexpression of miR-19b in mouse amygdala led to decreased cued freezing behavior, while knockdown of miR-19b led to the mirror phenotype, suggesting that miR-19-Adbr1 interaction regulates the circuits controlling learned fear.[74] Finally, it was elegantly shown that downregulation of miR-33 mediates the anxiolytic effects of a GABA agonist as measured in the contextual FC test. Viral hippocampal overexpression of miR-33 blocked the effects of the agonist, while using locked nucleic acid oligos to inhibit miR-33 mimicked the anxiolytic effects. The authors suggest these effects are mediated by several GABA-related genes that are regulated in vivo by miR-33.[36] Taken together, a large amount of data demonstrates the role of miRNAs in normal fear processing and dysregulation in pathological FC behavior (summarized in Table 11.1).

The Role of miRNAs in Chronic Stress-Induced Major Depression

Exposure to chronic psychological mild stress often precedes the onset of major depression disorder (MDD) episodes. This phenomenon is partially mediated by miRNAs. Screen studies using rodent models demonstrated an altered miRNAs' expression pattern in several brain sites following exposure to chronic stress paradigms aimed to induce depression-like behaviors. For example, maternal separation early life stress,[56,71] chronic restraint stress,[52,61] and repeated inescapable shocks,[68] all altered miRNAs' expression profiles in brain areas associated with MDD, such as the PFC, amygdala, and hippocampus. Similarly, postmortem studies analyzing the miRNome of depressed patients identified modified expression patterns in the PFC[49,51,67,69] and hippocampus.[39] Meta-analysis and computational efforts to combine the results from rodent and human studies is required. In addition, extending the screens to other brain areas associated with MDD, such as the raphe nucleus, may be informative.

A few in-depth studies have focused on specific miRNAs highlighted by screens and bioinformatic analysis, and associated with exposure to chronic stress, MDD, and antidepressant treatment. For example, a study showed that miR-16 facilitated antidepressant activity by reducing serotonin reuptake and inducing serotonergic characteristics in noradrenergic neurons.[2] Specifically, miR-16 targets the serotonin reuptake transporter (SERT) in normal conditions. MiR-16 is upregulated in the serotonergic raphe nucleus and downregulated in the noradrenergic locus coeruleus by selective serotonin reuptake inhibitor (SSRI) antidepressants. Apart from reducing serotonin reuptake in serotonergic neurons, Braudry et al. showed that miR-16 induces a serotonergic profile of neuroadrenergic neurons mediated by the neurotropic factor S100β. Finally, depression-like behaviors, induced by exposure to chronic stress, were reduced by infusion of miR-16 into the raphe nucleus or anti-miR-16 into the locus coeruleus, to the same extent as SSRI infusion to the raphe nucleus. These results offer a mechanistic understanding to the mode of action of serotonin-related antidepressant drugs.[2] Moreover, in a follow-up study, an additional role was assigned to miR-16 in the response to SSRI, molecularly mediating increased adult neurons neurogenesis in the hippocampus.[44]

Another study that examined the serotonergic circuitry identified an interaction between miR-135, the SERT, and the serotonin receptor 1a (HTR1A) in mediating stress-induced anxiety and depression and in the response to SSRIs.[35] In this study, miR-135 was identified in a screen, profiling the miRNA fingerprint of serotonergic neurons and bioinformatics. Further experimental work demonstrated that miR-135 regulates two key genes in the serotonergic system, SERT and HTR1A. Furthermore, it was shown that miR-135 levels increased in the serotonergic raphe nucleus of mice following SSRI administration. Mimicking this by increasing miR-135 level specifically in serotonergic neurons using transgenic mice induced behavioral resiliency to chronic stress along with adaption of the serotonergic tone. Furthermore, in human studies' samples it was reported that miR-135 levels are down-regulated in the serotonergic raphe nucleus of suicide victims with MDD and circulating levels of miR-135 were suggested as a potential biomarker for depression and response to treatment.[35] In summary, along with screening studies reporting an alternation in the miRNA's expression profile by antidepressant treatment in animal models[56] and human lymphoblastoid cell lines,[57,58] there are causal reports on a role for both miR-16 and miR-135 in regulating the serotonergic tone and mediating the response to antidepressants.

TRANSGENERATIONAL EFFECTS OF STRESS MEDIATED BY SPERM miRNAs

Recent studies reported intriguing results regarding altered behavioral, physiological, and epigenetic modifications in offspring of a stressed parent that were not directly exposed to stress themselves (reviewed in Ref. 24). There is evidence that this phenomenon is modulated by sperm miRNAs. Initially, a study by Rodgers et al.[62] showed that parental exposure to chronic variable stress altered the nonstressed offspring's HPA axis function, and the transcriptome in the paraventricular nucleus of the hypothalamus (PVN) and bed nucleus of the stria terminalis (BNST). In parallel, sperm miRNAs' expression pattern was altered in stressed mice and the authors suggested that this potentially mediated the modified phenotype observed in their offspring.[62] Similarly, a study by Gapp et al.[23] showed that parental exposure to early life stress altered anxiety, depression, and metabolism of the offspring, along with modifying their brain and sperm transcriptome, including the miRNAs' expression profiles. Furthermore, directly manipulating the oocytes by injecting RNA from sperm of stressed mice to embryos of naïve parents passed the stress-induced phenotype transgenerationally.[23] These findings demonstrate inheritance of acquired stress-induced traits and by that can be considered as evidence for Lamarckism. Epigenetic regulation by miRNAs emerges as a mechanistic molecular mediator of this phenomenon.

CIRCULATING miRNAs AS BIOMARKERS FOR STRESS-RELATED PSYCHOPATHOLOGIES

Blood levels of miRNAs can be utilized as potential biomarkers for diagnosis of stress-related psychiatric disorders and as a means to evaluate the response of patients to treatment. In the circulation, miRNAs are detectable in blood cells, plasma, or serum, either in particular membrane vesicles, exosomes, or bound to protein. Primarily, correlations have been reported between levels of circulating miRNAs and disease states, such as different types of cancers[65] and diabetes.[29] Recently, there is increasing evidence of altered pattern of circulating miRNAs associated with mental disorders.[59] The mechanism by which miRNAs enter the circulation is only partially understood; there is evidence that miRNAs are released from cells that undergo apoptosis, or may be actively secreted from living cells, via lipid structures, such as exosomes or in high-density lipoproteins (e.g., Refs. 40, 72, 76; reviewed in Ref. 14).

Two studies focused on the effects of stress-induced anxiety on the circulating miRNA's profile in healthy controls and characterized the blood miRNome at several time points before and after a stressful exam, and both highlighted upregulation of miR-16.[33,37] The circulating miRNA's profile of psychiatric patients was also tested, for example in patients with PTSD,[82] MDD,[3,4,20,75,79] schizophrenia,[27,43] and autism.[55] More specifically, one study suggests that miR-1202 levels in the circulation could be potentially used for assigning patients to different treatments.[49] Initially it was found that miR-1202 levels are downregulated in PFC postmortem brain tissue from depressed subjects. Next, in a prospective study of depressed patients treated with SSRIs, miR-1202 circulation levels could distinguish between responders and nonresponders, as the miR-1202 level is initially lower and upregulated upon drug administration, only in responders.[49] Furthermore, to complete the mouse and in vitro studies described above, we reported that miR-135a levels are lower in the blood of depressed subjects. In another cohort, miR-135 levels were increased by treatment.[35] Replication studies using bigger cohorts are needed to further validate these findings and identify additional relevant miRNAs. Taken together, psychiatric diagnosis and response to treatment might be reflected in fingerprints of circulating miRNAs that can potentially be used in the clinic for diagnosis and treatment assignment.

Alternately, patient-derived stable cell lines can be generated from noninvasive samples collected from psychiatric patients, for example, lymphoblast lines generated from blood lymphocytes or dermal-derived fibroblasts. These cells would represent the genetic landscape of the patient's disorder, and studying their miRNA expression profile at baseline or in response to treatment may indicate potential miRNAs that are relevant to the disorder of interest. Furthermore, cell lines derived from patients may serve as tools for screening response to drugs as part of a personalized medicinal approach. For example, the miRNAs' expression pattern was analyzed in human lymphoblastoid cell lines derived from healthy subjects treated with SSRIs[58] and in fibroblasts derived from MDD patients and controls.[25,26] Such studies highlighted a panel of miRNAs potentially associated either with MDD or the response to antidepressants that should be further tested for use as biomarkers.

OPEN QUESTIONS AND FUTURE DIRECTIONS

Reviewing the first era of research on the role of stress-related miRNAs in mental disorders raises novel questions requiring further study:

1. Not much is known regarding the molecular mechanism mediating the observed changes in miRNAs' levels in models of psychopathologies. This is not the case for all miRNA domains. In cancer research, for example, altered miRNA promoter methylation patterns can explain some of the modified expression patterns of miRNAs associated with the disease (reviewed in Ref. 19).

2. Comprehensively mapping of the spatial and temporal expression profile of endogenous miRNAs in the developing and adult brain, in specific neuronal cell types, in subcellular fractions (such as the synaptosomes), in health, and in different psychopathological conditions is greatly needed. Such knowledge is crucial for the understanding of the possible behavioral and physiological functions of these miRNAs, as there are reports of great cell type/organism specificity for the miRNome.[48] Some systematic efforts were made to profile miRNAs' expression patterns in human, mouse, and other organisms across different tissue types (http://www.mirz.unibas.ch/smiRNAdb), in the developing human[83] or zebra fish brain,[10] and in part of the adult mouse brain.[32] In addition, the interaction between miRNAs and their target genes has been tested in young mouse brain[12] and in a couple of human brain sites[6] using high-throughput sequencing of RNAs isolated by cross-linking immunoprecipitation (HITS-CLIP) in several brain regions. Extending such studies and mapping mouse and human miRNA's expression profiles equivalently to the mRNA Allen brain atlas (http://www.brain-map.org.) and miRNA–RNA interactions would be very informative.

3. The response to stress and the risk to develop stress-related psychopathologies varies between the sexes; particularly the increased risk of developing anxiety and depression in women. There is evidence that miRNAs can mediate the effects of prenatal stress on brain sexual dimorphic organization,[54] suggesting that more research is needed to explore the role of miRNAs in sex differences in the normal and the pathological response to stress.

4. Another field of interest that is relatively unexplored concerns the possible involvement of miRNAs in mediating individual differences related to resiliency or susceptibility to psychiatric disorders. Meaning, upon exposure to an abnormal stressor, some individuals will display susceptibility and develop psychopathologies while others will remain healthy and are defined as resilient. Such individual differences are mediated both by genetic and epigenetic mechanisms, and indeed there is first evidence from mouse models that miRNA can mediate this phenomena,[16] yet further studies are needed.

CONCLUDING REMARKS

miRNAs are emerging as pivotal modulators of normal and pathological responses to stress. The fact that more than half of the protein-coding genes are predicted to be regulated by miRNAs and that each miRNA can regulate hundreds of different genes positions these molecules as possible master regulators of many cellular processes. Furthermore, an miRNA may regulate the expression of several genes within a specific biological or cellular pathway. These unique features, together with rapidly increasing experimental data, encourage scientists to study the role of miRNAs in the regulation of stress-related psychopathologies. miRNAs have an established role in brain development, this, taken together with the fact that responses to stress have a neurodevelopmental origin and that most psychopathologies are considered to be diseases of multiple genes, provides further support to the possible involvement of miRNAs in different psychiatric disorders.

The accumulating evidence presented in this review suggests that miRNAs may function through several mechanisms to direct stress-related behavior. Some miRNAs' expression levels change following acute and chronic stress challenges, thus facilitating a subsequent change in the expression of target genes, which are putatively needed in order to direct certain behavioral outcomes. On the other hand, miRNAs may serve as "buffers" for keeping protein targets stable and avoid being upregulated to a pathological level following challenge and contribute to restoration of homeostasis. Shedding light on the role of miRNAs in stress-related psychopathologies may potentially enable a better understanding of the molecular pathways of these disorders and possibly promote the much needed development of new therapeutic and diagnostic approaches (Fig. 11.3).

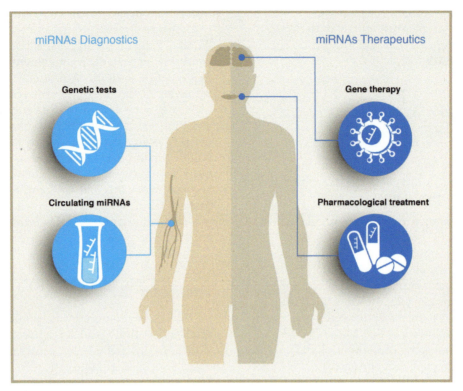

FIGURE 11.3   **Utilizing miRNA biology in psychiatry for diagnostics and therapeutics.** miRNA-related analysis can be used as noninvasive biomarkers in psychiatry for diagnostic purposes. Association studies linking genetic variation with susceptibility for a psychiatric disorder can potentially be used for diagnostics. Single nucleotide polymorphisms (SNPs) or chromosomal modifications, such as DNA duplication or deletion in miRNA genes are examples of such genetic markers. Similarly, the levels of specific miRNAs in, for example, different fractions of the blood, can be used for diagnosis and monitoring of response to treatment. Manipulating specific miRNA levels can potentially be used for therapeutics of psychiatric disorders. Viral vectors such as adeno-associated virus (AAV) can be used as part of gene therapy to overexpress or knockdown specific miRNA. Alternatively, drugs could be developed containing synthetic miRNA mimics or oligos designed to reduce a specific miRNA level, such as anti-miRNA or "miRNA sponges."

References

1. Bartel DP. MicroRNAs: target recognition and regulatory functions. *Cell.* 2009;136(2):215–233.
2. Baudry A, Mouillet-Richard S, Schneider B, Launay JM, Kellermann O. miR-16 targets the serotonin transporter: a new facet for adaptive responses to antidepressants. *Science.* 2010;329(5998):1537–1541.
3. Belzeaux R, Bergon A, Jeanjean V, et al. Responder and non-responder patients exhibit different peripheral transcriptional signatures during major depressive episode. *Transl Psychiatry.* 2012;2:e185. http://dx.doi.org/10.1038/tp.2012.112.
4. Bocchio-Chiavetto L, Maffioletti E, Bettinsoli P, et al. Blood microRNA changes in depressed patients during antidepressant treatment. *Eur Neuropsychopharmacol.* 2013;23(7):602–611. http://dx.doi.org/10.1016/j.euroneuro.2012.06.013.
5. Borchert GM, Lanier W, Davidson BL. RNA polymerase III transcribes human microRNAs. *Nat Struct Mol Biol.* 2006;13(12):1097–1101. http://dx.doi.org/10.1038/nsmb1167.
6. Boudreau RL, Jiang P, Gilmore BL, et al. Transcriptome-wide discovery of microRNA binding sites in human brain. [Research Support, N.I.H., Extramural Research Support, Non-U.S. Gov't] *Neuron.* 2014;81(2):294–305. http://dx.doi.org/10.1016/j.neuron.2013.10.062.
7. Bushati N, Cohen SM. MicroRNAs in neurodegeneration. *Curr Opin Neurobiol.* 2008;18(3):292–296. http://dx.doi.org/10.1016/j.conb.2008.07.001.
8. Carthew RW, Sontheimer EJ. Origins and Mechanisms of miRNAs and siRNAs. *Cell.* 2009;136(4):642–655. http://dx.doi.org/10.1016/j.cell.2009.01.035.
9. Caspi A, Sugden K, Moffitt TE, et al. Influence of life stress on depression: moderation by a polymorphism in the 5-HTT gene. *Science.* 2003;301(5631):386–389.
10. Chen PY, Manninga H, Slanchev K, et al. The developmental miRNA profiles of zebrafish as determined by small RNA cloning. *Genes Dev.* 2005;19(11):1288–1293. http://dx.doi.org/10.1101/gad.1310605.
11. Chendrimada TP, Gregory RI, Kumaraswamy E, et al. TRBP recruits the Dicer complex to Ago2 for microRNA processing and gene silencing. *Nature.* 2005;436(7051):740–744. http://dx.doi.org/10.1038/nature03868.
12. Chi SW, Zang JB, Mele A, Darnell RB. Argonaute HITS-CLIP decodes microRNA-mRNA interaction maps. [Research Support, N.I.H., Extramural Research Support, Non-U.S. Gov't] *Nature.* 2009;460(7254):479–486. http://dx.doi.org/10.1038/nature08170.
13. Corcoran C, Walker E, Huot R, et al. The stress cascade and schizophrenia: etiology and onset. *Schizophr Bull.* 2003;29(4):671–692.
14. Cortez MA, Bueso-Ramos C, Ferdin J, Lopez-Berestein G, Sood AK, Calin GA. MicroRNAs in body fluids–the mix of hormones and biomarkers. *Nat Rev Clin Oncol.* 2011;8(8):467–477. http://dx.doi.org/10.1038/nrclinonc.2011.76.
15. Dias BG, Goodman JV, Ahluwalia R, Easton AE, Andero R, Ressler KJ. Amygdala-dependent fear memory consolidation via miR-34a and Notch signaling. *Neuron.* 2014;83(4):906–918. http://dx.doi.org/10.1016/j.neuron.2014.07.019.

16. Dias C, Feng J, Sun H, et al. beta-catenin mediates stress resilience through Dicer1/microRNA regulation. [Research Support, N.I.H., Extramural Research Support, Non-U.S. Gov't] *Nature.* 2014;516(7529):51–55. http://dx.doi.org/10.1038/nature13976.

17. Eacker SM, Dawson TM, Dawson VL. Understanding microRNAs in neurodegeneration. *Nat Rev Neurosci.* 2009;10(12):837–841. http://dx.doi.org/10.1038/nrn2726.

18. Emde A, Hornstein E. miRNAs at the interface of cellular stress and disease. *EMBO J.* 2014;33(13):1428–1437. http://dx.doi.org/10.15252/embj.201488142.

19. Esteller M. Non-coding RNAs in human disease. *Nat Rev Genet.* 2011;12(12):861–874. http://dx.doi.org/10.1038/nrg3074.

20. Fan HM, Sun XY, Guo W, et al. Differential expression of microRNA in peripheral blood mononuclear cells as specific biomarker for major depressive disorder patients. *J Psychiatric Res.* 2014;59:45–52. http://dx.doi.org/10.1016/j.jpsychires.2014.08.007.

21. Friedman RC, Farh KK, Burge CB, Bartel DP. Most mammalian mRNAs are conserved targets of microRNAs. *Genome Res.* 2009;19(1):92–105.

22. Gao J, Wang WY, Mao YW, et al. A novel pathway regulates memory and plasticity via SIRT1 and miR-134. *Nature.* 2010;466(7310):1105–1109. http://dx.doi.org/10.1038/nature09271.

23. Gapp K, Jawaid A, Sarkies P, et al. Implication of sperm RNAs in transgenerational inheritance of the effects of early trauma in mice. *Nat Neurosci.* 2014;17(5):667–669. http://dx.doi.org/10.1038/nn.3695.

24. Gapp K, von Ziegler L, Tweedie-Cullen RY, Mansuy IM. Early life epigenetic programming and transmission of stress-induced traits in mammals: how and when can environmental factors influence traits and their transgenerational inheritance? *Bioessays.* 2014;36(5):491–502. http://dx.doi.org/10.1002/bies.201300116.

25. Garbett KA, Vereczkei A, Kalman S, et al. Coordinated messenger RNA/microRNA changes in fibroblasts of patients with major depression. *Biol Psychiatry.* 2014. http://dx.doi.org/10.1016/j.biopsych.2014.05.015.

26. Garbett KA, Vereczkei A, Kalman S, et al. Fibroblasts from patients with major depressive disorder show distinct transcriptional response to metabolic stressors. [Research Support, N.I.H., Extramural Research Support, Non-U.S. Gov't] *Transl Psychiatry.* 2015;5:e523. http://dx.doi.org/10.1038/tp.2015.14.

27. Gardiner E, Beveridge NJ, Wu JQ, et al. Imprinted DLK1-DIO3 region of 14q32 defines a schizophrenia-associated miRNA signature in peripheral blood mononuclear cells. *Mol Psychiatry.* 2012;17(8):827–840. http://dx.doi.org/10.1038/mp.2011.78.

28. Griggs EM, Young EJ, Rumbaugh G, Miller CA. MicroRNA-182 regulates amygdala-dependent memory formation. [Research Support, N.I.H., Extramural] *J Neurosci.* 2013;33(4):1734–1740. http://dx.doi.org/10.1523/JNEUROSCI.2873-12.2013.

29. Guay C, Regazzi R. Circulating microRNAs as novel biomarkers for diabetes mellitus. *Nat Rev Endocrinol.* 2013;9(9):513–521. http://dx.doi.org/10.1038/nrendo.2013.86.

30. Haramati S, Navon I, Issler O, et al. MicroRNA as repressors of stress-induced anxiety: the case of amygdalar miR-34. *J Neurosci.* 2011;31(40):14191–14203.

31. Hausser J, Zavolan M. Identification and consequences of miRNA-target interactions - beyond repression of gene expression. *Nat Rev Genet.* 2014;15(9):599–612. http://dx.doi.org/10.1038/nrg3765.

32. He M, Liu Y, Wang X, Zhang MQ, Hannon GJ, Huang ZJ. Cell-type-based analysis of microRNA profiles in the mouse brain. *Neuron.* 2012;73(1):35–48. http://dx.doi.org/10.1016/j.neuron.2011.11.010.

33. Honda M, Kuwano Y, Katsuura-Kamano S, et al. Chronic academic stress increases a group of microRNAs in peripheral blood. *PLoS One.* 2013;8(10):e75960. http://dx.doi.org/10.1371/journal.pone.0075960.

34. Hornstein E, Shomron N. Canalization of development by microRNAs. [Review] *Nat Genet.* 2006;38(suppl.):S20–S24. http://dx.doi.org/10.1038/ng1803.

35. Issler O, Haramati S, Paul ED, et al. MicroRNA 135 is essential for chronic stress resiliency, antidepressant efficacy, and intact serotonergic activity. *Neuron.* 2014;83(2):344–360. http://dx.doi.org/10.1016/j.neuron.2014.05.042.

36. Jovasevic V, Corcoran KA, Leaderbrand K, et al. GABAergic mechanisms regulated by miR-33 encode state-dependent fear. *Nat Neurosci.* 2015;18(9):1265–1271. http://dx.doi.org/10.1038/nn.4084.

37. Katsuura S, Kuwano Y, Yamagishi N, et al. MicroRNAs miR-144/144* and miR-16 in peripheral blood are potential biomarkers for naturalistic stress in healthy Japanese medical students. *Neurosci Lett.* 2012;516(1):79–84. http://dx.doi.org/10.1016/j.neulet.2012.03.062.

38. Kim VN, Han J, Siomi MC. Biogenesis of small RNAs in animals. *Nat Rev Mol Cell Biol.* 2009;10(2):126–139. http://dx.doi.org/10.1038/nrm2632.

39. Kohen R, Dobra A, Tracy JH, Haugen E. Transcriptome profiling of human hippocampus dentate gyrus granule cells in mental illness. *Transl Psychiatry.* 2014;4:e366. http://dx.doi.org/10.1038/tp.2014.9.

40. Kosaka N, Iguchi H, Yoshioka Y, Takeshita F, Matsuki Y, Ochiya T. Secretory mechanisms and intercellular transfer of microRNAs in living cells. *J Biol Chem.* 2010;285(23):17442–17452. http://dx.doi.org/10.1074/jbc.M110.107821.

41. Krol J, Loedige I, Filipowicz W. The widespread regulation of microRNA biogenesis, function and decay. *Nat Rev Genet.* 2010;11(9):597–610. http://dx.doi.org/10.1038/nrg2843.

42. Kye MJ, Neveu P, Lee YS, et al. NMDA mediated contextual conditioning changes miRNA expression. *PLoS One.* 2011;6(9):e24682.

43. Lai CY, Yu SL, Hsieh MH, et al. MicroRNA expression aberration as potential peripheral blood biomarkers for schizophrenia. *PLoS One.* 2011;6(6):e21635. http://dx.doi.org/10.1371/journal.pone.0021635.

44. Launay JM, Mouillet-Richard S, Baudry A, Pietri M, Kellermann O. Raphe-mediated signals control the hippocampal response to SRI antidepressants via miR-16. *Transl Psychiatry.* 2011;1:e56. http://dx.doi.org/10.1038/tp.2011.54.

45. Lee Y, Ahn C, Han J, et al. The nuclear RNase III Drosha initiates microRNA processing. *Nature.* 2003;425(6956):415–419.

46. Leung AK, Sharp PA. MicroRNA functions in stress responses. *Mol Cell.* 2010;40(2):205–215. http://dx.doi.org/10.1016/j.molcel.2010.09.027.

47. Lin Q, Wei W, Coelho CM, et al. The brain-specific microRNA miR-128b regulates the formation of fear-extinction memory. *Nat Neurosci.* 2011;14(9):1115–1117.

48. Londin E, Loher P, Telonis AG, et al. Analysis of 13 cell types reveals evidence for the expression of numerous novel primate- and tissue-specific microRNAs. [Research Support, N.I.H., Extramural Research Support, Non-U.S. Gov't Research Support, U.S. Gov't, Non-P.H.S.] *Proc Natl Acad Sci USA.* 2015;112(10):E1106–E1115. http://dx.doi.org/10.1073/pnas.1420955112.

49. Lopez JP, Lim R, Cruceanu C, et al. miR-1202 is a primate-specific and brain-enriched microRNA involved in major depression and antidepressant treatment. *Nat Med.* 2014;20(7):764–768. http://dx.doi.org/10.1038/nm.3582.

50. Mannironi C, Camon J, De Vito F, et al. Acute stress alters amygdala microRNA miR-135a and miR-124 expression: inferences for corticosteroid dependent stress response. *PLoS One.* 2013;8(9):e73385. http://dx.doi.org/10.1371/journal.pone.0073385.

51. Maussion G, Yang J, Yerko V, et al. Regulation of a truncated form of tropomyosin-related kinase B (TrkB) by Hsa-miR-185* in frontal cortex of suicide completers. *PLoS One.* 2012;7(6):e39301. http://dx.doi.org/10.1371/journal.pone.0039301.

52. Meerson A, Cacheaux L, Goosens KA, Sapolsky RM, Soreq H, Kaufer D. Changes in brain microRNAs contribute to cholinergic stress reactions. *J Mol Neurosci.* 2010;40(1–2):47–55. http://dx.doi.org/10.1007/s12031-009-9252-1.

53. Mendell JT, Olson EN. MicroRNAs in stress signaling and human disease. *Cell*. 2012;148(6):1172–1187. http://dx.doi.org/10.1016/j.cell.2012.02.005.

54. Morgan CP, Bale TL. Early prenatal stress epigenetically programs dysmasculinization in second-generation offspring via the paternal lineage. *J Neurosci*. 2011;31(33):11748–11755. http://dx.doi.org/10.1523/JNEUROSCI.1887-11.2011.

55. Mundalil Vasu M, Anitha A, Thanseem I, et al. Serum microRNA profiles in children with autism. *Mol Autism*. 2014;5:40. http://dx.doi.org/10.1186/2040-2392-5-40.

56. O'Connor RM, Grenham S, Dinan TG, Cryan JF. microRNAs as novel antidepressant targets: converging effects of ketamine and electroconvulsive shock therapy in the rat hippocampus. *Int J Neuropsychopharmacol*. 2013;16(8):1885–1892. http://dx.doi.org/10.1017/S1461145713000448.

57. Oved K, Morag A, Pasmanik-Chor M, et al. Genome-wide miRNA expression profiling of human lymphoblastoid cell lines identifies tentative SSRI antidepressant response biomarkers. *Pharmacogenomics*. 2012;13(10):1129–1139. http://dx.doi.org/10.2217/pgs.12.93.

58. Oved K, Morag A, Pasmanik-Chor M, Rehavi M, Shomron N, Gurwitz D. Genome-wide expression profiling of human lymphoblastoid cell lines implicates integrin beta-3 in the mode of action of antidepressants. *Transl Psychiatry*. 2013;3:e313. http://dx.doi.org/10.1038/tp.2013.86.

59. Rao P, Benito E, Fischer A. MicroRNAs as biomarkers for CNS disease. *Front Mol Neurosci*. 2013;6:39. http://dx.doi.org/10.3389/fnmol.2013.00039.

60. Rinaldi A, Vincenti S, De Vito F, et al. Stress induces region specific alterations in microRNAs expression in mice. *Behav Brain Res*. 2009;208(1):265–269.

61. Rinaldi A, Vincenti S, De Vito F, et al. Stress induces region specific alterations in microRNAs expression in mice. *Behav Brain Res*. 2010;208(1):265–269. http://dx.doi.org/10.1016/j.bbr.2009.11.012.

62. Rodgers AB, Morgan CP, Bronson SL, Revello S, Bale TL. Paternal stress exposure alters sperm microRNA content and reprograms offspring HPA stress axis regulation. *J Neurosci*. 2013;33(21):9003–9012. http://dx.doi.org/10.1523/JNEUROSCI.0914-13.2013.

63. Salta E, De Strooper B. Non-coding RNAs with essential roles in neurodegenerative disorders. *Lancet Neurol*. 2012;11(2):189–200. http://dx.doi.org/10.1016/S1474-4422(11)70286-1.

64. Schratt G. microRNAs at the synapse. *Nat Rev Neurosci*. 2009;10(12):842–849.

65. Schwarzenbach H, Hoon DS, Pantel K. Cell-free nucleic acids as biomarkers in cancer patients. *Nat Rev Cancer*. 2011;11(6):426–437. http://dx.doi.org/10.1038/nrc3066.

66. Serretti A, Kato M, De Ronchi D, Kinoshita T. Meta-analysis of serotonin transporter gene promoter polymorphism (5-HTTLPR) association with selective serotonin reuptake inhibitor efficacy in depressed patients. *Mol Psychiatry*. 2007;12(3):247–257. http://dx.doi.org/10.1038/sj.mp.4001926.

67. Smalheiser NR, Lugli G, Rizavi HS, Torvik VI, Turecki G, Dwivedi Y. MicroRNA expression is down-regulated and reorganized in prefrontal cortex of depressed suicide subjects. *PLoS One*. 2012;7(3):e33201.

68. Smalheiser NR, Lugli G, Rizavi HS, et al. MicroRNA expression in rat brain exposed to repeated inescapable shock: differential alterations in learned helplessness vs. non-learned helplessness. *Int J Neuropsychopharmacol*. 2011:1–11.

69. Smalheiser NR, Lugli G, Zhang H, Rizavi H, Cook EH, Dwivedi Y. Expression of microRNAs and other small RNAs in prefrontal cortex in schizophrenia, bipolar disorder and depressed subjects. *PLoS One*. 2014;9(1):e86469. http://dx.doi.org/10.1371/journal.pone.0086469.

70. Sun K, Lai EC. Adult-specific functions of animal microRNAs. *Nat Rev Genet*. 2013;14(8):535–548. http://dx.doi.org/10.1038/nrg3471.

71. Uchida S, Hara K, Kobayashi A, et al. Early life stress enhances behavioral vulnerability to stress through the activation of REST4-mediated gene transcription in the medial prefrontal cortex of rodents. *J Neurosci*. 2010;30(45):15007–15018. http://dx.doi.org/10.1523/JNEUROSCI.1436-10.2010.

72. Vickers KC, Palmisano BT, Shoucri BM, Shamburek RD, Remaley AT. MicroRNAs are transported in plasma and delivered to recipient cells by high-density lipoproteins. *Nat Cell Biol*. 2011;13(4):423–433. http://dx.doi.org/10.1038/ncb2210.

73. Vo NK, Cambronne XA, Goodman RH. MicroRNA pathways in neural development and plasticity. *Curr Opin Neurobiol*. 2010;20(4):457–465. http://dx.doi.org/10.1016/j.conb.2010.04.002.

74. Volk N, Paul ED, Haramati S, et al. MicroRNA-19b associates with Ago2 in the amygdala following chronic stress and regulates the adrenergic receptor beta 1. *J Neurosci*. 2014;34(45):15070–15082. http://dx.doi.org/10.1523/JNEUROSCI.0855-14.2014.

75. Wan Y, Liu Y, Wang X, et al. Identification of differential microRNAs in cerebrospinal fluid and serum of patients with major depressive disorder. [Research Support, Non-U.S. Gov't] *PLoS One*. 2015;10(3):e0121975. http://dx.doi.org/10.1371/journal.pone.0121975.

76. Wang K, Zhang S, Weber J, Baxter D, Galas DJ. Export of microRNAs and microRNA-protective protein by mammalian cells. *Nucleic Acids Res*. 2010;38(20):7248–7259. http://dx.doi.org/10.1093/nar/gkq601.

77. Wang RY, Phang RZ, Hsu PH, Wang WH, Huang HT, Liu IY. In vivo knockdown of hippocampal miR-132 expression impairs memory acquisition of trace fear conditioning. *Hippocampus*. 2013;23(7):625–633. http://dx.doi.org/10.1002/hipo.22123.

78. Wang W, Kwon EJ, Tsai LH. MicroRNAs in learning, memory, and neurological diseases. *Learn Mem*. 2012;19(9):359–368. http://dx.doi.org/10.1101/lm.026492.112.

79. Wang X, Sundquist K, Hedelius A, Palmer K, Memon AA, Sundquist J. Circulating microRNA-144-5p is associated with depressive disorders. *Clin Epigenetics*. 2015;7(1):69. http://dx.doi.org/10.1186/s13148-015-0099-8.

80. Wright P, Gill M, Murray RM. Schizophrenia: genetics and the maternal immune response to viral infection. *Am J Med Genet*. 1993;48(1):40–46. http://dx.doi.org/10.1002/ajmg.1320480110.

81. Yi R, Qin Y, Macara IG, Cullen BR. Exportin-5 mediates the nuclear export of pre-microRNAs and short hairpin RNAs. *Genes Dev*. 2003;17(24):3011–3016.

82. Zhou J, Nagarkatti P, Zhong Y, Ginsberg JP, Singh NP, Zhang J, et al. Dysregulation in microRNA expression is associated with alterations in immune functions in combat veterans with post-traumatic stress disorder. *PLoS One*. 2014;9(4):e94075. http://dx.doi.org/10.1371/journal.pone.0094075.

83. Ziats MN, Rennert OM. Identification of differentially expressed microRNAs across the developing human brain. *Mol Psychiatry*. 2014;19(7):848–852. http://dx.doi.org/10.1038/mp.2013.93.

84. Zoghbi HY. Postnatal neurodevelopmental disorders: meeting at the synapse? *Science*. 2003;302(5646):826–830. http://dx.doi.org/10.1126/science.1089071.

12

Stress Reactions Orchestrate Parasympathetic Functioning and Inflammation Under Diverse Cholinergic Manipulations

N. Mishra, H. Soreq

The Hebrew University of Jerusalem, Jerusalem, Israel

Abstract

Stress reactions impair parasympathetic functioning, elevate inflammation, and modify brain microRNAs (miRNA) via cholinergic genes and their targeting miRNAs (CholinomiRs), and newer technologies may provide novel pharmaceutic avenues to these issues. Here, we describe stress-inducible changes in the expression of cholinergic genes (muscarinic, nicotinic, cholinesterases, etc.), which result in rapid increases of cholinergic signaling, and demonstrate how surveillance by CholinomiRs provides fine-tuning of these changes in cholinergic signaling, retrieving homeostasis. This chapter focuses on miRNA-132 (an evolutionarily conserved acetylcholinesterase (AChE) targeting miRNA), and its role in stress and inflammation, both of which upregulate miRNA-132 in the mammalian brain, potentiate the cholinergic tone, and suppress inflammation by downregulating AChE. Future technologies, including the use of chemically protected antisense oligonucleotides (e.g., anti-miRNA-132) may open new ways to suppress the levels of miRNAs involved in stress and inflammation. The delicate balance between stress and inflammation and ways to retrieve it thus merits renewed discussion.

INTRODUCTION

Stress is the body's response to external demands or threats, but at excessive levels it may cause major damage to brain and body.[6,36] Stress can compromise the immune system[37] through altered cholinergic gene expression[11,21] leading to behavioral abnormalities, such as anxiety, posttraumatic stress disorder, and depression.[38] The parasympathetic nervous system functions to maintain homeostasis through multiple receptors, enzymes, and transporters, with the primary neurotransmitter and neuromodulator acetylcholine (ACh). Imbalanced ACh levels may give rise to many diseases also observed under conditions of various cholinergic insults, like smoking, myasthenia gravis, and organophosphate poisoning that may expose both brain and body to stress and inflammation insults.[27] Therefore, parasympathetic stresses, psychological insults, and inflammation are intertwined via a vicious cycle, with ACh being the common link.

MicroRNAs (miRNAs) are small noncoding RNAs that regulate posttranscriptional functioning of coding transcripts across different tissues. Many researchers including us have shown the importance of miR-NAs in neuronal-immune process,[4,29,40] cholinergic-inflammation signaling,[25] the neuronal consequences of stress,[19,35] Alzheimer disease (AD),[14] and brain-to-body communicator of inflammation.[18,34] These developments call for revisiting the current knowhow on these aspects. In this chapter, we highlight the cholinergic attributes of stress and inflammation, their miRNA regulation, and implication of newer RNA technologies in these issues.

BODY TO BRAIN SIGNALING OF STRESS AND INFLAMMATION

Stress and inflammation are well-known perturbations that impair parasympathetic functioning (or vice versa). For example, psychological stress increases the susceptibility to inflammation and alters cholinergic gene expression.[21] Similarly, bacterial or viral infection is often associated with "sickness behavior" and increases in ACh signaling.[43] Many studies were aimed at exploring the molecular mechanisms through which stress and inflammation relate to each other and at outlining the underlying neural and genetic mechanisms.[37] Studies of the neural mechanism of stress indicate that social stress, similar to parallel stresses that were addressed earlier in human evolution, prepares the body in advance from forthcoming risks like physical injury, pathogens, and social–environmental threats.[47] These experiences are processed in higher brain regions (the anterior insula and dorsal anterior cingulate cortex) and transmit this information to lower brain regions (such as the hypothalamus and brainstem) that simultaneously modulate three inflammatory pathways: (1) the hypothalamic-pituitary-adrenal axis, (2) the sympathetic nervous system, and (3) the efferent vagus nerve. These pathways release glucocorticoids, epinephrine, norepinephrine, and the ACh neurotransmitter. The glucocorticoids and ACh both act as antiinflammatory mediators, whereas epinephrine and norepinephrine inversely upregulate inflammatory gene expression. Specifically, the transcription factor NFkB mediates upregulation of proinflammatory gene expression, increasing the levels of cytokines that induce depression-like behavior; and ACh interacts with the α7 nicotinic AChR on the surface of hematopoietic cells to block this inflammatory reaction. As a consequence, ACh-regulated antiinflammatory mechanisms largely depend on the α7 nicotinic ACh receptor (α7nAChR), which inhibits the proinflammatory pathway of NF-κB nuclear translocation and suppresses cytokine release.[43] This further implies that reduced destruction of ACh may potentiate the blockade of inflammation.

At the psychology level, both early and more recent reports described direct connection between increased cholinergic signaling and psychological stress.[5,10,11,41] Stress-induced glucocorticoids release triggers ACh release in the hippocampus, prefrontal cortex, and the nucleus accumbens of rats.[43] To get further insights on the role of ACh in the etiology of depression, Mineur and coworkers employed viral-mediated knockdown of acetylcholinesterase (AChE) in the hippocampus. Both physostigmine treatment and AChE knockdown by shRNAs increased depression-like behavior which could be reversed by fluoxetine (an antidepressant drug),[22] which also elevates the production of the "readthrough" AChE-R variant. In addition, in a separate study fluoxetine was investigated for protection from excessive ACh signaling which suggested that fluoxetine acts via inhibiting the muscarinic ACh receptor.[44] A recent finding associated the nicotinic receptor as well with stress and alcoholism. The authors observed coexpression of the α4 and α7 nicotinic receptors with glucocorticoid receptors that together may induce both stress and stress-related alcoholism.[8] Thus, both elevated receptor functioning and suppressed ACh degradation support stress responses and reduce inflammation.

We investigated the involvement of the Ach-hydrolyzing enzymes AChE and butyrylcholinesterase (BChE) in inflammation. Our studies demonstrated a correlation between circulation cholinesterase activities and inflammation, both in the brain and in peripheral lymphocytes.[27] Correspondingly, carriers of polymorphisms in the AChE and BChE genes show altered association with inflammatory biomarkers. In addition, the different AChE splice variants exert distinct impacts on inflammatory reactions,[20,39] probably since the AChE-R protein is soluble and monomeric, whereas the "synaptic" AChE-S variant is membrane-bound and slow to be modified. Globally increased blockade of cholinesterase (ChEs) activities (e.g., through pharmaceutic or insecticide blockers) reduces the release of proinflammatory cytokines by macrophages, which in turn stimulates the cholinergic antiinflammatory pathway.

Importantly, multiple cholinergic genes (including cholinesterases, nicotinic, and muscarinic receptors) are involved in the cholinergic pathway and are profoundly modulated under stress and inflammation at several interactive levels. Those include epigenetic changes of the corresponding promoters, such as the AChE promoter and alternative splicing,[21] which expands the diversity of stress-promoted cholinergic gene involvement. These complexities were studied more conspicuously for one cholinergic gene, AChE at the level of alternative splicing,[20] epigenetic modifications of histone methylation,[33] and posttranscriptional modifications.[25] Apart from α7nAChR, the toll-like receptor TLR9 also regulates NFκB-mediated proinflammatory cytokine secretion and minimizes stress-associated behavioral consequences.[25,47] Thus, both the innate and the adaptive immune systems communicate with the cholinergic signaling pathway for modulating immune reactions. Checks and balances of cholinergic signaling are hence multileveled.

MICRORNAs-DRIVEN CHOLINERGIC MODULATION OF INFLAMMATION

miRNAs are short RNA regulators that target mRNA transcripts by interacting with short "seed" sequences, often in the 3'-untranslated regions of coding transcripts, leading to suppressed protein levels via translational arrest and/or induction of mRNA degradation. Single miRNAs may silence multiple mRNAs, and many different miRNAs can target one mRNA, thereby controlling entire biological pathways by a complex network of interactions that also affect downstream sequences. The functional role of miRNAs depends on the cell type, developmental stage, environmental factors, and evolutionary conservation of the miRNA-target pathway.[2] High-throughput microarrays, deep sequencing, and advanced bioinformatics studies show remarkable changes in the miRNA profiles with inflammation and stress. Here, we highlight the involvement of CholinomiRs (miRNAs targeting cholinergic genes) in stress disorders, such as inflammation, anxiety, and neurodegeneration.

The major neuronal CholinomiR, miRNA-132 provides a remarkable example of an evolutionarily conserved CholinomiR that targets AChE. This miRNA can attenuate inflammation by reducing AChE concentration, serving as regulator of the brain-to-body

resolution of inflammation.[34] Moreover, miRNA-132 levels increase sharply in the mouse hippocampus under acute psychological stress, a phenomenon that associates with decreases in AChE activity and increases in ACh signaling. This raised the question of which of the many targets of miRNA-132 was causally involved. To solve this question, we used viral infection in vivo to suppress AChE activity in the murine hippocampus, essentially mimicking the impact of miRNA-132 increases with regards to this target alone. Notably, suppression of hippocampal AChE activity sufficed to prevent both miRNA-132 increase and the behavioral malfunctioning following stressful insults, indicating that in this particular context, AChE is the causally involved target of miRNA-132.[35] Next, we addressed the question of inflammatory responses and evolutionary conservation of miRNA-132. Importantly, we found that miRNA-132 expression also increases in postoperative inflamed intestinal sections from patients with acute inflammatory bowel disease (IBD). Patient samples showed conspicuously higher levels of miRNA-132 than controls, and in specific patients, the inflamed intestinal sections showed much higher levels of miRNA-132 compared to the noninflamed edges from the same tissues. In addition, we observed a negative correlation between the intestinal AChE activities and colonic miRNA-132 levels, suggesting new homeostatic role of miRNA-132 in IBD, where it indirectly stimulates cholinergic signaling (decreasing AChE level) and promotes antiinflammatory processes.[18] These and other inflammatory conditions with elevated miRNA-132 expression and its relevant validated and/or predicted targets are summarized in Table 12.1.

That a single miRNA may target multiple mRNAs or vice versa makes the biological functioning of miRNAs more intriguing and complicates studies of their roles. Our early study addressed the association of the impact of miRNA activities with alternative splicing. In acutely stressed rats, we found increases in the expression of the amygdala miRNA-183 and miRNA-134, both of which predictably target the stress-inducible splicing factor SC35.[19] This splicing factor promotes the splicing of AChE mRNA from the synapse-associated isoform AChE-S to the normally rare, stress-induced soluble AChE-R protein,[20] possibly explaining part of the rapid stress-related changes in cholinergic signaling. Furthermore, the miRNA genes are considerably smaller than those of their coding mRNA targets, suggesting that the target genes are more readily susceptible for mutagenesis than the miRNA controllers of those genes, simply because these target genes are 100-fold larger. This raises an additional question, namely, what are the consequences of single nucleotide changes in the 3'-untranslated regions of the target genes of such miRNAs. Such mutations or polymorphisms would, until

TABLE 12.1   Involvement of miRNA-132 in Different Inflammatory Conditions and Its Relevant Targets

Inflammation	Tissue/Disease	Target(s)	References
Lipopolysaccharide	Bone marrow and splenocytes	Acetylcholinesterase	34
Lipopolysaccharide	Alveolar macrophages	Acetylcholinesterase	17
Inflammatory bowel disease	Intestinal tissue	Acetylcholinesterase	18
IgE-mediated activation	Murine and human mast cells	Heparin-binding EGF-like growth factor	24
Wound repair	Human skin wounds	Heparin-binding EGF-like growth factor	15
IL-12 stimulation	Human natural killer cells	Signal transducer and activator of transcription 4	9
Viral infection	Primary human lymphatic endothelial cells	P300	13
Multiple sclerosis	Human B cells	Sirtuin-1	23
Atherosclerosis	Human umbilical vein endothelial cells	Sirtuin-1	45
Myeloid-related protein-induced astrocyte activation	Astrocytes	Interleukin-1 receptor-associated kinase 4	12
Peptidoglycan	Primary macrophages	Interleukin-1 receptor-associated kinase 4	26

EGF, epidermal growth factor; *IgE*, Immunoglobulin E; *IL*, Interleukin; *P300*, protein-300.

recently, be considered biologically irrelevant as these genomic changes do not affect the protein product of those target genes; however, impairments in the miRNA interactions of such genes may be more profoundly effective as they can change the levels of the protein product due to weakened miRNA suppression. Moreover, leaving free miRNA chains "unoccupied" may cause further changes in other targets of this miRNA, which would be more excessively suppressed than under normal circumstances. The result of such imbalance could hence be expressed as changes in multiple coding genes.

The cholinergic system offers an excellent example of such a scenario. We have shown that the primate-specific miRNA-608 targets the 3'-untranslated region of the human AChE mRNA transcript. A common SNP impairs the interaction of miRNA-608 with AChE, and we employed Surface Plasmon Resonance to demonstrate that the rare rs17228616 AChE allele interacts with miR-608 at 15-fold weaker binding efficacy than the common allele. This difference is expressed as a significant 40% elevation in brain AChE activity of carriers of the rare allele that is accompanied by downstream reductions in the levels of other miRNA-608 targets [e.g., decrease in the transcript levels of the Rho GTPase CDC42 and the proinflammatory cytokine interleukin 6 (IL6)]. Analysis of over 600 genotyped individuals demonstrated an association of the rare allele with elevated trait, but not state anxiety, compatible with the assumption that this is an inherited phenotype; carriers of the rare allele further showed elevated blood pressure and plasma C-reactive protein, reflecting inflammation.[7] In addition,

miRNA-132 changed in human evolution in conjunction with its target genes.[1] These two studies together suggest that both inherited changes in miRNA functioning and in SNPs and stress-inducible alternative splicing modulation may alter the action of miRNAs, which increases the complexity of their action.

NEW TECHNOLOGIES: RNA-SEQUENCE PROFILING AND CHEMICALLY PROTECTED ANTISENSE OLIGONUCLEOTIDES

The current understanding of cholinergic participation in stress and inflammation requires an elaborate involvement of brain-to-body communication, as is reflected in ample interference or enhancement studies. Such modifications may be established at the cellular level, or by engineering diverse animal models with cholinergic perturbations, where pharmacological manipulations can demonstrate the impact of such genomic changes. In addition, gene profiling studies may complement these efforts to find cholinergic effects. To further understand the role of miRNAs in the stress-induced posttranscriptional changes, improved technologies are required which can expedite the studies of the cholinergic regulators. In the following we describe the corresponding miRNA technologies, applications, and clinical implications.

RNA sequencing is by far the leading technology in miRNA profiling studies. Different miRNA profiles reflect imbalanced functioning in many cases of comparison between health and disease and can enable studies of the role of miRNAs as predictors and biomarkers of various impairments in key biological processes. miRNA profiling is widely used for exploring mechanisms of gene regulation (cellular development, identifying novel miRNAs, elucidating miRNA–mRNA interactions, and miRNA–phenotype interactions). At the translational level, this is a key approach for identifying miRNA biomarkers in biological fluids as noninvasive diagnostic tools. Currently, the following techniques are often used[30] for studying the functions of miRNAs:

1. Quantitative reverse transcription PCR, a specific method used for validating the quantification of miRNA differences;
2. MicroRNA microarrays, used for high-throughput miRNA identification but with low specificity, no ability to identify novel miRNAs, and the limitation that only fold change differences between health and disease, but not the absolute expression levels of the tested miRNAs can be detected;
3. RNA sequencing, a high-throughput sequencing platform, which can efficiently distinguish between miRNAs, the complementary strand isomiRNAs, and novel miRNAs but depends on the quality of RNA, which may be an issue with human samples;
4. Smaller scale next-generation RNA sequencing, which may be less reliable in terms of depth and validity of the observed differences and requires different validation tests.
5. Single-cell RNA sequencing studies, which can determine absolute quantities of the identified miRNAs, but is expensive, and not widely used except for basic research.

These technologies further enhance the interest in miRNAs therapeutics.[16] To date, three main miRNA therapeutic approaches have been applied:

1. Expression vectors (miRNA sponges) that sequester endogenous miRNAs and allow expression of the target mRNA but are limited to transgenic animals;
2. Small-molecule inhibitors that specifically inhibit miRNAs and are limited by their high EC_{50} values, and the lack of direct target information;
3. Antisense oligonucleotides (ASOs), a technology that has received much attention recently and reached the level of clinical use; ASOs bind with high complementarity to the nucleic acids–protein RISC complexes where miRNAs perform their suppression activities (miRISCs), thereby blocking their binding to endogenous mRNA targets.

Anti-miRNA therapeutics is a developing biotechnology field, yet it involves many challenges that need to be solved before it can be established as a therapeutic strategy. A major difficulty with the use of antisense oligonucleotide blockers of miRNAs is their susceptibility to rapid in vivo degradation by nucleases. To avoid this demise, a variety of chemical modifications have been incorporated into anti-miRNA (AntagomiR) oligonucleotides. A most widely used modification is that of locked nucleic acid (LNA) that offers both enhanced binding affinity to miRNAs and satisfactory nuclease resistance. LNAs are further employed for in vivo delivery, and they may be either conjugated with cholesterol or delivered in liposome carriers, or antibody complexes or polymer-based nanoparticles, all of which may increase the functional efficiency of these small size oligonucleotides.[46] Key concerns regard those miRNAs with similar seed regions, which may interfere with each other; in addition, specific AntagomiRs and carrier proteins may lead to rejection by the immune system, and the delivery of such AntagomiRs to the tissue of interest may fail. The delivery of AntagomiRs to the brain is particularly complicated because of the hurdle of crossing the blood–brain barrier.[32] For research purposes, a number of approaches are devised to overcome this problem: (1) direct infection into the brain with viral vectors that may be modified for better tropism to neuronal cells[3]; (2) direct introduction of antisense oligonucleotides by intracerebroventricular (icv) injection. However, these approaches are mainly used in basic research, and each of them has its own limitations. In addition, these methods are unsuitable for use for the treatment of neurodegenerative diseases, such as Huntington, Creutzfeldt-Jakob disease, or AD, which are all diagnosed when many of the relevant cells have died and involve general failure in miRNA metabolism.[14] Novel solutions are hence sought for avoiding these technical limitations.

Acute life-risking crises enable attempts for using oligonucleotide therapeutics; in this context, an interesting putative application of miRNAs as therapeutic targets involves cases of nervous system poisoning, such as organophosphate poisoning.[38] We and others have validated the use of anti-miR-132 (AM132) against such toxicity[28] and showed suppressed inflammation as an outcome of such treatment,[34] which opens new possibilities for AM132 use in therapy. Clinical development of miRNA therapeutics is thus an area of intense interest to pharmaceutical purposes, as miRNA-targeting oligonucleotides have several advantages over traditional small-molecule drugs, most notably thanks to the ease with which oligonucleotides can be chemically modified to enhance their PK/PD profiles and due to the ability of miRNAs to target multiple genes simultaneously. Currently, miRNA-targeting therapies in preclinical and clinical development include use of antimiR therapies against various systemic disorders, e.g., miR-122[42] for liver disease, miR-208[31] for heart failure, etc. The new pharmaceutical use of oligonucleotide therapeutics for miRNA manipulations may therefore emerge first in the case of lethal risks with no alternative solutions.

SUMMARY AND FUTURE PROSPECTS

Stressful insults involve impaired cholinergic signaling and potentiate the cholinergic antiinflammatory pathway, which compromises the immune system through multi-leveled feedback mechanisms modulating the expanded family of cholinergic genes involved in various steps of ACh signaling. The relevant modifications span epigenetic changes, alternative splicing, and miRNA-mediated alterations. This newly emerged complexity of cholinergic signaling regulates brain–body communication in association with stress and inflammation and opens new avenues for cholinergic research that may span behavior, immune processes, and metabolic function and includes a novel pharmaceutics strategy by AntagomiR agents.

Acknowledgments

This research was supported by the German Research Foundation Trilateral Cooperation Program and The German Israeli Foundation Grant No. 1093–32.2/2010. N.M. was supported by The Planning and Budgeting Committee (PBC) and The Edmond and Lily Safra Center of Brain Science (ELSC) postdoctoral fellowships.

References

1. Barbash S, Shifman S, Soreq H. Global coevolution of human microRNAs and their target genes. *Mol Biol Evol.* 2014;31(5):1237–1247.
2. Bartel DP. MicroRNAs: target recognition and regulatory functions. *Cell.* 2009;136(2):215–233.
3. Berson A, Barbash S, Shaltiel G, et al. Cholinergic-associated loss of hnRNP-A/B in Alzheimer's disease impairs cortical splicing and cognitive function in mice. *EMBO Mol Med.* 2012;4(8):730–742.
4. Brain O, Owens BM, Pichulik T, et al. The intracellular sensor NOD2 induces microRNA-29 expression in human dendritic cells to limit IL-23 release. *Immunity.* 2013;39(3):521–536.
5. Charles HC, Lazeyras F, Krishnan KRR, Boyko OB, Payne M, Moore D. Brain choline in depression: in vivo detection of potential pharmacodynamic effects of antidepressant therapy using hydrogen localized spectroscopy. *Prog Neuropsychopharmacol Biol Psychiatry.* 1994;18(7):1121–1127.
6. Cohen S, Janicki-Deverts D, Doyle WJ, et al. Chronic stress, glucocorticoid receptor resistance, inflammation, and disease risk. *Proc Natl Acad Sci USA.* 2012;109(16):5995–5999.
7. Hanin G, Shenhar-Tsarfaty S, Yayon N, et al. Competing targets of microRNA-608 affect anxiety and hypertension. *Hum Mol Genet.* 2014. ddu170.
8. Holgate JY, Bartlett SE. Early life stress, nicotinic acetylcholine receptors and alcohol use disorders. *Brain Sci.* 2015;5(3):258–274.
9. Huang Y, Lei Y, Zhang H, Hou L, Zhang M, Dayton AI. MicroRNA regulation of STAT4 protein expression: rapid and sensitive modulation of IL-12 signaling in human natural killer cells. *Blood.* 2011;118(26):6793–6802.
10. Janowsky D, Davis J, El-Yousef MK, Sekerke HJ. A cholinergic-adrenergic hypothesis of mania and depression. *Lancet.* 1972; 300(7778):632–635.
11. Kaufer D, Friedman A, Seidman S, Soreq H. Acute stress facilitates long-lasting changes in cholinergic gene expression. *Nature.* 1998;393(6683):373–377.
12. Kong H, et al. The effect of miR-132, miR-146a, and miR-155 on MRP8/TLR4-induced astrocyte-related inflammation. *J Mol Neurosci.* 2015;57(1):28–37.
13. Lagos D, et al. miR-132 regulates antiviral innate immunity through suppression of the p300 transcriptional co-activator. *Nat Cell Biol.* 2010;12(5):513–519.
14. Lau P, Bossers K, Salta E, et al. Alteration of the microRNA network during the progression of Alzheimer's disease. *EMBO Mol Med.* 2013;5(10):1613–1634.
15. Li D, Wang A, Liu X, et al. MicroRNA-132 enhances transition from inflammation to proliferation during wound healing. *J Clin Invest.* 2015;125(8):3008–3026. http://dx.doi.org/10.1172/jci79052.
16. Li Z, Rana TM. Therapeutic targeting of microRNAs: current status and future challenges. *Nat Rev Drug Discov.* 2014;13(8): 622–638.
17. Liu F, et al. miR-132 inhibits lipopolysaccharide-induced inflammation in alveolar macrophages by the cholinergic anti-inflammatory pathway. *Exp Lung Res.* 2015;41(5):261–269.
18. Maharshak N, Shenhar-Tsarfaty S, Aroyo N, et al. MicroRNA-132 modulates cholinergic signaling and inflammation in human inflammatory bowel disease. *Inflamm Bowel Dis.* 2013;19(7): 1346–1353. http://dx.doi.org/10.1097/MIB.0b013e318281f47d.
19. Meerson A, Cacheaux L, Goosens KA, Sapolsky RM, Soreq H, Kaufer D. Changes in brain MicroRNAs contribute to cholinergic stress reactions. *J Mol Neurosci.* 2010;40(1–2):47–55.
20. Meshorer E, Soreq H. Virtues and woes of AChE alternative splicing in stress-related neuropathologies. *Trends Neurosci.* 2006;29(4):216–224.
21. Meshorer E, Erb C, Gazit R, et al. Alternative splicing and neuritic mRNA translocation under long-term neuronal hypersensitivity. *Science.* 2002;295(5554):508–512.
22. Mineur YS, Obayemi A, Wigestrand MB, et al. Cholinergic signaling in the hippocampus regulates social stress resilience and anxiety-and depression-like behavior. *Proc Natl Acad Sci USA.* 2013;110(9):3573–3578.
23. Miyazaki Y, et al. A novel microRNA-132-surtuin-1 axis underlies aberrant B-cell cytokine regulation in patients with relapsing-remitting multiple sclerosis. *PLoS One.* 2014;9(8):e105421.
24. Molnár V, Érsek B, Wiener Z, et al. MicroRNA-132 targets HB-EGF upon IgE-mediated activation in murine and human mast cells. *Cell Mol Life Sci.* 2012;69(5):793–808.
25. Nadorp B, Soreq H. Predicted overlapping microRNA regulators of acetylcholine packaging and degradation in neuroinflammation-related disorders. *Front Mol Neurosci.* 2014;7.
26. Nahid MA, Yao B, Dominguez-Gutierrez PR, Kesavalu L, Satoh M, Chan EK. Regulation of TLR2-mediated tolerance and cross-tolerance through IRAK4 modulation by miR-132 and miR-212. *J Immunol.* 2013;190(3):1250–1263.
27. Ofek K, Soreq H. Cholinergic involvement and manipulation approaches in multiple system disorders. *Chem Biol Interact.* 2013;203(1):113–119.
28. Ofek K, Hanin G, Gilboa-Geffen A, et al. Prophylactic oligonucleotide-mediated enhancement of host acetylcholinesterase protects from organophosphate poisoning. In: *Paper Presented at the Defense Science Research Conference and Expo (DSR).* ; 2011.
29. Ponomarev ED, Veremeyko T, Barteneva N, Krichevsky AM, Weiner HL. MicroRNA-124 promotes microglia quiescence and suppresses EAE by deactivating macrophages via the C/EBP-α-PU.1 pathway. *Nat Med.* 2011;17(1):64–70.
30. Pritchard CC, Cheng HH, Tewari M. MicroRNA profiling: approaches and considerations. *Nat Rev Genet.* 2012;13(5):358–369.
31. Rodino-Klapac LR. MicroRNA therapy of heart failure. *Heart Metab.* 2014;31.

32. Greenberg DS, Soreq H. MicroRNA therapeutics in neurological disease. *Curr Pharm Des*. 2014;20(38):6022–6027.

33. Sailaja BS, Cohen-Carmon D, Zimmerman G, Soreq H, Meshorer E. Stress-induced epigenetic transcriptional memory of acetylcholinesterase by HDAC4. *Proc Natl Acad Sci USA*. 2012;109(52):E3687–E3695.

34. Shaked I, Meerson A, Wolf Y, et al. MicroRNA-132 potentiates cholinergic anti-inflammatory signaling by targeting acetylcholinesterase. *Immunity*. 2009;31(6):965–973.

35. Shaltiel G, Hanan M, Wolf Y, et al. Hippocampal microRNA-132 mediates stress-inducible cognitive deficits through its acetylcholinesterase target. *Brain Struct Funct*. 2013;218(1):59–72.

36. Shenhar-Tsarfaty S, Berliner S, Bornstein NM, Soreq H. Cholinesterases as biomarkers for parasympathetic dysfunction and inflammation-related disease. *J Mol Neurosci*. 2014;53(3):298–305.

37. Slavich GM, Irwin MR. From stress to inflammation and major depressive disorder: a social signal transduction theory of depression. *Psychol Bull*. 2014;140(3):774.

38. Soreq H. Checks and balances on cholinergic signaling in brain and body function. *Trends Neurosci*. 2015;38(7):448–458.

39. Soreq H, Seidman S. Acetylcholinesterase—new roles for an old actor. *Nat Rev Neurosci*. 2001;2(4):294–302.

40. Soreq H, Wolf Y. NeurimmiRs: microRNAs in the neuroimmune interface. *Trends Mol Med*. 2011;17(10):548–555.

41. Steingard RJ, Yurgelun-Todd DA, Hennen J, et al. Increased orbito-frontal cortex levels of choline in depressed adolescents as detected by in vivo proton magnetic resonance spectroscopy. *Biol Psychiatry*. 2000;48(11):1053–1061.

42. Thakral S, Ghoshal K. miR-122 is a unique molecule with great potential in diagnosis, prognosis of liver disease, and therapy both as miRNA mimic and antimir. *Curr Gene Ther*. 2015;15(2):142.

43. Tracey KJ. Reflex control of immunity. *Nat Rev Immunol*. 2009;9(6):418–428.

44. Waiskopf N, Ofek K, Gilboa-Geffen A, et al. AChE and RACK1 promote the anti-inflammatory properties of fluoxetine. *J Mol Neurosci*. 2014;53(3):306–315. http://dx.doi.org/10.1007/s12031-013-0174-6.

45. Zhang L, et al. MiR-132 inhibits expression of SIRT1 and induces pro-inflammatory processes of vascular endothelial inflammation through blockade of the SREBP-1c metabolic pathway. *Cardiovasc Drugs Ther*. 2014;28(4):303–311.

46. Zhang Y, Wang Z, Gemeinhart RA. Progress in microRNA delivery. *J Control Release*. 2013;172(3):962–974.

47. Zimmerman G, Shaltiel G, Barbash S, et al. Post-traumatic anxiety associates with failure of the innate immune receptor TLR9 to evade the pro-inflammatory NFκB pathway. *Transl Psychiatry*. 2012;2(2):e78.

13

Early Life Stress- and Sex-Dependent Effects on Hippocampal Neurogenesis

P.J. Lucassen, A. Korosi, H.J. Krugers, C.A. Oomen

SILS-CNS, University of Amsterdam, Amsterdam, The Netherlands

Abstract

Neurogenesis refers to the birth of new neurons in an adult brain, a form of structural plasticity that has been implicated in cognition, mood, and anxiety, and is well regulated by environmental and hormonal factors. Exposure to stress (hormones) generally inhibits neurogenesis. Here, we discuss (sex-dependent) effects of stress on adult hippocampal neurogenesis, and focus on stress during the sensitive period of early life. While the effects of acute, mild stress are generally short-lasting and recover quickly, chronic or severe forms of stress can induce longer-lasting reductions in adult neurogenesis, especially when encountered during early life. Some of these inhibitory effects of early stress can normalize after appropriate recovery periods, exercise, drugs targeting the stress system, and some antidepressants. Early life stress may (re-)program hippocampal plasticity, thereby altering the overall composition of the hippocampal circuit. This may modify stress responsivity, hippocampal function, later cognition, and the risk for psychopathology.

STRESS, TIME DOMAINS, AND MEDIATORS OF THE STRESS RESPONSE

In daily life, environmental challenges and exposure to stressful experiences can often not be avoided. Stress could be defined as any environmental demand that exceeds the physiological regulatory capacity of an organism, particularly during unpredictable situations. Stress can be psychological in nature as during financial or work-related problems,[190] or also involve biological changes, such as metabolic crises or inflammation. In many instances, exposure to any (perceived or real) stressor elicits a stress response that enables the individual to respond appropriately and ultimately maintain or regain homeostasis.

- Stem cells are present in the brains of many different species including primates and humans. In a few brain regions, mainly olfactory bulb and hippocampus, these stem cells continue to produce new neurons in the adult brain.

- During this process of "adult neurogenesis," stem cells go through subsequent phases of proliferation, selection, fate specification, migration, and neuronal differentiation before they eventually form fully functional neurons that integrate into the adult neuronal circuit and contribute to brain functions, such as cognition, pattern separation, mood, and anxiety.

- Adult neurogenesis is regulated by many environmental and hormonal factors. Enriched environmental housing and voluntary physical exercise generally stimulate, whereas age and exposure to stress inhibit neurogenesis.

- While effects of acute and mild stress are generally short-lasting and recover quickly, chronic or severe forms of stress can induce lasting reductions in neurogenesis, especially when encountered during early life. Different paradigms and effects are discussed.

- Early life stress may (re-)program hippocampal plasticity, thereby altering the overall composition of the hippocampal circuit, which may modify later stress responsivity, hippocampal function, cognition, and the risk for psychopathology.

The perception and interpretation of different types of stress, and the individual's response to it, depends largely on genetic background, sex, coping strategies, and personality traits. Early life experiences, early nutrition, epigenetics, and gene–environmental interactions are also important in "programming" the adult response to stress, and from there, the risk for psychopathology.[14,70,72,80,101,103,117]

The endocrinological response to stress first involves adrenal epinephrine and norepinephrine release, hormones that elevate blood pressure and respiration, and increase blood flow to essential organs. Later, the hypothalamic–pituitary–adrenal (HPA) axis is activated as well, a classic neuroendocrine circuit that coordinates various emotional, cognitive, neuroendocrine, and autonomic inputs, and determines the specific behavioral, neural, and hormonal repertoire of an individual's response to stress.[69,72] Ultimately, the stress response helps to (re-)direct energy and focus attention on the most urgent elements of a challenge, whereas less urgent,

"maintenance" functions (e.g., digestion), are temporarily suppressed.[72]

Activation of the HPA-axis starts with corticotropin-releasing hormone (CRH) production in the paraventricular nucleus (PVN) that eventually releases glucocorticoids (GCs) from the adrenal. Negative feedback of the stress response occurs after binding of GCs to high-affinity mineralocorticoid (MR) and lower-affinity glucocorticoid receptors (GR) in brain.[29] The GR helps to maintain GC levels within physiological limits[38,84] and aberrant GR expression, or GR/MR variants, have for example been implicated in hypercortisolism, hippocampal changes, stress resistance, anxiety, and depression.[29,149,197,202,205] GCs act as transcription factors that bind to glucocorticoid responsive genes where they act in a slow, genomic manner, although also faster, nongenomic actions exist.[76,186] GC plasma levels are under circadian and ultradian control[96,147] that modifies stress sensitivity.[39,52,118,157–159]

Upon their release in the periphery, GCs affect energy, inflammatory responses, and lipid metabolism, among others. Hence, imbalances in GC regulation can have deleterious consequences,[29] which is particularly relevant for the brain, where a high density of GRs exists, especially in the hippocampus, which makes this structure highly responsive to stress.[29,181,203] Indeed, GCs can influence memory, fear, and attention in a negative manner, particularly when exposure to stress is chronic and uncontrollable. This may relate to altered MR/GR levels and/or balance,[29,146] which may alter HPA feedback and increase the risk for psychopathology. Positive effects of stress, such as enhanced memory formation, have been described too, which depend on the timing, type, and controllability of a stressor.[71,167] While functional changes after stress involve reductions in hippocampal excitability, long-term potentiation, and hippocampal memory, morphological consequences of stress include hippocampal volume reductions as well as a number of cellular changes, notably dendritic atrophy and a suppressed rate of adult neurogenesis (see later).[24,104,156]

ADULT NEUROGENESIS

Adult neurogenesis (AN) refers to the production of new neurons in the adult brain. These adult-generated neurons are derived from stem cells that go through stages of proliferation, fate specification and apoptosis, migration, and neuronal differentiation, before they eventually yield new, functional neurons that integrate into the preexisting, adult network of the hippocampus.[1,67,77,189,198,216] AN is dynamically regulated by various environmental factors and declines with age (e.g., Refs. 55,87). Adult cytogenesis and neurogenesis has also been reported in other brain structures, like the amygdala, striatum, hypothalamus, and neocortex, with differences between species, and often in response to specific challenges or injury.[45,112]

AN in the dentate gyrus (DG) is potently stimulated by exercise and environmental enrichment, notably parallel to improvements in hippocampal function.[78,199] Whereas rewarding experiences can stimulate AN,[182] aversive experiences like stress generally decrease AN.[9,102,105]

STRESS AND ADULT HIPPOCAMPAL NEUROGENESIS

Exposure to stress during adult life is one of the best known environmental suppressors of AN. Both psychosocial[26,46] and physical stressors[108,144,201] can inhibit one or more phases of the neurogenesis process.[102,105,120] In classical studies, rodents exposed to the odor of a predator generated a strong stress hormone response that was associated with significant parallel reductions in hippocampal proliferation. Both acute and chronic stressors generally suppress proliferation and many different types of stressors, including physical restraint, social defeat, inescapable foot shock, sleep deprivation, and mixed types of multiple, unpredictable or mild stressors, generally all decrease numbers of new neurons in the dentate gyrus.[26,34,46,54,55,65,66,90,122,142,144,165,172,212]

Notably, exceptions exist in stress inhibition of neurogenesis in that negative findings have also been reported.[27,50,51,106,134,141] These might depend on the type of stressor applied, or the species, sex, or strain.[50,51,73,165,207] Interindividual variation in the behavioral susceptibility to stress is also a relevant factor.[94] In some instances, increased AN has been reported after stress, but in these studies, the stressors were often predictable, controllable, and/or mild, may actually have enriched an otherwise boring environment and could have been perceived as rewarding experiences.[141,193,164]

AN in the hippocampus is further required for the beneficial effects of an enriched environment on recovery from stress-induced changes in behavior[161] where it is correlated with increased survival of newborn cells and AN.[161,185] Surprisingly, housing animals in an enriched environment that includes voluntary exercise, increases GCs,[200] suggesting that this rise in GC levels is essential for increased AN in the hippocampus.[154,161] When rats are adrenalectomized, admittedly a highly artificial condition, environmental enrichment-induced increases in AN are no longer apparent,[90] indicating that GCs can facilitate adult hippocampal neurogenesis under specific conditions. Also somewhat counterintuitively, exercise, which is considered a potent stimulus for AN,[195] stimulates GC levels, even though exercise per se reduces stress.[75] Cessation of voluntary exercise subsequently impairs AN and can increase anxietylike behavior[132] consistent with other studies that indicate that changes in AN often correlate with anxiety measures.[58,61,143,148,168,214]

When no other transmitter systems are altered and the stressor is unpredictable or uncontrollable and its nature severe, stress generally reduces AN.[34,56,65,66,90,122,142,144,165,172] In fact, this type of stress can reduce multiple stages of the neurogenic process, including the initial phase of proliferation of the neural stem cells and amplifying progenitor cells, as well as subsequent neuronal differentiation phase and dendritic expansion. Exposure to GCs per se was even shown to deplete the neural precursor pool.[213] Stress not only reduces proliferation and AN in many different species, it may also shift neural stem cells away from neuronal differentiation, and instead "redirect" them towards the generation of oligodendrocytes.[20] Although not studied in great detail yet, such (early) stress-induced fate shifts may have important functional consequences: e.g., for the myelination of axons and/or mossy fibers, and hence network connectivity, particularly when they occur during early development when cell division is massive.

Although different types of stress trigger different behavioral and functional responses, adrenal GCs (GCs; corticosterone in rodents; cortisol in man) are considered instrumental in mediating the suppressing effects of stress on AN.[165] The basis for this assertion is as follows. First, exogenous GC administration to animals has effects similar to those of stress on cell proliferation, neuronal differentiation, and cell survival,[62,115,211,213] as well as on the production of oligodendrocytes and microglia responses. Second, the reductions in AN after stress, and many of the molecular[28] and physiological[85] changes, can be prevented by blocking the GR, for a very short period[62,115,137] or by CRH antagonists.[5] Furthermore, in a transgenic mouse model of AN inhibition, a transient increase in the corticosterone response to stress occurs as well as an attenuated dexamethasone-induced suppression of corticosterone release.[175] This is indicative of a role for the newborn cells in regulating HPA axis activity. On the other hand, ablation of AN by irradiation did not impair basal HPA axis activity.[179,101,103]

Although general blockers of different parameters of the stress system are thus effective, the precise mechanism(s) by which GCs decrease the numbers of new neurons remains poorly understood. More information has become available on its molecular control.[6,119,166] NMDA receptors, GRs and MRs, all present on the new cells, albeit in different ratios over time, likely act in concert to mediate effects of stress on AN.[43,125,209,210] Notably, GR knockdown, selectively in newborn cells, accelerates their neuronal differentiation and migration, alters their dendritic complexity, parallel to impaired contextual freezing during fear conditioning. Hence, GR expression in the newborn hippocampal cells is important for structural as well as functional integration into the mature hippocampal circuits involved in fear memory.[40]

Furthermore, most precursors in the brain are located closely to blood vessels.[140] Although often not distinguished in quantitative analysis, this proximity makes this population particularly sensitive to stress hormones[57] and many other peripheral factors. Astrocytes are also of relevance as they closely align the vasculature, express GRs, support the survival of developing neurons, and are involved in their synaptic integration.[178] Notably, astrocytes are affected by some, but not all, types of stress.[10,25,136,191,203]

Stress further slows down neuronal differentiation of the adult-born cells, as evidenced by the upregulation of markers indicating cell cycle arrest[54] that may be induced by specific changes in DNA methylation.[15] Stress also reduces the survival of new neurons that were born already: i.e., prior to the actual stressful experience. A change in "corticoid environment"[211] is thought to be mediated by stress-induced reductions in neurotrophins and survival promoting factors, such as brain-derived neurotrophic factor (BDNF), insulinlike growth factor-1 (IGF-1) and vascular endothelial growth factor (VEGF) (e.g., Refs. 162,208). Reductions in newborn cell survival may involve microglia, which can phagocytose new neurons. Indeed, stress influences microglia, both number and responsivity, which may modulate their efficiency in clearing debris or dead neurons,[59,126,170] or their capacity to release neurotoxic cytokines.[48,83,97,169]

An important difference between studies on temporal aspects of stress and AN is whether GC levels remain elevated or not after the initial exposure to stress has ended. In some psychosocial stress models, GC levels remain elevated, which has stronger suppressive effects on AN than exposure to severe, but predictable, physical stressors, such as restraint.[209] Several examples exist of a persistent and lasting inhibition of AN after an initial exposure to stress, despite a later lowering of GC levels.[26,120] In contrast, GC levels can remain elevated after a psychosocial stressor, with AN being suppressed long term. In milder models of stress, GC levels generally normalize, yet AN remains reduced.[165,192] This suggests that while GCs are involved in the initial suppression of proliferation, they are not always necessary for maintaining this effect.

When studying effects of stress on AN in laboratory conditions, it is further important to realize that many variables influence the outcome.[12] These variables include interindividual genetic or gender differences in stress coping and resilience,[93] prior handling of the animals, time of day at sacrifice, and previous exposure to stressful learning tasks, e.g., the water maze, or exercise.[35,36,60,111] Anatomical differences exist, such as in projections to specific subregions of the hippocampus or in the larger networks, or neuromatrix.[176] Thus, stress

effects on AN might differ between the dorsal or the ventral hippocampus depending upon the stimulus.[133,184]

Many other factors may contribute to the stress-induced inhibition of AN, such as the stress-induced increase in glutamate release and NMDA receptor activation[46,128,165] or through stress effects on various neurotransmitter systems implicated in the regulation of AN, such as GABA,[44] serotonin,[32] noradrenaline,[68] acetylcholine,[16] dopamine,[33,182] cannabinoids, opioids, nitric oxide, and gonadal steroids.[9,42,107] Many antidepressant drugs that interfere with stress-related behavior in animals also modulate AN. The relation between stress, AN, antidepressants, and mental illnesses, such as major depression has been extensively discussed, but is beyond our current scope and we therefore refer to recent literature.[34,104,102,105,108,142,151,155,164,179,184]

In functional terms, AN has been linked to various cognitive measures, with AN being relevant for some, but not all forms of hippocampal dependent learning and memory.[2,22,135,152,153,160,215] Stress-induced suppression of AN is linked to an impaired performance on various hippocampal tasks, such as spatial navigation and object memory, as well as anxiety-like behaviors, as measured with the elevated plus maze, open field and novelty suppressed feeding tasks.[91,123,124,175,196]

LONG-LASTING EFFECTS OF EARLY, PERINATAL STRESS EXPOSURE; SEX DIFFERENCES

The brain is particularly sensitive to stressful experiences during the early postnatal period given the large numbers of dividing cells that form neuronal networks and eventually produce behavior. By interacting with inborn genetic risk factors, environmental factors such as early life stress (ELS) specifically target the development of brain structures and mechanisms involved in emotional and stress regulation.[30,174,194] Thus stress during early life has been implicated in many changes in later brain structure and function, and in an increased risk for psychopathology and age-related cognitive decline.[19,21,49,53,86,113,116,131,188] Mechanistically, exposure to early stress can induce epigenetic modifications of stress-related genes.[8,23,88,101,103] In addition to changes in the set point of HPA axis activity, e.g., GC feedback sensitivity, regulation of AN also appears to be modified by stress-exposure during the perinatal period.

In experimental conditions, ELS affects hippocampal, emotional and cognitive functions, and stress reactivity later in life.[4,11,17,63,139] AN is very sensitive to ELS and exposure to perinatal stress typically induces reductions in AN in adult offspring[82,91,100,121,129,180] although exceptions also occur.[187] Infection during pregnancy represents a stressor and reduces cell proliferation and AN in both juvenile and adult offspring.[95] Undernourished mothers similarly produce offspring with lower rates of cell proliferation and neuronal survival in adulthood.[114] Prenatal stress and maternal separation further decrease the size and complexity of adult born neurons.[92,183] The effects are generally brain region-specific: prenatal stress e.g., impaired AN in the DG but not in the olfactory bulb.[13]

While in utero exposure to stress almost invariably reduces AN in adulthood,[91] postnatal stress exposure yields more variable results, and is modified by maternal and paternal factors, sex, genetic background, and epigenetic changes, although suppression of AN prevails as well.[79,89,98,101,103]

During the early postnatal period, individuals are particularly dependent on parental care, which is important for emotional and cognitive development and for attachment styles.[11,19,31] In rodents, postnatal stress is often induced by disturbing this important mother–child interaction. Many models have been developed, including 24-h maternal deprivation at different days during the stress hyporesponsive period of the first 2 weeks of rodent life. When applied on postnatal day (PND) 3, this can be considered a model for maternal neglect.[110,145] Maternal separation protocols exist as well, where the mother is removed from her pups for shorter periods of time, but this is done repeatedly, e.g., for several days. Interestingly, this generally increases the amount of maternal care and is hence considered less stressful than maternal deprivation. Limiting the amount of nesting and bedding material during the first week of life has further become a popular animal model for ELS. The latter induces lasting effects on AN and cognition, often in a sex-dependent manner. In maternal separation or maternal deprivation paradigms generally reductions are seen in cell proliferation, AN, and the survival of new neurons in adult rodents in the dentate gyrus.[3,92,121,139] These effects often display a clear sex difference and depend on the moment when AN is studied.[89,138,139]

Alternative methods to study effects of stress include selected breeding that can result in animal lines that differ in specific traits, like stress responsivity or resilience, anxiety, or in different displays of maternal care. In such lines, neuronal survival was found to be decreased, and apoptosis increased, in offspring of low-caring mothers versus offspring of high-caring mothers, that differ in their stress response.[204] In addition, repeated maternal separation (MS) leads to lastingly decreased levels of proliferation,[121] without affecting neuronal survival in the DG.[47,121] In a similar model, a biphasic response was also found in hippocampal AN as well as in BDNF expression and cognition, suggesting that ELS may endow animals with a potential adaptive advantage in stressful environments on the short term but with increasing age, is associated with long-term deleterious effects.[180] MS

was also shown to alter the capacity of neural precursor cells to differentiate into neurons, which is mediated via methylation of the retinoic acid receptor gene promoter.[15,101,103] Furthermore, in rats selectively bred for inborn levels of anxiety, prenatal stress was reported to decrease the survival of newly-generated cells as well as AN in the hippocampus of high-anxiety breeders only.[100] Interestingly, this lower rate of AN was paralleled by an impaired integration of the newborn neurons when compared to normal rats, specifically in females.[150]

The maternal deprivation paradigm differs from maternal separation in that it is not repeated for several days, and is longer lasting, i.e., 24 h long instead of daily periods of a few hours. Maternal deprivation was found to transiently increase the numbers of immature (DCX-positive) neurons in male rats 3 weeks of age,[136] ultimately leading to reduced proliferation throughout the full rostrocaudal axis of the DG, and reduced differentiation in the caudal part of the DG in adult males.[139] Strikingly different effects of ELS on AN is seen in female rats. Whereas neurogenesis is enhanced at PND21 in male rats, a strong suppression was reported in females after maternal deprivation at PND3.[136] However, as opposed to males, early life adversity does not result in a decreased number of DCX-positive cells, but an overall lower total number of adult dentate gyrus granule cells in adult females.[138] Similarly in mice, chronic ELS affected cognitive function and rates of survival of adult born neurons more robustly in males when compared to females.[74,129]

The effects of prenatal stress on AN are often sex dependent.[107,109,163,206] Male rats show a brief period in adolescence during which neurogenesis, BDNF expression, and spatial learning are actually improved, possibly allowing the individual to temporarily compensate for the effects of early life adversity. Female rats do not show such a period of improved performance but rather show a very strong suppression of neurogenesis during the prepubertal period, which then subsides with age. The consequences of this period of suppressed neurogenesis in females may be long-lasting, and female rats exposed to 24 h of maternal deprivation at PND3 exhibited a lower total number of mature granule cells in adulthood, potentially limiting the number of synaptic contacts that can be established in this region. Finally, it is important to mention that AN can be permanently affected also by other early life stressors that are not necessarily related to the mother–infant interaction alone. For example, early life inflammation,[64,127] radiation therapy,[41,130] anesthesia,[217] stroke,[177] infection,[14a] and ethanol exposure[173] induce long-lasting effects on AN associated with late-onset cognitive impairment.

Despite the large variation in ELS models, species, age of testing, and outcome parameters, the majority of studies reports only mild behavioral changes in females after early life adversity, i.e., two-third of the experimental series in female rodents did not show a significant change in behavior after early life adversity. Possibly this number is lower than the actual situation, because the influence of the hormonal cycle, which could have added to variation in the behavioral outcome, is not always taken into account. On the other hand, more likely than not, the prevalence of significant effects, at least in animal studies, is overestimated due to a publication bias towards positive findings.[81] Thus, early life experiences during both pre- and postnatal development can bidirectionally alter hippocampal neuronal plasticity, which supports the possibility that these structural changes might be involved in affected cognition as shown recently by a novel causal statistical method demonstrating that cognitive impairments induced by ELS are largely AN dependent.[129]

When tested in adulthood or middle-age, cell proliferation and AN are usually found to be decreased after stress. Yet, at earlier stages, e.g., at P9[129] or PND 21,[180] AN in males is actually enhanced by ELS, as was BDNF expression and performance in a stressful version of the Morris water maze when studied at 2 months, but impaired at 15 months of age. Apparently, early life adversity can transiently improve the functionality of the dentate gyrus, possibly allowing the organism to survive in the adverse conditions. However, in the long run, this adaptation to early life adversity may deplete specific populations of stem cells and seems to program structural plasticity such that it may become a disadvantage later in life,[82,99,121] most notably under low to moderately stressful conditions. The ELS-induced reduction in neurogenic capacity later in life might be due to increased AN postnatally, that might in fact deplete the neurogenic pool. This phenomenon is consistent with other studies on cellular[213] and in vivo level[37,129,171,180] where effects of early stress or early antidepressant treatment and neuronal hyperactivity were studied.

Interestingly, when tested under stressful conditions, early maternal deprivation improved learning and memory of rats. In fact, contextual learning was enhanced in both contextual and cued fear conditioning task. Long-term potentiation, when measured in the presence of corticosterone, was facilitated in male, not female, MD rats.[138,139] Similar effects on physiology and dendritic structure were observed as a function of maternal care with animals receiving more care having better LTP and more complex dendrites and spines.[7,18] The behavioral phenotype in female rats appeared to be more subtle and confined to amygdala-dependent learning paradigms.[138] Literature on cognitive performance in adult females exposed early in life to these types of ELS is generally less extensive than literature on males, as reviewed recently.[98] These data suggest that early life

events, whether adverse or not, might increase the sensitivity of the hippocampus to future environments, and prepare the organism to respond optimally when a similar stress is experienced in adult life as the one encountered shortly after birth.[79,139]

CONCLUDING REMARKS

Stress and GCs interfere with one or more of the phases of the neurogenetic process. Their inhibitory effects can normalize after a recovery period, voluntary exercise, or antidepressant treatment. Adult neurogenesis has been implicated in cognitive functions, in the regulation of mood and anxiety, and in the therapeutic effects of antidepressant drugs. A reduced rate of AN may be indicative of impaired hippocampal plasticity. Lasting reductions in AN or turnover rate of DG granule cells, as programmed by early life events, will alter the overall composition of the DG cell population and can modify stress responsivity and thereby influence functioning of the adult hippocampal circuit and later cognition, as well as the risk for psychopathology.

Acknowledgments

PJL is supported by the HersenStichting Nederland, Alzheimer Nederland, ISAO and Amsterdam Brain & Cognition (ABC). AK and PJL are further supported by NWO and by Amsterdam Brain & Cognition (ABC).

References

1. Abrous DN, Koehl M, Le Moal M. Adult neurogenesis: from precursors to network and physiology. *Physiol Rev*. 2005;85(2):523–569.
2. Aimone JB, Deng W, Gage FH. Adult neurogenesis: integrating theories and separating functions. *Trends Cogn Sci*. 2010;14(7):325–337. doi:S1364-6613(10)00088-4. pii:10.1016/j.tics.2010.04.003.
3. Aisa B, Elizalde N, Tordera R, Lasheras B, Del Rio J, Ramirez MJ. Effects of neonatal stress on markers of synaptic plasticity in the hippocampus: implications for spatial memory. *Hippocampus*. 2009;19(12):1222–1231.
4. Aisa B, Tordera R, Lasheras B, Del Rio J, Ramirez MJ. Cognitive impairment associated to HPA axis hyperactivity after maternal separation in rats. *Psychoneuroendocrinology*. 2007;32(3):256–266.
5. Alonso R, Griebel G, Pavone G, Stemmelin J, Le Fur G, Soubrie P. Blockade of CRF(1) or V(1b) receptors reverses stress-induced suppression of neurogenesis in a mouse model of depression. *Mol Psychiatry*. 2004;9(3):278–286.
6. Anacker C, Cattaneo A, Luoni A, et al. Glucocorticoid-related molecular signaling pathways regulating hippocampal neurogenesis. *Neuropsychopharmacology*. 2013;38(5):872–883. http://dx.doi.org/10.1038/npp.2012.253.
7. Bagot RC, van Hasselt FN, Champagne DL, Meaney MJ, Krugers HJ, Joels M. Maternal care determines rapid effects of stress mediators on synaptic plasticity in adult rat hippocampal dentate gyrus. *Neurobiol Learn Mem*. 2009;92(3):292–300.
8. Bale TL, Baram TZ, Brown AS, et al. Early life programming and neurodevelopmental disorders. *Biol Psychiatry*. 2010;68(4):314–319. http://dx.doi.org/10.1016/j.biopsych.2010.05.028.

9. Balu DT, Lucki I. Adult hippocampal neurogenesis: regulation, functional implications, and contribution to disease pathology. *Neurosci Biobehav Rev.* 2009;33(3):232–252. http://dx.doi.org/10.1016/j.neubiorev.2008.08.007.

10. Banasr M, Duman RS. Regulation of neurogenesis and gliogenesis by stress and antidepressant treatment. *CNS Neurol Disord Drug Targets.* 2007;6(5):311–320.

11. Baram TZ, Davis EP, Obenaus A, et al. Fragmentation and unpredictability of early-life experience in mental disorders. *Am J Psychiatry.* 2012;169(9):907–915. http://dx.doi.org/10.1176/appi.ajp.2012.11091347.

12. Bekinschtein P, Oomen CA, Saksida LM, Bussey TJ. Effects of environmental enrichment and voluntary exercise on neurogenesis, learning and memory, and pattern separation: BDNF as a critical variable? *Semin Cell Dev Biol.* 2011;22(5):536–542. http://dx.doi.org/10.1016/j.semcdb.2011.07.002.

13. Belnoue L, Grosjean N, Ladeveze E, Abrous DN, Koehl M. Prenatal stress inhibits hippocampal neurogenesis but spares olfactory bulb neurogenesis. *PLoS One.* 2013;8(8):e72972. http://dx.doi.org/10.1371/journal.pone.0072972.

14. Binder EB, Bradley RG, Liu W, et al. Association of FKBP5 polymorphisms and childhood abuse with risk of posttraumatic stress disorder symptoms in adults. *JAMA.* 2008;299(11):1291–1305. http://dx.doi.org/10.1001/jama.299.11.1291.

14a. Bland ST, Beckley JT, Young S, et al. Enduring consequences of early-life infection on glial and neural cell genesis within cognitive regions of the brain. *Brain Behav Immun.* 2010;24:329–338.

15. Boku S, Toda H, Nakagawa S, et al. Neonatal maternal separation alters the capacity of adult neural precursor cells to differentiate into neurons via methylation of retinoic acid receptor gene promoter. *Biol Psychiatry.* 2015;77(4):335–344. http://dx.doi.org/10.1016/j.biopsych.2014.07.008.

16. Bruel-Jungerman E, Lucassen PJ, Francis F. Cholinergic influences on cortical development and adult neurogenesis. *Behav Brain Res.* 2011;221(2):379–388. http://dx.doi.org/10.1016/j.bbr.2011.01.021.

17. Brunson KL, Kramar E, Lin B, et al. Mechanisms of late-onset cognitive decline after early-life stress. *J Neurosci.* 2005;25(41):9328–9338.

18. Champagne DL, Bagot RC, van Hasselt F, et al. Maternal care and hippocampal plasticity: evidence for experience-dependent structural plasticity, altered synaptic functioning, and differential responsiveness to glucocorticoids and stress. *J Neurosci.* 2008;28(23):6037–6045.

19. Chen Y, Baram TZ. Toward understanding how early-life stress reprograms cognitive and emotional brain networks. *Neuropsychopharmacology.* 2016;41(1):197–206. http://dx.doi.org/10.1038/npp.2015.181.

20. Chetty S, Friedman AR, Taravosh-Lahn K, et al. Stress and glucocorticoids promote oligodendrogenesis in the adult hippocampus. *Mol Psychiatry.* 2014;19(12):1275–1283. http://dx.doi.org/10.1038/mp.2013.190.

21. Chu DA, Williams LM, Harris AW, Bryant RA, Gatt JM. Early life trauma predicts self-reported levels of depressive and anxiety symptoms in nonclinical community adults: relative contributions of early life stressor types and adult trauma exposure. *J Psychiatr Res.* 2013;47(1):23–32. http://dx.doi.org/10.1016/j.jpsychires.2012.08.006.

22. Clelland CD, Choi M, Romberg C, et al. A functional role for adult hippocampal neurogenesis in spatial pattern separation. *Science.* 2009;325(5937):210–213.

23. Cottrell EC, Seckl JR. Prenatal stress, glucocorticoids and the programming of adult disease. *Front Behav Neurosci.* 2009;3:19. http://dx.doi.org/10.3389/neuro.08.019.2009.

24. Czeh B, Lucassen PJ. What causes the hippocampal volume decrease in depression? Are neurogenesis, glial changes and apoptosis implicated? *Eur Arch Psychiatry Clin Neurosci.* 2007;257(5):250–260.

25. Czeh B, Simon M, Schmelting B, Hiemke C, Fuchs E. Astroglial plasticity in the hippocampus is affected by chronic psychosocial stress and concomitant fluoxetine treatment. *Neuropsychopharmacology.* 2006;31(8):1616–1626.

26. Czeh B, Welt T, Fischer AK, et al. Chronic psychosocial stress and concomitant repetitive transcranial magnetic stimulation: effects on stress hormone levels and adult hippocampal neurogenesis. *Biol Psychiatry.* 2002;52(11):1057–1065.

27. Dagyte G, Van der Zee EA, Postema F, et al. Chronic but not acute foot-shock stress leads to temporary suppression of cell proliferation in rat hippocampus. *Neuroscience.* 2009;162(4):904–913.

28. Datson NA, Speksnijder N, Mayer JL, et al. The transcriptional response to chronic stress and glucocorticoid receptor blockade in the hippocampal dentate gyrus. *Hippocampus.* 2012;22(2):359–371. http://dx.doi.org/10.1002/hipo.20905.

29. de Kloet ER, Joels M, Holsboer F. Stress and the brain: from adaptation to disease. *Nat Rev Neurosci.* 2005;6(6):463–475.

30. de Rooij SR, Veenendaal MV, Raikkonen K, Roseboom TJ. Personality and stress appraisal in adults prenatally exposed to the Dutch famine. *Early Hum Dev.* 2012;88(5):321–325. http://dx.doi.org/10.1016/j.earlhumdev.2011.09.002.

31. De Wolff MS, van Ijzendoorn MH. Sensitivity and attachment: a meta-analysis on parental antecedents of infant attachment. *Child Dev.* 1997;68(4):571–591.

32. Djavadian RL. Serotonin and neurogenesis in the hippocampal dentate gyrus of adult mammals. *Acta Neurobiol Exp (Wars).* 2004;64(2):189–200.

33. Dominguez-Escriba L, Hernandez-Rabaza V, Soriano-Navarro M, et al. Chronic cocaine exposure impairs progenitor proliferation but spares survival and maturation of neural precursors in adult rat dentate gyrus. *Eur J Neurosci.* 2006;24(2):586–594. http://dx.doi.org/10.1111/j.1460-9568.2006.04924.x.

34. Dranovsky A, Hen R. Hippocampal neurogenesis: regulation by stress and antidepressants. *Biol Psychiatry.* 2006;59(12):1136–1143.

35. Droste SK, Gesing A, Ulbricht S, Muller MB, Linthorst AC, Reul JM. Effects of long-term voluntary exercise on the mouse hypothalamic-pituitary-adrenocortical axis. *Endocrinology.* 2003;144(7):3012–3023.

36. Ehninger D, Kempermann G. Paradoxical effects of learning the Morris water maze on adult hippocampal neurogenesis in mice may be explained by a combination of stress and physical activity. *Genes Brain Behav.* 2006;5(1):29–39.

37. Encinas JM, Sierra A. Neural stem cell deforestation as the main force driving the age-related decline in adult hippocampal neurogenesis. *Behav Brain Res.* 2012;227(2):433–439. http://dx.doi.org/10.1016/j.bbr.2011.10.010.

38. Erdmann G, Berger S, Schutz G. Genetic dissection of glucocorticoid receptor function in the mouse brain. *J Neuroendocrinol.* 2008;20(6):655–659. http://dx.doi.org/10.1111/j.1365-2826.2008.01717.x.

39. Fitzsimons C, Herbert J, Schouten M, Meijer OC, Lucassen PJ, Lightman SL. Circadian and ultradian gluc effects of glucocorticoids on neural stem cells and adult hippoal stem cells and adult hippocampal neurogenesis. *Front Neuroendocrinol.* 2016;41:44–58. http://dx.doi.org/10.1016/j.yfrne.2016.05.001.

40. Fitzsimons CP, van Hooijdonk LW, Schouten M, et al. Knockdown of the glucocorticoid receptor alters functional integration of newborn neurons in the adult hippocampus and impairs fear-motivated behavior. *Mol Psychiatry.* 2013;18(9):993–1005. http://dx.doi.org/10.1038/mp.2012.123.

41. Fukuda A, Fukuda H, Swanpalmer J, et al. Age-dependent sensitivity of the developing brain to irradiation is correlated with the number and vulnerability of progenitor cells. *J Neurochem.* 2005;92(3):569–584. http://dx.doi.org/10.1111/j.1471-4159.2004.02894.x.

42. Galea LA. Gonadal hormone modulation of neurogenesis in the dentate gyrus of adult male and female rodents. *Brain Res Rev.* 2008;57(2):332–341.

43. Garcia A, Steiner B, Kronenberg G, Bick-Sander A, Kempermann G. Age-dependent expression of glucocorticoid- and mineralocorticoid receptors on neural precursor cell populations in the adult murine hippocampus. *Aging Cell*. 2004;3(6):363–371.

44. Ge S, Yang CH, Hsu KS, Ming GL, Song H. A critical period for enhanced synaptic plasticity in newly generated neurons of the adult brain. *Neuron*. 2007;54(4):559–566. doi:S0896-6273(07)00334-0 pii:10.1016/j.neuron.2007.05.002.

45. Gould E. How widespread is adult neurogenesis in mammals? *Nat Rev Neurosci*. 2007;8(6):481–488. http://dx.doi.org/10.1038/nrn2147.

46. Gould E, McEwen BS, Tanapat P, Galea LA, Fuchs E. Neurogenesis in the dentate gyrus of the adult tree shrew is regulated by psychosocial stress and NMDA receptor activation. *J Neurosci*. 1997;17(7):2492–2498.

47. Greisen MH, Altar CA, Bolwig TG, Whitehead R, Wortwein G. Increased adult hippocampal brain-derived neurotrophic factor and normal levels of neurogenesis in maternal separation rats. *J Neurosci Res*. 2005;79(6):772–778.

48. Guadagno J, Swan P, Shaikh R, Cregan SP. Microglia-derived IL-1beta triggers p53-mediated cell cycle arrest and apoptosis in neural precursor cells. *Cell Death Dis*. 2015;6:e1779. http://dx.doi.org/10.1038/cddis.2015.151.

49. Gupta A, Labus J, Kilpatrick LA, et al. Interactions of early adversity with stress-related gene polymorphisms impact regional brain structure in females. *Brain Struct Funct*. 2016;221(3):1667–1679. http://dx.doi.org/10.1007/s00429-015-0996-9.

50. Hanson ND, Owens MJ, Boss-Williams KA, Weiss JM, Nemeroff CB. Several stressors fail to reduce adult hippocampal neurogenesis. *Psychoneuroendocrinology*. 2011;36(10):1520–1529. http://dx.doi.org/10.1016/j.psyneuen.2011.04.006.

51. Hanson ND, Owens MJ, Nemeroff CB. Depression, antidepressants, and neurogenesis: a critical reappraisal. *Neuropsychopharmacology*. 2011;36(13):2589–2602. http://dx.doi.org/10.1038/npp.2011.220.

52. Harris AP, Holmes MC, de Kloet ER, Chapman KE, Seckl JR. Mineralocorticoid and glucocorticoid receptor balance in control of HPA axis and behaviour. *Psychoneuroendocrinology*. 2013;38(5):648–658. http://dx.doi.org/10.1016/j.psyneuen.2012.08.007.

53. Heim C, Newport DJ, Mletzko T, Miller AH, Nemeroff CB. The link between childhood trauma and depression: insights from HPA axis studies in humans. *Psychoneuroendocrinology*. 2008;33(6):693–710.

54. Heine VM, Maslam S, Joels M, Lucassen PJ. Increased P27KIP1 protein expression in the dentate gyrus of chronically stressed rats indicates G1 arrest involvement. *Neuroscience*. 2004;129(3):593–601.

55. Heine VM, Maslam S, Joels M, Lucassen PJ. Prominent decline of newborn cell proliferation, differentiation, and apoptosis in the aging dentate gyrus, in absence of an age-related hypothalamus-pituitary-adrenal axis activation. *Neurobiol Aging*. 2004;25(3):361–375.

56. Heine VM, Maslam S, Zareno J, Joels M, Lucassen PJ. Suppressed proliferation and apoptotic changes in the rat dentate gyrus after acute and chronic stress are reversible. *Eur J Neurosci*. 2004;19(1):131–144.

57. Heine VM, Zareno J, Maslam S, Joels M, Lucassen PJ. Chronic stress in the adult dentate gyrus reduces cell proliferation near the vasculature and VEGF and Flk-1 protein expression. *Eur J Neurosci*. 2005;21(5):1304–1314.

58. Hill AS, Sahay A, Hen R. Increasing adult hippocampal neurogenesis is sufficient to reduce anxiety and depression-like behaviors. *Neuropsychopharmacology*. 2015;40(10):2368–2378. http://dx.doi.org/10.1038/npp.2015.85.

59. Hinwood M, Morandini J, Day TA, Walker FR. Evidence that microglia mediate the neurobiological effects of chronic psychological stress on the medial prefrontal cortex. *Cereb Cortex*. 2012;22(6):1442–1454. http://dx.doi.org/10.1093/cercor/bhr229.

60. Holmes MM, Galea LA, Mistlberger RE, Kempermann G. Adult hippocampal neurogenesis and voluntary running activity: circadian and dose-dependent effects. *J Neurosci Res*. 2004;76(2):216–222.

61. Hu P, Liu J, Wang J, et al. Chronic retinoic acid treatment suppresses adult hippocampal neurogenesis, in close correlation with depressive-like behavior. *Hippocampus*. 2016;26(7):911–923. http://dx.doi.org/10.1002/hipo.22574.

62. Hu P, Oomen C, van Dam AM, et al. A single-day treatment with mifepristone is sufficient to normalize chronic glucocorticoid induced suppression of hippocampal cell proliferation. *PLoS One*. 2012;7(9):e46224. http://dx.doi.org/10.1371/journal.pone.0046224.

63. Ivy AS, Brunson KL, Sandman C, Baram TZ. Dysfunctional nurturing behavior in rat dams with limited access to nesting material: a clinically relevant model for early-life stress. *Neuroscience*. 2008;154(3):1132–1142.

64. Jakubs K, Bonde S, Iosif RE, et al. Inflammation regulates functional integration of neurons born in adult brain. *J Neurosci*. 2008;28(47):12477–12488. http://dx.doi.org/10.1523/JNEUROSCI.3240-08.2008.

65. Jayatissa MN, Bisgaard C, Tingstrom A, Papp M, Wiborg O. Hippocampal cytogenesis correlates to escitalopram-mediated recovery in a chronic mild stress rat model of depression. *Neuropsychopharmacology*. 2006;31(11):2395–2404.

66. Jayatissa MN, Henningsen K, West MJ, Wiborg O. Decreased cell proliferation in the dentate gyrus does not associate with development of anhedonic-like symptoms in rats. *Brain Res*. 2009;1290:133–141. http://dx.doi.org/10.1016/j.brainres.2009.07.001.

67. Jessberger S, Gage FH. Adult neurogenesis: bridging the gap between mice and humans. *Trends Cell Biol*. 2014;24(10):558–563. http://dx.doi.org/10.1016/j.tcb.2014.07.003.

68. Joca SR, Ferreira FR, Guimaraes FS. Modulation of stress consequences by hippocampal monoaminergic, glutamatergic and nitrergic neurotransmitter systems. *Stress*. 2007;10(3):227–249. http://dx.doi.org/10.1080/10253890701223130.

69. Joels M, Baram TZ. The neuro-symphony of stress. *Nat Rev Neurosci*. 2009;10(6):459–466.

70. Joels M, Karst H, Krugers HJ, Lucassen PJ. Chronic stress; implications for neuron morphology, function and neurogenesis. *Front Neuroendocrinol*. 2007;28(2–3):72–96.

71. Joels M, Pu Z, Wiegert O, Oitzl MS, Krugers HJ. Learning under stress: how does it work? *Trends Cogn Sci*. 2006;10(4):152–158.

72. Joels M, Sarabdjitsingh RA, Karst H. Unraveling the time domains of corticosteroid hormone influences on brain activity: rapid, slow, and chronic modes. *Pharmacol Rev*. 2012;64(4):901–938. http://dx.doi.org/10.1124/pr.112.005892.

73. Kanatsou S, Fearey BC, Kuil LE, et al. Overexpression of mineralocorticoid receptors partially prevents chronic stress-induced reductions in hippocampal memory and structural plasticity. *PLoS One*. 2015;10(11):e0142012. http://dx.doi.org/10.1371/journal.pone.0142012.

74. Kanatsou S, Ter Horst JP, Harris AP, Seckl J, Krugers HJ, Joels M. Effects of mineralocorticoid receptor overexpression on anxiety and memory after early life stress in female mice. *Front Behav Neurosci*. 2016;9.

75. Kannangara TS, Lucero MJ, Gil-Mohapel J, et al. Running reduces stress and enhances cell genesis in aged mice. *Neurobiol Aging*. 2011;32(12):2279–2286. http://dx.doi.org/10.1016/j.neurobiolaging.2009.12.025.

76. Karst H, Berger S, Erdmann G, Schutz G, Joels M. Metaplasticity of amygdalar responses to the stress hormone corticosterone. *Proc Natl Acad Sci USA*. 2010;107(32):14449–14454. http://dx.doi.org/10.1073/pnas.0914381107.

77. Kempermann G. New neurons for 'survival of the fittest'. *Nat Rev Neurosci*. 2012;13(10):727–736. http://dx.doi.org/10.1038/nrn3319.

78. Kempermann G, Fabel K, Ehninger D, et al. Why and how physical activity promotes experience-induced brain plasticity. *Front Neurosci.* 2010;4:189. http://dx.doi.org/10.3389/fnins.2010.00189.

79. Koehl M, van der Veen R, Gonzales D, Piazza PV, Abrous DN. Interplay of maternal care and genetic influences in programming adult hippocampal neurogenesis. *Biol Psychiatry.* 2012;72(4):282–289. http://dx.doi.org/10.1016/j.biopsych.2012.03.001.

80. Koolhaas JM, Bartolomucci A, Buwalda B, et al. Stress revisited: a critical evaluation of the stress concept. *Neurosci Biobehav Rev.* 2011;35(5):1291–1301. http://dx.doi.org/10.1016/j.neubiorev.2011.02.003.

81. Korevaar DA, Hooft L, ter Riet G. Systematic reviews and meta-analyses of preclinical studies: publication bias in laboratory animal experiments. *Lab Anim.* 2011;45(4):225–230. http://dx.doi.org/10.1258/la.2011.010121.

82. Korosi A, Naninck EF, Oomen CA, et al. Early-life stress mediated modulation of adult neurogenesis and behavior. *Behav Brain Res.* 2012;227(2):400–409. http://dx.doi.org/10.1016/j.bbr.2011.07.037.

83. Kreisel T, Frank MG, Licht T, et al. Dynamic microglial alterations underlie stress-induced depressive-like behavior and suppressed neurogenesis. *Mol Psychiatry.* 2014;19(6):699–709. http://dx.doi.org/10.1038/mp.2013.155.

84. Kretz O, Reichardt HM, Schutz G, Bock R. Corticotropin-releasing hormone expression is the major target for glucocorticoid feedback-control at the hypothalamic level. *Brain Res.* 1999;818(2):488–491.

85. Krugers HJ, Goltstein PM, van der Linden S, Joels M. Blockade of glucocorticoid receptors rapidly restores hippocampal CA1 synaptic plasticity after exposure to chronic stress. *Eur J Neurosci.* 2006;23(11):3051–3055.

86. Krugers HJ, Joels M. Long-lasting consequences of early life stress on brain structure, emotion and cognition. *Curr Top Behav Neurosci.* 2014;18:81–92. http://dx.doi.org/10.1007/7854_2014_289.

87. Kuhn HG, Dickinson-Anson H, Gage FH. Neurogenesis in the dentate gyrus of the adult rat: age-related decrease of neuronal progenitor proliferation. *J Neurosci.* 1996;16(6):2027–2033.

88. Kundakovic M, Champagne FA. Early-life experience, epigenetics, and the developing brain. *Neuropsychopharmacology.* 2015;40(1):141–153. http://dx.doi.org/10.1038/npp.2014.140.

89. Lajud N, Torner L. Early life stress and hippocampal neurogenesis in the neonate: sexual dimorphism, long term consequences and possible mediators. *Front Mol Neurosci.* 2015;8:3. http://dx.doi.org/10.3389/fnmol.2015.00003.

90. Lehmann ML, Brachman RA, Martinowich K, Schloesser RJ, Herkenham M. Glucocorticoids orchestrate divergent effects on mood through adult neurogenesis. *J Neurosci.* 2013;33(7):2961–2972. http://dx.doi.org/10.1523/JNEUROSCI.3878-12.2013.

91. Lemaire V, Koehl M, Le Moal M, Abrous DN. Prenatal stress produces learning deficits associated with an inhibition of neurogenesis in the hippocampus. *Proc Natl Acad Sci USA.* 2000;97(20):11032–11037.

92. Leslie AT, Akers KG, Krakowski AD, et al. Impact of early adverse experience on complexity of adult-generated neurons. *Transl Psychiatry.* 2011;1:e35. http://dx.doi.org/10.1038/tp.2011.38.

93. Levine S. Developmental determinants of sensitivity and resistance to stress. *Psychoneuroendocrinology.* 2005;30(10):939–946.

94. Levone BR, Cryan JF, O'Leary OF. Role of adult hippocampal neurogenesis in stress resilience. *Neurobiol Stress.* 2015;1(1):147–155.

95. Lin YL, Wang S. Prenatal lipopolysaccharide exposure increases depression-like behaviors and reduces hippocampal neurogenesis in adult rats. *Behav Brain Res.* 2014;259:24–34. http://dx.doi.org/10.1016/j.bbr.2013.10.034.

96. Liston C, Cichon JM, Jeanneteau F, Jia Z, Chao MV, Gan WB. Circadian glucocorticoid oscillations promote learning-dependent synapse formation and maintenance. *Nat Neurosci.* 2013;16(6):698–705. http://dx.doi.org/10.1038/nn.3387.

97. Llorens-Martin M, Jurado-Arjona J, Bolos M, Pallas-Bazarra N, Avila J. Forced swimming sabotages the morphological and synaptic maturation of newborn granule neurons and triggers a unique pro-inflammatory milieu in the hippocampus. *Brain Behav Immun.* 2016;53:242–254. http://dx.doi.org/10.1016/j.bbi.2015.12.019.

98. Loi M, Koricka S, Lucassen PJ, Joels M. Age- and sex-dependent effects of early life stress on hippocampal neurogenesis. *Front Endocrinol (Lausanne).* 2014;5:13. http://dx.doi.org/10.3389/fendo.2014.00013.

99. Loman MM, Gunnar MR, Early Experience, S., & Neurobehavioral Development, C. Early experience and the development of stress reactivity and regulation in children. *Neurosci Biobehav Rev.* 2010;34(6):867–876. http://dx.doi.org/10.1016/j.neubiorev.2009.05.007.

100. Lucassen PJ, Bosch OJ, Jousma E, et al. Prenatal stress reduces postnatal neurogenesis in rats selectively bred for high, but not low, anxiety: possible key role of placental 11beta-hydroxysteroid dehydrogenase type 2. *Eur J Neurosci.* 2009;29(1):97–103. http://dx.doi.org/10.1111/j.1460-9568.2008.06543.x.

101. Lucassen PJ, Fitzsimons CP, Korosi A, Joels M, Belzung C, Abrous DN. Stressing new neurons into depression? *Mol Psychiatry.* 2013;18(4):396–397. http://dx.doi.org/10.1038/mp.2012.39.

102. Lucassen PJ, Meerlo P, Naylor AS, et al. Regulation of adult neurogenesis by stress, sleep disruption, exercise and inflammation: implications for depression and antidepressant action. *Eur Neuropsychopharmacol.* 2010;20(1):1–17.

103. Lucassen PJ, Naninck EF, van Goudoever JB, Fitzsimons C, Joels M, Korosi A. Perinatal programming of adult hippocampal structure and function; emerging roles of stress, nutrition and epigenetics. *Trends Neurosci.* 2013;36(11):621–631. http://dx.doi.org/10.1016/j.tins.2013.08.002.

104. Lucassen PJ, Pruessner J, Sousa N, et al. Neuropathology of stress. *Acta Neuropathol.* 2014;127(1):109–135. http://dx.doi.org/10.1007/s00401-013-1223-5.

105. Lucassen PJ, Stumpel MW, Wang Q, Aronica E. Decreased numbers of progenitor cells but no response to antidepressant drugs in the hippocampus of elderly depressed patients. *Neuropharmacology.* 2010;58:940–949.

106. Lyons DM, Buckmaster PS, Lee AG, et al. Stress coping stimulates hippocampal neurogenesis in adult monkeys. *Proc Natl Acad Sci USA.* 2010;107(33):14823–14827. http://dx.doi.org/10.1073/pnas.0914568107.

107. Mahmoud R, Wainwright SR, Galea LA. Sex hormones and adult hippocampal neurogenesis: regulation, implications, and potential mechanisms. *Front Neuroendocrinol.* April 2016;41:129–152. http://dx.doi.org/10.1016/j.yfrne.2016.03.002.

108. Malberg JE, Duman RS. Cell proliferation in adult hippocampus is decreased by inescapable stress: reversal by fluoxetine treatment. *Neuropsychopharmacology.* 2003;28(9):1562–1571.

109. Mandyam CD, Crawford EF, Eisch AJ, Rivier CL, Richardson HN. Stress experienced in utero reduces sexual dichotomies in neurogenesis, microenvironment, and cell death in the adult rat hippocampus. *Dev Neurobiol.* 2008;68(5):575–589.

110. Marco EM, Llorente R, Lopez-Gallardo M, et al. The maternal deprivation animal model revisited. *Neurosci Biobehav Rev.* 2015;51:151–163. http://dx.doi.org/10.1016/j.neubiorev.2015.01.015.

111. Marlatt MW, Potter MC, Lucassen PJ, van Praag H. Running throughout middle-age improves memory function, hippocampal neurogenesis, and BDNF levels in female C57BL/6J mice. *Dev Neurobiol.* 2012;72(6):943–952. http://dx.doi.org/10.1002/dneu.22009.

112. Marlatt MW, Philippens I, Manders E, et al. Distinct structural plasticity in the hippocampus and amygdala of the middle-aged common marmoset (*Callithrix jacchus*). *Exp Neurol.* 2011;230:91–301.

113. Maselko J, Kubzansky L, Lipsitt L, Buka SL. Mother's affection at 8 months predicts emotional distress in adulthood. *J Epidemiol Community Health*. 2011;65(7):621–625. http://dx.doi.org/10.1136/jech.2009.097873.

114. Matos RJ, Orozco-Solis R, Lopes de Souza S, Manhaes-de-Castro R, Bolanos-Jimenez F. Nutrient restriction during early life reduces cell proliferation in the hippocampus at adulthood but does not impair the neuronal differentiation process of the new generated cells. *Neuroscience*. 2011;196:16–24. http://dx.doi.org/10.1016/j.neuroscience.2011.08.071.

115. Mayer JL, Klumpers L, Maslam S, de Kloet ER, Joels M, Lucassen PJ. Brief treatment with the glucocorticoid receptor antagonist mifepristone normalises the corticosterone-induced reduction of adult hippocampal neurogenesis. *J Neuroendocrinol*. 2006;18(8):629–631.

116. McGowan PO, Szyf M. The epigenetics of social adversity in early life: implications for mental health outcomes. *Neurobiol Dis*. 2010;39(1):66–72. http://dx.doi.org/10.1016/j.nbd.2009.12.026.

117. Meaney MJ, Szyf M, Seckl JR. Epigenetic mechanisms of perinatal programming of hypothalamic-pituitary-adrenal function and health. *Trends Mol Med*. 2007;13(7):269–277. http://dx.doi.org/10.1016/j.molmed.2007.05.003.

118. Medina A, Seasholtz AF, Sharma V, et al. Glucocorticoid and mineralocorticoid receptor expression in the human hippocampus in major depressive disorder. *J Psychiatr Res*. 2013;47(3):307–314. http://dx.doi.org/10.1016/j.jpsychires.2012.11.002.

119. Miller JA, Nathanson J, Franjic D, et al. Conserved molecular signatures of neurogenesis in the hippocampal subgranular zone of rodents and primates. *Development*. 2013;140(22):4633–4644. http://dx.doi.org/10.1242/dev.097212.

120. Mirescu C, Gould E. Stress and adult neurogenesis. *Hippocampus*. 2006;16(3):233–238.

121. Mirescu C, Peters JD, Gould E. Early life experience alters response of adult neurogenesis to stress. *Nat Neurosci*. 2004;7(8):841–846.

122. Mitra R, Sundlass K, Parker KJ, Schatzberg AF, Lyons DM. Social stress-related behavior affects hippocampal cell proliferation in mice. *Physiol Behav*. 2006;89(2):123–127.

123. Montaron MF, Drapeau E, Dupret D, et al. Lifelong corticosterone level determines age-related decline in neurogenesis and memory. *Neurobiol Aging*. 2006;27(4):645–654.

124. Montaron MF, Koehl M, Lemaire V, Drapeau E, Abrous DN, Le Moal M. Environmentally induced long-term structural changes: cues for functional orientation and vulnerabilities. *Neurotox Res*. 2004;6(7–8):571–580.

125. Montaron MF, Piazza PV, Aurousseau C, Urani A, Le Moal M, Abrous DN. Implication of corticosteroid receptors in the regulation of hippocampal structural plasticity. *Eur J Neurosci*. 2003;18(11):3105–3111.

126. Morris GP, Clark IA, Zinn R, Vissel B. Microglia: a new frontier for synaptic plasticity, learning and memory, and neurodegenerative disease research. *Neurobiol Learn Mem*. 2013;105:40–53. http://dx.doi.org/10.1016/j.nlm.2013.07.002.

127. Musaelyan K, Egeland M, Fernandes C, Pariante CM, Zunszain PA, Thuret S. Modulation of adult hippocampal neurogenesis by early-life environmental challenges triggering immune activation. *Neural Plast*. 2014;2014:194396. http://dx.doi.org/10.1155/2014/194396.

128. Nacher J, McEwen BS. The role of N-methyl-D-asparate receptors in neurogenesis. *Hippocampus*. 2006;16(3):267–270.

129. Naninck EF, Hoeijmakers L, Kakava-Georgiadou N, et al. Chronic early life stress alters developmental and adult neurogenesis and impairs cognitive function in mice. *Hippocampus*. 2015;25(3):309–328. http://dx.doi.org/10.1002/hipo.22374.

130. Naylor AS, Bull C, Nilsson MK, et al. Voluntary running rescues adult hippocampal neurogenesis after irradiation of the young mouse brain. *Proc Natl Acad Sci USA*. 2008;105(38):14632–14637. http://dx.doi.org/10.1073/pnas.0711128105.

131. Nemeroff CB. Paradise lost: the neurobiological and clinical consequences of child abuse and neglect. *Neuron*. 2016;89(5):892–909. http://dx.doi.org/10.1016/j.neuron.2016.01.019.

132. Nishijima T, Llorens-Martin M, Tejeda GS, et al. Cessation of voluntary wheel running increases anxiety-like behavior and impairs adult hippocampal neurogenesis in mice. *Behav Brain Res*. 2013;245:34–41. http://dx.doi.org/10.1016/j.bbr.2013.02.009.

133. O'Leary OF, Cryan JF. A ventral view on antidepressant action: roles for adult hippocampal neurogenesis along the dorsoventral axis. *Trends Pharmacol Sci*. 2014;35(12):675–687. http://dx.doi.org/10.1016/j.tips.2014.09.011.

134. O'Leary OF, O'Connor RM, Cryan JF. Lithium-induced effects on adult hippocampal neurogenesis are topographically segregated along the dorso-ventral axis of stressed mice. *Neuropharmacology*. 2012;62(1):247–255. http://dx.doi.org/10.1016/j.neuropharm.2011.07.015.

135. Oomen CA, Bekinschtein P, Kent BA, Saksida LM, Bussey TJ. Adult hippocampal neurogenesis and its role in cognition. *Wiley Interdiscip Rev Cogn Sci*. 2014;5(5):573–587. http://dx.doi.org/10.1002/wcs.1304.

136. Oomen CA, Girardi CE, Cahyadi R, et al. Opposite effects of early maternal deprivation on neurogenesis in male versus female rats. *PLoS One*. 2009;4(1):e3675.

137. Oomen CA, Mayer JL, de Kloet ER, Joels M, Lucassen PJ. Brief treatment with the glucocorticoid receptor antagonist mifepristone normalizes the reduction in neurogenesis after chronic stress. *Eur J Neurosci*. 2007;26(12):3395–3401.

138. Oomen CA, Soeters H, Audureau N, et al. Early maternal deprivation affects dentate gyrus structure and emotional learning in adult female rats. *Psychopharmacology Berl*. 2011;214(1):249–260. http://dx.doi.org/10.1007/s00213-010-1922-8.

139. Oomen CA, Soeters H, Audureau N, et al. Severe early life stress hampers spatial learning and neurogenesis, but improves hippocampal synaptic plasticity and emotional learning under high-stress conditions in adulthood. *J Neurosci*. 2010;30(19):6635–6645. http://dx.doi.org/10.1523/JNEUROSCI.0247-10.2010.

140. Palmer TD, Willhoite AR, Gage FH. Vascular niche for adult hippocampal neurogenesis. *J Comp Neurol*. 2000;425(4):479–494.

141. Parihar VK, Hattiangady B, Kuruba R, Shuai B, Shetty AK. Predictable chronic mild stress improves mood, hippocampal neurogenesis and memory. *Mol Psychiatry*. 2011;16(2):171–183. http://dx.doi.org/10.1038/mp.2009.130.

142. Perera TD, Dwork AJ, Keegan KA, et al. Necessity of hippocampal neurogenesis for the therapeutic action of antidepressants in adult nonhuman primates. *PLoS One*. 2011;6(4):e17600. http://dx.doi.org/10.1371/journal.pone.0017600.

143. Pham K, McEwen BS, Ledoux JE, Nader K. Fear learning transiently impairs hippocampal cell proliferation. *Neuroscience*. 2005;130(1):17–24.

144. Pham K, Nacher J, Hof PR, McEwen BS. Repeated restraint stress suppresses neurogenesis and induces biphasic PSA-NCAM expression in the adult rat dentate gyrus. *Eur J Neurosci*. 2003;17(4):879–886.

145. Pryce CR, Feldon J. Long-term neurobehavioural impact of the postnatal environment in rats: manipulations, effects and mediating mechanisms. *Neurosci Biobehav Rev*. 2003;27(1–2):57–71.

146. Qi XR, Kamphuis W, Wang S, et al. Aberrant stress hormone receptor balance in the human prefrontal cortex and hypothalamic paraventricular nucleus of depressed patients. *Psychoneuroendocrinology*. 2013;38(6):863–870. http://dx.doi.org/10.1016/j.psyneuen.2012.09.014.

147. Qian X, Droste SK, Lightman SL, Reul JM, Linthorst AC. Circadian and ultradian rhythms of free glucocorticoid hormone are highly synchronized between the blood, the subcutaneous tissue, and the brain. *Endocrinology*. 2012;153(9):4346–4353. http://dx.doi.org/10.1210/en.2012-1484.

148. Revest JM, Dupret D, Koehl M, et al. Adult hippocampal neurogenesis is involved in anxiety-related behaviors. *Mol Psychiatry*. 2009;14(10):959–967. http://dx.doi.org/10.1038/mp.2009.15.

149. Ridder S, Chourbaji S, Hellweg R, et al. Mice with genetically altered glucocorticoid receptor expression show altered sensitivity for stress-induced depressive reactions. *J Neurosci*. 2005;25(26):6243–6250. http://dx.doi.org/10.1523/JNEUROSCI.0736-05.2005.

150. Sah A, Schmuckermair C, Sartori SB, et al. Anxiety- rather than depression-like behavior is associated with adult neurogenesis in a female mouse model of higher trait anxiety- and comorbid depression-like behavior. *Transl Psychiatry*. 2012;2:e171. http://dx.doi.org/10.1038/tp.2012.94.

151. Sahay A, Hen R. Adult hippocampal neurogenesis in depression. *Nat Neurosci*. 2007;10(9):1110–1115.

152. Sahay A, Scobie KN, Hill AS, et al. Increasing adult hippocampal neurogenesis is sufficient to improve pattern separation. *Nature*. 2011;472(7344):466–470. http://dx.doi.org/10.1038/nature09817.

153. Sahay A, Wilson DA, Hen R. Pattern separation: a common function for new neurons in hippocampus and olfactory bulb. *Neuron*. 2011;70(4):582–588. http://dx.doi.org/10.1016/j.neuron.2011.05.012.

154. Sampedro-Piquero P, Begega A, Arias JL. Increase of glucocorticoid receptor expression after environmental enrichment: relations to spatial memory, exploration and anxiety-related behaviors. *Physiol Behav*. 2014;129:118–129. http://dx.doi.org/10.1016/j.physbeh.2014.02.048.

155. Santarelli L, Saxe M, Gross C, et al. Requirement of hippocampal neurogenesis for the behavioral effects of antidepressants. *Science*. 2003;301(5634):805–809.

156. Sapolsky RM, Uno H, Rebert CS, Finch CE. Hippocampal damage associated with prolonged glucocorticoid exposure in primates. *J Neurosci*. 1990;10(9):2897–2902.

157. Sarabdjitsingh RA, Conway-Campbell BL, Leggett JD, et al. Stress responsiveness varies over the ultradian glucocorticoid cycle in a brain-region-specific manner. *Endocrinology*. 2010;151(11):5369–5379. http://dx.doi.org/10.1210/en.2010-0832.

158. Sarabdjitsingh RA, Jezequel J, Pasricha N, et al. Ultradian corticosterone pulses balance glutamatergic transmission and synaptic plasticity. *Proc Natl Acad Sci USA*. 2014;111(39):14265–14270. http://dx.doi.org/10.1073/pnas.1411216111.

159. Sarabdjitsingh RA, Joels M, de Kloet ER. Glucocorticoid pulsatility and rapid corticosteroid actions in the central stress response. *Physiol Behav*. 2012;106(1):73–80. http://dx.doi.org/10.1016/j.physbeh.2011.09.017.

160. Saxe MD, Battaglia F, Wang JW, et al. Ablation of hippocampal neurogenesis impairs contextual fear conditioning and synaptic plasticity in the dentate gyrus. *Proc Natl Acad Sci USA*. 2006;103(46):17501–17506.

161. Schloesser RJ, Lehmann M, Martinowich K, Manji HK, Herkenham M. Environmental enrichment requires adult neurogenesis to facilitate the recovery from psychosocial stress. *Mol Psychiatry*. 2010;15(12):1152–1163. http://dx.doi.org/10.1038/mp.2010.34.

162. Schmidt HD, Duman RS. The role of neurotrophic factors in adult hippocampal neurogenesis, antidepressant treatments and animal models of depressive-like behavior. *Behav Pharmacol*. 2007;18(5–6):391–418. http://dx.doi.org/10.1097/FBP.0b013e3282ee2aa8.

163. Schmitz C, Rhodes ME, Bludau M, et al. Depression: reduced number of granule cells in the hippocampus of female, but not male, rats due to prenatal restraint stress. *Mol Psychiatry*. 2002;7(7):810–813.

164. Schoenfeld TJ, Cameron HA. Adult neurogenesis and mental illness. *Neuropsychopharmacology*. 2015;40(1):113–128. http://dx.doi.org/10.1038/npp.2014.230.

165. Schoenfeld TJ, Gould E. Differential effects of stress and glucocorticoids on adult neurogenesis. *Curr Top Behav Neurosci*. 2013;15:139–164. http://dx.doi.org/10.1007/7854_2012_233.

166. Schouten M, Buijink MR, Lucassen PJ, Fitzsimons CP. New neurons in aging brains: molecular control by small non-coding RNAs. *Front Neurosci*. 2012;6:25. http://dx.doi.org/10.3389/fnins.2012.00025.

167. Schwabe L, Joels M, Roozendaal B, Wolf OT, Oitzl MS. Stress effects on memory: an update and integration. *Neurosci Biobehav Rev*. 2012;36(7):1740–1749. http://dx.doi.org/10.1016/j.neubiorev.2011.07.002.

168. Seo DO, Carillo MA, Chih-Hsiung Lim S, Tanaka KF, Drew MR. Adult hippocampal neurogenesis modulates fear learning through associative and nonassociative mechanisms. *J Neurosci*. 2015;35(32):11330–11345. http://dx.doi.org/10.1523/JNEUROSCI.0483-15.2015.

169. Sierra A, Beccari S, Diaz-Aparicio I, Encinas JM, Comeau S, Tremblay ME. Surveillance, phagocytosis, and inflammation: how never-resting microglia influence adult hippocampal neurogenesis. *Neural Plast*. 2014;610343. http://dx.doi.org/10.1155/2014/610343.

170. Sierra A, Encinas JM, Deudero JJ, et al. Microglia shape adult hippocampal neurogenesis through apoptosis-coupled phagocytosis. *Cell Stem Cell*. 2010;7(4):483–495. http://dx.doi.org/10.1016/j.stem.2010.08.014.

171. Sierra A, Martin-Suarez S, Valcarcel-Martin R, et al. Neuronal hyperactivity accelerates depletion of neural stem cells and impairs hippocampal neurogenesis. *Cell Stem Cell*. 2015;16(5): 488–503. http://dx.doi.org/10.1016/j.stem.2015.04.003.

172. Simon M, Czeh B, Fuchs E. Age-dependent susceptibility of adult hippocampal cell proliferation to chronic psychosocial stress. *Brain Res*. 2005;1049(2):244–248.

173. Singh AK, Gupta S, Jiang Y, Younus M, Ramzan M. In vitro neurogenesis from neural progenitor cells isolated from the hippocampus region of the brain of adult rats exposed to ethanol during early development through their alcohol-drinking mothers. *Alcohol Alcohol*. 2009;44(2):185–198. http://dx.doi.org/10.1093/alcalc/agn109.

174. Singh-Taylor A, Korosi A, Molet J, Gunn BG, Baram TZ. Synaptic rewiring of stress-sensitive neurons by early-life experience: a mechanism for resilience? *Neurobiol Stress*. 2015;1:109–115. http://dx.doi.org/10.1016/j.ynstr.2014.10.007.

175. Snyder JS, Soumier A, Brewer M, Pickel J, Cameron HA. Adult hippocampal neurogenesis buffers stress responses and depressive behaviour. *Nature*. 2011;476(7361):458–461. http://dx.doi.org/10.1038/nature10287.

176. Sousa N. The dynamics of the stress neuromatrix. *Mol Psychiatry*. 2016;21:302–312.

177. Spadafora R, Gonzalez FF, Derugin N, Wendland M, Ferriero D, McQuillen P. Altered fate of subventricular zone progenitor cells and reduced neurogenesis following neonatal stroke. *Dev Neurosci*. 2010;32(2):101–113. http://dx.doi.org/10.1159/000279654.

178. Sultan S, Li L, Moss J, et al. Synaptic integration of adult-born hippocampal neurons is locally controlled by astrocytes. *Neuron*. 2015;88(5):957–972. http://dx.doi.org/10.1016/j.neuron.2015.10.037.

179. Surget A, Tanti A, Leonardo ED, et al. Antidepressants recruit new neurons to improve stress response regulation. *Mol Psychiatry*. 2011;16(12):1177–1188. http://dx.doi.org/10.1038/mp.2011.48.

180. Suri D, Veenit V, Sarkar A, et al. Early stress evokes age-dependent biphasic changes in hippocampal neurogenesis, BDNF expression, and cognition. *Biol Psychiatry*. 2013;73(7):658–666. http://dx.doi.org/10.1016/j.biopsych.2012.10.023.

181. Swaab DF, Bao AM, Lucassen PJ. The stress system in the human brain in depression and neurodegeneration. *Ageing Res Rev*. 2005;4(2):141–194.

182. Takamura N, Nakagawa S, Masuda T, et al. The effect of dopamine on adult hippocampal neurogenesis. *Prog Neuropsychopharmacol Biol Psychiatry*. 2014;50:116–124. http://dx.doi.org/10.1016/j.pnpbp.2013.12.011.

183. Tamura M, Sajo M, Kakita A, Matsuki N, Koyama R. Prenatal stress inhibits neuronal maturation through downregulation of mineralocorticoid receptors. *J Neurosci.* 2011;31(32):11505–11514. http://dx.doi.org/10.1523/JNEUROSCI.3447-10.2011.

184. Tanti A, Belzung C. Neurogenesis along the septo-temporal axis of the hippocampus: are depression and the action of antidepressants region-specific? *Neuroscience.* 2013;252:234–252. http://dx.doi.org/10.1016/j.neuroscience.2013.08.017.

185. Tanti A, Rainer Q, Minier F, Surget A, Belzung C. Differential environmental regulation of neurogenesis along the septo-temporal axis of the hippocampus. *Neuropharmacology.* 2012;63(3):374–384. http://dx.doi.org/10.1016/j.neuropharm.2012.04.022.

186. Tasker JG. Rapid glucocorticoid actions in the hypothalamus as a mechanism of homeostatic integration. *Obesity (Silver Spring).* 2006;14(suppl 5):259S–265S. http://dx.doi.org/10.1038/oby.2006.320.

187. Tauber SC, Bunkowski S, Schlumbohm C, et al. No long-term effect two years after intrauterine exposure to dexamethasone on dentate gyrus volume, neuronal proliferation and differentiation in common marmoset monkeys. *Brain Pathol.* 2008;18(4):497–503.

188. Teicher MH, Anderson CM, Polcari A. Childhood maltreatment is associated with reduced volume in the hippocampal subfields CA3, dentate gyrus, and subiculum. *Proc Natl Acad Sci USA.* 2012;109(9):E563–E572. http://dx.doi.org/10.1073/pnas.1115396109.

189. Toni N, Laplagne DA, Zhao C, et al. Neurons born in the adult dentate gyrus form functional synapses with target cells. *Nat Neurosci.* 2008;11(8):901–907.

190. Ursin H, Eriksen HR. The cognitive activation theory of stress. *Psychoneuroendocrinology.* 2004;29(5):567–592. http://dx.doi.org/10.1016/S0306-4530(03)00091-X.

191. Vallieres L, Campbell IL, Gage FH, Sawchenko PE. Reduced hippocampal neurogenesis in adult transgenic mice with chronic astrocytic production of interleukin-6. *J Neurosci.* 2002;22(2):486–492.

192. Van Bokhoven P, Oomen CA, Hoogendijk WJ, Smit AB, Lucassen PJ, Spijker S. Reduction in hippocampal neurogenesis after social defeat is long-lasting and responsive to late antidepressant treatment. *Eur J Neurosci.* 2011;33(10):1833–1840. http://dx.doi.org/10.1111/j.1460-9568.2011.07668.x.

193. Van der Borght K, Meerlo P, Luiten PG, Eggen BJ, Van der Zee EA. Effects of active shock avoidance learning on hippocampal neurogenesis and plasma levels of corticosterone. *Behav Brain Res.* 2005;157(1):23–30.

194. van der Doelen RH, Kozicz T, Homberg JR. Adaptive fitness; early life adversity improves adult stress coping in heterozygous serotonin transporter knockout rats. *Mol Psychiatry.* 2013;18(12):1244–1245. http://dx.doi.org/10.1038/mp.2012.186.

195. van Praag H, Kempermann G, Gage FH. Running increases cell proliferation and neurogenesis in the adult mouse dentate gyrus. *Nat Neurosci.* 1999;2(3):266–270.

196. Veena J, Srikumar BN, Mahati K, Bhagya V, Raju TR, Shankaranarayana Rao BS. Enriched environment restores hippocampal cell proliferation and ameliorates cognitive deficits in chronically stressed rats. *J Neurosci Res.* 2009;87(4):831–843.

197. Vinkers CH, Joels M, Milaneschi Y, Kahn RS, Penninx BW, Boks MP. Stress exposure across the life span cumulatively increases depression risk and is moderated by neuroticism. *Depress Anxiety.* 2014;31(9):737–745. http://dx.doi.org/10.1002/da.22262.

198. Vivar C, Potter MC, Choi J, et al. Monosynaptic inputs to new neurons in the dentate gyrus. *Nat Commun.* 2012;3:1107. http://dx.doi.org/10.1038/ncomms2101.

199. Vivar C, Potter MC, van Praag H. All about running: synaptic plasticity, growth factors and adult hippocampal neurogenesis. *Curr Top Behav Neurosci.* 2013;15:189–210. http://dx.doi.org/10.1007/7854_2012_220.

200. Vivinetto AL, Suarez MM, Rivarola MA. Neurobiological effects of neonatal maternal separation and post-weaning environmental enrichment. *Behav Brain Res.* 2013;240:110–118. http://dx.doi.org/10.1016/j.bbr.2012.11.014.

201. Vollmayr B, Simonis C, Weber S, Gass P, Henn F. Reduced cell proliferation in the dentate gyrus is not correlated with the development of learned helplessness. *Biol Psychiatry.* 2003;54(10):1035–1040.

202. Wang Q, Joels M, Swaab DF, Lucassen PJ. Hippocampal GR expression is increased in elderly depressed females. *Neuropharmacology.* 2012;62(1):527–533. http://dx.doi.org/10.1016/j.neuropharm.2011.09.014.

203. Wang Q, Van Heerikhuize J, Aronica E, et al. Glucocorticoid receptor protein expression in human hippocampus; stability with age. *Neurobiol Aging.* 2013;34(6):1662–1673. http://dx.doi.org/10.1016/j.neurobiolaging.2012.11.019.

204. Weaver IC, Szyf M, Meaney MJ. From maternal care to gene expression: DNA methylation and the maternal programming of stress responses. *Endocr Res.* 2002;28(4):699.

205. Wei Q, Hebda-Bauer EK, Pletsch A, et al. Overexpressing the glucocorticoid receptor in forebrain causes an aging-like neuroendocrine phenotype and mild cognitive dysfunction. *J Neurosci.* 2007;27(33):8836–8844. http://dx.doi.org/10.1523/JNEUROSCI.0910-07.2007.

206. Weinstock M. Sex-dependent changes induced by prenatal stress in cortical and hippocampal morphology and behaviour in rats: an update. *Stress.* 2011;14(6):604–613. http://dx.doi.org/10.3109/10253890.2011.588294.

207. Westenbroek C, Den Boer JA, Veenhuis M, Ter Horst GJ. Chronic stress and social housing differentially affect neurogenesis in male and female rats. *Brain Res Bull.* 2004;64(4):303–308.

208. Wilson CB, Ebenezer PJ, McLaughlin LD, Francis J. Predator exposure/psychosocial stress animal model of post-traumatic stress disorder modulates neurotransmitters in the rat hippocampus and prefrontal cortex. *PLoS One.* 2014;9(2):e89104. http://dx.doi.org/10.1371/journal.pone.0089104.

209. Wong EY, Herbert J. The corticoid environment: a determining factor for neural progenitors' survival in the adult hippocampus. *Eur J Neurosci.* 2004;20(10):2491–2498.

210. Wong EY, Herbert J. Roles of mineralocorticoid and glucocorticoid receptors in the regulation of progenitor proliferation in the adult hippocampus. *Eur J Neurosci.* 2005;22(4):785–792.

211. Wong EY, Herbert J. Raised circulating corticosterone inhibits neuronal differentiation of progenitor cells in the adult hippocampus. *Neuroscience.* 2006;137(1):83–92.

212. Wu MV, Shamy JL, Bedi G, et al. Impact of social status and antidepressant treatment on neurogenesis in the baboon hippocampus. *Neuropsychopharmacology.* 2014;39(8):1861–1871. http://dx.doi.org/10.1038/npp.2014.33.

213. Yu S, Patchev AV, Wu Y, et al. Depletion of the neural precursor cell pool by glucocorticoids. *Ann Neurol.* 2010;67(1):21–30. http://dx.doi.org/10.1002/ana.21812.

214. Yun S, Donovan MH, Ross MN, et al. Stress-induced anxiety- and depressive-like phenotype associated with transient reduction in neurogenesis in adult nestin-CreERT2/diphtheria toxin fragment a transgenic mice. *PLoS One.* 2016;11(1):e0147256. http://dx.doi.org/10.1371/journal.pone.0147256.

215. Zhang CL, Zou Y, He W, Gage FH, Evans RM. A role for adult TLX-positive neural stem cells in learning and behaviour. *Nature.* 2008;451(7181):1004–1007. http://dx.doi.org/10.1038/nature06562.

216. Zhao C, Deng W, Gage FH. Mechanisms and functional implications of adult neurogenesis. *Cell.* 2008;132(4):645–660. http://dx.doi.org/10.1016/j.cell.2008.01.033.

217. Zhu C, Gao J, Karlsson N, et al. Isoflurane anesthesia induced persistent, progressive memory impairment, caused a loss of neural stem cells, and reduced neurogenesis in young, but not adult, rodents. *J Cereb Blood Flow Metab.* 2010;30(5):1017–1030. http://dx.doi.org/10.1038/jcbfm.2009.274.

14

Stress, Alcohol and Epigenetic Transmission

D.K. Sarkar

Rutgers, The State University of New Jersey, New Brunswick, NJ, United States

Abstract

The stress system is a major neuroendocrine system with pivotal regulatory physiological functions. This system is regulated by intricate communication between the hypothalamus, pituitary, and adrenal glands. Stress axis function is often perturbed by alcohol abuse that lasts long and which is linked to molecular, neurophysiological, and behavioral changes in exposed individuals. The cellular mechanism involved in the development of aberrant stress response to alcohol is multifactorial and partly controlled by epigenetic modifications of various genes of the stress regulatory system. One of these targeted genes is the stress response inhibitory proopiomelanocortin gene, which is hypermethylated and its transcription reduced by alcohol. Recent evidence that the alcohol epigenetic changes on the proopiomelanocortin gene are inherited provides a plausible explanation for transmission of alcohol-induced complex behavioral traits including aberrant stress responses across generations.

THE STRESS RESPONSE SYSTEM

Stress is a biological and psychological response experienced on encountering a change in the body's internal or external environment. In response to a stress, the body initiates a cascade of physiological changes in the central nervous system (CNS) and periphery, that subsequently triggers the autonomic nervous system (ANS) and the hypothalamic–pituitary–adrenal axis (HPA). Stressful experiences activate components of the limbic system including the hippocampus and the amygdala that modulate the activity of hypothalamic and brainstem structures controlling the HPA axis and ANS activities. The hypothalamus secretes corticotrophin-releasing factor (CRF) and vasopressin from the paraventricular nucleus (PVN) of the hypothalamus, both of which activate the pituitary to release proopiomelanocortin (POMC)-derived peptides such as adrenocorticotropic hormone (ACTH) that stimulates the secretion of glucocorticoids from the adrenal cortex into the peripheral circulation. In addition to the activation of the HPA axis, stressful stimuli also stimulate the sympathetic nervous system (SNS) that mediates the "fight or flight" response. Like other parts of the nervous system, the SNS operates through a series of interconnected neurons, many of which have direct connections with the neurons of the PVN and also project to the sympathetic ganglia. From these ganglia, postganglionic fibers run to effector organs including spleen, lymphoid tissues, liver, and many other peripheral organs. The main neurotransmitter of the preganglionic sympathetic fibers is acetylcholine, a chemical messenger that binds and activates nicotinic acetylcholine receptors on postganglionic neurons. The principal neurotransmitter released by the postganglionic neurons

Stress: Neuroendocrinology and Neurobiology
http://dx.doi.org/10.1016/B978-0-12-802175-0.00014-0

is noradrenaline (norepinephrine), which act on beta-adrenergic receptors on cardiovascular system and many other peripheral tissues. The other division of the ANS is the parasympathetic nervous system, which is responsible for stimulation of "rest-and-digest" activi-

KEY POINTS

Moderate alcohol drinking is known to relieve body's stress response, but copious alcohol consumptions increase the body's stress response. This alcohol-altered stress response is caused by a disrupted feedback regulation of the various components of the hypothalamic-pituitary-adrenal (HPA) axis that often results in the development of alcohol use disorders. The HPA axis is very vulnerable to the detrimental effect of alcohol particularly during the developmental period. Alcohol exposure during the developmental period leads to an increased incidence of adult-onset stress-related diseases that is long-lasting and may carry over across generations. Alcohol effects on the developmental abnormalities are often contributed by epigenetic modifications of genes regulating cellular functions. Epigenetic changes of genes have been gaining traction for a mechanism of inheritance of certain behavioral traits. Recent studies using animal models of alcohol drinking have provided evidence for the transmission of alcohol epigenetic marks on the genes that govern the HPA and stress-related disorders possibly across generations.

ties that occur when the body is at rest and is normally functioning in opposition to the sympathetic neurons. The main neurotransmitter of the parasympathetic system is acetylcholine, which acts on two types of receptors, the nicotinic cholinergic at postganglionic neurons and muscarinic at the target organ.[31]

The HPA axis plays a key role in the body's stress response. Stress increases the production and secretion of the CRF peptide from a neuronal population in the PVN region of the hypothalamus. The CRF from these cells is released into the hypothalamic-pituitary portal vessels and thereby transmitted to the anterior pituitary gland, where it acts by way of the CRF receptor1 (CRFR1), on corticotropic cells to trigger the synthesis and release of ACTH into the systemic circulation. ACTH, in turn, stimulates the release of glucocorticoids (i.e., corticosterone in rats and cortisol in humans) from the cortex of the adrenal glands. Glucocorticoids act on numerous tissues throughout the organism to coordinate the body's stress response as described in detail, elsewhere in this

handbook. The CRF/CRFR1 system is also located in areas of the brain outside the HPA and plays a key role in the central stress response systems.[2]

The CRF–HPA system is controlled by several central neurotransmitters including opioid peptides (e.g., β-endorphin; β-EP) and their receptors. For example, in the rats, CRF-producing neurons in the PVN are collocated with the fibers of β-EP-releasing neurons. In the median eminence μ-opioid receptors (MOP-r) are located on the terminals of CRF-releasing neurons. Agents that stimulate the activity of MOP-r can inhibit neurotransmitter-stimulated CRF release from the hypothalamus in vitro. Likewise, β-EP infusion decreases CRF release into the hypothalamic-pituitary portal vessels, and pretreatment with the MOP-r agonist, morphine, prevents stress-induced HPA activation. Finally, transplantation of β-EP-producing cells into the PVN suppresses HPA activation and reduces stress responses to various stressors. Hence, endogenous opioids (and, by extension, opiate drugs) moderate the stress response.[25]

ALCOHOL (ETHANOL) AND THE STRESS RESPONSE

There is substantial evidence that alcohol consumption affects the stress response pathways and activates the HPA axis. In humans, plasma cortisol levels increase in healthy subjects at alcohol doses exceeding 100 mg/dL equivalent to 3–4 standard drinks at a session. Healthy men who were in the top percentile of self-reported alcohol consumption often show higher levels of excreted cortisol in urine. In addition, in heavy drinkers, the inhibitory control of the HPA axis was impaired.[30] The development of alcohol use disorders progresses from a heightened to a blunted HPA stress responsiveness. For example, numerous studies have reported that individuals with a family history of alcoholism have a higher risk of developing the disorder than those with no family history and that this risk seems to be associated with dysregulated HPA axis activity.[32] Alcohol dependence was associated with a decrease in Crf mRNA expression in the PVN as well as a reduced responsiveness of the pituitary to CRF.[23]

Similar findings were reported in animal studies. An acute administration of ethanol to rats increases plasma ACTH and corticosterone levels, through an enhanced release of CRF from the hypothalamus. Neutralization of circulating CRF with antibodies inhibits the stimulatory actions of ethanol on ACTH and corticosterone secretion.[24] Using an operant self-administration animal model of alcohol dependence, it was shown that the HPA response to several weeks of daily 30-min self-administration of alcohol was higher in "low-responding" nondependent animals (<0.2 mg/kg/session), intermediate in nondependent animals (~0.4 mg/kg/session), and most blunted in dependent animals (~1.0 mg/kg/session).[23] Attenuated cortisol response to alcohol was also detected in heavy social drinkers.[9] In an animal model, when ethanol was introduced gradually using a liquid diet paradigm for 4 weeks followed by 3 weeks withdrawal, there was a persistent reduction in the HPA activity.[20] These persistent alterations in the HPA axis activity in chronic alcohol drinking animal model are similar to the aberrant HPA function observed in abstinent alcoholics and sons of alcoholics, thereby suggesting a connection between HPA dysregulation and the vulnerability to relapse.[19]

Several studies have now connected the aberrant HPA function in the mechanisms involved in alcohol dependence. Studies have shown that low levels of CRF are often associated with more intense cravings and an increased probability of relapse.[29] In Crh knock-out mice, it was shown that ethanol exposure, in a continuous- or a limited-access paradigm, consumed twice as much ethanol as their wild-type counterparts.[15] These Crh-deficient mice exhibited a reduced sensitivity to the locomotor stimulant and rewarding effects of ethanol. In rhesus macaques, a single nucleotide polymorphism in the promoter region of Crh (−248C→T) conferred increased stress reactivity: in addition to higher levels of plasma ACTH and cortisol, these animals exhibited a suppression of environmental exploration (a behavioral response to social separation stress), and consumed more alcohol then controls.[3] Mice lacking a functional Crhr1 receptor, when subjected to repeated stress, progressively increased their ethanol intake, an effect that seemed to persist throughout life.[26] The Crhr2 receptor also contributes to the development of alcohol abuse behaviors: thus, since Crhr2 mRNA expression was found to be lower in alcohol-preferring (P) rats compared to nonpreferring (NP) rats.[34] P rats also showed polymorphisms in the coding region as well as the 3′-untranslated region of the Crhr2 promoter.[34] The Crhr2 promoter polymorphism was associated with reduced mRNA expression and reduced Crhr2 receptor density, particularly in the amygdala of P rats compared to NP rats. The authors also observed a decrease in social interactions and a higher corticosterone levels following 30-min restraint in P rats.[34]

Glucocorticoid signaling has also been implicated in the development and progression of alcohol use disorders. It is suggested that alcohol intoxication and withdrawal by activating the HPA axis and increasing circulating glucocorticoid levels induce the reinforcing effects of alcohol on drinking behavior. These effects of glucocorticoids are primarily activated by glucocorticoid receptors (GRs) and minimally by mineralocorticoid receptors (MRs). In a double-blind study on 56 alcohol-dependent human subjects who received the GR antagonist mifepristone exhibited a significant reduction in alcohol consumption during the 1 week treatment with mifepristone and the 1 week posttreatment.[33] The GR antagonist mifepristone

also dose-dependently decreased ethanol intake in rats exposed to ethanol using a limited-access model. On the other hand, the MR antagonist spironolactone had no effect on ethanol intake.[10] In an animal study, acute administration of ORG34517, a type II GR antagonist, during early withdrawal significantly reduced the ethanol-induced behavioral abnormalities, and although blood corticosterone levels were increased, no sedative-like effects were observed in these animals.[22] These studies identified an important role for glucocorticoids in the development of alcohol use disorders.

The stress response system influences alcohol use disorders by interacting with the reward pathways. For example, the CRF/CRFR1 system can activate mesolimbic dopaminergic pathways and increase dopamine-mediated signal transmission in various parts of the mesolimbic system, including the nucleus accumbens, amygdala, and medial prefrontal cortex. Corticosterone directly stimulates activity of the mesolimbic dopamine system, subsequently increasing drug-seeking behavior. Thus, stress, via activation of the CRH–HPA circuits and/or extrahypothalamic CRH circuits, increases mesolimbic dopamine that, in turn, increases drug seeking in drug-treated animals.[27]

Another aspect that adds to the complexity of the mechanisms leading to alcohol dependence is the relationship between the endogenous opioid system, the HPA axis, and the dopaminergic system.[16] The endogenous endorphin, β-EP, released from neurons in the arcuate nucleus of the hypothalamus, inhibits CRF release in the PVN and stimulates dopamine (DA) release in the nucleus accumbens.[5] Glucocorticoids, released during stress and following ethanol-induced activation of the HPA axis, can modulate the activities of the opioid, CRH, and mesolimbic dopaminergic systems and have been shown to interact with the rewarding properties of alcohol abuse. Several investigators have speculated that diminished opioid activity, which is either the result of alcoholism or genetically linked to the risk of alcoholism, could induce hypercortisolemia, alter mesolimbic DA production, and lead to abnormal ethanol reinforcement.[5,16] In fact, it was reported that individuals who are predisposed to alcoholism had a decreased synthesis and release of plasma β-EP and that administration of alcohol ameliorates this deficiency.[6] Recently, it has been shown that transplantation of β-EP-producing cells into the PVN suppresses HPA activation and anxiety-like behaviors and normalizes stress hyperresponse in prenatal alcohol-exposed adult rats and anxiety-like behavior and alcohol drinking in alcohol-preferring P rats.[12]

The 31-amino acid peptide β-EP, with a reasonably high affinity for the MOP-r and ∂-opioid receptors (DOP-r), is present in high concentrations in various areas of the brain. The β-EP-producing perikarya are located mainly in the ventromedial arcuate nucleus region, which projects to widespread brain structures, including many areas of the hypothalamus and the limbic system, where this opioid peptide has been proposed to function as a neurotransmitter or neuromodulator regulating a variety of brain functions. These brain functions include psychomotor stimulation; positive reinforcement; adaptive processes; drinking, eating, and sexual behaviors; pituitary function; thermoregulation; nociception; and mood. It is generally believed that the endogenous opioid system mediates some of the reinforcing properties of ethanol. Neurobiological studies indicate that alcohol alters opioid peptide systems. Opioid-receptor antagonists have been shown to decrease ethanol consumption. The extended amygdala is considered to be a site of the opioid action because a high proportion of cells in the medial nucleus of the amygdala and the bed nucleus of the stria terminalis express MOP-r and DOP-r. The extended amygdala is also a major brain area involved in excessive ethanol. These studies[7,21] clearly show that the opioid system, the HPA axis, and the reward system are intertwined, and that chronic alcohol-induced dysregulation of either of these systems play an important role in the reinforcing effects of alcohol on the development of alcohol use disorders.

ALCOHOL, EPIGENETIC AND STRESS RESPONSE

Epigenetic changes are now considered potential mechanisms for the long-term effects of toxicants or drugs of abuse including alcohol. The epigenetic mechanisms include DNA methylation, histone modifications, chromatin remodeling, and noncoding RNAs including microRNAs (miRNAs). Studies have demonstrated that ethanol exposure causes selective histone modifications in a tissue-specific manner and an alteration in neuronal gene expression and/or function.[13] Modulation in DNA methylation also has been linked to many neurological diseases or disorders including drug addiction.[14] However, only a limited number of studies have been conducted in determining the role of epigenetic mechanisms in alcohol-induced alteration of the HPA axis function.

By employing the animal model of fetal alcohol exposure (FAE), it was shown that alcohol fetal programming that causes lowering expression of the *Pomc* gene involves epigenetic mechanism. This is because fetal alcohol exposed animals showed increased methylation of several CpG dinucleotides in the proximal part of the *Pomc* gene promoter region.[8] CpG methylation in the promoter region of a gene most often correlates with silencing of its promoter activity. Measurements of the changes in protein and gene levels of histone-modifying proteins and DNMTs levels in POMC neurons provided plausible mechanisms by which alcohol programs histone modification and DNA methylation to increase *Pomc* gene methylation and decrease

gene expression. It was observed that FAE decreased the level of histone-modifying enzymes that methylate H3K4 and its associated gene Set7/9 and acetylate H3K9 and its associated gene CREB-binding protein (CBP). In addition, FAE increased the level of HDAC2, which is known to suppress H3K9 acetylation in the brain.[4] H3K4 methylation and H3K9 acetylation are known to activate gene expression. On the other hand, FAE increased the levels of histone-modifying enzymes that methylate H3K9 and its associated genes G9a and Setdb1. These are repressive marks for gene activation. Hence, increased H3K9 methylation/deacetylation and decreased H3K4 methylation might be the key modifications of the histone tail surrounding the POMC DNA in FAE animals (Fig. 14.1).

FAEs also induced some endophenotypes of the *Pomc* gene expression defect, including elevated basal and immune stimulus (lipopolysaccharide)-activated ACTH and corticosterone levels in plasma of both male and female offspring.[8] Noticeably, the suppression of histone deacetylation and DNA methylation by pharmacological agents normalized *Pomc* gene expression and POMC neuronal functional abnormalities (elevated corticosterone and ACTH responses to a stress challenge). These data suggest that FAE epigenetic mark on *Pomc* gene might

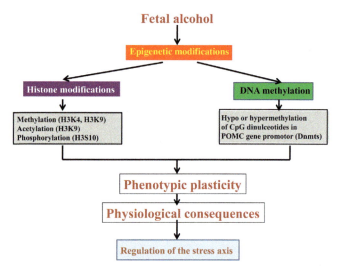

FIGURE 14.1 **Fetal alcohol marks on epigenetic machinery that may repress POMC gene expression.** It was observed that FAE decreases the level of histone-modifying enzymes that methylate H3 lysine 4 (H3K4) and acetylate H3 lysine 9 (H3K9). FAE also increases the levels of histone-modifying enzymes that methylate H3K9. Increased levels of DNA methyltransferases1 (Dnmt1), methyl CpG-binding protein (MeCP2), and methyl-binding proteins1 (MBD1) were also observed in POMC cells in fetal alcohol exposed offspring. It is postulated that histone modifications, such as H3 lysine 9 acetylation and H3K4 methylation, create a signal that regulates *Pomc* gene transcription. Histone deacetylases (HDACs) remove acetyl groups from H3 lysine residues making way for methylation. HDACs also activate DNA methyltransferases (DNMT) and methyl C-binding proteins (MBD1 and Mecp2) that aid in spreading the silencing signal. Reduction in *Pomc* gene transcription results in development of POMC neuronal plasticity and aberrant stress response.

lead to the abnormal production of stress hormones and hyperstress response for a prolonged period of time in the offspring. Pandey and colleagues[18] demonstrated that histone deacetylase inhibitor trichostatin A (TSA) treatment reversed the molecular changes that were observed in alcohol-withdrawn rats in the amygdala and attenuated some of the behavioral manifestations that were associated with alcohol withdrawal, such as anxiety. Collectively, these findings suggest that the components of the epigenetic machinery have potential medical applications for the treatment of some of the behavioral changes that are usually associated with alcohol exposure including the induction of an abnormal stress response. Behavioral tests should be performed in future studies to compare the stress response in control and fetal alcohol-exposed rats after inhibition of HDAC activity.

TRANSGENERATIONAL EFFECT OF ALCOHOL EPIGENETIC MARKS

In addition to the direct effects of early life exposure to toxicants or drugs of abuse on adult onset disease, it has been recently shown that they also affect subsequent generations. Transgenerational inheritance of epigenetic changes in the genome is an interesting area of research that is still subject to debate and not fully understood. It is now considered an additional molecular mechanism, along with classic induction of genetic mutations, for the germline transmission of environmentally induced phenotypic changes.[28]

Recently, it was shown that following in utero exposure to ethanol, regulatory regions of POMC in the hypothalamus of rats undergo epigenetic modifications: altered histone marks and DNA methylation of the proximal promoter. In addition, histone-modifying HDACs and DNA methyltransferases (DNMTs) were shown to be impacted, suggesting a causal relationship between alcohol and epigenetic changes. As a result, POMC neurons are impacted across at least three generations, perturbing the expression of key POMC-derived peptides, such as β-EP and affecting the production of its downstream messenger corticosterone leading to dysregulation of the HPA axis and an elevated response to stress in the adult offspring. This was the first demonstration of a true transgenerational epigenetic effect for prenatal alcohol exposure.[8] Interestingly, fetal alcohol effect was reversed through HDAC and DNA methylation inhibitors, providing additional support for this conclusion.[8] These results provide the first direct evidence that fetal alcohol effects on *Pomc* gene hypermethylation and stress axes abnormalities persist throughout adulthood and perpetuate into subsequent generations.

There are a number of studies in human that provide indirect support of the heritability of alcohol-related disorders in the literature. A recent study of Native American women who abused alcohol showed that the F2 generation offspring (that is the grandchildren) of an alcohol-abusing woman have a higher tendency to show fetal alcohol syndrome (FAS) than those F2 progeny of control women.[11] There is also evidence that hypomethylation occurs in the sperm of alcoholic men.[17] Transmission of the effects of alcohol through the paternal line has precedents in the literature for induction of symptoms like those found in Fetal Alcohol Spectrum Disorders (FASD). These include mental impairment, cardiac defects, low birth weight, and hyperactivity, compared with controls, as assessed in human epidemiological studies and backed up by animal studies.[1] This supports the findings of Govorko and colleagues[8] that factors that impact POMC and subsequently affect the HPA axis and FASD can be transmitted by males through the germline.

CONCLUSION

The stress response system is a key to homeostatic regulation in humans and other species during environmental challenges. Although alcohol (ethanol) is known to reduce the body's physiological stress response, copious alcohol drinking activates the stress response pathway and increases the HPA axis activity. Chronic alcohol drinking results in blunted HPA stress responsiveness and development of alcohol use disorders. The cellular mechanism involved in the development of blunted stress response to alcohol is multifactorial and may be controlled by epigenetic modification of various genes of the stress regulatory system including *Pomc* gene. Epigenetic inheritance has been gaining traction as a plausible explanation for transmission of complex behavioral traits across generations. Recent evidence from studies in laboratory rats showed transmission of alcohol epigenetic marks on the stress axis for three generations. As stress is a major risk factor for excessive and problematic alcohol drinking, this has significant implications for intergenerational alcohol drinking behavior.

References

1. Abel E. Paternal contribution to fetal alcohol syndrome. *Addict Biol.* 2004;9:127–133.
2. Bale TL, Vale WW. CRF and CRF receptors: role in stress responsivity and other behaviors. *Annu Rev Pharmacol Toxicol.* 2004;44:525–557.
3. Barr CS, Dvoskin RL, Gupte M, et al. Functional CRH variation increases stress-induced alcohol consumption in primates. *Proc Natl Acad Sci USA.* 2009;106:14593–14598.
4. Bekdash RA, Zhang C, Sarkar DK. Gestational choline supplementation normalized fetal alcohol-induced alterations in histone modifications, DNA methylation, and proopiomelanocortin (POMC) gene expression in β-endorphin-producing POMC neurons of the hypothalamus. *Alcohol Clin Exp Res.* 2013;37:1133–1142.

5. Gianoulakis C. Influence of the endogenous opioid system on high alcohol consumption and genetic predisposition to alcoholism. *J Psychiatry Neurosci.* 2001;26:304–318.

6. Gianoulakis C. Endogenous opioids and addiction to alcohol and other drugs of abuse. *Curr Top Med Chem.* 2004;4:39–50.

7. Gianoulakis C. Endogenous opioids and addiction to alcohol and other drugs of abuse. *Curr Top Med Chem.* 2009;9:999–1015.

8. Govorko D, Bekdash RA, Zhang C, Sarkar DK. Male germ-line transmits fetal alcohol adverse effect on hypothalamic proopiomelanocortin gene across generations. *Biol Psychiatry.* 2012;72:378–388.

9. King A, Munisamy G, de Wit H, Lin S. Attenuated cortisol response to alcohol in heavy social drinkers. *Int J Psychophysiol.* 2006;59:203–209.

10. Koenig HN, Olive MF. The glucocorticoid receptor antagonist mifepristone reduces ethanol intake in rats under limited access conditions. *Psychoneuroendocrinology.* 2004;29:999–1003.

11. Kvigne VL, Leonardson GR, Borzelleca J, Welty TK. Characteristics of grandmothers who have grandchildren with fetal alcohol syndrome or incomplete fetal alcohol syndrome. *Matern Child Health J.* 2008;12:760–765.

12. Logan RW, Wynne O, Maglakelidze G, et al. β-endorphin neuronal transplantation into the hypothalamus alters anxiety-like behaviors in prenatal alcohol-exposed rats and alcohol-non-preferring and alcohol-preferring rats. *Alcohol Clin Exp Res.* 2015;39:146–157.

13. Moonat S, Starkman BG, Sakharkar A, Pandey SC. Neuroscience of alcoholism: molecular and cellular mechanisms. *Cell Mol Life Sci.* 2010;67:73–88.

14. Nestler EJ. Epigenetic mechanisms of drug addiction. *Neuropharmacology.* 2014;76(Pt B):259–268.

15. Olive MF, Mehmert KK, Koenig HN, et al. A role for corticotropin releasing factor (CRF) in ethanol consumption, sensitivity, and reward as revealed by CRF-deficient mice. *Psychopharmacology.* 2003;165:181–187.

16. Oswald LM, Wand GS. Opioids and alcoholism. *Physiol Behav.* 2004;81:339–358.

17. Ouko LA, Shantikumar K, Knezovich J, Haycock P, Schnugh DJ, Ramsay M. Effect of alcohol consumption on CpG methylation in the differentially methylated regions of H19 and IG-DMR in male gametes: implications for fetal alcohol spectrum disorders. *Alcohol Clin Exp Res.* 2009;33:1615–1627.

18. Pandey SC, Ugale R, Zhang H, Tang L, Prakash A. Brain chromatin remodeling: a novel mechanism of alcoholism. *J Neurosci.* 2008;28:3729–3737.

19. Rasmussen DD, Boldt BM, Bryant CA, Mitton DR, Larsen SA, Wilkinson CW. Chronic daily ethanol and withdrawal: 1. Long-term changes in the hypothalamo-pituitary-adrenal axis. *Alcohol Clin Exp Res.* 2000;24:1836–1849.

20. Rasmussen DD, Mitton DR, Green J, Puchalski S. Chronic daily ethanol and withdrawal: 2. Behavioral changes during prolonged abstinence. *Alcohol Clin Exp Res.* 2001;25:999–1005.

21. Rasmussen DD, Boldt BM, Wilkinson CW, Mitton DR. Chronic daily ethanol and withdrawal: 3. Forebrain pro-opiomelanocortin gene expression and implications for dependence, relapse, and deprivation effect. *Alcohol Clin Exp Res.* 2002;26:535–546.

22. Reynolds AR, Saunders MA, Brewton HW, Winchester SR, Elgumati IS, Prendergast MA. Acute oral administration of the novel, competitive and selective glucocorticoid receptor antagonist ORG 34517 reduces the severity of ethanol withdrawal and related hypothalamic-pituitary-adrenal axis activation. *Drug Alcohol Depend.* 2015;154:100–104.

23. Richardson HN, Lee SY, O'Dell LE, Koob GF, Rivier CL. Alcohol self-administration acutely stimulates the hypothalamic-pituitary-adrenal axis, but alcohol dependence leads to a dampened neuro-endocrine state. *Eur J Neurosci.* 2008;28:1641–1653.

24. Rivier C, Lee S. Acute alcohol administration stimulates the activity of hypothalamic neurons that express corticotropin-releasing factor and vasopressin. *Brain Res.* 1996;726:1–10.

25. Sarkar DK, Zhang C. Beta-endorphin neuron regulates stress response and innate immunity to prevent breast cancer growth and progression. *Vitam Horm.* 2013;93:263–276.

26. Sillaber I, Rammes G, Zimmermann S, et al. Enhanced and delayed stress-induced alcohol drinking in mice lacking functional CRH1 receptors. *Science.* 2002;296:931–933.

27. Sinha R. Chronic stress, drug use, and vulnerability to addiction. *Ann N Y Acad Sci.* 2008;1141:105–130.

28. Skinner MK, Manikkam M, Guerrero-Bosagna C. Epigenetic transgenerational actions of environmental factors in disease etiology. *Trends Endocrinol Metab.* 2010;21:214–222.

29. Tartter MA, Ray LA. A prospective study of stress and alcohol craving in heavy drinkers. *Pharmacol Biochem Behav.* 2012;101:625–631.

30. Thayer JF, Hall M, Sollers 3rd JJ, Fischer JE. Alcohol use, urinary cortisol, and heart rate variability in apparently healthy men: evidence for impaired inhibitory control of the HPA axis in heavy drinkers. *Int J Psychophysiol.* 2006;59:244–250.

31. Tsigos C, Chrousos GP. Hypothalamic-pituitary-adrenal axis, neuroendocrine factors and stress. *J Psychosom Res.* 2002;53:865–871.

32. Uhart M, Oswald L, McCaul ME, Chong R, Wand GS. Hormonal responses to psychological stress and family history of alcoholism. *Neuropsychopharmacology.* 2006;31:2255–2263.

33. Vendruscolo LF, Estey D, Goodell V, et al. Glucocorticoid receptor antagonism decreases alcohol seeking in alcohol-dependent individuals. *J Clin Invest.* 2015;125:3193–3197.

34. Yong W, Spence JP, Eskay R. Alcohol-preferring rats show decreased corticotropin-releasing hormone-2 receptor expression and differences in HPA activation compared to alcohol-nonpreferring rats. *Alcohol Clin Exp Res.* 2014;38:1275–1283.

15

Stress, Panic, and Central Serotonergic Inhibition

J.E. Hassell Jr.[1], P.S.M. Yamashita[1], P.L. Johnson[2], H. Zangrossi Jr.[3], A. Shekhar[2], C.A. Lowry[1]

[1]University of Colorado Boulder, Boulder, CO, United States; [2]Indiana University School of Medicine, Indianapolis, IN, United States; [3]University of São Paulo, Ribeirão Preto, Brazil

Abstract

Panic disorder (PD) is an anxiety disorder associated with the occurrence of panic attacks, which arise suddenly without warning. Panic disorder represents a serious psychiatric condition and it can induce complications related to the fear of having subsequent panic attacks and avoidance behaviors. Given its importance, many studies have been conducted to elucidate the circuitry involved in this disorder. Clinical and preclinical studies suggest that PD can be modulated by a specific network of brain structures controlling emotional behaviors and autonomic responses. Using animal models that allow measurement of responses related to behavioral and autonomic symptoms of panic attacks in humans, it has been shown that the neuromodulator serotonin plays an inhibitory role in control of panic attacks associated with PD. Understanding the pathways through which serotonergic systems modulate panic-like responses is key to understanding the biological basis of panic attacks and PD, and, consequently, to establishing novel therapeutic strategies for treatment of PD.

INTRODUCTION

From Sigmund Freud's "anxiety neurosis" to hypothalamic disinhibition as a model of panic-like behavior,[34,52] the effort to understand the psychological and neurobiological basis of panic disorder (PD) has an extensive history. The conceptualization of the biological basis of PD has evolved partially due to advances in the understanding of the pharmacology of serotonin (5-hydroxytryptamine; 5-HT) and the discovery of the therapeutic potential of imipramine, a tricyclic antidepressant.[60] Donald Klein first documented that this tricyclic antidepressant has the ability to reduce panic attacks, without affecting anticipatory anxiety, in patients diagnosed with anxiety disorders. Further studies consolidated the distinction between panic attacks, a core symptom of PD, and anticipatory anxiety, which eventually led to the separation of PD from other anxiety disorders.[61] In 1980, PD was added to the third edition of the Diagnostic and Statistical Manual (DSM);[2] as a distinct anxiety disorder, partially because of the panic attack symptomology. According to the DSM 5th edition,[3] PD involves the occurrence of unexpected panic attacks with at least one of the attacks followed by persistent concern or worry about additional panic attacks or their consequences (e.g., losing control or going crazy) or significant maladaptive change in behavior related to the attacks. According to the DSM-5, panic attacks are defined as an abrupt surge of intense fear or intense discomfort that reaches a peak within minutes, and during which 4 or more of 13 defined symptoms develop: (1) palpitations, pounding heart, or accelerated heart rate, (2) sweating, (3) trembling or shaking, (4) sensations of shortness of breath or smothering, (5) feeling of choking, (6) chest pain or discomfort, (7) nausea or abdominal distress, (8) feeling dizzy, lightheaded, or faint, (9) chills or heat sensations, (10) paresthesias (numbness or tingling sensations), (11) derealization (feelings of unreality) or depersonalization (being detached from oneself), (12) fear of losing control or going crazy, and (13) fear of dying.

Since the clinical findings from Donald Klein,[60] advances have been made in understanding the biological basis of PD. A breakthrough in PD research was the ability to induce panic attacks in patients with PD in a laboratory setting. Sodium lactate, caffeine, and cholecystokinin tetrapeptide (CCK-4) infusions, and CO_2 inhalation, are among the most commonly used panicogens to study panic attacks.[63,73,86,92] Symptoms reported during the infusion/inhalation resemble those reported by the patients when they experience spontaneous panic attacks, namely fear of losing control, difficulty breathing, sweating, headache, and palpitations.[73,86,99] A key finding from studies of clinically induced panic attacks was the ability of imipramine, given acutely before sodium lactate infusion, to prevent symptoms of panic attacks induced by sodium lactate.[117] This finding and others set the stage for further studies of how these panicogens induce panic attacks and investigations of the neural substrates involved.

NEUROBIOLOGY OF PANIC DISORDER AND PANIC ATTACKS

Many theoretical frameworks have been proposed to explain the neurobiological basis of PD, from Clark's hypothesis of misinterpretation of bodily sensations[11] to Klein's false suffocation alarm hypothesis of PD.[28,62] In 1989, it was hypothesized by Gorman and colleagues that specific neural substrates were responsible for the clinical symptoms of PD.[38] Gorman and colleagues hypothesized that there was a panic network within the brain of individuals with PD that was hypersensitive to otherwise innocuous stimuli. Specifically, they hypothesized that (1) the brainstem had a role in the autonomic components of a panic attack, (2) the limbic lobe had a role in anticipatory anxiety associated with PD, and (3) the cortical areas of the brain had a role in phobic avoidance.[38]

Much work has been done since, with advances in imaging techniques that have allowed identification of neural correlates of panic attacks and PD in human subjects, pharmacological studies characterizing the neurochemical substrates in human and animal studies, and development of animal models for panic attacks and PD.[92,119] Key areas of the brain have been identified as having important roles in the possible etiology and/or maintenance of PD. These areas include the periaqueductal gray (PAG), dorsomedial/perifornical hypothalamus (DMH/PeF), and RVLM.[23,82] Conversely, the dorsal raphe nucleus (DR), a major source for serotonergic innervation of these panic-related structures, has been implicated in the inhibitory control of panic attacks.[23,82]

KEY POINTS

- Functional and anatomical evidence shows that serotonin plays an important role in the inhibition of panic-related physiologic and behavioral responses. The inhibitory influence of serotonergic systems on the autonomic and behavioral responses associated with panic attacks appear to be mediated by a subset of serotonergic neurons in the dorsal raphe nucleus projecting to a distributed system mediating panic-related responses, including periventricular structures such as the dorsal periaqueductal gray and dorsomedial hypothalamus as well as the rostral ventrolateral medulla (RVLM).

Here, we will discuss the involvement of the PAG, DMH/PeF, RVLM, and amygdala in panic attacks and PD; then we will discuss the influences of serotonergic systems arising from the DR and median raphe nucleus (MR) on this circuitry. Specifically, we will highlight evidence that serotonin plays an important role in the inhibition of panic-related physiology and behavior through actions in the PAG, DMH/PeF, RVLM, and amygdala.

Dorsal Periaqueductal Gray (DPAG), Panic Attacks, and Panic Disorder

Functional magnetic resonance imaging (fMRI) revealed that in healthy individuals, the PAG responds to emotionally arousing images or when a threat becomes increasingly closer to the individual, implicating the PAG in a defensive behavior network.[46,76] The PAG has been implicated in both clinical and preclinical studies as an important neural substrate mediating key features of panic attacks. In human studies, the PAG has been found to be activated in response to CO_2 inhalation, a substance that has been shown to induce panic attacks in PD patients.[27,39]

Behaviorally, one of the first reports of the PAGs' role in fear in humans was based on implantation of electrodes for chronic pain relief.[79] Individuals that received electrical stimulation of the PAG and other tegmental areas had increased autonomic responses and reported that they felt "fearful," "frightful," or "terrible" whenever the PAG was stimulated.[79] Consistent with these findings, electrical stimulation of the dorsal column of the PAG (DPAG) in rats induces defensive responses, such as freezing and escape behaviors and has been used as an animal model of some aspects of panic attacks.[66,67,90,91] Electrical stimulation of the DPAG as a model of panic attacks is sensitive to panic-potentiating agents, such as yohimbine, caffeine, as well as panic-attenuating drugs, like clonazepam and alprazolam.[49] In addition, electrical stimulation of the DPAG demonstrates behavioral responses in rodents that are reflective of self-reported physiological and emotional responses in humans experiencing panic attacks.[49]

In addition to electrical stimulation of the DPAG, studies using chemical stimulation of the DPAG in rodents have further supported the DPAG's involvement in panic-like behaviors. For example, microinjections of the glutamic acid decarboxylase inhibitor, L-allylglycine (L-AG) into the DPAG, which inhibits gamma-aminobutyric acid (GABA) synthesis and induces functional disinhibition of the DPAG, increases levels of freezing behavior and, at higher doses of L-AG, escape behavior, such as jumping.[15]

A limited, but expanding, number of behavioral tests involving assessment of escape behaviors have been proposed to have predictive validity for panic attacks.

The most commonly used test is the elevated T-maze (ETM). The ETM behavioral test is derived from the elevated plus-maze and it allows for the measurement, in the same rat, of anxiety- and panic-related defensive responses, inhibitory avoidance, and escape, respectively.[119] As predicted, the DPAG of rats is activated during behavioral escape tasks in the ETM.[103] Escape behaviors are assessed also in an open-field or most recently in a polygonal arena with a burrow.[8,45,119]

Dorsomedial/Perifornical Hypothalamus, Panic Attacks, and Panic Disorder

The dorsomedial regions of the posterior hypothalamus are important areas of the brain involved in sympathoexcitatory responses, control of the hypothalamic–pituitary–adrenal (HPA) axis, and escape behaviors.[31,50,56] For instance, deep brain stimulation of the perifornical area of the posterior hypothalamus of humans was done to alleviate chronic headaches, but, based on self-reports, induced panic and fear of dying, as well as tachycardia, pressor responses and tachypnea,[85,115] thermal sensations, and paresthesias.[115] In rodents, electrical stimulation of the rat dorsomedial hypothalamus (DMH) induces escape behavior.[18] Here, we define the DMH as defined by Fontes.[32] Briefly, we refer to DMH to indicate a region of the hypothalamus that includes the dorsomedial hypothalamic nucleus (DMN) but also adjoining areas, particularly dorsal and posterior to the nucleus itself as well as laterally including the medial part of the perifornical area. Of particular interest in this regard are orexinergic neurons, which are distributed throughout the DMH/PeF region, connected, via multisynaptic pathways, to autonomic control systems[58] and may play an important role in the neural mechanisms controlling panic attacks.[53]

Both the DMH/PeF and ventromedial (VMH) subregions of the medial hypothalamus have been implicated in control of panic-like behaviors.[50,108] For example, disinhibition of the DMH with bicuculline methiodide (BMI), a GABA_A receptor antagonist, in rats, increases heart rate and anxiety-like behavior as measured on the elevated plus-maze.[95] Intra-DMH or intra-VMH BMI increases defensive immobility and escape behavior.[33] In addition, when L-AG is continuously delivered into the DMH of rats, they become sensitive to intravenous infusions of sodium lactate, a compound that evokes panic attacks in PD patients, but not in healthy individuals.[65,96] This sensitivity to sodium lactate includes increases in heart rate and blood pressure as well as anxiety-like behavioral responses.[96] Anatomical studies support the idea of a network of brain nuclei, including the DMH and DPAG, that coordinates appropriate responses to panicogenic stimuli.[16,58,107]

Amygdala, Panic Attacks, and Panic Disorder

Symptoms associated with a panic attack involve fear, such as the fear of dying and fear of going crazy, while PD includes persistent anxiety or persistent maladaptive avoidance behaviors.[3] The amygdala is an important nucleus in the brain associated with the expression of fear, anxiety, and avoidance behavior and has been proposed to be involved in the etiology of PD and the focus of Gorman's neuroanatomical hypothesis of PD.[25] Amygdalar hypoactivation in patients with PD was found in response to emotional faces.[21] In addition, a significantly smaller amygdalar volume was found in patients with PD using quantitative magnetic resonance imaging (MRI).[74] Interestingly, three patients with Urbach–Wiethe disease, a disease characterized by focal bilateral lesions in the amygdala and reduced capacity for expression of fear, experienced fear and panic during an acute inhalation of 35% CO_2, a panicogenic substance.[29] In addition, these patients did not experience anticipatory anxiety but only fear and panic after the inhalation of 35% CO_2, which suggests that the amygdala is not necessary for interoceptive (e.g., CO_2-induced) panic attacks to occur,[29] which may be more representative of spontaneous panic attacks. In another study, a patient with Urbach–Wiethe disease experienced panic attacks and sought medical assistance.[114] Together, these findings support the notion that panic attacks can still occur in the absence of amygdala integrity[114] and that the amygdala may instead play a role in the inhibition of panic.[90] Yet, unlike healthy controls, patients with Urbach–Wiethe disease did not experience anticipatory anxiety before a subsequent CO_2 inhalation, which demonstrates that the amygdala does appear to be important for anticipatory anxiety that may be more closely associated with expected panic attacks. This is important to note, since patients with PD experience approximately equal numbers of spontaneous and situational/expected (e.g., in situations that are associated with previous panic attacks or stress) panic attacks.[97] Agoraphobia, which is associated with fear-associated learning, is also estimated to affect up to 50% of those with PD.[59]

Although the above findings suggest that the amygdala plays a minimal role in the induction of panic attacks, this does not preclude its role in panic attacks or the fear symptoms associated with PD. A preclinical study showed that disinhibition of the basolateral amygdala (BLA) following repeated infusions of BMI primes rats within 3–5 days to respond to BMI with exaggerated increases in heart rate, blood pressure, and anxiety as measured in the social interaction test.[89] In another study, priming of the BLA with BMI made rats sensitive to an intravenous infusion of sodium lactate, significantly increasing lactate-induced heart rate and blood pressure as well as anxiety in the social interaction test.[88]

These findings support the involvement of the amygdala in a neural network involved in PD. Further studies are required to resolve the apparent inconsistencies.

Autonomic Correlates of Panic Attacks and Panic Disorder

Many of the symptoms of panic attacks are related to autonomic functions, for example shortness of breath, palpitations, and increased heart rate, which has led to interest in the relationship between autonomic physiology and PD.[17] For example, sympathetic tone in patients diagnosed with PD is increased as evidenced by altered heart rate variability, a proxy for sympathetic tone, and altered breathing rates.[12] In addition, PD patients respond with increased sympathetic tone compared to healthy individuals when shown threatening pictures, suggesting dysfunctional autonomic regulation.[113] A single inhalation of 35% CO_2 in healthy individuals increases systolic blood pressure, lowers heart rate, and increases systemic norepinephrine concentrations, corroborating a relationship between panicogen administration and autonomic responses.[55]

Serotonin, Panic Attacks, and Panic Disorder

Serotonin is a monoaminergic neuromodulator that has been implicated in diverse physiological functions including thermoregulation, control of sleep–wake cycles, behavioral arousal, and circadian rhythms.[43,101] Serotonin is also implicated in the etiology and pathophysiology of anxiety disorders including PD.[26] Clinical studies have suggested that dysregulation of 5-HT$_{1A}$ receptors and the serotonin transporter (5-HTT) likely contribute to the etiology and maintenance of PD.[26,78] Studying patients with PD in the absence of panic attacks, Esler and coworkers[26] showed that these patients had about fourfold increases in brain serotonin turnover compared to healthy controls. The increase in brain serotonin turnover was highest in patients with the most severe symptoms. Esler and colleagues[26] assessed serotonin turnover by measurement of 5-hydroxyindoleacetic acid (5-HIAA), a metabolite of serotonin, in blood from jugular venous sampling from cortical and subcortical brain regions. They also identified the PD patients' genotypes for the polymorphic region (5-HTTLPR) of the 5-HTT gene promotor to identify individuals homozygous and heterozygous for the short allele, which is thought to confer impaired serotonin uptake.[64] PD patients homozygous or heterozygous for the short allele did not have higher serotonin turnover than patients without this short variant,[26] suggesting that the serotonin turnover increases in PD patients are not related to an impairment in serotonin reuptake. These findings clearly suggest that a serotonin dysregulation is present in PD and likely not because of a reduction in serotonin reuptake. Exaggerated activity of serotonergic mesolimbocortical systems projecting to the amygdala and prefrontal cortex is thought to mediate stress-induced increases in inhibitory avoidance and fear, and the increases in serotonin turnover in PD patients may reflect an overactivation of these mesolimbocortical systems. Studies using single photon emission computed tomography (SPECT) in PD subjects found lower binding of [^{123}I] nor-β-citalopram, a selective serotonin reuptake inhibitor, relative to controls, indicating lower densities of 5-HTT in the midbrain raphe nuclei compared to healthy volunteers,[72] an effect that would be consistent with the exaggerated turnover of serotonin in PD subjects, as observed by Esler and colleagues.[26]

Genetic allelic variance in the 5-HT$_{1A}$ receptor is associated with risk factors for frequency of panic attacks.[118] PD subjects with genetic variants of the *HTR1A* gene, which encodes for the 5-HT$_{1A}$ receptor, were found to show increased escape behaviors during a behavioral avoidance task.[106] In addition, it has been shown that PD subjects have reduced binding of the autoinhibitory 5-HT$_{1A}$ receptor in the midbrain raphe nuclei,[78] an effect that would be consistent with the exaggerated turnover of serotonin in PD subjects, as observed by Esler and colleagues.[26]

ANATOMICAL AND FUNCTIONAL HETEROGENEITY OF SEROTONERGIC SYSTEMS

The DR is a major source of serotonergic projections to limbic structures involved in the modulation of anxiety disorders.[4,22,110] DR serotonergic neurons are topographically organized into subregions (Fig. 15.1) including the rostral, caudal, dorsal, ventral, ventrolateral, and interfascicular regions.[44] These subregions differ in anatomy, hodology, and functional topography.

Since 1991, when Deakin and Graeff first proposed that serotonin plays a dual role in the modulation of defensive behaviors related to anxiety and panic,[20] several studies have shown that DR activation inhibits escape- or panic-like responses yet facilitates anxiety-like behavior.[84,94] According to Deakin and Graeff,[20] the DR serotonergic systems modulate defensive responses through two different serotonergic pathways. Deakin and Graeff hypothesized that the serotonergic dorsal raphe periventricular tract, consisting of projections to midline structures including the DPAG, inhibits escape expression, while the serotonergic dorsal raphe forebrain bundle tract, consisting of mesolimbocortical projections to forebrain limbic structures including the amygdala, facilitates the expression of inhibitory avoidance

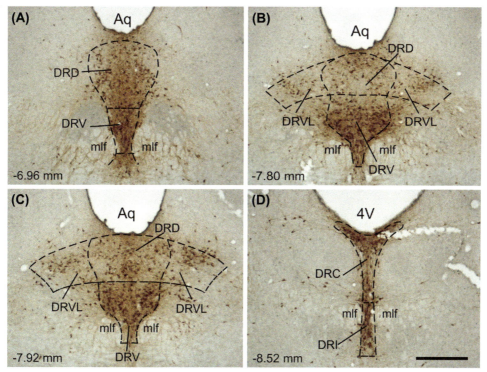

FIGURE 15.1 Photomicrographs showing tryptophan hydroxylase-immunoreactive neurons (orange/brown cytosolic staining) in the dorsal raphe nucleus at levels: −6.96 mm, −7.80 mm, −7.92 mm, and −8.52 mm from bregma. Tryptophan hydroxylase (TPH) is the rate-limiting enzyme in the biosynthesis of serotonin and is commonly used as a marker of serotonergic neurons. *Dashed lines* show the boundaries for the subregions of the dorsal raphe nucleus, according to Ref.83: dorsal raphe nucleus, dorsal part (DRD); dorsal raphe nucleus, ventral part (DRV); dorsal raphe nucleus, ventrolateral part (DRVL); dorsal raphe nucleus, caudal part (DRC); dorsal raphe nucleus, interfascicular part (DRI). Scale bar, 500 μm.

behaviors and risk assessment. In addition, according to the authors, dysregulation of these pathways would be involved in the neurobiology of generalized anxiety and panic disorders.[20,119] It is equally possible that both systems are involved in different symptoms of PD.

Consonant with the Deakin and Graeff hypothesis, anatomical as well as behavioral evidence suggest that specific subdivisions of the DR can modulate different behaviors related to anxiety.[1,30,69] Serotonergic neurons innervating brain structures implicated in escape- or panic-like behavioral responses (Fig. 15.2), including the DPAG and RVLM, an important structure mediating autonomic responses to aversive stimuli, are located in the dorsolateral part of the dorsal raphe nucleus (DRVL) and the adjacent ventrolateral periaqueductal gray (VLPAG).[82] Panicogenic stimuli such as exposure

to the open arms of the ETM,[103] exposure to high concentrations of CO_2,[40,51] and infusion of sodium lactate[52] increase c-Fos expression in serotonergic neurons within the DRVL/VLPAG (Table 15.1). As DRVL/VLPAG serotonergic neurons innervate the DPAG and RVLM, these serotonergic neurons are excellent candidates for the serotonergic dorsal raphe periventricular tract hypothesized by Deakin and Graeff that modulates escape- or panic-like behavior, more recently referred to as dorsal raphe arcuate tract.[70a]

In contrast, serotonergic neurons innervating brain structures implicated in inhibitory avoidance and fear responses, including the basolateral nucleus of the amygdala, are located in the dorsal part of the dorsal raphe nucleus (DRD). Exposing animals to anxiogenic stimuli, such as exposure to the inhibitory avoidance task in the

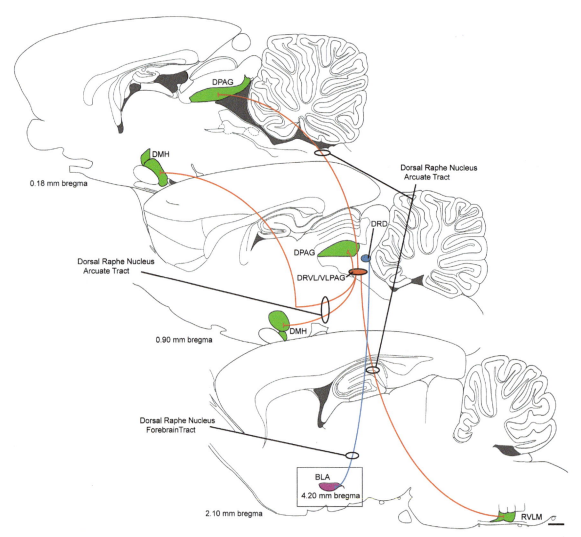

FIGURE 15.2   Hypothetical model of antipanic serotonergic systems illustrated using sagittal planes of the rat brain, 0.18 mm, 0.90 mm, and 2.10 mm lateral from bregma.[83] Serotonergic neurons located in the dorsal raphe nucleus, ventrolateral part (DRVL)/ventrolateral periaqueductal gray (VLPAG) give rise to the dorsal raphe nucleus arcuate tract to the dorsal periaqueductal gray (DPAG), dorsomedial hypothalamus (DMH)/perifornical region (PeF), and rostral ventrolateral medulla (RVLM). Serotonergic neurons located in the dorsal raphe nucleus, dorsal part (DRD) give rise to the dorsal raphe nucleus forebrain tract to the basolateral amygdala (BLA). Scale bar, 1 mm.

TABLE 15.1   Functional Dissociation of DRD and DRVL/VLPAG Serotonergic Neurons (*Conditions Associated With Adverse Early Life Experience/Autonomic/Panicogenic Stimuli*)

Stress/Arousal/Anxiogenic Stimulus	Effect on DRD	Effect on DRVL/VLPAG	Dependent Variable	References
Adolescent social isolation[71]	No effect	↓ DRVL/VLPAG	*tph2* mRNA	71
CO_2	↑ DRD	↑ DRVL/VLPAG	c-Fos	51
Cold swim (19°C, versus 25°C)	No effect	↑ DRVL/VLPAG	c-Fos	57
Contextual fear	No effect	↑ DRVL/VLPAG	c-Fos	102
Lipopolysaccharide, i.p.	No effect	↑ DRVL/VLPAG	c-Fos	47
Maternal separation × social defeat interaction	No effect	↑ DRVL/VLPAG	*slc6a4* mRNA	36
Maternal separation × social defeat interaction	No effect	↑ DRVL/VLPAG	*tph2* mRNA	37
Neonatal lipopolysaccharide	No effect	↑ DRVL/VLPAG	*tph2* mRNA	98
Sodium lactate (effect absent in panic model)	No effect (DRD)	↑ DRVL/VLPAG	c-Fos	52
Warm temperature	No effect (DRD)	↑ DRVL/VLPAG	c-Fos	43

CO_2, Carbon dioxide; *DRD*, dorsal raphe nucleus, dorsal part; *DRVL*, dorsal raphe nucleus, ventrolateral part; *i.p.*, intraperitoneal; *slc6a4*, solute carrier family 6 member 4, serotonin transporter gene; *tph2*, tryptophan hydroxylase 2; *VLPAG*, ventrolateral periaqueductal gray.

ETM,[103] high levels of illumination in the open-field[9] and administration of anxiogenic agents, such as the adenosine receptor antagonist caffeine, the 5-$HT_{2A/2C}$ receptor agonist *m*-chlorophenyl piperazine (mCPP), the $α_2$-adrenoreceptor antagonist yohimbine, and the partial inverse agonist at the benzodiazepine allosteric site on the $GABA_A$ receptor, N-methyl-beta-carboline-3-carboxamide (FG-7142),[1] increase c-Fos expression in serotonergic neurons within the DRD (Table 15.2). As DRD serotonergic neurons innervate the amygdala, including the BLA,[13,41] these serotonergic neurons are excellent candidates for the serotonergic DR forebrain tract proposed by Deakin and Graeff, which innervates forebrain limbic structures and modulates anxiety-like and fear responses. Tables 15.1 and 15.2 illustrate the double dissociation in conditions that involve activation of the DRVL/VLPAG versus DRD serotonergic systems.

DORSAL RAPHE NUCLEUS (DR) INNERVATION OF ESCAPE- OR PANIC-RELATED STRUCTURES

Inhibitory Role of Serotonin in the DPAG: Behavioral and Physiological Outcomes

Serotonin is involved in modulating defensive responses, such as inhibitory avoidance or escape behaviors in the ETM. Using this behavior test, it has been shown that activation of the DR induces anxiogenic- and panicolytic-like effects, likely by inducing serotonin release in the BLA and DPAG, respectively.[84,111,112] In these studies, the DR was activated using either intra-DR injections of the excitatory amino acid kainate or disinhibition of the DR using the 5-HT_{1A} receptor antagonist N-[2-[4-(2-methoxyphenyl)-1-piperazinyl]ethyl]-N-2-pyridinyl-cyclohexanecarboxamide maleate (WAY 100635), to block DR inhibitory presynaptic 5-HT_{1A} autoreceptors. In both experiments, latencies to leave the open and closed arms were increased,[84] suggesting panicolytic- and anxiogenic-like responses, respectively. The panicolytic-like effect was counteracted by intra-DPAG injections of 5-HT_{1A} and 5-$HT_{2A/2C}$ receptor antagonists.[84] To demonstrate that serotonin inhibits escape behavior on the ETM, it was shown that intra-DPAG serotonin, 5-HT_{1A} and 5-HT_{2A}, but not 5-HT_{2C}, receptor agonists, increase escape latencies to leave the open arm of the ETM.[19,116] These findings support not only an inhibitory role of serotonin on escape- or panic-like behaviors, but that the DPAG is one neural substrate where serotonergic systems can inhibit these behaviors (Fig. 15.3). Furthermore, using microdialysis, Viana and coworkers[111] showed that intra-DR kainate injections increase extracellular serotonin within the DPAG and BLA.[111] Intra-BLA injection of serotonin induces anxiogenic effects mediated by 5-HT_{2C} receptors,[112] while depletion of serotonin in the BLA reduces anxiety and disrupts fear conditioning.[54] Thus, these studies indicate that DR activation inhibits panic-like responses through serotonin actions in the DPAG and facilitates anxiety-like responses through serotonin actions in the BLA, supporting Deakin and Graeff's proposal. Inhibition of DR serotonergic systems by local microinjections of the 5-HT_{1A} receptor agonist 7-(dipropylamino)-5,6,7,8-tetrahydronaphthalen-1-ol (8-OH-DPAT) or serotonergic lesions using 5,7-dihydroxytryptamine (5,7-DHT) result in panicogenic- and anxiolytic-like effects in the ETM.[94] Overall, these studies support a role for DR serotonergic

TABLE 15.2   Functional Dissociation of DRD and DRVL/VLPAG Serotonergic Neurons (*Conditions Associated With Activation of DRD Serotonergic Neurons*)

Stress/Arousal/Anxiogenic Stimulus	Effect on DRD	Effect on DRVL/VLPAG	Dependent Variable	References
Acoustic startle	↑DRD	No effect	c-Fos	102
Anxiety due to intimate partner violence	↑DRD	No effect	c-Fos	14
Anxiogenic drug, caffeine	↑DRD	No effect	c-Fos	1
Anxiogenic drug, FG-7142	↑DRD	No effect	c-Fos	1
Anxiogenic drug, mCPP	↑DRD	No effect	c-Fos	1
Avoidance task on elevated T-maze	↑DRD	No effect	c-Fos	103
Crh overexpression (OE)	↑DRD	No effect	Crhr2 binding	100
Diurnal variation in Tph activity	↑DRD	No effect	Tph activity	24
Inescapable shock/learned helplessness	↑DRD	N.D.	↓5-HT$_{1A}$	87
Open-field	↑DRD	No effect	c-Fos	9
Repeated versus acute social defeat	↓DRD	No effect	c-Fos	81
Social defeat	↑DRD	No effect	c-Fos	35
Substance P in vivo	↑DRD	↓	Neuronal Firing rate	109
Ucn2, i.c.v.	↑DRD	No effect	c-Fos	42,105,104
Ucn3 overexpression (OE), 24 h post-restraint	↑DRD	No effect	5-HT, 5-HIAA	80

5-HT$_{1A}$, Serotonin receptor 1A subtype; *5-HT*, 5-hydroxytryptamine; *5-HIAA*, 5-hydroxyindoleacetic acid; *Crh*, corticotropin-releasing hormone; *Crhr2*, corticotropin-releasing hormone receptor 2; *DRD*, dorsal raphe nucleus, dorsal part; *DRVL*, dorsal raphe nucleus, ventrolateral part; *FG-7142*, N-methyl-beta-carboline-3-carboxamide; *i.c.v.*, intracerebroventricular; *mCPP*, m-chlorophenyl piperazine; *N.D.*, not determined; *Tph*, tryptophan hydroxylase; *Ucn2*, urocortin 2; *Ucn3*, urocortin 3; *VLPAG*, ventrolateral periaqueductal gray.

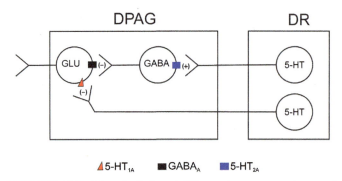

FIGURE 15.3   Hypothetical model illustrating serotonergic control of panic-like physiological and behavioral responses through inhibition of the dorsal periaqueductal gray (DPAG). Serotonin acts on inhibitory postsynaptic 5-HT$_{1A}$ receptors located on excitatory glutamatergic output neurons, while serotonin acts on excitatory 5-HT$_{2A}$ receptors located on local GABAergic interneurons in the DPAG, resulting in an overall inhibition of DPAG output and panic-like responses.

systems in inhibiting escape- or panic-like behavior and facilitating anxiety-like behavior.

The inhibitory role of serotonin in the DPAG also has been shown in other panic-like models, such as electrical stimulation of the DPAG, since local microinjections of serotonin increase the electrical current required to induce escape behavior.[10] It was further determined that serotonin within the DPAG inhibits panic-like responses through actions on 5-HT$_{1A}$ and 5-HT$_{2A}$ receptors.

Intra-DPAG microinjections of the 5-HT$_{1A}$ receptor agonist 8-OH-DPAT or DOI (2,5-dimethoxy-4-iodophenyl piperazine dihydrochloride), a preferential 5-HT$_{2A}$ receptor agonist, increase the effective threshold for defensive behavior expression.[10,93] Later, similar to the results found in the ETM, 5-HT$_{2C}$ receptor activation was shown to be ineffective in changing the electrical threshold required to evoke escape behavior, suggesting that 5-HT$_{2A}$ receptors, but not 5-HT$_{2C}$ receptors, are involved in the inhibitory role of serotonin in the DPAG.[116]

The role of DR serotonergic systems in inhibition of escape- or panic-like behavioral responses has further been elucidated in studies using chemical stimulation of the DPAG.[75] Microinjection of 3-morpholinylsydnonei-mine chloride (SIN-1), a nitric oxide donor, into the DPAG induced escape behavior and this effect was potentiated by prior intra-DR microinjection of 8-OH-DPAT.[75] This is in agreement with serotonin inhibition of DPAG-induced escape since intra-DR 8-OH-DPAT would have reduced serotonin within the DPAG, thus allowing escape behavior to occur.

Inhibitory Role of Serotonin in the DMH: Behavioral and Physiological Outcomes

Serotonergic neurons within the DRVL/VLPAG region project to the DMH/PeF,[68] as well as the

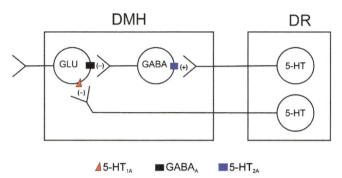

FIGURE 15.4 Hypothetical model illustrating serotonergic control of panic-like physiological and behavioral responses through inhibition of the dorsomedial hypothalamus (DMH)/perifornical region (PeF). Serotonin acts on inhibitory postsynaptic 5-HT$_{1A}$ receptors located on excitatory glutamatergic output neurons, while serotonin acts on excitatory 5-HT$_{2A}$ receptors located on local GABAergic interneurons in the DMH/PeF, resulting in an overall inhibition of DMH/PeF output and panic-like responses.

DPAG. The activation of 5-HT$_{1A}$ receptor in the DMH induces a panicolytic-like behavioral response in the ETM without affecting inhibitory avoidance.[77] In addition, inhibition of escape following intra-DMH 8-OH-DPAT was potentiated by 21 days of systemic treatment with the tricyclic antidepressant imipramine.[77]

The 5-HT$_{1A}$ receptors also have an inhibitory role on sympathoexcitation induced by disinhibition of the DMH with BMI.[48] Intracisternal injection of 8-OH-DPAT reverses the increase of heart rate and mean arterial blood pressure induced by intra-DMH BMI.[48] In addition, microinjection of BMI into the DMH also induces the expression of behaviors, such as alertness, immobility, escape, and backward movement.[7] In this study, lesions of DR serotonergic cells before the intra-DMH injection of BMI increased the frequencies of alertness, immobility, escape, and backward movement.[7] Escape behavior also is induced by electrical stimulation of the DMH and, as seen above, escape- or panic-like responses are inhibited by 5-HT$_{1A}$ and 5-HT$_{2A}$ receptor activation.[18] These data suggest that serotonergic neurotransmission within the DMH/PeF inhibits the expression of escape- or panic-related responses (Fig. 15.4).

Inhibitory Role of Serotonin in the RVLM: Physiological Outcomes

As is the case with the DPAG and DMH/PeF, the sympathoexcitatory region of the RVLM receives direct projections from DRVL/VLPAG serotonergic neurons.[6] Activation of neurons in the VLPAG decreases sympathetic nerve activity and decreases arterial blood pressure via activation of 5-HT$_{1A}$ receptors in the RVLM.[5,70]

CONCLUSION

Functional and anatomical evidence from preclinical and clinical studies show that dysfunction of serotonergic systems is present in PD. This neuromodulator plays an inhibitory role in physiological and behavioral responses relevant to the autonomic and behavioral components of panic attacks that are a defining symptom of PD. Serotonergic inhibition of these autonomic and behavioral correlates of panic attacks appears to be mediated by the DR serotonergic dorsal raphe arcuate tract projecting to the DPAG, DMH/PeF, and RVLM. In contrast, serotonin plays an excitatory role in anxiety and fear responses relevant to the persistent fear and avoidance behaviors in PD. Serotonergic exaggeration of anxiety and fear appears to be mediated by the serotonergic dorsal raphe forebrain bundle tract, consisting of mesolimbocortical projections to forebrain limbic structures including the amygdala. Much work still needs to be done for more complete understanding of the biological basis of PD and development of more effective therapeutic approaches to treatment of PD.

References

1. Abrams JK, Johnson PL, Hay-Schmidt A, Mikkelsen JD, Shekhar A, Lowry CA. Serotonergic systems associated with arousal and vigilance behaviors following administration of anxiogenic drugs. *Neuroscience*. 2005;133(4):983–997.
2. American Psychiatric Association. *Diagnostic and Statistical Manual of Mental Disorders*. 3rd ed. DSM-III; 1980.
3. American Psychiatric Association. *Diagnostic and Statistical Manual of Mental Disorders (DSM-5)*. American Psychiatric Pub; 2013.
4. Azmitia EC, Segal M. An autoradiographic analysis of the differential ascending projections of the dorsal and median raphe nuclei in the rat. *J Comp Neurol*. 1978;179(3):641–667.
5. Bago M, Dean C. Sympathoinhibition from ventrolateral periaqueductal gray mediated by 5-HT(1A) receptors in the RVLM. *Am J Physiol Regul Integr Comp Physiol*. 2001;280(4):R976–R984.
6. Bago M, Marson L, Dean C. Serotonergic projections to the rostroventrolateral medulla from midbrain and raphe nuclei. *Brain Res*. 2002;945(2):249–258.
7. Biagioni AF, de Freitas RL, da Silva JA, et al. Serotonergic neural links from the dorsal raphe nucleus modulate defensive behaviours organised by the dorsomedial hypothalamus and the elaboration of fear-induced antinociception via locus coeruleus pathways. *Neuropharmacology*. 2013;67:379–394.
8. Biagioni AF, Anjos-Garcia T, Ullah F, et al. Neuroethological validation of an experimental apparatus to evaluate oriented and non-oriented escape behaviours: comparison between the polygonal arena with a burrow and the circular enclosure of an open-field test. *Behav Brain Res*. 2015;298(Pt B):65–77.
9. Bouwknecht JA, Spiga F, Staub DR, Hale MW, Shekhar A, Lowry CA. Differential effects of exposure to low-light or high-light open-field on anxiety-related behaviors: relationship to c-Fos expression in serotonergic and non-serotonergic neurons in the dorsal raphe nucleus. *Brain Res Bull*. 2007;72(1):32–43.
10. Broiz AC, Oliveira LC, Brandao ML. Regulation of conditioned and unconditioned fear in rats by 5-HT1A receptors in the dorsal periaqueductal gray. *Pharmacol Biochem Behav*. 2008;89(1):76–84.

11. Clark DM. A cognitive approach to panic. *Behav Res Ther.* 1986;24(4):461–470.

12. Cohen H, Benjamin J, Geva AB, Matar MA, Kaplan Z, Kotler M. Autonomic dysregulation in panic disorder and in post-traumatic stress disorder: application of power spectrum analysis of heart rate variability at rest and in response to recollection of trauma or panic attacks. *Psychiatry Res.* 2000;96(1):1–13.

13. Commons KG, Connolley KR, Valentino RJ. A neurochemically distinct dorsal raphe-limbic circuit with a potential role in affective disorders. *Neuropsychopharmacology.* 2003;28(2):206–215.

14. Cordero MI, Poirier GL, Marquez C, et al. Evidence for biological roots in the transgenerational transmission of intimate partner violence. *Transl Psychiatry.* 2012;2:e106.

15. Cunha JM, Zanoveli JM, Ledvinka-Filho E, Brandao ML. L-allylglycine dissociates the neural substrates of fear in the periaqueductal gray of rats. *Brain Res Bull.* 2010;81(4–5):416–423.

16. da Silva LG, de Menezes RC, dos Santos RA, Campagnole-Santos MJ, Fontes MA. Role of periaqueductal gray on the cardiovascular response evoked by disinhibition of the dorsomedial hypothalamus. *Brain Res.* 2003;984(1–2):206–214.

17. Davies SJ, Lowry CA, Nutt DJ. Panic and hypertension: brothers in arms through 5-HT? *J Psychopharmacol.* 2007;21(6):563–566.

18. de Bortoli VC, Yamashita PS, Zangrossi Jr H. 5-HT1A and 5-HT2A receptor control of a panic-like defensive response in the rat dorsomedial hypothalamic nucleus. *J Psychopharmacol.* 2013;27(12):1116–1123.

19. de Paula Soares V, Zangrossi Jr H. Involvement of 5-HT1A and 5-HT2 receptors of the dorsal periaqueductal gray in the regulation of the defensive behaviors generated by the elevated T-maze. *Brain Res Bull.* 2004;64(2):181–188.

20. Deakin JF, Graeff FG. 5-HT and mechanisms of defence. *J Psychopharmacol.* 1991;5(4):305–315.

21. Demenescu LR, Kortekaas R, Cremers HR, et al. Amygdala activation and its functional connectivity during perception of emotional faces in social phobia and panic disorder. *J Psychiatr Res.* 2013;47(8):1024–1031.

22. Descarries L, Watkins KC, Garcia S, Beaudet A. The serotonin neurons in nucleus raphe dorsalis of adult rat: a light and electron microscope radioautographic study. *J Comp Neurol.* 1982;207(3):239–254.

23. Donner NC, Johnson PL, Fitz SD, Kellen KE, Shekhar A, Lowry CA. Elevated *tph2* mRNA expression in a rat model of chronic anxiety. *Depress Anxiety.* 2012;29(4):307–319.

24. Donner NC, Montoya CD, Lukkes JL, Lowry CA. Chronic non-invasive corticosterone administration abolishes the diurnal pattern of *tph2* expression. *Psychoneuroendocrinology.* 2012;37(5):645–661.

25. Dresler T, Guhn A, Tupak SV, et al. Revise the revised? New dimensions of the neuroanatomical hypothesis of panic disorder. *J Neural Transm.* 2013;120(1):3–29.

26. Esler M, Lambert E, Alvarenga M, et al. Increased brain serotonin turnover in panic disorder patients in the absence of a panic attack: reduction by a selective serotonin reuptake inhibitor. *Stress.* 2007;10(3):295–304.

27. Faull OK, Jenkinson M, Clare S, Pattinson KT. Functional subdivision of the human periaqueductal grey in respiratory control using 7 tesla fMRI. *NeuroImage.* 2015;113:356–364.

28. Fava L, Morton J. Causal modeling of panic disorder theories. *Clin Psychol Rev.* 2009;29(7):623–637.

29. Feinstein JS, Buzza C, Hurlemann R, et al. Fear and panic in humans with bilateral amygdala damage. *Nat Neurosci.* 2013;16(3):270–272.

30. Fernandez SP, Cauli B, Cabezas C, Muzerelle A, Poncer JC, Gaspar P. Multiscale single-cell analysis reveals unique phenotypes of raphe 5-HT neurons projecting to the forebrain. *Brain Struct Funct.* 2015, 1–19.

31. Fontes MA, Tagawa T, Polson JW, Cavanagh SJ, Dampney RA. Descending pathways mediating cardiovascular response from dorsomedial hypothalamic nucleus. *Am J Physiol Heart Circ Physiol.* 2001;280(6):H2891–H2901.

32. Fontes MA, Xavier CH, de Menezes RC, Dimicco JA. The dorsomedial hypothalamus and the central pathways involved in the cardiovascular response to emotional stress. *Neuroscience.* 2011;184:64–74.

33. Freitas RL, Uribe-Marino A, Castiblanco-Urbina MA, Elias-Filho DH, Coimbra NC. GABA(A) receptor blockade in dorsomedial and ventromedial nuclei of the hypothalamus evokes panic-like elaborated defensive behaviour followed by innate fear-induced antinociception. *Brain Res.* 2009;1305:118–131.

34. Freud S. The justification for detaching from neurasthenia a particular syndrome: the anxiety-neurosis. Collected Papers. 1894;1:76–106.

35. Gardner KL, Thrivikraman KV, Lightman SL, Plotsky PM, Lowry CA. Early life experience alters behavior during social defeat: focus on serotonergic systems. *Neuroscience.* 2005;136(1):181–191.

36. Gardner KL, Hale MW, Lightman SL, Plotsky PM, Lowry CA. Adverse early life experience and social stress during adulthood interact to increase serotonin transporter mRNA expression. *Brain Res.* 2009;1305:47–63.

37. Gardner KL, Hale MW, Oldfield S, Lightman SL, Plotsky PM, Lowry CA. Adverse experience during early life and adulthood interact to elevate *tph2* mRNA expression in serotonergic neurons within the dorsal raphe nucleus. *Neuroscience.* 2009;163(4):991–1001.

38. Gorman JM, Liebowitz MR, Fyer AJ, Stein J. A neuroanatomical hypothesis for panic disorder. *Am J Psychiatry.* 1989;146(2):148–161.

39. Gorman JM, Papp LA, Coplan JD, et al. Anxiogenic effects of CO_2 and hyperventilation in patients with panic disorder. *Am J Psychiatry.* 1994;151(4):547–553.

40. Griez EJ, Lousberg H, van den Hout MA, van der Molen GM. CO_2 vulnerability in panic disorder. *Psychiatry Res.* 1987;20(2):87–95.

41. Hale MW, Hay-Schmidt A, Mikkelsen JD, et al. Exposure to an open-field arena increases c-Fos expression in a subpopulation of neurons in the dorsal raphe nucleus, including neurons projecting to the basolateral amygdaloid complex. *Neuroscience.* 2008;157(4):733–748.

42. Hale MW, Stamper CE, Staub DR, Lowry CA. Urocortin 2 increases c-Fos expression in serotonergic neurons projecting to the ventricular/periventricular system. *Exp Neurol.* 2010;224(1):271–281.

43. Hale MW, Dady KF, Evans AK, Lowry CA. Evidence for in vivo thermosensitivity of serotonergic neurons in the rat dorsal raphe nucleus and raphe pallidus nucleus implicated in thermoregulatory cooling. *Exp Neurol.* 2011;227(2):264–278.

44. Hale MW, Shekhar A, Lowry CA. Stress-related serotonergic systems: implications for symptomatology of anxiety and affective disorders. *Cell Mol Neurobiol.* 2012;32(5):695–708.

45. Hall C. Emotional behavior in the rat. I. Defecation and urination as measures of individual differences in emotionality. *J Comp Psychol.* 1934;18(3):385–403.

46. Hermans EJ, Henckens MJ, Roelofs K, Fernandez G. Fear bradycardia and activation of the human periaqueductal grey. *NeuroImage.* 2013;66:278–287.

47. Hollis JH, Evans AK, Bruce KP, Lightman SL, Lowry CA. Lipopolysaccharide has indomethacin-sensitive actions on Fos expression in topographically organized subpopulations of serotonergic neurons. *Brain Behav Immun.* 2006;20(6):569–577.

48. Horiuchi J, McDowall LM, Dampney RA. Role of 5-HT(1A) receptors in the lower brainstem on the cardiovascular response to dorsomedial hypothalamus activation. *Auton Neurosci.* 2008;142(1–2):71–76.

49. Jenck F, Moreau JL, Martin JR. Dorsal periaqueductal gray-induced aversion as a simulation of panic anxiety: elements of face and predictive validity. *Psychiatry Res.* 1995;57(2):181–191.

50. Johnson PL, Shekhar A. An animal model of panic vulnerability with chronic disinhibition of the dorsomedial/perifornical hypothalamus. *Physiol Behav.* 2012;107(5):686–698.

51. Johnson PL, Hollis JH, Moratalla R, Lightman SL, Lowry CA. Acute hypercarbic gas exposure reveals functionally distinct subpopulations of serotonergic neurons in rats. *J Psychopharmacol.* 2005;19(4):327–341.

52. Johnson PL, Lowry C, Truitt W, Shekhar A. Disruption of GABAergic tone in the dorsomedial hypothalamus attenuates responses in a subset of serotonergic neurons in the dorsal raphe nucleus following lactate-induced panic. *J Psychopharmacol.* 2008;22(6):642–652.

53. Johnson PL, Truitt W, Fitz SD, et al. A key role for orexin in panic anxiety. *Nat Med.* 2010;16(1):111–115.

54. Johnson PL, Molosh A, Fitz SD, et al. Pharmacological depletion of serotonin in the basolateral amygdala complex reduces anxiety and disrupts fear conditioning. *Pharmacol Biochem Behav.* 2015;138:174–179.

55. Kaye J, Buchanan F, Kendrick A, et al. Acute carbon dioxide exposure in healthy adults: evaluation of a novel means of investigating the stress response. *J Neuroendocrinol.* 2004;16(3):256–264.

56. Keim SR, Shekhar A. The effects of GABAA receptor blockade in the dorsomedial hypothalamic nucleus on corticotrophin (ACTH) and corticosterone secretion in male rats. *Brain Res.* 1996;739(1–2):46–51.

57. Kelly KJ, Donner NC, Hale MW, Lowry CA. Swim stress activates serotonergic and nonserotonergic neurons in specific subdivisions of the rat dorsal raphe nucleus in a temperature-dependent manner. *Neuroscience.* 2011;197:251–268.

58. Kerman IA, Bernard R, Rosenthal D, Beals J, Akil H, Watson SJ. Distinct populations of presympathetic-premotor neurons express orexin or melanin-concentrating hormone in the rat lateral hypothalamus. *J Comp Neurol.* 2007;505(5):586–601.

59. Kessler RC, Chiu WT, Jin R, Ruscio AM, Shear K, Walters EE. The epidemiology of panic attacks, panic disorder, and agoraphobia in the national comorbidity survey replication. *Arch Gen Psychiatry.* 2006;63(4):415–424.

60. Klein DF. Delineation of two drug-responsive anxiety syndromes. *Psychopharmacologia.* 1964;5:397–408.

61. Klein DF. Anxiety reconceptualized. *Compr Psychiatry.* 1980;21(6):411–427.

62. Klein DF. False suffocation alarms, spontaneous panics, and related conditions. An integrative hypothesis. *Arch Gen Psychiatry.* 1993;50(4):306–317.

63. Leibold NK, Viechtbauer W, Goossens L, et al. Carbon dioxide inhalation as a human experimental model of panic: the relationship between emotions and cardiovascular physiology. *Biol Psychol.* 2013;94(2):331–340.

64. Lesch KP, Bengel D, Heils A, et al. Association of anxiety-related traits with a polymorphism in the serotonin transporter gene regulatory region. *Science.* 1996;274(5292):1527–1531.

65. Liebowitz MR, Gorman JM, Fyer A, Dillon D, Levitt M, Klein DF. Possible mechanisms for lactate's induction of panic. *Am J Psychiatry.* 1986;143(4):495–502.

66. Lim LW, Temel Y, Visser-Vandewalle V, Blokland A, Steinbusch H. Fos immunoreactivity in the rat forebrain induced by electrical stimulation of the dorsolateral periaqueductal gray matter. *J Chem Neuroanat.* 2009;38(2):83–96.

67. Lim LW, Blokland A, van Duinen M, et al. Increased plasma corticosterone levels after periaqueductal gray stimulation-induced escape reaction or panic attacks in rats. *Behav Brain Res.* 2011;218(2):301–307.

68. Ljubic-Thibal V, Morin A, Diksic M, Hamel E. Origin of the serotonergic innervation to the rat dorsolateral hypothalamus: retrograde transport of cholera toxin and upregulation of tryptophan hydroxylase mRNA expression following selective nerve terminals lesion. *Synapse.* 1999;32(3):177–186.

69. Lopes DA, Lemes JA, Melo-Thomas L, et al. Unpredictable chronic mild stress exerts anxiogenic-like effects and activates neurons in the dorsal and caudal region and in the lateral wings of the dorsal raphe nucleus. *Behav Brain Res.* 2016;297:180–186.

70. Lovick TA. Cardiovascular responses to 5-HT in the ventrolateral medulla of the rat. *J Auton Nerv Syst.* 1989;28(1):35–41.

70a. Lowry CA, Hale MW, Evans AK, Heerkens J, Staub DR, Gasser PJ, Shekhar A. Serotonergic systems, anxiety, and affective disorder: focus on the dorsomedial part of the dorsal raphe nucleus. *Ann N Y Acad Sci.* 2008;132:86–94.

71. Lukkes JL, Kopelman JM, Donner NC, Hale MW, Lowry CA. Development × environment interactions control *tph2* mRNA expression. *Neuroscience.* 2013;237:139–150.

72. Maron E, Kuikka JT, Shlik J, Vasar V, Vanninen E, Tiihonen J. Reduced brain serotonin transporter binding in patients with panic disorder. *Psychiatry Res.* 2004;132(2):173–181.

73. Masdrakis VG, Markianos M, Oulis P. Lack of specific association between panicogenic properties of caffeine and HPA-axis activation. A placebo-controlled study of caffeine challenge in patients with panic disorder. *Psychiatry Res.* 2015;229(1–2):75–81.

74. Massana G, Serra-Grabulosa JM, Salgado-Pineda P, et al. Amygdalar atrophy in panic disorder patients detected by volumetric magnetic resonance imaging. *NeuroImage.* 2003;19(1):80–90.

75. Miguel TL, Pobbe RL, Spiacci Junior A, Zangrossi Junior H. Dorsal raphe nucleus regulation of a panic-like defensive behavior evoked by chemical stimulation of the rat dorsal periaqueductal gray matter. *Behav Brain Res.* 2010;213(2):195–200.

76. Mobbs D, Petrovic P, Marchant JL, et al. When fear is near: threat imminence elicits prefrontal-periaqueductal gray shifts in humans. *Science.* 2007;317(5841):1079–1083.

77. Nascimento JO, Kikuchi LS, de Bortoli VC, Zangrossi Jr H, Viana MB. Dorsomedial hypothalamus serotonin 1A receptors mediate a panic-related response in the elevated T-maze. *Brain Res Bull.* 2014;109:39–45.

78. Nash JR, Sargent PA, Rabiner EA, et al. Serotonin 5-HT1A receptor binding in people with panic disorder: positron emission tomography study. *Br J Psychiatry.* 2008;193(3):229–234.

79. Nashold Jr BS, Wilson WP, Slaughter DG. Sensations evoked by stimulation in the midbrain of man. *J Neurosurg.* 1969;30(1):14–24.

80. Neufeld-Cohen A, Kelly PA, Paul ED, et al. Chronic activation of corticotropin-releasing factor type 2 receptors reveals a key role for 5-HT1A receptor responsiveness in mediating behavioral and serotonergic responses to stressful challenge. *Biol Psychiatry.* 2012;72(6):437–447.

81. Paul ED, Hale MW, Lukkes JL, Valentine MJ, Sarchet DM, Lowry CA. Repeated social defeat increases reactive emotional coping behavior and alters functional responses in serotonergic neurons in the rat dorsal raphe nucleus. *Physiol Behav.* 2011;104(2):272–282.

82. Paul ED, Johnson PL, Shekhar A, Lowry CA. The Deakin/Graeff hypothesis: focus on serotonergic inhibition of panic. *Neurosci Biobehav Rev.* 2014;46(Pt 3):379–396.

83. Paxinos G, Watson C. *The Rat Brain in Stereotaxic Coordinates.* Academic Press; 2007.

84. Pobbe RL, Zangrossi Jr H. 5-HT(1A) and 5-HT(2A) receptors in the rat dorsal periaqueductal gray mediate the anti-panic-like effect induced by the stimulation of serotonergic neurons in the dorsal raphe nucleus. *Psychopharmacol Berl.* 2005;183(3):314–321.

85. Rasche D, Foethke D, Gliemroth J, Tronnier VM. Deep brain stimulation in the posterior hypothalamus for chronic cluster headache. Case report and review of the literature. *Schmerz.* 2006;20(5):439–444.

86. Rifkin A. Panic disorder: response to sodium lactate, and treatment with antidepressants. *Psychopharmacol Bull.* 1983;19(3):432–434.

87. Rozeske RR, Evans AK, Frank MG, Watkins LR, Lowry CA, Maier SF. Uncontrollable, but not controllable, stress desensitizes 5-HT1A receptors in the dorsal raphe nucleus. *J Neurosci.* 2011;31(40):14107–14115.

88. Sajdyk TJ, Shekhar A. Sodium lactate elicits anxiety in rats after repeated GABA receptor blockade in the basolateral amygdala. *Eur J Pharmacol.* 2000;394(2–3):265–273.

89. Sanders SK, Morzorati SL, Shekhar A. Priming of experimental anxiety by repeated subthreshold GABA blockade in the rat amygdala. *Brain Res.* 1995;699(2):250–259.

90. Schenberg LC, Schimitel FG, Armini RS, et al. Translational approach to studying panic disorder in rats: hits and misses. *Neurosci Biobehav Rev.* 2014;46(Pt 3):472–496.

91. Schimitel FG, de Almeida GM, Pitol DN, Armini RS, Tufik S, Schenberg LC. Evidence of a suffocation alarm system within the periaqueductal gray matter of the rat. *Neuroscience.* 2012;200:59–73.

92. Schunck T, Erb G, Mathis A, et al. Functional magnetic resonance imaging characterization of CCK-4-induced panic attack and subsequent anticipatory anxiety. *NeuroImage.* 2006; 31(3):1197–1208.

93. Schutz MT, de Aguiar JC, Graeff FG. Anti-aversive role of serotonin in the dorsal periaqueductal grey matter. *Psychopharmacol Berl.* 1985;85(3):340–345.

94. Sena LM, Bueno C, Pobbe RL, Andrade TG, Zangrossi Jr H, Viana MB. The dorsal raphe nucleus exerts opposed control on generalized anxiety and panic-related defensive responses in rats. *Behav Brain Res.* 2003;142(1–2):125–133.

95. Shekhar A, Sims LS, Bowsher RR. GABA receptors in the region of the dorsomedial hypothalamus of rats regulate anxiety in the elevated plus-maze test. II. Physiological measures. *Brain Res.* 1993;627(1):17–24.

96. Shekhar A, Keim SR, Simon JR, McBride WJ. Dorsomedial hypothalamic GABA dysfunction produces physiological arousal following sodium lactate infusions. *Pharmacol Biochem Behav.* 1996;55(2):249–256.

97. Shulman ID, Cox BJ, Swinson RP, Kuch K, Reichman JT. Precipitating events, locations and reactions associated with initial unexpected panic attacks. *Behav Res Ther.* 1994;32(1):17–20.

98. Sidor MM, Amath A, MacQueen G, Foster JA. A developmental characterization of mesolimbocortical serotonergic gene expression changes following early immune challenge. *Neuroscience.* 2010;171(3):734–746.

99. Sinha SS, Coplan JD, Pine DS, Martinez JA, Klein DF, Gorman JM. Panic induced by carbon dioxide inhalation and lack of hypothalamic-pituitary-adrenal axis activation. *Psychiatry Res.* 1999;86(2):93–98.

100. Sink KS, Walker DL, Freeman SM, Flandreau EI, Ressler KJ, Davis M. Effects of continuously enhanced corticotropin releasing factor expression within the bed nucleus of the stria terminalis on conditioned and unconditioned anxiety. *Mol Psychiatry.* 2013;18(3):308–319.

101. Smith VM, Jeffers RT, Antle MC. Serotonergic enhancement of circadian responses to light: role of the raphe and intergeniculate leaflet. *Eur J Neurosci.* 2015;42(10):2805–2817.

102. Spannuth BM, Hale MW, Evans AK, Lukkes JL, Campeau S, Lowry CA. Investigation of a central nucleus of the amygdala/dorsal raphe nucleus serotonergic circuit implicated in fear-potentiated startle. *Neuroscience.* 2011;179:104–119.

103. Spiacci Jr A, Coimbra NC, Zangrossi Jr H. Differential involvement of dorsal raphe subnuclei in the regulation of anxiety- and panic-related defensive behaviors. *Neuroscience.* 2012;227:350–360.

104. Staub DR, Spiga F, Lowry CA. Urocortin 2 increases c-Fos expression in topographically organized subpopulations of serotonergic neurons in the rat dorsal raphe nucleus. *Brain Res.* 2005;1044(2):176–189.

105. Staub DR, Evans AK, Lowry CA. Evidence supporting a role for corticotropin-releasing factor type 2 (CRF2) receptors in the regulation of subpopulations of serotonergic neurons. *Brain Res.* 2006;1070(1):77–89.

106. Straube B, Reif A, Richter J, et al. The functional -1019C/G *HTR1A* polymorphism and mechanisms of fear. *Transl Psychiatry.* 2014;4.

107. Thompson RH, Canteras NS, Swanson LW. Organization of projections from the dorsomedial nucleus of the hypothalamus: a PHA-L study in the rat. *J Comp Neurol.* 1996;376(1):143–173.

108. Ullah F, dos Anjos-Garcia T, dos Santos IR, Biagioni AF, Coimbra NC. Relevance of dorsomedial hypothalamus, dorsomedial division of the ventromedial hypothalamus and the dorsal periaqueductal gray matter in the organization of freezing or oriented and non-oriented escape emotional behaviors. *Behav Brain Res.* 2015;293:143–152.

109. Valentino RJ, Bey V, Pernar L, Commons KG. Substance P acts through local circuits within the rat dorsal raphe nucleus to alter serotonergic neuronal activity. *J Neurosci.* 2003;23(18): 7155–7159.

110. Vertes RP. A PHA-L analysis of ascending projections of the dorsal raphe nucleus in the rat. *J Comp Neurol.* 1991;313(4):643–668.

111. Viana MB, Graeff FG, Loschmann PA. Kainate microinjection into the dorsal raphe nucleus induces 5-HT release in the amygdala and periaqueductal gray. *Pharmacol Biochem Behav.* 1997;58(1):167–172.

112. Vicente MA, Zangrossi H. Serotonin-2C receptors in the basolateral nucleus of the amygdala mediate the anxiogenic effect of acute imipramine and fluoxetine administration. *Int J Neuropsychopharmacol.* 2012;15(3):389–400.

113. Wang SM, Yeon B, Hwang S, et al. Threat-induced autonomic dysregulation in panic disorder evidenced by heart rate variability measures. *Gen Hosp Psychiatry.* 2013;35(5):497–501.

114. Wiest G, Lehner-Baumgartner E, Baumgartner C. Panic attacks in an individual with bilateral selective lesions of the amygdala. *Arch Neurol.* 2006;63(12):1798–1801.

115. Wilent WB, Oh MY, Buetefisch CM, et al. Induction of panic attack by stimulation of the ventromedial hypothalamus. *J Neurosurg.* 2010;112(6):1295–1298.

116. Yamashita PS, de Bortoli VC, Zangrossi Jr H. 5-HT2C receptor regulation of defensive responses in the rat dorsal periaqueductal gray. *Neuropharmacology.* 2011;60(2–3):216–222.

117. Yeragani VK, Pohl R, Balon R, Rainey JM, Berchou R, Ortiz A. Sodium lactate infusions after treatment with tricyclic antidepressants: behavioral and physiological findings. *Biol Psychiatry.* 1988;24(7):767–774.

118. Yevtushenko OO, Oros MM, Reynolds GP. Early response to selective serotonin reuptake inhibitors in panic disorder is associated with a functional 5-HT1A receptor gene polymorphism. *J Affect Disord.* 2010;123(1–3):308–311.

119. Zangrossi Jr H, Graeff FG. Serotonin in anxiety and panic: contributions of the elevated T-maze. *Neurosci Biobehav Rev.* 2014;46(Pt 3):397–406.

16

Neuroendocrinology of Posttraumatic Stress Disorder: Focus on the HPA Axis

M.E. Bowers[1], R. Yehuda[1,2]

[1]Icahn School of Medicine at Mount Sinai, New York, NY, United States; [2]James J. Peters Veterans Affairs Medical Center, Bronx, NY, United States

Abstract

The hypothalamic–pituitary–adrenal (HPA) axis is activated in response to stressor detection in order to mobilize resources critical for survival. Chronic or extreme stress precipitates dysregulation of the HPA axis concomitant with development of psychopathology, most notably in major depressive disorder. HPA axis alterations have similarly been noted in posttraumatic stress disorder (PTSD). Conversely, PTSD appears to be characterized by lower basal levels of cortisol and greater HPA axis negative feedback, which is thought to be mediated by greater sensitivity of the glucocorticoid receptor. Recent evidence calls into question whether HPA axis alterations represent a preexisting vulnerability or are altered with pathophysiology of the disease and, further, whether these alterations reflect trauma exposure, rather than PTSD per se. Rigorous comparison of participant trauma history, consistency in diagnostic criteria, and systems biology approaches may better inform questions related to an association between changes in the HPA axis and PTSD.

INTRODUCTION: ADAPATIVE AND MALADAPTIVE RESPONSES OF THE HYPOTHALAMIC–PITUITARY–ADRENAL AXIS

Stressors, including physical or psychological events that threaten survival and reproduction, activate the hypothalamic–pituitary–adrenal (HPA) axis [here, simplified as the signaling cascade composed of corticotropin-releasing hormone (CRH or CRF)-corticotropin (ACTH)-cortisol-glucocorticoid receptor (GR)] to mobilize a cache of processes to evade death and, eventually, to restore homeostasis. Stressor perception by the brain induces secretion of CRH by parvocellular neurons in the paraventricular nucleus (PVN) of the hypothalamus. Specifically, CRH is secreted into the portal vessel system, subsequently activating synthesis of proopiomelanocortin (POMC), which is cleaved to ACTH in the anterior pituitary. From the pituitary, ACTH accesses the adrenal cortex via the circulatory system, where it promotes secretion of cortisol, one of the major glucocorticoid hormones in humans. The glucocorticoids promote adaptive stress-responsive physiological processes and behaviors by mobilizing energy reserves, dampening inflammation, and modulating memory processing and storage. Stress-related increases in cortisol eventually result in inhibition of HPA axis, in part through activation of the GR, terminating the stress response and restoring homeostasis.[17,59,60,114]

While HPA axis stressor reactivity promotes survival by triggering specific physiological and behavioral responses, chronic or extreme stressor exposure and concomitant persistent activation of the HPA axis can have deleterious effects. Posttraumatic stress disorder can develop in response to an extreme, life-threatening event or set of events in a subset of vulnerable individuals and is characterized by persistent symptoms of avoidance, hyperarousal, avoidance, and reexperiencing (e.g., nightmares or "flashbacks").[2,45] The gradual conceptualization of PTSD as a failure to recover from a remitted stressor, along with advances in understanding of the neuroendocrinology of major depressive disorder (MDD), prompted investigation into neuroendocrine alterations, specifically cortisol and other aspects of the HPA axis, associated with PTSD. Here, we review evidence of neuroendocrine changes noted in subjects diagnosed with PTSD, particularly focusing on CRH, ACTH, cortisol, and GR. Alterations have also been noted with regard to the sympathetic-adrenal-medullary system, the "flight or fight" response, and PTSD, where greater levels of peripheral catecholamine levels (perhaps underlying a higher degree of arousal characteristic of the disorder) are observed in individuals with PTSD.[72,124,128] For review of the role of catecholamines in PTSD.[73,98]

KEY POINTS

- Exposure to trauma can precipitate posttraumatic stress disorder (PTSD).
- PTSD is associated with neuroendocrine changes, e.g., decreased cortisol.
- Variation in components of the hypothalamic–pituitary–adrenal (HPA)-axis is associated with risk and pathophysiology.
- Questions remain as to whether HPA axis changes are associated with trauma exposure or PTSD.

Several studies note higher levels of plasma and CSF CRH in individuals with PTSD compared to trauma-exposed and healthy controls, although whether these levels represent CRH of hypothalamic or extrahypothalamic origin is unknown.[6,11,15,87]

More studies have examined ACTH, at baseline and in response to perturbation, and although data suggest individuals with PTSD have lower levels of ACTH, similar to cortisol, there is some controversy over whether lower ACTH levels truly reflect psychopathology rather than trauma exposure. For instance, lower plasma ACTH is observed in formerly deployed Gulf War veterans compared to veterans with PTSD and healthy nondeployed subjects.[32] No differences in and higher ACTH levels between PTSD patients and controls have been noted as well.[15,107] In response to CRF, one study reports greater ACTH increases in individuals with PTSD compared to trauma-exposed and healthy controls, while another study observes a smaller ACTH response in individuals with PTSD compared to controls (although history of trauma in these controls is unclear).[30,94] In response to the cold pressor task (CPT), persons with PTSD exhibit a less robust ACTH response compared to responses in control subjects.[85] The HPA axis is able to self-terminate through GR-mediated negative feedback, e.g., via inhibition of ACTH secretion at the level of the anterior pituitary. Greater decline in ACTH to exogenous glucocorticoid administration in individuals with PTSD is thought to reflect greater negative feedback inhibition, potentially stemming from increased GR sensitivity, although others note no differences in ACTH in response to dexamethasone.[20,70,107,127]

Several studies have also examined ACTH in response to the dexamethasone/CRH test, which is thought to represent a more sensitive tool in detecting HPA axis dysregulation. Typically, dexamethasone is administered orally the evening before intravenous CRH administration and subsequent chronic blood sampling. Regarding ACTH response to the dexamethasone/CRH test, the findings are split. Some studies find lower ACTH response in PTSD patients compared to healthy and trauma controls, while others report no differences.[14,48,67,99]

Of the various components of the HPA axis, cortisol has been the most intensely studied. Given high comorbidity and symptom overlap between MDD and PTSD, and observed hypercortisolism in MDD, investigators initially hypothesized that PTSD would be associated with high levels of cortisol. Conversely, research consistently observes lower diurnal levels of urinary, plasma, and salivary cortisol across different populations with PTSD resulting from a variety of focal trauma.[10,56,83,101,112,118,121,125,126] According to Yehuda et al., low cortisol in persons with PTSD appears to result from a prolonged nadir (evening levels) and shorter duration of peak cortisol release (morning/cortisol awakening response). The cortisol awakening response (CAR) represents the surge in cortisol between awakening and approximately 30 min postawakening. Although the exact role of the CAR is still somewhat unclear, variation in the CAR is influenced by psychosocial and physical factors, as well as psychopathology.[52] Separate studies support altered morning and cortisol awakening responses in PTSD, where individuals with PTSD exhibit lower levels of cortisol at these time points.[9,12,26,54,75,82,92,93,108,111] However, it should be noted that other studies observe no differences in cortisol levels and, less often, higher levels of cortisol between individuals with PTSD and controls.[5,7,21,24,40,51,53,55,66,78,79,91,97,102,113,128–130]

As the field has evolved, investigators have begun to probe whether alterations in cortisol levels are associated with trauma exposure or development of PTSD in the aftermath of trauma exposure. Several studies observe lower basal cortisol levels in trauma-exposed (PTSD+ and PTSD−) subjects compared to nonexposed, healthy controls.[38,69,96] Still other studies find lower cortisol levels in participants with PTSD compared to both trauma-exposed and healthy, nonexposed subjects.[83,108,111,121] Several meta-analyses, however, conclude that low cortisol is associated with trauma exposure, rather than psychopathology.[48,62,64] Conflict in the data may stem from inconsistent comparison of severity of trauma, including index trauma and total trauma history, as well as trauma type and developmental timing of trauma exposure across studies. Current investigations are attempting to more systematically compare trauma-exposed (PTSD+ versus PTSD−) by appropriately matching trauma history, among other variables.

In further parsing the cortisol data, one meta-analysis observes that women with PTSD have significantly lower levels than healthy controls compared to no significant differences between men with and without PTSD, whereas a separate meta-analysis reports that lower P.M. cortisol is associated with a higher percent of males in trauma-exposed groups.[62,64] It is unclear, however, whether included studies accounted for phase of the menstrual cycle when assessing an effect of gender on cortisol levels. When controlling for time since trauma exposure, data suggest that time elapsed since focal trauma is negatively associated with P.M. cortisol for trauma-exposed groups. This data accords with other studies suggesting that cortisol is hypersecreted in the short term, whereas hyposecretion may develop subsequently in the long term.[64,93,106,110] Importantly, type of biological sample (plasma, saliva, urine) does not appear to mediate any significant effect on cortisol measures.[62] This is critical, as cortisol measured from plasma typically represents bound and unbound hormone, while salivary and urinary cortisol is entirely unbound. Cortisol bound to carrier proteins (cortisol-binding globulin, CBG) represents the proportion of hormone that is unable to act on target tissue. At least one study observes no difference in CBG levels between PTSD subjects and controls, while other data indicate increased CBG levels in individuals with PTSD.[16,43]

As PTSD is most often comorbid with MDD, and PTSD and MDD appear to be associated with divergent alterations in the HPA axis, the question of whether inconsistent cortisol data in populations with PTSD could be confounded by comorbid MDD has been specifically addressed by several studies. As expected, PTSD comorbid with MDD is associated with increased levels of basal cortisol, although other individual studies and meta-analyses observe no effect of comorbid diagnosis of MDD.[47,62,64,129]

With regard to conflict related to whether alterations in cortisol are reflective of trauma exposure or PTSD, it may be more instructive to compare cortisol levels against a continuum of PTSD symptoms, including specific PTSD subscales. These types of analyses could be particularly illuminating, as many PTSD subjects may have subthreshold or past PTSD. With evolving diagnostic criteria, in addition, distinction between PTSD+ and PTSD− is somewhat ambiguous. Although some studies find an association between cortisol and total number of PTSD symptoms,[10] correlations between cortisol and specific symptoms are more often reported. Findings of negative associations between number of hyperarousal symptoms, and, separately, number of avoidance-numbing symptoms, and levels of basal cortisol are more commonly observed, although some studies also report correlations between cortisol and hyperarousal and dissociation symptoms, additionally.[10,26,27,35,38,57,71,92,121]

NEGATIVE FEEDBACK/ GLUCOCORTICOID RECEPTOR

One of the critical features of the HPA axis is the ability to self-terminate via GR-mediated negative feedback. As a part of this feedback, increased cortisol synthesis is terminated to prevent deleterious effects of excess circulating cortisol. Data suggest that extreme or chronic stress can fundamentally impact specific features of the HPA axis, including negative feedback. For instance, MDD is associated with blunted negative feedback, where cortisol synthesis appears to be minimally perturbed in response to synthetic glucocorticoid challenge, as measured by the dexamethasone suppression test (DST). Response to the DST is typically measured by comparing baseline morning plasma cortisol to post-dexamethasone plasma cortisol on the following consecutive day. Dexamethasone is often administered orally the evening before the second blood collection at a dose of 0.5 mg, although doses of 0.25 and 1.0 mg are used as well. The observation that individuals with MDD exhibit blunted negative feedback prompted investigation into whether a similar phenomenon would be observed in patients with PTSD. Conversely, initial reports noted increased cortisol suppression in response to dexamethasone in individuals with PTSD with civilian, military, domestic violence, and child abuse trauma compared to healthy and, for some studies, trauma-exposed controls.[28,34,70,77,116,120] Increased sensitivity of GR and enhanced negative HPA axis feedback in individuals with PTSD may, in part, be mediated by increased number of GR. Several studies observe higher numbers of GR on lymphocytes from whole blood of individuals with PTSD compared to healthy controls and trauma-exposed controls.[33,116,122] Greater number of receptors could mediate increased negative feedback, particularly as PTSD has been associated with lower GR promoter methylation, but other mechanisms of increased GR sensitivity have yet to be explored, including increased binding capacity or changes in properties related to GR-targeted DNA transcription.[117]

Other studies, however, suggest that increased negative feedback/cortisol suppression is related to trauma exposure, rather than PTSD.[16,31,88] Still, other studies observe no differences in postdexamethasone cortisol between individuals with PTSD and healthy and trauma-exposed controls.[4,20,50,53,63,71,107]

In vitro tests may offer may direct test of GR sensitivity, clarifying the role of GR in PTSD, as they offer a more controlled experimental condition and allow for testing of multiple doses of dexamethasone. In these experiments, cells are incubated with a variety of doses of dexamethasone and then subsequently assayed for a specific outcome measure affected by GR activity (e.g., cytokine production). Data thus far suggest enhanced sensitivity of GR in whole blood or leukocytes isolated from whole blood of individuals with PTSD compared to healthy or trauma-exposed controls.[82,119]

Relatively less research has been dedicated to uncovering a potential role of the mineralocorticoid receptor (MR) in PTSD. MR has a higher affinity for glucocorticoids and is therefore thought to be activated by low cortisol levels, regulating key functions at baseline. Conversely, GR has a lower affinity for glucocorticoids and is activated when levels of glucocorticoids are high, e.g., during periods of stress.[68] The data thus far minimally implicate MR in PTSD, although one study reports lower levels of whole blood MR RNA expression in a subset of individuals with PTSD.[44,58,76,131]

Meta-analysis reveals greater cortisol suppression postdexamethasone in individuals with PTSD, PTSD co-morbid with MDD, and trauma controls compared to no trauma controls, suggesting that greater negative HPA axis feedback may be a feature of trauma exposure, rather than psychopathology per se.[64] Furthermore, postdexamethasone cortisol is moderated by age, sex, time since focal trauma, and development time window of trauma occurrence (e.g., childhood), consistent with findings from other meta-analyses.[47,64]

PERTURBATIONS OF THE HPA AXIS AS A VULNERABILITY FACTOR IN THE DEVELOPMENT OF PTSD

GR expression and sensitivity, rather than circulating cortisol levels, has been particularly implicated in vulnerability to development of PTSD before trauma exposure, although the data are limited. These studies have been exclusively conducted in populations at high risk for trauma exposure, including military, firefighter, and police samples, and these variables should be considered when interpreting the data. Peripheral blood mononuclear cell (PBMC) GR number, measured by binding occupancy, is predictive of subsequent development of PTSD in military personnel before deployment.[103,104] Interestingly, greater number of GRs in individuals who developed PTSD persisted months after deployment, suggesting that increased number of GRs is primarily a risk factor for PTSD, rather than a variable that is altered with development of the disorder.[104] Alternatively, there is limited evidence suggesting that diurnal cortisol, including CAR, before trauma exposure represents a preexisting vulnerability to PTSD, although a recent study finds that lower baseline hair cortisol concentration is predictive of later PTSD symptomatology upon trauma exposure during deployment in a German military population.[37,95,105] In addition, data suggest that a greater salivary CAR, assessed during academy training, is a prospective predictor of greater peritraumatic dissociation and Acute Stress Disorder (ASD) symptoms across the first 3 years of police service.[41] ASD reactions include acute dissociative symptoms, reexperiencing, avoidance, and anxiety and/or arousal after intense stressor exposure and are strong predictors of development of PTSD.[39]

More biological data exist with regard to changes in the aftermath (hours to weeks) after trauma and the relative predictive power of this data in relation to development of PTSD. In studies where cortisol was measured in the hours after trauma exposure, data seem to support the hypothesis that low cortisol (in urine and saliva) is predictive of subsequent development of PTSD or greater PTSD symptoms.[19,22,36,49,61,65,80] In rape survivors, previous sexual assault is associated with lower cortisol in the ER, and, separately, greater risk of developing PTSD.[81] A separate study independently supports this hypothesis, reporting that a history of prior assault in rape survivors is associated with lower serum cortisol at the ER, and, furthermore, acute cortisol levels are positively associated with PTSD symptoms at 6 weeks and negatively associated with PTSD symptoms at later time points.[109]

Some data also suggest that low cortisol measured in the days and weeks after trauma also predict PTSD symptoms.[1,3] It should be noted, however, that other studies do not observe a correlation between acute cortisol levels and subsequent development of PTSD.[123,91]

CONCLUSIONS AND FUTURE DIRECTIONS

Altogether, the data suggest that individuals with PTSD exhibit elevated levels of CRH, downregulation of ACTH and cortisol, as well as increased sensitivity of GR and, subsequently, greater negative feedback of the HPA axis. Questions persist as to whether neuroendocrine alterations are a consequence of trauma exposure or, specifically, the development of PTSD in response to trauma exposure. These questions will likely be answered upon completion of large-scale studies designed to explicitly address comparison of trauma exposure between PTSD and non-PTSD populations. Related is the issue of how methodology (frequency of biological sampling, sample size (N), type of biological sample e.g., urine, saliva, plasma—which is most meaningful?), somewhat addressed by the extant literature, contributes to differences in neuroendocrine findings. Studies should also continue to probe the contribution of current versus lifetime diagnoses of PTSD, number of traumatic exposures, developmental period of trauma exposure, gender, comorbid psychiatric disorders, and time elapsed since index trauma on neuroendocrine measures. Molecular biology and genetic studies, particularly encompassed within a systems biology approach, may help explain variance in the neuroendocrine data.[86,13] In addition, the frequent cooccurrence in military veterans of mild traumatic brain (mTBI) injury and PTSD, which are associated with diverging neuroendocrine phenotypes, may contribute to inconsistent cortisol data.[23] Opportunities to examine the neuroendocrine profile of persons before trauma exposure through the development of resultant psychopathology will elucidate whether perturbation of the HPA axis is reflective of some vulnerability factor or whether HPA axis changes are rooted in pathophysiology of PTSD.

Other evidence further underline the presence of neuroendocrine alterations in PTSD, including promising findings related to hydrocortisone, among other HPA axis-targeted drugs, for the treatment of PTSD. This includes secondary prevention, as well as chronic administration administered alone or in combination with behavioral therapy after development of PTSD.[18,29,89,90,100,115,132] Dexamethasone has also been shown to modulate intermediate phenotypes associated with PTSD, including increased fear-potentiated startle.[42] Conversely, behavioral therapies that ameliorate maladaptive PTSD-related symptoms and behaviors also appear to alter cortisol.[8,25,46,74,84]

In conclusion, evidence of lower cortisol in individuals with PTSD suggests a reconceptualization of PTSD reflecting a systemic failure of the stress response system. Lower levels of cortisol likely stems from increased GR sensitivity and, given the data, contributes to maladaptive behaviors and symptoms of PTSD. As more specific evidence regarding HPA axis dysfunction in PTSD accumulates, this will potentially lead to more targeted treatments and diagnostics.

Acknowledgments

The authors wish to acknowledge support from the US Department of Defense (DOD W81XWH-10-2-0072 and DOD W81XWH-13-1-0071) and from the Lightfighter Trust Foundation (LFT2009-02-1).

References

1. Aardal-Eriksson E, Eriksson TE, et al. Salivary cortisol, posttraumatic stress symptoms, and general health in the acute phase and during 9-month follow-up. *Biol Psychiatry.* 2001;50(12):986–993.
2. American Psychiatric Association and American Psychiatric Association DSM-5 Task Force. *Diagnostic and Statistical Manual of Mental Disorders: DSM-5.* Washington, DC: American Psychiatric Association; 2013.
3. Anisman H, Griffiths J, et al. Posttraumatic stress symptoms and salivary cortisol levels. *Am J Psychiatry.* 2001;158(9):1509–1511.
4. Bachmann AW, Sedgley TL, et al. Glucocorticoid receptor polymorphisms and post-traumatic stress disorder. *Psychoneuroendocrinology.* 2005;30(3):297–306.
5. Baker DG, Ekhator NN, et al. Higher levels of basal serial CSF cortisol in combat veterans with posttraumatic stress disorder. *Am J Psychiatry.* 2005;162(5):992–994.
6. Baker DG, West SA, et al. Serial CSF corticotropin-releasing hormone levels and adrenocortical activity in combat veterans with posttraumatic stress disorder. *Am J Psychiatry.* 1999;156(4):585–588.
7. Basu A, Levendosky AA, et al. Trauma sequelae and cortisol levels in women exposed to intimate partner violence. *Psychodyn Psychiatry.* 2013;41(2):247–275.
8. Bergen-Cico D, Possemato K, et al. Reductions in cortisol associated with primary care brief mindfulness program for veterans with PTSD. *Med Care.* 2014;52(12 suppl 5):S25–S31.
9. Bicanic IA, Postma RM, et al. Salivary cortisol and dehydroepiandrosterone sulfate in adolescent rape victims with post traumatic stress disorder. *Psychoneuroendocrinology.* 2013;38(3):408–415.
10. Bierer LM, Tischler L, et al. Clinical correlates of 24-h cortisol and norepinephrine excretion among subjects seeking treatment following the world trade center attacks on 9/11. *Ann N Y Acad Sci.* 2006;1071:514–520.
11. Bremner JD, Licinio J, et al. Elevated CSF corticotropin-releasing factor concentrations in posttraumatic stress disorder. *Am J Psychiatry.* 1997;154(5):624–629.
12. Chen T, Guo M, et al. A comparative study on the levels of serum cytokines and cortisol among post-traumatic stress disorder patients of Li and Han ethnicities in Hainan. *Chin Med J Engl.* 2014;127(15):2771–2774.
13. Daskalakis NP, Cohen H, et al. Expression profiling associates blood and brain glucocorticoid receptor signaling with trauma-related individual differences in both sexes. *Proc Natl Acad Sci USA.* 2014;111(37):13529–13534.

14. de Kloet C, Vermetten E, et al. Differences in the response to the combined DEX-CRH test between PTSD patients with and without co-morbid depressive disorder. *Psychoneuroendocrinology.* 2008;33(3):313–320.

15. de Kloet CS, Vermetten E, et al. Elevated plasma corticotrophin-releasing hormone levels in veterans with posttraumatic stress disorder. *Prog Brain Res.* 2008;167:287–291.

16. de Kloet CS, Vermetten E, et al. Enhanced cortisol suppression in response to dexamethasone administration in traumatized veterans with and without posttraumatic stress disorder. *Psychoneuroendocrinology.* 2007;32(3):215–226.

17. de Kloet ER, Joels M, et al. Stress and the brain: from adaptation to disease. *Nat Rev Neurosci.* 2005;6(6):463–475.

18. Delahanty DL, Gabert-Quillen C, et al. The efficacy of initial hydrocortisone administration at preventing posttraumatic distress in adult trauma patients: a randomized trial. *CNS Spectr.* 2013;18(2):103–111.

19. Delahanty DL, Raimonde AJ, et al. Initial posttraumatic urinary cortisol levels predict subsequent PTSD symptoms in motor vehicle accident victims. *Biol Psychiatry.* 2000;48(9):940–947.

20. Duval F, Crocq MA, et al. Increased adrenocorticotropin suppression following dexamethasone administration in sexually abused adolescents with posttraumatic stress disorder. *Psychoneuroendocrinology.* 2004;29(10):1281–1289.

21. Eckart C, Engler H, et al. No PTSD-related differences in diurnal cortisol profiles of genocide survivors. *Psychoneuroendocrinology.* 2009;34(4):523–531.

22. Ehring T, Ehlers A, et al. Do acute psychological and psychobiological responses to trauma predict subsequent symptom severities of PTSD and depression? *Psychiatry Res.* 2008;161(1):67–75.

23. Flory JD, Henn-Haase C, et al. Glucocorticoid functioning in male combat veterans with posttraumatic stress disorder and mild traumatic brain injury. *Biol Psychiatry.* 2015;78(3):e5–e6.

24. Friedman MJ, Jalowiec J, et al. Adult sexual abuse is associated with elevated neurohormone levels among women with PTSD due to childhood sexual abuse. *J Trauma Stress.* 2007;20(4):611–617.

25. Gerardi M, Rothbaum BO, et al. Cortisol response following exposure treatment for PTSD in rape victims. *J Aggress Maltreat Trauma.* 2010;19(4):349–356.

26. Gill J, Vythilingam M, et al. Low cortisol, high DHEA, and high levels of stimulated TNF-alpha, and IL-6 in women with PTSD. *J Trauma Stress.* 2008;21(6):530–539.

27. Goenjian AK, Pynoos RS, et al. Hypothalamic-pituitary-adrenal activity among Armenian adolescents with PTSD symptoms. *J Trauma Stress.* 2003;16(4):319–323.

28. Goenjian AK, Yehuda R, et al. Basal cortisol, dexamethasone suppression of cortisol, and MHPG in adolescents after the 1988 earthquake in Armenia. *Am J Psychiatry.* 1996;153(7):929–934.

29. Golier JA, Caramanica K, et al. A pilot study of mifepristone in combat-related PTSD. *Depression Res Treat.* 2012;2012:393251.

30. Golier JA, Caramanica K, et al. Neuroendocrine response to CRF stimulation in veterans with and without PTSD in consideration of war zone era. *Psychoneuroendocrinology.* 2012;37(3):350–357.

31. Golier JA, Schmeidler J, et al. Enhanced cortisol suppression to dexamethasone associated with Gulf War deployment. *Psychoneuroendocrinology.* 2006;31(10):1181–1189.

32. Golier JA, Schmeidler J, et al. Twenty-four hour plasma cortisol and adrenocorticotropic hormone in Gulf War veterans: relationships to posttraumatic stress disorder and health symptoms. *Biol Psychiatry.* 2007;62(10):1175–1178.

33. Gotovac K, Sabioncello A, et al. Flow cytometric determination of glucocorticoid receptor (GCR) expression in lymphocyte subpopulations: lower quantity of GCR in patients with posttraumatic stress disorder (PTSD). *Clin Exp Immunol.* 2003;131(2):335–339.

34. Griffin MG, Resick PA, et al. Enhanced cortisol suppression following dexamethasone administration in domestic violence survivors. *Am J Psychiatry.* 2005;162(6):1192–1199.

35. Habersaat S, Borghini A, et al. Posttraumatic stress symptoms and cortisol regulation in mothers of very preterm infants. *Stress Health.* 2014;30(2):134–141.

36. Hawk LW, Dougall AL, et al. Urinary catecholamines and cortisol in recent-onset posttraumatic stress disorder after motor vehicle accidents. *Psychosom Med.* 2000;62(3):423–434.

37. Heinrichs M, Wagner D, et al. Predicting posttraumatic stress symptoms from pretraumatic risk factors: a 2-year prospective follow-up study in firefighters. *Am J Psychiatry.* 2005;162(12):2276–2286.

38. Horn CA, Pietrzak RH, et al. Linking plasma cortisol levels to phenotypic heterogeneity of posttraumatic stress symptomatology. *Psychoneuroendocrinology.* 2014;39:88–93.

39. Howlett JR, Stein MB. Prevention of trauma and stressor-related disorders: a review. *Neuropsychopharmacology.* 2015;41(1):357–369.

40. Inslicht SS, Marmar CR, et al. Increased cortisol in women with intimate partner violence-related posttraumatic stress disorder. *Ann N Y Acad Sci.* 2006;1071:428–429.

41. Inslicht SS, Otte C, et al. Cortisol awakening response prospectively predicts peritraumatic and acute stress reactions in police officers. *Biol Psychiatry.* 2011;70(11):1055–1062.

42. Jovanovic T, Phifer JE, et al. Cortisol suppression by dexamethasone reduces exaggerated fear responses in posttraumatic stress disorder. *Psychoneuroendocrinology.* 2011;36(10):1540–1552.

43. Kanter ED, Wilkinson CW, et al. Glucocorticoid feedback sensitivity and adrenocortical responsiveness in posttraumatic stress disorder. *Biol Psychiatry.* 2001;50(4):238–245.

44. Kellner M, Baker DG, et al. Mineralocorticoid receptor function in patients with posttraumatic stress disorder. *Am J Psychiatry.* 2002;159(11):1938–1940.

45. Kessler RC, Sonnega A, et al. Posttraumatic stress disorder in the National Comorbidity Survey. *Arch Gen Psychiatry.* 1995;52(12):1048–1060.

46. Kim SH, Schneider SM, et al. PTSD symptom reduction with mindfulness-based stretching and deep breathing exercise: randomized controlled clinical trial of efficacy. *J Clin Endocrinol Metab.* 2013;98(7):2984–2992.

47. Klaassens ER, Giltay EJ, et al. Adulthood trauma and HPA-axis functioning in healthy subjects and PTSD patients: a meta-analysis. *Psychoneuroendocrinology.* 2012;37(3):317–331.

48. Klaassens ER, Giltay EJ, et al. Trauma exposure in relation to basal salivary cortisol and the hormone response to the dexamethasone/CRH test in male railway employees without lifetime psychopathology. *Psychoneuroendocrinology.* 2010;35(6):878–886.

49. Kobayashi I, Delahanty DL. Awake/sleep cortisol levels and the development of posttraumatic stress disorder in injury patients with peritraumatic dissociation. *Psychol Trauma.* 2014;6(5):449–456.

50. Kosten TR, Wahby V, et al. The dexamethasone suppression test and thyrotropin-releasing hormone stimulation test in posttraumatic stress disorder. *Biol Psychiatry.* 1990;28(8):657–664.

51. Laudenslager ML, Noonan C, et al. Salivary cortisol among American Indians with and without posttraumatic stress disorder (PTSD): gender and alcohol influences. *Brain Behav Immun.* 2009;23(5):658–662.

52. Law R, Hucklebridge F, et al. State variation in the cortisol awakening response. *Stress.* 2013;16(5):483–492.

53. Lindley SE, Carlson EB, et al. Basal and dexamethasone suppressed salivary cortisol concentrations in a community sample of patients with posttraumatic stress disorder. *Biol Psychiatry.* 2004;55(9):940–945.

54. Luecken LJ, Dausch B, et al. Alterations in morning cortisol associated with PTSD in women with breast cancer. *J Psychosom Res.* 2004;56(1):13–15.

55. Maes M, Lin A, et al. Increased 24-hour urinary cortisol excretion in patients with post-traumatic stress disorder and patients with major depression, but not in patients with fibromyalgia. *Acta Psychiatr Scand.* 1998;98(4):328–335.

56. Mason JW, Giller EL, et al. Urinary free-cortisol levels in posttraumatic stress disorder patients. *J Nerv Ment Dis.* 1986;174(3):145–149.

57. Mason JW, Wang S, et al. Psychogenic lowering of urinary cortisol levels linked to increased emotional numbing and a shame-depressive syndrome in combat-related posttraumatic stress disorder. *Psychosom Med.* 2001;63(3):387–401.

58. Matic G, Vojnovic Milutinovic D, et al. Mineralocorticoid receptor and heat shock protein expression levels in peripheral lymphocytes from war trauma-exposed men with and without PTSD. *Psychiatry Res.* 2014;215(2):379–385.

59. McEwen BS. Allostasis and allostatic load: implications for neuropsychopharmacology. *Neuropsychopharmacology.* 2000;22(2):108–124.

60. McEwen BS. Physiology and neurobiology of stress and adaptation: central role of the brain. *Physiol Rev.* 2007;87(3):873–904.

61. McFarlane AC, Barton CA, et al. Cortisol response to acute trauma and risk of posttraumatic stress disorder. *Psychoneuroendocrinology.* 2011;36(5):720–727.

62. Meewisse ML, Reitsma JB, et al. Cortisol and post-traumatic stress disorder in adults: systematic review and meta-analysis. *Br J Psychiatry.* 2007;191:387–392.

63. Metzger LJ, Carson MA, et al. Basal and suppressed salivary cortisol in female Vietnam nurse veterans with and without PTSD. *Psychiatry Res.* 2008;161(3):330–335.

64. Morris MC, Compas BE, et al. Relations among posttraumatic stress disorder, comorbid major depression, and HPA function: a systematic review and meta-analysis. *Clin Psychol Rev.* 2012;32(4):301–315.

65. Mouthaan J, Sijbrandij M, et al. The role of acute cortisol and DHEAS in predicting acute and chronic PTSD symptoms. *Psychoneuroendocrinology.* 2014;45:179–186.

66. Muhtz C, Godemann K, et al. Effects of chronic posttraumatic stress disorder on metabolic risk, quality of life, and stress hormones in aging former refugee children. *J Nerv Ment Dis.* 2011;199(9):646–652.

67. Muhtz C, Wester M, et al. A combined dexamethasone/corticotropin-releasing hormone test in patients with chronic PTSD–first preliminary results. *J Psychiatr Res.* 2008;42(8):689–693.

68. Myers B, McKlveen JM, et al. Glucocorticoid actions on synapses, circuits, and behavior: implications for the energetics of stress. *Front Neuroendocrinol.* 2014;35(2):180–196.

69. Neumeister A, Normandin MD, et al. Elevated brain cannabinoid CB1 receptor availability in post-traumatic stress disorder: a positron emission tomography study. *Mol Psychiatry.* 2013;18(9):1034–1040.

70. Newport DJ, Heim C, et al. Pituitary-adrenal responses to standard and low-dose dexamethasone suppression tests in adult survivors of child abuse. *Biol Psychiatry.* 2004;55(1):10–20.

71. Neylan TC, Brunet A, et al. PTSD symptoms predict waking salivary cortisol levels in police officers. *Psychoneuroendocrinology.* 2005;30(4):373–381.

72. Nicholson EL, Bryant RA, et al. Interaction of noradrenaline and cortisol predicts negative intrusive memories in posttraumatic stress disorder. *Neurobiol Learn Mem.* 2014;112:204–211.

73. O'Donnell T, Hegadoren KM, et al. Noradrenergic mechanisms in the pathophysiology of post-traumatic stress disorder. *Neuropsychobiology.* 2004;50(4):273–283.

74. Olff M, de Vries GJ, et al. Changes in cortisol and DHEA plasma levels after psychotherapy for PTSD. *Psychoneuroendocrinology.* 2007;32(6):619–626.

75. Olff M, Guzelcan Y, et al. HPA- and HPT-axis alterations in chronic posttraumatic stress disorder. *Psychoneuroendocrinology.* 2006;31(10):1220–1230.

76. Otte C, Muhtz C, et al. Mineralocorticoid receptor function in posttraumatic stress disorder after pretreatment with metyrapone. *Biol Psychiatry.* 2006;60(7):784–787.

77. Pfeffer CR, Altemus M, et al. Salivary cortisol and psychopathology in adults bereaved by the September 11, 2001 terror attacks. *Int J Psychiatry Med.* 2009;39(3):215–226.

78. Pico-Alfonso MA, Garcia-Linares MI, et al. Changes in cortisol and dehydroepiandrosterone in women victims of physical and psychological intimate partner violence. *Biol Psychiatry.* 2004;56(4):233–240.

79. Pitman RK, Orr SP. Twenty-four hour urinary cortisol and catecholamine excretion in combat-related posttraumatic stress disorder. *Biol Psychiatry.* 1990;27(2):245–247.

80. Price M, Kearns M, et al. Emergency department predictors of posttraumatic stress reduction for trauma-exposed individuals with and without an early intervention. *J Consult Clin Psychol.* 2014;82(2):336–341.

81. Resnick HS, Yehuda R, et al. Effect of previous trauma on acute plasma cortisol level following rape. *Am J Psychiatry.* 1995;152(11):1675–1677.

82. Rohleder N, Joksimovic L, et al. Hypocortisolism and increased glucocorticoid sensitivity of pro-inflammatory cytokine production in Bosnian war refugees with posttraumatic stress disorder. *Biol Psychiatry.* 2004;55(7):745–751.

83. Roth G, Ekblad S, et al. A longitudinal study of PTSD in a sample of adult mass-evacuated Kosovars, some of whom returned to their home country. *Eur Psychiatry.* 2006;21(3):152–159.

84. Rothbaum BO, Price M, et al. A randomized, double-blind evaluation of D-cycloserine or alprazolam combined with virtual reality exposure therapy for posttraumatic stress disorder in Iraq and Afghanistan war veterans. *Am J Psychiatry.* 2014;171(6):640–648.

85. Santa Ana EJ, Saladin ME, et al. PTSD and the HPA axis: differences in response to the cold pressor task among individuals with child vs. adult trauma. *Psychoneuroendocrinology.* 2006;31(4):501–509.

86. Sarapas C, Cai G, et al. Genetic markers for PTSD risk and resilience among survivors of the World Trade Center attacks. *Dis Markers.* 2011;30(2–3):101–110.

87. Sautter FJ, Bissette G, et al. Corticotropin-releasing factor in posttraumatic stress disorder (PTSD) with secondary psychotic symptoms, nonpsychotic PTSD, and healthy control subjects. *Biol Psychiatry.* 2003;54(12):1382–1388.

88. Savic D, Knezevic G, et al. Is there a biological difference between trauma-related depression and PTSD? DST says 'NO'. *Psychoneuroendocrinology.* 2012;37(9):1516–1520.

89. Schelling G, Kilger E, et al. Stress doses of hydrocortisone, traumatic memories, and symptoms of posttraumatic stress disorder in patients after cardiac surgery: a randomized study. *Biol Psychiatry.* 2004;55(6):627–633.

90. Schelling G, Roozendaal B, et al. Efficacy of hydrocortisone in preventing posttraumatic stress disorder following critical illness and major surgery. *Ann N Y Acad Sci.* 2006;1071:46–53.

91. Shalev AY, Videlock EJ, et al. Stress hormones and post-traumatic stress disorder in civilian trauma victims: a longitudinal study. Part I: HPA axis responses. *Int J Neuropsychopharmacol.* 2008;11(3):365–372.

92. Simeon D, Yehuda R, et al. Dissociation versus posttraumatic stress: cortisol and physiological correlates in adults highly exposed to the World Trade Center attack on 9/11. *Psychiatry Res.* 2008;161(3):325–329.

93. Simsek S, Uysal C, et al. BDNF and cortisol levels in children with or without post-traumatic stress disorder after sustaining sexual abuse. *Psychoneuroendocrinology.* 2015;56:45–51.

94. Smith MA, Davidson J, et al. The corticotropin-releasing hormone test in patients with posttraumatic stress disorder. *Biol Psychiatry.* 1989;26(4):349–355.

95. Steudte-Schmiedgen S, Stalder T, et al. Hair cortisol concentrations and cortisol stress reactivity predict PTSD symptom increase after trauma exposure during military deployment. *Psychoneuroendocrinology*. 2015;59:123–133.

96. Steudte S, Kirschbaum C, et al. Hair cortisol as a biomarker of traumatization in healthy individuals and posttraumatic stress disorder patients. *Biol Psychiatry*. 2013;74(9):639–646.

97. Steudte S, Kolassa IT, et al. Increased cortisol concentrations in hair of severely traumatized Ugandan individuals with PTSD. *Psychoneuroendocrinology*. 2011;36(8):1193–1200.

98. Strawn JR, Geracioti Jr TD. Noradrenergic dysfunction and the psychopharmacology of posttraumatic stress disorder. *Depress Anxiety*. 2008;25(3):260–271.

99. Strohle A, Scheel M, et al. Blunted ACTH response to dexamethasone suppression-CRH stimulation in posttraumatic stress disorder. *J Psychiatr Res*. 2008;42(14):1185–1188.

100. Suris A, North C, et al. Effects of exogenous glucocorticoid on combat-related PTSD symptoms. *Ann Clin Psychiatry*. 2010;22(4):274–279.

101. Thomas KS, Bower JE, et al. Post-traumatic disorder symptoms and blunted diurnal cortisol production in partners of prostate cancer patients. *Psychoneuroendocrinology*. 2012;37(8):1181–1190.

102. van der Hal-Van Raalte EA, Bakermans-Kranenburg MJ, et al. Diurnal cortisol patterns and stress reactivity in child Holocaust survivors reaching old age. *Aging Ment Health*. 2008;12(5):630–638.

103. van Zuiden M, Geuze E, et al. Glucocorticoid receptor pathway components predict posttraumatic stress disorder symptom development: a prospective study. *Biol Psychiatry*. 2012;71(4):309–316.

104. van Zuiden M, Geuze E, et al. Pre-existing high glucocorticoid receptor number predicting development of posttraumatic stress symptoms after military deployment. *Am J Psychiatry*. 2011;168(1):89–96.

105. van Zuiden M, Kavelaars A, et al. A prospective study on personality and the cortisol awakening response to predict posttraumatic stress symptoms in response to military deployment. *J Psychiatr Res*. 2011;45(6):713–719.

106. Vidovic A, Gotovac K, et al. Repeated assessments of endocrine- and immune-related changes in posttraumatic stress disorder. *Neuroimmunomodulation*. 2011;18(4):199–211.

107. Vythilingam M, Gill JM, et al. Low early morning plasma cortisol in posttraumatic stress disorder is associated with co-morbid depression but not with enhanced glucocorticoid feedback inhibition. *Psychoneuroendocrinology*. 2010;35(3):442–450.

108. Wahbeh H, Oken BS. Salivary cortisol lower in posttraumatic stress disorder. *J Trauma Stress*. 2013;26(2):241–248.

109. Walsh K, Nugent NR, et al. Cortisol at the emergency room rape visit as a predictor of PTSD and depression symptoms over time. *Psychoneuroendocrinology*. 2013;38(11):2520–2528.

110. Weems CF, Carrion VG. The association between PTSD symptoms and salivary cortisol in youth: the role of time since the trauma. *J Trauma Stress*. 2007;20(5):903–907.

111. Wessa M, Rohleder N, et al. Altered cortisol awakening response in posttraumatic stress disorder. *Psychoneuroendocrinology*. 2006;31(2):209–215.

112. Wingenfeld K, Whooley MA, et al. Effect of current and lifetime posttraumatic stress disorder on 24-h urinary catecholamines and cortisol: results from the Mind Your Heart Study. *Psychoneuroendocrinology*. 2015;52:83–91.

113. Witteveen AB, Huizink AC, et al. Associations of cortisol with posttraumatic stress symptoms and negative life events: a study of police officers and firefighters. *Psychoneuroendocrinology*. 2010;35(7):1113–1118.

114. Yehuda R. Post-traumatic stress disorder. *N Engl J Med*. 2002;346(2):108–114.

115. Yehuda R, Bierer LM, et al. Cortisol augmentation of a psychological treatment for warfighters with posttraumatic stress disorder: randomized trial showing improved treatment retention and outcome. *Psychoneuroendocrinology*. 2015;51:589–597.

116. Yehuda R, Boisoneau D, et al. Dose-response changes in plasma cortisol and lymphocyte glucocorticoid receptors following dexamethasone administration in combat veterans with and without posttraumatic stress disorder. *Arch Gen Psychiatry*. 1995;52(7):583–593.

117. Yehuda R, Flory JD, et al. Lower methylation of glucocorticoid receptor gene promoter 1F in peripheral blood of veterans with posttraumatic stress disorder. *Biol Psychiatry*. 2015;77(4):356–364.

118. Yehuda R, Golier JA, et al. Circadian rhythm of salivary cortisol in Holocaust survivors with and without PTSD. *Am J Psychiatry*. 2005;162(5):998–1000.

119. Yehuda R, Golier JA, et al. Enhanced sensitivity to glucocorticoids in peripheral mononuclear leukocytes in posttraumatic stress disorder. *Biol Psychiatry*. 2004;55(11):1110–1116.

120. Yehuda R, Halligan SL, et al. Effects of trauma exposure on the cortisol response to dexamethasone administration in PTSD and major depressive disorder. *Psychoneuroendocrinology*. 2004;29(3):389–404.

121. Yehuda R, Kahana B, et al. Low urinary cortisol excretion in Holocaust survivors with posttraumatic stress disorder. *Am J Psychiatry*. 1995;152(7):982–986.

122. Yehuda R, Lowy MT, et al. Lymphocyte glucocorticoid receptor number in posttraumatic stress disorder. *Am J Psychiatry*. 1991;148(4):499–504.

123. Yehuda R, Resnick HS, et al. Predictors of cortisol and 3-methoxy-4-hydroxyphenylglycol responses in the acute aftermath of rape. *Biol Psychiatry*. 1998;43(11):855–859.

124. Yehuda R, Siever LJ, et al. Plasma norepinephrine and 3-methoxy-4-hydroxyphenylglycol concentrations and severity of depression in combat posttraumatic stress disorder and major depressive disorder. *Biol Psychiatry*. 1998;44(1):56–63.

125. Yehuda R, Southwick SM, et al. Low urinary cortisol excretion in patients with posttraumatic stress disorder. *J Nerv Ment Dis*. 1990;178(6):366–369.

126. Yehuda R, Teicher MH, et al. Cortisol regulation in posttraumatic stress disorder and major depression: a chronobiological analysis. *Biol Psychiatry*. 1996;40(2):79–88.

127. Yehuda R, Yang RK, et al. Alterations in cortisol negative feedback inhibition as examined using the ACTH response to cortisol administration in PTSD. *Psychoneuroendocrinology*. 2006;31(4):447–451.

128. Young EA, Breslau N. Cortisol and catecholamines in posttraumatic stress disorder: an epidemiologic community study. *Arch Gen Psychiatry*. 2004;61(4):394–401.

129. Young EA, Breslau N. Saliva cortisol in posttraumatic stress disorder: a community epidemiologic study. *Biol Psychiatry*. 2004;56(3):205–209.

130. Young EA, Tolman R, et al. Salivary cortisol and posttraumatic stress disorder in a low-income community sample of women. *Biol Psychiatry*. 2004;55(6):621–626.

131. Zaba M, Kirmeier T, et al. Identification and characterization of HPA-axis reactivity endophenotypes in a cohort of female PTSD patients. *Psychoneuroendocrinology*. 2015;55:102–115.

132. Zohar J, Yahalom H, et al. High dose hydrocortisone immediately after trauma may alter the trajectory of PTSD: interplay between clinical and animal studies. *Eur Neuropsychopharmacol*. 2011;21(11):796–809.

17

Stress and Major Depression: Neuroendocrine and Biopsychosocial Mechanisms

A. Roy[1], R.N. Roy[2]

[1]University of Calgary, Calgary, AB, Canada; [2]Saba University School of Medicine, Saba, Netherlands-Antilles

Abstract

Depression is a prevalent and debilitating mental health disorder, and the leading cause of disability globally. Chronic psychosocial stress is central to depression etiology. The stress system involves the hypothalamic–pituitary–adrenal (HPA) axis and the locus coeruleus-norepinephrine/sympathetic system, which operate through stress mediators (monoamines, neuropeptides, corticosteroids) to mount an adaptive response. Normally, there is negative feedback control of the HPA axis. With chronic stress, this feedback control can be lost, leading to HPA axis dysregulation and consequent depression. Antidepressants increase synaptic levels of monoamines by blocking their reuptake or metabolism; long-term treatment leads to enhanced expression of neurotrophic factors, such as brain-derived neurotrophic factor (BDNF). Activation of BDNF and other neurotrophic factors leads to increased hippocampal neurogenesis, and improved structural plasticity and neurotransmission. As psychosocial stress, from life events and circumstances, is among the most potent triggers of HPA axis dysregulation and depression, a biopsychosocial approach is warranted for effective treatment and ultimate prevention.

INTRODUCTION

The World Health Organization has estimated that 350 million people suffer from depression worldwide, making it a highly prevalent disorder associated with considerable disability, morbidity, and mortality; it is the leading cause of disability globally.[91] Diagnosed clinically based on the American Psychiatric Association's Diagnostic and Statistical Manual of Mental Disorders (DSM 5) criteria,[2] symptoms of depression can include persistent depressed mood, lack of interest or pleasure, fatigue, problems in concentration and memory, problems in sleeping, changes in appetite, and changes in weight. Diverse hypotheses and theories have been proposed to explain the development of depression. In virtually all of these theories, be they biologically or psychosocially focused, psychosocial stress plays a major role.[69] An increasing emphasis is accordingly

Stress: Neuroendocrinology and Neurobiology
http://dx.doi.org/10.1016/B978-0-12-802175-0.00017-6

KEY POINTS

- Depression is a prevalent and debilitating mental health disorder, and is the leading cause of disability globally. Chronic psychosocial stress is central to depression etiology.

- To respond to physical and psychological stressors, animals (including humans) are endowed with a stress system, composed of the hypothalamic–pituitary–adrenal (HPA) axis and the locus coeruleus-norepinephrine (LC-NE)/sympathetic system, which is activated to maintain homeostatic stability in the face of challenging stimuli. The stress system provides stress mediators, which include monoamines (norepinephrine, serotonin, and dopamine), neuropeptides (corticotropin-releasing hormone and related peptides, urocortins, and vasopressin), and corticosteroids (glucocorticoids: corticosterone, and cortisol).

- On activation of the HPA axis and the LC-NE/sympathetic system, stress mediators, acting through their respective receptors, concertedly mount an adaptive physiological and behavioral response to stress. Physiologically, energy-consuming vegetative functions, such as feeding, digestion, growth, reproduction, and immune functions, are temporarily suspended; cardiovascular tone, respiratory rate, and intermediate metabolism (gluconeogenesis and lipolysis) are increased, and vital nutrients and oxygen are redirected to the central nervous system to help with the adaptive stress response.

- In the initial phase, when stress hormone levels are high, the response to stress is fast; it is mediated mainly by catecholamines and CRH/CRHR1, and is geared toward a "fight or flight" response, leading to increased vigilance, alertness, attention, and arousal. As the stress hormone levels begin to decline, gradually the slow phase of gene-mediated corticosteroid action via mineralocorticoid receptor (MR) and glucocorticoid receptor (GR) takes over. MR is mostly involved in maintenance of stress-related neural circuits. GR is important for restoration of homeostasis and storage of information for future use.

- To minimize the damaging catabolic and immune-suppressive effects, the stress system has a built-in mechanism for shutting down the stress response through negative feedback control of the HPA axis. This is accomplished through GR-mediated glucocorticoid actions, via both fast nongenomic and delayed gene-mediated mechanisms.

- With exposure to recurrent or chronic stress, prolonged stress response can lead to loss of the negative feedback control of the HPA axis; this loss can occur particularly in susceptible individuals with biological predisposition due to genetic background, adverse early-life events, or prenatal epigenomic changes. Depressive and anxiety disorders are consequences of the above. Hypercortisolemia in these individuals is associated with altered functions of serotonin receptors and the ratio of MR/GR occupancy. In the hippocampus, these changes lead to reduced neurogenesis, altered synaptic plasticity and decreased long-term potentiation (LTP), and a consequent decline in memory formation.

- Antidepressants act to reverse these damaging effects by increasing synaptic levels of monoamines by blocking their reuptake or metabolism; subsequently, long-term treatment leads to receptor activation to trigger intracellular signaling pathways downstream that regulate transcription and enhance the expression of neurotrophic factors, such as brain-derived neurotrophic factor (BDNF). Activation of BDNF and other neurotrophic factors leads to increased hippocampal neurogenesis, and improved structural plasticity and neurotransmission, consequently increasing LTP and improving memory.

- Biopsychosocial or diathesis-stress-based approaches to depression point to the interaction between various vulnerabilities (biological, cognitive, emotive, environmental, social) and stressors, such as stressful life events. Psychosocial stress from life events is among the most potent triggers of depressive episodes, via HPA axis hyperactivity and dysregulation, and interrelates with other risk and protective factors to yield depressive symptoms. Maladaptive patterns of thought, lack of social support, and social disadvantage are key risk factors for depression, which must be addressed for effective treatment and ultimate prevention.

being placed on a biopsychosocial, diathesis-stress-based framework for understanding depression, which points to a combination of biological predisposition and psychosocial stress stemming from social circumstances to explain depression etiology.[30,35,75] Correspondingly, the hypothalamic-pituitary-adrenal (HPA) axis plays a central role in mechanistic explanations for depression pathophysiology.[69]

This chapter seeks to present the key neuroendocrine mechanisms for stress and depression. After defining key terms and outlining key brain regions and peripheral structures of interest, a summary of the main components of a normal physiological stress response will be conveyed. Next, the neuroendocrine features and processes that define depression as an abnormal stress response will be described, followed by a summary of the mechanisms of antidepressant action. The chapter will then discuss the link between psychosocial stress and depression from a biopsychosocial perspective, concluding with reflections about the value of such an approach for advancing treatment and prevention.

NEUROENDOCRINE MECHANISMS OF NORMAL STRESS RESPONSE

Organisms strive to survive in changing environment by maintaining a dynamic equilibrium, termed *homeostasis*. This equilibrium is threatened from time to time by adverse physical stimuli (e.g., physical trauma, blood loss) or emotional (e.g., psychosocial) stimuli, known as *stressors*. The subjective sense of adverse changes in the environment, perceived by individuals, is called *stress*. For responding to stress, organisms have developed, through evolution, an array of molecular arsenals, called *stress mediators*, which include monoamines (norepinephrine; epinephrine; dopamine; serotonin), neuropeptides (corticotropin-releasing hormone (CRH); urocortin I, II, III; vasopressin; orexin; dynorphin; ghrelin; leptin), and corticosteroids (corticosterone; cortisol). Through their concerted actions, stress mediators influence various aspects of adaption to stress, at the physiological, immunological, neuronal, and behavioral levels. The various processes underlying the adaptive stress response are collectively referred to as *allostasis*. The term allostasis implies that the organism adapts to adverse situations by constantly adjusting the set points of physiological parameters, relating to the autonomic nervous system as well as the endocrine (HPA axis), cardiovascular, metabolic, and immune systems, to maintain stability. However, the effects of continual allostatic response, accrued over time (*allostatic load*), has negative consequences to the brain and the body; it may, in extreme cases (*allostatic*

overload), lead to pathological outcomes. Aside from repeated stress response, allostatic load may arise from inadequate, excessive, or prolonged stress response[58,67]; neuropsychological disorders such as depression and anxiety may thus be viewed as the negative outcome of prolonged neurochemical imbalance.

The brain regions primarily involved in the stress response include limbic structures (hippocampus, amygdala, hypothalamus), as well as the prefrontal cortex and the pituitary. In addition, brainstem nuclei contribute to the stress response process by providing important monoaminergic stress mediators: norepinephrine (locus coeruleus), serotonin (raphe nuclei), and dopamine (mesocorticolimbic pathway). Physical stressors (e.g., physical trauma) promptly stimulate neuronal populations in the hypothalamus and brainstem. Psychological stressors, on the other hand, activate brain regions that mediate emotion (amygdala), learning and memory (hippocampus), and decision making (prefrontal cortex).[25,59,46]

Aside from these specific areas of the brain, which constitute the central limb of the stress system, adaptive stress response requires participation of peripheral structures, namely the adrenal gland and the sympathetic/adrenomedullary system. The hypothalamus, pituitary, and adrenal gland act in concert via the HPA axis, the key mediator of stress response. In parallel, stress also activates the locus coeruleus (LC)-norepinephrine (NE)/sympathetic system. The HPA axis, together with the LC-NE/sympathetic system, constitutes the stress system that mediates the adaptive response to stress. The stress response involves both behavioral and physiological changes. Physiological adaptations include increased cardiovascular tone, respiratory rate, and intermediate metabolism; general vegetative functions, such as feeding, digestion, growth, reproduction, and immune functions, are inhibited.

Hypothalamic–Pituitary–Adrenal Axis

Stress triggers the release of a variety of neurotransmitters and neuropeptides in the brain. Among them, CRH, a 41-amino-acid peptide, undoubtedly plays a major role as a key effector of stress response. CRH is synthesized in the hypophysiotropic neurons, in the parvocellular subdivision of the hypothalamic paraventricular nuclei (PVN). The internal and external stimuli of stress are integrated at the hypothalamic level, stimulating the PVN to release CRH into the hypophyseal portal circulation that subsequently reaches the anterior pituitary. By binding to its receptor (CRHR1) on pituitary corticotrophs, CRH induces the release of adrenocorticotropic hormone (ACTH) into the systemic circulation. The effect of CRH on ACTH release is synergized by arginine vasopressin (AVP), a nonapeptide synthesized in the parvocellular neurons of PVN and released into the portal circulation, by binding to V_{1b} receptors on pituitary corticotrophs. ACTH is derived from the prohormone proopiomelanocortin (POMC), which also provides a number of other processed bioactive peptides, including β-endorphin, β-lipotropin hormone, and melanocortins. Circulating ACTH induces the zona fasciculata of the adrenal cortex, by binding to the type 2 melanocortin receptor (MC2-R), leading to synthesis of glucocorticoids (corticosterone and cortisol) and their release into the systemic circulation. CRH and AVP from the hypothalamic PVN, ACTH from the anterior pituitary, and glucocorticoids from the adrenal cortex are the principal effectors mediating the stress response; the anatomic structures contributing these effectors collectively constitute the HPA axis.[78] As discussed below, glucocorticoids modulate the expression of a wide array of gene products essential for mediating the stress response, by activating the expression of some and inhibiting others.

Corticotropin-Releasing Hormone in Stress Response

CRH action is mediated via two different G protein-coupled receptors (GPCRs), CRHR1 and CRHR2. The two receptors bear 70% homology in their amino acid sequences, with divergent ligand-binding N-terminals. Aside from CRH, the two receptors also bind other CRH-like ligands, namely, urocortin I, II, and III, with different affinities. CRH binds CRHR1 with high affinity and has poor affinity for CRHR2 receptor. CRHR2 is a high-affinity receptor for both urocortin II and III. Urocortin I, on the other hand, binds both receptors with similar affinities.[39,42,44]

CRH and its receptors have also been localized at various extrahypothalamic areas of the brain.[68,34] The two receptor types are differentially expressed; whereas CRHR2 is localized in discrete areas (lateral septum, ventromedial hypothalamus, cortical nucleus of the amygdala), CRHR1 is expressed more extensively, such as neocortical areas, lateral dorsal tegmentum, hypothalamic nuclei, pedunculopontine tegmental nucleus, basolateral and medial nuclei of the amygdala, anterior pituitary, and cerebellar Purkinje cells. At extrahypothalamic sites, CRH functions as a neuromodulator to integrate complex humoral and behavioral responses to stress. As suggested by various studies involving animal models[77,83,60a], as well as clinical studies in humans,[42a] CRH action in stress response is mediated essentially by CRHR1 receptor. Studies with knockout mice deficient for CRHR1[77,83] and CRHR2[7,50] receptors have clarified the roles played by these receptors in the stress response. Clearly, CRHR1 plays a key role in the functioning of the HPA axis; whereas the HPA axis is disrupted, and ACTH and corticosterone release

reduced in mice deficient for CRHR1 receptor, the HPA axis is unaffected in mice lacking CRHR2. In addition, the two receptors mediate contrasting behavioral responses to stress. CRHR1 mediates anxiogenic response, since CRHR1-deficient mice exhibit markedly reduced anxiety. Mice lacking CRHR2 receptor, on the other hand, display enhanced anxiety, suggesting that this receptor mediates anxiolytic response. Consistent with such conclusions, rats displayed anxiolytic-like behavior when urocortin II and, particularly, urocortin III—the agonists which have high affinities for CRHR2—were injected intracerebroventricularly.[85,86]

Several observations suggest the hypothesis that the anxiogenic effect of CRH is mediated via its action on the LC-NE system. There is anatomic evidence of direct synaptic contacts between CRH axon terminals and dendrites of NE cells in the LC.[88] Moreover, CRH-like immunoreactivity has been shown to be increased in the LC in animals, in both acute and chronic stress.[15] In addition, in humans, fear and anxiety are associated with increased release of NE[16] (reviewed in Ref. 4).

CRH mediates its action via different signal transduction pathways, depending on the intracellular context of various cell types. In most cell types, ligand-activated CRH receptors activate adenylate cyclase, following Gαs coupling, to generate cAMP as a second messenger, leading to activation of protein kinase A (PKA). PKA in turn phosphorylates cytosolic and nuclear targets downstream, such as the transcription factor cAMP response element-binding protein (CREB), thus inducing gene expression. Like other GPCRs, CRH receptors are also able to interact with multiple other G proteins linked to several intracellular signaling pathways, including mitogen-activated protein kinase (MAPK). MAPK, which is also known as extracellular regulated kinase (ERK), plays a key role in the global cellular response to CRH in neuroendocrine tissues (reviewed in Ref. 10).

In pituitary corticotrophs, besides activating CREB via cAMP/PKA pathway, CRH/CRHR1 signaling also activates other transcription factors, such as orphan nuclear receptors Nur77/NGF1-B, Nurr1, and NOR1 via the ERK/MAPK pathway to regulate the expression of POMC gene. This is accomplished through PKA further activating two other pathways downstream, one of which is dependent on the entry of calcium through voltage-gated Ca channels on the plasma membrane, mediated by calmodulin kinase II (CaMKII); the other pathway is calcium-independent. Both pathways involve activation of Rap1 (a small G protein of Ras family), which, in turn, activates a protein kinase, B-Raf, the first in a cascade of three protein kinases, B-Raf, MEK1/2 (mitogen-activated ERK-activating kinase), and ERK1/2, transmitting the signal via phosphorylation. The activated ERK1/2 (pERK1/2) modulates the induction and activity of orphan transcription factors Nur77 and Nurr1 involved in POMC expression (reviewed in Ref. 10).

Although ERK1/2 and CRHRs are widely expressed throughout the brain, CRH-induced, CRHR1-dependent ERK1/2-MAPK activation appears to be restricted to hippocampal CA1–CA3 areas and the basolateral amygdalar complex. These limbic structures are considered to be involved in learning and memory, the processing of environmental stimuli and stress-related behaviors. Other brain structures, including hypothalamic nuclei and the central nuclei of the amygdala, which are involved in the processing of ascending visceral information and neuroendocrine-autonomic response of stress, do not appear to employ MAPK pathway.[5]

Locus Coeruleus-Norepinephrine and Sympathetic System in Stress

Many of the wide array of stressful stimuli that trigger the HPA axis also activate the LC-NE system. The LC harbors the largest group of noradrenergic neurons in the brain. The LC-NE, along with catecholaminergic neurons of the brainstem nucleus of the solitary tract (NTS), is part of the central sympathetic system; the efferent adrenomedullary system constitutes the peripheral sympathetic counterpart. Reciprocal reverberatory neural connections that exist between PVN and LC allow CRH and NE to stimulate the secretion of each other through CRHR1 and α1 adrenergic receptors, respectively.[17,90,40] In addition, the parvocellular subdivision of the PVN is densely innervated by stress-receptive neurons from the NTS which induces, via α1 adrenergic receptors, the expression and release of CRH from the PVN.[66]

The parallel activation of the HPA axis and LC-NE system by stressful stimuli helps ensure coordination between the endocrine and the cognitive limbs of the stress response. The activity of the LC correlates with the state of arousal. By modulating its two modes of release, namely, phasic and tonic, the LC maintains a level of arousal appropriate for optimal processing of sensory information in the environment. In phasic mode, LC neurons discharge into the pericoerulear region in a synchronous manner (en masse), essentially by coupling through gap junctions between dendrites. Phasic discharge is mediated by the excitatory amino acid glutamate and is characterized by a brief period of excitation followed by a longer duration of inhibition. Behaviorally, phasic activity facilitates focused attention and engagement in specific behavioral tasks. On the other hand, tonic discharge occurs in a desynchronized manner and is mediated by CRH via CRHR1 receptor. It is characterized by heightened arousal, scanning attention, and searching for alternative tasks that are adaptive in a threatening environment. The primary source of CRH to

the LC is the central nucleus of the amygdala; its release within the LC is normally restrained by the basal level of corticosteroids. The tonic activity is opposed by endogenous opioids (β-endorphin, enkephalin), acting via μ opioid receptors, and is aimed at shifting the activity toward phasic mode.[87] Several other regulatory mechanisms exist which either stimulate or inhibit the central stress system. Several studies in rats, involving serotonin precursor, agonists and reuptake inhibitors, have indicated that the neurotransmitter serotonin (5-HT) stimulates the HPA axis at the hypothalamic level, mainly by indirect activation of CRH neurons in the PVN. The use of receptor-specific agonists and antagonists has also suggested that 5-HT exerts its action on CRH neurons via activation of 5-HT1a, 5-HT1b, 5-HT2a, and 5-HT2c receptors.[12,32,31,49]

Immune System and Stress Response

Inflammatory cytokines (TNF-α, IL-1, IL-2, IL-6, IFN-γ, platelet-activating factor (PAF)), produced from inflammatory sites, are potent activators of CRH; therefore, they stimulate the HPA axis. Among these cytokines, three, namely TNF-α, IL-1, and IL-6, are the most important and are produced in a cascade-like fashion, stimulating their own production; whereas IL-1β and TNF-α stimulate the production of IL-6, IL-6 inhibits TNF-α and IL-1 secretion. Glucocorticoids, the end product of the HPA axis, inhibit the function of essentially all inflammatory cells (lymphocytes, monocytes, neutrophils, basophils, and eosinophils). Glucocorticoids mediate this inhibition by affecting the transcription of these cytokine genes, altering the stability of their mRNA, as well as by inhibiting the production of inflammatory factors derived from arachidonic acid, such as leukotrienes and prostaglandins. Glucocorticoids may mediate immune response either by directly decreasing transcription of genes for cytokines (IL-6, IL-1β) or indirectly through inhibition of proinflammatory transcription factors, such as NF-κB and AP-1. NF-κB is a heterodimer (consisting of monomeric proteins p50 and p65) which is localized in the cytoplasm, complexed with IκBα and IκBβ, being prevented from entering the nucleus.[22] Many stressors (cytokines, antigens, viral infection, oxidants) activate NF-κB, releasing it from IκB, thus enabling it to enter the nucleus, where it binds to specific DNA sequences in the promoter regions of target genes. NF-κB upregulates the expression of genes for many cytokines, enzymes, and adhesion molecules involved in inflammatory diseases. Many of these proinflammatory cytokines (e.g., TNF-α, IL-1), in turn, activate NF-κB to perpetuate inflammatory response. As opposed to its central antiinflammatory role, CRH exerts proinflammatory effects at peripheral inflammatory sites.[63]

Glucocorticoids and the Stress System

Corticosteroids are released in a pulsatile manner, with circadian variation commensurate with the activity level of the organism; pulse amplitudes are low early in the inactive period, and high amplitude pulses begin just before the onset of the active period. Aside from circadian rhythm, additional amounts of corticosteroids are secreted in response to perceived stressful situation.[25,52]

Glucocorticoids (corticosterone in rodents, and primarily cortisol in humans), the final effectors of the HPA system, exert their action by binding to two types of receptors, namely, the type I, high-affinity or MR and the type II, low-affinity or GR. Corticosterone/cortisol binds to MR with affinity ten-fold higher than it does to GR.[47] In the brain, MRs are largely localized in the limbic structures of the hippocampus, septum, septohippocampal nucleus, and amygdala and are envisaged to mediate basal HPA activity. GRs are distributed throughout the brain, with high densities in the limbic structures of the hippocampus and septum, and in the parvocellular neurons of the PVN. GRs bind corticosterone/cortisol mostly during stress, when the levels of these steroids increase as much as 100-fold. The function of these low-affinity receptors is to suppress the hyperactivity of the HPA axis induced by stress, through their action at the level of the PVN, the anterior pituitary, and the hippocampus.[23]

Type I (MR) and type II (GR) corticosteroid receptors belong to the superfamily of nuclear hormone receptors.[56] The binding of the corticosteroid to the ligand-binding domain at the carboxy terminal of the cytoplasmic MR/GR receptor protein leads to the dissociation of associated heat shock protein (hsp), inducing a conformational change of the receptor, which enables its translocation to the nucleus. The homo- or heterodimers (GR–GR/MR–MR or GR-MR) of receptor–steroid complex then bind to specific DNA sequences, termed glucocorticoid-responsive elements (GREs) and interact with either coactivators or corepressors to either facilitate or inhibit transcription of the target genes. Steroid-bound GR monomers may also affect gene expression by interacting with stress-induced transcription factors such as nuclear factor-κB (NF-κB) and activator protein 1 (AP1) or other proteins to inhibit the transcription of target genes.[22a] The effects of glucocorticoids through genomic action, requiring translocation of ligand-bound receptors to the nucleus and regulation of gene transcription, require hours to days to develop. Apart from slow-acting genomic action, membrane-associated MRs can participate in rapid nongenomic glucocorticoid signaling within minutes.[49a,81a]

Under various experimental designs, aimed at creating differential MR/GR occupation, involving mice mutants and receptor agonists and antagonists, large-scale expression profiles (DNA microarray, serial analysis of gene expression) have been used to identify genes responsive to MR, GR, or both receptor types. The studies[25,46] have indicated that glucocorticoids are involved in coordinating various aspects of cellular structure, metabolism, and synaptic transmission. Some of the genes identified include known targets of GR, such as growth factors, cell adhesion factors, enzymes, and receptors for biogenic amines and neuropeptides.[73,41,70,71] MR and GR also modulate membrane properties by regulating the transcription of genes for GPCRs, ionotropic receptors, ion channels, and ion pumps.[25]

Time Course of Stress Response

The stress response takes place in a time scale spanning milliseconds to days. Classically, two phases of actions of stress mediators have been recognized. The early phase of stress response begins within seconds after the onset of the stressor, with activation of the HPA axis and a rise in corticosteroid levels, when the flight-or-fight response occurs, mediated by catecholamines, neuropeptides (CRH/AVP/CRHR1), and the fast nongenomic corticosteroid action via MR. At the cellular level, it is characterized by increased excitability and leads to behavioral display of vigilance, alertness, arousal, and attention. As the hormone level begins declining (usually normalizing by 2h), gradually the second phase of slow gene-mediated corticosteroid action via MR and GR takes hold. MR- and GR-mediated transcription of specific genes in cells that carry these receptors, such as CA1 hippocampal cells, affect structural integrity and excitability. MR is envisaged to mediate the maintenance of stress-related neural circuits and excitability; behaviorally it is implicated in the appraisal of sensory information. GR, on the other hand, is involved in normalizing homeostasis and storing information for future use. CRHR2-mediated anxiolytic effect likely has a contributory role in this slow termination phase of the stress response.[25]

In recognition of recent studies, three temporal domains of the stress response have been suggested to replace the classic two-phase understanding. In this model, in addition to rapid synaptic effects, receptor activation by monoamines and neuropeptides regulate transcription factors (such as CREB and AP-1) rapidly within seconds to minutes. The rapid activation of these transcription factors may lead to sustained genomic changes which may help the organism to respond to subsequent or recurrent stress.[25,46,48]

Feedback Control of the HPA Axis in Stress

The catabolic and immune-suppressive effects of the adaptive stress response are homeostatic, without damaging sequelae, if the stress response is of limited duration. Therefore, the adaptive function of the HPA axis is highly dependent on the proper functioning of the negative feedback control of CRH and ACTH secretion exerted by cortisol/corticosterone via type II GR. The mechanism involves repression of transcription of POMC in the anterior pituitary by GR, by interacting with orphan nuclear receptor NGF1/Nur77. Other nuclear proteins such as Brg1 and HDAC2 (histone deacetylase 2) also participate in the process of transrepression by GR, by scaffolding the repression complex. These proteins are frequently misexpressed in ACTH-secreting adenomas (Cushing disease), in which the negative feedback action of GR is lost.[57,9,10]

Apart from this delayed feedback through genomic action, there is an additional fast nongenomic negative feedback of the HPA axis. This is mediated by binding of glucocorticoid to a G protein-linked glucocorticoid membrane receptor, leading to a retrograde release of an endocannabinoid that results in suppression of glutamate release onto PVN CRH neurons.[26]

The mechanism for the loss of negative feedback of HPA axis observed in chronic stress is not known. However, the hippocampus exerts an inhibitory influence on the activity of the amygdala, PVN CRH, and LC-NE/sympathetic system, as well as on the HPA stress response. Since chronic stress leads to damage or atrophy of hippocampus, it has been hypothesized (glucocorticoid cascade hypothesis) that the prolonged HPA response is the consequence of impaired hippocampal function.[59]

NEUROENDOCRINE ABNORMALITIES IN CHRONIC STRESS AND DEPRESSION

Prolonged elevation of corticosteroid levels in chronic stress, likely due to loss of negative feedback control of the HPA axis, is envisaged to be among the most significant biological associated features of depression in humans. Studies in animals[49b] have indicated that long-term exposure to high levels of corticosterone leads to reduced expression of corticosteroid receptors (CR), especially the high-affinity MRs. In addition, there is attenuation of the functional serotonin response in CA1 pyramidal cells of the hippocampus, an area considered important in the etiology of depression. Various studies on stress, glucocorticoids, and monoamines (reviewed in Ref. 48) appear to suggest that chronic stress leads to reduced 5-HT metabolism, downregulation of 5-HT1A receptors, and upregulation of 5-HT2 receptors.

The relative occupancy and activation of MR and GR have profound effects on neurotransmission, as well as on learning ability and memory formation. At low corticosterone concentration, neural excitability is predominantly governed by MR, which is opposed by increasingly activated GR as the hormone level increases.

Strengthening of synaptic contacts by repeated stimulation is termed as LTP; this is accomplished when glutaminergic afferents in the CA1 area of the hippocampus are repeatedly stimulated. LTP correlates with capacity to learn and to retrieve memorized information. It is most pronounced under resting conditions, under basal levels of corticosteroids, when the ratio of ligand-activated MR/GR is high, as most MR but fewer GR are fully occupied. Such conditions favor a facilitated adaptation to stressful situations. Under chronic stress, when corticosteroid levels are high, the MR/GR ratio is reduced, as a result of increased GR occupation; such conditions are associated with reduced LTP and memory impairment, encompassing both consolidation and retrieval of memory.[43]

Studies in animals and clinical studies in humans have demonstrated that chronic stress and depression lead to reduction of hippocampal volume, with accompanying loss and atrophy of neurons. The dentate gyrus (DG)-CA3 system of the hippocampus is involved in memory formation. The intricate neural network comprising the DG-CA3 system, although delicately balanced with excitatory and inhibitory neurons, is nonetheless vulnerable to damage. It manifests adaptive structural plasticity in response to acute and chronic stress. Throughout adult life, new neurons are produced in the DG, arising from precursors of granule cells located in the subgranular zone. Besides DG neurogenesis, structural plasticity affects CA3 cells with remodeling of dendrites. Dendritic remodeling, involving dendritic length and spine density, also occurs in the amygdala and prefrontal cortex, leading to alterations in synaptic connections. Hippocampal structural plasticity is mediated by adrenal steroids in interactions with various neurochemical modulators, which include MR and GR, glutamate release, NMDA receptor activation, 5-HT, NE, GABA_B-benzodiazepine receptor, endogenous opioids, and growth factors BDNF and IGF-1. Chronic stress suppresses neurogenesis and reduces dendritic architecture.[59]

Hippocampal volume loss has been reported in several clinical studies involving brain imaging of depressed patients and postmortem investigations. Consistent with such clinical observations of hippocampal atrophy, many studies in animals have demonstrated that various types of stressors decrease the proliferation of new neurons in the subgranular zone of the hippocampus (reviewed in Ref. 89).

Adult neurogenesis in hippocampal DG is decreased when exposed to elevated levels of adrenal glucocorticoids, such as in acute or chronic stress. Both repeated stress and glucocorticoid administration also affect CA3 pyramidal cells adversely, leading to atrophy of dendritic branches and reduced spine density on apical dendrites. Stress and glucocorticoid administration also reduce the expressions of neurotrophic factors, such as brain-derived neurotrophic factor (BDNF). In contrast, long-term treatment with antidepressants reverses dendritic atrophy, increases hippocampal neurogenesis, and also upregulates the expression of BDNF and other neurotrophic factors. As emphasized in the neurotrophic hypothesis, the downregulation of the expression of BDNF and other neurotrophic factors in stress could contribute to adverse effects on hippocampal structural plasticity, associated with depression and mood disorders (reviewed in Ref. 89).

Glucocorticoids influence neurogenesis in the DG differentially via MR and GR. Whereas activation of GR induces apoptosis of DG granule cells by increasing the ratio of the proapoptotic factor Bax relative to the antiapoptotic factors Bcl-2 or Bcl-x, activation of MR has the opposite effect.[1]

The role of BDNF as a trophic factor modulating neurogenesis and synaptic plasticity, however, appears to be dependent on interactions with glucocorticoids and other factors in the brain.[38] BDNF exerts its action by binding to the extracellular domain of tropomyosin-related kinase B (TrkB) receptor, which initiates dimerization of the receptor and autophosphorylation of tyrosine residues within the intracellular domain, leading to activation of three intracellular signaling cascades (MAPK/ERK, PLCγ, and PI3K). Phosphorylation of TrkB is induced by antidepressant drugs. It has been reported recently that glucocorticoids can also induce phosphorylation of TrkB in neuronal cells, independently of neurotrophins.[45] Glucocorticoids seem to have a role as trophic factors, as removal of glucocorticoids by adrenalectomy in rats leads to reduced dendritic branching and death of DG granule neurons[37]; this is despite that administration of glucocorticoids over extended periods causes hippocampal cell death. The activity of BDNF is further dependent on the cleavage of proBDNF by plasmin, which is activated by the conversion of plasminogen to plasmin by tissue plasminogen activator (tPA), a serine protease. Since binding of proBDNF to p75[NTR] receptor induces apoptosis, tPA likely plays an important role in the balance between pro- and mature BDNF.[51] Alongside BDNF, glucocorticoids participate in mediating neuronal LTP; BDNF is important for glutamatergic and GABAergic synaptic maturation, and glucocorticoids enhance surface expression of NMDA and AMPA receptor subunits (reviewed in Ref. 36).

MECHANISMS OF ANTIDEPRESSANT ACTION

The mode of action of antidepressants that block the reuptake of 5-HT and NE by presynaptic neurons, thereby increasing postsynaptic concentration of these neurotransmitters, was thought to be consistent with the classic monoamine hypothesis of biogenic amine deficiency as the cause of mood disorders. Although reuptake inhibition occurs immediately, the classic monoamine hypothesis does not explain why it takes weeks and months for antidepressants to be effective. Therefore, alternative modes of action have been proposed to explain antidepressant effects.

The cAMP/CREB/BDNF hypothesis, advanced by Duman and colleagues,[27] is centered on the observation that the stress-induced decrease of hippocampal BDNF is blocked by antidepressants, thereby enhancing the survival and growth of neurons in the hippocampus. Since the BDNF gene contains a cAMP response element (CRE) to which phosphorylated binding protein (pCREB) binds to enhance BDNF transcription, the hypothesis suggests that antidepressants activate cAMP-mediated CREB phosphorylation via PKA to increase BDNF expression, by increasing amine-binding at the G protein-linked receptors at the cell membrane.

The CR hypothesis[43] emphasizes on the concept that the CR signaling is impaired in major depression and antidepressants variously improve CR (MR or GR) function to reduce HPA overactivity. For instance, many antidepressants upregulate MR, including 5-HT reuptake inhibitors and enhancers, such as selective serotonin reuptake inhibitors (SSRIs), monoamine oxidase (MAO) inhibitors, and NE reuptake inhibitors. Since MR has an inhibitory effect on the HPA axis, upregulation of MR leads to suppression of the HPA axis and activation of cAMP-mediated CREB phosphorylation. GR, on the other hand, abolishes CREB phosphorylation by binding to CREB. The antidepressant venlafaxine impairs GR signaling by interacting with associated chaperones to reduce overactivity of the HPA axis. The antidepressant tranylcypromin, an irreversible nonselective MAO inhibitor, induces AP-1 complex to enhance AP-1 activity, so as to counter the negative modulation of AP-1 by GR.[43] Thus, antidepressants appear to function by modifying the set point of the HPA system, so that its negative feedback capacity is improved.[43]

PSYCHOSOCIAL STRESS AND DEPRESSION: A BIOPSYCHOSOCIAL PERSPECTIVE

Mechanistically, HPA axis stress physiology and associated neuroendocrine mechanisms are central to understanding the biopsychosocial interface of depression etiology.[69] Virtually all environmental and genetic risk factors for depression show a link with increased HPA axis activity, with treatment or remission yielding a reversion to normal activity levels.[81,62] Psychosocial stress from life events is among the most potent triggers of depressive episodes and interrelates with other risk and protective factors to yield depressive symptoms. Life stress can be in the form of acute stressors (such as the death of a loved one) or of chronic strain (such as ongoing financial problems).[6] Both categories are reflected in the main animal models for depression (chronic mild stress model, social defeat stress model and learned helplessness model),[92] and both feature as causal predictors of depression in research in humans.[6,55,61]

Of course, not everyone who faces life stress goes on to develop depression. There is considerable individual variability in the degree of HPA axis activation to stress; this variability may be because of inherent physiological differences between individuals, and/or differences in type and perception of stress. Biopsychosocial or diathesis stress-based approaches to understanding depression and other affective disorders indicate that depression stems from the interaction between various vulnerabilities (stemming from biological, cognitive, emotive, environmental, and social factors, which either predispose or protect against distress) and stressors, such as stressful life events.[35,75]

An individual's initial HPA axis activity set point is genotypically programmed. However, it can be altered by in utero exposures and by negative events early in life; such experiences appear to predispose individuals to depression and/or anxiety disorders later in life, likely via programmed, permanent hyperactivity of the HPA axis and by altering proper neurodevelopment of pertinent brain regions.[3,18,24,81] Normally, the fetus is protected from high maternal glucocorticoid levels by the inactivating function of placental 11β-HSD2 enzyme. However, the enzyme can become saturated during severe maternal stress during pregnancy, exposing the fetus to increased cortisol levels and affecting the HPA axis of the offspring postnatally. The established association between severe maternal stress during pregnancy and subsequent neuropsychiatric disorders in offspring seems to suggest prenatal programming, involving epigenetic changes (such as DNA methylation) of specific genes,[76] though evidence for epigenetic mechanisms has not been conclusive.[60] Research in psychology, epidemiology, and neurobiology similarly shows compelling evidence for the role of adverse childhood experiences in influencing physical and mental health throughout the lifespan.[3,8] The Adverse Childhood Experiences (ACE) Study found a strong, dose–response relationship between childhood experiences of abuse or household dysfunction and adult health outcomes, including depression and other mental health issues[30a,3]; these findings have been echoed in numerous other studies to date.[14]

Biological predisposition to depression may be conferred in a number of ways. Factors that alter the physiological systems involved in depression—such as nutritional deficiencies, comorbid health conditions, and

polymorphisms of certain relevant genes—have been implicated.[69] In regards to the latter, individuals carrying polymorphisms of genes including human BDNF gene (Val66Met),[28] human FKBP5 gene (five SNP alleles),[93] and serotonin transporter gene (5-HTT) (s allele)[13] may be more susceptible to depression when faced with stressful life events (i.e., in gene–environment interactions).

The specific nature of stressful events and the subjective perception of stressful situations can lead to differential activation of the HPA axis[19]; such variables also form the basis of various theories stemming from the behaviorist, cognitive, cognitive-behaviorist, and humanistic perspectives in psychology. While these various theories propose different explanatory pathways or mechanisms, most involve or implicate experiences of stress. Overall, it is clear that a positive outlook on one's self and one's life, as well as a strong network of social support (intimate partner, family, friends, community) can buffer the emotional and biological impacts of stressful life situations. By contrast, maladaptive patterns of thought[74] and a lack of meaningful social support[72] can exacerbate the vulnerability of developing depression.[79] Helplessness, hopelessness, and inability to effectively cope in stressful situations all have been shown to contribute to HPA axis activation.[84]

In population health research, the importance is well established of social determinants of health, such as income, employment, education, social and physical environments, and social support[20]—all of which have been found to be associated with depression.[11,53,54,65] Similarly, there is also growing recognition of the individual-level and population-level health impacts of social marginalization from racism, sexism, and other axes of oppression.[21,64,80] While the etiological pathways linking these social and societal factors with health outcomes are likely complex and multifaceted,[82] the chronic life stress generated from social disadvantage and marginalization can serve as a perpetual activator of the HPA axis, leading to dysregulation and pathophysiology.[69] Moreover, social disadvantage and marginalization can impede access to protective factors which can buffer against HPA axis dysregulation and hyperactivity, such as social support, health and social services, and other resources for healthy living and positive coping.[33,21,29]

CONCLUSION

The complexity of the neuroendocrine mechanisms involved in stress and major depression likely underlie the striking diversity in symptoms, comorbid health conditions, and treatment outcomes among individuals living with depression. Further biomedical and clinical research may help to advance clinical diagnosis and pharmacologic treatment options.

A biopsychosocial engagement with depression etiology additionally points to the importance of addressing the predominantly psychosocial and societal causes and moderators of stress. Such an approach points to the value of nonpharmacologic treatment options such as psychotherapy. It also points to the need for primary prevention via population-level interventions (health and social policies, programs, services) that address the upstream societal causes of chronic psychosocial stress. Moreover, a biopsychosocial approach offers a meaningful framework for cross-disciplinary dialogue between biomedical and social scientists—and between researchers, practitioners, and policymakers—for further research and action on this prevalent and debilitating mental health disorder.

References

1. Almeida FX, Condé GL, Crochemore C, et al. Subtle shifts in the ratio between pro- and antiapoptotic molecules after activation of corticosteroid receptors decide neuronal fate. *FASEB J*. 2000;14:779–790.
2. American Psychiatric Association (APA). *Diagnostic and Statistical Manual of Mental Disorders*. 5th ed. Washington, DC: American Psychiatric Association; 2013.
3. Anda RF, Felitti VJ, Bremner JD, et al. The enduring effects of abuse and related adverse experiences in childhood. A convergence of evidence from neurobiology and epidemiology. *Eur Arch Psychiatry Clin Neurosci*. 2006;256(3):174–186.
4. Arborelius L, Owens MJ, Plotsky PM, Nemeroff CB. The role of corticotropin-releasing factor in depression and anxiety disorders. *J Endocrinol*. 1999;160:1–12.
5. Arzt E, Holsboer F. CRF signaling: molecular specificity for drug targeting in the CNS. *Trends Pharmacol Sci*. 2006;27:531–538.
6. Avison WR, Turner RJ. Stressful life events and depressive symptoms: disaggregating the effects of acute stressors and chronic strain. *J Health Soc Behav*. 1988;29:253–264.
7. Bale TI, Contarino A, Smith GW, et al. Mice deficient for corticotropin-releasing hormone receptor-2 display anxiety-like behaviour and are hypersensitive to stress. *Nat Genet*. 2000;24:410–414.
8. Beatson J, Taryan S. Predisposition to depression: the role of attachment. *Aust NZ J Psychiatry*. 2003;37:219–225.
9. Bilodeau S, Vallette-Kasic S, Gauthier Y, et al. Role of Brg1 and HDAC2 in GR transrepression of pituitary POMC gene and misexpression in Cushing disease. *Genes Dev*. 2006;20:2871–2886.
10. Bonfiglio JJ, Inda C, Refoso D, Holsboer F, Arzt E, Silberstein S. The corticotropin-releasing hormone network and the hypothalamic-pituitary-adrenal axis: molecular and cellular mechanisms involved. *Neuroendocrinology*. 2011;94:12–20.
11. Bowen A, Muhajarine N. Antenatal depression. *Can Nurse*. 2006;102:26–30.
12. Calogero AE, Bagdy G, Szemeredi K, Tartaglia ME, Gold PW, Chrousos GP. Mechanisms of serotonin receptor agonist-induced activation of the hypothalamic-pituitary-adrenal axis in the rat. *Endocrinology*. 1990;126:1888–1894.
13. Caspi A, Sugden K, Moffitt TE, et al. Influence of life stress on depression: moderation by a polymorphism in the 5-HTT gene. *Science*. 2003;301:386–389.
14. Chapman D, Dube S, Anda R. Adverse childhood events as risk factors for negative mental health outcomes. *Psychiatr Ann*. 2007;37(5):359–364.
15. Chappell PB, Smith MA, Kilts CD, et al. Alterations in corticotropin-releasing factor-like immunoreactivity in discrete rat brain regions after acute and chronic stress. *J Neurosci*. 1986;6:2908–2914.
16. Charney DS, Bremner JD, Redmond DE. Noradrenergic neural substrates for anxiety and fear: clinical associations based on preclinical research. In: Bloom FE, Kupfer DJ, eds. *Psychopharmacology: The Fourth Generation of Progress*. New York: Raven Press; 1995:387–395.
17. Chrousos GP. Regulation and dysregulation of the hypothalamic-pituitary-adrenal axis. The corticotropin-releasing hormone perspective. *Endocrinol Metab Clin North Am*. 1992;21:833–858.
18. Clark PM. Programming of the hypothalamo-pituitary-adrenal axis and the fetal origins of adult disease hypothesis. *Eur J Pediatr*. 1998;157(suppl 1):S7–S10.
19. Croes S, Merz P, Netter P. Cortisol reaction in success and failure condition in endogenous depressed patients and controls. *Psychoneuroendocrinology*. 1993;18:23–35.
20. CSDH. *Closing the Gap in a Generation: Health Equity through Action on the Social Determinants of Health. Final Report of the Commission on Social Determinants of Health*. Geneva: World Health Organization; 2008.
21. Cudd AE. *Analyzing Oppression*. New York: Oxford University Press; 2006.
22. De Bosscher K, Vanden Berghe W, Haegeman G. The interplay between the glucocorticoid receptor and nuclear factor-κB or activator protein-1: molecular mechanisms for gene repression. *Endocr Rev*. 2003;24:488–522.
22a. De Bosscher K, Vanden Berghe W, Vermeulen L, Plaisance S, Boone E, Haegeman G. Glucocorticoids repress NF-kappaB-driven genes by disturbing the interaction of p65 with the basal transcription machinery, irrespective of coactivator levels in the cell. *Proc Natl Acad Sci USA*. 2000;97:3919–3924.
23. de kloet ER. Brain corticosteroid receptor balance and homeostatic control. *Front Neuroendocrinol*. 1991;12:95–164.
24. de Kloet ER. Stress in the brain. *Eur J Pharmacol*. 2000;405:187–198.
25. de Kloet ER, Joël M, Holsboer F. Stress and the brain: from adaptation to disease. *Nat Rev Neurosci*. 2005;6:463–475.
26. Di S, Malcher-Lopes R, Halmos KC, Tasker JG. Nongenomic glucocorticoid inhibition via endocannabinoid release in the hypothalamus: a fast feedback mechanism. *J Neurosci*. 2003;23:4850–4857.
27. Duman RS, Heninger GR, Nestler EJ. A molecular and cellular theory of depression. *Arch Gen Psychiatry*. 1997;54:597–606.
28. Egan MF, Kojima M, Callicott JH, et al. The BDNF val66met polymorphism affects activity-dependent secretion of BDNF and human memory and hippocampal function. *Cell*. 2003;112:257–269.
29. Elo I. Social class differentials in health and mortality: patterns and explanations in comparative perspective. *Annu Rev Sociol*. 2009;35:553–572.
30. Engel GL. The need for a new medical model: a challenge for biomedicine. *Science*. 1977;196:129–136.
30a. Felitti VJ, Anda RF, Nordenberg D, et al. Relationship of childhood abuse and household dysfunction to many of the leading causes of death in adults: the adverse childhood experiences (ACE) study. *Am J Prev Med*. 1998;14(4):245–258.
31. Fuller RW. The involvement of serotonin in regulation of pituitary-adrenocortical function. *Front Neuroendocrinol*. 1992;13:250–270.
32. Fuller RW, Snoddy HD. Serotonin receptor subtypes involved in the elevation of serum corticosterone concentration in rats by direct and indirect-acting serotonin agonists. *Neuroendocrinology*. 1990;52:206–210.
33. Galabuzi G, Labonte R. *Social Inclusion as a Determinant of Health*. Ottawa: Health Canada; 2003.
34. Gallagher JP, Orozco-Cabal LF, Liu J, Shinnick-Gallagher P. Synaptic physiology of central CRH system. *Eur J Pharmacol*. 2008;583:215–225.
35. Garcia-Toro M, Aguirre I. Biopsychosocial model in depression revisited. *Med Hypotheses*. 2007;68:683–691.

36. Gottmann K, Mittmann T, Lessmann V. BDNF signaling in the formation, maturation and plasticity of glutamatergic and GABAergic synapses. *Exp Brain Res.* 2009;199:203–234.

37. Gould E, Woolley CS, McEwen BS. Short-term glucocorticoid manipulation affect neuronal morphology and survival in the adult dentate gyrus. *Neuroscience.* 1990;37:367–375.

38. Gray JD, Milner TA, McEwen BS. Dynamic plasticity: the role of glucocorticoids, brain-derived neurotrophic factor and other trophic factors. *Neuroscience.* 2013;239:214–227.

39. Grigoriadis DE. The corticotropin-releasing factor receptor: a novel target for the treatment of depression and anxiety-related disorders. *Expert Opin Ther Targets.* 2005;9:651–684.

40. Habib KE, Gold PW, Chrousos GP. Neuroendocrinology of stress. *Endocrinol Metab Clin North Am.* 2001;30:695–728.

41. Hansson AC, Cintra A, Belluardo N, et al. Gluco- and mineralo-corticoid receptor-mediated regulation of neurotrophic factor gene expression in the dorsal hippocampus and the neocortex of the rat. *Eur J Neurosci.* 2000;12:2918–2934.

42. Hauger RL, Risbrough V, Brauns O, Dautzenberg FM. Corticotropin releasing factor (CRF) receptor signaling in the central nervous system: new molecular targets. *CNS Neurol Disord Drug Targets.* 2006;5:453–479.

42a. Holsboer F. The rationale for corticotropin-releasing hormone receptor (CRH-R) antagonists to treat depression and anxiety. *J Psychiatr Res.* 1999;33:181–214.

43. Holsboer F. The corticosteroid receptor hypothesis of depression. *Neuropsychopharmacology.* 2000;23:477–501.

44. Holsboer F, Ising M. Central CRH system in depression and anxiety – evidence from clinical studies with CRH1 receptor antagonists. *Eur J Pharmacol.* 2008;583:350–357.

45. Jeanneteau F, Garabedian MJ, Chao MV. Activation of Trk neurotrophin receptors by glucocorticoids provides a neuroprotective effect. *Proc Natl Acad Sci USA.* 2008;105:4862–4867.

46. Joëls M, Baram T. The neuro-symphony of stress. *Nat Rev Neurosci.* 2009;10:459–466.

47. Joëls M, de Kloet E. Mineralocorticoid and glucocorticoid receptors in the brain. Implications for ion permeability and transmitter systems. *Prog Neurobiol.* 1994;43:1–36.

48. Joëls M, de Kloet E, Karst H. Corticosteroid actions on neurotransmission. In: Fink G, Pfaff DW, Levine J, eds. *Handbook of Neuroendocrinology.* San Diego: Academic Press; 2012:415–431.

49. Jørgensen H, Knigge U, Kjær A, Møller M, Warberg J. Serotonergic stimulation of corticotropin-releasing hormone and pro-opiomelanocortin gene expression. *J Neuroendocrinol.* 2002;14:788–795.

49a. Karst H, Berger S, Turiault M, Tronche F, Schütz G, Joëls M. Mineralocorticoid receptors are indispensable for nongenomic modulation of hippocampal glutamate transmission by corticosterone. *Proc Natl Acad Sci USA.* 2005;102:19204–19207.

49b. Karten YJ, Nair SM, van Essen L, Sibug R, Joëls M. Long-term exposure to high corticosterone levels attenuates serotonin responses in rat hippocampal CA1 neurons. *Proc Natl Acad Sci USA.* 1999;96:13456–13461.

50. Kishimoto T, Radulovic M, Lin CR, et al. Deletion of CRHR2 reveals an anxiolytic role for corticotropin-releasing hormone receptor-2. *Nat Genet.* 2000;24:415–419.

51. Lee R, Kermani P, Teng KK, Hempstead BL. Regulation of cell survival by secreted proneurotrophins. *Science.* 2001;294:1945–1948.

52. Lightman SL, Wiles CC, Atkinson HC, et al. The significance of glucocorticoid pulsatility. *Eur J Pharmacol.* 2008;583:255–262.

53. Lorant V, Deliège D, Eaton W, Robert A, Philippot P, Ansseau M. Socioeconomic inequalities in depression: a meta-analysis. *Am J Epidemiol.* 2003;157(2):98–112.

54. Lorant V, Croux C, Weich S, Deliège D, Mackenbach J, Ansseau M. Depression and socio-economic risk factors: 7-year longitudinal population study. *Br J Psychiatry.* 2007;190:293–298.

55. Kendler KS, Karkowski LM, Prescott CA. Causal relationship between stressful life events and the onset of major depression. *Am J Psychiatry.* 1999;156:837–841.

56. Mangelsdorf DJ, Thummel C, Beato M, et al. The nuclear receptor superfamily: the second decade. *Cell.* 1995;83:835–839.

57. Martens C, Bilodeau S, Maira M, Gauthier Y, Drouin J. Protein-protein interactions and transcriptional antagonism between the subfamily of NGFI-B/Nur77 orphan nuclear receptors and glucocorticoid receptor. *Mol Endocrinol.* 2005;19(4):885–897.

58. McEwen BS. Protective and damaging effects of stress mediators. *N Engl J Med.* 1998;338:171–179.

59. McEwen BS. Physiology and neurobiology of stress and adaptation: central role of the brain. *Physiol Rev.* 2007;87:873–904.

60. Miller G. Epigenetics. The seductive allure of behavioral epigenetics. *Science.* 2010;329:24–27.

60a. Muller MB, Zimmermann S, Sillaber I, et al. Limbic corticotropin-releasing hormone receptor 1 mediates anxiety-related behavior and hormonal adaptation to stress. *Nat Neurosci.* 2003;6:1100–1107.

61. Muscatell K, Slavich G, Monroe S, Gotlib I. Stressful life events, chronic difficulties, and the symptoms of clinical depression. *J Nerv Ment Dis.* 2009;197:154–160.

62. Nemeroff CB. The corticotropin-releasing factor (CRF) hypothesis of depression: new findings and new directions. *Mol Psychiatry.* 1996;1:336–342.

63. O'Connor TM, O'Halloran DJ, Shanahan F. The stress response and the hypothalamic-pituitary axis: from molecule to melancholia. *QJM.* 2000;93:323–333.

64. Paradies Y. A systematic review of empirical research on self-reported racism and health. *Int J Epidemiol.* 2006;35:888–901.

65. Patten SB, Juby H. *A Profile of Clinical Depression in Canada. Research Synthesis Series: #1. Research Data Centre Network;* 2007. https://dspace.ucalgary.ca/bitstream/1880/46327/6/Patten_RSS1.pdf.

66. Plotsky PM, Cunningham Jr ET, Widmaier EP. Catecholaminergic modulation of corticotropin-releasing factor and adrenocorticotropin secretion. *Endocr Rev.* 1989;10:437–458.

67. Read S, Grundy E. *Allostatic Load – A Challenge to Measure Multisystem Physiological Dysregulation. National Centre for Research Methods Working Paper 04/12;* 2012. http://eprints.ncrm.ac.uk/2879/.

68. Reul JM, Holsboer F. Corticotropin-releasing factor receptors 1 and 2 in anxiety and depression. *Curr Opin Pharmacol.* 2002;2:23–33.

69. Roy A, Campbell MK. A unifying framework for depression: bridging the major biological and psychosocial theories through stress. *Clin Invest Med.* 2013;36(4):E170–E190.

70. Sabban EL, Kvetnansky R. Stress-triggered activation of gene expression in catecholaminergic systems: dynamics of transcriptional events. *Trends Neurosci.* 2001;24:91–98.

71. Sandi C. Stress, cognitive impairment and cell adhesion molecules. *Nat Rev Neurosci.* 2004;5:917–930.

72. Santini Z, Koyanagi A, Tyrovolas S, Mason C, Haro J. The association between social relationships and depression: a systematic review. *J Affect Disord.* 2015;175:53–65.

73. Schaaf MJ, Hoetelmans RW, de Kloet ER, Vreugdenhil E. Corticosterone regulates expression of BDNF and trkB but not NT-3 and trkC mRNA in the rat hippocampus. *J Neurosci Res.* 1997;48:334–341.

74. Scher CD, Ingram RE, Segal ZV. Cognitive reactivity and vulnerability: empirical evaluation of construct activation and cognitive diatheses in unipolar depression. *Clin Psychol Rev.* 2005;25:487–510.

75. Schotte CK, Van Den Bossche B, De Doncker D, Claes S, Cosyns P. A biopsychosocial model as a guide for psychoeducation and treatment of depression. *Depress Anxiety.* 2006;23:312–324.

76. Seckl JR, Holmes MC. Mechanisms of disease: glucocorticoids, their placental metabolism and fetal 'programming' of adult pathophysiology. *Nat Clin Pract Endocrin Metabol.* 2006;3(6):479–488.

77. Smith GW, Aubry JM, Dellu F, et al. Corticotropin-releasing factor receptor 1-deficient mice display decreased anxiety, impaired stress response, and aberrant neuroendocrine development. *Neuron*. 1998;20:1093–1102.

78. Smith SM, Vale WW. The role of the hypothalamic-pituitary-adrenal axis in neuroendocrine responses to stress. *Dialogues Clin Neurosci*. 2006;8:383–395.

79. Southwick SM, Vythilingam M, Charney DS. The psychobiology of depression and resilience to stress: implications for prevention and treatment. *Annu Rev Clin Psychol*. 2005;1:255–291.

80. Sue DW. *Microaggressions in Everyday Life: Race, Gender and Sexual Orientation*. New Jersey: John Wiley & Sons; 2010.

81. Swaab DF, Bao AM, Lucassen PJ. The stress system in the human brain in depression and neurodegeneration. *Ageing Res Rev*. 2005;4:141–194.

81a. Tasker JG, Di S, Malcher-Lopes R. Minireview: rapid glucocorticoid signaling via membrane-associated receptors. *Endocrinology*. 2006;147:5549–5556.

82. Taylor SE, Repetti RL, Seeman T. Health psychology: what is an unhealthy environment and how does it get under the skin? *Annu Rev Psychol*. 1997;48:411–447.

83. Timpl P, Spanagel R, Sillaber I, et al. Impaired stress response and reduced anxiety in mice lacking a functional corticotropin-releasing hormone receptor 1. *Nat Genet*. 1998;19:162–166.

84. Ursin H, Eriksen HR. Cognitive activation theory of stress (CATS). *Neurosci Biobehav Rev*. 2010;34:877–881.

85. Valdez GR, Inoue K, Koob GF, Rivier J, Vale W, Zorrilla EP. Human urocortin II: mild locomotor suppressive and delayed anxiolytic-like effects of a novel corticotropin-releasing factor related peptide. *Brain Res*. 2002;943:142–150.

86. Valdez GR, Zorrilla EP, Rivier J, Vale WW, Koob GF. Locomotor suppressive and anxiolytic-like effects of urocortin 3, a highly selective type 2 corticotropin-releasing factor agonist. *Brain Res*. 2003;980:206–212.

87. Valentino RJ, Bockstaele EV. Convergent regulation of locus coeruleus activity as an adaptive response to stress. *Eur J Pharmacol*. 2008;583:194–203.

88. Van Bockstaele EJ, Colago EEO, Valentino RJ. Corticotropin-releasing factor-containing axon terminals synapse onto catecholamine dendrites and may presynaptically modulate other afferents in the rostral pole of the nucleus locus coeruleus in the rat brain. *J Comp Neurol*. 1996;364:523–534.

89. Warner-Schmidt JL, Duman RS. Hippocampal neurogenesis: opposing effects of stress and antidepressant treatment. *Hippocampus*. 2006;16:239–249.

90. Whitnall MH. Regulation of the hypothalamic corticotropin-releasing hormone neurosecretory system. *Prog Neurobiol*. 1993;40:573–629.

91. World Health Organization (WHO). *Depression*; 2015. http://www.who.int/mediacentre/factsheets/fs369/en/.

92. Yan HC, Cao X, Das M, Zhu XH, Gao TM. Behavioral animal models of depression. *Neurosci Bull*. 2010;26:327–337.

93. Zimmermann P, Bruckl T, Nocon A, et al. Interaction of FKBP5 gene variants and adverse life events in predicting depression onset: results from a 10-year prospective community study. *Am J Psychiatry*. 2011;168:1107–1116.

CHAPTER

18

Telomeres and Early Life Stress

K.K. Ridout[1,2], S.J. Ridout[1,2], K. Goonan[1], A.R. Tyrka[1,2], L.H. Price[1,2]

[1]Butler Hospital, Providence, RI, United States; [2]Alpert Medical School of Brown University,
Providence, RI, United States

Abstract

Telomeres are structures at the ends of chromosomes that preserve the encoding DNA during chromosome replication. The telomere itself shortens with each cell division, reflecting the age of a cell and the time until senescence. Telomere shortening, and changes in levels of telomerase, the enzyme that maintains telomeres, occurs in the context of certain somatic and psychiatric diseases. Emerging evidence indicates that telomeres shorten with exposure to multiple forms of stress experienced early in life, suggesting that telomere shortening might be a useful biomarker indicating the overall stress response of an organism to various pathogenic conditions. Thus, telomeres could serve as a unifying biomarker of stress response that transcends brain/body distinctions. The possibility that telomere shortening can be slowed or reversed by psychiatric and psychosocial interventions could represent an opportunity for developing novel preventative and therapeutic approaches.

INTRODUCTION

Experiencing early life stress (ELS) increases individual risk for a broad range of poor physical and mental health outcomes. Childhood maltreatment is associated with a greater risk for psychiatric disorders, including depression, anxiety, posttraumatic stress and substance use disorders,[44,80] and medical disorders,[90] including diabetes,[79] asthma,[7] cancer,[54] and heart disease.[53,93] Other forms of childhood adversity, including loss of a parent, poverty, and removal from the parental home, carry similarly increased risk for these disorders, which are in turn often associated with aging and increased mortality risk. These findings have led to interest in how external experiences could cause such physical and psychiatric manifestations.

The cellular aging hypothesis posits that the characteristic physical and mental declines seen with biological aging are driven by molecular alterations at the cellular level. Telomere attrition is one of the molecular changes associated with aging and seems to be accelerated in a number of disorders. Telomere shortening ultimately leads to cellular senescence, apoptosis, and mutations, while triggering proinflammatory pathways, and ultimately contributing to organ dysfunction and risk for such age-related conditions as diabetes and

Stress: Neuroendocrinology and Neurobiology
http://dx.doi.org/10.1016/B978-0-12-802175-0.00018-8

185

cardiovascular disease. A growing body of literature provides evidence that individuals exposed to ELS experience accelerated telomere shortening compared to individuals with no ELS exposure. This chapter will focus on the potential role of telomere biology in mediating some of the physical and psychiatric manifestations of ELS. Furthermore, we will address potential mechanisms linking ELS to accelerated telomere attrition. Finally, we will identify key areas deserving further research in this area.

KEY POINTS

- Early life stress is associated with a number of poor physical and mental health outcomes.
- Telomeres are structures that protect the ends of chromosomes that are associated with aging and cellular senescence but may also have a role in the pathogenesis and progression of disease.
- There is a large amount of evidence linking telomere length to early life stress.
- Future work should focus on identifying mechanisms underlying the association between early life stress and telomere length.

AN INTRODUCTION TO TELOMERE BIOLOGY

Human DNA exists as a double helical structure through which genetic information is passed along to preserve the integrity of and propagate the species. It is organized into specific sequences that contain this information and further organized into chromosomes. Maintaining the specific sequencing of DNA is vital to human health and reproduction. Normal DNA replication involves a mechanism of discontinuous synthesis on the lagging strand, which would lead to progressive shortening of chromosome ends with each cell replication and the eventual loss of genetic information. To prevent this harmful shortening, chromosome ends contain a specific sequence of nucleotides consisting of tandem TTAGGG repeats ranging from a few to 15 kilobases in length, called telomeres.[9] Telomeres are associated with a number of proteins that further help preserve the ends of chromosomes from degradation or recognition by DNA repair enzymes, which can trigger repair mechanisms, apoptosis, or cellular senescence cascades.[17] Maintenance of telomere function depends on both a minimal length of TTAGGG repeats and telomere-binding proteins.[8]

Telomere length is maintained by telomerase, a ribonucleoprotein reverse transcriptase heavily expressed in stem cells, germ cells, and regenerating tissues, which adds TTAGGG nucleotides to the ends of telomeres using its own template RNA. Despite this machinery, telomeres shorten with each cell division in most somatic tissues, as there is insufficient telomerase to indefinitely maintain telomere length.[101] Consequently, telomeres shorten by about 30–50 base pairs with each cell replication in most somatic tissues and, as the number of cell replication events increases over time, telomere shortening is correlated with aging.[101] As such, telomere length can serve as a marker of biological age.[2]

FACTORS IMPACTING TELOMERE LENGTH

Telomere length is impacted by a complex system of cellular processes and exposures, including oxidative damage, environmental exposures, DNA replication stress, epigenetic changes, and genetic polymorphisms. Due to their low reduction potential,[55] the guanine (G) nucleotides in telomeric repeats are particularly sensitive to oxidative damage secondary to increased reactive oxygen species (ROS). This damage induces telomere shortening, cell senescence, and cell death independent of oxidative stress elsewhere in the cell.[76] Exposure to ionizing radiation or carcinogens can lead to telomere DNA damage,[36] triggering DNA excision repair processes, including base excision repair (BER), nucleotide excision repair (NER), and mismatch repair (MMR),[36] to help restore the integrity of the telomere, albeit at the potential cost of shortening telomeres.

Emerging research suggests that epigenetic modifications may impact telomere length. Epigenetic modifications allow for elaboration of the genome beyond what is determined by the DNA and are carried forward during cellular division, but do not change the DNA sequence. Methylation is one of the most common forms of epigenetic modification.[38] Methylation can alter the chromatin state of telomeres, with greater methylation associated with a closed chromatin state[24] that is protective of the underlying DNA sequence. Decreased methylation of telomeres can lead to inappropriate signaling of DNA repair pathways, potential sequence error and shortening of the telomere DNA.[24]

Recently, genome-wide association studies (GWAS) have identified genetic loci that might affect variation in telomere length.[42,50] In a study of subjects with a history of familial longevity, Lee et al.[42] found that three loci, 4q25, 17q23.2, and 10q11.21, were associated with telomere length. Mangino et al.[49] have also identified a locus on chromosome 18q12.2 associated with telomere length. A recent meta-analysis[48] confirmed the association between leukocyte telomere length and loci 3q26.2 and 10q24.33 that had been previously reported,[14,43,50] and identified novel genomic regions associated with telomere length variation (17p13.1 and 19p12). Specific single nucleotide polymorphisms (SNPs) in genes that produce telomerase have been associated with telomere length and longevity.[45,74] Supporting the importance of genetic polymorphisms in regulating telomere length in relationship to psychopathology risk, inherited telomere syndromes due to inherited genetic mutations of telomere maintenance machinery are associated with characteristic pathologies, including neuropsychiatric sequelae.[8] A recent study of 2026 individuals showed that homozygotes for a risk SNP in the gene coding for human telomerase reverse transcriptase (hTERT) linked to shorter telomeres had a higher risk of depression compared to heterozygote carriers or homozygotes for the protective allele. However, this association only occurred in subjects without a history of ELS, suggesting that the environmental effects of early stress on telomere length may be large enough to conceal such genetic influences.[92]

TELOMERES IN HUMAN HEALTH AND DISEASE

Telomere shortening has been shown to accompany physical disease states that are associated with aging and stress exposure, including diabetes mellitus, obesity, heart disease, chronic obstructive pulmonary disease (COPD), asthma, as well as psychiatric illnesses, such as depression, anxiety, posttraumatic stress disorder (PTSD), bipolar disorder, and schizophrenia.[61,63] There is, as yet, no definitive unitary mechanism that accounts for telomere shortening across these various conditions. One common factor, however, is that they all entail physiologic stress. GWAS have identified sites associated with telomere shortening associated with several disease states (which are a source of physiologic stress) in various patient populations.[15,30,67,72] Furthermore, some studies have demonstrated relationships between the stress response system, oxidative stress, and telomere shortening in these diseases.[63] These are first steps toward a mechanistic understanding of how telomere shortening can be viewed as a common outcome of physiologic disequilibrium due to multiple etiologies. While telomere length in these conditions could be merely a disease marker (i.e., an indicator of ongoing disease), it is possible that telomere shortening primes the cell for the development of disease pathology. In a study of healthy older adults, Cawthon et al.[12] found telomere length to be predictive of eventual mortality, even though cause of death was variable. Other studies implicate telomere length as a risk marker for cancer[48] and hypertension.[97] Reports of reduced telomere length in association with smoking,[52] obesity,[86] and alcohol

abuse[58] are consistent with these conditions as risk factors for increased mortality.

Telomere dysfunction can play a causal role in disease. Telomerase deficiency underlies the pathology seen with several genetic disorders, including dyskeratosis congenita, familial idiopathic pulmonary fibrosis, and familial bone marrow failure syndromes.[2] Progeroid syndromes, characterized by clinical manifestations of accelerated aging and molecular evidence of defective DNA repair, may also reflect causal involvement of telomeres.[2]

EARLY LIFE STRESS AND TELOMERES

Current evidence shows that telomere length is maximum at birth and decreases progressively with age.[32] Furthermore, large differences in telomere length develop in the first few years after birth but are relatively stable after that[63a]. As such, early exposure to stress may be particularly detrimental to telomere length, resulting in faster age-related deterioration, earlier onset of cellular senescence, and age-related diseases.[59,62,63] Tyrka et al.[85] were the first to report a link between ELS and shortened telomeres in a study of healthy adults with ($n = 10$) or without ($n = 21$) a reported history of childhood maltreatment. These results were recently replicated in a sample of 290 adults, where parental loss and childhood maltreatment were negatively associated with telomere length.[84] A number of other studies have reported a negative association between telomere length and ELS. Kananen et al.[39] confirmed a negative association between telomere length and increasing number of reported childhood adverse life events, even absent a relationship with current psychological distress or DSM-IV anxiety disorder diagnosis. Similarly, in a study of 215 older adults recalling ELS, Savolainen et al.[66] found that although temporary separation from parents or a self-reported history of trauma were not associated with telomere length, participants reporting both experiences had shorter telomeres, suggesting an additive effect of stress and traumatic events on telomere length. Zalli et al.[99] reported that in a study of 333 healthy men and women aged 54–76, shorter telomeres and higher telomerase activity were associated with a history of ELS. In a study of older (mean age 70.6 years) subjects, Schaakxs et al.[68] found that having experienced any childhood adverse event was weakly but significantly negatively associated with telomere length. Kiecolt-Glaser et al.[40] reported that shorter telomeres were associated with multiple childhood adversities in a study comprising family caregivers of those with dementia ($n = 58$) and control subjects ($n = 74$). Surtees et al.,[77] studying 4441 women in the United Kingdom European Prospective Investigation into Cancer-Norfolk database, found that shorter telomeres correlated with increased reported childhood adversity experiences, although not with current social adversity or emotional health. In a sample of 11,443 adult women, Cai et al.[11] showed that telomere length was significantly shorter in those who had experienced more stressful life events and in those reporting childhood sexual abuse; the relationship between ELS and telomere length was mediated through a history of major depressive disorder. O'Donovan et al.[56] observed reduced telomere length in adults with chronic PTSD ($n=43$) versus healthy control subjects ($n=47$); this was accounted for by those PTSD subjects reporting multiple categories of childhood trauma ($n=18$). Similarly, Bersani et al.[6] reported that, in combat-exposed male veterans, a history of ELS was negatively associated with telomere length, as was psychopathology severity and perceived stress.

Not all studies have confirmed the negative association between ELS and telomere length. In the Twins United Kingdom study, no difference in telomere length was detected between individuals who endorsed childhood sexual ($n=34$) or physical abuse ($n=20$) and those who did not ($n=516$ and $n=520$, respectively).[26] Similarly, Jodczyk et al.[37] reported that, in participants aged 28–30 ($n=677$), telomere length did not correlate with 26 measures of life course adversity or stress which occurred before 25 years of age, or with the average number of life events reported from age 16 to 25 for each developmental domain. Verhoeven et al.[88] reported an association between telomere length and stressors in the preceding 5 years but found no association with ELS in a sample of 2936 male and female adult patients from the Netherlands. These divergent findings could be due to differences in study design and methodology, subject age, severity of ELS, or genetic factors.

In the first study to show effects of early adversity on telomere length in children, Drury et al.[20] reported that in 100 children aged 6–10 years, greater time spent in institutional care correlated with reduced telomere length obtained from buccal swab. Furthermore, in a sample of children aged 5–15 ($n=80$), Drury et al.[19] reported that telomere length in cells obtained from buccal swabs was significantly shorter in those with higher exposure to family violence and disruption and that witnessing family violence had an especially potent impact on telomere length in girls but not boys. Entringer et al.[21] demonstrated that maternal experience of severe psychosocial stress during pregnancy was associated with shorter telomeres in young adult offspring ($n=45$) versus control subjects ($n=49$), suggesting that telomere vulnerability to ELS could extend even to the antenatal period. In the first prospective longitudinal study of telomere length and ELS, Shalev et al.[71] found, in a sample of 236 children tested at age 5 and again at age 10 years, buccal

cell telomere shortening in those exposed to more than two forms of violence ($n=39$) compared with those who were unexposed ($n=128$) or less exposed ($n=69$).

Other aspects of the social environment have also been shown to have an effect on telomere length. Theall et al.,[82] in a study of neighborhood stress, found that salivary telomere length in 99 children aged 4–14 was significantly shorter in those exposed to highly disordered neighborhood environments. Similarly, Mitchell et al.[51] found reduced telomere length in 40 9-year-old African-American boys associated with low income, low maternal education, unstable family structure, and harsh parenting. This effect was moderated by genetic variants in serotonergic and dopaminergic pathways, suggesting interplay between telomere length and genetic sensitivity with environmental exposure. Recent findings raise the possibility that early intervention with at-risk children might have a beneficial impact on telomere length. Asok et al.,[1] in a study of 89 children, found that parental responsiveness to the child's cues moderated the association between ELS and telomere length, resulting in longer telomeres in children with exposure to ELS and high parental responsiveness compared to children with ELS and low parental responsiveness. These findings suggest the possibility of interventions that might provide protective effects if implemented early. Taken together, these studies support a relationship between ELS and reduced telomere length and suggest that this effect is dose-dependent and might be susceptible to amelioration or reversal with appropriate intervention.

POTENTIAL MECHANISMS UNDERLYING THE ASSOCIATION BETWEEN EARLY LIFE STRESS AND TELOMERE LENGTH

Stress refers to the organism's response to environmental demands or pressures. Neuroendocrine mechanisms of the stress response are well conserved across vertebrates and are designed to mobilize energy and prioritize expenditure of this energy toward necessary biological functions. A number of studies examining the relationship between the hypothalamic-pituitary-adrenal (HPA) axis and telomere length reported decreased telomere length with prolonged cortisol exposure in adults.[3] Furthermore, both autonomic[41,57] and adrenocortical activity have been linked to shorter telomere length in children.[28,41] However, in certain stress-related disease states, shortened telomeres have been associated with a hypocortisolemic state,[94] and, in older adults, telomere length was not associated with HPA axis function.[65] In primary human lymphocytes maintained in vitro, cortisol administration was not associated with changes in telomere length, but did negatively affect cell mitosis as evidenced by reduced cell division and cell growth.[10]

In addition, cortisol decreases telomerase activity and reduces hTERT expression in human cultured lymphocytes,[13] suggesting that cortisol may exert effects on telomeres via effects on telomerase. Consistent with this, women exposed to chronic stress had decreased telomerase expression and greater leukocyte telomere shortening.[22] However, in rats, experimental stress exposure resulted in elevated telomerase expression in leucocytes.[5] These contrasting findings suggest complex regulatory processes that may differ depending on stress exposure severity and duration, among other factors.[8]

It is possible that the HPA axis potentiates its effects on telomere length through inflammation or oxidative stress. Glucocorticoids inhibit the production of the proinflammatory cytokines interleukin (IL)-6 and IL-1 by immune cells, while adrenocorticotropic hormone (ACTH) and corticotropin-releasing hormone (CRH) have proinflammatory properties. Furthermore, IL-1, IL-6, and tumor necrosis factor (TNF)-α activate the HPA axis.[46,73] Sympathetic neurons of the autonomic nervous system (ANS) secrete proinflammatory and anti-inflammatory neuropeptides,[46] providing an additional mechanism through which the stress response interacts with systemic inflammation. Oxidative damage and antioxidant mechanisms are responsive to HPA axis function.[16,75] Oxidative stress has been associated with telomere shortening in aging[98,100] and in a number of disorders,[31,69,83] and in vitro studies show that telomeres are shortened in cells grown under hypoxia conditions or in the presence of antioxidant.[64,81] Thus, environmental stress may be linked to telomere attrition through physiologic stress response systems or via interactions with inflammatory or oxidative pathways.

METHODOLOGICAL ISSUES IN STUDIES OF TELOMERES AND EARLY LIFE STRESS

Two main methodological issues arise in studies of telomere length and ELS: the source tissue in which telomere length is measured and the type of clinical study methodology used to assess the relationship between ELS and telomeres.

Source Tissue and Telomere Length

Most studies have used peripheral blood leukocytes to measure telomere length, but some have used saliva, buccal cells, and brain tissues. In animal models, there is an association between length of telomeres in blood and in other tissues, including brain.[60] In humans, telomere length in peripheral blood leukocytes has been correlated with telomere length in skin,[23] synovium,[23] and the vasculature.[95] In a study of humans with and without Alzheimer's disease, peripheral blood leukocyte telomere length was significantly correlated with brain cerebellum telomere length,[47] suggesting that peripheral measures may be reflective of central processes. In another study comparing 12 tissue types obtained from cadavers with no previous history of tumors, metabolic disease (with the exception of diabetes), or systemic inflammation, there was a significant correlation between telomere length in peripheral blood leukocytes and that in muscle and liver. The association between telomere length in leukocytes and that in brain and spleen tissues approached significance.[18] In this study, histories of other exposures that might impact telomere length were not available and there was high variability in leukocyte telomere length, which limited their ability to detect associations. Interestingly, telomeres were the longest in leukocytes compared to the other tissue types examined, likely reflecting the lifespan of different cell types. Taken together, these studies suggest that telomere measurement in easily accessible tissues, such as blood, could serve as a surrogate parameter for relative telomere length in other tissues.

Aside from correlations with telomere length in other tissues, leukocyte telomere length could be an important marker of the systemic influences affecting telomere length, such as stress signaling and metabolic hormones. In addition, shortened leukocyte telomere length is a marker for the senescent status of circulating immune cells.[8] A role for inflammation has been documented in the pathogenesis of many conditions, including diabetes, cardiovascular disease, and psychiatric disorders.[59,62] Immune cell senescence is proinflammatory; as such, telomere shortening in leukocytes may have relevance for the pathogenesis of these conditions.[8]

Clinical Study Design: Cross-Section Versus Longitudinal Measurements

The majority of studies examining ELS and telomere length have been cross-sectional in their design. Several facts regarding telomere biology limit the robustness of this approach. Starting at birth, telomere length varies greatly between individuals,[4,78] and although telomere length is equal between the sexes at birth, there is evidence that age-associated telomere shortening may occur more rapidly in males than females and rates may also differ between ethnic groups.[25] In addition to careful controls for age and sex, cross-sectional studies require large sample sizes because of the marked variability of telomere length, which may reflect variation in telomere length from birth. Aviv et al.[4] have estimated that longitudinal studies, measuring actual telomere erosion rates within individuals over time, would require five times fewer subjects than cross-sectional studies. Moreover, longitudinal studies are more powerful than cross-sectional designs for inferring causal effects.

Very few longitudinal telomere studies have been conducted, particularly in psychiatric or stress-related conditions.

EFFECTS OF TELOMERES ON CELLULAR BIOLOGY

As new evidence emerges regarding telomere biology, it has become evident that telomere length is not only a marker of aging, exposure to environmental stressors, or systemic disease but may have a role in cellular signaling and replication.[8] Telomeres and associated telomeric binding proteins are part of extensive networks linking telomerase, DNA damage response proteins, cell cycle regulation, and shelterin structures.[24] Shelterin inhibits specific DNA damage response and DNA recombination pathways[70] that would result in cell cycle arrest.[27] Proteins of the shelterin complex recruit and regulate telomerase activity at telomeres.[35] In addition, synthesis of telomeric repeats by telomerase is temporally regulated by cell cycle-specific pathways and by cellular factors that determine its intracellular localization.[35]

Deletion of the telomerase gene leads to heterogeneity in cell cycle stage durations, an increased time spent in the G2/M checkpoint (also known as the DNA damage checkpoint), and accelerated cell aging.[96] Cancers that are caused by dysregulated telomerase function demonstrate the effects of telomerase regulation on cell cycle. In humans, a germline mutation activating transcription of TERT expression is sufficient to cause fully penetrant melanoma.[33] In addition, common germline variants in telomere maintenance genes associated with longer telomeres raise the risk of developing melanomas,[34] nonsmokers' lung cancer,[91] and gliomas.[89] Telomere dysfunction is linked to perturbation of other cellular processes that include the Wnt/β-catenin signaling network,[24] which controls many aspects of organism development, cell proliferation, and differentiation.[87] Thus, changes in telomere biology can signal cellular pathways that halt cell division and tissue replenishment, and critically shortened or damaged telomeres can trigger cellular DNA repair pathways, leading to cellular senescence, apoptosis, or carcinogenesis, thus promoting or advancing systemic disease.[29]

FUTURE DIRECTIONS

Many pivotal issues linking ELS and telomere length remain to be addressed. First, how the relationship between ELS and telomere length might be moderated by intervention is unknown. The timing at which a potential psychosocial or pharmacologic intervention might best be implemented, with regard to development, age, and proximity to a stressor, is an area warranting further attention as well. Viewing telomere length as a biomarker proxy for cellular aging or cumulative stress exposure might be helpful in quantifying the medical and psychiatric disease burden associated with ELS. From this perspective, future studies determining factors associated with telomere length preservation after ELS would be a crucial step toward linking this global measure of cellular functioning and senescence with functional disease outcomes.

A good understanding of the mechanistic linkage between ELS and telomere length is still lacking. While several potential mechanisms have been suggested, it is still unclear how psychological stress is translated into cumulative cellular stress exposure as may be indexed by telomere length. As defining such mechanisms is also a critical step in the identification of potential intervention points, both pharmacologic and psychosocial, this remains a critical area of research into the true nature of how ELS may be translated into cellular aging.

Finally, it may be that specific types of ELS trigger a more robust response in terms of telomere shortening. There may be a dose–response relationship between ELS and telomere length. To date, these questions are unanswered and will likely require very large population data to be meaningfully addressed. The importance of such research lies in the potential for deciphering more "critical" types or timing of stress exposures that might have especially profound effects on telomere biology and greater potential for sequelae that impact the health of the patient throughout the life cycle.

References

1. Asok A, Bernard K, Roth TL, Rosen JB, Dozier M. Parental responsiveness moderates the association between early-life stress and reduced telomere length. *Dev Psychopathol.* 2013;25(3):577–585. http://dx.doi.org/10.1017/S0954579413000011.
2. Aubert G, Lansdorp PM. Telomeres and aging. *Physiol Rev.* 2008;88(2):557–579. http://dx.doi.org/10.1152/physrev.00026.2007. [pii]:88/2/557.
3. Aulinas A, Ramirez MJ, Barahona MJ, et al. Telomeres and endocrine dysfunction of the adrenal and GH/IGF-1 axes. *Clin Endocrinol (Oxf).* 2013;79(6):751–759. http://dx.doi.org/10.1111/cen.12310.
4. Aviv A, Valdes AM, Spector TD. Human telomere biology: pitfalls of moving from the laboratory to epidemiology. *Int J Epidemiol.* 2006;35(6):1424–1429. http://dx.doi.org/10.1093/ije/dyl169. [pii]:dyl169.
5. Beery AK, Lin J, Biddle JS, Francis DD, Blackburn EH, Epel ES. Chronic stress elevates telomerase activity in rats. *Biol Lett.* 2012;8(6):1063–1066. http://dx.doi.org/10.1098/rsbl.2012.0747.
6. Bersani FS, Lindqvist D, Mellon SH, et al. Association of dimensional psychological health measures with telomere length in male war veterans. *J Affect Disord.* 2016;190:537–542. http://dx.doi.org/10.1016/j.jad.2015.10.037.
7. Bhan N, Glymour MM, Kawachi I, Subramanian SV. Childhood adversity and asthma prevalence: evidence from 10 US states (2009–2011). *BMJ Open Respir Res.* 2014;1(1):e000016. http://dx.doi.org/10.1136/bmjresp-2013-000016.

8. Blackburn EH, Epel ES, Lin J. Human telomere biology: a contributory and interactive factor in aging, disease risks, and protection. *Science*. 2015;350(6265):1193–1198.

9. Blackburn EH, Greider CW, Szostak JW. Telomeres and telomerase: the path from maize, Tetrahymena and yeast to human cancer and aging. *Nat Med*. 2006;12(10):1133–1138. http://dx.doi.org/10.1038/nm1006-1133.

10. Bull C, Christensen H, Fenech M. Cortisol is not associated with telomere shortening or chromosomal instability in human lymphocytes cultured under low and high folate conditions. *PLoS One*. 2015;10(3):e0119367. http://dx.doi.org/10.1371/journal.pone.0119367.

11. Cai N, Chang S, Li Y, et al. Molecular signatures of major depression. *Curr Biol*. 2015;25(9):1146–1156. http://dx.doi.org/10.1016/j.cub.2015.03.008.

12. Cawthon RM, Smith KR, O'Brien E, Sivatchenko A, Kerber RA. Association between telomere length in blood and mortality in people aged 60 years or older. *Lancet*. 2003;361(9355):393–395. Retrieved from: http://www.ncbi.nlm.nih.gov/entrez/query.fcgi?cmd=Retrieve&db=PubMed&dopt=Citation&list_uids=12573379.

13. Choi J, Fauce SR, Effros RB. Reduced telomerase activity in human T lymphocytes exposed to cortisol. *Brain Behav Immun*. 2008;22(4):600–605. http://dx.doi.org/10.1016/j.bbi.2007.12.004.

14. Codd V, Mangino M, van der Harst P, et al. Common variants near TERC are associated with mean telomere length. *Nat Genet*. 2010;42(3):197–199. http://dx.doi.org/10.1038/ng.532.

15. Codd V, Nelson CP, Albrecht E, et al. Identification of seven loci affecting mean telomere length and their association with disease. *Nat Genet*. 2013;45(4):422–427. http://dx.doi.org/10.1038/ng.2528. 427e421–422.

16. Costantini D, Marasco V, Moller AP. A meta-analysis of glucocorticoids as modulators of oxidative stress in vertebrates. *J Comp Physiol B*. 2011;181(4):447–456. http://dx.doi.org/10.1007/s00360-011-0566-2.

17. de Lange T. Shelterin: the protein complex that shapes and safeguards human telomeres. *Genes Dev*. 2005;19(18):2100–2110. Retrieved from: http://www.ncbi.nlm.nih.gov/entrez/query.fcgi?cmd=Retrieve&db=PubMed&dopt=Citation&list_uids=16166375.

18. Dlouha D, Maluskova J, Kralova Lesna I, Lanska V, Hubacek JA. Comparison of the relative telomere length measured in leukocytes and eleven different human tissues. *Physiol Res*. 2014;63(suppl 3):S343–S350. Retrieved from: http://www.ncbi.nlm.nih.gov/pubmed/25428739.

19. Drury SS, Mabile E, Brett ZH, et al. The association of telomere length with family violence and disruption. *Pediatrics*. 2014;134(1):e128–137. http://dx.doi.org/10.1542/peds.2013-3415.

20. Drury SS, Theall K, Gleason MM, et al. Telomere length and early severe social deprivation: linking early adversity and cellular aging. [published online ahead of print May 18] *Mol Psychiatry*. 2011. http://dx.doi.org/10.1038/mp.2011.53. pii:mp201153.

21. Entringer S, Epel ES, Kumsta R, et al. Stress exposure in intrauterine life is associated with shorter telomere length in young adulthood. *Proc Natl Acad Sci USA*. 2011;108(33):E513–E518. http://dx.doi.org/10.1073/pnas.1107759108. [pii]:1107759108.

22. Epel ES, Blackburn EH, Lin J, et al. Accelerated telomere shortening in response to life stress. *Proc Natl Acad Sci USA*. 2004;101(49):17312–17315. http://dx.doi.org/10.1073/pnas.0407162101. [pii]:0407162101.

23. Friedrich U, Griese E, Schwab M, Fritz P, Thon K, Klotz U. Telomere length in different tissues of elderly patients. *Mech Ageing Dev*. 2000;119(3):89–99. Retrieved from: http://www.ncbi.nlm.nih.gov/pubmed/11080530.

24. Gardano L, Pucci F, Christian L, Le Bihan T, Harrington L. Telomeres, a busy platform for cell signaling. *Front Oncol*. 2013;3:146. http://dx.doi.org/10.3389/fonc.2013.00146.

25. Geronimus AT, Hicken MT, Pearson JA, Seashols SJ, Brown KL, Cruz TD. Do US black women experience stress-related accelerated biological aging?: a novel theory and first population-based test of black–white differences in telomere length. *Hum Nat*. 2010;21(1):19–38. http://dx.doi.org/10.1007/s12110-010-9078-0.

26. Glass D, Parts L, Knowles D, Aviv A, Spector TD. No correlation between childhood maltreatment and telomere length. *Biol Psychiatry*. 2010;68(6):e21–22. http://dx.doi.org/10.1016/j.biopsych.2010.02.026. author reply e23–24. [pii]: S0006-3223(10)00579-2.

27. Goodarzi AA, Block WD, Lees-Miller SP. The role of ATM and ATR in DNA damage-induced cell cycle control. *Prog Cell Cycle Res*. 2003;5:393–411. Retrieved from: http://www.ncbi.nlm.nih.gov/pubmed/14593734.

28. Gotlib IH, LeMoult J, Colich NL, et al. Telomere length and cortisol reactivity in children of depressed mothers. *Mol Psychiatry*. 2015;20(5):615–620. http://dx.doi.org/10.1038/mp.2014.119.

29. Gramatges MM, Bertuch AA. Short telomeres: from dyskeratosis congenita to sporadic aplastic anemia and malignancy. *Transl Res*. 2013;162(6):353–363. http://dx.doi.org/10.1016/j.trsl.2013.05.003.

30. Gu J, Chen M, Shete S, et al. A genome-wide association study identifies a locus on chromosome 14q21 as a predictor of leukocyte telomere length and as a marker of susceptibility for bladder cancer. *Cancer Prev Res (Phila)*. 2011;4(4):514–521. http://dx.doi.org/10.1158/1940-6207.CAPR-11-0063.

31. Guan JZ, Guan WP, Maeda T, Guoqing X, GuangZhi W, Makino N. Patients with multiple sclerosis show increased oxidative stress markers and somatic telomere length shortening. *Mol Cell Biochem*. 2015;400(1–2):183–187. http://dx.doi.org/10.1007/s11010-014-2274-1.

32. Holohan B, De Meyer T, Batten K, et al. Decreasing initial telomere length in humans intergenerationally understates age-associated telomere shortening. *Aging Cell*. 2015;14(4):669–677. http://dx.doi.org/10.1111/acel.12347.

33. Horn S, Figl A, Rachakonda PS, et al. TERT promoter mutations in familial and sporadic melanoma. *Science*. 2013;339(6122):959–961. http://dx.doi.org/10.1126/science.1230062.

34. Huang FW, Hodis E, Xu MJ, Kryukov GV, Chin L, Garraway LA. Highly recurrent TERT promoter mutations in human melanoma. *Science*. 2013;339(6122):957–959. http://dx.doi.org/10.1126/science.1229259.

35. Hukezalie KR, Wong JM. Structure-function relationship and biogenesis regulation of the human telomerase holoenzyme. *FEBS J*. 2013;280(14):3194–3204. http://dx.doi.org/10.1111/febs.12272.

36. Jia P, Her C, Chai W. DNA excision repair at telomeres. *DNA Repair (Amst)*. 2015. http://dx.doi.org/10.1016/j.dnarep.2015.09.017.

37. Jodczyk S, Fergusson DM, Horwood LJ, Pearson JF, Kennedy MA. No association between mean telomere length and life stress observed in a 30 year birth cohort. *PLoS One*. 2014;9(5):e97102. http://dx.doi.org/10.1371/journal.pone.0097102.

38. Jones PA, Takai D. The role of DNA methylation in mammalian epigenetics. *Science*. 2001;293(5532):1068–1070. http://dx.doi.org/10.1126/science.1063852.

39. Kananen L, Surakka I, Pirkola S, et al. Childhood adversities are associated with shorter telomere length at adult age both in individuals with an anxiety disorder and controls. *PLoS One*. 2010;5(5):e10826. http://dx.doi.org/10.1371/journal.pone.0010826.

40. Kiecolt-Glaser JK, Gouin JP, Weng NP, Malarkey WB, Beversdorf DQ, Glaser R. Childhood adversity heightens the impact of later-life caregiving stress on telomere length and inflammation. *Psychosom Med*. 2011;73(1):16–22. http://dx.doi.org/10.1097/PSY.0b013e31820573b6. [pii]:PSY.0b013e31820573b6.

41. Kroenke CH, Epel E, Adler N, et al. Autonomic and adrenocortical reactivity and buccal cell telomere length in kindergarten children. *Psychosom Med.* 2011;73(7):533–540. http://dx.doi.org/10.1097/PSY.0b013e318229acfc.

42. Lee JH, Cheng R, Honig LS, et al. Genome wide association and linkage analyses identified three loci-4q25, 17q23.2, and 10q11.21-associated with variation in leukocyte telomere length: the long life family study. *Front Genet.* 2013;4:310. http://dx.doi.org/10.3389/fgene.2013.00310.

43. Levy D, Neuhausen SL, Hunt SC, et al. Genome-wide association identifies OBFC1 as a locus involved in human leukocyte telomere biology. *Proc Natl Acad Sci USA.* 2010;107(20):9293–9298. http://dx.doi.org/10.1073/pnas.0911494107.

44. Li M, D'Arcy C, Meng X. Maltreatment in childhood substantially increases the risk of adult depression and anxiety in prospective cohort studies: systematic review, meta-analysis, and proportional attributable fractions. *Psychol Med.* 2015:1–14. http://dx.doi.org/10.1017/S0033291715002743.

45. Liu Y, Cao L, Li Z, et al. A genome-wide association study identifies a locus on TERT for mean telomere length in Han Chinese. *PLoS One.* 2014;9(1):e85043. http://dx.doi.org/10.1371/journal.pone.0085043.

46. Louati K, Berenbaum F. Fatigue in chronic inflammation—a link to pain pathways. *Arthritis Res Ther.* 2015;17:254. http://dx.doi.org/10.1186/s13075-015-0784-1.

47. Lukens JN, Van Deerlin V, Clark CM, Xie SX, Johnson FB. Comparisons of telomere lengths in peripheral blood and cerebellum in Alzheimer's disease. *Alzheimers Dement.* 2009;5(6):463–469. http://dx.doi.org/10.1016/j.jalz.2009.05.666.

48. Ma H, Zhou Z, Wei S, et al. Shortened telomere length is associated with increased risk of cancer: a meta-analysis. *PLoS One.* 2011;6(6):e20466. http://dx.doi.org/10.1371/journal.pone.0020466. [pii]:PONE-D-11-04747.

49. Mangino M, Hwang SJ, Spector TD, et al. Genome-wide meta-analysis points to CTC1 and ZNF676 as genes regulating telomere homeostasis in humans. *Hum Mol Genet.* 2012;21(24):5385–5394. http://dx.doi.org/10.1093/hmg/dds382.

50. Mangino M, Richards JB, Soranzo N, et al. A genome-wide association study identifies a novel locus on chromosome 18q12.2 influencing white cell telomere length. *J Med Genet.* 2009;46(7):451–454. http://dx.doi.org/10.1136/jmg.2008.064956.

51. Mitchell C, Hobcraft J, McLanahan SS, et al. Social disadvantage, genetic sensitivity, and children's telomere length. *Proc Natl Acad Sci USA.* 2014;111(16):5944–5949. http://dx.doi.org/10.1073/pnas.1404293111.

52. Morla M, Busquets X, Pons J, Sauleda J, MacNee W, Agusti AG. Telomere shortening in smokers with and without COPD. *Eur Respir J.* 2006;27(3):525–528. http://dx.doi.org/10.1183/09031936.06.00087005. [pii]:27/3/525.

53. Morton PM, Mustillo SA, Ferraro KF. Does childhood misfortune raise the risk of acute myocardial infarction in adulthood? *Soc Sci Med.* 2014;104:133–141. http://dx.doi.org/10.1016/j.socscimed.2013.11.026.

54. Morton PM, Schafer MH, Ferraro KF. Does childhood misfortune increase cancer risk in adulthood? *J Aging Health.* 2012;24(6):948–984. http://dx.doi.org/10.1177/0898264312449184.

55. Neeley WL, Essigmann JM. Mechanisms of formation, genotoxicity, and mutation of guanine oxidation products. *Chem Res Toxicol.* 2006;19(4):491–505. http://dx.doi.org/10.1021/tx0600043.

56. O'Donovan A, Epel E, Lin J, et al. Childhood trauma associated with short leukocyte telomere length in posttraumatic stress disorder. *Biol Psychiatry.* 2011;70(5):465–471. http://dx.doi.org/10.1016/j.biopsych.2011.01.035. [pii]:S0006-3223(11)00129-6.

57. Parks CG, Miller DB, McCanlies EC, et al. Telomere length, current perceived stress, and urinary stress hormones in women. *Cancer Epidemiol Biomarkers Prev.* 2009;18(2):551–560. http://dx.doi.org/10.1158/1055-9965.EPI-08-0614.

58. Pavanello S, Hoxha M, Dioni L, et al. Shortened telomeres in individuals with abuse in alcohol consumption. *Int J Cancer.* 2011;129(4):983–992. http://dx.doi.org/10.1002/ijc.25999.

59. Price LH, Kao HT, Burgers DE, Carpenter LL, Tyrka AR. Telomeres and early-life stress: an overview. *Biol Psychiatry.* 2013;73(1):15–23. http://dx.doi.org/10.1016/j.biopsych.2012.06.025.

60. Reichert S, Criscuolo F, Verinaud E, Zahn S, Massemin S. Telomere length correlations among somatic tissues in adult zebra finches. *PLoS One.* 2013;8(12):e81496. http://dx.doi.org/10.1371/journal.pone.0081496.

61. Ridout KK, Ridout SJ, Price LH, Sen S, Tyrka AR. Depression and telomere length: a meta-analysis. *J Affect Disord.* 2015;191:237–247. http://dx.doi.org/10.1016/j.jad.2015.11.052.

62. Ridout SJ, Ridout KK, Kao HT, et al. Telomeres, early-life stress and mental illness. *Adv Psychosom Med.* 2015;34:92–108. http://dx.doi.org/10.1159/000369088.

63. Rizvi S, Raza ST, Mahdi F. Telomere length variations in aging and age-related diseases. *Curr Aging Sci.* 2014;7(3):161–167. Retrieved from: http://www.ncbi.nlm.nih.gov/pubmed/25612739.

63a. Rufer N, Brümmendorf TH, Kolvraa S, et al. Telomere fluorescence measurements in granulocytes and T lymphocyte subsets point to a high turnover of hematopoietic stem cells and memory T cells in early childhood. *J Exp Med.* 1999;190(2):157–167. PMID: 10432279.

64. Saretzki G. Extra-telomeric functions of human telomerase: cancer, mitochondria and oxidative stress. *Curr Pharm Des.* 2014;20(41):6386–6403. Retrieved from: http://www.ncbi.nlm.nih.gov/pubmed/24975608.

65. Savolainen K, Eriksson JG, Kajantie E, Lahti J, Raikkonen K. Telomere length and hypothalamic-pituitary-adrenal axis response to stress in elderly adults. *Psychoneuroendocrinology.* 2015;53:179–184. http://dx.doi.org/10.1016/j.psyneuen.2014.12.020.

66. Savolainen K, Eriksson JG, Kananen L, et al. Associations between early life stress, self-reported traumatic experiences across the lifespan and leukocyte telomere length in elderly adults. *Biol Psychol.* 2014;97:35–42. http://dx.doi.org/10.1016/j.biopsycho.2014.02.002.

67. Saxena R, Bjonnes A, Prescott J, et al. Genome-wide association study identifies variants in casein kinase II (CSNK2A2) to be associated with leukocyte telomere length in a Punjabi Sikh diabetic cohort. *Circ Cardiovasc Genet.* 2014;7(3):287–295. http://dx.doi.org/10.1161/CIRCGENETICS.113.000412.

68. Schaakxs R, Wielaard I, Verhoeven JE, Beekman AT, Penninx BW, Comijs HC. Early and recent psychosocial stress and telomere length in older adults. *Int Psychogeriatr.* 2015:1–9. http://dx.doi.org/10.1017/S1041610215001155.

69. Sekoguchi S, Nakajima T, Moriguchi M, et al. Role of cell-cycle turnover and oxidative stress in telomere shortening and cellular senescence in patients with chronic hepatitis C. *J Gastroenterol Hepatol.* 2007;22(2):182–190. http://dx.doi.org/10.1111/j.1440-1746.2006.04454.x.

70. Sfeir A, de Lange T. Removal of shelterin reveals the telomere end-protection problem. *Science.* 2012;336(6081):593–597. http://dx.doi.org/10.1126/science.1218498.

71. Shalev I, Moffitt TE, Sugden K, et al. Exposure to violence during childhood is associated with telomere erosion from 5 to 10 years of age: a longitudinal study. *Mol Psychiatry.* 2012. http://dx.doi.org/10.1038/mp.2012.32. [pii]:mp201232.

72. Shi J, Sun F, Peng L, et al. Leukocyte telomere length-related genetic variants in 1p34.2 and 14q21 loci contribute to the risk of esophageal squamous cell carcinoma. *Int J Cancer.* 2013;132(12):2799–2807. http://dx.doi.org/10.1002/ijc.27959.

73. Silverman MN, Sternberg EM. Glucocorticoid regulation of inflammation and its functional correlates: from HPA axis to glucocorticoid receptor dysfunction. *Ann N. Y Acad Sci.* 2012;1261:55–63. http://dx.doi.org/10.1111/j.1749-6632.2012.06633.x.

74. Soerensen M, Thinggaard M, Nygaard M, et al. Genetic variation in TERT and TERC and human leukocyte telomere length and longevity: a cross-sectional and longitudinal analysis. *Aging Cell.* 2012;11(2):223–227. http://dx.doi.org/10.1111/j.1474-9726.2011.00775.x.

75. Spiers JG, Chen HJ, Sernia C, Lavidis NA. Activation of the hypothalamic-pituitary-adrenal stress axis induces cellular oxidative stress. *Front Neurosci.* 2014;8:456. http://dx.doi.org/10.3389/fnins.2014.00456.

76. Sun L, Tan R, Xu J, et al. Targeted DNA damage at individual telomeres disrupts their integrity and triggers cell death. *Nucleic Acids Res.* 2015;43(13):6334–6347. http://dx.doi.org/10.1093/nar/gkv598.

77. Surtees PG, Wainwright NW, Pooley KA, et al. Life stress, emotional health, and mean telomere length in the European Prospective Investigation into Cancer (EPIC)-Norfolk population study. *J Gerontol A Biol Sci Med Sci.* 2011;66(11):1152–1162. http://dx.doi.org/10.1093/gerona/glr112. [pii]:glr112.

78. Takubo K, Izumiyama-Shimomura N, Honma N, et al. Telomere lengths are characteristic in each human individual. *Exp Gerontol.* 2002;37(4):523–531. Retrieved from: http://www.ncbi.nlm.nih.gov/entrez/query.fcgi?cmd=Retrieve&db=PubMed&dopt=Citation&list_uids=11830355.

79. Tamayo T, Christian H, Rathmann W. Impact of early psychosocial factors (childhood socioeconomic factors and adversities) on future risk of type 2 diabetes, metabolic disturbances and obesity: a systematic review. *BMC Public Health.* 2010;10:525. http://dx.doi.org/10.1186/1471-2458-10-525.

80. Teicher MH, Samson JA. Childhood maltreatment and psychopathology: a case for ecophenotypic variants as clinically and neurobiologically distinct subtypes. *Am J Psychiatry.* 2013;170(10):1114–1133. http://dx.doi.org/10.1176/appi.ajp.2013.12070957.

81. Thanan R, Oikawa S, Hiraku Y, et al. Oxidative stress and its significant roles in neurodegenerative diseases and cancer. *Int J Mol Sci.* 2015;16(1):193–217. http://dx.doi.org/10.3390/ijms16010193.

82. Theall KP, Brett ZH, Shirtcliff EA, Dunn EC, Drury SS. Neighborhood disorder and telomeres: connecting children's exposure to community level stress and cellular response. *Soc Sci Med.* 2013;85:50–58. http://dx.doi.org/10.1016/j.socscimed.2013.02.030.

83. Tyrka AR, Carpenter LL, Kao HT, et al. Association of telomere length and mitochondrial DNA copy number in a community sample of healthy adults. *Exp Gerontol.* 2015;66:17–20. http://dx.doi.org/10.1016/j.exger.2015.04.002.

84. Tyrka AR, Parade SH, Price LH, et al. Alterations of mitochondrial DNA copy number and telomere length with early adversity and psychopathology. *Biol Psychiatry.* 2016;79(2):78–86. http://dx.doi.org/10.1016/j.biopsych.2014.12.025.

85. Tyrka AR, Price LH, Kao HT, Porton B, Marsella SA, Carpenter LL. Childhood maltreatment and telomere shortening: preliminary support for an effect of early stress on cellular aging. *Biol Psychiatry.* 2010;67(6):531–534. http://dx.doi.org/10.1016/j.biopsych.2009.08.014. [pii]:S0006-3223(09)01013-0.

86. Valdes AM, Andrew T, Gardner JP, et al. Obesity, cigarette smoking, and telomere length in women. *Lancet.* 2005;366(9486):662–664. http://dx.doi.org/10.1016/S0140-6736(05)66630-5. [pii]:S0140-6736(05)66630-5.

87. Valenta T, Hausmann G, Basler K. The many faces and functions of beta-catenin. *EMBO J.* 2012;31(12):2714–2736. http://dx.doi.org/10.1038/emboj.2012.150.

88. Verhoeven JE, van Oppen P, Puterman E, Elzinga B, Penninx BW. The association of early and recent psychosocial life stress with leukocyte telomere length. *Psychosom Med.* 2015;77(8):882–891. http://dx.doi.org/10.1097/PSY.0000000000000226.

89. Walsh KM, Wiencke JK, Lachance DH, et al. Telomere maintenance and the etiology of adult glioma. *Neuro Oncol.* 2015;17(11):1445–1452. http://dx.doi.org/10.1093/neuonc/nov082.

90. Wegman HL, Stetler C. A meta-analytic review of the effects of childhood abuse on medical outcomes in adulthood. *Psychosom Med.* 2009;71(8):805–812. http://dx.doi.org/10.1097/PSY.0b013e3181bb2b46.

91. Wei R, DeVilbiss FT, Liu W. Genetic polymorphism, telomere biology and non-small lung cancer risk. *J Genet Genomics.* 2015;42(10):549–561. http://dx.doi.org/10.1016/j.jgg.2015.08.005.

92. Wei YB, Martinsson L, Liu JJ, et al. hTERT genetic variation in depression. *J Affect Disord.* 2015;189:62–69. http://dx.doi.org/10.1016/j.jad.2015.09.025.

93. Wickrama KK, O'Neal CW, Lee TK, Wickrama T. Early socioeconomic adversity, youth positive development, and young adults' cardio-metabolic disease risk. *Health Psychol.* 2015;34(9):905–914. http://dx.doi.org/10.1037/hea0000208.

94. Wikgren M, Maripuu M, Karlsson T, et al. Short telomeres in depression and the general population are associated with a hypocortisolemic state. *Biol Psychiatry.* 2012;71(4):294–300. Retrieved from: http://www.ncbi.nlm.nih.gov/entrez/query.fcgi?cmd=Retrieve&db=PubMed&dopt=Citation&list_uids=22055018.

95. Wilson WR, Herbert KE, Mistry Y, et al. Blood leucocyte telomere DNA content predicts vascular telomere DNA content in humans with and without vascular disease. *Eur Heart J.* 2008;29(21):2689–2694. http://dx.doi.org/10.1093/eurheartj/ehn386.

96. Xie Z, Jay KA, Smith DL, et al. Early telomerase inactivation accelerates aging independently of telomere length. *Cell.* 2015;160(5):928–939. http://dx.doi.org/10.1016/j.cell.2015.02.002.

97. Yang Z, Huang X, Jiang H, et al. Short telomeres and prognosis of hypertension in a Chinese population. *Hypertension.* 2009;53(4):639–645. http://dx.doi.org/10.1161/HYPERTENSIONAHA.108.123752. [pii]:HYPERTENSIONAHA.108.123752.

98. Yokoo S, Furumoto K, Hiyama E, Miwa N. Slow-down of age-dependent telomere shortening is executed in human skin keratinocytes by hormesis-like-effects of trace hydrogen peroxide or by anti-oxidative effects of pro-vitamin C in common concurrently with reduction of intracellular oxidative stress. *J Cell Biochem.* 2004;93(3):588–597. http://dx.doi.org/10.1002/jcb.20208.

99. Zalli A, Carvalho LA, Lin J, et al. Shorter telomeres with high telomerase activity are associated with raised allostatic load and impoverished psychosocial resources. *Proc Natl Acad Sci USA.* 2014;111(12):4519–4524. http://dx.doi.org/10.1073/pnas.1322145111.

100. Zhang J, Rane G, Dai X, et al. Ageing and the telomere connection: an intimate relationship with inflammation. *Ageing Res Rev.* 2016;25:55–69. http://dx.doi.org/10.1016/j.arr.2015.11.006.

101. Zhao Z, Pan X, Liu L, Liu N. Telomere length maintenance, shortening, and lengthening. *J Cell Physiol.* 2014;229(10):1323–1329. http://dx.doi.org/10.1002/jcp.24537.

ENDOCRINE SYSTEMS AND MECHANISMS IN STRESS CONTROL

CHAPTER

19

Stress, Glucocorticoids, and Brain Development in Rodent Models

C.M. McCormick, T.E. Hodges

Brock University, St. Catharines, ON, Canada

Abstract

Organ systems are sensitive to malleability by environmental stressors during times of rapid development and biological transitions. For the developing nervous system, there are three windows of heightened sensitivity to environmental stressors, the prenatal, neonatal, and adolescent periods, although the brain maintains some sensitivity throughout the lifespan. Glucocorticoid hormones are the signaling molecules that underlie the effects of environmental stressors, able to affect all aspects of neuronal development from cell birth to cell death. This chapter provides an overview of how environmental stressors result in long-lasting changes during the different windows of brain development. The extent to which stress-induced developmental changes are adaptive or maladaptive depends on the neural region, the severity and duration of the stressor exposure, the sex and genotype of the individual, and the quality of the individual's environment in life stages beyond the stress exposures.

INTRODUCTION

Rats and mice are the species used most in studies of neurodevelopment; thus this chapter focuses on evidence from studies in rats. Despite marked species differences, the timing and sequencing of neural development is highly conserved across mammals.[12] CNS development starts with the induction of ectodermal embryonic tissue by the notochord to form the neural plate, from which the neural tube (precursor cells of the CNS) is formed, and the formation of the neural crest (includes precursors of Schwann cells, nerve ganglia, meninges) (Fig. 19.1). The developmental processes of proliferation, migration, aggregation, differentiation, synaptogenesis, apoptosis, and myelination unfold in precise sequences and timelines, the interruption of which can result in abnormalities (Fig. 19.2).[34] Moreover, because the CNS has a protracted development and maintains some plasticity throughout the lifespan (a necessity for the brain's learning and memory functions), it is highly susceptible to environmental experiences. Environmental stressors that result in elevations of glucocorticoid hormones, however, can have a more widespread influence on the brain in periods of greater maturational change. The effects of stressors are thus intimately linked to the ontogenetic stage of the organism.

KEY POINTS

- Environmental stressors influence the development of the brain.
- Stressors have their greatest effects when the brain is undergoing rapid change.
- Fetal, neonatal, and adolescence are windows of greater susceptibility to stressors.
- Stressors have long-lasting effects on emotional, cognitive, and social behavior.
- Stress-induced elevations of glucocorticoid hormones underlie the effects on brain development and function.
- Effects of stressors vary depending on severity of the stressor and duration, genotype, brain region, and stage of ontogeny.

STRESSORS, GLUCOCORTICOIDS, AND THE HYPOTHALAMIC–PITUITARY–ADRENAL AXIS

Environmental stressors influence brain development and function primarily through effects on glucocorticoids (GCs, corticosterone in rats, cortisol in humans). There are receptors for GCs in almost all cell types of the body, thus GCs have widespread effects on many functions, including metabolism, endocrine, and immune function. The production and release of GCs is controlled by the hypothalamic–pituitary–adrenal (HPA) axis (Fig. 19.3). Under basal conditions, there is circadian cyclicity in the release of GCs, with peak and nadir concentrations that map on to wake–sleep cycles. The perception of stressors initiates neural signaling that is integrated in the paraventricular nucleus (PVN) in the hypothalamus, initiating the release of corticotropin-releasing factor (CRF) and arginine vasopressin (AVP), which act to release adrenocorticotropic hormone (ACTH) from the pituitary into blood circulation. ACTH then activates the production and release of GCs from the adrenal cortex into circulation, where the majority of GCs are bound to corticosteroid-binding globulin (CBG), which limits their entry into cells. Although HPA stress responses are adaptive, they can become dysfunctional when confronted by chronic stressors.

As lipophilic hormones, unbound GCs readily traverse cell membranes to reach their cognate receptors,

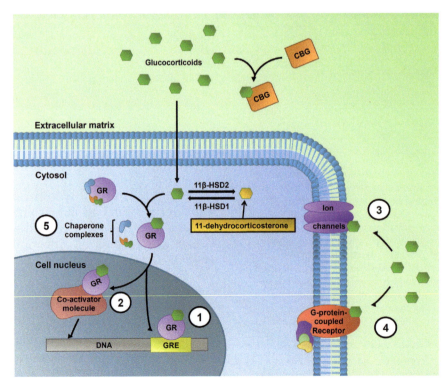

FIGURE 19.1 Genomic and nongenomic actions of glucocorticoids (GCs). GCs diffuse across the cell membrane, bind to cytoplasmic glucocorticoid receptors (GR), and translocate into the nucleus to influence gene expression either by (1) binding directly to glucocorticoid response elements found on DNA or (2) binding with coactivator molecules. Several nongenomic actions have been identified that alter intracellular signaling pathways and cell excitability, including (3) ion channels (4) G protein-coupled receptors, and (5) the dissociation of chaperone complexes from bound GR.

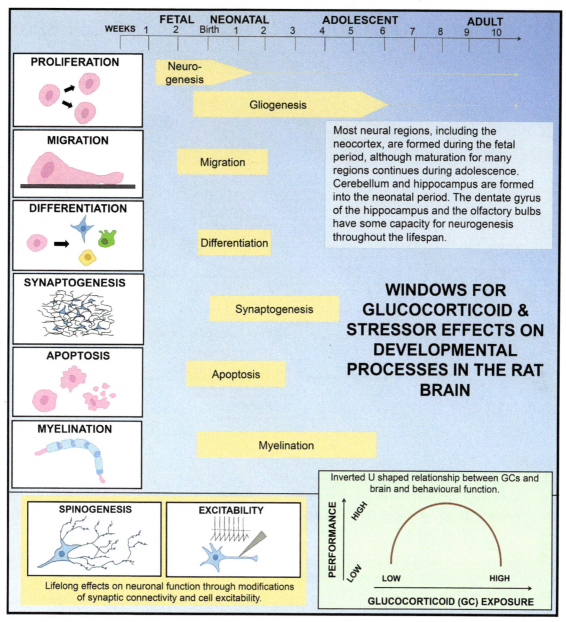

FIGURE 19.2 The timeframe for processes of brain development indicating when in development each process is most modifiable by stressors and glucocorticoids. *Adapted from Rice D, Barone Jr S. Critical periods of vulnerability for the developing nervous system: evidence from humans and animal models.* Environ Health Perspect. 2000;108(suppl 3):511–533.

the mineralocorticoid and glucocorticoid receptors (MR and GR), which are located primarily in the cytosol, but also in the nucleus, synaptic terminals, and mitochondria. MR are limited to limbic regions, with most expression in the hippocampus. GR are expressed more widely, with their greatest expression found in the hippocampus, medial prefrontal cortex, amygdala, and PVN. MR has a 10-fold higher affinity for corticosterone than does GR, and MR is near maximally bound under basal concentrations of hormone. Thus, GR signaling is a main mechanism of the effects of stressors on brain development.

When bound by GCs, cytosolic GRs translocate to the nucleus and exert cell-type and context-specific genomic effects by forming dimers and binding at glucocorticoid response elements or other transcription factors to enhance or repress target gene expression. GRs also have rapid nongenomic effects involving actions at membrane-bound receptors and through other mechanisms.[20] Through the multiplicity of GR mechanisms, GCs have widespread effects on neuronal development and function. For example, GCs influence regulators of actin and microtubules, and thereby have effects on neuronal migration, cytoarchitecture, and

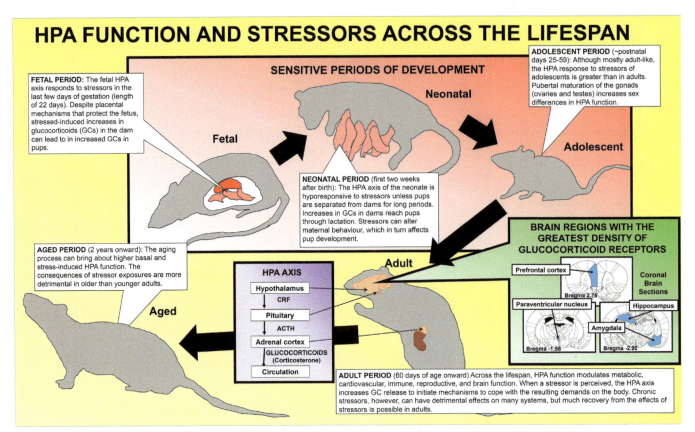

FIGURE 19.3   The hypothalamic–pituitary–adrenal axis and its function at different periods of development.

ultimately on synaptic development and plasticity. In addition, GCs affect numerous cell death and survival signaling pathways. Through nongenomic actions, GCs influence many aspects of neurotransmission (e.g., AMPA receptor trafficking, glutamate release) and have rapid effects on neuronal excitability through actions on ion channels. GCs also influence intracellular signaling pathways through the dissociation of bound GRs from cytosolic chaperone complexes. The effects of acute GC exposures on neural function differ from those of chronic exposures and vary according to the developmental stage. Many effects of chronic stressors experienced in adulthood dissipate with time, whereas effects at other periods of life are longer-lasting and can occur after shorter durations and severities of exposures. The fetal, neonatal, and adolescent periods, and time of increased brain development are particularly sensitive to stressors.

FETAL PERIOD

Gestational length in rats is ~22 days. The fetal rat brain is comparable to the first two trimesters of humans, and brain development in the neonatal rat is comparable to the last trimester in humans.[12] Neurogenesis, migration,

and differentiation are the main neural developmental events in the fetal rat and begin at about gestational day (GD) 9.5, with most proliferation occurring between GD13–18. Myelination begins before birth and declines in mid-adolescence.[34] The effects of prenatal stressors on the CNS, however, are more widespread.

Development of the Hypothalamic–Pituitary–Adrenal Axis

By the third week of gestation, the fetal HPA axis is becoming functional.[39] Corticosterone is detectable in the adrenal by GD16–17, although its rhythm of release is set by that of the dam. The brain expresses GR and MR mRNA and protein as early as GD12–15. The neurons of the PVN develop in the last week of gestation, and the main secretagogues of ACTH, CRF, and AVP, are detectable by about GD17–18. The fetal pituitary secretes ACTH in response to PVN secretagogues as early as GD17. In addition, the HPA axis responds to physical stressors in the days before parturition.[39] The fetus, however, is exposed to maternal GCs long before the development of its HPA function, and GCs are important for normal development. GR is widely expressed in many fetal tissues other than the brain, and a rise in GCs is necessary for the transition from fetus to neonate, with the maturation of many organs dependent on this rise. Nevertheless, inappropriately timed or excessive exposure to GCs can disrupt development.

Exposure to Maternal Glucocorticoids

The placenta provides a barrier between maternal and fetal blood supplies while allowing the exchange of nutrients and the removal of waste. It also offers protection to the fetus from overexposure to the higher GC concentrations of the dam through expression of 11β-hydroxysteroid dehydrogenase 2 (11β-HSD2), which converts corticosterone to inert 11-dehydrocorticosterone. The placenta, however, has increased expression of 11β-hydroxysteroid dehydrogenase 1 (11β-HSD1), which converts 11-dehydrocorticosterone to corticosterone, from about GD16, likely to allow the increased GC exposure required for the organ maturation that is critical for transition from fetus to neonate. P-glycoprotein in the placenta and in the blood–brain barrier may also serve to protect the developing nervous system by extruding GCs. Further, in the last week of gestation, the HPA response to acute stressors is attenuated in the dam compared with other times of gestation or with nonpregnant females, which limits the effects of stressors on both dams and fetuses in the last week.[7]

The protective mechanisms against GCs may be overcome by stress-induced elevations in the dam; for example, chronic stress limits the ability of placental 11β-HSD2 to respond effectively, and plasma concentrations of corticosterone in fetuses are correlated with those of dams after a stressor. Exposing dams to stressors increased the expression of immediate early genes in the PVN of both dams and fetuses, and expression was graded in terms of the severity of the stressor.[17] Further, epigenetic modifications in genes controlling GC signaling are found in the placenta in response to maternal stressors.[30]

Sex Differences in the Fetal Environment

There is little evidence of sex differences in the HPA axis in fetal rats.[39] Nevertheless, there are several mechanisms for sex differences in the effects of gestational stressors. In humans, maternal HPA function differs depending on the sex of the fetus.[18] In rats, female fetuses have higher basal corticosterone concentrations than do male fetuses in the absence of differences in adrenal content of corticosterone, which has been attributed to differential placental transport of corticosterone from the dam to males and females. Sex differences in gene expression in the placenta were found for both X chromosome and autosomal genes that also may result in sex differences in the effects of maternal stressors on the fetus.[30] There also is evidence that stressors influence

gene expression in the placenta of females to a greater extent than in males.[15]

Effects of Maternal Gestational Stress on the CNS

Immediate effects of excessive GCs in the dam on fetal brain development include reductions in neurogenesis in the hippocampus, a retardation of the migration of neurons in the developing cerebral cortex, reduced proliferation and altered migration of telencephalic GABA neurons, and changes in MR and GR densities in limbic regions, among other effects.[4,40] There is more research, however, on the programming effects of maternal gestational stress. GC signaling is one of the main means of environmental programming of the HPA axis in development and involves epigenetic modifications that influence gene expression, including microRNA expression, histone modifications, and DNA methylation.[2] For example, through epigenetic mechanisms, embryonic treatment of neural stem cells with the synthetic GC, dexamethasone, altered the expression of genes critical for the regulation of cell proliferation, differentiation, migration, cellular senescence, DNA methylation, ion channels, mitochondrial function, and oxidative stress in neural stem cells (Fig. 19.4).[5]

Increased HPA axis responsiveness to stressors and decreased hippocampal neurogenesis are commonly reported in adult offspring of stressed dams. There are effects of maternal stress on dendritic architecture; the neural structures typically affected are the prefrontal cortex, amygdala, and hippocampus. Broad changes in gene expression, including epigenetic regulators, were found in the hippocampus and prefrontal cortex in adult offspring of stressed dams.[38] Changes in cognitive and emotional behavior in adulthood are reported, with reductions in performance on hippocampal-dependent tasks and heightened anxiety-like behavior the most typical effects of maternal stress.[4,40] Furthermore, changes in motivational and reward systems are evident. Prospective studies in humans found maternal stress to be associated with attention-deficit hyperactive disorder and schizophrenia in male offspring, anxiety and depression in female offspring, and some evidence of negative consequences on cognitive performance in both sexes.[40] Elevations of GCs in the dam are implicated in the effects of maternal stress in offspring. Some of the effects of maternal stress can be obtained by exogenous administration of corticosterone to dams and can be attenuated by maternal adrenalectomy.[40] Although care of neonates is altered in dams exposed to stressors during gestation, changes in maternal behavior do not account for all the effects in offspring. Neural region, stressor procedure and intensity, and sex of offspring, however, moderate the extent and directionality of the effects of maternal stress.

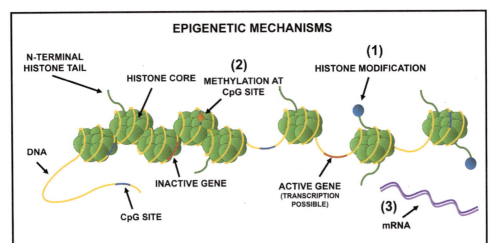

EPIGENETIC MECHANISMS

Epigenetics mechanisms change gene expression without changing the DNA sequence. There are three main mechanisms: **(1) Histone modifications** involve changes to the N-terminal tails of histone protein cores that form the chromatin structure that DNA is wrapped around. Numerous histone modifications have been identified (e.g., acetylation, methylation, phosphorylation) that alter the interaction of chromatin and DNA to either facilitate or repress transcription. **(2) DNA methylation** of cytosines at CpG sites (sites where cytosines are adjacent to guanines) typically leads to gene silencing. **(3) Non-coding RNA** (e.g., microRNAs) have post-transcriptional effects through the regulation of mRNA and chromatin structure.
Many of the genes for the enzymes required for these epigenetic mechanisms (e.g., DNA methyltransferases, histone acetyltransferases, demethylases, protein kinases) are known to be regulated by glucocorticoids (GCs), and thus provide a mechanism for GCs to produce epigenetic modifications. In turn, many of these epigenetic mechanisms are known to regulate expression of the glucocorticoid receptor, and thus the effects of GCs. Epigenetic modifications can be short-lived or long-lasting in the individual; in germ cells, the modifications may persist into subsequent generations.

FIGURE 19.4   Epigenetic mechanisms.

Elevated GCs in the dam can be secondary to other environmental factors. For example, low protein diets reduce 11β-HSD2, thereby increasing the exposure of the fetus to GCs, and maternal malnutrition produces many effects in offspring similar to those of maternal stress.[13] In turn, the effects of GCs on brain development may involve indirect mechanisms; for example, maternal stress or administration of exogenous GCs suppress the surge in testosterone involved in the sexual differentiation of the fetal brain. A decrease in plasma testosterone concentrations in maternally stressed males compared with controls is evident in adulthood, thus maternal stress may also dampen the secondary period of sexual differentiation of the brain that occurs in adolescence.[22] Finally, exposure to synthetic GCs may have greater detrimental and long-lasting effects on the developing fetus than exposure to natural GCs.[27]

NEONATAL PERIOD

The first two weeks after birth constitute the neonatal period in rats. The dam regulates many of the physiological systems of the neonate, including HPA function. The newborn is poikilothermic, unable to see or hear, and requires anogenital stimulation by the dam for waste elimination. Eyes open at about postnatal day (PD) 14, pups begin to chew on solids at about PD16, and thermoregulate by the third week. Weaning is completed by PD25, although in many labs pups are housed separately from dams at PD21. Neurogenesis has declined significantly by birth, whereas gliogenesis remains relatively high until weaning. Differentiation of the cerebral cortex, cerebellum, and hippocampus continues after birth, and synaptogenesis occurs mostly in the first three weeks.[34]

Neonatal Hypothalamic–Pituitary–Adrenal Axis Function

Corticosterone concentrations in the neonate decline after birth, and a circadian rhythm in plasma corticosterone is detectable about 2 weeks later. MR binding capacity in the hippocampus is low at birth and adult-like by the end of the first week, whereas GR continues to increase until the third week. From about PD4 to PD12, the HPA axis of rat pups is resistant to activation by a number of stressors, and this period is known as the stress hyporesponsive period (SHRP).[39] CRF and ACTH concentrations are also low during this time. Social buffering accounts for some of the hyporesponsiveness; removal of the dam for a minimum of 12h increases HPA responsiveness to stressors in the neonate, and removal of all familiar cues (isolating pup from littermates, nest) increases pup's corticosterone concentrations within 30 min. A reduced exposure to GCs may be required to ensure that the high rate of development that occurs during that time is not compromised. Some exposure to GCs, however, is necessary for normal brain development; adrenalectomy of pups toward the end of the SHRP led to increased gliogenesis throughout the brain and increased neurogenesis in the hippocampus and cerebellum leading to adult brains that were about 14% larger than that of control rats.[26]

Maternal Hypothalamic–Pituitary–Adrenal Axis Function and the Neonate

Although increased basal and suckling-induced HPA activity is required to support lactation, lactating dams have a dampened HPA axis response to stressors.[6] Nevertheless, corticosterone concentrations in milk rise after stress exposures, although they do not reach the peak concentrations of those in plasma. Adding corticosterone to the drinking water of dams or injections of corticosterone to dams results in elevated corticosterone concentrations in pups. Small elevations in corticosterone may have significant actions because concentrations of CBG are low in neonates.[39]

Neonatal Stressors and Brain Development

The timing of the application of stress procedures in neonates is an important factor in the extent and direction of outcomes of stress exposures. Although there are significant differences in neonatal stress procedures, they typically involve repeated separations of dam from pups from periods of 15 min to several hours across a number of days, and the extent to which these procedures result in heightened GCs in pups varies. Shorter periods of maternal separation result in efficient HPA responses to stressors and reduced emotionality in offspring as adults, whereas longer periods result in prolonged HPA activation to stressors and greater behavioral dysfunction. As adults, pups of dams provided with corticosterone in drinking water (which produced moderate elevations of corticosterone in pups) have reduced HPA responses to stressors, reduced anxiety-like behavior, and improved performance on hippocampal-dependent learning and memory tasks.[11] Studies in mice found that higher doses of corticosterone in the drinking water produce the opposite pattern of effects of lower doses on pup development, in keeping with other evidence of inverted U-shaped effects of GCs.

Further, the effects of stressors experienced by the dam on neonates may be secondary to a rise in her GCs. Stressors alter maternal behavior, and maternal behavior in the neonatal period induces long-lasting epigenetic alterations that shape the function of the HPA and hypothalamic–pituitary–gonadal axes of adult offspring.[14] Compared with rats that received high neonatal licking and grooming (LG) by dams, pups that received low LG had prolonged HPA activation in response stressors, greater anxiety-like behavior, and reduced learning and memory performance. The behavioral changes in low LG compared with high LG pups in adulthood are accompanied by changes in many parameters of the nucleus accumbens (e.g., reductions in dopamine receptors in females), hippocampus (e.g., reduced GR and MR, reductions in markers of synaptic plasticity in males), amygdala (e.g., changes in GABA receptor subunits in males), and medial preoptic area (e.g., reduced estrogen receptor and oxytocin receptor expression in females).[14]

The importance of LG for development is demonstrated by pups reared artificially without dams for which many of the resulting deficits can be prevented by stroking with a paintbrush, simulating the tactile stimulation typically received from the dam.[23] Within-litter variation in the maternal attention a pup receives is thus a source of individual differences. Furthermore, males receive more LG than do females, which contributes to sexual differentiation of the nervous system and behavior. Adding corticosterone to dams' drinking water increases their LG of pups.[11] It is thus difficult to separate effects of a rise in GCs from changes to the dam's behavior on the brain development of pups.

ADOLESCENCE

There is no agreed upon definition of the timespan of adolescence in either rodents or humans; a liberal definition for rats has adolescence spanning from weaning until ~PD59. Adolescence involves prepubertal and postpubertal stages, with the onset of puberty (indexed by physical markers) occurring earlier in females (vaginal opening, ~PD35) than in males (balanopreputial separation, ~PD42). Activational effects of gonadal hormones, which rise in circulation with pubertal maturation, lead to greater sex differences in brain function and behavior in adolescence. Sex differences in HPA function also emerge, and the HPA axis becomes sensitive to regulation by gonadal steroids (e.g., estradiol tends to increase and testosterone tends to decrease HPA responses to stressors). Further, adolescence is a second period of sexual differentiation than that which occurs perinatally, whereby gonadal hormones shape developing neural circuitry in a relatively permanent manner.[22] There is much brain development in adolescence, however, that is independent of gonadal maturation.

Although neurogenesis has decreased by birth, hippocampal neurogenesis remains about five times higher in adolescence than in adulthood. There is cell proliferation in other structures in adolescence, including the amygdala, locus coeruleus, nucleus accumbens, and prefrontal cortex. Some of the most significant changes are found in neurotransmitter systems, most of which exhibit a peak in receptor expression in mid-adolescence and then decline to adult levels. There is extensive synaptic pruning and reorganization in adolescence. Of note for considering stressor effects, brain regions rich in GR expression are continuing to develop in adolescence.[36]

Behavioral measures highlight that the adolescent brain is qualitatively different from the adult brain, in part because of ongoing maturation, but also because the brain is tuned to meet requirements that are specific to adolescence. For example, dispersal from the nest requires the animal to have a heightened interest in novelty-seeking, and the social restructuring involved in dispersal means the adolescent brain should be geared for social learning. Indeed, increased risk-taking and novelty-seeking are behavioral hallmarks of adolescence, a time when social play is highest and social interactions carry greater reward value. There are differences in cognitive performance in adolescence. For example, adolescents have enhanced contextual fear conditioning (hippocampal-reliant) and impaired working memory (prefrontal cortical-reliant) compared with adults. Further, the same level of performance in adolescents and adults sometimes involves different neural mechanisms in each.

Hypothalamic–Pituitary–Adrenal Axis Function in Adolescence

Most features of the HPA axis are adultlike in adolescence, including morphology of the PVN, and expression of MR mRNA and GR mRNA and protein in the PVN and hippocampus, basal ACTH, corticosterone, and CBG concentrations. Nevertheless, adolescents often show greater or more prolonged release of corticosterone in response to a variety of stressors than do adults, which may involve differences in the neural circuits projecting to the PVN (e.g., greater CRF receptor binding in adolescents than adults in amygdala).[16] Several studies report a prolonged expression of immediate early genes in several brain regions in both pre- and postpubertal adolescents compared with adults.[21]

Consequences of Adolescent Stressors

Malleability in adolescence is not as great as in earlier periods. Nevertheless, adolescence is an opportunity for remodeling and recovery from earlier stressor exposures.

This same plasticity, however, renders the trajectory of adolescent brain development vulnerable to environmental stressors. The stress procedures used in investigations of adolescence are numerous (e.g., predator stress exposures, social defeat, social isolation housing, social instability, restraint) and vary as to the ages administered, which lead to variability in the effects observed in addition to sex-specific effects. Nevertheless, some general conclusions can be drawn. Changes in baseline and stress-induced HPA function are reported after chronic variable stress procedures experienced in early adolescence. The potential for permanent alterations in HPA function may be limited to a timeframe before PD30; across most studies, any change in basal or acute stress-induced HPA function from stressors experience in later adolescence is no longer present in adulthood. Differences in HPA function, however, might reemerge after repeated stress exposures in adulthood.

Most studies of chronic stress in adolescence report changes in emotional behavior (typically heightened anxiety in adulthood).[25] Similar to findings for stress exposures in neonates, however, mild chronic stress exposures in adolescence may render the animal more resilient, with studies reporting less anxiety- and depressive-like behavior as adults after such exposures compared with nonadolescent-stressed controls, and some evidence that mild stress exposures in adolescence may buffer against chronic stress exposures in adulthood.[9] Effects on motivational systems as measured by behavioral responses to drugs of abuse are mixed, although adolescent-stressed rats typically show greater locomotor activity when given psychostimulants as adults compared with control rats. Further, there is much evidence for adolescent stress-induced changes in the mesocorticolimbic dopamine system and dopamine receptor expression that could underlie some of the effects on motivational behavior.[8]

The main conclusion from studies of cognition is that hippocampal-dependent tasks are particularly vulnerable to adolescent stress exposures, and the effects tend to be greater when tested in adulthood than when tested in adolescence.[19] This delay suggests that stressors in adolescence perturb ongoing hippocampal development, such that functional effects are only observed with time; differences are found in hippocampal neuroanatomy and neurogenesis in adulthood after adolescent stress. Although these changes may underlie differences in hippocampal-dependent learning and memory tasks, they may also be relevant for the effects of adolescent stress on emotionality, given the role of the hippocampus in anxiety and depressive behavior that is distinct from its function in learning and memory.

There are fewer studies of prefrontal cortical function; the available studies suggest that for some prefrontal cortical measures, effects of adolescent stress exposures were attenuated after several weeks of stress-free recovery.[29] The better recovery of function in the prefrontal cortex may be because its development is protracted and does not stabilize until the fourth month. Thus, the critical period for stress exposures on prefrontal cortical function may involve a different window than typically used in studies of adolescents. Studies in humans, however, clearly implicate a role of stress exposures on the development of the prefrontal cortex.[10]

CONCLUSIONS AND NEW DIRECTIONS

Through the actions of GC hormones, particularly during times of rapid development and biological transitions (e.g., fetus to neonate, adolescent, and pubertal), exposure to environmental stressors alters the trajectory of brain development, serving to make the individual more or less resilient to stressors in the future, and producing long-lasting effects on emotional, cognitive, and social behaviors. Recent research efforts are focused on uncovering epigenetic mechanisms. Indeed, increased DNA methylation of a promoter of the GR gene is proving to be a highly consistent long-lasting effect of early life stressors across studies in both animal models and humans.[37] There is increasing evidence that the effects of chronic stressors and exposures to GCs lead to intergenerational effects. In rodents, in which the father provides no paternal care, father's stress history influences the development of offspring through epigenetic modifications in the sperm.[3] Epigenetic mechanisms that underlie the sensitivity of the adolescent period are also being investigated.[28] In the adult, stress-induced rises in GCs drive epigenetic processes and changes in dendritic arborization that shape learning and memory, which are functions governed by the brain structures rich in GR.[33] In aging, the animal may again have a heightened sensitivity to stressors; for example, elevations in corticosterone led to greater loss of prefrontal cortical dendritic spines and impaired working memory in aged rats than in young adult rats.[1]

Investigations of individual differences are elucidating the factors underlying variation in the susceptibility to chronic stressors. For example, studies of selective breeding for high or low anxiety (or high or low responses to novelty) have shown that the effects of maternal stress[24] and of adolescent stress[32] are moderated by genotype. There is limited understanding of one of the largest individual difference factors, that of sex, because most studies have been in males. In studies that included females, effects of stressors are sex-specific.

A remaining question is why is brain development so malleable by stressors? A leading hypothesis is that GCs allow for environmental tuning of development, initiating adaptive processes that prepare for the extent to which the organism will face an adverse environment. This proposal has led to evidence for a "mismatch" hypothesis, whereby an animal will appear more or less resilient when facing stressors depending on whether stressful experiences early in life prepared it for these challenges.[35] Thus, whether or not the influences of chronic stressors on brain development are adaptive or maladaptive depends not only on stage of ontogeny but also on the quality of the environment in which the individual lives.

Acknowledgments

The research program is supported by a Natural Sciences and Engineering Research Council (NSERC) Discovery Grant to CMM and an NSERC Graduate Scholarship to TEH.

References

1. Anderson RM, Birnie AK, Koblesky NK, Romig-Martin SA, Radley JJ. Adrenocortical status predicts the degree of age-related deficits in prefrontal structural plasticity and working memory. *J Neurosci.* 2014;34:8387–8397.
2. Babenko O, Kovalchuk I, Metz GA. Stress-induced perinatal and transgenerational epigenetic programming of brain development and mental health. *Neurosci Biobehav Rev.* 2015;48:70–91.
3. Bale T. Lifetime stress experience: transgenerational epigenetics and germ cell programming. *Dialogues Clin Neurosci.* 2014;16:297–305.
4. Bock J, Wainstock T, Braun K, Segal M. Stress in utero: prenatal programming of brain plasticity and cognition. *Biol Psychiatry.* 2015;78:315–326.
5. Bose R, Spulber S, Kilian P, et al. Tet3 mediates stable glucocorticoid-induced alterations in DNA methylation and Dnmt3a/Dkk1 expression in neural progenitors. *Cell Death Dis.* 2015;6:e1793.
6. Brunton PJ, Russell JA, Douglas AJ. Adaptive responses of the maternal hypothalamic–pituitary–adrenal axis during pregnancy and lactation. *J Neuroendocrinol.* 2008;20:764–776.
7. Brunton PJ, Russell JA, Hirst JJ. Allopregnanolone in the brain: protecting pregnancy and birth outcomes. *Prog Neurobiol.* 2014;113:106–136.
8. Burke AR, Miczek KA. Stress in adolescence and drugs of abuse in rodent models: role of dopamine, CRF, and HPA axis. *Psychopharmacology.* 2014;231:1557–1580.
9. Buwalda B, Stubbendorff C, Zickert N, Koolhaas JM. Adolescent social stress does not necessarily lead to a compromised adaptive capacity during adulthood: a study on the consequences of social stress in rats. *Neuroscience.* 2013;249:258–270.
10. Casey BJ. Beyond simple models of self-control to circuit-based accounts of adolescent behavior. *Annu Rev Psychol.* 2015;66:295–319.
11. Catalani A, Alemà GS, Cinque C, Zuena AR, Casolini P. Maternal corticosterone effects on hypothalamus–pituitary–adrenal axis regulation and behavior of the offspring in rodents. *Neurosci Biobehav Rev.* 2011;35:1502–1517.
12. Clancy B, Finlay BL, Darlington RB, Anand KJS. Extrapolating brain development from experimental species to humans. *Neurotoxicology.* 2007;28:931–937.
13. Cottrell EC, Holmes MC, Livingstone DE, Kenyon CJ, Seckl JR. Reconciling the nutritional and glucocorticoid hypotheses of fetal programming. *FASEB J.* 2012;26:1866–1874.

14. Curley JP, Champagne FA. Influence of maternal care on the developing brain: mechanisms, temporal dynamics and sensitive periods. *Front Neuroendocrinol.* 2016;40:52–66.

15. Davis EP, Pfaff DW. Sexually dimorphic responses to early adversity: implications for affective problems and autism spectrum disorder. *Psychoneuroendocrinology.* 2014;49:11–25.

16. Eiland L, Romeo RD. Stress and the developing adolescent brain. *Neuroscience.* 2013;249:162–171.

17. Fujioka T, Fujioka A, Endoh H, Sakata Y, Furukawa S, Nakamura S. Materno-fetal coordination of stress-induced Fos expression in the hypothalamic paraventricular nucleus during pregnancy. *Neuroscience.* 2003;118:409–415.

18. Giesbrecht GF, Campbell T, Letourneau N, Team AS. Sexually dimorphic adaptations in basal maternal stress physiology during pregnancy and implications for fetal development. *Psychoneuroendocrinology.* 2015;56:168–178.

19. Green MR, McCormick CM. Effects of stressors in adolescence on learning and memory in rodent models. *Hormones Behav.* 2013;64:364–379.

20. Groeneweg FL, Karst H, de Kloet ER, Joëls M. Rapid non-genomic effects of corticosteroids and their role in the central stress response. *J Endocrinol.* 2011;209:153–167.

21. Hodges TE, McCormick CM. Adolescent and adult rats habituate to repeated isolation, but only adolescents sensitize to partner unfamiliarity. *Hormones Behav.* 2015;69:16–30.

22. Juraska JM, Sisk CL, DonCarlos LL. Sexual differentiation of the adolescent rodent brain: hormonal influences and developmental mechanisms. *Hormones Behav.* 2013;64:203–210.

23. Lomanowska AM, Melo AI. Deconstructing the function of maternal stimulation in offspring development: Insights from the artificial rearing model in rats. *Hormones Behav.* 2016;77:224–236.

24. Lucassen PJ, Bosch OJ, Jousma E, et al. Prenatal stress reduces postnatal neurogenesis in rats selectively bred for high, but not low, anxiety: possible key role of placental 11β-hydroxysteroid dehydrogenase type 2. *Eur J Neurosci.* 2009;29:97–103.

25. McCormick CM, Green MR. From the stressed adolescent to the anxious and depressed adult: investigations in rodent models. *Neuroscience.* 2013;249:242–257.

26. Meyer J. Early adrenalectomy stimulates subsequent growth and development of the rat brain. *Exp Neurol.* 1983;82:432–446.

27. Moisiadis VG, Matthews SG. Glucocorticoids and fetal programming part 1: outcomes. *Nat Rev Endocrinol.* 2014;10:391–402.

28. Morrison KE, Rodgers AB, Morgan CP, Bale TL. Epigenetic mechanisms in pubertal brain maturation. *Neuroscience.* 2014;264:17–27.

29. Negrón-Oyarzo I, Dagnino-Subiabre A, Carvajal PM. Synaptic impairment in layer 1 of the prefrontal cortex induced by repeated stress during adolescence is reversed in adulthood. *Front Cell Neurosci.* 2015;9:442.

30. Nugent BM, Bale TL. The omniscient placenta: Metabolic and epigenetic regulation of fetal programming. *Front Neuroendocrinol.* 2015;39:28–37.

31. Deleted in review.

32. Rana S, Nam H, Glover ME, et al. Protective effects of chronic mild stress during adolescence in the low-novelty responder rat. *Stress.* 2016;19(1):133–138.

33. Reul JM. Making memories of stressful events: a journey along epigenetic, gene transcription, and signaling pathways. *Front Psychiatry.* 2014;5:5.

34. Rice D, Barone Jr S. Critical periods of vulnerability for the developing nervous system: evidence from humans and animal models. *Environ Health Perspect.* 2000;108(suppl 3):511–533.

35. Schmidt MV. Animal models for depression and the mismatch hypothesis of disease. *Psychoneuroendocrinology.* 2011;36:330–338.

36. Spear LP. Adolescent neurodevelopment. *J Adolesc Health.* 2013;52:S7–S13.

37. Turecki G, Meaney MJ. Effects of the social environment and stress on glucocorticoid receptor gene methylation: a systematic review. *Biol Psychiatry.* 2016;79:87–96.

38. Van den Hove DLA, Kenis G, Brass A, et al. Vulnerability versus resilience to prenatal stress in male and female rats; implications from gene expression profiles in the hippocampus and frontal cortex. *Eur Neuropsychopharmacol.* 2013;23:1226–1246.

39. Walker CD, McCormick CM. Development of the stress axis: maternal and environmental influences. In: Arnold A, Pfaff D, Etgen A, Farbach S, Rubin R, eds. *Hormones, Brain, and Behavior.* 2nd ed. New York: Elsevier; 2009:1931–1973.

40. Weinstock M. Sex-dependent changes induced by prenatal stress in cortical and hippocampal morphology and behaviour in rats: an update. *Stress.* 2011;14:604–613.

20

Aging and Adrenocortical Factors

J.C. Pruessner

McGill University, Montreal, QC, Canada

Abstract

Some of the most interesting factors associated with aging is the significant interindividual variability in mental and physical changes. Identifying factors that can explain this heterogeneity is thus of great interest to many research groups around the world. One factor that is often investigated for its role in the aging process is stress, and the effect it has on the body. The hypothalamic-pituitary-adrenal (HPA) axis is the major neuroendocrine system that regulates the body's response to stress, by changing its activity, resulting in a release of a cascade of hormones, with cortisol being the final product in humans. A number of theories propose that high amounts of stress lead to a change of regulation of this system over time, which can then become a risk factor for psychological and physical disease in older age. These theories include the general adaptation syndrome, the allostatic load model, and the glucocorticoid cascade hypothesis. They argue that with high amounts of stress over time individuals become more prone to disease, especially with advanced age. There is some evidence for mechanisms relating to the HPA axis and stress-related disease in aging in accordance with these models; however, findings from especially more recent studies question the strength of these effects. It appears that other factors, especially events early in life can have a lifelong, programming effect on the regulation of the HPA axis, sometimes stronger than age-related changes. Newer theories such as life history theory consider these effects, and can complement previous models to better explain the available data. Currently, studies are underway to determine how well these theories can be combined to explain age-related disease.

INTRODUCTION

No one wants to die. Even people who want to go to heaven don't want to die to get there. And yet, death is the destination we all share. No one has ever escaped it. And that is as it should be, since death is very likely the single best invention of life. It is life's change agent, it clears out the old to make way for the new. *Steve Jobs, in his commencement speech to graduates of Stanford University, in 2005.*

Regardless of whether the individual believes in concepts like evolution or intelligent design, the fact that we all are going to die is undisputed. And it likely serves a purpose so important that it is deeply hardwired into our DNA with indeed no exceptions. The process that leads to death is called aging, and it begins at conception (how *old* is the fetus?). From there on, it continues gradually until we reach our personal, individual expiry date. While the eventual end of life is not under discussion, the paths the individual can take to get there show an astonishing amount of variability.

KEY POINTS

- Aging is not associated with a decline in mental and physical abilities per se, but shows great variation across subjects.

- Stress exposure and stress coping abilities have been discussed as factors contributing to the significant interindividual variation.

- The current chapter reviews the theories that have emerged to explain the role of stress in mediating age-related decline.

- These include the general adaptation syndrome, the allostatic load model, the glucocorticoid cascade hypothesis, and life history theory.

- Little or no scientific evidence is available to support the general adaption syndrome and the glucocorticoid cascade hypothesis.

- In contrast, both the allostatic load model and life history theory are being supported by numerous epidemiological studies and basic research in animals and humans.

- While the allostatic load model concentrates on changes in the organism as a result of cumulative stress exposure over the life span that facilitate disease, life history theory highlights developmental effects that might arise from specific early life environments that contribute to specific stress reactivity patterns.

- These models are thus complementary and together might be better able to explain existing patterns and findings.

If you read the beautiful book "Still Alice" by Lisa Genova,[23] you followed the story of a female Harvard Professor who possesses a mutation in a certain gene sequence referred to as PSEN 1. As a result, it encodes the protein Presenilin 1 in such a way that amyloid-beta is allowed to accumulate in certain parts of the central nervous system, rather than being cleared away. The consequence is that Alice, who lives a successful life as a Harvard psychology professor, experiences a sudden onset of severe memory loss at the age of 50 which signals the onset of rapidly progressing dementia. Loss of cognitive control and executive function, increased impulsivity and irritability follow, before she becomes completely disabled within the next 2 years. This is when

the book ends, but we know that subjects diagnosed with the hereditary form of Alzheimer's dementia typically die within 5 years from the initial diagnosis.

But even in the absence of dramatic genetic predispositions, life expectancy and health in old age vary drastically among individuals. On the one hand, you find individuals like Joy Johnson who at the age of 86 have enough muscle strength, bone substance, and cardiovascular fitness to participate in the 2013 New York Marathon. On the other hand, more than 35% of women, and more than 30% of men in the age range of 65–74 are obese, more than 25% of women of that age range suffer from osteoporosis, and more than half the elderly population in that age range show signs of metabolic syndrome with hypertension and poor cardiovascular health. More than 20% of men and women over 75 are so brittle that they can't walk more than 1 km.[1]

If one could choose, I am certain the vast majority of us would vote to run the marathon at the age of 86. But of course, this is not so much a choice rather than a myriad of factors which all come together to contribute to the way our health and fitness evolve as we age. The burden which is associated with declining cardiovascular and physical fitness is also creating considerable costs for the health care system making this a problem not only for the individual but also for society. With the low birth rate and the general population aging considerably especially in the western developed countries, it becomes a pressing economical burden to determine factors that contribute to the premature decline of health and fitness in old age. The hope of health researchers is that these factors—once identified—can be changed in such a way that we can all age successfully, with high quality of life and all of our capacities intact, for much of our natural life expectancy.

One of the factors that has received significant attention as a potential moderator of successful aging in the health field is stress. The guiding and intuitive idea is that stress is a burden for health and that with cumulative exposure to vast amounts of stress over the course of the lifetime the individual is more prone to develop age-related disease and negative health outcomes as compared to an individual who has led a relatively stress-free life. Several theories have been developed over the years, and numerous studies have been conducted to determine if and how stress contributes to impaired health and onset of disease in old age. These theories all look at the physiological and endocrinological consequences of stress—the changes that our body undergoes as a consequence of stress exposure, and here one system has especially been in the focus of researchers—the major endocrine stress response system, that secretes a cascade of hormones which change metabolism and energy availability across the organism.

THE HYPOTHALAMIC–PITUITARY–ADRENAL AXIS

The hypothalamic–pituitary–adrenal (HPA) axis is this neuroendocrine system that consists of a number of distinct structures both in and outside of the central nervous system, which work together to release a series of hormones in response to varying energy demands from the organism. These energy demands are either the product of varying internal metabolic states, or occur as a response to perceived external challenges, both referred to as stressors. The distinction is important as it suggests that an inflammation that occurs in part of the body can be similarly "stressful" as an important deadline at work.

If an internal or external challenge is processed and perceived as stressful, first corticotropin-releasing hormone (CRH) is released from a specific subnucleus of the hypothalamus, called the paraventricular nucleus. CRH is using specific capillary vessels to connect to the anterior pituitary, where it binds to target receptors and stimulates production and release of adrenocorticotropic hormone (ACTH). From the pituitary, ACTH is released into blood circulation and travels freely through the body to eventually reach receptors in the adrenal cortex, on top of the adrenals. The adrenal cortex then produces and secretes cortisol into blood circulation.[66] It can be assumed that many if not all cells of the human body are targets for these glucocorticoids, and they are reported to have various genomic and nongenomic effects. By far the best-known effects of cortisol have to do with metabolic regulation, changing the activity of the liver (stimulating gluconeogenesis), suppressing the effects of insulin to increase the available amounts of glucose in the body, and showing various effects on the cardiovascular system (typically stimulating), the immune system (typically suppressive), which can all be summarized as aiming to increase the available amounts of energy during times of high demand.[70]

Further, cortisol like all steroid hormones is liposoluble, allowing it to cross the blood–brain barrier and find targets inside the central nervous system. There, it binds to mineralocorticoid (type I) receptors, predominantly in the hippocampus, amygdala, and prefrontal cortex, and glucocorticoid receptors (type II), diffusely located across the cerebrum.[13]

These two receptor types also have distinct affinities for cortisol: mineralocorticoid receptors bind cortisol at 10 times lower concentrations than glucocorticoid receptors, which means that during periods of low HPA axis activity, only amygdala, hippocampus, and prefrontal cortex are aware of circulating cortisol levels.[48] When levels are high, however, for example after stress, the majority of structures and systems in the brain are stimulated by glucocorticoids.

Besides their difference in affinity for cortisol, these receptor subtypes also play different roles in regulating activity of the HPA axis. While type I receptors are believed to mildly stimulate further activity of the axis (a positive feedback mechanism), type II receptors are believed to exert a strong negative feedback on further activity of the axis, i.e., when activated they suppress subsequent activity.[49] Thus, after the initial reaction of the HPA axis to a stressor, there is a period of attenuated activity until cortisol levels have returned close to baseline, similar to the refractory period observed in action potential activity of neurons in the central nervous system.

Various periodic patterns of activity combine to determine the baseline levels of cortisol. The ultradian rhythm shows the highest frequency, with secretory pulses that occur every 60–90 min. These lead to rhythmic increases and decreases of cortisol levels at all times. Superimposed on that is a strong circadian rhythm, with the ultradian pulses leading to the greatest amounts of cortisol being released in the early morning hours; this is followed by declining pulses throughout the day and lowest levels of cortisol (the nadir) in the first few hours of night rest, typically around midnight.[26] As a consequence of the strong circadian variation, the type I and type II receptors for cortisol are differentially occupied in the morning versus the afternoon and evening. During peak baseline activity of the HPA axis, almost all type I receptors are occupied, and more than two-thirds of type II. In contrast, in the afternoon and evening, still the majority of type I receptors are occupied, but only around 10% of type II.[72]

In addition to the circadian rhythm, there are reports in the literature that describe a circavigintan (approximately every 20 days) and circatrigintan (every 30 days) rhythm of HPA axis activity, with smaller cyclic variations of HPA axis activity peaking every 20 or 30 days. The importance of these rhythms, or the mechanisms controlling these pulsatory cycles, is at this point unclear. Finally, as the last of these infradian types of HPA axis rhythms, there are reports of seasonal rhythms of HPA axis activity with overall lower levels of HPA axis activity during the winter months compared to summer. Fig. 20.1 illustrates how these various rhythms combine to create a multicyclic activity of cortisol over time.

Both the circadian and the seasonal rhythm of cortisol are linked to the suprachiasmatic nucleus (SCN) of the hypothalamus, often referred to as the "master clock" of the central nervous system. The SCN is directly stimulated by retinal fibers of the visual system, which signal changes in light and dark.[59] The SCN is in constant cross talk with the PVN, thus enabling light-dependent regulation of the HPA axis.

In addition to these varying rhythms governing baseline activation of HPA axis activity, changes or even *perceived changes* in the environment can cause an additional activation of the HPA axis at almost any time (with the exception of the acute refractory period). One

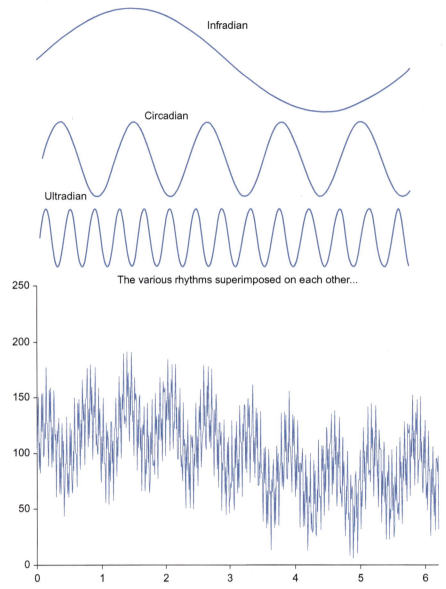

FIGURE 20.1 The various infradian, circadian, and ultradian rhythms interact to create the typical multicyclic pattern of hypothalamic–pituitary–adrenal axis activity.

example for a distinct change in the environment which leads to a strong and persistent activation of HPA axis is the cortisol awakening response (CAR). Following the first 30–45 min after awakening, cortisol levels typically double in concentration to reach the highest baseline levels of the day.[47] A lot of research has gone into investigating possible causes of the CAR, with some evidence pointing to a cognitive mechanism, possibly linked to an orientation response after awakening.[71]

Perhaps the best example for a perceived change in the environment that can cause acute activation of the HPA axis is the presence of a psychosocial stressor that can lead to a doubling or tripling of cortisol levels within 20–30 min. It is one of the most interesting aspects of HPA axis activity, as it causes the strongest changes in

activity of this endocrine system to something as benign as a thought. The fact that anticipatory stressors (e.g., the thought of having to give a talk in front of an audience, like in the "Trier Social Stress Test")[28] can lead to the same robust HPA axis activation and cortisol increase like a marathon run[42] can serve as an example to that effect.

THE HYPOTHALAMIC–PITUITARY–ADRENAL AXIS AND NORMAL AGING

Aging causes significant changes in a variety of physiological, metabolic, and endocrine systems, including the HPA axis. These changes occur with respect to baseline as well as stress-reactive aspects of HPA axis regulation.

A number of questions are especially important to address in the context of aging—first, are these changes signs of a pathological process, or changes that occur normally and that have to do with an attenuation of metabolic and physiological activity which happens naturally with advancing age. Second, if they are related to pathological processes, are they causally related to these processes (and if so, are they cause or consequence?), or a correlate of another factor which is primarily associated with a specific pathology? And finally, can these changes be observed during baseline, stress reactivity, or both? In order to address these questions, it is useful to first quickly review age associations of HPA axis activity at baseline and after activation, and then take a look at HPA axis activity associated with age-related pathology.

Baseline (Unstimulated) Cortisol Levels

Studies have shown inconsistent results over the years as far as baseline changes of cortisol with age are concerned. Some studies suggest that older adults show higher baseline cortisol levels as compared to younger adults,[67] while other studies have shown the opposite, i.e., lower baseline cortisol levels especially in the morning with advancing age, leading to an overall flatter diurnal profile.[58] These findings are contrasted by yet other studies which have shown stable cortisol levels across the life span.[69]

A number of factors can contribute to explain these different findings across studies. For one, variations in study design, sampling methods, and time points can lead to these different results. In addition, differences within the populations under investigation can certainly contribute to different findings as well. Elderly populations often suffer from a number of minor ailments (e.g., hypertension, arthritis, overweight), which if excluded would lead to overly restrictive recruitment criteria and difficulties in finding enough subjects for a particular study. However, each of these factors could by itself affect activity and regulation of the HPA axis and could thus explain interindividual differences in cortisol levels. Sometimes, these differences can be observed within the same study population, if researchers are looking for it. For example, Lupien et al.[32] in a longitudinal approach observed subgroups of healthy older individuals with high levels of cortisol which were either annually increasing, or moderate levels that were annually increasing, or moderate levels that were annually decreasing. These findings point to the importance of follow-up measures to better understand cortisol dynamics and also at the importance of taking individual factors into account to explain variations in hormonal activity.

Although the findings in the available studies point to an inconsistent view of age-related changes of HPA axis activity, the predominant belief continues to be that changes in baseline cortisol levels can be a marker of pathological aging and might be associated with mental and physical impairment.[31] It might, however, be a safer conclusion for us at this point in time to argue that interindividual variations are greater than a general age effect.

Reactive Cortisol

There is also only mixed evidence to suggest that cortisol levels after stimulation are changing with age. First, a number of studies have observed changes both in response to psychological (stress) or pharmacological (drug) challenges.[44] Here, it seems to be particularly the later aspects of the cortisol response dynamics, the return to baseline after stimulation. This could point to a change in type II-related mechanisms of HPA axis regulation. For example, Boscaro and colleagues[7] could show that after pharmacological challenge, older adults were delayed in returning cortisol levels back to baseline. Also pointing to a central mechanism is the fact that the delay in returning to baseline is associated with memory impairments, suggesting that perhaps the hippocampus, a critical structure involved in negative feedback regulation of HPA axis regulation, is affected, since it is also critical for memory. Indirect evidence for this hypothesis comes from a study by Buchanan showing a complete absence of cortisol stress reactivity in subjects with hippocampal lesions.[8] Psychological stressors have also been used to demonstrate a change in responsivity in elderly individuals. Seeman and colleagues, for example, used a driving simulation challenge to test young and older subjects and could demonstrate a greater responsivity in the older participants.[55] The difficulty with psychological challenges like these, however, is to standardize the subjective experience to allow accurate interpretation of variations in endocrine responsivity observed across age groups. It is conceivable that elderly subjects experienced the driving simulation as more stressful to begin with, thus leading to greater amounts of psychological stress. This is a problem that can be avoided by pharmacological challenges although other aspects have to be taken into account (reduced metabolism in elderly individuals might lead to longer drug exposure, for example). The public speaking and mental arithmetic in the "Trier Social Stress Test" is also not without its problems, as the test is typically presented and conducted by younger individuals and as a result, elderly individuals might simply show a different response because of the age gap. In fact, when controlling experimentally for the age of the presenters and the age of the test group, Sindi and colleagues have observed effects of both.[62]

Studies showing effects of age on the cortisol stress response are contrasted by studies showing no such effects. Here, a carefully conducted study by Kudielka employing the Trier Social Stress Test (TSST) in children, young, and old adults while simultaneously measuring ACTH, plasma, and salivary cortisol can be used to exemplarily illustrate this point.[30] The authors of this review could first show that ACTH levels are highest in younger men, followed by younger women, who were showing the second highest levels overall. This was then followed by the older men, with older women being the lowest ACTH group. This suggests a stronger effect of age as compared to gender (young always higher than old, with men within each age group higher than women) and could point to a central effect. This picture changed dramatically, however, when taking the total plasma cortisol levels into account; here, it was the older women who presented with the highest levels, followed by the old men, followed by younger men, with the younger women being the lowest group. Thus, age as a main factor was now reversed, with gender no longer being a main effect, but if anything showing an interaction with age (women in older age higher than men, and in younger age lower than men). Finally, this picture changed again when looking at salivary (free) cortisol levels (this time also including children): here, older men showed by far the highest levels, with no observable differences among older women, younger men or women, or boys and girls. It is difficult to conclude anything specific from this set of finding; what stands out perhaps is that despite having low ACTH levels, older men have the highest free salivary cortisol levels, which would point to an increased sensitivity of ACTH receptors at the level of the adrenal cortex, resulting in larger cortisol output. This would match with the higher levels of total plasma cortisol in the elderly women if we knew their estrogen status; as estrogens are known to stimulate production of cortisol-binding globulins (CBG), higher amounts of estrogens might lead to higher CBG levels and thus the effect of greater ACTH receptor sensitivity would only be visible in the plasma cortisol, and not the free cortisol. Unfortunately, the authors did not disclose the exact number of their elderly female subjects being menopausal, thus one can only speculate about this being the explanation.

Taken together, however, this study provides some signs of a lower ACTH production in the elderly, complemented by signs of higher glucocorticoid production. What has to be taken into account, however, is that these studies did not screen, or contrast, factors associated with healthy ageing with those of age-related disease. It is well possible that overall, the aged population does not present with significant differences in HPA axis activity but that systematic effects emerge once age-related disease or impairment is taken into account.

THEORIES RELATING TO AGE-RELATED CHANGE

To approach the data on age-related disease and adrenocortical activity, it makes the most sense to combine this with the presentation of theories which have been formulated over the years suggesting a role of adrenocortical factors in age-related disease. Here, a number of opposing theories have been formulated to suggest mechanisms by which various forms of lifelong stress can have these permanent effects on the regulation of the HPA axis, including the general adaption syndrome,[56] the glucocorticoid cascade hypothesis,[52] and allostatic load models.[53] Finally, there is the more recent adaptive calibration model,[15] as part of life history theory approaches.[17] All of these theories can be employed to formulate how aging, or in the case of life history theory, development across the life span, can affect HPA axis regulation. In the following, we will review these models and some of the available data to support or refute them.

The General Adaptation Syndrome

In 1936, Hans Selye initiated the era of modern stress research by formulating the "general adaptation syndrome" (GAS)—the notion that chronic stress can make you ill.[56] Selye—himself a pioneer in stress research—developed this model following the work of Walter Cannon, another pioneer whose theories on the autonomous nervous system and emotion are still taught in today's psychology classes. Cannon in 1932 had coined the term homeostasis[9] to describe changes in an organism's external environment that disturb the internal balance of its body (such as extreme heat or cold) and require it to adapt; in this case, by a change in metabolism to keep the body's temperature stable. Selye was inspired by this view in formulating his theory on stress and argued that the body needs to adapt similarly when exposed to threatening and stressful stimuli. In this view, the stress response is defined as the body's attempt to restore its homeostatic balance while providing the organism with the necessary energy to cope with the threatening stimuli. In Selye's studies on chronic stress he could observe that the organism's stress system—and the HPA axis is the body's predominant stress system—would show first signs of increased load [for example, an enlargement (hyperplasia) of the cortisol producing adrenal cortex] but that with a persistence of the chronic stress the system would eventually break down, and the body would fail. The basic idea is illustrated in Fig. 20.2.

While entirely intuitive, evidence in humans supporting the validity of the general adaptation syndrome has been limited. Importantly, the GAS implies an invariable

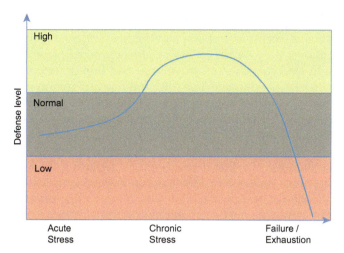

FIGURE 20.2 The general adaptation syndrome as originally formulated by Hans Selye. The individual can temporarily increase its stress resistance during periods of high demand (stress) through internal or external causes. The initial increase in stress resistance (the period of adaptation to the increased demand) is referred to as the alarm stage, where stress resistance is increasing. This is followed by the period of the resistance stage, where the stress resistance is maximized and remains on high levels. If stressors persist for too long, the body cannot keep up the increased stress resistance levels, at which point defenses break down and illness can occur.

order of events—first, the defenses are increased, then they remain steady on higher level, before they eventually decrease and fail, which coincides with the onset of (stress-related) illness. When linked to stress-related disease with aging, it would imply that one should always see an increase in markers of stress defense before observing a decrease. In cases where one only observes a decrease and the study type is cross-sectional, this would only make sense under the assumption that those subjects who are currently decreasing are in the last phase of the adaptation, i.e., at the stage shortly before the systems fail, and the individual becomes ill. This would work in the example mentioned above where elderly subjects had been investigated, and subgroups with stable, increasing, or decreasing cortisol levels were observed.[32] Here, one would have to assume that those subjects with stable cortisol levels were not currently experiencing an adaptation, those with increasing/high levels were currently experiencing an adaptation because of chronic stress, and those with decreasing/high levels would be in the last stage of the syndrome, shortly before systems fail and the subjects fall ill. On the other hand, it would be difficult to argue that the general adaptation syndrome could still apply in cases where a blunting of the stress response is occurring very early in development, or—in case of a longitudinal study—following directly from what appears to be a normal stress response. Examples for the former pattern are frequently observed,[45] while longitudinal studies are

notoriously more difficult to perform, and thus to find in the literature. In addition, while numerous studies have reported a blunting of the stress response, the blunting per se does not immediately seem to lead to higher incidence of mental or physical disease, thus putting into question one of the key assumptions of this model.

Taken together, at this point in time a general validity of this model is unlikely. It is conceivable that a pattern of breaking down of stress system defenses can be observed in the presence of extreme stressors (as Selye himself exposed his animals to extreme stressors to observe the stress effects, both physically and psychologically), or only in conjunction with other risk factors present in the individual.

The Glucocorticoid Cascade Hypothesis

Another model that has generated quite a bit of interest in the context of aging and HPA axis regulation is the "glucocorticoid cascade hypothesis."[52] This theory basically claimed that periods of excessive and chronic stress could lead to damage in those areas of the brain that are rich in receptors for cortisol—prominently among them, the hippocampus. Because the hippocampus is further involved in the inhibition of subsequent activity of the HPA axis (through glucocorticoid receptors on its surface it becomes aware that cortisol has been released and signals to the hypothalamus to shut down further HPA axis activity—the *negative feedback* process mentioned earlier), a damaged hippocampus would be impaired in its ability to relay the negative feedback signal, and thus the HPA axis would not shut down as efficiently, leading to subsequently even higher circulating levels of cortisol, with even greater damage to the hippocampus. This model would explain why particularly lower hippocampal volume is associated with higher cortisol stress responses (the hippocampus is less capable of shutting down HPA axis activity), and it could further explain why chronic stress could lead to memory impairment, which is associated with increased age (the smaller hippocampus would be not as good as a bigger hippocampus in executing its memory-related functions). This hypothesis has been extremely influential, in part because its central message was so intuitive: stress is bad for you! If researchers can now show that stress leads to damage of essential structures in the brain, then there is ample justification to not overdo one's work pressures to avoid stress-related brain damage!

A number of questions immediately arose when researchers first started to test this hypothesis: since it is excessive stress that needs to be present to lead to the damaging effects, what exactly defines "excessive"—is a one-time traumatic stressor sufficient, or is a period of chronic stress required? Can many small stressors over the course of a lifetime (an accumulation of daily hassles, for example) do damage, or are only individual big stressors to blame? While the early studies by Sapolsky et al.[52] in rodents could produce evidence for hippocampal damage after periods of chronic stress, studies in humans that have been performed since have produced rather inconsistent results.

This is not a consequence of too few studies; the question of glucocorticoid toxicity in the human brain has by now been addressed by a considerable number of studies, and it would be beyond the scope of the current contribution to try and even list all the studies that have tried to address this question. However, there is one piece of evidence that is particular compelling in suggesting that a strict model of toxicity might not work as suggested, at least in humans. This evidence comes from patients diagnosed with Cushing's syndrome, a condition with excessive glucocorticoid production often caused by a tumor in the pituitary, who also present with cognitive disturbances at the time of diagnosis. Moreover, upon assessment of their brain function and integrity, these patients show signs of reduced hippocampal volumes.[65] This looks like fine evidence to support the Sapolsky model! However, once the underlying reason for their excessive glucocorticoid production is found and they are successfully treated (for example, by surgery to remove the tumor from their brain), Cushing patients' cognitive function returns to a normal level. Even more striking, when they undergo another brain scan, their hippocampal volume is no longer reduced in comparison to a control group. When following patients diagnosed with Cushing's syndrome into old age, their incidence of neurodegeneration and dementia in old age is not higher than in the general population, suggesting that actually, no permanent damage has been done. And all of this occurs despite the fact that the disease has led to excessive glucocorticoid production which sometimes lasted for years. Since their HPA axis is—because of the tumor—completely out of control, with the highest possible glucocorticoid exposure going on sometimes for years, it casts more than a shadow of a doubt on the idea that glucocorticoids cause permanent damage. If anyone should suffer from glucocorticoid neurotoxicity, this population should certainly be affected! In the light of these findings, some authors have suggested that glucocorticoid exposure represents an insult from which people can recover.[39] These authors discussed the observed lower hippocampal volumes during the presence of high cortisol levels instead as a consequence of transient intracellular changes in water and electrolyte content.[65]

But if it is not glucocorticoid toxicity, what could cause the association between hippocampal volumes and cortisol regulation that we[46] and others[33] have observed, that looked like cortisol might damage the brain? One argument could simply be the idea that cortisol is an insult to the brain (or rather, those areas of the brain rich in glucocorticoid receptors) but one that the neurons can recover from over time. While excessive glucocorticoid presence might lead to reduction in volume and loss of function that looks like damage, in reality it is just an impairment that the brain can recover from. The notion of transient changes in water and electrolyte content that Swaab and colleagues made would fit well with that explanation. Another possibility is that the hippocampus really did take damage, but that it recovered by replacing the lost cells. Since Eriksson in 1998 has contributed compelling evidence for the notion that also in the adult human brain, new cells can be generated in significant numbers, one can speculate that even lost tissue might be replaced over time.[21]

Allostatic Load Models

The allostatic load models are also at least in part inspired by the "homeostasis" concept introduced by Cannon.[9] Following from the work of Selye, however, these models look at the long-term cost of the individual to maintain homeostasis. The basic idea here is that the individual organism depends on surplus energy (typically from endogenous stores—fat, glycogen, and proteins that have been secured from exogenous sources) for the emergency responses to take place in times of increased demand. During times of chronic increased demand, an organism would then be in danger of using up these surplus energy stores, and must be concerned about not depleting them in order to survive. This ability to maintain these emergency responses especially during times of chronic demand has been called "allostasis,"[63] which implies that an additional adaptive process has to take place to avoid depletion.[36] Allostasis can be understood as "stability through change" thus denoting this adaptive process. Similar to Selye who followed in the footsteps of Cannon with homeostasis of physiological systems, allostasis was first described by Sterling and Eyer in 1988 referring to the cardiovascular system, which adjusts during resting and active states. From there, allostasis was then generalized to other systems, especially the HPA axis and cortisol.[36]

Like homeostasis, allostasis is a necessary process to ensure the survival of the organism. Unlike homeostasis, it changes physiological parameters that allow homeostasis to continue. Thus, while homeostasis keeps "set points" and describes the processes to maintain these (such as body temperature), allostasis describes the modification of other parameters to keep these set points. For example, the increase in food intake in response to an increased metabolism when exposed to a harsh environment would be the allostatic change that would allow the body temperature to remain stable when the organism is out in the cold. Thus, while homeostasis involves keeping functions and systems stable, allostasis describes the changes in the function of these systems in response to internal or external challenges that allows for the required stability.

This can be further exemplified by looking at various examples in the context of various systems across the body. For example, to maintain the necessary amounts of oxygen and nutrient supply to the muscles during exercise, catecholamines have to be released through the sympathetic nervous system to adjust heart rate and blood pressure. The increase in catecholamines, and the subsequent increase in heart rate and blood pressure, would be an allostatic process. Within the endocrine system, the increase in glucocorticoids, and the subsequent increase in food intake after exposure to a psychological or physical stressor, would be an example for an allostatic process to keep energy levels stable, which itself is a homeostatic process. In a sense, allostasis and homeostasis are part of the same process, but instead of focusing on the aspects which remain stable, the focus is on the aspects that are changing for the system to remain stable.

Once the focus shifts from the system that overall remains stable to the parameters which are changing, it becomes easy to realize that there might be a price to pay for the stability; the "wearing out" of those parameters that undergo constant change to ensure the stability of the overall system. If allostasis refers to the necessary "constant change" of specific parameters to ensure overall stability, then the question arises what long-term costs might be associated with this constant adaption. This is the most critical component of this theory, which has also provided the name for this line of models: the concept of allostatic load. This has been a source of intense research for almost two decades now, and one that carries particular importance for the aging individual, as it can be rightfully argued that this wearing out process is taking place over time, and becomes more prevalent with advancing age.[37,50] One example that can serve to illustrate the long-term cost for the stability of the system is the so-called metabolic syndrome, a condition that affects an increasing number of people in the developed countries and that shows a strong association with age. It describes a cluster of risk factors for physical disease, including abdominal obesity, elevated blood pressure, elevated glucose levels, and high blood triglyceride levels, among others. It is associated with a sharply reduced life expectancy, increased risk for diabetes, and atherosclerosis.

A case can be made that stress plays a significant role in the development of the metabolic syndrome, through mechanisms described in the allostatic load model. The key to understand the role of stress in the development of the metabolic syndrome with increasing age is to understand the interaction between the stress systems and how it might change with age. Recently, our group has begun to systematically examine the interaction between both the HPA and SNS stress systems on the psychological, physiological, and endocrine level during an acute stress response. To this end, we looked at the cross-correlation in peak response between SNS and HPA[20] and the ratio of markers of HPA and SNS and their association with behavioral variables.[2] Further, we combined the dexamethasone suppression test with a psychosocial stress test consisting of public speaking and mental arithmetic, the TSST, to study the effects of stress in the absence of the HPA stress system. This study was the first attempt to systematically investigate the effect of manipulating the stress systems, and thus deserves a somewhat more detailed description: here, we exposed 30 healthy young men to psychosocial stress and measured salivary cortisol, salivary alpha-amylase, heart rate, blood pressure, and subjective stress. As the main experimental manipulation, half of the subjects received a standard dose of dexamethasone (DEX; 2mg) the night before testing, resulting in elevated feedback at the level of the pituitary and a central hypocorticoid state since DEX does not cross the blood–brain barrier,[12,14] resulting in increased CRH levels but rendering the pituitary unable to respond to this CRH stimulation after psychosocial stress. This can be compared to a state where the HPA axis is dysfunctional and unable to respond to stress, as discussed for states like burnout and chronic fatigue syndrome. As a result of this manipulation, subjects who had received DEX demonstrated a higher increase in subjective stress post-TSST and a significantly higher heart rate throughout the protocol, when compared to the placebo (PLC) group. This suggested an active compensation between the two systems, where SNS activity may be elevated in the presence of a suppressed HPA axis response.[3] While we were the first group to combine an acute psychosocial stressor with dexamethasone to investigate this cross talk, others have also reported such an association, and a number of hypotheses have been proposed to explain the increased SNS activity following dexamethasone administration. First, the dexamethasone-induced hypocorticoid state in the brain could have caused an elevated heart rate via a CRH surge (due to the lack of negative feedback), by way of a PVN and LC/NE connection.[73] In addition, Chrousos and Gold[10] found that NE potentiates the release of CRH, creating a feed forward mechanism between the systems; thus a central SNS mechanism could have been initiated to augment the HPA axis response, a mechanism also discussed in Ref. 64. Consequently, the increased SNS activity found in a population where cortisol output of the HPA is suppressed may be due to these factors individually and/or in combination.[3] Since the regulation of SNS and HPA overlaps at several points in the brain, notably also the hypothalamus, *it can be speculated that this is an active compensatory mechanism, such that the absence of the HPA response leads to the increase in the SNS activity, compensating for the lack of a stress response in one system with increased activity in the other, to keep the organism in an equilibrium, and allow allostasis.*

Thus, we have via experimental manipulation of the stress systems demonstrated a potential pathway for the development of cardiovascular disease, metabolic syndrome, and diabetes.[11,51,54,57] This points to a potential disease mechanism where either a blunted HPA axis activity or an increased SNS activity might lead to these types of disease.

We next conducted a study to investigate the opposite effect, by using propranolol (PROP) in combination with the TSST. PROP is a nonselective beta-blocker mainly used in the treatment of hypertension, and also to treat social anxiety and tremors.[22,24] It is highly lipophilic, thus achieving high concentrations in the brain. It blocks the action of epinephrine and norepinephrine on β1 and β2 receptors in the central nervous system, i.e., it suppresses the activation of the SNS.[38] While it is routinely used in memory research, there are only a few studies where PROP has been combined with a psychosocial stress task to investigate the effects on the acute stress response,[34] showing higher cortisol levels in response to stress. PROP is rapidly and completely absorbed, with peak plasma levels occurring approximately 60–120 min after ingestion. It has a half-life of 3–4 h.

When administering subjects PROP 1h before the onset of the TSST, thereby blocking an SNS response, we observed signs of a compensatory increase in HPA axis activity in response to stress.[4] As expected, SNS activation was strongly suppressed in the PROP group. Heart rate, salivary alpha-amylase, and systolic blood pressure levels all showed very small or no increases in response to stress, in the PROP group. In contrast, subjective stress and diastolic blood pressure were not different between the two groups. The main result that indicated a significant cross talk between the stress systems was observed for the cortisol levels, which were found to be significantly higher in the experimental group. The finding of higher HPA axis activity after SNS suppression has also been reported by a number of other laboratories,[5,29,35,43,60] *suggesting that the absence of SNS response leads to an increase in HPA axis activity, compensating for the lack of a stress response with an increase in the complementary stress system, to keep the organism in an equilibrium.* These results are quite intriguing, as one could have expected a lower cortisol response in the absence of a physiological stress response; after all, no physiological arousal is signaling to the brain a state of stress, as would be expected from past research on emotion, and emotion processing.[16] The fact that a suppression of the SNS leads to higher cortisol stress responses strongly suggests that at a central level the lack of a physiological response is responded to by a stronger activation of the HPA axis.

This mechanism that results in higher cortisol responses to stress has potential psychopathological effects as well, since central hyperactivity of the HPA (i.e., CRF secretion) is associated with depression and mood disorders. Some of the most prominent theories on depression suggest that it is the central effect of CRH that is associated with depressive symptomatology and mood dysfunction, through the effects of CRH in core limbic structures related to mood and anxiety, e.g., the amygdala and the prefrontal cortex.[6,25,40,41] This is in line with the proposed neuroanatomical mechanisms to explain the compensatory effect between the HPA and the SNS. First, there is the possibility of an increase in adrenaline production due to propranolol's blocking noradrenergic binding to β2-adrenoceptors. Adrenaline could then in turn trigger increased CRH release from the hypothalamus, with subsequent increase in cortisol[68]—this would represent a central mechanism. Second, there is also the possibility of a direct inhibitory effect of SNS activation on the adrenal cortex. In cases where the SNS system is inhibited, a disinhibition of the HPA axis could result.[68] This would present a compensatory mechanism at the level of the adrenal cortex. These two mechanisms could further complement each other, making the HPA/SNS interaction more potent. While the exact routes of action still await further empirical confirmation, the available evidence points to a negative interaction between the SNS and HPA after they have been triggered by central nervous system components, in the presence of a stimulus that has been perceived as stressful. Taken together, these preliminary pharmacological manipulations point toward mechanisms by which the wear and tear on the stress systems can lead to cardiovascular disease and metabolic syndrome with aging. Especially when combined with a sedentary lifestyle and physical inactivity, an increase in food consumption, insulin resistance, abdominal obesity, and hypertension could potentially follow from these effects.

Life History Theory

The final model to cover in the context of aging and adrenocortical factors falls into the broader context of "life history theory,"[18] and it complements the previous theories by arguing that some individuals might be especially prone to age-related changes because of specific environmental cues during critical development periods which might have contributed to a specific "programming" of adrenocortical regulation. Life history theory in this context talks about slow versus fast "life history strategies," which among other aspects are characterized by high or low adrenocortical responsivity. The main idea is that environmental cues during early life indicative of high stress environments trigger a "fast life history strategy." The authors of this line of theories argue that a number of physiological and metabolic consequences arise from this environment which can be summarized to make the subject more stress resistant in the short-term, at the cost of longevity. The short life expectancy that ensues would then also be associated with a greater emphasis on reproductive efforts. In contrast, safe environments are believed to cause "slow life history strategy," where more moderate acute stress resilience comes at the benefit of longevity, or to reuse the terminology associated with the allostatic load models, less wear and tear. Most life history authors, like Simpson,[61] suspect that these reactivity patterns become entrained during critical development periods early in life and then are stable throughout life.

There are by now a substantial number of studies which support the idea that distinct events during critical development periods early in life can shape the developing brain, and in turn, the stress response. Studies looking at prenatal and postnatal adversity like malnutrition, toxic exposure, physical or sexual abuse, parental neglect, etc. have shown that these factors can be additive in affecting the volume of key structures in the brain, including the hippocampus, the anterior cingulate, and the precuneus, all structures involved in personality, emotion, and stress regulation.[19,27] In these studies, we could demonstrate partial support for a neurodevelopmental model by establishing a link between early life adversity (in the form of self-reported early life maternal care, or reports of childhood trauma), and morphological changes in the brain (either on cortical thickness of the neocortex, or the hippocampus). In the case of the 2010 study, we were then able to also show how early life adversity was associated with a changed stress regulation. To identify the directionality of these results, we then performed a mediation analysis to determine whether the correlation between early life adversity and cortisol regulation was mediated by the hippocampal volume of the participants. In other words, if early life adversity has an effect on the morphology of the hippocampus in the developing brain, and the hippocampus in turn determines the stress responsivity of the individual, then this could shed light on the mechanisms by which the variations in early life can have the programming effect on adrenocortical regulation. And indeed, this was exactly what we were able to observe—the correlation between markers of early life adversity on stress regulation in adulthood, which were significant in isolation, lost their effect when testing the mediation through hippocampus volume, suggesting that this is the neurodevelopmental link between the two.[19] The subsequent study showing an effect of childhood trauma on the thickness of specific areas of the neocortex suggests that hippocampal volume is but one target among many to be affected by variations in early life adversity.

SUMMARY AND CONCLUSION

The current chapter has looked at some of the available studies investigating changes of the adrenocortical system with ageing. While there are some studies suggesting a systematic change, the overall evidence is inconsistent, thus there does not appear to be a strong general effect with ageing. Interindividual differences seem to mask any systematic age-related effects. The various models that can be employed to make a case for stress playing a role in age-related disease take this into account—they all would argue that individual factors have to be considered when investigating age-related disease. The first of these models, the general adaption syndrome, provided a framework in which to conceptualize these age-related changes. Evidence for this model is sparse, however; although intuitive, aside from some animal studies and some extreme stressors not much points to a complete breakdown of the stress systems over time. The glucocorticoid cascade hypothesis, the idea that chronic stress leads to an eventual destruction of brain structures involved in the regulation of the adrenocortical system, has also been very popular for a considerable amount of time in the stress field, but at least in humans does not seem to be strongly supported by evidence, either.

The last two models reviewed in this chapter seem to find more and consistent support from studies in animals and humans. First, the allostatic load model—shifting the emphasis from homeostasis to allostasis and looking at the wear and tear of the associated systems—can be recruited to explain the age-related increase in metabolic syndrome, diabetes, hypertension, when looking at the interaction among the stress systems in combination with age. While still not capable of explaining individual age-related disease, it provides a framework that in its core elements is supported by evidence. The growing popularity of this model would further support its validity.

Finally, the neurodevelopmental model shifts the focus on events early in the developmental process—both brain structures and adrenocortical regulation are considered as a consequence of exposure to adverse environmental factors early in life. This is to some extent reminiscent of psychoanalytical theories and models. However, there is also quite a bit of evidence in favor of life history theory, thus it is likely to play an important role as one of the individual factors determining age-related changes.

Taken together, this chapter has reviewed the models which in the past or presence have guided stress researchers in trying to understand age-related changes that might be influenced by stress. From the current line of evidence, a picture emerges where events early in life "shape" to some extent the brain, which in turn has a significant effect on adrenocortical regulation. This might lead to a more short-term stress resilience, which comes at the price of greater long-term wear and tear, and thus increased risk for those physical and mental diseases which are stress- and age-related—metabolic syndrome, diabetes, depression, burnout, to name a few. Future studies will have to continue gathering evidence for these models and identify further factors which are contributing to age-related disease processes in addition.

References

1. Ahmadi SF, Streja E, Zahmatkesh G, et al. Reverse epidemiology of traditional cardiovascular risk factors in the geriatric population. *J Am Med Dir Assoc.* 2015;16:933–939.
2. Ali N, Pruessner JC. The salivary alpha amylase over cortisol ratio as a marker to assess dysregulations of the stress systems. *Physiol Behav.* 2011;106(1):65–72.
3. Andrews J, D'Aguiar C, Pruessner JC. The combined dexamethasone/TSST paradigm – a new method for psychoneuroendocrinology. *PLoS One.* 2012;7:e38994.
4. Andrews J, Pruessner JC. The combined propranolol/TSST paradigm–a new method for psychoneuroendocrinology. *PLoS One.* 2013;8:e57567.
5. Benschop RJ, Jacobs R, Sommer B, et al. Modulation of the immunologic response to acute stress in humans by beta-blockade or benzodiazepines. *FASEB J.* 1996;10:517–524.
6. Binder EB, Nemeroff CB. The CRF system, stress, depression and anxiety-insights from human genetic studies. *Mol Psychiatry.* 2010;15:574–588.
7. Boscaro M, Paoletta A, Scarpa E, et al. Age-related changes in glucocorticoid fast feedback inhibition of adrenocorticotropin in man. *J Clin Endocrinol Metab.* 1998;83:1380–1383.
8. Buchanan TW, Tranel D, Kirschbaum C. Hippocampal damage abolishes the cortisol response to psychosocial stress in humans. *Horm Behav.* 2009;56:44–50.
9. Cannon W. *The Wisdom of the Body.* New York, NY: W W Norton; 1932.
10. Chrousos GP, Gold PW. The concepts of stress and stress system disorders. *J Am Med Assoc.* 1992;267:1244–1252.
11. Crews DE. Composite estimates of physiological stress, age, and diabetes in American Samoans. *Am J Phys Anthropol.* 2007;133:1028–1034.
12. de Kloet ER, van der Vies J, de Wied D. The site of the suppressive action of dexamethasone on pituitary-adrenal activity. *Endocrinology.* 1974;94:61–73.
13. De Kloet ER, Vreugdenhil E, Oitzl MS, Joëls M. Glucocorticoid feedback resistance. *Trends Endocrinol Metab.* 1997;8:26–33.
14. De Kloet R, Wallach G, McEwen BS. Differences in corticosterone and dexamethasone binding to rat brain and pituitary. *Endocrinology.* 1975;96:598–609.
15. Del Giudice M, Ellis BJ, Shirtcliff EA. The adaptive calibration model of stress responsivity. *Neurosci Biobehav Rev.* 2011;35:1562–1592.
16. Dutton DG, Aron AP. Some evidence for heightened sexual attraction under conditions of high anxiety. *J Pers Soc Psychol.* 1974;30:510–517.
17. Ellis BJ, Del Giudice M. Beyond allostatic load: rethinking the role of stress in regulating human development. *Dev Psychopathol.* 2014;26:1–20.
18. Ellis BJ, Essex MJ, Boyce WT. Biological sensitivity to context: II. Empirical explorations of an evolutionary-developmental theory. *Dev Psychopathol.* 2005;17:303–328.
19. Engert V, Buss C, Khalili-Mahani N, Wadiwalla M, Dedovic K, Pruessner JC. Investigating the association between early life parental care and stress responsivity in adulthood. *Dev Neuropsychol.* 2010;35:570–581.
20. Engert V, Vogel S, Efanov SI, et al. Investigation into the cross-correlation of salivary cortisol and alpha-amylase responses to psychological stress. *Psychoneuroendocrinology.* 2011;36(9):1294–1302.
21. Eriksson PS, Perfilieva E, Bjoerk-Eriksson T, et al. Neurogenesis in the adult human hippocampus. *Nat Med.* 1998;4:1313–1317.
22. Fonte RJ, Stevenson JM. The use of propranolol in the treatment of anxiety disorders. *Hillside J Clin Psychiatry.* 1985;7:54–62.
23 Genova L. New York: Gallery Books; 2008.
24. Hansson L. The use of propranolol in hypertension: a review. *Postgrad Med J.* 1976;52(suppl 4):77–80.
25. Hauger RL, Risbrough V, Oakley RH, Olivares-Reyes JA, Dautzenberg FM. Role of CRF receptor signaling in stress vulnerability, anxiety, and depression. *Ann N Y Acad Sci.* 2009;1179: 120–143.
26. Haus E, Touitou Y. Principles of clinical chronobiology. In: Touitou Y, Haus E, eds. *Biologic Rhythms in Clinical and Laboratory Medicine.* Berlin: Springer Verlag; 1994:6–33.
27. Heim CM, Mayberg HS, Mletzko T, Nemeroff CB, Pruessner JC. Decreased cortical representation of genital somatosensory field after childhood sexual abuse. *Am J Psychiatry.* 2013;170:616–623.
28. Kirschbaum C, Pirke KM, Hellhammer DH. The 'Trier Social Stress Test' – a tool for investigating psychobiological stress responses in a laboratory setting. *Neuropsychobiology.* 1993;28:76–81.
29. Kizildere S, Gluck T, Zietz B, Scholmerich J, Straub RH. During a corticotropin-releasing hormone test in healthy subjects, administration of a beta-adrenergic antagonist induced secretion of cortisol and dehydroepiandrosterone sulfate and inhibited secretion of ACTH. *Eur J Endocrinol.* 2003;148:45–53.
30. Kudielka BM, Buske-Kirschbaum A, Hellhammer DH, Kirschbaum C. HPA axis responses to laboratory psychosocial stress in healthy elderly adults, younger adults, and children: impact of age and gender. *Psychoneuroendocrinology.* 2004;29:83–98.
31. Lee BK, Glass TA, McAtee MJ, et al. Associations of salivary cortisol with cognitive function in the Baltimore memory study. *Arch Gen Psychiatry.* 2007;64:810–818.
32. Lupien S, Lecours AR, Schwartz G, et al. Longitudinal study of basal cortisol levels in healthy elderly subjects: evidence for subgroups. *Neurobiol Aging.* 1996;17:95–105.
33. Lupien SJ, de Leon M, de Santi S, et al. Cortisol levels during human aging predict hippocampal atrophy and memory deficits [see comments]. *Nat Neurosci.* 1998;1:69–73.
34. Maheu FS, Joober R, Beaulieu S, Lupien SJ. Differential effects of adrenergic and corticosteroid hormonal systems on human short- and long-term declarative memory for emotionally arousing material. *Behav Neurosci.* 2004;118:420–428.
35. Maheu FS, Joober R, Lupien SJ. Declarative memory after stress in humans: differential involvement of the beta-adrenergic and corticosteroid systems. *J Clin Endocrinol Metab.* 2005;90:1697–1704.
36. McEwen BS. Stress, adaptation, and disease. Allostasis and allostatic load. *Ann N Y Acad Sci.* 1998;840:33–44.
37. McEwen BS. Sex, stress and the hippocampus: allostasis, allostatic load and the aging process. *Neurobiol Aging.* 2002;23:921–939.
38. McGaugh JL. The amygdala modulates the consolidation of memories of emotionally arousing experiences. *Annu Rev Neurosci.* 2004;27:1–28.
39. Muller MB, Lucassen PJ, Yassouridis A, Hoogendijk WJ, Holsboer F, Swaab DF. Neither major depression nor glucocorticoid treatment affects the cellular integrity of the human hippocampus. *Eur J Neurosci.* 2001;14:1603–1612.
40. Nemeroff CB. The corticotropin-releasing factor (CRF) hypothesis of depression: new findings and new directions. *Mol Psychiatry.* 1996;1:336–342.

41. Nemeroff CB. The neurobiology of depression. *Sci Am.* 1998;278: 42–49.

42. O'Leary CB, Hackney AC. Acute and chronic effects of resistance exercise on the testosterone and cortisol responses in obese males: a systematic review. *Physiol Res.* 2014;63:693–704.

43. Oei NY, Tollenaar MS, Elzinga BM, Spinhoven P. Propranolol reduces emotional distraction in working memory: a partial mediating role of propranolol-induced cortisol increases? *Neurobiol Learn Mem.* 2010;93:388–395.

44. Otte C, Yassouridis A, Jahn H, et al. Mineralocorticoid receptor-mediated inhibition of the hypothalamic-pituitary-adrenal axis in aged humans. *J Gerontol A Biol Sci Med Sci.* 2003;58:B900–B905.

45. Ouellet-Morin I, Odgers CL, Danese A, et al. Blunted cortisol responses to stress signal social and behavioral problems among maltreated/bullied 12-year-old children. *Biol Psychiatry.* 2011;70:1016–1023.

46. Pruessner JC, Baldwin MW, Dedovic K, et al. Self-esteem, locus of control, hippocampal volume, and cortisol regulation in young and old adulthood. *Neuroimage.* 2005;28:815–826.

47. Pruessner JC, Wolf OT, Hellhammer DH, et al. Free cortisol levels after awakening: a reliable biological marker for the assessment of adrenocortical activity. *Life Sci.* 1997;61:2539–2549.

48. Reul JM, de Kloet ER. Two receptor systems for corticosterone in rat brain: microdistribution and differential occupation. *Endocrinology.* 1985;117:2505–2511.

49. Reul JM, van den Bosch FR, de Kloet ER. Relative occupation of type-I and type-II corticosteroid receptors in rat brain following stress and dexamethasone treatment: functional implications. *J Endocrinol.* 1987;115:459–467.

50. Robertson T, Watts E. The importance of age, sex and place in understanding socioeconomic inequalities in allostatic load: evidence from the Scottish Health Survey (2008–2011). *BMC Public Health.* 2016;16:126.

51. Sabbah W, Watt RG, Sheiham A, Tsakos G. Effects of allostatic load on the social gradient in ischaemic heart disease and periodontal disease: evidence from the Third National Health and Nutrition Examination Survey. *J Epidemiol Community Health.* 2008;62:415–420.

52. Sapolsky RM, Krey LC, McEwen BS. The neuroendocrinology of stress and aging: the glucocorticoid cascade hypothesis. *Endocr Rev.* 1986;7:284–301.

53. Schulkin J, McEwen B, Gold PW. Allostasis, amygdala, and anticipatory angst. *Neurosci Biobehav Rev.* 1994;18:385–396.

54. Seeman TE, McEwen BS, Rowe JW, Singer BH. Allostatic load as a marker of cumulative biological risk: MacArthur studies of successful aging. *Proc Natl Acad Sci USA.* 2001;98:4770–4775.

55. Seeman TE, Robbins RJ. Aging and hypothalamic-pituitary-adrenal response to challenge in humans. *Endocr Rev.* 1994;15:233–260.

56. Selye H. A syndrome produced by diverse noxious agents. *Nature.* 1936;138.

57. Selye H. The evolution of the stress concept. Stress and cardiovascular disease. *Am J Cardiol.* 1970;26:289–299.

58. Sharma M, Palacios-Bois J, Schwartz G, et al. Circadian rhythms of melatonin and cortisol in aging. *Biol Psychiatry.* 1989;25:305–319.

59. Silver R, LeSauter J, Tresco PA, Lehman MN. A diffusible coupling signal from the transplanted suprachiasmatic nucleus controlling circadian locomotor rhythms. *Nature.* 1996;382:810–813.

60. Simeckova M, Jansky L, Lesna I, Vybiral S, Sramek P. Role of beta adrenoceptors in metabolic and cardiovascular responses of cold exposed humans. *J Therm Biol.* 2000;25:437–442.

61. Simpson EA, Sclafani V, Paukner A, et al. Inhaled oxytocin increases positive social behaviors in newborn macaques. *Proc Natl Acad Sci USA.* 2014;111:6922–6927.

62. Sindi S, Fiocco AJ, Juster RP, Lord C, Pruessner J, Lupien SJ. Now you see it, now you don't: testing environments modulate the association between hippocampal volume and cortisol levels in young and older adults. *Hippocampus.* 2014;24:1623–1632.

63. Sterling P, Eyer J. Allostasis: a new paradigm to explain arousal pathology. In: Fisher S, Reason J, eds. *Handbook of Life Stress, Cognition and Health.* New York: Wiley & Sons; 1988:629–649.

64. Suzuki T, Nakamura Y, Moriya T, Sasano H. Effects of steroid hormones on vascular functions. *Microsc Res Tech.* 2003;60:76–84.

65. Swaab DF, Bao AM, Lucassen PJ. The stress system in the human brain in depression and neurodegeneration. *Ageing Res Rev.* 2005; 4:141–194.

66. Ulrich-Lai YM, Herman JP. Neural regulation of endocrine and autonomic stress responses. *Nat Rev Neurosci.* 2009;10(6):397–409.

67. Van Cauter E, Leproult R, Kupfer DJ. Effects of gender and age on the levels and circadian rhythmicity of plasma cortisol. *J Clin Endocrinol Metab.* 1996;81:2468–2473.

68. Viru A, Viru M, Karelson K, et al. Adrenergic effects on adrenocortical cortisol response to incremental exercise to exhaustion. *Eur J Appl Physiol.* 2007;100:241–245.

69. Waltman C, Blackman MR, Chrousos GP, Riemann C, Harman SM. Spontaneous and glucocorticoid-inhibited adrenocorticotropic hormone and cortisol secretion are similar in healthy young and old men. *J Clin Endocrinol Metab.* 1991;73:495–502.

70. Whitworth JA, Saines D, Scoggins BA. Blood pressure and metabolic effects of cortisol and deoxycorticosterone in man. *Clin Exp Hypertens A.* 1984;6:795–809.

71. Wilhelm I, Born J, Kudielka BM, Schlotz W, Wust S. Is the cortisol awakening rise a response to awakening? *Psychoneuroendocrinology.* 2007;32:358–366.

72. Xu RB, Liu ZM, Zhao Y. A study on the circadian rhythm of glucocorticoid receptor. *Neuroendocrinology.* 1991;53(suppl 1):31–36.

73. Yamaguchi N, Okada S. Cyclooxygenase-1 and -2 in spinally projecting neurons are involved in CRF-induced sympathetic activation. *Auton Neurosci.* 2009;151:82–89.

21

Aldosterone and Mineralocorticoid Receptors[1]

J.W. Funder[1,2]

[1]Hudson Institute of Medical Research (Formerly Prince Henry's Institute of Medical Research), Clayton, VIC, Australia; [2]Monash University, Clayton, VIC, Australia

Abstract

Physiologically, aldosterone is a homeostatic hormone, rising in response to sodium deficiency, potassium loading, and volume depletion to restore the status quo ante: it is also raised acutely but not chronically by adrenocorticotropic hormone (ACTH). Pathophysiologically, in primary aldosteronism (PA) aldosterone levels are inappropriately high for the subject's sodium/potassium/volume status, thus outside the normal feedback loop, and no longer homeostatic but causing widespread deleterious cardiovascular effects. Long thought to be a rare and relatively benign form of hypertension, patients with PA have much higher risk factors for cardiovascular disease than age-, sex-, and BP-matched essential hypertensives. The role of ACTH in aldosterone secretion has not received major attention, although recent studies suggest a major involvement in stress-related rather than homeostatic secretion, playing an additional, as yet unrecognized, causative role in PA.

BACKGROUND

Classically, aldosterone is the physiological mineralocorticoid hormone uniquely secreted from the zona glomerulosa of the adrenal cortex, and acting via intracellular mineralocorticoid receptors in epithelial target tissues to promote unidirectional transepithelial sodium transport. While this remains the case, over the past decade there has been mounting evidence for additional complexity in terms of aldosterone secretion and action, and of MR activation by glucocorticoids. For aldosterone, this ranges from possible extraadrenal aldosterone synthesis, to major pathophysiological actions via nonepithelial mineralocorticoid receptors, to rapid nongenomic actions via a range of receptor mechanisms.

---

[1]This article is a revision of the previous edition articles by J.W. Funder, volume 1, pp. 141–144, © Elsevier Inc, and volume 2. pp. 132–135.

- Classically, aldosterone is considered to have primarily epithelial actions, to retain sodium and excrete potassium: in fact, aldosterone also has physiological actions on blood vessels and the central nervous system.

- Mineralocorticoid receptors (MR)—often misnamed "aldosterone receptors"—have equal, high affinity for cortisol and aldosterone, with most MR normally occupied but not activated by cortisol.

- Primary aldosteronism (PA) reflects inappropriate aldosterone secretion for the patient's sodium status, at least in part outside the normal feedback mechanisms.

- Adrenocorticotropic hormone (ACTH) is well recognized as a stress hormone, but its role as an acute potent stimulus to aldosterone secretion, equivalent to that of angiotensin II and plasma [K$^+$] but outside the normal feedback loop, poorly explored.

- Recently, potential roles for inappropriate ACTH secretion and action in the genesis of PA have been reported.

ALDOSTERONE SECRETION: ADRENAL AND EXTRAADRENAL

Aldosterone was isolated in 1953. It was initially called electrocortin, reflecting its effects on electrolyte transport, but was given the definitive name of aldosterone in recognition of its unique aldehyde (—CHO) group at carbon 18, in contrast with the methyl group in other steroids. Formation of this aldehyde group, which is crucial for the ability of aldosterone to access mineralocorticoid receptors in epithelia (vide infra), is achieved by a three-step process catalyzed by the enzyme aldosterone synthase (CYP11B2), expression of which in the adrenal cortex is confined to the outermost cell layer, the zona glomerulosa. While possible extraadrenal synthesis does not appear to contribute to circulating hormone levels, the extent to which it may occur to support local tissue effects of aldosterone in vivo remains to be determined.

REGULATION OF ALDOSTERONE SECRETION

In vivo aldosterone secretion from the zona glomerulosa is predominantly and independently regulated by angiotensin II and plasma potassium concentrations.[1] Adrenocorticotropic hormone (ACTH) can also raise aldosterone secretion, but unlike the other two secretagogues, its action is not sustained. While a myriad of other possible modulators of aldosterone secretion have been reported, including endothelin, nitric oxide, and uncharacterized factors from the pituitary and adipose tissue, the relative importance of such inputs to pathophysiological control in vivo has not been established. Both experimentally and clinically, changes in angiotensin II and/or plasma potassium appear sufficient to explain changes in aldosterone secretion rates from nonneoplastic adrenals.

MINERALOCORTICOID RECEPTORS: CLONING

Classical mineralocorticoid receptors binding aldosterone with high affinity and blocked by the antagonist spironolactone have been studied since the mid-1970s, with the human MR cloned and expressed by the late 1980s. The human MR comprises 984 amino acids and, like other members of the steroid/thyroid/retinoid/orphan receptor superfamily, has a domain structure that includes a ligand-binding domain and a DNA-binding domain. Like other members of the receptor superfamily, mineralocorticoid receptors (MR) act by regulating gene expression; serum- and glucocorticoid-inducible protein kinase expression in A6 cells is actinomycin inhibitable, and aldosterone-induced sodium transport is both actinomycin and puromycin inhibitable in toad bladder models of mineralocorticoid action. MR are members of a subfamily with the receptors for glucocorticoids, progestins, and androgens, sharing ≥50% amino acid identity in the ligand-binding domain and ≥90% in the DNA-binding domain. Recent studies have shown MR to be the first of the four receptors to branch off from the common ancestral sequence, millions of years before the enzymes responsible for the synthesis of aldosterone;[2] their putative ligand in species such as cartilaginous and bony fish is thus cortisol.

MINERALOCORTICOID RECEPTORS: CHARACTERIZATION

The initial MR cloning and expression studies also highlighted two questions regarding aldosterone action, questions that had been previously addressed by studies on MR in tissue extracts. First, when mRNA from a variety of rat tissues was probed with MR cDNA, the highest levels of MR expression were found in hippocampus rather than classic epithelial aldosterone target tissues. Second, when the affinity of various steroids was determined by their ability to compete for [^{3}H] aldosterone binding to expressed human mineralocorticoid receptors, cortisol was shown to have high affinity for MR, equivalent to that of aldosterone, and corticosterone (the physiological glucocorticoid in rat and mouse) slightly higher affinity.[3] There are thus at least two operational questions, of which the first is the physiological role(s) of MR in nonepithelial tissues, such as hippocampus (and other brain areas and heart, for example). The second is how aldosterone can selectively activate MR, in epithelia or in nonepithelial tissues, in the face of at least equivalent receptor affinity and the much higher circulating levels of the physiological glucocorticoids.

ALDOSTERONE SPECIFICITY-CONFERRING MECHANISMS: TRANSCORTIN, 11βHSD2

Differential plasma binding of steroids has long been recognized as one factor favoring aldosterone occupancy of MR. The physiological glucocorticoids cortisol and corticosterone circulate at least 95% protein bound, both to the specific globulin transcortin and to albumin; aldosterone has negligible affinity for transcortin and is only ~50% albumin bound in plasma, giving it an order of magnitude advantage in terms of plasma free to total concentrations. Second, the enzyme 11βHSD2 is abundantly expressed in epithelial aldosterone target tissues. 11βHSD2 has a very low K_m (high affinity) for cortisol and corticosterone, acting essentially unidirectionally to convert them to cortisone and 11-dehydrocorticosterone.[4] These 11-keto congeners have very low affinity for MR, and when they do occupy MR, act as antagonists. Aldosterone, in contrast, is not so metabolized; its 11β-OH group cyclizes in solution with the unique and very active aldehyde group at C18, forming an 11,18-hemiacetal and rendering aldosterone not a substrate for 11βHSD2.

APPARENT MINERALOCORTICOID EXCESS

The crucial role of 11βHSD2 in conferring aldosterone specificity in otherwise nonselective epithelial MR is underlined by the clinical syndrome of apparent mineralocorticoid excess. In this rare autosomal recessive disorder, patients present with high blood pressure and marked sodium retention, despite low plasma levels of renin and aldosterone.[5] The syndrome can be diagnosed by an abnormally high ratio of urinary free cortisol to cortisone, reflecting loss-of-function mutation of the sequence coding for 11βHSD2. Under such circumstances the mutant enzyme shows poor or absent ability to convert cortisol to cortisone, and cortisol inappropriately activates MR, leading to uncontrolled sodium retention and hypertension. A milder version of the syndrome can be produced by licorice abuse, as the active principle of licorice (glycyrrhetinic acid) blocks 11βHSD2 activity, as does the erstwhile antipeptic ulcer drug carbenoxolone, the hemisuccinate of glycyrrhetinic acid.

EXTRAEPITHELIAL ACTIONS OF ALDOSTERONE

The physiological roles of extraepithelial MR and the extraepithelial effects of aldosterone have been elucidated only in part. Hippocampal MR appear involved in integration of baseline circadian ACTH secretion in response to stress; given the high reflection coefficient of aldosterone at the blood–brain barrier, it appears improbable that these receptors, unprotected by 11βHSD2, are ever occupied by aldosterone. Other brain MR, in contrast, are clearly occupied and at least to some extent activated by aldosterone, being protected by 11βHSD2. The amygdala, for example, express both MR and 11βHSD2, and have been shown to be responsible for the aldosterone-induced increase in salt appetite in sodium deficiency or after sodium loss; interestingly, the amygdala appear within the blood–brain barrier, with no evidence for facilitated aldosterone access to MR. Activation of MR in the anteroventral third ventricle (AV3V) region (which is outside the blood–brain barrier) is necessary for mineralocorticoid-salt hypertension: intracerebroventricular infusion of MR antagonists abolishes the elevation of blood pressure following peripheral infusion of aldosterone to rats, or feeding a high-salt diet to salt-sensitive strains of rats. This has generally been interpreted as a direct role for aldosterone in modulating blood pressure via MR activation in the AV3V region; physiologically, however, this appears unlikely, and the MR activation putatively reflects glucocorticoid occupancy of the receptors under particular conditions.

ALDOSTERONE, CARDIAC HYPERTROPHY, AND CARDIAC FIBROSIS

Aldosterone occupancy of cardiac MR has similarly been shown in a variety of experimental models to produce cardiac hypertrophy and fibrosis. These effects are independent of hemodynamic changes, as clamping blood pressure by the concomitant infusion of intracerebroventricular MR antagonist does not alter the effects of peripherally infused aldosterone on cardiac hypertrophy and fibrosis. They are also crucially dependent on an aldosterone-salt imbalance; infusion of the same dose of aldosterone to rats fed a low-salt diet is not followed by increases in blood pressure or in cardiac hypertrophy and cardiac fibrosis.[6] Cardiac MR, like those in the CNS, are normally unprotected by 11βHSD2; the cardiac effects seem to require only a minority of MR to be occupied by mineralocorticoid, as shown in studies of transgenic mice expressing 11βHSD2 specifically in cardiomyocytes.[7] Importantly, under normal circumstances, occupancy of extraepithelial MR by glucocorticoids does not mimic the effect of aldosterone. Also in contrast with the epithelial effects of aldosterone via MR, which have a time course of hours, the extraepithelial effects of aldosterone may take days or weeks to be manifest: vascular inflammatory responses are seen after 1–2 days, increases in collagen type I and type III mRNA are seen by the third week of continuous aldosterone infusion, and plateau levels of blood pressure elevation are seen at 4–6 weeks. The mechanisms underlying these epithelial/extraepithelial differences in the action of aldosterone via identical MR are not understood, but presumably reflect at least in part tissue-specific differences in the coactivators, corepressors, and response elements involved.

ALDOSTERONE: A STRESS HORMONE?

Aldosterone is a stress hormone depending on the definition of stress. Psychological or physical stressors that raise ACTH and adrenal catecholamine output in most subjects lead to acute but not sustained elevations in aldosterone levels.[1] Physiologically, aldosterone is raised in response to upright posture to defend intravascular volume; longer-term stressors, in terms of fluid and electrolyte balance (e.g., diarrhea, florid sweating, pregnancy, and lactation on a low-sodium diet) raise aldosterone to act in a tight feedback loop to retain sodium and water. Our renin-angiotensin-aldosterone system appears to have evolved in a context of sodium deficiency, which is often more of an issue for herbivores than for omnivores, and we have much less well developed mechanisms for suppressing aldosterone secretion than for elevating it. On contemporary salt intakes, our inability to lower aldosterone secretion commensurately may thus in a counterintuitive sense be in itself a stressor for the organism, which requires 2–5% of the normal modern salt intake except in response to the stress of reproduction, gastrointestinal infection, or environmental extremes. The evolutionary craving for salt, and its routine addition to a range of foodstuffs, may thus constitute a stress in terms of the cardiovascular system, most noticeably hypertension, but with increasing evidence for direct effects of MR activation on the heart. The RALES trial, in which adding low-dose spironolactone to conventional ACE inhibitor/loop diuretic/digitalis therapy for heart failure substantially lowered mortality,[8] would also seem to constitute evidence for MR activation operating as a stress mechanism in this sense.

MR ACTIVATION AND STRESS

As noted previously, 11βHSD2 acts in epithelial tissues to allow aldosterone to selectively activate MR. In such tissues, however, the enzyme debulks intracellular glucocorticoid levels by 90%, rather than completely remove cortisol/corticosterone, so that their levels are still ~10 times those of aldosterone, and thus most MR in such tissues are normally occupied but not activated by glucocorticoids when the enzyme is operating. The key to this distinction appears to be the stoichiometric generation of NADH, which alters the redox state of the cell so that MR-glucocorticoid complexes are inactive; when 11βHSD2 is blocked, levels of NADH fall, and the MR-glucocorticoid complexes are activated.[9] This bivalent effect of cortisol can be reproduced in Langendorf ischemia reperfusion studies, in which cortisol activates MR in the context of tissue damage, thus mimicking aldosterone.[10] In pathophysiological terms, this suggests that the MR in nonepithelial tissues (e.g., cardiomyocytes and most neurons) that are always essentially occupied by glucocorticoids may be activated by the metabolic or redox state of the cell, in particular, the generation of reactive oxygen species. If this is the case, then metabolic stress may be mediated via MR through a novel mechanism of activating a hitherto quiescent steroid–receptor complex, rather than by increasing the secretion rate and blood levels of a stress hormone, such as cortisol.

PRIMARY ALDOSTERONISM

Primary aldosteronism (PA) is the inappropriate oversecretion of aldosterone at least in part outside the influence of its physiological secretagogues (angiotensin II, plasma [K+], ACTH). Long thought (and taught) to be a rare and relatively benign cause of hypertension, we now know that neither of these is the case.[11] Current consensus is that PA represents 5–13% of all hypertension, and recent evidence has suggested that inappropriate hypersecretion of aldosterone may play a part in elevating blood pressure in ~50% of all hypertensives.

Hypersecretion of aldosterone may be unilateral or bilateral. Unilateral hypersecretion is usually from an aldosterone producing adenoma (APA), comprising approximately one-third of all cases; bilateral adrenal hyperplasia (BAH) accounts for approximately two-thirds of PA. Occasionally bilateral APA, unilateral hyperplasia or (very rarely) adrenal carcinoma may cause aldosterone excess. Over 60% of APA has now been shown to reflect a somatic mutation in a potassium or calcium channel, or Na+/K+or Ca++ATPases, confined to the adrenal cortex.[12] The drivers of BAH are currently unknown, but the subject of intense investigation.

Screening for PA is by determination of the aldosterone to renin ratio (ARR). Patients with a high ARR, and a plasma aldosterone concentration (PAC) in the higher levels of the normal range or beyond, then undergo one of a series of confirmatory/exclusion tests, of which the most elaborate, expensive, and arguably most discriminatory is the fludrocortisone suppression test (FST). Patients receive six hourly fludrocortisone and 200 mEq/day sodium for four days: if this combination fails to suppress PAC to very low levels the diagnosis of PA is confirmed.

Two recent studies from Greece have modified the FST by giving patients 2 mg dexamethasone at midnight on the last day, a so-called dexamethasone enhanced FST, or FDST. When 80 normotensive controls underwent an FDST, the upper limit of PAC was 3 ng/dl; when a cohort of hypertensives underwent an FDST, 29% had PAC above the normal limit.[13] The logical interpretation of these data is that absent dexamethasone the "normal" levels of PAC are elevated by ACTH, with two-thirds of PA patients falling into this elevated "normal" range, thus avoiding confirmation as PA.

In a subsequent study the same group took hypertensives negative for PA by FDST and subjected them to an ultralow ACTH infusion, measuring PAC and cortisol in response.[14] The subjects segregated into two tight groups: most had a very muted response (in cortisol and aldosterone) to the ACTH infusion, indistinguishable from that in normotensive controls. In contrast, 27% of the hypertensive group showed an equivalent cortisol response, but a pronounced elevation in PAC. When the same three groups of subjects were treadmill exercised, these 27% were clearly hyperresponsive in terms of PAC, with no differences in plasma ACTH or cortisol between groups.

If these studies are replicated and validated they open up a new chapter in the pathophysiology of aldosterone.[15] For a start "inappropriate" aldosterone secretion would now appear to cause, or contribute to, elevated blood pressure in ~50% of hypertensives. In addition, they point to a hitherto poorly recognized role for ACTH as a potent, acute secretagogue of aldosterone, firmly establishing its role as a true "stress-induced" hormone. The physiological—as well as the pathophysiological—implications of this expanded role remain to be systematically explored.

Glossary

Aldosterone The classic salt-retaining steroid hormone secreted from the zona glomerulosa of the adrenal cortex.

Aldosterone synthase/CYP11B2 The enzyme abundantly expressed in adrenal zona glomerulosa and reported in PCR studies on heart, blood vessels, and brain, which, by a sequential three-step process, converts the methyl ($—CH_3$) group at carbon 18 to an aldehyde ($—CHO$) group.

Apparent mineralocorticoid excess (AME) An autosomal recessive syndrome of inactivating mutations in the gene coding for 11β-hydroxysteroid dehydrogenase type 2, producing juvenile hypertension, marked salt retention reflecting inappropriate cortisol activation of epithelial mineralocorticoid receptors, and premature death.

11β-Hydrozysteroid dehydrogenase type 2 (11βHSD2) The enzyme expressed in very high abundance (3–4 million copies per cell) in epithelial aldosterone target cells that converts cortisol and corticosterone (which have affinity for mineralocorticoid receptors similar to that of aldosterone) to receptor-inactive metabolites, and NAD to NADH, thus allowing aldosterone to activate the relatively nonselective mineralocorticoid receptors.

Mineralocorticoid receptors (MR) The intracellular receptors that have equal high affinity for aldosterone, corticosterone, and cortisol and via which aldosterone acts in epithelial tissues to promote unidirectional transepithelial sodium transport.

Acknowledgment

This work is supported by the Victorian Government's Operational Infrastructure Support Program.

References

1. Hattangady NG, Olala LO, Bollag WB, Rainey WE. Acute and chronic regulation of aldosterone production. *Mol Cell Endocrinol*. 2012;350:151–162.
2. Kassahn KS, Ragan MA, Funder JW. Mineralocorticoid receptors: evolutionary and pathophysiological considerations. *Endocrinology*. 2011;152:1883–1890.
3. Arriza JL, Weinberger C, Cerèlli G, et al. Cloning of human nimeralocorticoid receptor complementary DNA: structural and functional kinship with the glucocorticoid receptor. *Science*. 1987;237:268–275.
4. Funder JW, Pearce PT, Smith AI. Mineralocorticoid action: target tissue specificity is enzyme, not receptor, mediated. *Science*. 1988;242:583–586.
5. Wilson RC, Krozowski ZS, Li K, et al. A mutation in the HSD11B2 gene in a family with apparent mineralocorticoid excess. *J Clin Endocrinol Metab*. 1995;80:2263–2266.
6. Brilla CG, Weber KT. Mineralocorticoid excess, dietary sodium, and myocardial fibrosis. *J Lab Clin Med*. 1992;120:893–901.
7. Qin W, Rudolph AE, Bond BR, et al. Transgenic model of aldosterone-driven cardiac hypertrophy and heart failure. *Circ Res*. 2003;93:69–76.
8. Pitt B, Zannad F, Remme WJ, Cody R, et al. The effect of spironolactone on morbidity and mortality in patients with severe heart failure. Randomized aldactone evaluation study investigators. *N. Engl J Med*. 1999;341:709–717.
9. Funder JW. RALES, EPHESUS and redox. *J Steroid Biochem Mol Biol*. 2005;93:121–125.
10. Mihailidou AS, Le TYL, Mardini M, Funder JW. Glucocorticoids activate cardiac mineralocorticoid receptors during experimental myocardial infarction. *Hypertension*. 2009;54:1306–1312.
11. Funder JW. Primary aldosteronism and salt. *Pflugers Arch*. 2015;467:587–594.
12. Fernandes-Rosa FL, Giscos-Douriez I, Amar L, et al. Different somatic mutations in multinodular adrenals with aldosterone-producing adenoma. *Hypertension*. 2015;66:1014–1022.
13. Gouli A, Kaltsas G, Tzonou A, et al. High prevalence of autonomous aldosterone secretion among patients with essential hypertension. *Eur J Clin Invest*. 2011;41:1227–1236.
14. Markou A, Sertedak A, Kaltsas G, et al. Stress-induced aldosterone hypersecretion in a substantial subset of patients with essential hypertension. *J Clin Endocrinol Metab*. 2015. http://dx.doi.org/10.1210/jc.2015-1268. [Epub ahead of print].
15. Funder JW. Primary aldosteronism: seismic shifts. *J Clin Endocrinol Metab*. 2015;100:2853–2855.

22

Androgen Action and Stress

M.A. Holschbach, R.J. Handa

Colorado State University, Fort Collins, CO, United States

Abstract

Androgens belong to a class of steroid hormones that circulate at high levels in males. However, androgens should not be thought of as exclusively male hormones since they are also produced in females, though at much lower levels. The predominant mode of androgen action is through binding in target tissues to a specific receptor termed the androgen receptor. This receptor can regulate the expression of a set of androgen-responsive genes. The synthesis, transport, and release of androgens to the bloodstream, their actions on target tissues, and their effects on stressor related neuroendocrine function and behaviors are addressed here.

ANDROGENS MAJOR FORMS OF ANDROGENS

The term "androgen" is derived from the Greek roots andro (man) and gennan (to produce). The biological definition of an androgen is any substance that specifically promotes the growth of the male gonads, although they also masculinize the brain, as described in Box 22.1.

There are four major forms of circulating androgens. Testosterone (T) is the predominant form in males and is of gonadal origin. Other common forms of androgens are dihydrotestosterone (DHT); androstenedione, a precursor of T; and dehydroepiandrosterone (DHEA), and its sulfonylated derivative, DHEAS, the latter being primarily of adrenal origin. Androstenedione and DHEA/S are less potent than T or DHT, but in addition to weakly activating androgen receptors (AR) themselves, can also be metabolized into the more potent androgens.[13]

Both males and females produce all forms of androgens, though in differing amounts. In men, the preponderance of circulating androgens is T, which circulates at 10-fold higher levels than DHT. In women, T is also present at higher levels than DHT, but the predominant forms of circulating androgens are DHEA/S.[6] Moreover, T levels are roughly 10-fold higher in men than in women. In addition to differing levels in males and females, androgens also undergo a modest daily variation, which is controlled by sleep rather than the circadian cycle. Highest levels of circulating T are found around the time

Stress: Neuroendocrinology and Neurobiology
http://dx.doi.org/10.1016/B978-0-12-802175-0.00022-X

of waking and then decline over the course of the day in men.[24] By contrast, males of nocturnal species, such as rodents, experience peak T levels in the first few hours after nightfall. Moreover, in women, cyclic variations in T have been reported with peak levels occurring during the preovulatory phase of the menstrual cycle and both sexes experience declining androgens as they age.[18]

Androgen Origins

In males, most T is produced by Leydig cells in the testes. In females, T and androstenedione are produced by thecal cells in the ovaries. In addition to the gonadally derived T, which circulates at high levels in males, a small proportion is synthesized in target tissues from the T precursors androstenedione and DHEA/S. These often circulate at relatively high levels in men and women but are very weak activators of AR on their own. In contrast, the majority of circulating DHT, an extremely potent androgen, is locally converted from T in androgen target tissues, such as the brain, prostate, liver, and skin, but it is also synthesized to a limited extent in the gonads. Androstenedione and DHEA/S are largely produced by the adrenal cortex of men and women but little DHEA is produced by the adrenal cortex of rats and mice. Recent studies have also demonstrated that the brain has the appropriate enzymes to produce androgens de novo. The extent to which this synthesis affects physiology remains to be determined. Androgen synthesis and metabolism can occur anywhere the enzymes that catalyze such reactions are present. Though the primary tissues of origin are presented earlier, precursors and metabolites are produced in many tissues.

BOX 22.1

HOW DO WE KNOW THAT ANDROGENS ORGANIZE THE BRAIN?

Sex differences in the brain have been a rich area for research. In the rodent brain, many of these differences are set up during a perinatal critical period, either before or shortly after birth. Differences could be due to masculinization pushing the brain and body toward a male phenotype, or feminization pushing the brain and body toward a female phenotype. The process of masculinization is often coupled with a similar process, defeminization, which removes feminine behavioral and morphological characteristics of the brain and behavior. Because females have very low levels of sex hormones during this time, these brain sex differences are often thought to be organized by androgens present only in the males. To test this, scientists have removed or blocked the actions of androgens in males to see if masculinization is prevented or supplied androgens to female fetuses or embryos to see if they are sufficient to cause a male phenotype. These types of experiments have shown that testosterone is sufficient for masculinization of the sexually dimorphic nucleus of the hypothalamus, as shown in the following figure (figure adapted from Ref. 5). This nucleus is remarkable in that it contains over two fold greater density of neurons in males than females and this difference is established perinatally through the direct and indirect actions of androgens acting to maintain neuron numbers in the male.

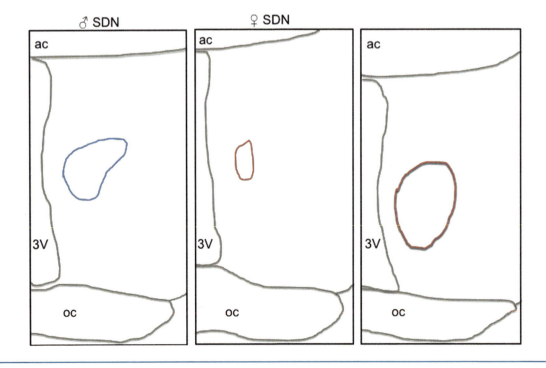

KEY POINTS

- Androgens work through binding to androgen receptors
- Metabolites of testosterone are important players in regulating androgen action
- Androgens can act through nongenomic (rapid) mechanisms that bypass classical transcriptional mechanisms
- Androgens are important for sexual differentiation of the brain
- Stress can influence androgen levels and actions
- Androgens can influence neuroendocrine and behavioral responses to stress

Androgen Synthesis Pathways

The biosynthetic pathway responsible for producing endogenous androgens is common to all steroid hormones and begins with the conversion of cholesterol to pregnenolone in the mitochondria of steroid-synthesizing cells.[10] Cholesterol is composed of 27 carbon atoms, including a 17-carbon sterol backbone with an 8-carbon side chain, but the major androgens, including T, have no side chain and are composed of only 19 carbon atoms. The regulation of steroid synthesis begins at the level of cholesterol transport from cellular storage sites to the inner mitochondrial membrane by the steroidogenic acute regulatory protein. Once in the mitochondrion, the P450 side-chain cleavage enzyme (P450scc or CYP11A) catalyzes the oxidation and removal of the six carbon atoms that comprise the majority of the side chain on cholesterol. The product, pregnenolone, then diffuses passively out of mitochondrion to the smooth endoplasmic reticulum. Next, the P450cl7 (CYP17) enzyme further shortens the side chain, removing all carbons and leaving only a ketone at the C17 position. The final two steps in T biosynthesis are catalyzed by dehydrogenases that have specificity for the C3-and C17-bound oxygens, respectively. The dehydrogenase that converts the C3 ketone to a hydroxyl group also isomerizes the C5 double bond to a C4 double bond. In addition to this brief description, the androgen synthesis pathway is detailed in Fig. 22.1.

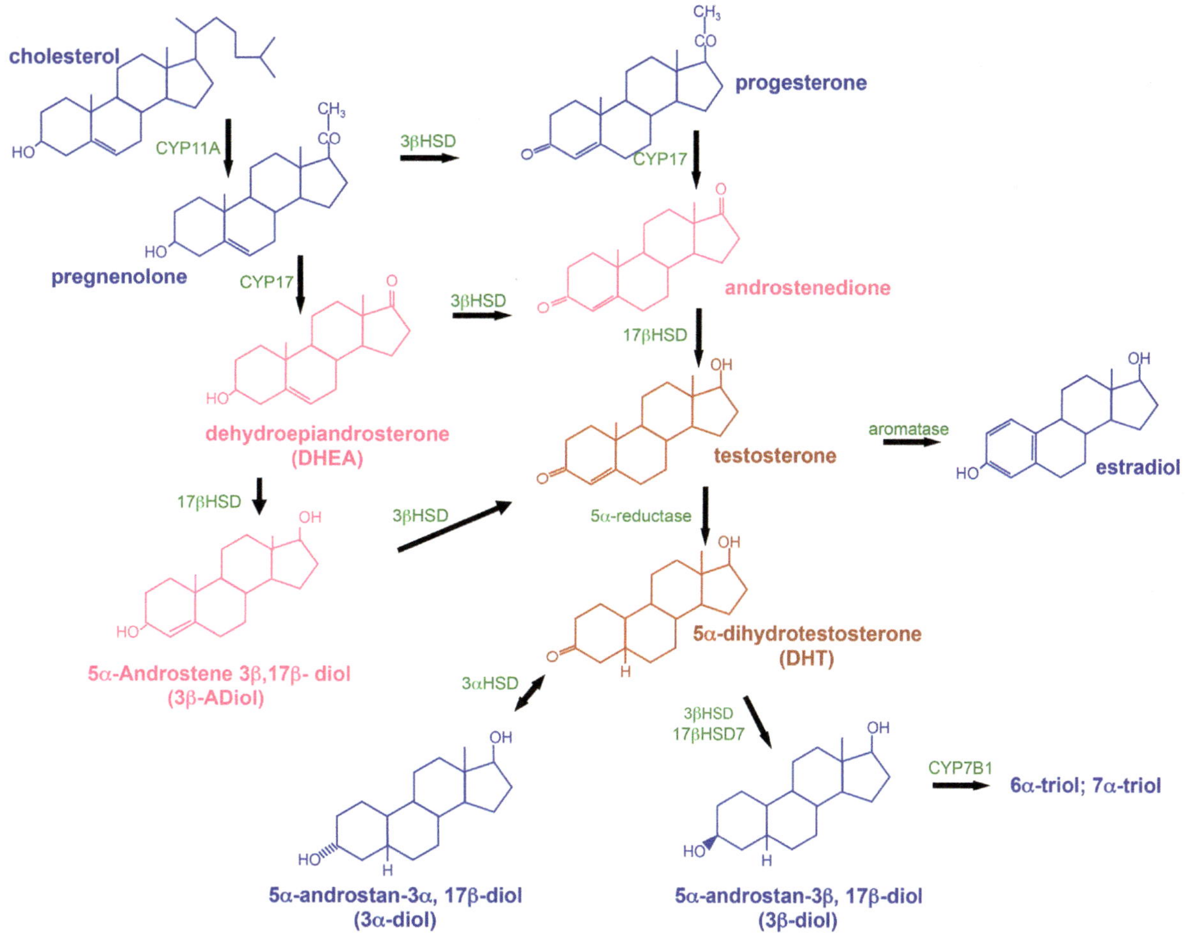

FIGURE 22.1 Schematic diagram depicting the biosynthetic pathway of androgens from cholesterol. Strong androgens are depicted in *red*, weak androgens are depicted in *pink*, non-androgenic precursors and metabolites are depicted in *blue*, and enzymes are designated by *green* text.

Metabolites of Testosterone

Androgens can be converted to estrogens by further enzyme catalysis in some target tissues and estrogen-synthesizing tissues, such as the granulosa cells of the ovaries. Aromatase is the enzyme responsible for converting androstenedione or T to estrone or estradiol, respectively, by the oxidoreduction of the C3 ketone and removal of the C10 methyl group, which allows for aromatization of the A-ring in the sterol backbone. This leads to estrogenic products that have very different effects than the parent compound.

T can also be converted by the enzyme 5α-reductase to DHT, which is a more potent androgen than T.[23] The pharmacological inhibition of 5α-reductase with compounds, such as finasteride and dutasteride, has proven to be an effective therapeutic for inhibiting androgen action in cases where excess androgen signaling leads to pathology, such as androgen-sensitive cancers. As discussed earlier, DHT is synthesized locally in many peripheral tissues and brain regions. This local metabolism of T to DHT allows the androgenic signal to be amplified in a tissue-specific manner. For some tissues, growth is largely dependent upon the production of DHT. For example, excessive production of or sensitivity to DHT has been shown to be a factor in prostate abnormalities, such as benign prostatic hyperplasia. DHT concentrations in androgen-dependent tissues are maintained through the direct formation of DHT from circulating T and from the interconversion of DHT and 5α-androstane 3α,17β-diol (3α diol), a primary metabolite of DHT. This bidirectional catalysis is accomplished by the actions of the oxidoreductase 3α-hydroxysteroid dehydrogenase (3αHSD). A variety of enzymes also convert DHT to 5α-androstane 3β,17β-diol (3β-diol), including 3βHSD, 3αHSD, and 17βHSD7; however this process is irreversible.

While DHT is a pure androgen, binding and activating only AR, the metabolites 3α-diol and 3β-diol have very low affinity for AR, and 3β-diol in particular preferentially binds the beta form of estrogen receptors. In the prostate, 3β-diol has been implicated in the regulation of epithelial cell proliferation by augmenting apoptosis and thereby counteracting the actions of DHT. 3β-diol is ultimately converted to the inactive compounds 6α- or 7α-triol and excreted in the urine. Similarly, the adrenal androgen, androstenediol can also be converted to a metabolite that binds estrogen receptor β. This metabolite, 5α-androstene 3β,17β-diol (3β-ADiol) has been shown to be a potent anti-inflammatory steroid, along with 3β-diol.

Secretion and Transport

Steroid hormones are secreted along a concentration gradient from synthetic cells to the circulating plasma and do not utilize a vesicular membrane fusion pathway. Consequently, circulating levels of androgens accurately reflect rates of synthesis. Steroid hormones are lipophilic and thus, are usually transported in the plasma bound to a serum binding protein, such as albumin- or sex hormone–binding globulin (SHBG). These binding proteins protect the steroid from degradation, which would otherwise shorten their half-life, and also inhibit renal excretion. Only free, unbound steroid is biologically active, so once at a target tissue, steroid hormones are released from the binding protein and because of their lipophilic nature, are able to easily enter cells by diffusing across the plasma membrane. Inside the cell, steroid hormones are bound by intracellular receptors. In the case of androgens, such as T and DHT, the specific receptor has been termed the AR.

THE ANDROGEN RECEPTOR

Molecular Organization

ARs are dynamic proteins that have been classified as members of a nuclear receptor superfamily.[11] This superfamily of proteins has a common structure composed of functional domains, depicted schematically in Fig. 22.2A. For steroid hormone receptors, the domains include a hormone-binding domain near the carboxyl terminus and a DNA-binding domain near the center of the linearized protein. The other domains include a hinge region to facilitate conformational change and a transactivation domain, which is important for interaction of the steroid receptor with transcriptional machinery. In the absence of hormone, the AR is maintained in an inactive state by a complex of chaperone proteins called heat-shock proteins. Additional information regarding the AR and all the other members of the nuclear receptor superfamily can be found at the Nuclear Receptor Signaling Atlas Website: www.nursa.org.

Function

Classically, the AR is a nuclear transcription factor and modulates gene expression in a hormone-dependent manner. In this respect, hormone binding induces a conformational change that dissociates a chaperone protein, such as heat-shock protein, thus activating the receptor. The classical activation pathway of ARs by T is depicted in Fig. 22.2B. The conformational change sequesters the ligand-binding domain along with the bound androgen into a hydrophobic pocket and reveals the DNA-binding domain, which has high affinity for certain DNA sequences called hormone response elements (HRE). Upon activation, other regions are revealed, which allow dimerization with another hormone-bound AR. The activated dimers then are able to bind to the HRE, a palindromic sequence

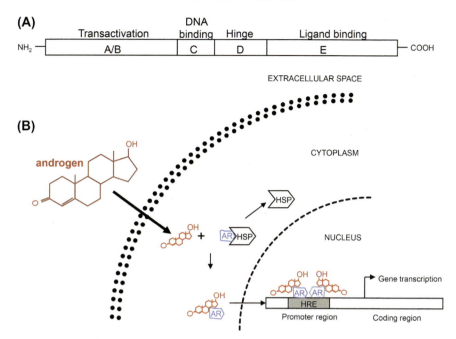

FIGURE 22.2 Schematic diagram of (A) androgen receptor (AR) functional domains and (B) pathway for activation of gene transcription. *HRE*, Hormone response element; *HSP*, heat-shock protein. Cell membrane is depicted by two layers of circular bars separating the cytoplasm from extracellular space. *Dashed line* indicates the nuclear membrane separating the nucleus from the cytoplasm. DNA of androgen-responsive gene is shown as a rectangle in the nucleus.

of nucleotides in DNA. These HREs are not present in the promoter regions of all genes, but only in those that are androgen responsive. Examples of such genes are those that encode the cytokine, interleukin-6, prostate-specific antigen, and AR itself.

Coregulatory Proteins

The control of gene expression by ARs is also dependent on factors such as the gene promoter context and the cell-specific presence of appropriate coregulatory proteins. Once the activated AR is bound to the HRE, gene expression changes can occur only if the receptor can communicate on a molecular level with the RNA polymerase holoenzyme, which occupies the transcription initiation site. It is generally accepted that this contact is not direct, but occurs through the attraction of several coregulatory proteins that do not bind to DNA directly but act through protein–protein contacts with the receptor at one level and the basal transcriptional machinery on another.

Nongenomic Androgen Actions

Hormone-dependent transcription is the primary mechanism of action for ARs; however, a growing body of evidence suggests that androgens can act rapidly, within seconds to minutes, to affect cell function suggesting the presence of a nongenomic pathway. These nongenomic effects involve the activation of second messenger signaling cascades that result in the phosphorylation of a variety

of transcription factors. Historically, ARs were thought to be inactive prior to localization in the nucleus; however, the observed rapid effects of androgens have forced the reevaluation of this hypothesis (Fig. 22.3).

One possibility is that the AR, following ligand binding and dissociation of heat-shock proteins, is able to associate with the plasma membrane through interactions with G-protein coupled receptors.[17] Many of AR's nongenomic actions have been attributed to cytoplasmic activation of the nonreceptor tyrosine kinase, Src, which stimulates cell proliferation and represents a target for therapy of androgen-sensitive cancers. In addition, androgens can bind directly to the SHBG receptor, also localized in the plasma membrane, and activate the mitogen-activated kinase second messenger system. Approximately 5% of ARs in frog oocytes are localized on the cell membrane. Androgens are potent activators of oocyte maturation and, surprisingly, BSA-bound androgens, which are unable to cross the cell membrane, are just as capable as free steroid to induce this process. Moreover, steroid-induced oocyte maturation does not require gene transcription.[12] These results suggest that it is membrane ARs, rather than nuclear ARs that promote oocyte maturation. Androgens also promote oocyte maturation in mammals by activating ARs, but a true membrane AR has not yet been identified in mammals.

Still, there is compelling evidence that some T activity is conferred by a membrane receptor. For example, T increases intracellular calcium in both oocytes and muscle cells, and these effects are not prevented by AR

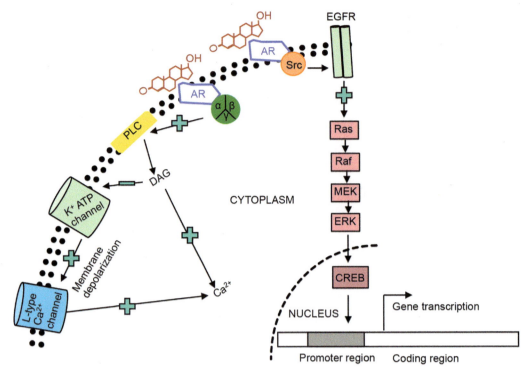

FIGURE 22.3　Schematic diagram of pathway for nongenomic androgen actions. *AR*, Membrane associated androgen receptor; *CREB*, cAMP response element binding protein; *DAG*, diacylglycerol; *EGFR*, epidermal growth factor receptor; *ER*, endoplasmic reticulum; *GPCR*, G-protein–coupled receptor; *PLC*, phospholipase C; *SRC*, Src kinase; *RAS*, Ras GTPase protein; *RAF*, Raf kinase.

antagonists, suggesting that T can alter G-protein signaling independently of known ARs, perhaps through an unidentified membrane receptor.[7] Although membrane ARs have yet to be identified in mammals, T binds to classical ARs at or near the plasma membrane of Sertoli cells within the testes to activate Src, which, in turn, phosphorylates the epidermal growth factor receptor to begin the process of spermatogenesis.[8] This extranuclear action of androgens is similar to that of growth factors, which also involve extranuclear kinases and binding of the transcriptional coactivator, cAMP response element binding protein, to cAMP response elements. In addition, this extranuclear androgen activation also leads to rapid, transient calcium influx that induces various secondary messenger cascades both in Sertoli cells and in muscle.

of facial and axillary hair, activation of sebaceous glands, coarsening of the skin, and growth of long bones and muscle. Androgens also regulate spermatogenesis, as discussed earlier, a process that is initiated as testicular size increases during puberty. Spermatogenesis persists throughout the life of the male but begins to decline as circulating T levels decline. The effects of androgens are not limited to the adult animal and effects of androgens and its metabolites on developmental processes are well known. In brain, androgens can program the brain during critical periods of development resulting in long-lasting changes in behavior and physiology in adulthood (See Box 22.1). Similar effects have been described in peripheral tissues with androgens being responsible for the sexual differentiation of the genital tract and external genitalia of males.

PHYSIOLOGICAL ACTIONS OF ANDROGENS

The specific set of genes that is regulated by androgens controls diverse physiological functions. In general, growth-associated processes are involved, such as those in prostate, hair, skin, muscle, and bone. Typical examples of androgen-dependent functions are those associated with puberty in males. Increasing levels of T during the midteen years lead to a deepening of the voice, growth

ANDROGEN-ASSOCIATED PATHOLOGY

As noted earlier, androgens are involved in many anabolic processes. When these growth processes go awry in the testes or prostate, cancerous pathology can result. One of the most common treatments for prostate cancer is to block AR function with antagonist drugs.[14]

Some of the developmental defects associated with malfunctioning androgen signaling have led to the discovery of specific androgen functions. The most common syndrome associated with a nonfunctional AR is called androgen insensitivity syndrome or tfm (testicular feminization mutation). This rare disorder in genetic males results in a phenotypically female individual with breast development, a lack of pubic hair, and formation of feminized genitalia.[3,25] Another syndrome with symptoms of hermaphroditism is a defect in the 5α-reductase gene. Individuals with this defect cannot produce DHT, the most potent androgen, and thus have ambiguous genitalia until puberty when T levels rise and partially compensate for the DHT deficiency. Conversely, a group of recessive genetic disorders referred to as congenital adrenal hyperplasia leads to high synthesis of adrenal androgens and masculinization of female fetuses. These individuals are often born with fused labioscrotal folds and a common urogenital sinus instead of separate urethra and vagina. Prenatal treatment with dexamethasone, a synthetic glucocorticoid that is bioavailable to the fetus, often prevents or ameliorates these changes by inhibiting adrenal function.[13a] Unfortunately, these treatments must be delivered during early prenatal development, before the sex or congenital adrenal hyperplasia (CAH) diagnosis could occur, so common treatment plans often included unnecessary treatment of 7/8 at-risk fetuses (all males and ¾ females). Thus, it is important to consider potential adverse reactions to prenatal dexamethasone treatment to mother and fetus.[9a] CAH represents an excellent example of the interactions of androgens and stress hormones.

INTERACTIONS BETWEEN STRESS AND ANDROGENS

Studies using animal models have shown that stress can affect circulating levels of T (Box 22.2), and conversely, T can influence behavioral and physiological responses to stress (Box 22.3).

BOX 22.2

HOW DO WE KNOW THAT ANDROGENS ARE AFFECTED BY STRESS?

Stress during critical periods can program HPG axis function. For example, prenatal stress can have lasting effects on male offspring's T levels. In rodents, prenatal stress feminizes (i.e., shortens) the distance between male newborns' anus and genitalia, a common readout of prenatal androgen exposure. Even as adults, male offspring whose mothers were stressed during pregnancy have significantly lower T levels and fewer AR-expressing neurons in some brain regions.[15] By contrast, prenatal stress does not strongly affect anogenital distance in newborn boys but masculinizes (i.e., increases) the anogenital distance of newborn girls and increases adolescent girls' testosterone levels.[1,1a,15a] This is not to say that boys are less vulnerable to the effects of prenatal stress. In fact, the birth ratio of boys to girls drops following catastrophic events such as terrorism or environmental disasters, suggesting that vulnerable male fetuses are less likely to survive severe prenatal stress.[4] Prenatal stress can also lead to persistent behavioral changes in offspring. Girls who experienced prenatal stress tend to engage in more male-typical play behaviors.[2] In male rodents, prenatal stress increases anxietylike behavior in adult male offspring. Intriguingly, early postnatal stress reversed many of these outcomes perhaps suggesting that alignment of the environment (i.e., similar stress levels during fetal and neonatal development) can rescue effects of challenging fetal development.[9]

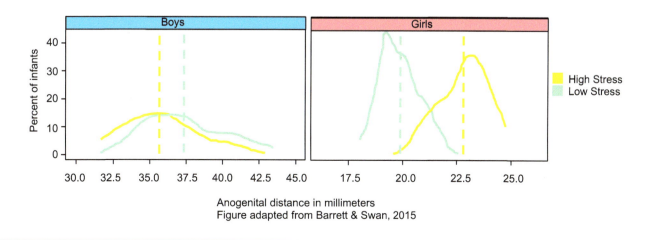

Anogenital distance in millimeters
Figure adapted from Barrett & Swan, 2015

BOX 22.3

HOW DO WE KNOW THAT ANDROGENS AFFECT BEHAVIORAL AND ENDOCRINE RESPONSES TO STRESS?

Androgens, acting through AR, can affect the neuroendocrine response to stressors and/or stress-related behaviors such as anxiety and depression. In fact, hypogonadal men are at risk for depression and that depression is alleviated by testosterone treatments. Similarly, stress causes more severe depressionlike behaviors in gonadectomized male rats compared to intact males.[20] Moreover, testosterone treatment alleviates depression-related symptoms in rats that were exposed to chronic, unpredictable stress and enhances antidepressant effects of tricyclic antidepressant treatments.[21]

Testosterone may alleviate depression in this population by providing negative feedback for the HPA axis through AR activation.[19] Depressed patients tend to have high levels of stress hormones, indicating an overactive HPA axis. In addition to causing depression-related behaviors, chronic unpredictable stress doubles the level of circulating corticosterone in male rats. Not only does testosterone treatment reduce the behavioral effects of chronic, unpredictable stress, it also prevents the increase in corticosterone, indicating that it normalized the HPA axis.

There are many interesting relationships between androgens and stress. Observations of baboons in their natural habitat have shown that in response to a stressful stimulus, T levels respond differently depending on the animal's rank in a dominance hierarchy. Dominant males show an increase in T levels following stress, whereas subordinate males' T levels decrease.[16] The reason for this differential response appears to be related to the animal's normal glucocorticoid levels. The subordinate males are in a state of chronic stress characterized by an elevated basal level of circulating glucocorticoids. In contrast, dominant males have low basal levels of circulating glucocorticoids. In response to an acute stressor, both dominant and subordinate males show rapid increases in circulating glucocorticoid levels and decreases in LH secretion. However, dominant males appear to be less sensitive to the inhibitory effects of glucocorticoids on T production. This differential response of T levels to stressful stimuli promotes the reproductive fitness of the dominant males while reducing that of subordinate males.

SUMMARY

The androgens are best known for their functions in reproduction and anabolic growth, though they have been shown to be important for other physiological functions such as regulating stress responses and guiding neural development. Ongoing research is attempting to more precisely identify the role of androgens in cognitive function, stress reactivity, and mood.

References

1. Barrett ES, Parlett LE, Sathyanarayana S, et al. Prenatal exposure to stressful life events is associated with masculinized anogenital distance (AGD) in female infants. *Physiol Behav*. 2013;114–115:14–20.
1a. Barrett ES, Parlett LE, Sathyanarayana S, Redmon JB, Nguyen RH, Swan SH. Prenatal stress as a modifier of associations between phthalate exposure and reproductive development: results from a multicentre pregnancy cohort study. *Paediatr Perinat Epidemiol*. 2016;30(2):105–114.
2. Barrett ES, Redmon JB, Wang C, Sparks A, Swan SH. Exposure to prenatal life events stress is associated with masculinized play behavior in girls. *Neurotoxicology*. 2014;41:C20–C27.
3. Brown TR. Androgen receptor dysfunction in human androgen insensitivity. *Trends Endocrinol Metab*. 1995;6:170–175.
4. Bruckner T, Catalano R. The sex ratio and age-specific male mortality: evidence for culling in utero. *Am J Hum Biol*. 2007;19:763–773.
5. Döhler KD, Coquelin A, Davis F, Hines M, Shryne JE, Gorski RA. Differentiation of the sexually dimorphic nucleus in the preoptic area of the rat brain is determined by the perinatal hormone environment. *Neurosci Lett*. 1982;33(3):295–298.
6. Donovan JL, DeVane CL, Lewis JG, et al. Effects of St John's Wort (*Hypericum perforatum* L.) extract on plasma androgen concentrations in healthy men and women: a pilot study. *Phytother Res*. 2005;19:901–906.
7. Estrada M, Espinosa A, Muller M, Jaimovich E. Testosterone stimulates intracellular calcium release and mitogen-activated protein kinases via a G protein-coupled receptor in skeletal muscle cells. *Endocrinology*. 2003;144(8):3586–3597.
8. Fix C, Jordan C, Cano P, Walker WH. Testosterone activates mitogen-activated protein kinase and the cAMP response element binding protein transcription factor in Sertoli cells. *Proc Natl Acad Sci USA*. 2004;101(30):10919–10924.
9. He FQ, Fang G, Wang B, Guo XJ, Guo CL. Perinatal stress effects on later anxiety and hormone secretion in male mandarin voles. *Behav Neurosci*. 2015;129(6):789–800.
9a. Heland S, Hewitt JK, McGillivray G, Walker SP. Preventing female virilisation in congenital adrenal hyperplasia: the controversial role of antenatal dexamethasone. *Aust N Z J Obstet Gynaecol*. 2016;56(3):225–232.

10. Hiipakka RA, Liao S. Molecular mechanism of androgen action. *Trends Endocrinol Metab*. 1998;9:317–324.

11. Lindzey J, Kumar MV, Grossman M, Young C, Tindall DJ. Molecular mechanisms of androgen action. *Vitam Worm*. 1994;49:383–432.

12. Lutz LB, Jamnongjit M, Yang WH, Jahani D, Gill A, Hammes SR. Selective modulation of genomic and nongenomic androgen responses by androgen receptor ligands. *Mol Endocrinol*. 2003;17(6): 1106–1116.

13. Luu-The V. Assessment of steroidogenesis and steroidogenic enzyme functions. *J Steroid Biochem Mol Biol*. 2013;137:176–182.

13a. Miller WL, Witchel SF. Prenatal treatment of congenital adrenal hyperplasia: risks outweigh benefits. *Am J Obstet Gynecol*. 2013;208(5):354–359.

14. Narayanan S, Srinivas S, Feldman D. Androgen-glucocorticoid interactions in the era of novel prostate cancer therapy. *Nat Rev Urol*. 2015;13(1):47–60.

15. Pallarés ME, Adrover E, Baier CJ, et al. Prenatal maternal restraint stress exposure alters the reproductive hormone profile and testis development of the rat male offspring. *Stress*. 2013;16:429–440.

15a. Sandman CA, Glynn LM, Davis EP. Is there a viability-vulnerability tradeoff? Sex differences in fetal programming. *J Psychosom Res*. 2013;75(4):327–335.

16. Sapolsky RM. Stress in the wild. *Sci Am*. 1990;262:106–113.

17. Sen A, Prizant H, Hammes SR. Understanding extranuclear (non-genomic) androgen signaling: what a frog oocyte can tell us about human biology. *Steroids*. 2011;76(9):822–828.

18. Ukkola O1, Gagnon J, Rankinen T, et al. Age, body mass index, race and other determinants of steroid hormone variability: the HERITAGE family study. *Eur J Endocrinol*. 2001;145(1):1–9.

19. Viau V. Functional cross-talk between the hypothalamic–pituitary gonadal and–adrenal axes. *J Neuroendocrinol*. 2002;14:506–513.

20. Wainwright SR, Lieblich SE, Galea LA. Hypogonadism predisposes males to the development of behavioural and neuroplastic depressive phenotypes. *Psychoneuroendocrinology*. 2011;36:1327–1341.

21. Wainwright SR, Workman JL, Tehani A, et al. Testosterone has antidepressant-like efficacy and facilitates imipramine-induced neuroplasticity in male rats exposed to chronic unpredictable stress. *Horm Behav*. 2016;79:58–69.

22. Deleted in review.

23. Wilson JD. Role of dihydrotestosterone in androgen action. *Prostate Suppl*. 1996;6:88–92.

24. Wittert G. The relationship between sleep disorders and testosterone in men. *Asian J Androl*. 2014;16(2):262–265.

25. Zuloaga DG, Puts DA, Jordan CL, Breedlove SM. The role of androgen receptors in the masculinization of brain and behavior: what we've learned from the testicular feminization mutation. *Horm Behav*. 2008;53(5):613–626.

CHAPTER

23

Angiotensin—Encyclopedia of Stress

L.A. Campos[1], M. Bader[2], O.C. Baltatu[1]

[1]Anhembi Morumbi University – Laureate International Universities, Sao Paulo, Brazil; [2]Max-Delbruck-Center for Molecular Medicine, Berlin-Buch, Germany

Abstract

The renin–angiotensin system (RAS) is classically known as an endocrine system that regulates cardiovascular and fluid–electrolyte balance. In addition, there are local RASs in a variety of organs, such as heart, vasculature, kidney, adrenal glands, and brain. These local RASs contribute to the endocrine RAS functions and also have organ-specific functions. Both physical and emotional stressors induce activation of endocrine and local RASs designed to preserve cardiovascular homeostasis.

COMPONENTS OF THE RENIN–ANGIOTENSIN SYSTEM

Angiotensinogen

Angiotensinogen (AOGEN) is the only known precursor molecule to be converted by the renin–angiotensin system (RAS) into active angiotensins (Fig. 23.1). It is a glycoprotein with a molecular weight between 61 and 65 kDa, and it shows close structural similarities to the serine protease inhibitor family. Originally found to be synthesized in the liver, AOGEN is known to be present in several other organs, especially in glial cells of the brain.

Renin

Renin is a very specific aspartyl proteinase whose only known substrate is AOGEN. Renin cleaves the decapeptide angiotensin I (Ang I) from the N-terminus of mature AOGEN (Fig. 23.1). Renin is produced primarily by the kidney and released into the circulation but is also produced in other tissues. Renin synthesis is stimulated by the sympathetic nervous system and by loss of volume and sodium.

Angiotensin-Converting Enzyme

Angiotensin-converting enzyme (ACE) is a dipeptidyl-carboxypeptidase that cleaves the C-terminal histidine–leucine of Ang I to generate angiotensin II (Ang II) (Fig. 23.1). It is found in high concentrations in the lung, kidney, and heart and is distributed throughout the brain. High amounts are detected in areas lacking the blood–brain barrier, such as the diencephalon, pituitary, and pineal gland. ACE is also able to hydrolyze other substrates, such as bradykinin, opioid peptides, and substance P. Thus, ACE is additionally involved in physiological mechanisms beyond the RAS.

ALTERNATIVE ENZYMES

There is evidence that proteases other than renin and ACE can produce bioactive angiotensin species. Such proteases include cathepsins, tonin, tissue plasminogen activator, chymase, ACE 2, and other neutral peptidases (Fig. 23.1).

Active Angiotensins

The best-studied peptide of the system is Ang II, and the functions of the RAS are largely attributed to this active metabolite.

Besides Ang II, additional peptides of the RAS have been found to be biologically active, such as Ang III,[14] Ang IV,[9] and Ang II(1–7)[11] (Fig. 23.1). In fact, Ang III may be even more potent than Ang II in the brain when acting on a common receptor. Ang II(1–7) can be formed via an ACE-independent mechanism and has the same potency as Ang II in vasopressin release.

Studies on Ang IV suggest a biological role for this peptide as well.

Receptors

The use of highly specific ligands has led to the description and cloning of two main Ang II receptors, namely AT_1 and AT_2. Most of the known classical cardiovascular effects of Ang II are attributed to the AT_1 receptor. In contrast to humans, two different subtypes have been found for this receptor in rodents: AT_{1a} and AT_{1b}. Whereas AT_{1a} has been localized in blood vessels, kidney, lung, liver, and specific brain areas that are involved in the control of blood pressure and fluid homeostasis, AT_{1b} receptors were originally described in glandular tissues, such as the anterior pituitary and adrenal gland.

ENDOCRINE RENIN–ANGIOTENSIN SYSTEM

The endocrine RAS is dependent on renin released by the kidney into the blood, so that AOGEN is cleaved in the circulation to liberate the inactive decapeptide Ang I. ACE then cleaves Ang I to yield the octapeptide Ang II, the active molecule of the system. Circulating Ang II is one of the most potent vasoconstrictors and acts on specific receptors to elicit the following cardiovascular and renal effects:

- constriction of the vasculature,
- reabsorption of sodium and water in the kidney,
- release of aldosterone and catecholamine from the adrenal glands,

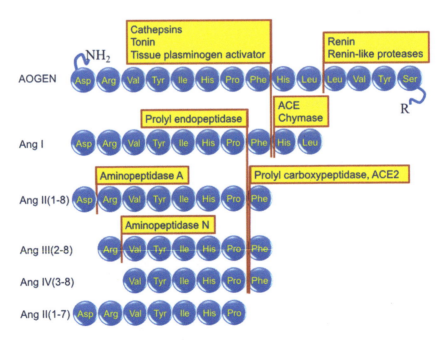

FIGURE 23.1   AOGEN is the only known precursor molecule to be converted by the RAS into active angiotensins.

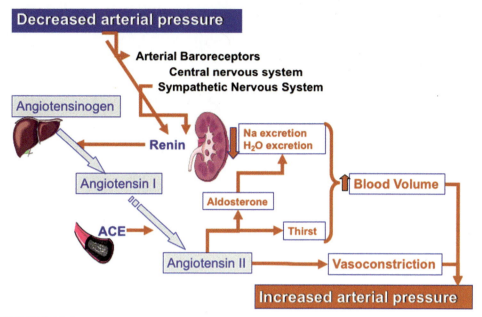

FIGURE 23.2 Summary of the RAS mechanisms in cardiovascular and fluid–electrolyte regulation.

- induction of inotropic and arrhythmogenic effects in the heart,
- induction of thirst and increased sympathetic discharge at specific brain areas outside the blood–brain barrier,
- enhancement of peripheral noradrenergic neurotransmission.

Fig. 23.2 summarizes the RAS mechanisms in cardiovascular and fluid–electrolyte regulation.

LOCAL RENIN–ANGIOTENSIN SYSTEMS

In addition to the circulating RAS, a local formation and action of Ang II has been observed at the level of various organs.[1,7]

The components of the RAS have been detected in several tissues including the heart, vasculature, kidney, adrenal glands, and brain. The local formation and action of Ang II in peripheral organs can therefore occur independent from the circulating RAS. However, RAS components required for angiotensin synthesis, such as renin and AOGEN, can be also taken up from the circulating blood. The latter is possible only for organs accessible to blood proteins, such as the kidney, heart, adrenal glands, and brain areas situated outside the blood–brain barrier. Nevertheless, there are areas in the brain that are inaccessible to the circulating Ang II because of the blood–brain barrier but express functional angiotensin receptors. In these brain regions, only locally formed angiotensins can act.

The tissue RASs seem to regulate long-term organ function and may be responsible for pathological structural changes.

RENIN–ANGIOTENSIN SYSTEM AND STRESS

Both physical and emotional stressors induce peripheral and central responses designed to preserve cardiovascular homeostasis. As one of the main regulators of cardiovascular homeostasis, the RAS is stimulated by stressors that lead to a lowering in the blood pressure, such as hemorrhage, loss of blood volume and sodium, or increased activity of the sympathetic nervous system. The RAS acts synergistically with the sympathetic nervous system, increasing blood pressure and maintaining organ perfusion, and electrolyte and volume balance.[5,7] Recently, it has been demonstrated that Ang II is also an important mediator of inflammation by stimulating oxidative stress.[10]

The physical and behavioral responses to stress are integrated at the level of the central nervous system, including areas known to regulate the homeostasis of the milieu interieur (Fig. 23.3). These areas include hypothalamic (organum vasculosum of the lamina terminalis, OVLT; subfornical organ, SFO; the paraventricular nucleus of the hypothalamus, PVN; supraoptic nucleus, SON) and brainstem nuclei (ventrolateral medulla, VLM; area postrema, AP; dorsal motor nucleus of the vagus, DMV; nucleus tractus solitarius, NTS). AOGEN and the angiotensin receptors are present in these areas that are responsible for the maintenance of homeostasis and participate in the stress response.[2,5–7] The components of the RAS are present in the brain:

- outside the blood–brain barrier—subfornical organ, organum vasculosum of lamina terminalis, area postrema (circumventricular organs), pineal and pituitary glands, which are accessible to angiotensins from circulation and locally produced;

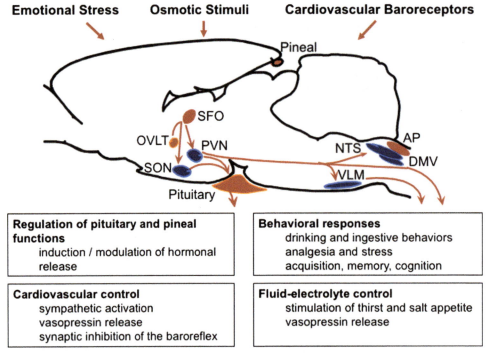

FIGURE 23.3   The physical and behavioral responses to stress are integrated at the level of the central nervous system, including areas known to regulate the homeostasis of the milieu interieur.

- inside the blood–brain barrier—supraoptic and paraventricular nuclei, nucleus tractus solitarius, and ventrolateral medulla, which are accessible exclusively to the brain-born locally produced angiotensins.

The sympathetic nervous system, the component of the autonomic nervous system that is essentially involved in the fight-or-flight reaction, is modulated by the RAS not only in the periphery but also in central areas of autonomic control. During the reaction to stress, there is an increase in blood levels of renin in addition to epinephrine. In the central nervous system, an interrelationship between the activity of the brain RAS and the sympathetic system and other central autonomic pathways has also been shown.[3,4] Stress-induced brain RAS activation may induce endocrine RAS activation through sympathetic pathways that induce renin release from the kidney[6] (Fig. 23.4). Consequently, an increase of plasma Ang II levels can act in brain sites situated outside the blood–brain barrier, interfering with the brain RAS.

Various stressor agents activate the hypothalamic–pituitary axis.[8] Ang II, together with other neurotransmitters and neuromodulators, modulates the activity of the hypothalamic–pituitary axis and stimulates the release of vasopressin and corticotropin-releasing factor (CRF).[2,7] In addition, it has been proposed that the brain

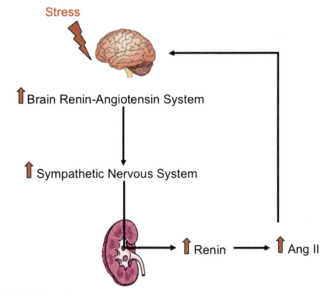

FIGURE 23.4   Stress-induced brain RAS activation may induce endocrine RAS activation through sympathetic pathways that induce renin release from the kidney.

RAS is involved in behavioral responses, such as cognition, memory, and central analgesia.[5,12,13]

Prolonged overactivation of the RAS may lead to a disease state. Drugs that block the actions of the RAS, such as ACE or renin inhibitors and AT1 antagonists, are widely used for the treatment of various cardiovascular diseases, and their effectiveness may be partially due to the inhibition of the brain RAS and thereby the attenuation of the sympathetic output and the stress response.[1,7]

References

1. Bader M. Tissue renin-angiotensin-aldosterone systems: targets for pharmacological therapy. *Annu Rev Pharmacol Toxicol.* 2010;50:439–465.
2. Bader M, Peters J, Baltatu O, Muller DN, Luft FC, Ganten D. Tissue renin-angiotensin systems: new insights from experimental animal models in hypertension research. *J Mol Med Berl.* 2001;79:76–102.
3. Bali A, Jaggi AS. Angiotensin as stress mediator: role of its receptor and interrelationships among other stress mediators and receptors. *Pharmacol Res.* 2013;76:49–57.
4. Bali A, Singh N, Jaggi AS. Renin-angiotensin system in pain: existing in a double life? *J Renin Angiotensin Aldosterone Syst.* 2014;15:329–340.
5. Baltatu O, Bader M. Brain renin-angiotensin system. Lessons from functional genomics. *Neuroendocrinology.* 2003;78:253–259.
6. Baltatu O, Campos LA, Bader M. Genetic targeting of the brain renin-angiotensin system in transgenic rats: impact on stress-induced renin release. *Acta Physiol Scand.* 2004;181:579–584.
7. Baltatu OC, Campos LA, Bader M. Local renin-angiotensin system and the brain–a continuous quest for knowledge. *Peptides.* 2011;32:1083–1086.
8. Grippo AJ, Johnson AK. Stress, depression and cardiovascular dysregulation: a review of neurobiological mechanisms and the integration of research from preclinical disease models. *Stress.* 2009;12:1–21.
9. Hallberg M. Targeting the insulin-regulated aminopeptidase/AT4 receptor for cognitive disorders. *Drug News Perspect.* 2009;22:133–139.
10. Husain K, Hernandez W, Ansari RA, Ferder L. Inflammation, oxidative stress and renin angiotensin system in atherosclerosis. *World J Biol Chem.* 2015;6:209–217.
11. Ohishi M, Yamamoto K, Rakugi H. Angiotensin (1-7) and other angiotensin peptides. *Curr Pharm Des.* 2013;19:3060–3064.
12. von Bohlen und Halbach O, Albrecht D. The CNS renin-angiotensin system. *Cell Tissue Res.* 2006;326:599–616.
13. Wright JW, Harding JW. The brain renin-angiotensin system: a diversity of functions and implications for CNS diseases. *Pflugers Arch.* 2013;465:133–151.
14. Yugandhar VG, Clark MA. Angiotensin III: a physiological relevant peptide of the renin angiotensin system. *Peptides.* 2013;46:26–32.

24

Stress, Angiotensin, and Cognate Receptors

G. Aguilera

National Institutes of Health, Bethesda, MD, United States

Abstract

Angiotensin II (AngII), the end product of the renin–angiotensin system (RAS), regulates cardiovascular homeostasis, acting upon type 1 AngII receptors (AT1R) coupled to calcium-/phospholipid-dependent pathways. In addition, AngII acts as neuropeptide in the brain, mediating the neuroendocrine regulation of blood pressure; sympathetic outflow; water and energy homeostasis; and modulating behavioral, autonomic, and hormonal responses to stress. The neuroendocrine effects depend on AT1R activation in brain areas outside the blood-brain barrier, which stimulate angiotensinergic pathways projecting to the hypothalamic paraventricular nucleus, median preoptic area, and other nuclei-containing AT1R. AngII also activates subtype 2 receptors (AT2), which counteracts AT1R activity. Other angiotensin peptides (Ang(1-7) and Ang IV), and their cognate receptors (Mas and AT4-IRAP), also oppose AT1R actions. These peptides and receptors are critical for balancing the effects of AngII/AT1R in homeostatic control and have relevance for clinical effects of drugs targeting the RAS.

INTRODUCTION

The octapeptide, angiotensin II (AngII), the final product of the renin–angiotensin system (RAS), is well recognized as a pressor agent and regulator of aldosterone secretion.[37] Initially considered a peripheral hormonal system targeting vascular smooth muscle and the adrenal zona glomerulosa, it has become evident that AngII can be formed locally in tissues, including the brain, independently from the circulating components.[5] In the brain, AngII acts as a neuropeptide, having dramatic effects on central cardiovascular control, fluid and electrolyte balance, and stress responses, such as activation of the hypothalamic–pituitary–adrenal (HPA) axis and the sympathoadrenal system (Fig. 24.1).

KEY POINTS

- Angiotensin II (AngII), the end product of the classical endocrine renin–angiotensin system (RAS), is a major regulator of blood pressure and aldosterone secretion.

- AngII, also produced by local RASs in the kidney, adipose tissue, other peripheral organs, and the brain, contributes to systemic and central control of blood pressure, water and salt balance, sympathetic and hypothalamic–pituitary–adrenal axis activity, behavior, and energy homeostasis.

- AngII exerts its actions by binding to two G protein–coupled receptors, type 1 (AT1R), which mediates the classic effects of the peptide, and type 2 (AT2R), which opposes most actions of AT1Rs.

- Previously thought inactive angiotensin peptides such as Ang(1-7) and Ang IV, formed in the brain and other tissues interact with cognate receptors (Mas, and AT4-IRAP, respectively), also counteracting the actions AT1Rs.

- Stress is a potent stimulator of the RAS and increases circulating Ang II levels.

- Circulating AngII interacts with AT1Rs outside the blood–brain barrier, and activates angiotensinergic pathways projecting to hypothalamic and limbic nuclei responsible for regulation of blood pressure; sympathetic outflow; water and energy homeostasis; and modulating behavioral, autonomic, and hormonal responses to stress.

- The balanced interaction of circulating- and tissue-produced active angiotensin peptides acting upon multiple receptors with diverse signaling mechanisms, mediate the pleotropic actions of the renin–angiotensin systems in cardiovascular homeostasis and stress adaptation.

The components of the RAS include the enzymes responsible of AngII formation from the substrate, angiotensinogen (Agt), as well as the receptors mediating the pleotropic effects of the peptide. Adding to the complexity of the RAS, research during the last four decades has uncovered biological effects for AngII fragments previously thought inactive. These peptides activate novel receptors and signaling pathways, many of them counteracting the classical effects of AngII. The coordinated action of the classical and novel angiotensin systems is critical for cardiovascular regulation and neuroendocrine control of homeostasis.

Following a description of the classical and nonclassical components of the RAS, this chapter will focus on the neuroendocrine effects of angiotensin peptides and discuss the participation of different angiotensin peptides and receptors in maintaining homeostasis.

COMPONENTS OF THE CLASSICAL RAS

The RAS comprises the enzymes and substrates required for the formation of the active octapeptide, AngII, including (a) the precursor protein, Agt, (b) the enzyme, renin, (c) the decapeptide angiotensin I (Ang I), and (d) angiotensin-converting enzyme (ACE), as well as the specific receptors that recognize AngII in the target tissues (Fig. 24.2).

Renin

Renin, an aspartyl protease produced by the juxtaglomerular apparatus of the kidney nephron, cleaves the decapeptide angiotensin I (Ang I), the direct precursor of AngII, from the amino terminus of the substrate Agt. The renin precursor, preprorenin, undergoes cleavage in the rough endoplasmic reticulum to yield the still inactive 60 kDa protein, prorenin. Prorenin stored in the cells is cleaved to active renin and secreted to the circulation in response to specific stimuli. In addition, a splice variant lacking the signal peptide remains intracellular, but it is active and capable of generating AngII in tissues such as the brain.[15]

Prorenin is also constitutively secreted (nonregulated) and circulates at levels higher than those of active renin (up to 100-fold). Prorenin is inactive, with the 43 amino terminus amino acids folded over preventing contact of the enzymatic cleft with Agt. However, prorenin becomes active through conformational changes following interaction with specific receptors in the periphery and brain. There is evidence that the prorenin receptor can generate Ang I from Agt and that it may be part tissue RASs. However, its role as part of the RAS has been disputed mainly because of its intracellular localization. In addition, the prorenin receptor is a subunit of the vacuolar H^+-ATPase, acts as a scaffold component required for the Wnt/β-catenin pathways and has critical roles in cell function independently of prorenin binding.[7] The exact role of prorenin receptors as part of the RAS is still controversial but its high expression in areas involved in blood pressure and water balance suggest an involvement in central cardiovascular control.

The kidney is the source of renin secreted to the circulation, and the production of renin from prorenin and its secretion into the blood is the rate-limiting step in the activation of the peripheral RAS. In addition, renin mRNA is present in the adrenal, brain, and other tissues, which may contribute to local AngII production. Consistent with the role of the RAS in cardiovascular

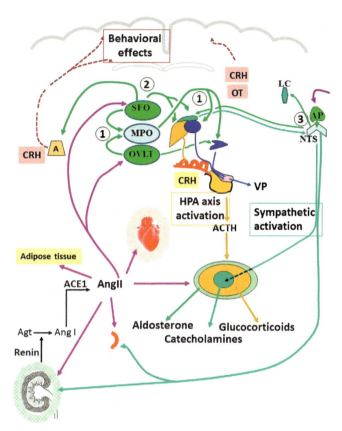

FIGURE 24.1 **Hormonal and neuropeptide effects of angiotensin II (AngII).** The hormonal effects of circulating AngII [generated by activation of the RAS (*black arrows*)], including adrenal aldosterone production, vasoconstriction, renal sodium reabsorption, cardiomyocyte contractility and growth, adipose tissue metabolism, and activation of brain areas outside the blood–brain barrier are indicated by the solid *purple arrows*. The patterned surfaces surrounding the kidney, heart, and adipose tissue indicate the additional effect of locally produced AngII by tissue RAS. Binding of circulating AngII to AT1R in the subfornical organ (SFO), vascular organ of the lamina terminalis (OVLT), and area postrema (AP) (green) activate angiotensinergic pathways responsible for the neuroendocrine effects of AngII. This include (1) projections from the SFO and OVLT to the median preoptic nucleus (MPO), essential relay nucleus for stimulation of vasopressin (VP) secretion from magnocellular neurons in the paraventricular and SON (blue) and dipsogenic responses (not shown). (2) Direct projections from the SFO to CRH neurons in the amygdala (A) and parvocellular hypophysiotropic (dark yellow) and autonomic (turquoise) CRH neurons in the PVN, responsible for HPA axis stimulation and sympathetic activation, respectively. The AP conveys chemoreceptor and baroreceptor signals to the nucleus of the solitary tract (NTS), which relays the information to autonomic CRH neurons in the PVN, the locus coeruleus (LC), and other noradrenergic areas (not shown) responsible for sympathetic activation and catecholaminergic transmission in the brain. The behavioral effects of Ang II promoting fear memory and anorexia are at least in part mediated by CRH and CRH and oxytocin (OT), respectively.

and water/sodium homeostasis, regulation of renin release depends on changes in blood pressure, plasma volume, and sodium balance.[37]

Angiotensinogen

Agt, a large alpha 2 globulin, is the sole substrate of renin in the formation of Ang I, the immediate precursor of AngII. In addition, Agt can be hydrolyzed by other enzymes to generate Ang(1-7) and desaspartic 1 AngII (or AngIII). Formation of Ang I from Atg is the rate limiting step in AngII generation. Circulating levels of Agt are sufficiently close to the Michaelis constant of renin for Atg that small variations in its level can influence the activity of the RAS. In this regard, genetic manipulations in mice or polymorphisms in the Agt gene leading to increased plasma Atg levels are associated with hypertension.[29]

The regulation of Atg production is multifactorial and occurs at the transcriptional and posttranscriptional levels. Several factors influence Atg production including AngII itself, glucocorticoids, estrogens, thyroxine, growth hormone, and various cytokines. Physiological and pathological conditions, such as pregnancy, use of oral contraceptives, increased glucocorticoid secretion (due to adrenal hyperfunction or chronic stress), and chronic inflammation can lead to increased Agt production. Conversely, pituitary, thyroid, gonads, or adrenals deficiency is associated with low plasma Agt levels. The main source of circulating Agt is the liver hepatocyte, which secretes Agt through constitutive pathways, leading to relatively stable plasma levels, irrespective of acute stress or cardiovascular changes. In addition to the liver, tissues such as adipocytes and epithelial cells of the renal proximal tubule, and the brain express Agt, and production in

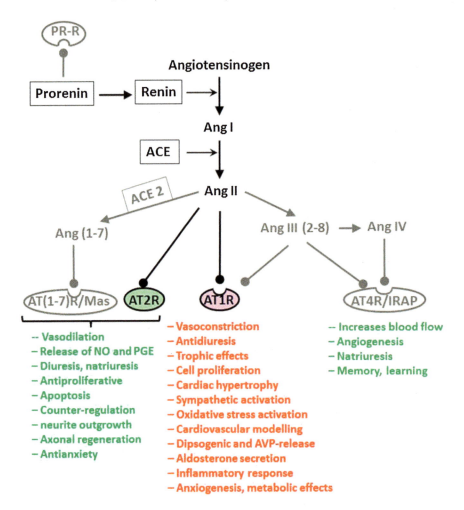

FIGURE 24.2 **Schematic depiction of the classical (black) and nonclassical (gray) renin–angiotensin systems.** Renin derived from its precursor, prorenin, cleaves the amino terminus decapeptide angiotensin I from the precursor angiotensinogen (Agt), which is converted to the active octapeptide angiotensin II (AngII) by the carboxy-peptidase, angiotensin-converting enzyme 1 (ACE1). AngII acts by binding to type 1 (AT1) and type 2 (AT2) receptors in the target tissues. AngII is converted to Ang(1-7) by ACE2, which interacts with its receptor, Mas. Other peptidases convert AngII to heptapeptide, Ang 2-8 (also known as AngII or des-aspartic AngII), which also interact with AT1R and the hexapeptide Ang IV, which interacts with the AngIV receptor or insulin-regulated aminopeptidase (IRAP). Prorenin can be activated in tissues by binding to the prorenin receptor. The biological consequences of activation of AT2R, Ang(1-7)R/Mas and AT4R/IRAP (green) are in general opposed to the effects of AT1R (red).

these tissues contributes to activation of local RASs. In addition, evidence of a positive correlation between BMI and circulating Atg levels suggests that Atg production in adipocytes contributes to circulating Atg levels.

In the brain, Agt is broadly distributed but with higher levels in areas related to blood pressure control. At the cellular level, expression occurs predominantly in astrocytes but also in neurons. Evidence from transgenic mice and rats suggests that Agt expression in the brain is part of a local RAS responsible for AngII production within the brain.[10,28]

Angiotensin-Converting Enzyme

ACE is a zinc metalloprotease that removes the C-terminal histidyl-leucine fragment from inactive AngI to form the active octapeptide, AngII [Ang-(1-8)]. This ACE, also known as angiotensin-converting enzyme 1 (ACE1), is located on the plasma membranes of various cell types, including the luminal surface of vascular endothelial cells, brush border epithelial cells in the renal proximal tubule cells, and discrete areas in the brain. ACE with characteristics identical to those of the membrane enzyme is also present in the circulation, and it is probably derived from clearance of membrane-bound enzyme. As a dipeptidyl carboxypeptidase, ACE hydrolyzes and inactivates a wide range of substrates in addition to Atg. These include bradykinin, met-enkephalin, dynorphin, substance P, neurotensin, and gonadotropin-releasing hormone.[19]

In the brain, the presence of ACE in areas also expressing AngII receptors, such as the circumventricular organs (including the subfornical organ (SFO) and area postrema) and choroid plexus suggest that ACE is part of a tissue RAS, which contributes to local production of AngII in the brain.[9] The expression of abundant ACE in extrapyramidal areas lacking of Ang receptors, such as the striatum and substantia nigra, suggesting that in these regions ACE plays a role unrelated to the RAS, probably associated with its ability to degrade other neuropeptides.[46]

NONCLASSICAL ANGIOTENSIN PEPTIDES

In addition to AngII, alternative pathways lead to the formation of biologically active peptides from AngI or AngII.[13,42] These include AngII fragments, initially thought to be inactive breakdown products of AngII, such as des-aspartic 1 AngII (AngIII), Ang (3-8) of Ang IV, and Ang(1-7). The major alternative pathways include the following: (a) conversion of AngII to the heptapeptide, AngIII (or des-aspartic AngII) through cleavage of the N-terminus aspartic acid by glutamyl aminopeptidase A. AngIII interacts with the same receptors as AngII and has actions similar to AngII. (b) Conversion of AngIII to the hexapeptide, Ang IV through cleavage of the arginine residue by the membrane alanyl aminopeptidase N. By interacting with specific receptors (see section Ang IV Receptor), the hexapeptide opposes the cardiovascular actions of AngII and has important functions in the brain related to memory and cognition. (c) Conversion of AngII to the heptapeptide, Ang(1-7), by removal of the carboxy-terminus, phenylalanine by an exopeptidase called angiotensin-converting enzyme 2 (ACE2). The enzyme is present in the endothelium of blood vessels and in several tissues including the brain. The physiological effects of Ang(1-7) oppose the pressor, profibrotic, prothrombotic, and anxiogenic effects of AngII. Administration of the peptide in rodents induces vasodilation, natriuresis, and inhibits AngII-induced angiogenesis and cell growth. The putative receptor for Ang(1-7) is the seven-transmembrane protein Mas, first identified as an oncogene (see section Ang(1-7) and Mas).

Other bioactive angiotensin peptides identified by mass spectrometry in blood and tissues, include the octapeptide, Ang A, and the heptapeptide, almandine, in which the amino terminus aspartic acid is substituted by alanine, from carboxylation of the aspartate residue in the amino terminus. These less understood peptides act through binding to a Mas-related G protein–coupled receptor (MrgD), causing vasodilation and could have a role in pain perception in the brain.[42]

RECEPTORS FOR ANGIOTENSIN PEPTIDES

Angiotensin II regulates cellular function by interacting with specific plasma membrane receptors in the target tissues.[17] These receptors have two major functions: first,

detection and recognition of the peptide, and second, transmission of information (signaling) to intracellular compartments leading to modification of cell function. The binding properties of AngII receptors were originally characterized using radiolabeled angiotensin analogs in membrane rich fractions or dispersed cells from adrenal cortex and smooth muscle. The affinity of AngII to its receptor is in the nanomolar range, and there is a close correlation between the dissociation constants and the biological actions of the peptide in vitro. Modified angiotensin peptides, or analogs, capable of binding to the receptor can either activate the receptor acting as agonists or are unable to couple to transduction mechanisms, thus acting as receptor antagonists.

Two major AngII receptor subtypes have been cloned, type 1 (AT1) and type 2 (AT2) receptors, which are indistinguishable by their binding affinities to AngII and peptides analogs. However, they differ pharmacologically by their sensitivity to disulfide bond reducing agents and ability to bind two classes of nonpeptide AngII receptor antagonists. Structurally, both receptor subtypes belong to the G protein–coupled receptor family, but they share only 32%–34% homology and differ in their tissue distribution and actions.[25] In addition to the classical AT1 and AT2 receptors, more recent studies have identified receptors for nonclassical components of the RAS. These include the Ang (1–7) Mas receptor, the Mas-related G protein–coupled receptor (MrgD) (cited in section Components of the Classical RAS), the Ang IV receptor (AT4R), later identified as the insulin-regulated aminopeptidase (IRAP), and the prorenin receptor (discussed in section Renin).

Type 1 AngII Receptors (AT1R)

The structural characteristics of AT1Rs are similar across all species studied. It contains 359 amino acids, two disulphide bridges, is glycosylated, and has several phosphorylation sites. The structure exhibits seven-transmembrane domains with the characteristics of a G protein–coupled receptor. Rodents have two closely related forms of AT1R, designated AT1a and AT1b. They exhibit high amino acid homology in their coding regions and almost identical binding properties and signaling mechanisms, but their noncoding regions differ significantly. Both subtypes of AT1R are present in most AngII-responsive tissues but their relative abundance varies at different sites. Although humans have a single AT1R, the existence of two receptors in rodents should be taken into consideration when using AT1R knockout mouse or rat models.[33,25]

Consistent with their role mediating the classic actions of AngII on aldosterone production and blood pressure, AT1Rs are most abundant in the adrenal zona glomerulosa, vascular smooth muscle, and kidney, and in areas of the brain involved in cardiovascular regulation and catecholamine release. AT1Rs in the brain and in the anterior pituitary mediate the neuroendocrine effects of AngII. In addition, AT1Rs are present in a number of peripheral tissues, including liver, gonads, and urogenital tract, where their physiological role is not fully understood.

The AT1R activates multiple signaling pathways dependently or independently of their coupling to heterotrimeric guanyl nucleotide binding proteins (G proteins)[22,25]: First, the predominant pathway involves coupling to Gq/11 with activation of phospholipase C, leading to increased inositol triphosphate and diacylglycerol generation and consequent increases in intracellular calcium and protein kinase C (PKC) activity. Calcium and PKC regulate a number of cell processes leading to hormone secretion, vascular, or cardiomyocyte contraction. In addition, these messenger systems activate additional pathways including (a) calcium-mediated activation of calcium/calmodulin-dependent kinase II (CamK II), which mediates the effects of AngII on cell growth; (b) delayed activation of phospholipases D, which contributes to sustained vascular muscle contraction; (c) activation of phospholipase A, leading to increases in arachidonic acid and its metabolites, which may serve as a feedback mechanism by counteracting vasoconstriction; (d) transactivation of growth factor receptors and consequent activation of mitogen-stimulated protein (MAP) kinase; (e) activation of small G proteins, such as Ras, Rho, and Rac, leading to activation of the MAP kinase pathway. Second, signaling through the release of Gβγ subunits following coupling of the AT1R with Gα, leads to stimulation of the Jak-STAT and activation of the mitogen-activated protein kinase pathways. Third, AT1Rs can couple to Gi leading to a decrease in adenylate cyclase activity. Fourth, activation of AT1Rs can elicit delayed effects by interacting with a number of scaffolding proteins, such as β-arrestin and AT1R-interacting scaffold proteins. It is well known that β-arrestin allows AT1R phosphorylation by G-protein receptor kinases leading to homologous desensitization of the receptor. In addition to terminating receptor activation, β-arrestin can cause delayed ERK-MAP kinase activation confined to the cytoplasm. The role of cytoplasmic phosphorylated ERK is not clear but it is not involved in transcriptional regulation. Activation of these multiple signaling mechanisms modulate ion channels, membrane potential, and the transcription of a number of genes in the cell, and is responsible for the wide range of actions of AngII mediated by AT1R.

Mapping of AT1Rs in the brain of a number of species by in vitro autoradiography and in situ hybridization histochemistry shows a highly conserved distribution across mammalian species.[4] In general, there is a good correlation between the presence of receptor protein (binding autoradiography) and AT1R mRNA (in situ hybridization). Mismatch can occur when receptors synthesized in cell bodies are transported to presynaptic terminals in distal sites, such as the nucleus of the solitary tract, the median eminence, and bed nucleus of the stria terminalis.

The highest levels of AT1R in the brain occurs in areas located outside the blood–brain barrier and are therefore responsive to circulating AngII, such as the SFO, vascular organ of the lamina terminalis (OVLT), area postrema, median eminence, and choroid plexus. These highly vascularized structures serve as a linkage between the peripheral circulation and the central nervous system, and mediate effects of changes in circulating AngII on water and salt appetite, blood pressure, and pituitary hormone release. AT1Rs are also present in a number of areas inside the blood–brain barrier, including (a) hypothalamic nuclei such as the paraventricular nucleus (PVN), together with the median eminence important in the regulation of pituitary function, and the median preoptic nucleus; (b) sites in the midbrain and hindbrain, such as the lateral parabrachial nucleus, a region implicated in the regulation of fluid balance, nucleus of the solitary tract, dorsal motor nucleus of the vagus, intermediate reticular nucleus, and rostral and caudal ventrolateral medulla, all areas involved in autonomic activity and cardiovascular reflexes.

Type 2 Angiotensin Receptors

The AT2R has a molecular weight ranging from 70 to 113 kDa depending on the degree of glycosylation. It contains two disulfide bridges, which are responsible for characteristic ligand binding potentiation by disulfide bond reducing agents, such as dithiothreitol. The sequence of rat and mouse AT2Rs is almost identical, and they share about 70% homology with the human receptor. As for the AT1R, the AT2R has seven-transmembrane domains and the structural characteristics of a G protein–coupled receptor, but its signaling mechanisms are atypical and appear to vary in different tissues.[23] Consistent with a lack of evidence of coupling to stimulatory G-proteins, AT2Rs fail to activate classical signaling pathways, such as cyclic AMP, inositol triphosphate production, or intracellular calcium levels. In contrast to AT1, AT2Rs are not regulated through phosphorylation or undergo internalization and recycling.

A number of signaling mechanisms have been implicated in the actions of AT2Rs, including (a) inhibition of inositol phosphates generation, induction of nitric oxide synthase and, inhibition of Na-ATPase in the proximal tubule in the kidney; (b) stimulation of vanadate-sensitive tyrosine protein phosphatase and Ser/Thr phosphatase activity, either in a G protein–dependent or independent manner, according to the tissue. The resultant protein dephosphorylation may limit kinase activation and mediate the antagonistic action of AT2Rs on AngII and growth factor–induced cell growth; (c) G protein–independent signaling through receptor coupling to scaffolding proteins, such as the AT2R interacting protein, which inhibits growth factor–stimulated kinase. In addition, interaction with the transcription factor, promyelocytic zinc finger protein, promotes internalization and perinuclear localization of the receptor; (d) nitric oxide–mediated stimulation of soluble guanylyl cyclase activity mediates beneficial effects of AT2Rs in the cardiovascular system and in the brain; (e) coupling to ion channels, including T-type calcium channels and the Kv2.1 delayed rectifier potassium channel, which have impact on cell growth, differentiation, and apoptosis.[25,26]

Rodents and primate AT2Rs are abundantly expressed in the fetus and levels decrease markedly after birth.[14] In the adult, AT2R binding is present only in a few tissues including the nonpregnant uterus, ovarian granulosa cells, adrenal zona glomerulosa, Langerhans islets, and in selected areas of the brain. In situ hybridization studies in transgenic AT2R -eGFP reporter mice revealed localization in neurons within nuclei that regulate blood pressure, metabolism, and fluid balance, as well as limbic and cortical areas involved in stress responses and mood.[46] Usually, AT1 and AT2Rs colocalize within a nucleus but rarely in the same neuron, with AT2Rs being presynaptic and AT1R usually postsynaptic. Within the nucleus of the solitary tract, AT2R are in glutamatergic neurons but not in tyrosine hydroxylase expressing neurons. In the PVN, they are confined to afferent terminals and GABAergic interneurons surrounding the AT1R-rich PVN. This suggests that the opposing effects of AT1 and AT2Rs in the brain depend on functional interaction between neurons rather than on activation of different signaling pathways within a neuron.

The high expression of AT2Rs in the fetus and situations involving tissue remodeling in adults, such as wound healing, suggest an important role in growth and differentiation. However, the role of the receptor during development is not clear since AT2R knockout mice are developmentally normal. Consistent with opposing the effects of AT1Rs, these mice show enhanced pressor responses to AngII, in addition to anxietylike behavior and reduced exploratory behavior. There is also evidence that AT2R activation stimulates neurite outgrowth in neuronal cultures in vitro.[25,26]

Ang(1-7) and Mas

Angiotensin (1-7) (Ang(1-7)), initially believed to be an inactive metabolite of AngII, was later found to be a ligand for the orphan G protein–coupled receptor, Mas, and to induce signaling in cells transfected with Mas. Tissues of Mas knockout mice lack of Ang(1-7) binding and show blunted responses to the Ang(1-7) in vitro. Similarly, the diuretic effects of the peptide are absent in Mas-null mice. Although the above evidence supports the role of Mas as the Ang(1-7) receptor, there are few radioligand binding and pharmacological data confirming a direct interaction between Ang(1-7) and MAS.[40]

In spite of having the characteristics of a G protein–coupled receptor, there is no evidence that ANG(1-7) activates conventional G protein signaling, such as calcium, IP3, and cAMP in cells transfected with Mas.[36] Several reports have shown activation G

protein–independent pathways in cells transfected with Mas, including, (a) calcium-independent activation of protein kinase B (or AKT) with subsequent activation of nitric oxide synthase and NO production, and (b) activation of phospholipase A and arachidonic acid production. While activation of these pathways by Ang(1-7) may mediate some biological effects of the peptide, several aspects of Mas signaling remain unclear. Discrepancies between Ang(1-7) responses of cells expressing endogenous Mas and transfected cells raise the possibility that Mas coupling to Gs requires the interaction with other proteins absent in COS or CHO cells. In addition, Mas overexpression in cardiomyocytes elicits ligand-independent effects, suggesting that Mas has ligand-independent constitutive activity and that Ang(1-7) may act by antagonizing Mas endogenous activity rather than as an agonist. Lastly, several natural and synthetic ligands, including neuropeptide F and several peptide and non-peptide ligands elicit G protein–dependent signaling, suggesting that Ang(1-7) is not the only ligand for Mas.

The Mas gene is widely expressed in the body, predominantly in vascular endothelial cells, with the highest expression in neurons of the brain and in Leydig and Sertoli cells in the testis. In the brain, Mas expression is high in the hippocampus, and number of areas in the medulla and hypothalamus related with cardiovascular and neuroendocrine regulation, including the nucleus of the solitary tract, caudal and rostral ventrolateral medulla, inferior olive, parvo and magnocellular portions of the paraventricular hypothalamic nucleus, supraoptic nucleus (SON), and lateral preoptic area.[8] These receptors are likely to mediate the opposing effects of Ang(1-7) on cardiovascular and neuroendocrine regulation, as well as the neuroprotective effects of the peptide.[6]

Ang IV Receptor

The Ang IV receptor, originally defined as a specific, high-affinity binding site for the hexapeptide angiotensin 3-8 (Ang IV), was later identified as the transmembrane enzyme, IRAP. IRAP is a 160–190 kDa membrane spanning protein shown to homodimerize and act in a G protein–independent manner. Ang IV binds AT4R-IRAP (nanomolar affinity) inhibiting its peptidase activity. In addition, AT4-IRAP interacts with other peptides including the globin fragment, LVV-hemorphin-7, oxytocin and vasopressin, and IRAP degrade neuropeptides including oxytocin.[41]

AT4-IRAP is widely expressed in the body and is predominantly found in GLUT4 vesicles in insulin responsive cells. In the brain it is located in areas associated with cognition and sensory function. Central administration of either Ang IV or its analogs, or LVV-hemorphin-7, induces c-fos expression in brain areas expressing AT4/AT4R-IRAP, as well as enhancing learning and memory in normal rodents and reversing memory deficits in animal models of amnesia.[45,46]

It is clear that AT4R-IRAP binds Ang IV, and the fact that deletion of the AT4R-IRAP gene in mice abolishes Ang IV binding confirms that the peptidase acts as the Ang IV receptor. It is also well established that central Ang IV administration has neuroprotective and cognitive effects but it is unclear whether AT4R-IRAP is the sole mediator of the effects of Ang IV. The signaling mechanisms of AT4R-IRAP are not fully understood. Ang IV can activate different signaling pathways, such as cytosolic calcium, MAP kinase, cyclic GMP production, and NOS, in cells expressing endogenous receptors, but it is inactive in CHO cells transfected with AT4R-IRAP, indicating that AT4R-IRAP alone cannot mediate the effects of Ang IV. In addition, there is no existing information concerning the ability of Ang IV to enhance cognitive function in AT4R-IRAP null mice. Thus, further studies are needed to elucidate its signaling mechanisms and to fully validate the physiological actions of the AngIV/IRAP system.

NEUROENDOCRINE ACTIONS OF ANGIOTENSIN

Angiotensin peptides can regulate neuronal function by acting as a neurotransmitter or as a hormone (Fig. 24.1). In general, all components of the RAS including the cognate receptors in target neurons exist in the brain. In spite of conflicting information concerning colocalization of the substrate and processing enzymes in the same cell, there is good evidence that angiotensin can be produced and released in neurons. Based on immunocytochemical studies, in several brain nuclei neurons containing AngII-like immunoreactivity transport the peptide through axonal projections to nerve endings, where the peptide is utilized as a neurotransmitter. This is clear for the SFO in which angiotensinergic neurons send AngII immunoreactive projections to neurons in the median preoptic nucleus, SON, and PVN, all areas containing angiotensin receptors.[12,32]

In addition to AngII produced locally in the brain, circulating AngII can affect central activity acting as a hormone. While unable to cross the blood–brain barrier, AngII modifies central activity by acting upon AT1Rs located in areas outside the blood–brain barrier in the circumventricular organs. These structures are an important component of the neuroendocrine effects of AngII on cardiovascular homeostasis and stress adaptation by interfacing the peripheral and central RASs.[16,32]

Angiotensin and Hypothalamic–Pituitary–Adrenal Axis Activity

Activation of the HPA axis, the main hormonal response to stress, is initiated by rapid release of corticotropin releasing hormone (CRH) from parvocellular neurons of the PVN into the portal circulation, with subsequent release of ACTH from the anterior pituitary and glucocorticoids from the adrenal zona fasciculata. The main mechanism by which AngII influences HPA axis activity involves regulation of the CRH neuron. Both the AngII peptide and AT1Rs are present in the PVN. Immunoreactive AngII is present in nerve endings originating in the SFO and other circumventricular organs are found in close proximity to CRH neurons. Activation of this angiotensinergic pathway by peripheral AngII plays an important role in the stimulation of the HPA axis observed during decreases in blood volume or exogenous AngII administration. In addition, there is colocalization of immunoreactive AngII in CRH neurons suggesting that the peptide could act as an autocrine factor.

Together with the circumventricular organs, the PVN is one of brain nuclei containing the highest concentration of AT1Rs, and these receptors are located in parvocellular CRH neurons.[3] Stress causes At1R upregulation due to the increased glucocorticoid levels, as receptor mRNA and ligand binding decrease following adrenalectomy and increase after glucocorticoid administration.[1] Central injection of a small dose of angiotensin II increases ACTH secretion and CRH mRNA in the PVN, while injection of the AT1 antagonist, losartan, reduces acute stress-induced CRH mRNA and sympathetic activation but not the increases in plasma ACTH and corticosterone.[1] In contrast, peripheral administration of the nonpeptide AT1R antagonist, candesartan, abolishes HPA responses to isolation stress, including elevations in plasma ACTH and corticosterone, the increase of AT1R expression in the PVN, and the decreases in CRH mRNA and CRH immunoreactivity observed during isolation stress.[39] This suggests that AT1Rs are required for maintaining HPA axis responses during chronic stress. In addition, it is likely that in the absence of active AT1Rs, the unbalanced action of AngII (and probably other Ang peptides) upon other receptor subtypes, such as AT2 or Mas, contribute to the long-term effects of the AT1 antagonist. Being a proinflammatory peptide, AngII can induce CRH expression by stimulating TNF and other cytokines from astroglia.[39]

At the pituitary level, several in vitro studies in rodents and primates have shown that AngII has the potential to directly stimulate ACTH secretion. Pituitary corticotrophs of rats and nonhuman primates express AT1Rs, and these receptors are exposed to peripheral AngII. Angiotensin II alone is a weak stimulator of ACTH secretion, but the peptide potentiates the stimulatory effect of CRH. Injection of AngII increases plasma ACTH in vivo, but the fact that a CRH antiserum prevents this effect suggests that the stimulation is primarily mediated by CRH release.[38] Although it is clear that AngII can directly modulate ACTH secretion in the pituitary, the physiological relevance of this effect during stress is unclear. Thus, it is possible that AngII has other roles in the pituitary, such as influencing corticotroph growth and differentiation, as shown for vasopressin.

At the adrenal level, the functional involvement of AngII in the zona fasciculata may be species-dependent. Although some species, such as bovine and primates, express AngII receptors in the adrenal zona fasciculata and AngII increases cortisol production in vitro, the physiological significance of such receptors is unclear.

Angiotensin and Regulation of Sympathetic Activity

The major autonomic response to stress is activation of the sympathetic system, which is closely related to the RAS. First, the rapid stimulation of sympathetic activity during stress stimulates beta adrenergic receptors in the juxtaglomerular apparatus increasing the secretion of renin and consequent increase in AngII formation. In its turn, AngII stimulates the sympathetic nervous system acting either as a hormone in the circumventricular organs or as a neurotransmitter within the brain.[35] A single peripheral or central injection of AngII induces transient increases in catecholamine secretion, mostly epinephrine from the adrenal medulla, and norepinephrine from nerve endings. In contrast, central administration of the AT1 antagonist, losartan, reduces the increase in blood catecholamine in response to immobilization stress in rats.[24] Likewise, chronic peripheral administration of the AT1 antagonist, candesartan, reduces the increase in urinary catecholamine in response to isolation stress.[39]

Circulating AngII stimulates catecholamine secretion through AT1Rs in the SFO and area postrema, since lesions of these areas in experimental animals attenuate catecholamine responses to AngII injection. AT1R in the SFO and area postrema activate angiotensinergic pathways, with consequent release of AngII from nerve endings synapsing with sympathetic CRH neurons expressing AT1R in the PVN. Studies in mice with selective deletion of AT1Rs in the PVN showing a reduction of catecholamine responses to stress emphasize the importance of the AT1Rs in the PVN mediating the sympathetic activation.[11]

Catecholaminergic neurons in the brain and the adrenal medulla express AT1Rs, and there is evidence that activation of AT1Rs directly stimulate catecholamine production. Activation of AT1Rs in neuronal cultures induces tyrosine hydroxylase gene transcription. In addition to AT1Rs, the locus coeruleus and the adrenal medulla express high levels of AT2Rs but the role of these receptors in the regulation of sympathetic activity during stress is unknown.[39]

The Renin–Angiotensin System and Behavior

Studies using blockers of the RAS or genetic models of hyper- or hypoactivity of the RAS indicate that AngII has effects on anxiety, cognition, memory, and feeding behavior. It is well accepted that reducing AngII formation with ACE inhibitors or blocking AT1R with receptor subtype specific antagonists reduces anxiety and improves cognition and memory acquisition. Although this suggests that AngII is anxiogenic, divergent information in the literature indicates that the behavioral effects of AngII are complex. For example, transgenic rats or mice models of either hyperactivity (overexpression of renin or Agt), or reduced activity (expression of antisense Agt) of the RAS in the brain, display similar increases in anxiety behavior in the elevated plus maze and other tests. Thus, under different experimental or physiological conditions, AngII may affect behavior depending on the concentration and mode of exposure of specific brain areas to the peptide, with consequent differential activation of receptor subtypes and signaling pathways.

A major mechanism for the proanxiety effects of AngII is AT1R-dependent stimulation of the anxiogenic peptide, CRH.[1,3] Both endogenous release of AngII from angiotensinergic pathways during stress and central Ang II injection increase CRH expression in the PVN and amygdala by activating AT1Rs in CRH neurons. The fact that AT1 antagonists prevent the increase in CRH mRNA in rats subjected to isolation stress supports the role of CRH in mediating the proanxiety actions of AngII.

The participation of AT2Rs and other angiotensin peptide receptors, such as the Ang(1-7)/Mas and Ang IV/AT4-IRAP adds complexity to the behavioral effects of AngII, and to the mechanisms of the anxiolytic actions of AT1R antagonists. Experimental evidence showing increases in AT2Rs in the locus coeruleus following administration of AT1 antagonists suggest that upregulation of this receptor subtype contributes to the antianxiety effects of AT1 antagonism. The fact that AT2R knockout mice develop anxietylike behavior also supports an anxiolytic role of these receptors in normal conditions.[20,21]

The mechanism of the anxiolytic effect of AT2Rs involves adrenergic receptors, since AT2R knockout mice show alpha1-adrenergic receptor binding downregulation in the amygdala without alteration in CRH receptor levels, and alpha-1-adrenergic but not CRH antagonists reverse the anxiety behavior in these mice.[34] This indicates that AT1 and AT2Rs regulate sympathetic outflow through different mechanisms, with the anxiogenic effects of AT1Rs depending on CRH, and the anxiolytic effects of AT2 mediated by noradrenergic regulation in the amygdala.

There is also published evidence that the ACE2/Ang(1-7)Mas and the Ang IV/IRAP systems have anxiolytic actions. The anxiolytic effect of central injection of the Mas ligand, Ang(1-7), correlates with reduced levels of oxidative stress markers in the amygdala, a nucleus involved in anxiety and bear behavior.[6,44] Mas knockout mice display anxiety, as well as impairment of object recognition memory. The mechanisms of the behavioral effects may include Mas-induced NO generation. Concerning Ang IV/IRAP, the effects of Ang IV injection resemble those of oxytocin and they can be reversed by oxytocin antagonists, thus the antianxiety effects of the peptide may involve oxytocin secretion.

Additional factors implicated in the behavioral effects of AngII are serotonin and the benzodiazepine receptor, which is a subunit of the GABA A receptor complex. There is evidence that AngII modulates serotonin production and the expression of tryptophan hydroxylase in the brain, and mice with disruption of Agt in the brain is associated with decreased serotoninergic activity and more pronounced anxiogenic responses to a serotonin antagonist. Thus, modulation of serotonin content in the brain may contribute to the anxiogenic effects of AngII.[43] Concerning the benzodiazepine receptor, several stressors decrease benzodiazepine receptor binding in the frontal cortex, and AT1 antagonists reverse the effect. Thus, the consequent decrease in GABAergic activity probably contributes to the development of anxiety.[18]

The Renin–Angiotensin System and Memory Consolidation

The AT1R interacts with a number of signaling pathways involved in memory consolidation, and the brain renin–angiotensin system has been implicated in both synaptic plasticity and some forms of memory consolidation. In addition to AT1, specific ligands for AT2 and AT4R-IRAP receptors facilitate memory consolidation.[6,20,45] The molecular mechanisms mediating the effect of angiotensin peptides on memory are not clear, but for the AT1R may involve phospholipase C–dependent increases in intracellular calcium, which is known to facilitate synaptic plasticity. AT4R-IRAP activation has been implicated in spatial memory formation and may contribute to fear memory consolidation.

While the various angiotensin peptides and receptors are involved in memory, the demonstration that selective deletion of AT1Rs in CRH neurons reduces conditioned fear provides evidence for essential role of AT1R-induced CRH regulation on fear memory consolidation.[30] The key role of the amygdala in fear memory generation, and the presence of AT1Rs in CRH neurons of this nucleus, suggests that CRH mediates the effects of AT1R on fear memory. However, the involvement of AT1R-expressing CRH neurons in other areas, such as the stria terminalis and the PVN, cannot be ruled out.

The Renin–Angiotensin System and Energy Balance

Much evidence indicates that AngII has marked effects on feeding behavior and energy homeostasis in the brain and in the periphery. Elevations in central Ang-II by exogenously administered peptide, or transgenic overexpression of human renin and Agt, reduces body weight and fat deposition due to decreased food intake and increased sympathetic nerve activity. Conversely, decreasing central RAS activity by expression of antisense Agt, increases food intake and body weight and reduces sympathetic output. This indicates that in the brain AngII reduces feeding behavior and increases energy expenditure by increasing sympathetic outflow.

There is strong evidence based on the effects of deletion of AT1Rs exclusively in the PVN, indicating that AT1Rs in the PVN mediate increases in food intake and decreases in energy expenditure, during exposure to high fat diet.[11] The lack of AT1Rs reduces the expression of CRH and oxytocin mRNAs in the PVN, without altering the expression vasopressin in the PVN and other neuropeptides in the arcuate nucleus. Therefore, a decrease in the anorexigenic peptides, CRH, and oxytocin are likely to mediate the increased feeding behavior. Since CRH is critical for sympathetic activation, a decrease in CRH can also explain the decreased sympathetic outflow, which is responsible for the reduced energy expenditure in these animals.

In contrast to mice with targeted deletion of AT1Rs in the PVN, whole body AT1R knockout exhibited attenuation of body weight gain and adiposity induced by high fat diet, due to increased sympathetic activation and energy expenditure.[27] Adipose tissue expresses Agt and can be a source of AngII production in obesity. The divergence between the central and peripheral RAS actions suggests the presence of a negative feedback pathway from adipose tissue–derived AngII to the brain, resulting in decrease in food intake and increase in energy expenditure.

Angiotensin II and Water Homeostasis

In addition to its effects on blood pressure and sodium balance, AngII modulates water homeostasis through coordinated actions in the brain and kidney.[31] In the brain, AngII acts as a potent dipsogen, and stimulates the synthesis and release of vasopressin, which promotes water conservation in the kidney. Studies using specific receptor subtype analogs have shown that effects of AngII on drinking behavior and vasopressin secretion are mediated by AT1Rs.

Peripheral signals such as increase in plasma osmolarity and decreases in blood volume markedly stimulate renin production and circulating AngII, which is sensed by AT1Rs in circumventricular organs, including the SFO and OVLT. Excitation of these neurons, mainly the SFO, activates angiotensinergic projections to the median preoptic nucleus, which have projections to the cortical areas responsible for thirst.[31]

Angiotensin II also participates in the central control of water homeostasis by regulating the secretion of the antidiuretic hormone, vasopressin.[31] Peripheral or central administration of AngII increases plasma vasopressin concentrations. The stimulatory effects of AngII on vasopressin secretion are mediated by AT1Rs in the SFO, since its ablation prevents the effects of peripheral AngII. As for drinking behavior, the median preoptic area is an important relay nucleus, part of the multisynaptic network transmitting signals to magnocellular neurons. In contrast to the predominantly direct effects of AngII on parvocellular neurons, AngII stimulates magnocellular neurons mostly indirectly by inhibiting GABAergic input resulting in activation of glutamate interneurons. However, direct effects cannot be ruled out since electrophysiological studies have shown that AngII modulates transient potassium conductance in magnocellular neurons.

Renin–Angiotensin Aldosterone System and Stress

Activation of AT1Rs in the adrenal zona glomerulosa by AngII is the main stimulator of aldosterone production in response to changes in volume and electrolyte balance. Aldosterone can influence neuroendocrine function and central control of blood pressure and sodium appetite by acting upon mineralocorticoid receptors in the hippocampus and brain areas involved cardiovascular control. On the other hand, neuroendocrine signals unrelated to cardiovascular homeostasis, such as stress, markedly stimulate aldosterone secretion through sympathetic stimulation of the RAS, as well as HPA axis activation and direct effect of ACTH on the adrenal glomerulosa cell.[2] While ACTH is a potent stimulus of aldosterone secretion in acute conditions, long-term administration of ACTH results in marked inhibition of basal and stimulated aldosterone secretion, due to downregulation of adrenal AT1Rs, and the activity of the rate limiting enzyme in the biosynthetic pathway, aldosterone synthase. For this reason, chronic stress leads to reduced basal and stimulated plasma aldosterone levels despite normal or increased plasma renin activity. While stimulation of the renin–angiotensin–aldosterone system is positive for adaptation to acute stress, alterations of this system during chronic stress may contribute

to cardiovascular pathology often associated with this condition.

CONCLUSIONS AND PERSPECTIVES

Since angiotensin II (AngII) was first discovered as the pressor substance generated by renin in 1940, much progress has been made in characterizing the components of the renin–angiotensin system, including AngII receptors, additional bioactive fragments of AngII, and their receptors. In addition to its roles in the regulation of blood pressure and aldosterone secretion as a circulating hormone, locally produced AngII by RASs in the kidney, adipose tissue, other peripheral organs, and in the brain contributes to systemic and central control of blood pressure, water and salt balance, sympathetic and HPA axis activity behavior, and energy homeostasis. The effects of AngII depend on binding to two G protein–coupled receptors, AT1R, which mediates most classic effects of the peptide, and AT2R, which oppose the pressor, profibrosis, autonomic, and proanxiety actions of AT1Rs. Angiotensin receptors located in brain areas outside the blood–brain barrier (circumventricular organs) play a key role in transmitting information from the periphery, and activate angiotensinergic pathways to the hypothalamic paraventricular nucleus and other areas responsible for the neuroendocrine actions of AngII. It is now clear that Ang II fragments formed in the blood or tissues and their cognate receptors are part of additional beneficial RASs, with actions opposing the effects of AT1R activation by Ang II. These novel axes include the ACE2/Ang(1-7)/Mas, and the Ang IV/Ang IVR-IRAP systems are not completely characterized but together with the AT2R are likely to be critical in counteracting potential damaging effects of excessive AT1R activity.

The added complexity of the RASs with endocrine, paracrine, and autocrine effects extending beyond cardiovascular regulation expands range of new perspective for the pharmacological targeting of cardiovascular disorders. Inhibition of ACE1 and AT1R is one of the most common and effective approaches for the pharmacological treatment of cardiovascular disease, especially hypertension. Blocking the predominant AngII/AT1R axis promotes the beneficial RASs by providing substrate for the synthesis of other peptides and increasing their relative receptor activity. Therefore, it is likely that increased activity of AT2R and the ACE2/Ang(1-7)/Mas and AngIV/IRAP axes contribute to the beneficial effects of blocking Ang formation and AT1R. A better knowledge of the physiology and signaling of the RASs will contribute to the understanding of the pathophysiology of cardiovascular and neuroendocrine disorders and development of new therapeutic tools for these conditions.

Acknowledgments

This work was supported by the Intramural Research Program of the Eunice Kennedy Shiver National Institute of Child Health and Human Development, National Institutes of Health.

References

1. Aguilera G, Kiss A, Luo X, Sunar-Akbasak B. The renin angiotensin system and the stress response. *Ann NY Acad Sci.* 1995;771:173–186.
2. Aguilera G, Kiss A, Sunar-Akbasak B. Hyperreninemic hypoaldosteronism after chronic stress in the rat. *J Clin Invest.* 1995;96:1512–1519.
3. Aguilera G, Young WS, Kiss A, Bathia A. Direct regulation of hypothalamic corticotropin-releasing-hormone neurons by angiotensin II. *Neuroendocrinology.* 1995;61:437–444.
4. Allen AM, Zhuo J, Mendelsohn FAO. Localization and function of angiotensin AT1 receptors. *Am J Hypertens.* 2000;13:S31–S38.
5. Bader M, Ganten D. Update on tissue renin-angiotensin systems. *J Mol Med.* 2008;86:615–621.
6. Bader M. ACE2, angiotensin-(1–7), and Mas: the other side of the coin. *Pflügers Archiv Eur J Physiol.* 2013;465:79–85.
7. Batenburg W, Krop M, Garrelds I, et al. Prorenin is the endogenous agonist of the (pro)renin receptor. Binding kinetics of renin and prorenin in rat vascular smooth muscle cells overexpressing the human (pro)renin receptor. *J Hypertens.* 2007;25:2441–2453.
8. Bunnemann B, Fuxe K, Metzger R, et al. Autoradiographic localization of mas proto-oncogene mRNA in adult rat brain using in situ hybridization. *Neurosci Lett.* 1990;114(2):147–153.
9. Correa FM, Plunkett LM, Saavedra JM. Quantitative distribution of angiotensin-converting enzyme (kininase II) in discrete areas of the rat brain by autoradiography with computerized microdensitometry. *Brain Res.* 1986;375:259–266.
10. Darby IA, Sernia C. In situ hybridization and immunohistochemistry of renal angiotensinogen in neonatal and adult rat kidneys. *Cell Tissue Res.* 1995;281:197–206.
11. de Kloet A, Pati D, Wang L, et al. Angiotensin type 1a receptors in the paraventricular nucleus of the hypothalamus protect against diet-induced obesity. *J Neurosci.* 2013;33:4825–4833.
12. Ferguson AV, Washburn DL, Latchford KJ. Hormonal and neurotransmitter roles for angiotensin in the regulation of central autonomic function. *Exp Biol Med.* 2001;226:85–96.
13. Ferrario CM, Chappell MC. What's new in the renin-angiotensin system? *Cell Mol Life Sci.* 2014;61:2720–2727.
14. Feuillan PP, Millan MA, Aguilera G. Angiotensin II binding sites in the rat fetus: characterization of receptor subtypes and interaction with guanyl nucleotides. *Regul Pept.* 1993;44:159–169.
15. Friis U, Madsen K, Stubbe J, et al. Regulation of renin secretion by renal juxtaglomerular cells. *Pflügers Archiv Eur J Physiol.* 2013;465:25–37.
16. Ganong WF. Circumventricular organs: definition and role in the regulation of endocrine and autonomic function. *Clin Exp Pharmacol Physiol.* 2000;27:422–427.
17. Glossmann H, Baukal AJ, Catt KJ. Properties of angiotensin II receptors in the bovine and rat adrenal cortex. *J Biol Chem.* 1974;249:825–834.
18. Gomes da Silva AQ, Xavier CH, Campagnole-Santos MJ, et al. Cardiovascular responses evoked by activation or blockade of GABA(A) receptors in the hypothalamic PVN are attenuated in transgenic rats with low brain angiotensinogen. *Brain Res.* 2012;1448:101–110.
19. Gonzalez Villalobos R, Shen X, Bernstein E, et al. Rediscovering ACE: novel insights into the many roles of the angiotensin-converting enzyme. *J Mol Med.* 2013;91:1143–1154.
20. Guimond MO, Gallo-Payet N. The angiotensin II type 2 receptor in brain functions: an update. *Int J Hypertens.* 2012;2012:351758.
21. Hein L, Barsh GS, Pratt RE, Dzau V, Kobilka BK. Behavioural and cardiovascular effects of disrupting the angiotensin II type-2 receptor in mice. *Nature.* 1995;377:744–747.
22. Higuchi S, Ohtsu H, Suzuki H, Shirai H, Frank G, Eguchi S. Angiotensin II signal transduction through the AT1 receptor: novel insights into mechanisms and pathophysiology. *Clin Sci.* 2007;112:417–428.
23. Ichiki T, Kambayashi Y, Inagami T. Molecular cloning and expression of angiotensin II type 2 receptor gene. *Adv Exp Med Biol.* 1996;396:145–152.
24. Jezova D, Ochedalski T, Kiss A, Aguilera G. Brain angiotensin II modulates sympathoadrenal and hypothalamic pituitary adrenocortical activation during stress. *J Neuroendocrinol.* 1998;10:67–72.
25. Karnik S, Unal H, Kemp J, et al. International union of basic and clinical pharmacology. XCIX. Angiotensin receptors: interpreters of pathophysiological angiotensinergic stimuli. *Pharmacol Rev.* 2015;67:754–819.
26. Kaschina E, Unger T. Angiotensin AT1/AT2 receptors: regulation, signalling and function. *Blood Press.* 2003;12:70–88.
27. Kouyama R, Suganami T, Nishida J, et al. Attenuation of diet-induced weight gain and adiposity through increased energy expenditure in mice lacking angiotensin II type 1a receptor. *Endocrinology.* 2005;146:3481–3489.
28. Lavoie J, Sigmund C. Minireview: overview of the renin-angiotensin system–an endocrine and paracrine system. *Endocrinology.* 2003;144(6):2179–2183.
29. Lu H, Cassis LA, Kooi CW, Daugherty A. Structure and functions of angiotensinogen. *Hypertens Res.* 2016;39. [Epub ahead of print].
30. Marvar P, Goodman J, Fuchs S, Choi D, Banerjee S, Ressler K. Angiotensin type 1 receptor inhibition enhances the extinction of fear memory. *Biol Psychiatry.* 2014;75:864–872.
31. McKinley MJ, Allen AM, Mathai ML, et al. Brain angiotensin and body fluid homeostasis. *Jpn J Physiol.* 2001;51:281–289.
32. McKinley MJ, Allen AM, May CN, et al. Neural pathways from the lamina terminalis influencing cardiovascular and body fluid homeostasis. *Clin Exp Pharmacol Physiol.* 2001;28:990–992.
33. Murphy TJ, Alexander RW, Griendling KK, Runge MS, Bernstein KE. Isolation of a cDNA encoding the vascular type-1 angiotensin II receptor. *Nature.* 1991;351:233–236.
34. Okuyama S, Sakagawa T, Chaki S, Imagawa Y, Ichiki T, Inagami T. Anxiety-like behavior in mice lacking the angiotensin II type-2 receptor. *Brain Res.* 1999;821(1):150–159.
35. Osborn J, Fink G, Kuroki M. Neural mechanisms of angiotensin II-salt hypertension: implications for therapies targeting neural control of the splanchnic circulation. *Curr Hypertens Rep.* 2011;13:221–228.
36. Pernomian L, Permonian L, Baraldi Araújo Restini C. Counter-regulatory effects played by the ACE – Ang II – AT1 and ACE2-Ang-(1-7) – Mas axes on the reactive oxygen species-mediated control of vascular function: perspectives to pharmacological approaches in controlling vascular complications. *VASA.* 2014;43:404–414.
37. Reid IA, Morris BJ, Ganong WF. The renin-angiotensin system. *Annu Rev Physiol.* 1978;40:377–410.
38. Rivier C, Vale W. Effect of angiotensin II on ACTH release in vivo: role of corticotropin-releasing factor. *Regul Pept.* 1983;7:253–258.
39. Saavedra J, Benicky J. Brain and peripheral angiotensin II play a major role in stress. *Stress.* 2007;10:185–193.
40. Santos RAS, Simoes e Silva AC, Maric C, et al. Angiotensin-(1-7) is an endogenous ligand for the G protein-coupled receptor Mas. *Proc Natl Acad Sci USA.* 2003;100:8258–8263.
41. Vanderheyden PML. From angiotensin IV binding site to AT4 receptor. *Mol Cell Endocrinol.* 2009;302(2):159–166.

42. Villela D, Passos Silva D, Santos R. Alamandine: a new member of the angiotensin family. *Curr Opin Nephrol Hypertens.* 2014;23:130–134.

43. Voigt J-P, Hörtnagl H, Rex A, van Hove L, Bader M, Fink M. Brain angiotensin and anxiety-related behavior: the transgenic rat TGR(ASrAOGEN)680. *Brain Res.* 2005;1046:145–156.

44. Wang L, de Kloet AD, Pati D, et al. Increasing brain angiotensin converting enzyme 2 activity decreases anxiety-like behavior in male mice by activating central Mas receptors. *Neuropharmacology.* 2016;105:114–123.

45. Wright JW, Krebs LT, Stobb JW, Harding JW. The angiotensin IV system: functional implications. *Front Neuroendocrinol.* 1995;16:23–52.

46. Zhuo J, Moeller I, Jenkins T, et al. Mapping tissue angiotensin-converting enzyme and angiotensin AT1, AT2 and AT4 receptors. *J Hypertens.* 1998;16:2027–2037.

25

Annexin A1

J.C. Buckingham[1], R.J. Flower[2]

[1]Brunel University London, Uxbridge, Middlesex, United Kingdom; [2]Queen Mary University of London, London,
United Kingdom

Abstract

Annexin A1 is a 37 kDa monomeric protein with selective tissue distribution. It is abundant in some key cell types of the innate and adaptive immune systems and the neuroendocrine system. The significance of this protein to the stress response is that its synthesis and release is regulated by glucocorticoids (GCs). Once released from target cells, the extracellular protein can act in an autocrine or paracrine fashion utilizing cell surface receptors of the formylpeptide family to bring about its biological actions. Many of the acute effects of GCs in the immune and neuroendocrine systems can be accounted for by the operation of this pathway.

INTRODUCTION

Glucocorticoids (GCs) exert diverse actions in the body maintaining and regulating homeostasis as well as controlling and moderating the activity of the innate and adaptive immune systems. Their complex effects are mediated through binding to the glucocorticoid receptor (GR), a member of the nuclear receptor superfamily.

When liganded, GR can activate cytosolic signaling pathways in the target cell and translocate to the nucleus where, acting as a transcription factor, it regulates the expression of key target genes. Amongst many significant genes upregulated or suppressed by GCs is the Annexin A1 (ANX-A1) gene. Utilizing both cytosolic and nuclear mechanisms, GCs regulate the synthesis and cellular disposition of the 37 kDa protein transcript of this gene. Amongst the most striking effects of ANX-A1, and relevant to this volume, is a prominent role in mediating the regulatory actions of the steroids on the host defense and neuroendocrine systems.

STRUCTURE OF ANX-A1

The "annexins" are an evolutionarily ancient family of monomeric proteins. They are widely distributed throughout eukaryotic phyla—specifically the animal, plant and fungal kingdoms—but are largely absent from

Stress: Neuroendocrinology and Neurobiology
http://dx.doi.org/10.1016/B978-0-12-802175-0.00025-5

prokaryotes and yeasts.[25] A characteristic feature of the family is the presence of an annexin core domain, which generally comprises four (occasionally more) repeating subunits of approximately 70 amino acids (the "annexin" repeat). These subunits usually harbor "type 2" calcium binding sites (although these are substituted in some annexins, with other motifs). These sites facilitate the binding of annexins to negatively charged phospholipids in the presence of calcium.[17]

KEY POINTS

- Annexin-A1 is a 37 kDa protein that is regulated and released by glucocorticoids.

- When released, the protein acts on cell surface G protein–coupled receptors of the formylpeptide family.

- The biological effects this bring about are key to the operation of the host defense response. These include:
  ○ A general suppression of the activity of the innate immune system and stimulation of proresolving actions.
  ○ Modulation of T-cell proliferation and differentiation.
  ○ Mediation of the acute feedback effects of glucocorticoids on the HPA axis.

- The biologically active domain of the protein is the N-terminal region and peptides derived from this region mimic the effects of the full length protein but are less potent.

Well over 100 annexins have been identified in over 50 species and 12 are found in humans. These are conventionally referred to as ANX-A1-13 (the ANX-A12 gene being unassigned), the descriptor "A" denoting their vertebrate origin. ANX-A13 is thought to be the ancestral gene from which all the other mammalian genes were ultimately derived.[20]

In addition to the characteristic core domain, individual vertebrate annexins have a unique N-terminal domain of variable length. This region of the protein harbors motifs that can recognize and bind to other intracellular or extracellular protein partners, including cell surface receptors (see later) and intracellular members of the S100 family of proteins. This N-terminal domain is a rapidly evolving component of these molecules. It often contains residues that can be modified by posttranslational processing, including phosphorylation and is probably responsible for the diversity of functions found within the family.

The crystal structure of human ANX-A1 has been determined at the 1.8 Å level.[34] It appears that the protein assumes a roughly lenticular "doughnut" configuration when it is folded. The four repeating domains are packed around the circumference whilst the N-terminal domain folds into the "hole" in the concave surface of the protein although, in the presence of calcium, this may "flip" out so as to interact with other binding partners.[35]

THE ANX-A1 GENE

ANX-A1 is encoded by a single gene in mammals and in the chicken, but two anx-A1 genes have been identified in the pigeon. The rat, mouse, and human genes show a high degree of homology; each comprises 13 exons, the first of which is noncoding. A TATA motif 30 bp upstream of the transcription start site is consistent with an inducible and tissue-specific mode of regulation. The ANX-A1 promoter has been studied in some detail by several groups[4,37]: surprisingly, for a protein that is strongly regulated by GCs in many tissues, it contains few GC response elements (in humans, there is one half-site in intron 1).

However, numerous other potential regulatory sites have been discovered including an AP1 site, a nuclear factor interleukin-6 site, and four GATA-binding protein-3 sites. Studies with mutant forms of the ANX-A1 promoter suggest that at least some of the positive GC actions on the synthesis of this protein rely on a transactivation mechanism utilizing a factor that binds to CCAT enhancer-binding protein in the upstream region.

It is clear that the regulation of this protein is subtle and that there are many factors that can influence its synthesis and disposition. Apart from GCs, the addition of cytokines, lipopolysaccharide, and possibly other substances that compromise cell integrity, such as heavy metal ions or thermal stress, can upregulate expression of the protein. In this respect, at least in some instances, it seems to act like a cellular "acute-phase" protein.[37]

Little is known about any process of posttranscriptional control over ANX-A1 synthesis.

DISTRIBUTION OF ANX-A1

The annexin family is generally widely distributed in different cell types[17] but not all annexins are coexpressed. ANX-A1 is found in many organs and tissues of the body, but is often localized in discrete populations of differentiated cells, particularly epithelia. It is particularly abundant in the lung, thymus, spleen, placenta, and seminal fluid. In addition, many cells involved in the inflammatory response (e.g., macrophages, polymorphonuclear neutrophils, and monocytes) express very substantial amounts of ANX-A1, as do certain cells of the neuroendocrine system.[5,6]

Within the pituitary gland ANX-A1 is expressed mainly by nonsecretory folliculostellate cells but also by endocrine cells, whereas in the hypothalamus it is expressed most strongly in the median eminence and paraventricular nucleus (i.e., areas associated with the regulation of anterior pituitary function). ANX-A1 is also expressed in lesser amounts elsewhere in the brain where it appears to be confined mainly to microglial cells, although some neuronal expression may also occur.

The subcellular distribution of ANX-A1 is unusual; it is abundant in the cytoplasm, but a small proportion, which in many cases represents the key biologically active pool, is also found on the external surface of the cell membrane, where it attaches via a Ca^{2+}-dependent linkage to binding proteins or to specific receptors (see later). ANX-A1 is also found attached to the inner leaflet of the plasma membrane, and cell fractionation studies have identified an integral membrane pool of ANX-A1.[28,36]

REGULATION OF THE EXPRESSION AND CELLULAR DISPOSITION OF ANX-A1

Glucocorticoids

ANX-A1 (formerly, "macrcortin," "lipocortin 1") was first identified as a GC-inducible protein in extracts of GC-conditioned media from rat peritoneal macrophage preparations[2,3] and the protein was cloned in 1986.[39] Subsequent studies showed that the expression of ANX-A1 mRNA and protein in experimental animals was reduced substantially by adrenalectomy but that it was restored by maintenance doses of GCs. The administration of the GC-receptor antagonist mifepristone (RU486) also lowered the resting ANX-A1 levels in rodents, thus confirming a role for the GCs in the regulation of ANX-A1 expression.[28] In man, the blood or salivary ANX-A1 concentration is broadly correlated with the blood cortisol.[26]

In addition to inducing the synthesis of ANX-A1, GCs also modify the cellular disposition of the protein in such a way that intracellular (cytoplasmic) ANX-A1 is exported to the outer surface of the cell. This process of externalization is an important facet of ANX-A1 biology because it enables the protein to interact with specific binding sites on the outer surface of cells that are critical to its biological action. Externalization, which is evident within minutes of GC challenge, is triggered by activation of protein kinase C (PKC), which phosphorylates ANX-A1 on Ser 27 (and other sites).[13,21]

This is apparently a signal for translocation of the protein to the plasma membrane from where, despite the fact that it lacks a classical secretory signal sequence, it is moved across the membrane using a mechanism that may involve ABC transporters.[33] Any agent that activates PKC may potentially promote ANX-A1 phosphorylation and externalization, but within the context of this volume, it should be noted that liganding of the GR is a potent stimulus for this event. The signaling mechanism regulating this may involve an intermediary action of PI3 kinase.[38]

Following externalization, intracellular (cytoplasmic) stores of the protein may fall and during this period, cells may become temporarily refractory to GC stimulation.[7] Eventually, the positive effects of the GCs on ANX-A1 transcription and synthesis replenishes the depleted stores. The effects of GCs on the synthesis and cellular disposition of ANX-A1 are unique, and other steroids (androgens, estrogens, progestogens, mineralocorticoids, and cholesterol) are without effect.

Other Factors

Several factors other than GCs are implicated in the regulation of ANX-A1 expression, including interleukin 6 (IL-6) and phorbol esters, which exert positive influences.[37] The apparently complementary action of IL-6 and the steroids is interesting because there are other examples of these substances acting cooperatively, for example, in the generation of acute-phase proteins in the liver. ANX-A1 in the brain is upregulated readily by tissue injury and excitotoxins, such as the excitatory amino acids glutamate or kainate, but the underlying mechanisms are unknown.

THE ANX-A1 RECEPTOR

Early observations, using immune-precipitation, fluorescence-activated cell analysis and other techniques, identified high-affinity saturable ANX-A1–binding sites on the surface of a variety of cell types including human and rodent peripheral blood leukocytes, rat anterior pituitary cells, and various cell lines (e.g., the rat pituitary GH3 line).[9,19] Confocal imaging revealed that the distribution of the surface-binding sites was punctate and variable, consistent with the widespread phenomenon of ligand-induced receptor clustering, a process essential for receptor internalization and signaling.

There is now compelling evidence that these binding sites are in fact receptors belonging to the formyl peptide receptor family, a family of GPCRs involved in the regulation of many cellular functions.[40] Many of the striking biological actions of ANX-A1 are now known to be mediated through activation of these receptors (see Refs. 31,32).

There are three FPR isoforms in humans, FPR1, FPR2, and FPR3. The FPR2 isoform (also known as ALX) seems to be the most important for transduction of ANX-A1 effects. Discrete binding domains have been identified that are essential for ANX-A1 action.[1] There is also evidence that this receptor family can homodimerize and heterodimerize. Homodimerization of FPR2 seems to be critical for the signaling of the antiinflammatory effects of ANX-A1 (see later)[12]: ligand-induced heterodimerizaion of FPR1 and FPR2 gives rise to a different spectrum of biological activity. FPR2 null mice are unable to respond to the acute inhibitory actions of GC in experimental models of inflammation.[16] Of interest is the fact that FPR2 is the receptor that also mediates the antiinflammatory/pro-resolving and other effects of lipid mediators, such as Lipoxin A4 and resolvin D2.[8]

FUNCTIONS OF ANX-A1

Overwhelming experimental evidence now supports the notion that ANX-A1 is crucial to many of the acute actions of GCs in several systems relevant to the stress response, including the innate and adaptive immune systems (see Refs. 15,29; for more comprehensive reviews) and the HPA axis (see Refs. 5,6).

Control of Inflammatory Resolution

Studies using ANX-A1 null mice, neutralizing anti ANX-A1 antibodies, or antisense constructs have firmly established that ANX-A1 exhibits significant antiinflammatory or pro-resolving activity in a number of experimental models of inflammation and it is now a well-recognized endogenous regulator of the inflammatory response as well as a mediator of GC action.[15,29] Some of the experimental systems in which these effects have been observed are listed in Table 25.1.

In general, in the innate immune system, the function of this protein, acting through the FPR receptor system, is to exert a general suppressive tone, reducing the amounts of proinflammatory eicosanoids generated, the release of histamine and preformed cytokines from mast cells, and so on. In contrast, ANX-A1 may trigger the release of pro-resolving mediators, such as IL-10.

In terms of cell migration, ANX-A1 potently down-regulates polymorphonuclear leukocytes (PMN) migration into inflammatory sites and accelerates their apoptosis. PMN contain abundant ANX-A1 and a significant proportion is contained within gelatinase granules.[30] When PMN are activated, this ANX-A1 is released from the granules and acts in a negative regulatory fashion on the PMN FPR cell surface receptors

TABLE 25.1 Some Actions of ANX-A1 in the Innate and Adaptive Immune Systems

Experimental Model		Innate Immune System	Adaptive Immune System
In vitro	Functions inhibited	• cPLA2 activation • Eicosanoid production • Superoxide generation • Phagocytosis • Mast cell degranulation	• Th2 cell differentiation
	Functions increased	• L-selectin shedding • PMN apoptosis • Phagocytosis of apoptotic PMN	• T-cell proliferation • TCR downstream signaling • Th1 cell differentiation
In vivo	Functions inhibited	• PMN trafficking • PMN adherence • Histamine release • Cytokine generation • Acute inflammation (some models) • Chronic inflammation (some models) • Hyperalgesia/ nociception • Fever (some pyrogens) • NMDA induced neuronal damage • Endotoxemia and sepsis	
	Functions increased		• T-cell dependent inflammatory responses

PMN, polymorphonuclear leukocyte; TCR, T-cell receptor.
Adapted and modified from D'Acquisto F, Perretti M, Flower RJ. Annexin-A1: a pivotal regulator of the innate and adaptive immune systems. Br J Pharmacol. 2008;155(2):152–169. http://dx.doi.org/10.1038/bjp.2008.252.

to reduce further release. PMN isolated from ANX-A1 null mice are "hyperactive," releasing excessive superoxide and other mediators. In contrast, ANX-A1 promotes the migration of monocytes into inflammatory sites that eventually transform the microenvironment into a resolving lesion.

In vivo, the time course and intensity of acute and chronic inflammation is greatly exaggerated in the absence of ANX-A1 (or its principal receptor, FPR2) whether this is bought about by gene deletion or immunoneutralization. Furthermore, the acute antiinflammatory actions of GCs are often abrogated, or greatly reduced, by the same procedures. Conversely, human recombinant ANX-A1 (hu-r-ANX-A1) and peptides derived from the N-terminal domain (see later) can rescue this phenotype and exert potent antiinflammatory and proresolution effects in many models of inflammation.

Interestingly, autoantibodies to ANX-A1 have been identified in the plasma of some patients with chronic inflammatory disease, such as rheumatoid arthritis and systemic lupus erythematosus.[18] Moreover, there is evidence of a causal relationship between the antibody titers and the incidence of certain types of steroid resistance, particularly in cases in which higher doses of GCs are required to repress the symptoms of the disease.

However, not all the antiinflammatory actions of GCs can be attributed to ANX-A1 and, in some models, GCs are effective but ANX-A1 is not; for example, steroids inhibit the generation of both carrageenin- and histamine-induced paw edema in the rat, but ANX-A1 is effective only against the carrageenan-induced lesion.[10]

Interestingly, the effect of the ability of GCs to regulate the induction and release of the protein varies between cells of the innate and adaptive immune systems. Whilst the effect of the protein acting through FPR2 is generally inhibitory on the actions of the innate immune system, its effect on the adaptive system is more complex.

Early observations[24] established that ANX-A1 is much less abundant in the cells of the adaptive immune system when compared to their counterparts in the innate system. The regulation of ANX-A1 synthesis and disposition between these two types of cells also varies.

Studies on T cells in vitro have established that whilst ANX-A1 has little effect on resting T cells, when they are activated by addition of anti-CD3 and anti-CD-28, both the intracellular ANX-A1 and FPR2 were translocated to the plasma membrane, where they could interact. Addition of ANX-A1 to T cells so activated, promoted T-cell proliferation.[14] It therefore seems possible that the presence of ANX-A1 in the inflammatory microenvironment regulates of T-cell signaling and activation.

Further studies on T-cell differentiation in the presence of ANX-A1 revealed that the protein skewed the differentiation of T0 cells towards a Th1 phenotype: conversely, activation of ANX-A1–deficient T0 cells differentiate almost exclusively to the Th2 phenotype.

When the effect of GCs on T-cell ANX-A1 was examined, it was found, paradoxically, to reduce ANX-A1 levels rather than to enhance them as was usually the case with cells of the innate immune system. This apparently counter-intuitive finding suggested a mechanism through which GCs could bring about a predominately Th2 phenotype.

In summary, the role and effect of ANX-A1 in the innate immune system is generally suppressive and pro-resolution whereas in the adaptive system however, it is generally stimulatory to T-cell signaling and proliferation. However, because of the contrasting effects of GCs on the regulation of the protein in the different cell types, the overall effect of the protein is to suppress the action of the innate system, promote resolution, and drive the adaptive system towards a Th2 response.

ANX-A1 and the Neuroendocrine System

ANX-A1 fulfills a significant role as a mediator of GC action in the hypothalamic-pituitary-adrenocortical (HPA) axis, acting mainly at the levels of the hypothalamus and anterior pituitary gland but also possibly at other sites in the brain (e.g., the hippocampus) to affect the negative feedback actions of the steroids on the axis.[5,22] In vivo central or peripheral administration of anti-ANX-A1 antisera effectively negates the ability of GCs (corticosterone or dexamethasone) to block the overt increases in adrenocorticotropic hormone (ACTH) secretion induced by IL-1β, whereas central injections of hu-r-ANX-A1 abolish the rises in GC secretion induced by either IL-1β or IL-6. Similarly, the inhibitory actions of GCs on the evoked release of the corticotropin releasing hormone from the hypothalamus and ACTH from the anterior pituitary gland in vitro are mimicked by hu-r-ANX-A1 and are reversed specifically by anti-ANX-A1 antisera and ANX-A1 antisense oligodeoxynucleotides directed against sequences specific to the ANX-A1 gene.

ANX-A1 also contributes to the regulation of steroidogenesis in the adrenal gland and plays a critical role in the manifestation of adrenal insufficiency in a murine model of endotoxaemia.[6a]

ANX-A1 also contributes to other aspects of GC action within the neuroendocrine system. In particular, it acts at the pituitary level to mediate the inhibitory actions of the steroids on the secretion of prolactin by the lactotrophs; in addition, it acts within the hypothalamus to mediate the acute stimulatory actions of the steroids on the secretion of GH. Many of the effects of ANX-A1 in the neuroendocrine system can be mimicked by its N-terminal peptides (see later).

As in other cell types, there is convincing evidence that the actions of ANX-A1 in the brain and neuroendocrine system are mediated by the FPR receptor system, specifically FPR2.[23]

ANX-A1–DERIVED PEPTIDES

An interesting and important observation concerns that location of the pharmacologically active moiety of ANX-A1. At least some facets of the biological activity of ANX-A1 can be attributed to sequences in the N-terminal of ANX-A1 of the molecule.[11] For example, a peptide comprising residues N-acetyl 2–26 has a profile of antiinflammatory and antiproliferative activity similar to that of the full-length recombinant molecule, although it is approximately two orders of magnitude less potent. N-acetyl 2–26 also mimics the effects of the full-length molecule on the neuroendocrine system, although once again it is less potent. Even shorter peptides from this domain may also be active. These peptides have the advantage that they are easier to work with than the full-length protein and inexpensive to produce; they thus provide valuable tools for research and may ultimately provide the base from which ANX-A1 peptides or nonpeptide analogs may be developed for therapeutic use.

However, there are some differences in the pharmacology of these peptides and the native protein. These have been traced to changes in the relative activation of FPR subtypes by the polypeptide as opposed to the full-length protein. In addition, the N-terminal peptide sequences do not mimic the effects of ANX-A1 on vesicular aggregation, as requires the presence of Ca^{2+} sequences in the core domain.[12]

The N-terminal domain also contains sensitive proteolytic sites. Since much of the biological activity of the protein is dependent upon the N-terminal structure, these probably regulate the inactivation of the protein in the extracellular environment. Interestingly, inhibition of these proteases, or the use of an ANX-A1 construct where the PR3 proteolytic cleavage site have been removed, greatly increases the antiinflammatory activity of the protein.[27]

SUMMARY

GCs are an integral and indispensable facet of the host response to stress, injury, or illness. As such, they have far reaching effects on the functioning of the immune system as well as many metabolic and other biochemical pathways. It is estimated that approximately 1% of all genes are regulated directly or indirectly by these hormones.

With such pluripotent actions then it is not surprising that the manifold pathways through which GCs bring about their effects are complex and, in many cases, continue to defy analysis.

The ability of the GCs to regulate the synthesis and release of ANX-A1 is only one such pathway but increasing weight of evidence suggests that it is crucial to GC actions in both the adaptive and innate immune systems as well as the neuroendocrine system that controls the release of these hormones.

References

1. Bena S, Brancaleone V, Wang JM, Perretti M, Flower RJ. Annexin A1 interaction with the FPR2/ALX receptor: identification of distinct domains and downstream associated signaling. *J Biol Chem*. 2012;287(29):24690–24697. http://dx.doi.org/10.1074/jbc.M112.377101.

2. Blackwell GJ, Carnuccio R, Di Rosa M, Flower RJ, Parente L, Persico P. Macrocortin: a polypeptide causing the anti-phospholipase effect of glucocorticoids. *Nature*. 1980;287(5778):147–149.

3. Blackwell GJ, Carnuccio R, Di Rosa M, et al. Glucocorticoids induce the formation and release of anti-inflammatory and antiphospholipase proteins into the peritoneal cavity of the rat. *Br J Pharmacol*. 1982;76(1):185–194.

4. Browning JL, Ward MP, Wallner BP, Pepinsky RB. Studies on the structural properties of lipocortin-1 and the regulation of its synthesis by steroids. *Prog Clin Biol Res*. 1990;349:27–45.

5. Buckingham JC, John CD, Solito E, et al. Annexin 1, glucocorticoids, and the neuroendocrine-immune interface. *Ann NY Acad Sci*. 2006;1088:396–409. http://dx.doi.org/10.1196/annals.1366.002.

6. Buckingham JC. Fifteenth Gaddum Memorial Lecture December 1994. Stress and the neuroendocrine-immune axis: the pivotal role of glucocorticoids and lipocortin 1. *Br J Pharmacol*. 1996;118(1):1–19.

6a. Buss NA, Gavins FN, Cover PO, Terron A, Buckingham JC. Targeting the annexin 1-formyl peptide receptor 2/ALX pathway affords protection afford protection against bacterial LPS-induced pathologic changes in the murine adrenal cortex. *FASEB J* 2015;29(7):2930–2942. http://dx.doi.org/10.1096/fj.14-268375.

7. Carnuccio R, Di Rosa M, Flower RJ, Pinto A. The inhibition by hydrocortisone of prostaglandin biosynthesis in rat peritoneal leucocytes is correlated with intracellular macrocortin levels. *Br J Pharmacol*. 1981;74(2):322–324.

8. Chiang N, Fierro IM, Gronert K, Serhan CN. Activation of lipoxin A(4) receptors by aspirin-triggered lipoxins and select peptides evokes ligand-specific responses in inflammation. *J Exp Med*. 2000;191(7):1197–1208.

9. Christian HC, Taylor AD, Flower RJ, Morris JF, Buckingham JC. Characterization and localization of lipocortin 1-binding sites on rat anterior pituitary cells by fluorescence-activated cell analysis/sorting and electron microscopy. *Endocrinology*. 1997;138(12):5341–5351. http://dx.doi.org/10.1210/endo.138.12.5593.

10. Cirino G, Peers SH, Flower RJ, Browning JL, Pepinsky RB. Human recombinant lipocortin 1 has acute local anti-inflammatory properties in the rat paw edema test. *Proc Natl Acad Sci USA*. 1989;86(9):3428–3432.

11. Cirino G, Cicala C, Sorrentino L, et al. Anti-inflammatory actions of an N-terminal peptide from human lipocortin 1. *Br J Pharmacol*. 1993;108(3):573–574.

12. Cooray SN, Gobbetti T, Montero-Melendez T, et al. Ligand-specific conformational change of the G-protein-coupled receptor ALX/FPR2 determines proresolving functional responses. *Proc Natl Acad Sci USA*. 2013;110(45):18232–18237. http://dx.doi.org/10.1073/pnas.1308253110.

13. Croxtall JD, Choudhury Q, Flower RJ. Glucocorticoids act within minutes to inhibit recruitment of signalling factors to activated EGF receptors through a receptor-dependent, transcription-independent mechanism. *Br J Pharmacol*. 2000;130(2):289–298. http://dx.doi.org/10.1038/sj.bjp.0703272.

14. D'Acquisto F, Merghani A, Lecona E, et al. Annexin-1 modulates T-cell activation and differentiation. *Blood*. 2007;109(3):1095–1102. http://dx.doi.org/10.1182/blood-2006-05-022798.

15. D'Acquisto F, Perretti M, Flower RJ. Annexin-A1: a pivotal regulator of the innate and adaptive immune systems. *Br J Pharmacol*. 2008;155(2):152–169. http://dx.doi.org/10.1038/bjp.2008.252.

16. Dufton N, Hannon R, Brancaleone V, et al. Anti-inflammatory role of the murine formyl-peptide receptor 2: ligand-specific effects on leukocyte responses and experimental inflammation. *J Immunol*. 2010;184(5):2611–2619. http://dx.doi.org/10.4049/jimmunol.0903526.

17. Gerke V, Moss SE. Annexins: from structure to function. *Physiol Rev*. 2002;82(2):331–371. http://dx.doi.org/10.1152/physrev.00030.2001.

18. Goulding NJ, Podgorski MR, Hall ND, et al. Autoantibodies to recombinant lipocortin-1 in rheumatoid arthritis and systemic lupus erythematosus. *Ann Rheum Dis.* 1989;48(10):843–850.

19. Goulding NJ, Pan L, Wardwell K, Guyre VC, Guyre PM. Evidence for specific annexin I-binding proteins on human monocytes. *Biochem J.* 1996;316(Pt 2):593–597.

20. Iglesias JM, Morgan RO, Jenkins NA, Copeland NG, Gilbert DJ, Fernandez MP. Comparative genetics and evolution of annexin A13 as the founder gene of vertebrate annexins. *Mol Biol Evol.* 2002;19(5):608–618.

21. John CD, Christian HC, Morris JF, Flower RJ, Solito E, Buckingham JC. Kinase-dependent regulation of the secretion of thyrotrophin and luteinizing hormone by glucocorticoids and annexin 1 peptides. *J Neuroendocrinol.* 2003;15(10):946–957.

22. John CD, Christian HC, Morris JF, Flower RJ, Solito E, Buckingham JC. Annexin 1 and the regulation of endocrine function. *Trends Endocrinol Metab.* 2004;15(3):103–109. http://dx.doi.org/10.1016/j.tem.2004.02.001.

23. John CD, Gavins FN, Buss NA, Cover PO, Buckingham JC. Annexin A1 and the formyl peptide receptor family: neuroendocrine and metabolic aspects. *Curr Opin Pharmacol.* 2008;8(6):765–776. http://dx.doi.org/10.1016/j.coph.2008.09.005.

24. Morand EF, Hutchinson P, Hargreaves A, Goulding NJ, Boyce NW, Holdsworth SR. Detection of intracellular lipocortin 1 in human leukocyte subsets. *Clin Immunol Immunopathol.* 1995;76(2):195–202.

25. Moss SE, Morgan RO. The annexins. *Genome Biol.* 2004;5(4):219. http://dx.doi.org/10.1186/gb-2004-5-4-219.

26. Mulla A, Leroux C, Solito E, Buckingham JC. Correlation between the antiinflammatory protein annexin 1 (lipocortin 1) and serum cortisol in subjects with normal and dysregulated adrenal function. *J Clin Endocrinol Metab.* 2005;90(1):557–562. http://dx.doi.org/10.1210/jc.2004-1230.

27. Pederzoli-Ribeil M, Maione F, Cooper D, et al. Design and characterization of a cleavage-resistant Annexin A1 mutant to control inflammation in the microvasculature. *Blood.* 2010;116(20):4288–4296. http://dx.doi.org/10.1182/blood-2010-02-270520.

28. Peers SH, Smillie F, Elderfield AJ, Flower RJ. Glucocorticoid-and non-glucocorticoid induction of lipocortins (annexins) 1 and 2 in rat peritoneal leucocytes in vivo. *Br J Pharmacol.* 1993;108(1):66–72.

29. Perretti M, D'Acquisto F. Annexin A1 and glucocorticoids as effectors of the resolution of inflammation. *Nat Rev Immunol.* 2009;9(1):62–70. http://dx.doi.org/10.1038/nri2470.

30. Perretti M, Christian H, Wheller SK, et al. Annexin I is stored within gelatinase granules of human neutrophil and mobilized on the cell surface upon adhesion but not phagocytosis. *Cell Biol Int.* 2000;24(3):163–174. http://dx.doi.org/10.1006/cbir.1999.0468.

31. Perretti M, Getting SJ, Solito E, Murphy PM, Gao JL. Involvement of the receptor for formylated peptides in the in vivo anti-migratory actions of annexin 1 and its mimetics. *Am J Pathol.* 2001;158(6):1969–1973. http://dx.doi.org/10.1016/S0002-9440(10)64667-6.

32. Perretti M, Chiang N, La M, et al. Endogenous lipid- and peptide-derived anti-inflammatory pathways generated with glucocorticoid and aspirin treatment activate the lipoxin A4 receptor. *Nat Med.* 2002;8(11):1296–1302. http://dx.doi.org/10.1038/nm786.

33. Philip JG, Flower RJ, Buckingham JC. Blockade of the classical pathway of protein secretion does not affect the cellular exportation of lipocortin 1. *Regul Pept.* 1998;73(2):133–139.

34. Rosengarth A, Gerke V, Luecke H. X-ray structure of full-length annexin 1 and implications for membrane aggregation. *J Mol Biol.* 2001;306(3):489–498. http://dx.doi.org/10.1006/jmbi.2000.4423.

35. Rosengarth A, Rosgen J, Hinz HJ, Gerke V. Folding energetics of ligand binding proteins II. Cooperative binding of Ca^{2+} to annexin I. *J Mol Biol.* 2001;306(4):825–835. http://dx.doi.org/10.1006/jmbi.2000.4358.

36. Seemann J, Weber K, Gerke V. Annexin I targets S100C to early endosomes. *FEBS Lett.* 1997;413(1):185–190.

37. Solito E, de Coupade C, Parente L, Flower RJ, Russo-Marie F. IL-6 stimulates annexin 1 expression and translocation and suggests a new biological role as class II acute phase protein. *Cytokine.* 1998;10(7):514–521. http://dx.doi.org/10.1006/cyto.1997.0325.

38. Solito E, Mulla A, Morris JF, Christian HC, Flower RJ, Buckingham JC. Dexamethasone induces rapid serine-phosphorylation and membrane translocation of annexin 1 in a human folliculostellate cell line via a novel nongenomic mechanism involving the glucocorticoid receptor, protein kinase C, phosphatidylinositol 3-kinase, and mitogen-activated protein kinase. *Endocrinology.* 2003;144(4):1164–1174. http://dx.doi.org/10.1210/en.2002-220592.

39. Wallner BP, Mattaliano RJ, Hession C, et al. Cloning and expression of human lipocortin, a phospholipase A2 inhibitor with potential anti-inflammatory activity. *Nature.* 1986;320(6057):77–81. http://dx.doi.org/10.1038/320077a0.

40. Walther A, Riehemann K, Gerke V. A novel ligand of the formyl peptide receptor: annexin I regulates neutrophil extravasation by interacting with the FPR. *Mol Cell.* 2000;5(5):831–840.

26

Corticotropin-Releasing Factor Receptor Antagonists

E.P. Zorrilla

The Scripps Research Institute, La Jolla, CA, United States

Abstract

Corticotropin releasing factor (CRF) and related peptides play key roles in stress-related physiological processes via widely distributed CRF$_1$ and CRF$_2$ receptors. CRF receptor antagonists have effects that reflect the functions of CRF and urocortins in activating hormonal, autonomic, behavioral, and immunological responses to stress, and may provide novel treatment approaches for stress-related disorders.

Corticotropin-releasing factor (CRF) system receptor antagonists have been sought since the isolation of the eponymous stress-related adrenocorticotropin-releasing peptide in 1981 by Vale and colleagues.[84] Discovered for its hypophysiotropic function in the HPA-axis, CRF, and, later, urocortins (Ucn) 1, 2, and 3, paralogs of CRF[32,48,72,89] were subsequently found also to play roles in diverse homeostatic stress responses, via extrahypothalamic brain sites and peripheral tissues.[1,18] CRF receptor antagonists have potential therapeutic applications in cases where these responses are inappropriate in magnitude, timing, or context.

PHARMACOLOGY OF CRF/UROCORTIN RECEPTOR SYSTEMS

CRF-related peptides interact with two known mammalian CRF receptor subtypes, CRF$_1$ and CRF$_2$, both belonging to the class B1 ("secretinlike") subfamily of G protein–coupled receptors for polypeptide hormones. The CRF$_1$ receptor has been cloned in human, mouse, rat, and tree shrew, among other species, and exists in multiple isoforms (e.g., CRF$_{1a}$-CRF$_{1h}$), with the best known and functional isoform being the $CRF_{1(a)}$ subtype. The CRF_2 receptor has three known functional membrane-associated subtypes in humans, $CRF_{2(a)}$, $CRF_{2(b)}$, and $CRF_{2(c)}$, as well as a ligand-sequestering, soluble $CRF_{2(a)}$ isoform discovered in mouse. CRF_2 and CRF_1 receptors have ~70% sequence identity. Whereas CRF has high, preferential affinity for CRF_1 vs CRF_2, receptors, Ucn 1 is a high-affinity agonist at both receptors and the type 2 urocortins, Ucn 2 and Ucn 3 are more selective for membrane CRF_2 receptors. Because of this pharmacological profile, references to "anti-CRF" treatments most often refer to functional or competitive antagonists of CRF_1 receptor mediated effects, whereas "anti-type 2 Ucn" treatments might target CRF_2 receptors. The biological actions of CRF, Ucn 1, and Ucn 2 in rodents also are modulated by a CRF-binding protein (CRF-BP), a 37-kDa secreted glycoprotein that binds and putatively immunosequesters CRF and Ucn 1 with equal or greater affinity than do CRF receptors. Structural requirements for binding to CRF receptors and the CRF-BP differ, so many (if not most) CRF receptor antagonists do not interact with the CRF-BP.[1,18,107]

KEY POINTS

- Corticotropin releasing factor (CRF) and related peptides play key roles in stress-related physiological processes via widely distributed CRF_1 and CRF_2 receptors.

- CRF receptor antagonists may have therapeutic potential to oppose pathophysiological consequences of CRF and/or urocortins, which elicit hormonal, autonomic, behavioral, and immunological responses to stress via CRF receptors.

- Both peptide and small molecule CRF receptor antagonists have been developed, with varying degrees of receptor subtype selectivity.

- Blood–brain barrier–penetrating, small molecule CRF_1 antagonists have received special interest for their therapeutic potential in regulating diverse stress-related conditions, including endocrine disorders that involve centrally driven hypercortisolemia, sympathoadrenomedullary responses, emotional disorders, gastrointestinal disorders, urologic disorders, premature labor, and several inflammatory conditions.

- Sustained medicinal chemistry efforts have now yielded drug-like CRF_1 receptor antagonists with good pharmacokinetics and acceptable safety profiles, but no compound has passed a Phase III efficacy trial to date.

NONNEUROENDOCRINE DISTRIBUTION OF CRF_1 RECEPTORS

CRF_1 receptors mediate not only the HPA-axis neuroendocrine response to stress, but also other aspects of organism stress responses. In brain, the distribution of CRF_1 receptors is highly conserved in stress-responsive brain regions, including neocortex, the extended central amygdala, medial septum, hippocampus, hypothalamus, thalamus, cerebellum, and autonomic midbrain and hindbrain nuclei.[70,88] This receptor distribution, concordant with that of its natural ligands (CRF, Ucn 1), is consistent with the proposed role for extrahypothalamic CRF_1 receptors in behavioral and autonomic stress responses.[108] Accordingly, excessive CRF_1 signaling putatively has a pathophysiologic role in several stress-related disorders, including anxiety and depressive disorders, substance dependence, and irritable bowel syndrome.[107] Hereafter, the identity and potential therapeutic indications for CRF_1 receptor antagonists are summarized.

SELECTED CRF RECEPTOR ANTAGONISTS

Nonselective Peptide CRF Receptor Antagonists

N-terminally truncated and substituted analogs of CRF can act as competitive partial agonists or full receptor antagonists (Table 26.1). Examples of these, in chronological order of discovery, include the partial agonist [Met[18], Lys[23], Glu[27,29,40], Ala[32,41], Leu[33,36,38]] r/hCRF$_{9-41}$; a.k.a. α-helical CRF$_{9-41}$; and the full receptor antagonists [D-Phe[12], Nle[21,38]CαMeLeu[37]] r/hCRF$_{12-41}$; a.k.a. D-Phe-CRF$_{12-41}$; [cyclo(30–33)[D-Phe[12],Nle[21,38],Glu[30],Lys[33]] r/hCRF$_{12-41}$; a.k.a. astressin; and [cyclo(30–33)[D-Phe[12], Nle[21], CαMeLeu[27],Glu[30],Lys[33],Nle[38], CαMeLeu[40]]Ac-r/hCRF$_{9-41}$; a.k.a. astressin-B.[3,67,75] These peptide ligands have roughly the same order of binding and antagonist potency at CRF_1 vs CRF_2 receptors and do not cross the blood–brain barrier. Thus, they are subtype nonselective, peripherally acting CRF-receptor antagonists.[25,57,74]

Subtype-Selective Peptide CRF Receptor Antagonists

Antisauvagine-30 (ASVG-30) was reported as the first preferential CRF_2 antagonist. Followings its development, several subsequent series of CRF_2 receptor peptide antagonists have been identified, each which is more potent, selective, and enzymatic degradation resistant than anti-sauvagine-30. These antagonists included the ASVG-30 analogs K31440 and K41498; [Tyr[11],His[12],Nle[17]]SVG$_{10-40}$, which has 2360-fold

TABLE 26.1   Alphabetical Selection of Peptide CRF Receptor Antagonists

| Familiar Name | CRF Receptor Binding Affinity (IC_{50} or Ki) | | Chemical Structure |
	CRF_1 (nM)	CRF_2 (nM)	
α-Hel-CRF_{9-41}	19	1.1	[Met^{18},Lys^{23},Glu^{27,29,40},Ala^{32,41},Leu^{33,36,38}]-rat/human corticotropin-releasing factor$_{9-41}$
Antisauvagine-30	400	1.1	[D-Phe^{11},His^{12}]-sauvagine$_{11-40}$
Astressin	13.2	1.5	cyclo(30–33)[D-Phe^{12},Nle^{21,38},Glu^{30},Lys^{33}]-rat/human corticotropin-releasing factor $_{12-41}$
Astressin-B	High affinity	Moderate affinity	cyclo(30–33)[D-Phe^{12},Nle^{21},CαMeLeu^{27},Glu^{30},Lys^{33},Nle^{38}, CαMeLeu^{40}] acetyl-rat/human corticotropin-releasing factor$_{9-41}$
Astressin$_2$-B	>500	1.3	cyclo(31–34)[D-Phe^{11},His^{12},CαMeLeu^{13,39},Nle^{17},Glu^{31},Lys^{34}] acetyl-sauvagine$_{8-40}$
D-Phe-CRF_{12-41}	19.2	4.4	[D-Phe^{12},Nle^{21,38},CαMeLeu^{37}]-rat/human corticotropin-releasing factor $_{12-41}$
K31440	288	1.5	Acetyl-[D-Tyr^{11},His^{12},Nle^{17}]-sauvagine$_{11-40}$
K41498	425	0.7	[D-Phe^{11},His^{12},Nle^{17}]-sauvagine$_{11-40}$

selectivity and subnanomolar affinity for the CRF_2 receptor; and astressin$_2$-B, a long-acting, potent and selective CRF_2 antagonist (Table 26.1). Although these antagonists do not oppose "classic" stress-related effects of CRF reviewed above, they do reverse stress-induced CRF_2-mediated endpoints, some of which are recruited by urocortins and some by CRF. Such endpoints include, for example, stress-induced gastric stasis and, possibly a component of stress-induced anorexia.

Preferential CRF_1 peptide antagonists also appear to have been identified in the course of seeking minimal peptide fragments with CRF receptor antagonist activity. For example, Yamada and colleagues followed upon a patent application (W01/29086) of the Solvay pharmaceutical group, which described a peptide comprising the 12 C-terminal residues of astressin as a potent antagonist of CRF receptors. Through amino acid substitution, of Nle38 with the bulkier, lipophilic cyclohexylalanine residue and Ala31 with an unnatural residue (D-Ala), they identified a metabolically stable, high-affinity (Ki~3 nM) CRF_1 antagonist that potently (i.v. 0.1 mg/kg) reduced ACTH secretion in a rat model of sepsis.[103] This peptide may be a CRF_1-preferring antagonist because the Solvay group concurrently published their findings on the patented 12-residue N-terminal truncated astressin derivative on which Yamada and colleagues began their investigation. The disclosed fragment not only retained CRF_1 affinity, but was "inactive" at the $CRF_{2(a)}$ receptor. Thus, it appears that lactam bridge-constrained N-terminally truncated astressin derivatives of 12–15 residue length may be preferential CRF_1 receptor peptide antagonists.[73] Such compounds may be useful for the treatment of pathologies that involve only peripheral CRF_1 hyperactivity.

Small Molecule CRF_1 Receptor Antagonists

Pharmacophore and Selectivity

Blood–brain barrier penetrating small molecules with high and selective CRF_1 (vs CRF_2) affinity also have been identified (Table 26.2) and reviewed in considerable detail recently.[31,46,106–108] With few exceptions, small molecule CRF_1 antagonists follow a common pharmacophore. Prototypical compounds (see Fig. 26.1) share one or two lipophilic top/side units, a central mono-, bi-, or tricyclic ring core, and a conformation-stabilizing di- or tri-substituted aromatic bottom group. Each ring core contains a putative proton-accepting ring nitrogen, and the hydrogen-bond accepting core is thought to interact with histidine-199 of the CRF_1 receptor, a polar amino acid in the third transmembrane domain that is not shared in the CRF_2 receptor or CRF-BP sequences. Therefore, CRF_1 antagonists with this pharmacophore are highly selective (CRF_1 vs. CRF_2 and CRF-BP). Most known small molecule CRF_1 receptor antagonists also show at least 1000-fold binding selectivity relative to other receptors, ion channels, and reuptake sites screened to date. However, this is not universal as one exception, R278995/CRA0450 actually exhibits 50-fold greater affinity for the σ_1 receptor than for the CRF_1 receptor.

This binding site of the classic small molecule antagonist pharmacophore also differs from the agonist binding site, whose confirmation is allosterically modified by CRF_1 antagonists. Thus, most small molecules are noncompetitive antagonists of the peptide ligands. Some exceptions to the pharmacophore include oxo-7H-benzo[e]perimidine-4-carboxylic acid derivatives, subtype nonselective CRF receptor antagonists discovered

TABLE 26.2 Alphabetical Selection of Small Molecule CRF$_1$ Receptor Antagonists

Familiar Name	CAS Registry#	CRF Receptor Binding Affinity		Chemical Structure
		CRF$_1$ (nM)	CRF$_2$ (nM)	
Antalarmin	157284-96-3	1	>10,000	N-butyl-N-ethyl[2,5,6-trimethyl-7-(2,4,6-trimethylphenyl)-7H-pyrrolo[2,3-d]pyrimidin-4-yl]-amine
CC 2064460	438552-89-7	35	ND	Propanimidic acid, N-(3-ethynylphenyl)-2-oxo-, 2-(2-chlorophenyl) hydrazide (9CI)
CP-154,526	157286-86-7	2.7	>10,000	N-butyl-N-[2,5-dimethyl-7-(2,4,6-trimethylphenyl)-7H-pyrrolo[2,3-d]pyrimidin-4-yl]-N-ethylamine
CRA0165	402751-10-4	10	>10,000	4-[5,6-Dihydro-3-(2-methylphenyl)-1(2H)-pyridinyl]-N-ethyl-6-methyl-N-[4-(1-methylethyl)-2-(methylthio)phenyl]
CRA1000	226948-11-4	16–21	>10,000	N-ethyl-4-[4-(3-fluorophenyl)-1,2,3,6-tetrahydro-1-pyridinyl]-N-[4-isopropyl-2-(methylsulfanyl) phenyl]-6-methylpyrimidin-2-amine
CRA1001	229346-94-5	19–22	>10,000	2-[N-(2-bromo-4-isopropylphenyl)-N-ethylamino]-4-[4-(3-fluorophenyl)-1,2,3,6-tetrahydropyridin-1-yl]-6-methylpyrimidine
DMP695	354994-31-3	3.3	>10,000	N-(2-chloro-4,6-dimethylphenyl)-1-[1-methoxymethyl-(2-methoxyethyl]-6-methyl-1H-1,2,3-triazolo[4,5-c]pyridin-4-amine mesylate
DMP696	202578-52-7	1.7	>10,000	4-(1,3-Dimethoxyprop-2-ylamine)-2,7-dimethyl-8-(2,4-dichlorophenyl)-pyrazolo[1,5-a]-1,3,5-triazine
DMP904	202579-74-6	1.0	>10,000	N-(1-ethylpropyl)-3-(4-methoxy-2-methylphenyl)-2,5-dimethyl-pyrazolo[1,5-a]pyrimidin-7-amine
LWH-63	276890-57-4	0.7	>10,000	4-ethyl[2,5,6-trimethyl-7-(2,4,6-trimethylphenyl)-7H-pyrrolo[2,3-d]pyrimidin-4-yl]amino]-1-butanol
MJL-1-109-2	596107-16-3	1.9	ND	pyrazolo[1,5-a]-1,3,5-triazin-4-amine, 8-[4-(bromo)-2-chlorophenyl]-N,N-bis(2-methoxyethyl)-2,7-dimethyl- (9CI)
NBI 27914	184241-44-9	2	>10,000	5-chloro-N-(cyclopropylmethyl)-2-methyl-N-propyl-N'-(2,4,6-trichlorophenyl)-4,6-pyrimidinediamine
NBI 30545	195054-99-0	2.8	ND	pyrazolo[1,5-a]pyrimidin-7-amine, 3-(2,4-dimethoxyphenyl)-N-(2-methoxyethyl)-2,5-dimethyl-N-propyl-(9CI)
NBI 35965	354999-74-9	4	>10,000	ND
R121919/ NBI30775	195055-03-9	3.5	>10,000	3-[6-(dimethylamino)-4-methyl-pyrid-3-yl]-2,5-dimethyl-N,N-dipropyl-pyrazolo[2,3-a]pyrimidin-7-amine
R121920	195055-01-7	4	ND	3-[6-(dimethylamino)-3-pyridinyl]-2,5-dimethyl-N,N-dipropyl-pyrazolo[1,5-a]pyrimidin-7-amine
R278995/ CRA0450	N/A	54	>10,000	1-[8-(2,4-dichlorophenyl)-2-methylquinolin-4-yl]-1,2,3,6-tetrahydropyridine-4-carboxamide benzenesulfonate
SC241	392336-72-0	5	>10,000	3-(2-bromo-4-isopropylphenyl)-5-methyl- 3H-[1,2,3]triazolo[4,5-d]pyrimidin-7-yl]-bis-(2-methoxy-ethyl)-amine
SN003	197801-88-0	6.8	ND	(+/−)-N-[2-methyl-4-methoxyphenyl]-1-(1-(methoxymethyl)propyl)-6-methyl-1H-1,2,3-triazolo[4,5-c]pyridin-4-amine
SSR125543A	321839-75-2	2	>10,000	4-(2-Chloro-4-methoxy-5-methylphenyl)-N-[2-cyclopropyl-1(S)-(3-fluoro-4-methylphenyl)ethyl]-5-methyl-N-(2-propynyl)thiazol-2-amine hydrochloride

ND, not determined.

by Alanex (10 reported in Ref. 52); CC 2064460, a moderately potent arylamidrazone CRF$_1$ antagonist that lacks a central ring core with the customary hydrogen-bond accepting nitrogen (compound 51 in Ref. 95); and stereospecific N-phenylphenylglycines, which also lack a ring core but were identified through computational screening based on a classic pharmacophore training set.[59]

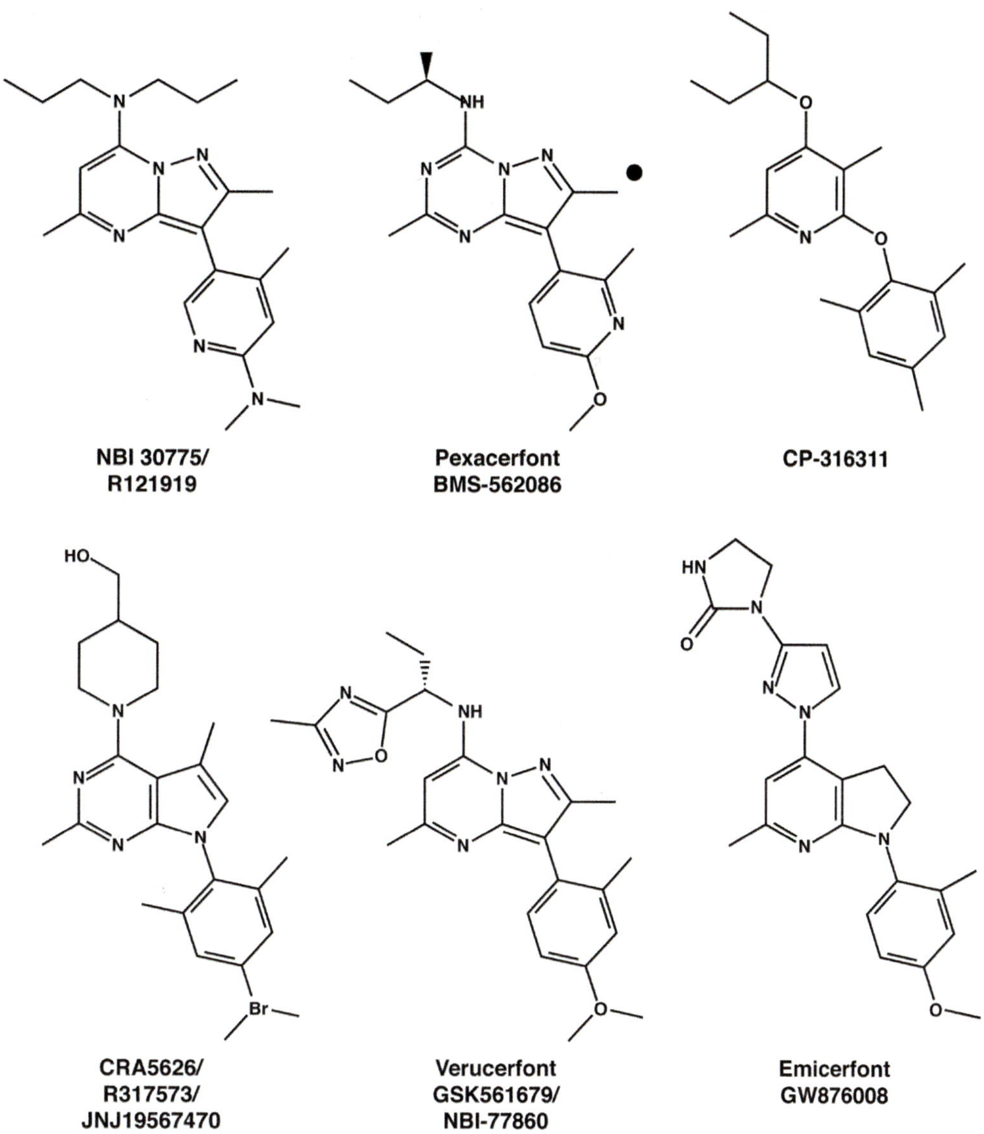

FIGURE 26.1 Representative selection of heterocyclic, small-molecule selective CRF_1 receptor antagonists that conform to the classic CRF_1 antagonist pharmacophore. See text for description of pharmacophore structural features.

Lipophilicity

An obstacle in the clinical development of CRF_1 antagonists was that many preclinical structures were excessively lipophilic with poor water solubility, bioavailability, and pharmacokinetic properties. Because of their high lipophilicity, many CRF_1 receptor antagonists have very high estimated bioconcentration factors at physiological pHs, a potential index of toxicity liability. Most early CRF_1 antagonists also failed Lipinski's "rule of 5" criteria for drug candidates, derived from review of physiochemical properties of compounds within the United States Adopted Names database, due to excessive lipophilicity (cLogP>5). Due to unfavorable pharmacokinetics, this degree of lipophilicity is rare for effective central nervous system-acting agents. Of more than 100 drugs marketed as of 1992 for CNS indications, not one had a logD >4; most (~85%) had a logD of 0–3.[49] Much preclinical discovery therefore focused on identifying less hydrophobic CRF_1 antagonists, with published discoveries accelerating ca. 2003–2005.[10,15,16,20,24,26,27,33–36,50,80,92,109]

CRF_1 antagonists as a class do not share some of the drawbacks of benzodiazepines, such as tolerance, sedative-hypnotic or amnestic effects, or abuse liability. They also do not lead to HPA-axis insufficiency at preclinically therapeutic doses,[106] but isolated reports of liver enzyme elevations led to the discontinued clinical development of R121919, which had shown dose-related efficacy in an open label Phase IIa clinical trial of patients with major depression, as well as of PF-00572778.[46]

THERAPEUTIC POTENTIAL OF CRF$_1$ ANTAGONISTS

Endocrine Disorders

CRF released into the portal blood from the median eminence stimulates ACTH release from the anterior pituitary into systemic circulation, which elicits the synthesis and release of glucocorticoids from the adrenal cortex, the classic HPA-axis stress response. HPA-axis dysregulation can result in endocrine disorders of hypercortisolemia, including Cushing syndrome. Most cases of Cushing syndrome are iatrogenic (e.g., from synthetic glucocorticoids), or due to ACTH or cortisol hypersecreting tumors and hyperplasias. However, central hypersecretion of "neuroendocrine" CRF leads to a Cushingoid state. For example, clinical syndromes with hypothalamic CRF-driven chronic hypercortisolemia, including severe alcohol dependence, visceral obesity, melancholic major depression, and anorexia nervosa, are marked by Cushingoid somatic and endocrine features that remit when cortisol levels normalize. Peripherally acting, long-acting peptide CRF receptor antagonists, such as astressin-B, have been especially effective in opposing CRF-evoked ACTH release[74] and therefore may have a therapeutic role in managing Cushingoid consequences of these disorders, if not also the disorders themselves. Consistent with this view, a recent phase Ib, single-blind, placebo-controlled, fixed-sequence, single-dose trial of verucerfont (NBI-77860/GSK561679; a CRF$_1$ antagonist) found efficacy to reduce ACTH levels in women with 21-hydroxylase deficiency.[83]

Sympathoadrenomedullary Response

Intracerebroventricular CRF administration evokes the sympathoadrenomedullary stress response in animal models, as it elevates blood pressure and heart rate, increases circulating levels of catecholamines, glucagon, and glucose; increases body temperature; and raises energy expenditure. Conversely, in animal models, CRF receptor antagonists, in particular those for the CRF$_1$ subtype, such as CP-154,526, attenuate both CRF and stress-induced tachycardia, hypertension, and hyperthermia.[5,61,62,65]

CNS-Related Disorders

Anxiety Disorders

Consistent with autonomic findings, clinical data suggest that brain CRF is hypersecreted in some pathological anxious conditions. Cerebrospinal (CSF) CRF levels, which mostly reflect nonneuroendocrine brain CRF secretion, are high in some patients with posttraumatic stress disorder, underweight patients with anorexia nervosa, ethanol withdrawn alcoholics, and some adults with obsessive-compulsive disorder, all conditions marked by high anxiety levels. Conversely, effective treatment of juveniles with obsessive-compulsive disorder with the tricyclic clomipramine decreased their CSF CRF levels.

The anxiolytic-like actions of nonselective peptide and small-molecule CRF$_1$ receptor antagonists, CRF$_1$ knockout, and CRF$_1$ antisense knockdown all collectively emphasize an endogenous role for extrahypothalamic, limbic CRF$_1$ neurotransmission in hyperarousal and anxiety (for review, see Refs. 107,108). For example, i.c.v. α-helical CRF$_{9-41}$, a nonselective CRF receptor partial agonist, reversed the ability of restraint, footshock, or social isolation to wake rodents prematurely from pentobarbital-induced sleep, and small molecule CRF$_1$ antagonists such as CRA1000 and R121919 also mitigate insomnia-like effects of stress on sleep. Stress or CRF administration also increase locomotor activity in familiar environments, a sign of behavioral hyperarousal blocked by pretreatment with CRF$_1$ receptor antagonists and absent in CRF$_1$-deficient mice.

Nonselective or CRF$_1$ receptor antagonists also reduce the avoidance of unfamiliar, exposed areas in the open field, elevated plus-maze, defensive withdrawal or light/dark box tests by rodents that were stressed by restraint, swim stress, social defeat, conspecific stress, ethanol withdrawal, conditioned effects of psychostimulants, or novelty (but these actions are not observed in unstressed rodents). Knockout of CRF$_1$, but not CRF$_2$, receptors in mice or antisense knockdown of CRF$_1$, but not CRF$_2$, receptors in rats likewise produces anxiolytic-like effects in exploration-based models of anxiety. Anxiolytic-like actions of nonselective or CRF$_1$ receptor antagonists are not limited to explorationbased models as they also were seen in rodent models of footshock or conditioned freezing, fear- or light-potentiated acoustic startle, shock-induced defensive-burying, social interaction, neonatal isolation-induced pup ultrasonic vocalization, and in unconditioned conflict models, such as punished drinking and punished crossing. Peptide nonselective or nonpeptide CRF$_1$ antagonists also reduced stress-induced anorexia in rats and unconditioned submissive/defensive behaviors in social defeat models. In nonhuman primates, CRF$_1$ exhibited oral anxiolytic-like action as well. For example, in rhesus monkeys, DMP904 (ED$_{50}$=21 mg/kg) reduced stereotypical mouthing of rhesus monkeys in a human intruder test, and antalarmin (20 mg/kg) attenuated behavioral signs of anxiety, normalized exploratory and sexual activity, and decreased HPA-axis and sympathoadrenomedullary hormone responses. Finally, depressed patients who participated in the open-label clinical trial with R121919 reported a reduction of anxious symptoms (and improved sleep quality) in relation to escalating doses of the small molecule CRF$_1$ antagonist.

Anxiolytic-like efficacy of antagonists is evident once they occupy 50%–60% of brain CRF$_1$ receptors in the rat, with further efficacy seen with incremental brain receptor occupancy. Acute anxiolytic-like efficacy of CRF$_1$ antagonists does not depend on actions at the HPA axis, as least six compounds—antalarmin, CP 154,526, CRA1000, CRA1001, DMP904 and R121919—reduced anxiety-like behavior at doses that did not reduce ACTH or corticosterone responses to stress. The importance of central as opposed to pituitary CRF$_1$ receptors in anxiety-like behavior is highlighted by the finding that limbic-selective deletion of CRF$_1$ receptors in mice is sufficient to produce anxiolytic-like effects. Anxiogenic-like CRF signaling may be mediated by mitogen-activated protein kinase extracellular signal-related kinase 1/2 (ERK1/2) pathways in the forebrain. Consistent with these findings and sites of anxiogenic-like action by exogenous CRF, the locus coeruleus, bed nucleus of the stria terminalis, and amygdala are sensitive to site-specific anxiolytic-like CRF$_1$ antagonist administration.[107,108]

On the other hand, other recent Phase II/III trials have not shown efficacy of CRF$_1$ antagonists to reduce anxiety in psychopathological populations (for review, see Refs. 46,106). Pexacerfont (BMS-562086) did not relieve generalized anxiety disorder symptoms[11] or suicidal ideation in anxious patients.[12] Trials of verucerfont and emicerfont for social anxiety disorder were completed years ago with undisclosed results (NCT00555139).

In more recent trials, GlaxoSmithKline and NIH evaluated verucerfont against startle in healthy women (NCT01059227), in women with PTSD (NCT01018992), and against stress-induced alcohol craving in anxious women (NCT01187511). A trial for pexacerfont has likewise been initiated in anxious alcoholics by Bristol Myers Squibb and NIAAA (NCT01227980). Results may be disclosed soon.

Depression

Evidence for increased extrahypothalamic CRF activity also has been observed in affective disorders and may contribute to depressive symptomatology. Patients with major depression exhibit signs of chronic, hypothalamically driven HPA axis hyperactivation, including hypercortisolemia, blunted ACTH responses to CRF infusion, nonsuppression in the dexamethasone suppression test, and increased numbers of PVN CRF neurons postmortem.[21,64,69] The resulting excess glucocorticoid levels exert positive feedback on extrahypothalamic CRF systems in the amygdala and bed nucleus of the stria terminalis, and thereby may lead to behavioral symptoms of depression.[30,60,98,99] Accordingly, abnormally high CSF CRF levels are seen in severely depressed patients,

who frequently exhibit comorbid pathological anxiety. Finally, very high CSF CRF levels and downregulated cortical CRF receptor binding and mRNA also were seen post-mortem in patients who committed suicide.[29,58]

Despite initial positive results, however, small-molecule CRF$_1$ antagonists have not consistently shown efficacy in animal models that predict antidepressant activity (see Ref. 107, for a review). Regarding positive findings, subchronic administration of DMP696 and R121919 reduced forced swim immobility in mice, and chronic SSR125543 treatment increased swimming in Flinder Sensitive Line rats, a putative genetic model of depression. Acute antalarmin treatment likewise reduced forced swim immobility in CRF$_2$-receptor-null mutant mice, and antalarmin, SSR125543A, LWH234, and CRA1000 acutely reduced immobility in several studies of outbred rats. R278995 also reduced hyper-emotionality of olfactory bulbectomized rats, a putative model of depression. Finally, chronic treatment with antalarmin or SSR125543A improved coat appearance and reversed reductions in hippocampal neurogenesis in a chronic mild stress model.

Regarding negative findings (see Ref. 107), R121919, CP-154,526, and R278995 failed to reduce forced swim immobility in rats. Antalarmin, CP-154,526, DMP904, R121919, and DMP696 all failed to reduce forced swim immobility in mice after acute, subchronic, or chronic (16 days) dosing. Furthermore, antalarmin, CP-154526, DMP904, R121919, DMP696, and R278995 were all inactive in the tail suspension test with acute dosing. Although acute treatment with CP-154526 was initially reported to produce antidepressant-like effects in the learned helplessness paradigm, a subsequent study with CP-154526 failed to replicate this finding, and DMP904, DMP696, CRA1000, and R2789995 likewise were unable to reduce the expression of learned helplessness after acute dosing. R278995 also did not produce antidepressant-like effects in the rat differential-reinforcement-of-low-rate 72-s model.

The clinical literature similarly indicates unrealized promise of CRF$_1$ antagonist treatment of major depression. Whereas an initial open-label study found that escalating doses of R121919 reduced depressive symptoms[105] and showed normalization of sleep EEG,[28] at least four subsequent double-blind, placebo-controlled studies have failed to observe antidepressant efficacy, including those for ONO-2333Ms (NCT00514865), SSR125543 (EU Clinical Trial DFI5687), CP-316311, and verucerfont (GSK561679) (see Ref. 46).

Gastrointestinal Disorders

Still, CRF$_1$ receptor antagonists may be promising tools for the treatment of functional gastrointestinal disorders, including irritable bowel syndrome (see Refs.

81,66 for reviews). In animal models, diverse stressors delay gastric emptying and stimulate colonic motor function, and endogenous CRF, through both brain and peripheral receptors, mediates these stress-induced changes in gastrocolonic motility. The colonic stimulating effects of stress are observed as colonic hypermotility, decreased transit time, defecation, and watery diarrhea and are mimicked by i.c.v. administration of CRF or urocortin 1 in rats, mice, and Mongolian gerbils. Central or peripheral administration of selective CRF_1, but not CRF_2, agonists stimulates colonic motility, and selective CRF_1, but not CRF_2, antagonists attenuate stress- or CRF/Ucn 1-induced stimulation of colonic motor function. For example, selective CRF_1 receptor antagonists, such as CP-154,526, CRA-1000, NBI-35965, NBI-27914, and antalarmin, injected peripherally or i.c.v. blunted the acceleration of colonic transit induced by restraint; the defecation induced by water avoidance stress, restraint, or social stress; and the diarrhea induced by morphine withdrawal. Likely substrates for CRF_1-mediated stimulation of colonic motor function include, centrally, the PVN and locus coerueus/Barrington nuclei that activate sacral parasympathetic nervous system activity, and, peripherally, the colonic myenteric nervous system.

CRF_1 receptor signaling also is involved in visceral hyperalgesia.[81] For example, both stress and i.c.v. CRF induce visceral hyperalgesia to colorectal distention in rats, and these actions of stress can be blocked by i.c.v. α-helical CRF_{9-41}. Systemic administration of NBI-35965, a nonpeptide CRF_1 antagonist, was found to block water avoidance stress-induced hyperalgesia to colorectal distention in adult offspring of maternal separation. Antalarmin (i.p.) likewise reduced i.c.v. CRF-induced hypersensitivity to colorectal distention and also blunted the visceral hyperalgesia of high-anxiety strain of rats. In addition, α-helical CRF_{9-41} attenuated electrical stimulation-induced visceral symptoms in patients with diarrhea-predominant irritable bowel syndrome. Although an early clinical trial in IBS-diarrhea predominant patients did not find significant efficacy of oral pexacerfont administration on irritable bowel symptoms, a dose-related trend to reduce visceral pain was seen. Furthermore, dosing may have been suboptimal.

CRF_1 receptors are involved in the early phase of postoperative gastric ileus. Peripheral CP154526 administration prevented abdominal surgery-induced slowing of gastric emptying. Similarly, CRF_1 KO mice did not show postoperative gastric ileus, whereas abdominal surgery inhibited postoperative gastric emptying 75% in wild-type mice.[51]

CRF_2 receptor antagonists might be effective treatments to relieve stress-induced gastric stasis and perhaps other adverse gastroparesis. Stress or central infusion of CRF receptor agonists, through vagal effectors, delay gastric emptying, reduce antral gastric motility, inhibit high amplitude gastric contractions, and shift duodenal activity from fasted to fed motor patterns, actions reversed by central astressin$_2$-B (CRF_2), but not NBI 27914 (CRF_1 antagonist) treatment. Complementing the central CRF_2 pathway for slowing gastric emptying, peripheral (i.v. or i.p.) injection of high affinity and/or selective CRF_2 agonists (Ucn 1, Ucn 2, or Ucn 3) also delays gastric emptying more potently than CRF. Conversely, systemic pretreatment with CRF_2 (e.g., ASVG-30 or astressin$_2$-B), but not CRF_1 (e.g., CP-154,526, DMP904, NBI 27914) antagonists blocks the gastromotor inhibiting effects of urocortins as well as restraint stress-induced gastric stasis. Thus, stress releases CRF-like peptides that inhibit gastric motility through brain-gut CRF_2 receptor systems. Likely substrates include, in brain, the PVN and dorsal vagal complex and, peripherally, myenteric fibers of the enteric nervous system (see Ref. 66).

Urologic Disorders

Epidemiologic and clinical findings indicate comorbidity of anxiety and stress with several urologic problems, including overactive bladder, interstitial cystitis, and incontinence.[4,44,76] Accumulating data implicate CRF-CRF1 systems in this relation (see Ref. 87 for a review). This is unsurprising given the overlap in neuroanatomical substrates of CRF-related emotional responding and of micturition and pelvic viscerosensation.[23,38,43,47] Shared substrates include Barrington nucleus, amygdala, hippocampus, and prefrontal cortex. Accordingly, in feline models of interstitial cystitis, CSF CRF levels are increased and subjects show exaggerated acoustic startle reactivity.[6] CRF also appears to play a role in the control of micturition and bladder sensation via the spinal cord.[71,90] In particular, CRF immunoreactivity is present in the intermediolateral cell column, suggesting a role in urinary motor function, and in sensory processing areas, including the dorsal horn, medial and lateral collateral pathways, dorsal intermediate gray, and laminae VII and X. Repeated social defeat induces bladder dysfunction in rodents in association with increased CRF neuron count and CRF mRNA expression in Barrington nucleus.[101] Furthermore, in rats, viral vector-mediated overexpression of CRF in Barrington nucleus leads to bladder dysfunction and urinary retention in association with increased burying, an anxiety-like behavior.[56] Conversely, CRF receptor antagonists have reversed bladder dysfunction in stress-induced animal models.[102]

Pregnancy and Parturition

Changes in circulating CRF, Ucn 1 and CRF-BP levels putatively help coordinate the events of pregnancy and parturition. CRF is produced in the hypothalamus and placenta during pregnancy, and placental CRF accesses both maternal and fetal circulations.[42] During pregnancy, placental CRF secretion increases exponentially, peaking at birth.[8] Accordingly, maternal plasma CRF levels increase from the second trimester exponentially towards term, rise dramatically during labor, and fall quickly after birth.[22,39,84] Unbound ("free") CRF levels are regulated by CRF-BP levels, which normally decrease near term (36 weeks gestation), resulting in a dramatic increase in bioavailable maternal plasma CRF. Women with preterm labor show high CRF and low CRF-BP levels, suggesting an involvement of high bioavailability of maternal CRF in the onset of parturition.[17,78] Consistent with this perspective, maternal "free" CRF measurement predicts outcome of certain at-risk pregnancies,[96,97] and the regulation of the length of gestation by CRF has been termed a "placental clock." For example, the highest levels of maternal CRF at 20 weeks gestation are observed in high-risk populations that ultimately deliver before 34 weeks.[63] A prospective study of 232 pregnancies likewise observed that elevated maternal plasma CRF levels at 33 weeks gestation carried a 3.3-fold greater risk of preterm birth, whereas women who delivered post-term had lower circulating CRF levels at that time than mothers who delivered at term.[91] Similarly, a prospective study of 282 pregnancies found that maternal plasma CRF levels at 18–20 weeks and 28–30 weeks gestation were higher in women who ultimately delivered preterm.[54] Thus, CRF_1 antagonists might delay premature birth. Supporting this hypothesis, CRF_1 receptor antagonists delayed the onset of labor in sheep model studies, though similar effects were not seen in rats.[19]

Inflammatory States

CRF is a proinflammatory peptide via direct autocrine/paracrine actions in immune and immune lineage cells. In brain, CRF and/or its receptors are localized in microglia, astrocytes, and infiltrating monocytes/macrophages. In periphery, CRF receptors are found in macrophages, lymphocytes, and mast cells.[7,79,94] "Immune" CRF and Ucn 1 are expressed by human cord blood-derived cultured mast cells by the 10th gestational week and also are produced by human leukemic mast cell lines. In addition to mast cells,[82] Ucn 1 also is seen in lymphocytes, macrophages, and fibroblasts,[2] a cellular distribution that underlies moderate overall expression in thymus, spleen, and skin.[68]

Supporting proinflammatory properties for CRF and Ucn 1, both peptides provoke resting monocytes to secrete IL-1β and IL-6 in vitro via a CRF_1-mediated mechanism, and CRF triggers lymphocyte proliferation.[45,100] CRF activates microglia, stimulates astrocytosis and monocyte migration, and augments macrophage production of free radicals and arachidonic acid. CRF receptor ligands also activate mast cells in vitro,[82] an action that can be blocked by the CRF_1 antagonist antalarmin, but not the CRF_2 antagonist astressin$_2$-B,[7] and which leads to release of vascular endothelial growth factor, elevated in arthritis, psoriasis, and other chronic inflammatory conditions. The findings suggest a potential antiinflammatory role for CRF_1 antagonists in some tissues (see Ref. 37).

Correspondingly, CRF immunoreactivity is elevated in local immune accessory cells in experimental models of arthritis and uveitis,[14,40,55,93] and CRF or Ucn 1 immunoreactivity also are evident in *human* inflamed tissue, including joints of patients with rheumatoid arthritis, lamina propria macrophages of patients with ulcerative colitis, reproductive tissue of patients with endometriosis, and thyroid glands of patients with Hashimoto thyroiditis.[13,41,77,85] The number of Ucn 1-positive cells in synovia of arthritic patients correlates strongly with inflammation severity.[85] Conversely, chronic systemic antalarmin treatment reduced joint inflammation in an adjuvant-induced arthritis model in LEW/N rats.[93] In addition, CRF_1 agonists (i.p.) increase intestinal mucosal permeability to macromolecules,[77] and chronic psychosocial stress reduces intestinal host defense and initiates intestinal inflammation through a mast cell CRF_1 receptor-dependent mechanism.

Stress is also known to worsen cutaneous diseases, such as psoriasis and atopic dermatitis. Recent evidence suggests that CRF/Ucn 1 crosstalk between mast cells, neurons, and keratinocytes underlie such exacerbation (see Ref. 104 for a review). Skin mast cells, which play a role in allergy and inflammation by releasing histamine, proteases, proteoglycans, prostaglandin D2, and leukotriene C4, secrete CRF and Ucn 1 in response to psychosocial stress. Accordingly, acute psychoscial stress increases skin CRF levels and triggers mast cell-dependent vascular permeability, which also results from intradermal CRF application.[53] CRA1000, a CRF_1 antagonist, blocks the actions of repeated stress to thicken the epidermis and increases mast cell numbers in dermis, suggesting a role for CRF_1 receptors in chronic contact dermatitis. Mast cells also release CRF and Ucn 1 in response to immunoglobulin E receptor crosslinking,[82] suggesting a general role for CRF_1 activity in atopic disorders. Consistent with this possibility, CRF_1 expression is increased in the skin of patients with psoriasis in direct relation to the intensity of clinical symptoms.[9]

Glossary

Adrenocorticotrophic hormone (ACTH) A hormone released into systemic circulation by the anterior pituitary as part of the HPA-stress response. ACTH, or corticotropin, induces the release of glucocorticoids from the adrenal cortex.

cLogP/cLogD Calculated partition coefficients that express a compound's lipophilicity. A partition coefficient expresses the differential solubility of a compound in two solvents, with the log ratio of the concentrations of a neural molecule partitioned between two solvents, classically octanol and water, called LogP. LogD expresses the partition coefficient at a specified pH, thereby accounting for charged forms of the molecule at a physiologically relevant pH (most typically that of blood, pH = 7.4). Drug-like substances have lipophilicities within a limited range, associated with favorable absorption, bioavailability, hydrophobic drug–receptor interactions, metabolism, and toxicity.

Corticotropin-Releasing Factor (CRF) Isolated by Vale and colleagues in 1981, the first hypothalamic ACTH secretagogue to be identified. In addition to its hypothalamic endocrine function, extrahypothalamic CRF is important in behavioral and autonomic stress responses.

Hypothalamic–pituitary–adrenal (HPA) axis The neuroendocrine system which, during stress, amplifies a neural signal of physiologic or psychologic stress into a systemic, endocrine response. Hormones secreted as part of the HPA stress response cascade equip the organism to mobilize and utilize energy resources more effectively and, more generally, respond to the stressor.

Paralog A gene (or gene product) related to a second gene (or gene product) by descent from an ancestral genetic duplication event. Although descended from common ancestral DNA sequences, paralogs generally evolve new, distinct functions, though often related to that of the original sequence/gene product.

Pharmacophore Term coined by Paul Ehrlich in 1909 as "a molecular framework that carries (*phoros*) the essential features responsible for a drug's (*pharmacon's*) biological activity." Updated by Peter Gund in 1977 to describe "a set of structural features in a molecule that is recognized at a receptor site and is responsible for that molecule's biological activity."

Urocortins Structurally related mammalian paralogs of corticotropin-releasing factor. Though these three known peptides (urocortins 1, 2, and 3) share sequence identity with corticotropin-releasing factor, each CRF family peptide is coded by a different gene and has different pharmacological properties, tissue distributions, and regulatory pathways, leading to their diverse roles in stress responses to homeostatic threats. Urocortins, unlike CRF, exhibit high affinity for CRF$_2$ receptors.

Acknowledgments

Financial support was received from the Pearson Center for Alcoholism and Addiction Research and National Institutes of Health grant AA06420 from the National Institute on Alcohol Abuse and Alcoholism. The content is solely the responsibility of the author and does not necessarily represent the official views of the National Institutes of Health or the National Institute of Alcohol Abuse and Alcoholism. The author thanks Michael Arends for editorial assistance.

References

1. Bale TL, Vale WW. CRF and CRF receptors: role in stress responsivity and other behaviors. *Annu Rev Pharmacol Toxicol.* 2004;44:525–557.
2. Bamberger CM, Wald M, Bamberger AM, Ergün S, Beil FU, Schulte HM. Human lymphocytes produce urocortin, but not corticotropin-releasing hormone. *J Clin Endocrinol Metab.* February 1998;83(2):708–711.
3. Barquist E, Zinner M, Rivier J, Tache Y. Abdominal surgery-induced delayed gastric emptying in rats: role of CRF and sensory neurons. *Am J Physiol.* 1992;262(4 Pt 1):G616–G620.
4. Bogner HR, Gallo JJ, Sammel MD, Ford DE, Armenian HK, Eaton WW. Urinary incontinence and psychological distress in community-dwelling older adults. *J Am Geriatr Soc.* March 2002;50(3):489–495.
5. Brown MR, Gray TS, Fisher LA. Corticotropin-releasing factor receptor antagonist: effects on the autonomic nervous system and cardiovascular function. *Regul Pept.* December 30, 1986;16(3–4):321–329.
6. Buffington CA. Comorbidity of interstitial cystitis with other unexplained clinical conditions. *J Urol.* October 2004;172(4 Pt 1):1242–1248.
7. Cao J, Papadopoulou N, Kempuraj D, et al. Human mast cells express corticotropin-releasing hormone (CRH) receptors and CRH leads to selective secretion of vascular endothelial growth factor. *J Immunol.* June 15, 2005;174(12):7665–7675.
8. Caritis SN. Treatment of preterm labour. A review of the therapeutic options. *Drugs.* September 1983;26(3):243–261.
9. Cemil BC, Canpolat F, Yilmazer D, Eskioğlu F, Alper M. The association of PASI scores with CRH-R1 expression in patients with psoriasis. *Arch Dermatol Res.* March 2012;304(2):127–132.
10. Chen C, Wilcoxen KM, Huang CQ, et al. Design of 2,5-dimethyl-3-(6-dimethyl-4-methylpyridin-3-yl)-7-dipropyl-aminopyrazolo[1,5-a]pyrimidine (NBI 30775/R121919) and structure-activity relationships of a series of potent and orally active corticotropin-releasing factor receptor antagonists. *J Med Chem.* 2004;47:4787–4798.
11. Coric V, Feldman HH, Oren DA, et al. Multicenter, randomized, double-blind, active comparator and placebo-controlled trial of a corticotropin-releasing factor receptor-1 antagonist in generalized anxiety disorder. *Depress Anxiety.* 2010;27:417–425.
12. Coric V, Stock EG, Pultz J, Marcus R, Sheehan DV. Sheehan Suicidality Tracking Scale (Sheehan-STS): preliminary results from a multicenter clinical trial in generalized anxiety disorder. *Psychiatry (Edgmont).* 2009;6:26–31.
13. Crofford LJ, Sano H, Karalis K, et al. Corticotropin-releasing hormone in synovial fluids and tissues of patients with rheumatoid arthritis and osteoarthritis. *J Immunol.* August 1, 1993;151(3):1587–1596.
14. Crofford LJ, Sano H, Karalis K, et al. Local secretion of corticotropin-releasing hormone in the joints of Lewis rats with inflammatory arthritis. *J Clin Invest.* December 1992;90(6):2555–2564.
15. Dyck B, Grigoriadis DE, Gross RS, et al. Potent, orally active corticotropin-releasing factor receptor-1 antagonists containing a tricyclic pyrrolopyridine or pyrazolopyridine core. *J Med Chem.* 2005;48:4100–4110.
16. Dzierba CD, Takvorian AG, Rafalski M, et al. Synthesis, structure-activity relationships, and in vivo properties of 3,4-dihydro-1H-pyrido[2,3-b]pyrazin-2-ones as corticotropin-releasing factor-1 receptor antagonists. *J Med Chem.* November 4, 2004;47(23):5783–5790.
17. Fadalti M, Pezzani I, Cobellis L, et al. Placental corticotropin-releasing factor. An update. *Ann N. Y Acad Sci.* 2000;900:89–94.
18. Fekete EM, Zorrilla EP. Physiology, pharmacology, and therapeutic relevance of urocortins in mammals: ancient CRF paralogs. *Front Neuroendocrinol.* 2007;28:1–27.
19. Funai EF, O'Neill LM, Davidson A, Roqué H, Finlay TH. A corticotropin releasing hormone receptor antagonist does not delay parturition in rats. *J Perinat Med.* 2000;28(4):294–297.
20. Gilligan PJ, Folmer BK, Hartz RA, et al. Pyrazolo-[1,5-a]-1,3,5-triazine corticotropin-releasing factor (CRF) receptor ligands. *Bioorg Med Chem.* September 1, 2003;11(18):4093–4102.
21. Gold PW, Chrousos GP. The endocrinology of melancholic and atypical depression: relation to neurocircuitry and somatic consequences. *Proc Assoc Am Physicians.* 1999;111:22–34.
22. Grammatopoulos DK, Hillhouse EW. Role of corticotropin-releasing hormone in onset of labour. *Lancet.* October 30, 1999;354(9189):1546–1549.
23. Grill WM, Bhadra N, Wang B. Bladder and urethral pressures evoked by microstimulation of the sacral spinal cord in cats. *Brain Res.* July 31, 1999;836(1–2):19–30.

24. Gross RS, Guo Z, Dyck B, et al. Design and synthesis of tricyclic corticotropin-releasing factor-1 antagonists. *J Med Chem.* 2005;48:5780–5793.

25. Gulyas J, Rivier C, Perrin M, et al. Potent, structurally constrained agonists and competitive antagonists of corticotropin-releasing factor. *Proc Natl Acad Sci USA.* 1995;92:10575–10579.

26. Guo Z, Tellew JE, Gross RS, et al. Design and synthesis of tricyclic imidazo[4,5-b]pyridin-2-ones as corticotropin-releasing factor-1 antagonists. *J Med Chem.* 2005;48:5104–5107.

27. Hartz RA, Nanda KK, Ingalls CL, et al. Design, synthesis, and biological evaluation of 1,2,3,7-tetrahydro-6h-purin-6-one and 3,7-dihydro-1h-purine-2,6-dione derivatives as corticotropin-releasing factor(1) receptor antagonists. *J Med Chem.* September 9, 2004;47(19):4741–4754.

28. Held K, Kunzel H, Ising M, et al. Treatment with the CRH1-receptor-antagonist R121919 improves sleep-EEG in patients with depression. *J Psychiatric Res.* 2004;38:129–136.

29. Holsboer F. The corticosteroid receptor hypothesis of depression. *Neuropsychopharmacology.* November 2000;23(5):477–501.

30. Holsboer F, Barden N. Antidepressants and hypothalamic-pituitary-adrenocortical regulation. *Endocr Rev.* April 1996;17(2):187–205.

31. Holsboer F, Ising M. Central CRH system in depression and anxiety: evidence from clinical studies with CRH1 receptor antagonists. *Eur J Pharmacol.* 2008;583:350–357.

32. Hsu SY, Hsueh AJ. Human stresscopin and stresscopin-related peptide are selective ligands for the type 2 corticotropin-releasing hormone receptor. *Nat Med.* 2001;7:605–611.

33. Huang CQ, Grigoriadis DE, Liu Z, et al. Design, synthesis, and SAR of 2-dialkylamino-4-arylpyrimidines as potent and selective corticotropin-releasing factor$_1$ (CRF$_1$) receptor antagonists. *Bioorg Med Chem Lett.* May 3, 2004a;14(9):2083–2086.

34. Huang CQ, Wilcoxen K, McCarthy JR, Haddach M, Grigoriadis D, Chen C. Synthesis of 1-methyl-3-phenylpyrazolo[4,3-b]pyridines via a methylation of 4-phthalimino-3-phenylpyrazoles and optimization toward highly potent corticotropin-releasing factor type-1 antagonists. *Bioorg Med Chem Lett.* October 6, 2003a;13(19):3371–3374.

35. Huang CQ, Wilcoxen K, McCarthy JR, et al. Synthesis and SAR of 8-arylquinolines as potent corticotropin-releasing factor$_1$ (CRF$_1$) receptor antagonists. *Bioorg Med Chem Lett.* October 6, 2003b;13(19):3375–3379.

36. Huang CQ, Wilcoxen KM, Grigoriadis DE, McCarthy JR, Chen C. Design and synthesis of 3-(2-pyridyl)pyrazolo[1,5-a]pyrimidines as potent CRF$_1$ receptor antagonists. *Bioorg Med Chem Lett.* August 2, 2004b;14(15):3943–3947.

37. Im E. Multi-facets of corticotropin-releasing factor in modulating inflammation and angiogenesis. *J Neurogastroenterol Motil.* January 1, 2015;21(1):25–32.

38. Imaki T, Nahan JL, Rivier C, Sawchenko PE, Vale W. Differential regulation of corticotropin-releasing factor mRNA in rat brain regions by glucocorticoids and stress. *J Neurosci.* 1991;11:585–599.

39. Karalis K, Goodwin G, Majzoub JA. Cortisol blockade of progesterone: a possible molecular mechanism involved in the initiation of human labor. *Nat Med.* May 1996;2(5):556–560.

40. Karalis K, Sano H, Redwine J, Listwak S, Wilder RL, Chrousos GP. Autocrine or paracrine inflammatory actions of corticotropin-releasing hormone in vivo. *Science.* October 18, 1991;254(5030):421–423.

41. Kawahito Y, Sano H, Mukai S, et al. Corticotropin releasing hormone in colonic mucosa in patients with ulcerative colitis. *Gut.* October 1995;37(4):544–551.

42. Keller PA, Elfick L, Garner J, Morgan J, McCluskey A. Corticotropin releasing hormone: therapeutic implications and medicinal chemistry developments. *Bioorg Med Chem.* June 2000;8(6):1213–1223.

43. Kergozien S, Ménétrey D. Environmental influences on viscero(noci)ceptive brain activities: the effects of sheltering. *Brain Res Cogn Brain Res.* September 2000;10(1–2):111–117.

44. Klausner AP, Steers WD. Corticotropin releasing factor: a mediator of emotional influences on bladder function. *J Urol.* December 2004;172(6 Pt 2):2570–2573.

45. Kohno M, Kawahito Y, Tsubouchi Y, et al. Urocortin expression in synovium of patients with rheumatoid arthritis and osteoarthritis: relation to inflammatory activity. *J Clin Endocrinol Metab.* September 2001;86(9):4344–4352.

46. Koob GF, Zorrilla EP. Update on corticotropin-releasing factor pharmacotherapy for psychiatric disorders: a revisionist view. *Neuropsychopharmacol Rev.* 2012;37:308–309.

47. Koyama K. Effects of amygdaloid and olfactory tubercle stimulation on efferent activities of the vesical branch of the pelvic nerve and the urethral branch of the pudendal nerve in dogs. *Urol Int.* 1991;47(suppl 1):23–30.

48. Lewis K, Li C, Perrin MH, et al. Identification of urocortin III, an additional member of the corticotropin-releasing factor (CRF) family with high affinity for the CRF2 receptor. *Proc Natl Acad Sci USA.* 2001;98:7570–7575.

49. Lin JH, Lu AY. Role of pharmacokinetics and metabolism in drug discovery and development. *Pharmacol Rev.* December 1997;49(4):403–449.

50. Lowe RF, Nelson J, Dang TN, et al. Rational design, synthesis, and structure-activity relationships of aryltriazoles as novel corticotropin-releasing factor-1 receptor antagonists. *J Med Chem.* 2005;48:1540–1549.

51. Luckey A, Wang L, Jamieson PM, et al. Corticotropin-releasing factor receptor 1-deficient mice do not develop postoperative gastric ileus. *Gastroenterology.* September 2003;125(3):654–659.

52. Luthin DR, Rabinovich AK, Bhumralkar DR, et al. Synthesis and biological activity of oxo-7H-benzo[e]perimidine-4-carboxylic acid derivatives as potent, nonpeptide corticotropin releasing factor (CRF) receptor antagonists. *Bioorg Med Chem Lett.* March 8, 1999;9(5):765–770.

53. Lytinas M, Kempuraj D, Huang M, Boucher W, Esposito P, Theoharides TC. Acute stress results in skin corticotropin-releasing hormone secretion, mast cell activation and vascular permeability, an effect mimicked by intradermal corticotropin-releasing hormone and inhibited by histamine-1 receptor antagonists. *Int Arch Allergy Immunol.* March 2003;130(3):224–231.

54. Mancuso RA, Schetter CD, Rini CM, Roesch SC, Hobel CJ. Maternal prenatal anxiety and corticotropin-releasing hormone associated with timing of delivery. *Psychosom Med.* September–October, 2004;66(5):762–769.

55. Mastorakos G, Bouzas EA, Silver PB, et al. Immune corticotropin-releasing hormone is present in the eyes of and promotes experimental autoimmune uveoretinitis in rodents. *Endocrinology.* October 1995;136(10):4650–4658.

56. McFadden K, Griffin TA, Levy V, Wolfe JH, Valentino RJ. Overexpression of corticotropin-releasing factor in Barrington's nucleus neurons by adeno-associated viral transduction: effects on bladder function and behavior. *Eur J Neurosci.* November 2012;36(10):3356–3364.

57. Menzaghi F, Howard RL, Heinrichs SC, Vale W, Rivier J, Koob GF. Characterization of a novel and potent corticotropin-releasing factor antagonist in rats. *J Pharmacol Exp Ther.* 1994;269:564–572.

58. Mitchell AJ. The role of corticotropin releasing factor in depressive illness: a critical review. *Neurosci Biobehav Rev.* September 1998;22(5):635–651.

59. Molteni V, Penzotti J, Wilson DM, et al. N-phenylphenylglycines as novel corticotropin releasing factor receptor antagonists. *J Med Chem.* May 6, 2004;47(10):2426–2429.

60. Murphy BE. Antiglucocorticoid therapies in major depression: a review. *Psychoneuroendocrinology.* 1997;22(suppl 1):S125–S132.

61. Nalivaiko E, Blessing WW. CRF1-receptor antagonist CP-154526 reduces alerting-related cutaneous vasoconstriction in conscious rabbits. *Neuroscience*. 2003;117(1):129–138.

62. Nalivaiko E, Blessing WW. CRF1 receptor antagonist CP-154,526 reduces cardiovascular responses during acute psychological stress in rabbits. *Brain Res*. August 13, 2004;1017(1–2):234–237.

63. Nappi RE, Petraglia F, Luisi S, Polatti F, Farina C, Genazzani AR. Serum allopregnanolone in women with postpartum "blues.". *Obstetrics Gynecol*. 2001;97:77–80.

64. Nemeroff CB. The neurobiology of depression. *Sci Am*. 1998;278:42–49.

65. Nijsen MJ, Croiset G, Stam R, et al. The role of the CRH type 1 receptor in autonomic responses to corticotropin-releasing hormone in the rat. *Neuropsychopharmacology*. April 2000;22(4):388–399.

66. Nozu T, Okumura T. Corticotropin-releasing factor receptor type 1 and type 2 interaction in irritable bowel syndrome. *J Gastroenterol*. August 2015;50(8):819–830.

67. Nozu T, Martinez V, Rivier J, Taché Y. Peripheral urocortin delays gastric emptying: role of CRF receptor 2. *Am J Physiol*. April 1999;276(4 Pt 1):G867–G874.

68. Pisarchik A, Slominski A. Molecular and functional characterization of novel CRFR1 isoforms from the skin. *Eur J Biochem*. July 2004;271(13):2821–2830.

69. Plotsky PM, Owens MJ, Nemeroff CB. Psychoneuroendocrinology of depression. Hypothalamic-pituitary-adrenal axis. *Psychiatr Clin North Am*. June 1998;21(2):293–307.

70. Primus RJ, Yevich E, Baltazar C, Gallager DW. Autoradiographic localization of CRF1 and CRF2 binding sites in adult rat brain. *Neuropsychopharmacology*. 1997;17:308–316.

71. Puder BA, Papka RE. Distribution and origin of corticotropin-releasing factor-immunoreactive axons in the female rat lumbosacral spinal cord. *J Neurosci Res*. December 15, 2001;66(6):1217–1225.

72. Reyes TM, Lewis K, Perrin MH, et al. Urocortin II: a member of the corticotropin-releasing factor (CRF) neuropeptide family that is selectively bound by type 2 CRF receptors. *Proc Natl Acad Sci USA*. 2001;98:2843–2848.

73. Rijkers DT, Kruijtzer JA, van Oostenbrugge M, Ronken E, den Hartog JA, Liskamp RM. Structure-activity studies on the corticotropin releasing factor antagonist astressin, leading to a minimal sequence necessary for antagonistic activity. *Chembiochem*. March 5, 2004;5(3):340–348.

74. Rivier J, Gulyas J, Corrigan A, et al. Astressin analogues (corticotropin-releasing factor antagonists) with extended duration of action in the rat. *J Med Chem*. December 3, 1998;41(25):5012–5019.

75. Rivier J, Rivier C, Vale W. Synthetic competitive antagonists of corticotropin-releasing factor: effect on ACTH secretion in the rat. *Science*. 1984;224:889–891.

76. Rothrock NE, Lutgendorf SK, Kreder KJ, Ratliff T, Zimmerman B. Stress and symptoms in patients with interstitial cystitis: a life stress model. *Urology*. March 2001;57(3):422–427.

77. Saruta M, Takahashi K, Suzuki T, Torii A, Kawakami M, Sasano H. Urocortin 1 in colonic mucosa in patients with ulcerative colitis. *J Clin Endocrinol Metab*. November 2004;89(11):5352–5361.

78. Sasaki A, Sato S, Murakami O, et al. Immunoreactive corticotropin-releasing hormone present in human plasma may be derived from both hypothalamic and extrahypothalamic sources. *J Clin Endocrinol Metab*. July 1987;65(1):176–182.

79. Singh VK. Stimulatory effect of corticotropin-releasing neurohormone on human lymphocyte proliferation and interleukin-2 receptor expression. *J Neuroimmunol*. August 1989;23(3):257–262.

80. St Denis Y, Di Fabio R, Bernasconi G, et al. Substituted tetraazaacenaphthylenes as potent CRF1 receptor antagonists for the treatment of depression and anxiety. *Bioorg Med Chem Lett*. 2005;15:3713–3716.

81. Taché Y, Million M. Role of corticotropin-releasing factor signaling in stress-related alterations of colonic motility and hyperalgesia. *J Neurogastroenterol Motil*. January 31, 2015;21(1):8–24.

82. Theoharides TC, Donelan JM, Papadopoulou N, Cao J, Kempuraj D, Conti P. Mast cells as targets of corticotropin-releasing factor and related peptides. *Trends Pharmacol Sci*. November 2004;25(11):563–568.

83. Turcu AF, Spencer-Segal JL, Farber RH, et al. Single-dose study of a corticotropin-releasing factor receptor-1 antagonist in women with 21-hydroxylase deficiency. *J Clin Endocrinol Metab*. March 2016;101(3):1174–1180.

84. Ur E, Grossman A. Corticotropin-releasing hormone in health and disease: an update. *Acta Endocrinol (Copenh)*. September 1992;127(3):193–199.

85. Uzuki M, Sasano H, Muramatsu Y, et al. Urocortin in the synovial tissue of patients with rheumatoid arthritis. *Clin Sci (Lond)*. June 2001;100(6):577–589.

86. Vale W, Spiess J, Rivier C, Rivier J. Characterization of a 41-residue ovine hypothalamic peptide that stimulates the secretion of corticotropin and -endorphin. *Science*. 1981;213:1394–1397.

87. Valentino RJ1, Wood SK, Wein AJ, Zderic SA. The bladder-brain connection: putative role of corticotropin-releasing factor. *Nat Rev Urol*. January 2011;8(1):19–28.

88. Van Pett K, Viau V, Bittencourt JC, et al. Distribution of mRNAs encoding CRF receptors in brain and pituitary of rat and mouse. *J Comp Neurology*. 2000;428:191–212.

89. Vaughan J, Donaldson C, Bittencourt J, et al. Urocortin, a mammalian neuropeptide related to fish urotensin I and to corticotropin-releasing factor. *Nature*. 1995;378:287–292.

90. Vincent SR, Satoh K. Corticotropin-releasing factor (CRF) immunoreactivity in the dorsolateral pontine tegmentum: further studies on the micturition reflex system. *Brain Res*. August 13, 1984;308(2):387–391.

91. Wadhwa PD, Garite TJ, Porto M, et al. Placental corticotropin-releasing hormone (CRH), spontaneous preterm birth, and fetal growth restriction: a prospective investigation. *Am J Obstet Gynecol*. October 2004;191(4):1063–1069.

92. Webb TR, Moran T, Huang CQ, McCarthy JR, Grigoriadis DE, Chen C. Synthesis of benzoylpyrimidines as antagonists of the corticotropin-releasing factor-1 receptor. *Bioorg Med Chem Lett*. August 2, 2004;14(15):3869–3873.

93. Webster EL, Barrientos RM, Contoreggi C, et al. Corticotropin releasing hormone (CRH) antagonist attenuates adjuvant induced arthritis: role of CRH in peripheral inflammation. *J Rheumatol*. June 2002;29(6):1252–1261.

94. Webster EL, Tracey DE, Jutila MA, Wolfe Jr SA, De Souza EB. Corticotropin-releasing factor receptors in mouse spleen: identification of receptor-bearing cells as resident macrophages. *Endocrinology*. July 1990;127(1):440–452.

95. Wilson DM, Termin AP, Mao L, et al. Arylamidrazones as novel corticotropin releasing factor receptor antagonists. *J Med Chem*. May 23, 2002;45(11):2123–2126.

96. Wolfe CD, Patel SP, Campbell EA, et al. Plasma corticotrophin-releasing factor (CRF) in normal pregnancy. *Br J Obstet Gynaecol*. October 1988a;95(10):997–1002.

97. Wolfe CD, Patel SP, Linton EA, et al. Plasma corticotrophin-releasing factor (CRF) in abnormal pregnancy. *Br J Obstet Gynaecol*. October 1988b;95(10):1003–1006.

98. Wolkowitz OM, Reus VI. Treatment of depression with antiglucocorticoid drugs. *Psychosom Med*. 1999 Sep-Oct;61(5):698–711.

99. Wolkowitz OM, Reus VI, Chan T, et al. Antiglucocorticoid treatment of depression: double-blind ketoconazole. *Biol Psychiatry*. April 15, 1999;45(8):1070–1074.

100. Woloski BM, Smith EM, Meyer 3rd WJ, Fuller GM, Blalock JE. Corticotropin-releasing activity of monokines. *Science*. November 29, 1985;230(4729):1035–1037.

101. Wood SK, Baez MA, Bhatnagar S, Valentino RJ. Social stress-induced bladder dysfunction: potential role of corticotropin-releasing factor. *Am J Physiol Regul Integr Comp Physiol*. May 2009;296(5):R1671–R1678.

102. Wood SK1, McFadden K, Griffin T, Wolfe JH, Zderic S, Valentino RJ. A corticotropin-releasing factor receptor antagonist improves urodynamic dysfunction produced by social stress or partial bladder outlet obstruction in male rats. *Am J Physiol Regul Integr Comp Physiol*. June 1, 2013;304(11):R940–R950.

103. Yamada Y, Mizutani K, Mizusawa Y, et al. New class of corticotropin-releasing factor (CRF) antagonists: small peptides having high binding affinity for CRF receptor. *J Med Chem*. 2004;47:1075–1078.

104. Zhou C, Yu X, Cai D, Liu C, Li C. Role of corticotropin-releasing hormone and receptor in the pathogenesis of psoriasis. *Med Hypotheses*. October 2009;73(4):513–515.

105. Zobel AW, Nickel T, Kunzel HE, et al. Effects of the high-affinity corticotropin-releasing hormone receptor 1 antagonist R121919 in major depression: the first 20 patients treated. *J Psychiatric Res*. 2000;34:171–181.

106. Zorrilla EP, Heilig M, de Wit H, Shaham Y. Behavioral, biological, and chemical perspectives on targeting CRF_1 receptor antagonists to treat alcoholism. *Drug Alcohol Depend*. March 1, 2013;128(3):175–186.

107. Zorrilla EP, Koob GF. Progress in corticotropin-releasing factor-1 antagonist development. *Drug Discov Today*. 2010;15:371–383.

108. Zorrilla EP, Koob GF. The therapeutic potential of CRF1 antagonists for anxiety. *Expert Opin Invest Drugs*. 2004;13:799–828.

109. Zuev D, Michne JA, Pin SS, Zhang J, Taber MT, Dubowchik GM. Optimization of CRF1R binding affinity of 2-(2,4,6-trichlorophenyl)-4-trifluoromethyl-5-aminomethylthiazoles through rapid and selective parallel synthesis. *Bioorg Med Chem Lett*. January 17, 2005;15(2):431–434.

Further Reading

1. Alonso R, Griebel G, Pavone G, Stemmelin J, Le Fur G, Soubrie P. Blockade of CRF1 or V1b receptors reverses stress-induced suppression of neurogenesis in a mouse model of depression. *Mol Psychiatry*. 2004;9:278–286. 224.

2. Bale TL, Picetti R, Contarino A, Koob GF, Vale WW, Lee KF. Mice deficient for both corticotropin-releasing factor receptor 1 (CRFR1) and CRFR2 have an impaired stress response and display sexually dichotomous anxiolytic-like behavior. *J Neurosci*. 2002;22:193–199.

3. Bale TL, Vale WW. Increased depression-like behaviors in corticotropin-releasing factor receptor-2-deficient mice: sexually dichotomous responses. *J Neurosci*. 2003;23:5295–5301.

4. Chaki S, Nakazato A, Kennis L, et al. Anxiolytic- and antidepressant-like profile of a new CRF1 receptor antagonist, R278995/CRA0450. *Eur J Pharmacol*. 2004;485:145–158.

5. Chan EC, Falconer J, Madsen G, et al. A corticotropin-releasing hormone type I receptor antagonist delays parturition in sheep. *Endocrinology*. July 1998;139(7):3357–3360.

6. Gilligan PJ, Baldauf C, Cocuzza A, et al. The discovery of 4-(3-pentylamino)-2,7-dimethyl-8-(2-methyl-4-methoxyphenyl)-pyrazolo-[1,5-a]-pyrimidine: a corticotropin-releasing factor (hCRF1) antagonist. *Bioorg Med Chem*. 2000;8:181–189.

7. Griebel G, Simiand J, Steinberg R, et al. 4-(2-Chloro-4-methoxy-5-methylphenyl)-N-[(1S)-2-cyclopropyl-1-(3-fluoro-4-methylphenyl)ethyl]5-methyl-N-(2-propynyl)-1,3-thiazol-2-amine hydrochloride (SSR125543A), a potent and selective corticotrophin-releasing factor1 receptor antagonist: II. Characterization in rodent models of stress-related disorders. *J Pharmacol Exp Ther*. 2002;301:333–345.

8. Harro J, Tonissaar M, Eller M. The effects of CRA 1000, a nonpeptide antagonist of corticotropin-releasing factor receptor type 1, on adaptive behaviour in the rat. *Neuropeptides*. 2001;35:100–109.

9. Jutkiewicz EM, Wood SK, Houshyar H, Hsin LW, Rice KC, Woods JH. The effects of CRF antagonists, antalarmin, CP154,526, LWH234, and R121919, in the forced swim test and on swim-induced increases in adrenocorticotropin in rats. *Psychopharmacology*. 2005;180:215–223.

10. Li YW, Fitzgerald L, Wong H, et al. The pharmacology of DMP696 and DMP904, non-peptidergic CRF1 receptor antagonists. *CNS Drug Rev*. 2005;11:21–52.

11. Licinio J, O'Kirwan F, Irizarry K, et al. Association of a corticotropin-releasing hormone receptor 1 haplotype and antidepressant treatment response in Mexican-Americans. *Mol Psychiatry*. December 2004;9(12):1075–1082.

12. Liu X, Peprah D, Gershenfeld HK. Tail-suspension induced hyperthermia: a new measure of stress reactivity. *J Psychiatric Res*. 2003;37:249–259.

13. Mansbach RS, Brooks EN, Chen YL. Antidepressant-like effects of CP-154,526, a selective CRF1 receptor antagonist. *Eur J Pharmacol*. 1997;323:21–26.

14. Nielsen DM, Carey GJ, Gold LH. Antidepressant-like activity of corticotropin-releasing factor type-1 receptor antagonists in mice. *Eur J Pharmacol*. 2004;499:135–146.

15. Oshima A, Flachskamm C, Reul JM, Holsboer F, Linthorst AC. Altered serotonergic neurotransmission but normal hypothalamic-pituitary-adrenocortical axis activity in mice chronically treated with the corticotropin-releasing hormone receptor type 1 antagonist NBI 30775. *Neuropsychopharmacology*. 2003;28:2148–2159.

16. Overstreet DH, Griebel G. Antidepressant-like effects of CRF1 receptor antagonist SSR125543 in an animal model of depression. *Eur J Pharmacol*. 2004;497:49–53.

17. Takamori K, Kawashima N, Chaki S, Nakazato A, Kameo K. Involvement of corticotropin-releasing factor subtype 1 receptor in the acquisition phase of learned helplessness in rats. *Life Sci*. 2001b;69:1241–1248.

18. Takamori K, Kawashima N, Chaki S, Nakazato A, Kameo K. Involvement of the hypothalamus-pituitary-adrenal axis in antidepressant activity of corticotropin-releasing factor subtype 1 receptor antagonists in the rat learned helplessness test. *Pharmacol Biochem Behav*. 2001a;69:445–449.

19. Yamano M, Yuki H, Yasuda S, Miyata K. Corticotropin-releasing hormone receptors mediate consensus interferon- YM643-induced depression-like behavior in mice. *J Pharmacol Exp Ther*. 2000;292:181–187.

CHAPTER

27

Antidepressant Actions on Glucocorticoid Receptors

N. Nikkheslat[1], P.A. Zunszain[1], L.A. Carvalho[2], C. Anacker[3], C.M. Pariante[1]

[1]King's College London, United Kingdom; [2]University College London, United Kingdom; [3]Columbia University, New York, NY, United States

Abstract

Effective antidepressant treatment has been shown to resolve the hypothalamic–pituitary–adrenal (HPA) axis dysregulation implicated in pathogenesis of depression. The hyperactivity of the HPA axis in depressed patients appears to reflect an altered function of glucocorticoid receptors (GR), which play a crucial part in the feedback regulation of the axis. Evidence suggests that antidepressants exert their effect directly on GR through increasing expression, promoting translocation, and enhancing function of the receptor in the brain and other target tissues, and hence normalizing HPA axis abnormalities. This chapter focuses on the HPA axis and on the GR and describes how effective antidepressant medications restore the neuroendocrine-immune balance through modulation of the GR. Investigation of the molecular pathways underlying the effects of antidepressants on regulation of the GR function is crucial for enhancing our understanding of the therapeutic efficacy of these medications and provides mechanistic insight for developing new antidepressant agents.

INTRODUCTION

Over the last decades, there have been mounting lines of investigations trying to understand the role of the hypothalamic–pituitary–adrenal (HPA) axis in the pathogenesis of major depressive disorders (MDD). In a nutshell, depressed individuals display alterations in stress responses mediated by the neuroendocrine system, indicating an association between HPA axis hyperactivity and cortisol hypersecretion with the development of depression.[48,61] Moreover, patients with depression tend to exhibit dysfunction of the immune system characterized by an overactivation of inflammatory responses as well as a failure of the regulatory control by the HPA axis.[32,69] Therefore, recent studies in depression have been trying to understand how these two main biological factors, HPA axis and inflammation, interact in the pathogenesis of this disorder. This chapter focuses on the glucocorticoid receptor (GR), the receptor for cortisol and corticosterone, which plays a crucial part in the regulation of both these systems, and on how effective antidepressant medications restore the neuroendocrine-immune balance through direct effects on the GR. The chapter updates and extends our previous reviews on these topics.[4,13]

HPA AXIS AND GLUCOCORTICOID RECEPTORS IN DEPRESSION

The HPA Axis

Involvement of the HPA axis in depression has been reported consistently by various studies.[48] HPA axis is a central physiological circuit, which connects the brain with the endocrine system and plays a fundamental role as a regulatory system in stress responses. In response to psychological, physical, and/or environmental stressful stimuli, the HPA system is activated, through the activation of the parvocellular neurons of the hypothalamic paraventricular nucleus secreting corticotropin-releasing hormone (CRH) and arginine vasopressin. These neuropeptides exert their effect on the pituitary gland promoting the release of adrenocorticotropic hormone (ACTH) from the anterior lobe of the gland into the bloodstream. Circulating ACTH in turn triggers the biosynthesis and secretion of glucocorticoids from the adrenal cortex.[48]

The Glucocorticoids

Glucocorticoids, cortisol (in human and primates) and corticosterone (in rodents), are the final product of the HPA axis. These steroid hormones are synthesized from cholesterol and exhibit a critical role in restoring and maintaining bodily stress-related homeostasis, modulating neuroendocrine and immune responses, regulating energy metabolism and inflammatory reactions, and influencing cardiovascular function. The physiological function of the endogenous glucocorticoids, as both central and peripheral effect, is mediated upon their interaction with specific intracellular receptors expressed in target tissues, including the HPA axis itself where they play a critical role in maintaining the intrinsic homeostasis of the axis activity through a glucocorticoid-mediated negative feedback mechanism inhibiting the secretion of ACTH from the pituitary as well as CRH from the hypothalamus.[15]

The Glucocorticoid Receptors

There are two distinct corticosteroid receptor subtypes: type I or mineralocorticoid receptor (MR) and type II or GR. These specialized intracellular ligand-binding receptors are the members of the superfamily of nuclear transcription factors, and are involved in modulating the transcription of target genes. Under normal physiological conditions, MR is expressed in heart, intestine, and renal tissue as well as limbic brain regions, while GR is found in almost all tissues and organs in the human body.[4] Compared with MR, GR has lower affinity for endogenous corticosteroids, and it is activated when the glucocorticoid levels are high. The GR also shows high affinity for dexamethasone which is a synthetic analog of cortisol that binds with a very high affinity to the GR and virtually not to MR. While MR is mainly associated with circadian regulation of cortisol, GR plays a role in the modulation of peak morning response. In addition, GR activity appears to be more crucial in regulating stress-related responses when specifically there are elevated levels of endogenous glucocorticoids.[48]

Inactivated GR primarily resides in the cytoplasm. The unbound cytosolic receptor is stabilized within an assembly of chaperone proteins including immunophilins (FKBP5, Cyp44, PP5) and heat shock proteins (HSP90, HSP70, HSP56, HSP23). GR is a ligand-dependent transcription factor and is activated following binding to glucocorticoids, which passively diffuse across the cell membrane. Upon activation, GR undergoes conformational changes, dissociates from its protein complex, and translocates to the nucleus where it regulates gene expression. In fact, GR positively or negatively alters gene transcription either directly by interaction with DNA-binding domains (in particular the promoter sequence of the DNA known as glucocorticoid responsive elements), or indirectly via binding with other transcription factors (protein–protein interactions), which in turn bind to the DNA response elements. GR sensitivity to glucocorticoids is vital in order to produce an appropriate response, and it is determined by the number, affinity, and function of the receptor, including the ability of GR to bind the ligand, to translocate from cytoplasm into the nucleus, and to interact with other signal transduction pathways.[4,43]

Activation of gene expression by GR is known as transactivation in which the GR stimulates transcription rate of a respective target gene, while negative alteration of gene expression by GR is called transrepression, in which GR suppresses the activity of other transcription factors, such as cyclic AMP (cAMP) response element-binding protein, nuclear factor kappa B, interferon regulatory factor 3, and activating protein-1. The important immunosuppressive and antiinflammatory roles of glucocorticoid are mediated through GR-dependent transrepression, which targets the genes associated with proinflammatory cytokines.[4] Pharmacologically, the gene-repressing properties of GR are targeted for treatment of inflammatory and autoimmune disorders using exogenous glucocorticoids.[14] However, the desired therapeutic use of these steroid agents is usually followed by adverse side effects, mainly associated with gene-activating properties of GR.[59]

KEY POINTS

- Depression is associated with dysregulation of HPA axis activity and alteration of glucocorticoid receptor function.
- Effective antidepressant medications restore the neuroendocrine-immune imbalance associated with depression.
- Antidepressants normalize the HPA axis hyperactivity through direct modulation of the glucocorticoid receptors.
- Antidepressants exert their effects on expression, phosphorylation, translocation, and function of the glucocorticoid receptors.
- Glucocorticoid receptor is a novel therapeutic target for development of new antidepressant medications and treatment of depressive symptoms.

There are two homologous isoforms associated with human GR that are encoded by the GR gene: GRα and GRβ. GRα is the cytoplasmic ligand-binding isoform, which is the classic GR and regulates gene transcription. In contrast, GRβ, which is the nuclear localized isoform, does not seem to bind to any ligand, but is known to affect GRα as a dominant-negative inhibitor attenuating its transcriptional activity through formation of inactive GRα/GRβ heterodimers and competition for DNA response elements binding and transcriptional coregulators. Elevation in inflammatory responses, and specifically in pro-inflammatory cytokines, increase expression of the GRβ leading to overexpression of this isoform that result in decreased sensitivity of GRα.[10,38]

Alteration in HPA Axis and Glucocorticoid Receptors in Depression

HPA axis dysfunctionality in psychosomatic and psychiatric disorders includes both hyperactivity and hypoactivity of the system, leading to dysregulation of stress-related responses and associated changes in GR sensitivity.[26,57] Most clinical studies point to hyperactivity of the HPA axis, in association with reduced GR sensitivity, as the most prevalent abnormality (see below for further discussion of HPA axis hypoactivity).

Elevated levels of glucocorticoids are observed in a significant proportion of patients with major depression and are believed to be involved in both etiology and pathogenesis of the disease. Indeed, almost half of the depressed patients show increased levels of cortisol in cerebrospinal fluid, plasma, and urine, and enlargement of pituitary and adrenal glands.[50] The hyperactivity of the HPA axis in depressed patients appears to reflect an impaired ability of glucocorticoid

hormones to exert their physiological effects: that is, a reduced GR sensitivity, also called "glucocorticoid resistance," which not only leads to an impairment of the negative feedback regulation of the axis but also diminishes glucocorticoid antiinflammatory actions at the periphery. Glucocorticoid resistance is also reflected by reduced GR function in peripheral blood cells[35] and in the skin,[19] and by the lack of Cushing-like stigmata in hypercortisolemic depressed patients,[34] suggesting that GR dysfunction is not confined to the HPA axis tissues. Indeed, a variety of studies report that depressed patients show HPA axis hyperactivity in the presence of GR resistance and reduced expression and receptor number, as discussed extensively before.[48,50]

Although the mechanisms leading to glucocorticoid resistance are not yet clear, some studies suggest that prolonged inflammation has a direct effect in reducing GR sensitivity through the interaction of cytokine signaling pathway with GR signaling pathway.[35,68] Proinflammatory cytokines may generate glucocorticoid resistance by directly affecting functional capacity of the GR at multiple levels. Apart from the above-mentioned overexpression of GRβ modulating the sensitivity of GRα, cytokines trigger glucocorticoid resistance by reducing GR ligand and DNA-binding capacities, inhibiting GR translocation to the nucleus and influencing GR protein–protein interactions; for example, activation of the mitogen-activated protein kinase (MAPK) signaling pathway in cytoplasm leads to phosphorylation of the GR, thus diminishing its transcriptional activity.[10,42,53]

Despite the extensive evidence supporting HPA axis hyperactivity and overproduction of glucocorticoids in depression, insufficient glucocorticoid signaling through hypoactivity of the HPA axis is also believed to contribute to the pathogenesis of depression, in at least some subgroups of patients. The evidence suggests that inadequate signaling capacity of glucocorticoids could be due not only to impaired GR-mediated signal transduction, but also to reduced glucocorticoid bioavailability itself.[35,57] For example, the study by Nikkheslat et al.[35] has recently shown reduced cortisol awakening response together with impaired GR sensitivity in elderly depressed patients with coronary heart disease.[35] In atypical depression, HPA hypoactivity probably contributes to the fatigue and reversed neurovegetative symptoms presented in this specific subtype.[36]

ANTIDEPRESSANTS EFFECTS ON GLUCOCORTICOID RECEPTORS

Effective antidepressant treatment has been shown to resolve the HPA axis dysregulation. Evidence suggests that the mechanism by which antidepressants normalize the HPA axis hyperactivity is at least partially independent of the effect on neurotransmitter systems, the classic mechanism of antidepressant action. More specifically, as shown by both in vitro and in vivo studies, some of these medications exert their effect directly on GR through increasing expression, promoting translocation, and enhancing function in the brain and other target tissues. Thus, direct effects of antidepressants on GR may not only normalize HPA axis abnormalities but also mediate a host of other antidepressant-related effects, such as reducing inflammation and increasing neurogenesis.[2,3,5,12,11] The rest of this chapter will focus specifically on studies examining the effects of antidepressants on the GR in vitro, as these allow dissection of the molecular mechanisms involved without the confounding effects of stress or depression (as is seen in animal or clinical models), and of the indirect actions of antidepressants on the HPA axis via other biological systems. For example, antidepressants are well known to acutely activate the HPA axis via direct serotonergic and noradrenergic stimulation, an effect that is independent (and virtually opposite) to that on the GR.[51] Moreover, some antidepressants may indirectly affect the GR through reducing levels of inflammatory biomarkers[25] and of inducible nitric oxide synthase, leading to reduced inflammatory reactions and oxidative stress.[28,37] Again, in vitro studies may partially allow us to separate direct effects of antidepressants on the GR from the indirect effects on inflammation.

The Effects of Antidepressants on GR Expression

In the late 1980s it was shown for the first time that monoamine reuptake inhibitors, irrespective of their ability to block serotonin or norepinephrine reuptake, increase GR messenger RNA (mRNA) concentrations in primary neuronal cell cultures of the hypothalamus, the amygdala, and the cerebral cortex[55]—and most studies so far have indeed evaluated GR mRNA rather than protein. An increase in GR-binding capacity and GR mRNA expression has also been described in rat hypothalamic and hippocampal neurons following treatment with tricyclic antidepressants (TCAs).[39,54] The findings were further supported by studies showing that the TCAs and selective serotonin reuptake inhibitors antidepressants (SSRIs) increase GR expression in neuronal cell cultures[24,27] and in fibroblasts.[56]

The data on the effects of antidepressants on the GR in immune cells are less consistent. An antidepressant-induced increase in GR mRNA level was reported in human immune cells by various studies.[20,21,63] Heiske and colleagues reported that antidepressants exert differential effects on GR mRNA levels in primary human leukocytes and monocytic cells: for example,

imipramine and maprotiline do not affect the GR, while the levels of GR mRNA decrease when cells are treated with desipramine and mirtazapine. However, these effects were found to be time dependent, with an initial GR upregulation after a short-term treatment (2.5 h) with mirtazapine, and a downregulation after longer periods of incubation.[22] In contrast, the data from another study has detected increased GRα mRNA expression following treatment with desipramine and paroxetine after 48 h as well as 7 days.[20] Similarly, others have found a significant increase mRNA levels of GRα in human monocytic blood cells after 24 h treatment with different classes of antidepressants, including desipramine, imipramine, maprotiline, and mirtazapine, all in the presence of corticosteroid treatment.[21] This last finding is particularly interesting in view of the functional data, described below, showing that corticosteroids are required for the GR-mediated effects of antidepressants on neurogenesis.[5] In addition, treatment of human monocytic cells with hypericum perforatum, a putative antidepressant strategy, was found to increase GRα mRNA levels and decrease GRβ mRNA levels after 16 h.[18]

The effects of antidepressants on GR protein have been also contradictory. While some studies observed an increase in GR protein levels[9,27] others showed a decrease.[52,46] The inconsistency in the findings on the in vitro effect of antidepressants on the GR mRNA levels appears to be due to the differences in the cell type, treatment time points, as well as concentration of antidepressants used.

The Effects of Antidepressants on GR Function

Considering the complex mechanisms of actions of GR, more integrative approaches in relation to functional properties of the receptor are required to better understand the effects of antidepressants on the receptor, rather than simple quantification of the GR protein and expression.

Differential phosphorylation of the GR by different kinases may modulate GR nuclear translocation or transactivation leading to alteration in glucocorticoid responsiveness.[1,4] Evidence suggests that antidepressants also enhance GR sensitivity by modifying the phosphorylation, facilitating nuclear translocation as well as gene transcription ability of the receptor.[5,20,23,52,46,47] On the other hand, some studies have reported inhibition of GR-induced gene transcription by antidepressants.[6,8,40] Indeed, antidepressants have been found to both increase and decrease GR-mediated gene transcription depending on the cell type and the incubation time.[12,49,52,46,47]

Pepin and colleagues in their pivotal study observed that in rat fibroblasts, TCA desipramine increases GR function after 24h (measured as GR-mediated gene transcription) and upregulates GR protein after 72h. They proposed that antidepressants directly augment GR function and the effect occurred before and in the absence of GR upregulation.[56] In addition, 24h treatment with desipramine in mouse fibroblasts was shown to enhance GR function by facilitating GR translocation in the absence of glucocorticoids.[52] The findings were further confirmed by other studies. Indeed, in primary human leukocytes and monocytic cells, TCAs imipramine was found to induce a translocation of GR to the nucleus, with and without presence of dexamethasone after 24 and 48h of treatment, respectively.[22] In vitro treatment of human lymphocytes with different types of antidepressants (desipramine, clomipramine, fluoxetine, milnacipran, and clorgyline) also induces translocation of GR and inhibits dexamethasone-induced GR-mediated transcription.[40] Another study reported an enhanced dexamethasone-induced GR nuclear translocation by treatment with clomipramine and desipramine in SY5Y cells, but the effect was not observed in the absence of dexamethasone.[20]

In human whole blood cells, clomipramine has been observed to reduce GR function in vitro in healthy individuals, while this effect was not present in treatment resistance depressed patients.[12] These findings were interpreted as showing that healthy controls exhibit clomipramine-induced GR translocation and thus reduced GR expression, resulting in reduced GR function, whereas treatment resistance patients show GR resistance in vitro to the effect of the antidepressant.[12,44] Further investigation on the effect of antidepressants on human blood cells has revealed the modulation of GR function by various classes of antidepressants (clomipramine, amitriptyline, sertraline, paroxetine, and venlafaxine) but not by the antipsychotics (haloperidol and risperidone).[11]

Of note, the effects of antidepressants on GR function may be relevant to brain function. For example, antidepressant treatment modulates GR in the brain of healthy human subject as investigated by examining the effect of citalopram on reducing the brain electric and cognitive responses to cortisol administration.[45] Moreover, the ability of antidepressants to enhance hippocampal neurogenesis, proposed by studies in animals as well as in humans,[7,17,64] seems to be mediated by GR-dependent mechanisms associated with modulation of GR phosphorylation and transcription[5,2].

The effects of antidepressants on GR function have been also examined in other psychiatric disorders. For example, in patients with posttraumatic stress disorder exhibiting hypersensitivity of GR (opposite to depressed patients who show reduced GR sensitivity), sertraline

was found to reduce GR function as measured in vitro on mononuclear leukocytes.[66,67]

Mechanisms of Antidepressants Actions on GR

To date, several molecular pathways have been involved in the effects of antidepressants on regulation of the GR function. Independent of the chemical class, antidepressants have been shown to directly inhibit membrane steroid transporters, most notably the multidrug resistance P-glycoprotein (MDR PGP), which expel glucocorticoids from the intracellular space and regulate the hormone bioavailability. Therefore, antidepressant-induced alterations in MDR PGP activity may lead to varying concentrations of available glucocorticoids and thus contribute to glucocorticoid resistance in different cells and tissues.[13] By blocking these transporters, antidepressants, such as clomipramine and fluoxetine, have been shown to increase the intracellular levels of glucocorticoids as investigated in rat cortical neurons and mice fibroblasts,[46,47] consequently leading to increased glucocorticoid responsiveness. Indeed, the inhibiting effects of antidepressants on membrane steroid transporters have been reported for various classes of antidepressants.[62,65] This molecular pathway may explain the time-dependent differences in GR expression upon antidepressant treatment described before, as inhibiting MDR PGP increases the intracellular glucocorticoid level that might subsequently lead to increased GR nuclear translocation and an initial downregulation of GR expression during the initial hours of treatment. This is followed by GR upregulation after days, possibly caused by MDR PGP downregulation and/or as a compensatory mechanism following the initial GR downregulation.[4,13] It is of note that this concept has, however, been challenged by animal studies reporting that PGP does not tightly regulate corticosteroid entry through the blood–brain barrier and suggesting the involvement of other transporters.[31,30]

It has been shown that antidepressants can also influence GR through mechanisms involving the phosphorylation of the receptor by cAMP-dependent protein kinase A (PKA), one of the signaling transduction pathways associated with regulation of GR function and inflammation. PKA has been shown to increase GR function by altering its DNA-binding domain and enhancing DNA-binding activity, thus upregulating GR transactivation.[58] Treatment of cells with the phosphodiesterases type 4 inhibitor rolipram that prevents cAMP breakdown, increases PKA activity and induces GR nuclear translocation and function leading to enhanced GR-mediated gene transcription in vitro. In addition, rolipram was found to potentiate the effect of antidepressant desipramine on GR function enhancement.[33] The findings were extended by observations on the effect of antidepressants on enhancing hippocampal neurogenesis via GR-dependent mechanisms and the involvement of PKA signaling leading to modulation of GR phosphorylation and transcription.[5] Involvement of PKA in the regulation of GR transcriptional activity and function were also described by other studies.[6,41,60] In vitro evaluation of GR function in human blood cells has revealed involvement of cAMP signaling transduction pathway as a mechanism of antidepressant inhibition of GR function, an effect that was augmented by rolipram and abolished by a GR antagonist.[11]

Antidepressants can also affect phospholipase C (PLC) and other intracellular protein kinases, including calcium/calmodulin-dependent kinase and protein kinase C (PKC), leading to inhibition of corticosterone-induced gene transcription in a time-dependent and dose-dependent manner, as investigated in fibroblast cells. One study reported a decrease in binding of corticosterone–GR complex to DNA by imipramine and that the inhibitory effect of the antidepressant on GR-mediated gene transcription depends partly on the inhibition of the PLC/PKC pathway.[8]

Although the role of MDR PGP and PKA/PLC as potential mechanisms investigated on the actions of antidepressants on GR have been discussed, how exactly antidepressants of different chemical classes act on these targets is poorly understood. It has been suggested that antidepressant action could be mediated, at least in part, by changes in membrane-associated lipid rafts and induction of G-protein signaling, which could ultimately activate PKA.[16]

CONCLUSIONS

Extensive lines of evidence have been discussed in relation to the therapeutic mechanism of action of antidepressants through direct effects on the GR. Antidepressants, independent of their structure and chemical class, appear to commonly target GR at various cellular levels. Considering the critical role in the regulation of HPA axis and inflammation, the involvement in the pathophysiology of depression, and the direct contribution to the mechanisms of antidepressants actions, targeting the GR more efficiently remains a novel therapeutic strategy in improving treatment of depression. Investigation of the molecular pathways underlying the effects of antidepressants on regulation of the GR function is crucial for enhancing our understanding of the therapeutic efficacy of these medications and provides mechanistic insight for developing new antidepressant agents in order to achieve successful clinical treatment of depression.

References

1. Adzic M, Djordjevic J, Djordjevic A, et al. Acute or chronic stress induce cell compartment-specific phosphorylation of glucocorticoid receptor and alter its transcriptional activity in Wistar rat brain. *J Endocrinol.* 2009;202:87–97.
2. Anacker C, Cattaneo A, Luoni A, et al. Glucocorticoid-related molecular signaling pathways regulating hippocampal neurogenesis. *Neuropsychopharmacology.* 2013;38:872–883.
3. Anacker C, Cattaneo A, Musaelyan K, et al. Role for the kinase SGK1 in stress, depression, and glucocorticoid effects on hippocampal neurogenesis. *Proc Natl Acad Sci USA.* 2013;110:8708–8713.
4. Anacker C, Zunszain PA, Carvalho LA, Pariante CM. The glucocorticoid receptor: pivot of depression and of antidepressant treatment? *Psychoneuroendocrinology.* 2011;36:415–425.
5. Anacker C, Zunszain PA, Cattaneo A, et al. Antidepressants increase human hippocampal neurogenesis by activating the glucocorticoid receptor. *Mol Psychiatry.* 2011;16:738–750.
6. Augustyn M, Otczyk M, Budziszewska B, et al. Effects of some new antidepressant drugs on the glucocorticoid receptor-mediated gene transcription in fibroblast cells. *Pharmacol Rep.* 2005;57:766–773.
7. Boldrini M, Underwood MD, Hen R, et al. Antidepressants increase neural progenitor cells in the human hippocampus. *Neuropsychopharmacology.* 2009;34:2376–2389.
8. Budziszewska B, Jaworska-Feil L, Kajta M, Lason W. Antidepressant drugs inhibit glucocorticoid receptor-mediated gene transcription – a possible mechanism. *Br J Pharmacol.* 2000;130:1385–1393.
9. Cai W, Khaoustov VI, Xie Q, Pan T, Le W, Yoffe B. Interferon-alpha-induced modulation of glucocorticoid and serotonin receptors as a mechanism of depression. *J Hepatol.* 2005;42:880–887.
10. Carvalho LA, Bergink V, Sumaski L, et al. Inflammatory activation is associated with a reduced glucocorticoid receptor alpha/beta expression ratio in monocytes of inpatients with melancholic major depressive disorder. *Transl Psychiatry.* 2014;4:e344.
11. Carvalho LA, Garner BA, Dew T, Fazakerley H, Pariante CM. Antidepressants, but not antipsychotics, modulate GR function in human whole blood: an insight into molecular mechanisms. *Eur Neuropsychopharmacol.* 2010;20:379–387.
12. Carvalho LA, Juruena MF, Papadopoulos AS, et al. Clomipramine in vitro reduces glucocorticoid receptor function in healthy subjects but not in patients with major depression. *Neuropsychopharmacology.* 2008;33:3182–3189.
13. Carvalho LA, Pariante CM. In vitro modulation of the glucocorticoid receptor by antidepressants. *Stress.* 2008;11:411–424.
14. De Bosscher K, Haegeman G. Minireview: latest perspectives on antiinflammatory actions of glucocorticoids. *Mol Endocrinol.* 2009;23:281–291.
15. de Kloet ER. From receptor balance to rational glucocorticoid therapy. *Endocrinology.* 2014;155:2754–2769.
16. Donati RJ, Schappi J, Czysz AH, Jackson A, Rasenick MM. Differential effects of antidepressants escitalopram versus lithium on Gs alpha membrane relocalization. *BMC Neurosci.* 2015;16:40.
17. Duman RS. Role of neurotrophic factors in the etiology and treatment of mood disorders. *Neuromol Med.* 2004;5:11–25.
18. Enning F, Murck H, Krieg JC, Vedder H. *Hypericum perforatum* differentially affects corticosteroid receptor-mRNA expression in human monocytic U-937 cells. *J Psychiatr Res.* 2011;45:1170–1177.
19. Fitzgerald P, O'Brien SM, Scully P, Rijkers K, Scott LV, Dinan TG. Cutaneous glucocorticoid receptor sensitivity and pro-inflammatory cytokine levels in antidepressant-resistant depression. *Psychol Med.* 2006;36:37–43.
20. Funato H, Kobayashi A, Watanabe Y. Differential effects of antidepressants on dexamethasone-induced nuclear translocation and expression of glucocorticoid receptor. *Brain Res.* 2006;1117:125–134.

21. Gebhardt S, Heiser P, Fischer S, Schneyer T, Krieg JC, Vedder H. Relationships among endocrine and signaling-related responses to antidepressants in human monocytic U-937 blood cells: analysis of factors and response patterns. *Prog Neuropsychopharmacol Biol Psychiatry*. 2008;32:1682–1687.

22. Heiske A, Jesberg J, Krieg JC, Vedder H. Differential effects of antidepressants on glucocorticoid receptors in human primary blood cells and human monocytic U-937 cells. *Neuropsychopharmacology*. 2003;28:807–817.

23. Herr AS, Tsolakidou AF, Yassouridis A, Holsboer F, Rein T. Antidepressants differentially influence the transcriptional activity of the glucocorticoid receptor in vitro. *Neuroendocrinology*. 2003;78:12–22.

24. Hery M, Semont A, Fache MP, Faudon M, Hery F. The effects of serotonin on glucocorticoid receptor binding in rat raphe nuclei and hippocampal cells in culture. *J Neurochem*. 2000;74:406–413.

25. Horowitz MA, Wertz J, Zhu D, et al. Antidepressant compounds can be both pro- and anti-inflammatory in human hippocampal cells. *Int J Neuropsychopharmacol*. 2015;18.

26. Kudielka BM, Bellingrath S, Hellhammer DH. Cortisol in burnout and vital exhaustion: an overview. *G Ital Med Lav Ergon*. 2006;28:34–42.

27. Lai M, McCormick JA, Chapman KE, Kelly PA, Seckl JR, Yau JL. Differential regulation of corticosteroid receptors by monoamine neurotransmitters and antidepressant drugs in primary hippocampal culture. *Neuroscience*. 2003;118:975–984.

28. Lu DY, Tsao YY, Leung YM, Su KP. Docosahexaenoic acid suppresses neuroinflammatory responses and induces heme oxygenase-1 expression in BV-2 microglia: implications of antidepressant effects for omega-3 fatty acids. *Neuropsychopharmacology*. 2010;35:2238–2248.

29. Deleted in review.

30. Mason BL, Pariante CM, Jamel S, Thomas SA. Central nervous system (CNS) delivery of glucocorticoids is fine-tuned by saturable transporters at the blood-CNS barriers and nonbarrier regions. *Endocrinology*. 2010;151:5294–5305.

31. Mason BL, Pariante CM, Thomas SA. A revised role for P-glycoprotein in the brain distribution of dexamethasone, cortisol, and corticosterone in wild-type and ABCB1A/B-deficient mice. *Endocrinology*. 2008;149:5244–5253.

32. Miller AH, Raison CL. The role of inflammation in depression: from evolutionary imperative to modern treatment target. *Nat Rev Immunol*. 2015;16:22–34.

33. Miller AH, Vogt GJ, Pearce BD. The phosphodiesterase type 4 inhibitor, rolipram, enhances glucocorticoid receptor function. *Neuropsychopharmacology*. 2002;27:939–948.

34. Murphy BE. Steroids and depression. *J Steroid Biochem Mol Biol*. 1991;38:537–559.

35. Nikkheslat N, Zunszain PA, Horowitz MA, et al. Insufficient glucocorticoid signaling and elevated inflammation in coronary heart disease patients with comorbid depression. *Brain Behav Immun*. 2015;48:8–18.

36. O'Keane V, Frodl T, Dinan TG. A review of atypical depression in relation to the course of depression and changes in HPA axis organization. *Psychoneuroendocrinology*. 2012;37:1589–1599.

37. O'Sullivan JB, Ryan KM, Curtin NM, Harkin A, Connor TJ. Noradrenaline reuptake inhibitors limit neuroinflammation in rat cortex following a systemic inflammatory challenge: implications for depression and neurodegeneration. *Int J Neuropsychopharmacol*. 2009;12:687–699.

38. Oakley RH, Cidlowski JA. Cellular processing of the glucocorticoid receptor gene and protein: new mechanisms for generating tissue-specific actions of glucocorticoids. *J Biol Chem*. 2011;286:3177–3184.

39. Okugawa G, Omori K, Suzukawa J, Fujiseki Y, Kinoshita T, Inagaki C. Long-term treatment with antidepressants increases glucocorticoid receptor binding and gene expression in cultured rat hippocampal neurones. *J Neuroendocrinol*. 1999;11:887–895.

40. Okuyama-Tamura M, Mikuni M, Kojima I. Modulation of the human glucocorticoid receptor function by antidepressive compounds. *Neurosci Lett*. 2003;342:206–210.

41. Otczyk M, Mulik K, Budziszewska B, et al. Effect of some antidepressants on the low corticosterone concentration-induced gene transcription in LMCAT fibroblast cells. *J Physiol Pharmacol*. 2008;59:153–162.

42. Pace TW, Hu F, Miller AH. Cytokine-effects on glucocorticoid receptor function: relevance to glucocorticoid resistance and the pathophysiology and treatment of major depression. *Brain Behav Immun*. 2007;21:9–19.

43. Pace TW, Miller AH. Cytokines and glucocorticoid receptor signaling. Relevance to major depression. *Ann NY Acad Sci*. 2009;1179:86–105.

44. Pariante CM. Risk factors for development of depression and psychosis. Glucocorticoid receptors and pituitary implications for treatment with antidepressant and glucocorticoids. *Ann NY Acad Sci*. 2009;1179:144–152.

45. Pariante CM, Alhaj HA, Arulnathan VE, et al. Central glucocorticoid receptor-mediated effects of the antidepressant, citalopram, in humans: a study using EEG and cognitive testing. *Psychoneuroendocrinology*. 2012;37:618–628.

46. Pariante CM, Hye A, Williamson R, Makoff A, Lovestone S, Kerwin RW. The antidepressant clomipramine regulates cortisol intracellular concentrations and glucocorticoid receptor expression in fibroblasts and rat primary neurones. *Neuropsychopharmacology*. 2003;28:1553–1561.

47. Pariante CM, Kim RB, Makoff A, Kerwin RW. Antidepressant fluoxetine enhances glucocorticoid receptor function in vitro by modulating membrane steroid transporters. *Br J Pharmacol*. 2003;139:1111–1118.

48. Pariante CM, Lightman SL. The HPA axis in major depression: classical theories and new developments. *Trends Neurosci*. 2008;31:464–468.

49. Pariante CM, Miller AH. Glucocorticoid receptors in major depression: relevance to pathophysiology and treatment. *Biol Psychiatry*. 2001;49:391–404.

50. Pariante CM, Nemeroff CB. Unipolar depression. *Handb Clin Neurol*. 2012;106:239–249.

51. Pariante CM, Papadopoulos AS, Poon L, et al. Four days of citalopram increase suppression of cortisol secretion by prednisolone in healthy volunteers. *Psychopharmacology (Berl)*. 2004;177:200–206.

52. Pariante CM, Pearce BD, Pisell TL, Owens MJ, Miller AH. Steroid-independent translocation of the glucocorticoid receptor by the antidepressant desipramine. *Mol Pharmacol*. 1997;52:571–581.

53. Pariante CM, Pearce BD, Pisell TL, et al. The proinflammatory cytokine, interleukin-1alpha, reduces glucocorticoid receptor translocation and function. *Endocrinology*. 1999;140:4359–4366.

54. Peiffer A, Veilleux S, Barden N. Antidepressant and other centrally acting drugs regulate glucocorticoid receptor messenger RNA levels in rat brain. *Psychoneuroendocrinology*. 1991;16:505–515.

55. Pepin MC, Beaulieu S, Barden N. Antidepressants regulate glucocorticoid receptor messenger RNA concentrations in primary neuronal cultures. *Brain Res Mol Brain Res*. 1989;6:77–83.

56. Pepin MC, Govindan MV, Barden N. Increased glucocorticoid receptor gene promoter activity after antidepressant treatment. *Mol Pharmacol*. 1992;41:1016–1022.

57. Raison CL, Miller AH. When not enough is too much: the role of insufficient glucocorticoid signaling in the pathophysiology of stress-related disorders. *Am J Psychiatry*. 2003;160:1554–1565.

58. Rangarajan PN, Umesono K, Evans RM. Modulation of glucocorticoid receptor function by protein kinase A. *Mol Endocrinol*. 1992;6:1451–1457.

59. Schacke H, Docke WD, Asadullah K. Mechanisms involved in the side effects of glucocorticoids. *Pharmacol Ther*. 2002;96:23–43.

60. Skuza G, Szymanska M, Budziszewska B, Abate C, Berardi F. Effects of PB190 and PB212, new sigma receptor ligands, on glucocorticoid receptor-mediated gene transcription in LMCAT cells. *Pharmacol Rep*. 2011;63:1564–1568.

61. Stetler C, Miller GE. Depression and hypothalamic-pituitary-adrenal activation: a quantitative summary of four decades of research. *Psychosom Med*. 2011;73(2):114–126.

62. Varga A, Nugel H, Baehr R, et al. Reversal of multidrug resistance by amitriptyline in vitro. *Anticancer Res*. 1996;16:209–211.

63. Vedder H, Bening-Abu-Shach U, Lanquillon S, Krieg JC. Regulation of glucocorticoid receptor-mRNA in human blood cells by amitriptyline and dexamethasone. *J Psychiatr Res*. 1999;33:303–308.

64. Wang JW, David DJ, Monckton JE, Battaglia F, Hen R. Chronic fluoxetine stimulates maturation and synaptic plasticity of adult-born hippocampal granule cells. *J Neurosci*. 2008;28:1374–1384.

65. Weiss J, Dormann SM, Martin-Facklam M, Kerpen CJ, Ketabi-Kiyanvash N, Haefeli WE. Inhibition of P-glycoprotein by newer antidepressants. *J Pharmacol Exp Ther*. 2003;305:197–204.

66. Yehuda R, Golier JA, Yang RK, Tischler L. Enhanced sensitivity to glucocorticoids in peripheral mononuclear leukocytes in posttraumatic stress disorder. *Biol Psychiatry*. 2004;55:1110–1116.

67. Yehuda R, Yang RK, Golier JA, Grossman RA, Bierer LM, Tischler L. Effect of sertraline on glucocorticoid sensitivity of mononuclear leukocytes in post-traumatic stress disorder. *Neuropsychopharmacology*. 2006;31:189–196.

68. Zunszain PA, Anacker C, Cattaneo A, Carvalho LA, Pariante CM. Glucocorticoids, cytokines and brain abnormalities in depression. *Prog Neuropsychopharmacol Biol Psychiatry*. 2011;35:722–729.

69. Zunszain PA, Hepgul N, Pariante CM. Inflammation and depression. In: *Behavioral Neurobiology of Depression and Its Treatment*. Berlin, Heidelberg: Springer; 2012:135–151.

28

Lipids and Lipoproteins

C.M. Stoney

National Institutes of Health, Bethesda, MD, United States

Abstract

Cholesterol, other lipids, and lipoproteins play essential physiological functions, and data show reasonably consistent elevations in these parameters during both chronic and acute psychological stress. Among healthy individuals at risk for future cardiovascular disease, elevations are often exaggerated. In addition, those with excessive lipid reactivity to stress are at subsequently elevated risk. Several potential mechanisms for these elevations have been explored, and it is likely that they are multifactorial and operate differently for chronic versus acute stress. Understanding these mechanisms is essential for explicating the potential clinical significance of stress-associated lipid elevations.

INTRODUCTION

Physiological responses to psychological stressors have been noted throughout the history of biomedical research and are generally biologically adaptive. However, in some individuals during some situations, the physiological responses to stress are exaggerated, leading to speculation regarding the potential clinical significance of such reactivity. Cholesterol and lipoproteins play important roles in cell formation, intracellular transport, cell signaling, nerve conduction, and other functions. High circulating concentrations signal elevated risk for cardiovascular disease. Understanding the mechanisms for stress-related elevations will shed light on the potential clinical significance and inform future clinical interventions.

STRUCTURE, FUNCTION, AND STABILITY OF LIPIDS AND LIPOPROTEINS

Circulating lipoproteins are large structures that contain varying concentrations of cholesterol, proteins, triglycerides, and phospholipids and serve to transport cholesterol throughout the body. They are differentiated from each other by the concentrations and type of each particle. For example, low-density lipoproteins (LDL) consist of mainly cholesterol, while high-density lipoproteins (HDL) consist primarily of proteins. In addition, the type of protein contained in LDL (apolipoprotein B-100) differs from the protein in HDL (apolipoprotein A-I and A-II). Cholesterol is important for the formation and permeability of cell membranes. Phospholipids are also critical for the formation of cellular membranes and are transported in lipoproteins. Phospholipids are formed in every cell of the body. Triglycerides are an important energy source and are synthesized in the liver and in adipose tissue. In addition to serving important physiological functions, lipoproteins serve as a marker for increased risk of disease. Elevated circulating concentrations of low-density lipoprotein-cholesterol (LDL-c) and low circulating concentrations of high-density lipoprotein-cholesterol (HDL-c) are associated with an increased risk for cardiovascular disease (CVD) and stroke. Clinical trials have demonstrated that reducing elevated total cholesterol and LDL-c levels, by pharmacologic means and/or lifestyle changes, reduces subsequent CVD risk. Thus, understanding factors that contribute to lipid and lipoprotein fluctuations is critical.

KEY POINTS

- Lipids and lipoproteins play essential physiological roles in optimal health.
- Elevations in the atherogenic lipids and lipoproteins are associated with increased risk of cardiovascular disease and stroke.
- Elevations in the atherogenic lipids occur under conditions of acute, episodic, and chronic stress, and individual differences in these responses are apparent.
- The exact mechanisms driving stress-related lipid elevations are not known, but many environmental, psychosocial, and physiological factors may be operating.
- The clinical significance of stress-related elevations in the atherogenic lipids is not clear, but cross-sectional data suggest that further study is warranted.

Circulating levels of blood cholesterol concentrations are widely and erroneously believed to be stable, with only modest increases with aging and transient, predictable increases in triglycerides and related lipids postprandially. These latter increases have also been thought to be clinically insignificant and relatively similar across individuals. However, it is clear now that there are significant variations and fluctuations in plasma cholesterol, triglycerides, and LDL-c concentrations both within individuals over time and between individuals. The extent of variation should not be underestimated; within individuals, the variability can be between 10% and 20% over the course of a single day or week and can remain high even under fairly standard conditions. The variations between individuals are as large or larger.

An understanding of the factors responsible for the fluctuations in lipids and lipoproteins are important both clinically and scientifically. Treatment decisions based on lipid concentrations must be informed by an understanding of such variations. Factors influencing lipid variations will also inform scientific exploration of cholesterol regulation. The discussion below focuses on environmental, psychological, and behavioral influences on lipids and lipoproteins.

ENVIRONMENTAL AND BEHAVIORAL INFLUENCES ON LIPID CONCENTRATIONS AND LIPID VARIABILITY

The robust individual differences in circulating lipid levels are in part a function of constitutional factors, such as age, gender, ethnicity, and genetic factors. Individual differences, as well as fluctuations within individuals, also occur as a function of a wide range of behavioral and environmental factors. For example, individuals robustly differ in their lipid responses to acute dietary challenges, as well as their typical dietary intake behaviors.[19] Although some evidence indicates there are genetic components to explain the variability in response to challenge, there are many conflicting studies and a need to examine multiple genes in well-controlled environments. When lipid responses to dietary challenge are exaggerated, an increased risk of CVD is often present. Independent of the relationship between diet and lipids is the effect of body weight, specifically obesity, on lipid concentrations. Obese and overweight adults and children have a higher risk lipid and lipoprotein profile than their normal weight counterparts,[43] and the relationship between degree of obesity and LDL-c may be reasonably linear.[32]

Several other behavioral and environmental factors significantly influence blood lipid levels. Aerobic exercise has a small but important impact on lipid levels, operating as a cardioprotective behavior primarily by increasing HDL-c concentrations. This effect occurs gradually with regular exercise, although short and irregular bouts of aerobic exercise can also induce small, transient changes in HDL-c levels.

Changes in posture significantly and acutely impacts lipid concentrations, primarily as a function of posture-induced acute decreases in intravascular plasma volume.

Postural changes result in hemoconcentration of blood cells and a resulting hemoconcentration of serum lipids. The potential impact of hemoconcentration of blood cells on stress-related lipid concentrations is addressed in greater detail later in this chapter.

When considering environmental factors, the effects of temperature and seasonality can have small but important effects, particularly in the research context when it is desirable to examine lipid levels longitudinally. Seasonal variations in lipids have been periodically noted[22] and are postulated to be related to changes in behaviors, such as diet and exercise, hemoconcentration during the coldest months, or measurement error. A recent large population-based study of 227,359 adult men and women reported significant increases in LDL-c during the winter months, and significant HDL-c and triglyceride elevations during the summer months,[20] confirming that seasonal variations should be carefully considered when examining lipid levels longitudinally. Other studies have failed to explain the seasonal variations by changes in diet or ambient temperature. Lipid levels change significantly during passive (e.g., not exercise-induced) heat stress,[42] suggesting that heat exposure may induce changes in lipid metabolism.

Because many of the above-noted variations in lipid levels are in the range associated with elevated cardiovascular risk, investigation of the role of these and other behavioral and psychological factors on lipid concentrations has potentially important public health impact. One of these factors is psychological stress, which is experienced episodically by most individuals over the lifespan and by some individuals is experienced chronically and persistently. The risk of cardiovascular events increases during stress,[11,12,16] suggesting the potential clinical importance of stress-related elevations in lipids during psychological stress.

CHRONIC PSYCHOLOGICAL STRESS

Naturalistic studies have provided unique insight regarding the role of chronic psychological stress on lipid levels. Real-world stressors such as major disasters in the form of earthquakes and war, and major stressors in the form of job loss, loss of income, and bereavement, have all been studied with regard to their impact on blood lipid levels, and been found to be related to elevations in the atherogenic lipids (e.g., total cholesterol, triglycerides, and/or LDL-c). One early study examined the effects of the 1980 major earthquake in southern Italy, which interrupted an on-going population-based longitudinal study of cardiovascular risk factors.[36] One of the follow-up examinations was interrupted by the earthquake, giving investigators the opportunity to compare lipid levels among those tested before (lower chronic stress) and those tested after (higher chronic stress) the earthquake. Participants tested after the earthquake had significantly higher total cholesterol and triglyceride levels than those tested prior to the disaster, and these differences were independent of baseline values. The differences did not persist at the subsequent screening 7 years later, suggesting the stress-related elevations may not have had long-term health consequences. The short-term consequences on health were not, however, studied. Findings such as these have been replicated in a variety of other studies examining effects of other naturalistic stressors, such as the Katrina disaster in the United States,[8,34] the effects of disaster relief staff on lipids,[3] and the effects on children exposed to the World Trade Center attacks,[35] all demonstrating increases of one or more of the atherogenic lipids.

A recent study examined a large sample ($n = 6714$) of men and women who reported either high levels of childhood psychological stress which resolved in adulthood; elevated levels of adult distress without high childhood distress; high levels of sustained lifetime psychological stress; or no significant stress. One significant strength of the study is the prospective design; participants reported on stress levels at six different time periods, when they were between the ages of 7 and 42 years. Outcomes at adulthood (age 42 years) were a comprehensive index of cardiometabolic risk, including nine biomarkers related to cardiovascular health. Importantly, total cholesterol, HDL-c, and triglycerides were included in the index of cardiometabolic health. Those reporting childhood only, adulthood only, or sustained lifetime stress had significantly higher cardiometabolic risk scores, relative to those reporting low levels of stress throughout the lifespan.[39] Although the outcome is a composite measure, these findings provide additional evidence that stress, particularly lifetime chronic stress, operates through one or more mechanisms to increase metabolic risk. These findings are similar to those found in the Whitehall II study, where chronic stress was also found to be associated with dyslipidemia (elevated triglycerides in concert with lower HDL-c).[15]

Occupational stress, another form of chronic stress, is associated with increased health risks, including cancer and cardiovascular disease, and understanding the pathophysiological mechanisms is an area of increased research. Individuals in some high-stress occupations have elevated lipids and lipoproteins, relative to individuals in lower-stress occupations who are otherwise matched. For example, several studies have reported that police officers and bus, taxi, and truck drivers report higher psychological stress than others and have elevated cholesterol, triglycerides and LDL-c, and lower HDL-c,[38] although null results regarding job strain and lipids have also been reported.[10,13,14] An innovative study examining lipid profiles of government workers assisting in relief efforts after the 2004 earthquake in Niigata Prefecture, Japan found elevated cholesterol among those male workers with high workloads.[3] Taken together, this body of research suggests that chronic stress in a variety of context is associated with elevated lipid levels. The mechanisms for such associations, however, not well established.

ACUTE AND EPISODIC PSYCHOLOGICAL STRESS

Short-term or episodic psychological stress has been linked with generally brief elevations in lipid concentrations in research studies spanning several decades. One of the earliest examples of stress-related lipid elevations comes from a study of race-car drivers, whose blood was tested at rest prior to a competitive race, and then immediately after the race.[33] Triglycerides increased over 100% from baseline and peaked a full hour after completion of the race. These and similar results point to the need to understand the timeframe of stress-related lipid elevations in order to ensure accurate assessments.

Early studies such as this one were conducted before a more complete lipid profile was accurately measurable and before subfractions had been fully recognized and characterized. More recent studies have supported the notion that lipids are elevated during acutely stressful experiences; total cholesterol, triglycerides, LDL, and other lipids have been shown to be significantly and acute altered by stressors such as examination stress,[18] short-term laboratory stressors,[26] and social stressors in human and nonhuman primates.[17,21] Individual differences in such metabolic changes can be striking; higher acute lipid responses have been noted in men who have a rigid (as opposed to flexible) way of expressing anger[7]; men have generally higher lipid responses than do women[28]; and healthy offspring of cardiac patients have higher lipid responses to stress than do those without a positive family history.[27] Interestingly, these individual difference factors are all associated with elevated CVD risk themselves, leading to speculation that acute stress responses may potentially explain the elevated risk. Other psychological stressors that have been examined in relationship to acute lipid responsiveness include those examining responses during conditions of social support and social isolation, which affect responsivity. For example, Steptoe and colleagues reported that stress-related cholesterol was elevated significantly more in men who reported being more socially isolated[9] relative to more socially integrated men.

MECHANISMS OF LIPID VARIABILITY DURING STRESS

The mechanisms by which metabolic parameters, in particular lipids, are elevated among adults with stress are not clear, but several pathways may be operating, either separately or together. Because the pattern of responses, as well as the timescale differs between acute and chronic stress, it is likely that the mechanisms driving stress-related lipid elevations are also different.[29] Several putative mechanisms have been tested in both human and animal studies.

Behavioral Factors

Individuals under chronic psychological stress often engage in more health-damaging behaviors such as increased smoking and tobacco use, increased sedentary lifestyle, and poor dietary habits[6,1] and each of these might result in chronically elevated lipid and other metabolic parameters. For example, there is some indication that individuals who respond to stress by increasing caloric intake have higher atherogenic lipid responses to stress. However, such indirect lipid consequences of changes in health behaviors are unlikely to fully explain the stress-related elevations in lipids, since most chronic stress studies find significant lipid elevations even after controlling for these behaviors[39] and are brief but are generally quite well controlled so that even minor behavioral changes such as posture are controlled and cannot therefore explain the change in lipids. A striking illustration of this is a very early study that examined daily fluctuations in lipid levels among a small number of men and women cardiac patients ($n = 4$) who were studied while hospitalized for between 4 weeks and 5 months. Frequent measures were taken of emotional experiences, social factors, stress experiences (among other parameters), and both diet and exercise were standardized. Periodic blood samples were taken for the measures of total cholesterol, and a "stress interview" was given to elicit experiences of psychological stress. This was an observational study which imposed extraordinary control of diet and physical activity, to understand the role of stress on lipid variations across time. In all patients, total cholesterol and triglycerides showed significant fluctuations and, in particular, rapid and acute elevations during the stressful interview and during reports (from patients and/or staff) of stressful days.[41] Although observational, the study underscores in this very small sample that cholesterol and triglyceride concentrations tend to rise with increased psychosocial stress and that these fluctuations cannot be attributed to fluctuations in diet and exercise behaviors.

Physiological and Biological Factors

Physiological mechanisms are likely to also be operating, separately or together with behavioral ones, and understanding these mechanisms will inform our understanding of the potential clinical significance of stress-related lipid elevations. For example, activation of the sympathetic nervous system during psychological stress can impact catecholamines, glucocorticoids such as cortisol, and free fatty acids, which can have resulting effects on lipid metabolism. Epinephrine stimulates the release of free fatty acids into circulation through the process of lipolysis. Increased free fatty acids can in term stimulate the production of very low-density lipoprotein (VLDL) catabolism, which over time results in increased circulating levels of LDL-c. Norepinephrine can decrease lipoprotein lipase activity by stimulating adipose beta receptors. A decrease in lipoprotein lipase activity would subsequently decrease triglyceride clearance, lower HDL, and increased circulating VLDL and LDL. Norepinephrine also has important effects on decreasing hepatic triglyceride lipase, which result in increased circulating HDL and triglyceride-rich lipoproteins. Cortisol activates the hypothalamic-pituitary-adrenocortical (HPA) axis to impact lipid metabolism. Cortisol, in concert with free fatty acids, increases the secretion of VLDL and hepatic triglycerides and inhibits insulin secretion. This latter effect is critical because decreased insulin during stress results in inhibition of triglyceride regulation and delays LDL clearance by suppressing hepatic LDL receptors and catabolism.

These putative physiological pathways have some, albeit not complete support from both animal and human data. The sympathetic-adrenal-medullary axis and HPA axis play key roles in stress reactivity and may modulate lipid and lipoprotein metabolism during psychological stress. For example, a study of hypertensive individuals found that acute stress-related norepinephrine responses predicted lipid responsivity, suggesting a potential mechanism for elevated CVD risk among hypertensive individuals.[40] An early study demonstrated that cynomolgus monkeys given chronic epinephrine infusions have subsequently elevated cholesterol.[5] Cortisol sometimes but not always is elevated in concert with LDL-c and total cholesterol in humans, and some studies have found that cortisol changes predict lipid elevations.[31] However, a study of pharmacologic blockade of sympathetic nervous system activity while participants were stressed in a series of laboratory challenges failed to change stress-related total cholesterol or LDL-c elevations, although blockade did prevent elevations in free fatty acids and triglycerides.[4] An important conclusion from these data is that stress-related elevations in various lipid parameters are likely to be operating through different physiological mechanisms, in addition to behavioral and environmental mechanisms. Overall, most investigations that have investigated these physiologic systems in concert with lipid changes have not found the lipid elevations to be fully explained.

Stress-Induced Metabolism

Earlier studies had indeed specifically and directly examined the role of short-term stress on triglyceride metabolism by studying the efficiency with which individuals metabolized a high-fat load during stress.[25] In these studies, a standard intravenous infusion of intralipid, providing a very high triglyceride load, was administered to a group of healthy adults under both high stress and rest. Individuals were tested a few days apart while fasting; all reported that the stress condition was experienced as stressful. Blood samples for the evaluation of triglyceride levels were taken before the intralipid infusion and then serially in order to calculate the disappearance rate constant (K_2) for each individual during both stress and rest conditions. K_2 defines the clearance rat of the exogenous fat solution; high K_2 values indicate a more efficient clearance of triglycerides from the circulation. During acute stress, K_2 values decreased significantly relative to a nonstress period, signaling diminished triglyceride clearance during stress. These data strongly suggest that stress-induced triglyceride concentrations are elevated during stress due to stress-induced changes in triglyceride clearance.

Hemoconcentration

During acutely stressful situations, there is a rapid and temporary decrease in blood plasma volume which results in a higher concentration of red blood cells and other cellular components (hemoconcentration). Hemoconcentration also occurs during postural shifts, dehydration, and a variety of pathophysiological conditions. During stress, hemoconcentration might occur by increased red blood cell mass, by acute changes in kidney function, or through hemodynamically induced fluid shifts (as might be expected with stress-related blood pressure increases). The resulting hemoconcentration leads to elevated lipid concentrations in the blood. There is good evidence for hemoconcentration during acute stress[2]; the increased lipid concentrations during acute stressors are likely to be partly, but not fully, accounted for by this mechanism, especially during stressors than induce significant changes in not only lipids, but catecholamines, cortisol, and blood pressure.

Multiple Mechanisms

Several pieces of evidence point to multiple mechanisms operating to cause acute stress-related lipid and lipoprotein concentrations, in addition to or instead of hemoconcentration. First, in many but not all studies, the total cholesterol, LDL-c, and/or triglyceride elevations during acute stress remain significant after correction for transvascular plasma volume shifts.[26] Second, some lipoproteins are consistently elevated during acute stress (e.g., LDL-c) while others are often not (e.g., HDL-c). If transvascular plasma volume shifts were fully responsible for stress-related lipid elevations, similar magnitude changes in concentration would be seen in each of these measures. Thus, it is highly likely that additional mechanisms for the stress-related elevations in lipids are operating during stress, and future studies would do well to simultaneously test multiple putative mechanisms. During chronic stress in particular, it is important to measure both behavioral and biological putative mechanisms, in order to begin to tease apart the complex interactions driving lipid reactivity. Understanding the mechanisms responsible to stress-related lipid elevations will help to delineate the potential clinical significance of these changes and inform later clinical trials if warranted.

INTERVENTIONS TO REDUCE STRESS-RELATED INCREASES IN LIPID CONCENTRATIONS

Because there is not definitive evidence that stress-related increases in lipids and lipoproteins are significant risk factors for CVD, there have been no systematic trials specifically testing interventions to modify such stress-related elevations. However, a few interventions have been empirically tested to reduce perceptions of stress and a few of these have tested whether such strategies might diminish stress-related lipid elevations. For example, individuals in very highly demanding jobs who spent a 3-week respite at a spa had significantly decreased LDL-c levels, relative to those who also participated in the 3-week spa visit but who reported less-demanding jobs.[30] Further investigation of the mechanisms by which psychological stress increases circulating levels of atherogenic lipids may suggest more targeted interventions for lipid reductions during stress.

CLINICAL IMPLICATIONS OF VARIATIONS IN LIPID CONCENTRATIONS

The data implicating a direct relationship between elevated blood lipid levels and risk for CVD is strong.[24] For example, chronically elevated LDL-c and total cholesterol levels are a risk for CVD, and lowering LDL-c through statin use and/or behavioral changes in diet and exercise produces a clinically significant decrease in risk, cardiovascular events, and even a regression of coronary atherosclerosis in some individuals. High triglycerides and low HDL-c are also correlated positively with CVD risk, although it is not clear to what degree lowering triglycerides or raising HDL-c levels subsequently reduces CVD risk.[37]

More relevant to the current discussion is the extent to which stress-associated elevations in lipids are clinically relevant. To date, there are limited prospective studies establishing lipid variations as a CVD risk factor, primarily because such day-to-day or even minute-to-minute variations are typically not studied in longitudinal investigations. However, a small but important and highly relevant study tested lipid concentrations among healthy participants engaging in an acutely stressful paradigm, and followed up three years later to assessing resting lipid levels. Importantly, individuals with exaggerated stress-induced lipid concentrations had, three years later, elevated baseline levels independent of a host of other risk factors.[23] These data suggest that acute lipid responses to acute psychological stress may be precursors to or indicate a vulnerability to later elevated lipid levels.

In addition, findings from several cross-sectional and laboratory studies provide an indirect suggestion that some lipid fluctuations may be clinically important. For example, laboratory studies of healthy individuals at elevated future risk for CVD show that these individuals have larger, short-term elevations in blood lipid concentrations during stress, compared with healthy age-matched individuals not at elevated future CVD risk.[27] Men, postmenopausal women, and healthy offspring of parents with documented myocardial infarction all have greater elevations in total cholesterol and LDL-c during acute psychological stress, relative to women, age-matched premenopausal women, and healthy offspring of parents with no cardiovascular disease, respectively. Other individual differences have been found in lipid responses to acute stressors. Individuals with psychosocial CVD risk factors such as elevated hostility, anxiety disorders, and personality characteristics[17] also have been shown to have transient elevations in total cholesterol and LDL-c during acute stress, as have hypertensive individuals.[17] It is an open question as to whether the increased risk of these individuals is related to the elevated lipid concentrations during stress.

FUTURE

Although there are no longitudinal studies definitively establishing lipid reactivity as a risk factor for subsequent CVD, available cross-sectional and animal data suggests that stress-related elevations in the atherogenic lipids may have clinical significance. Future prospective studies designed to explicate the mechanisms and clinical relevance of both acute and chronic stress-related lipid reactivity are warranted.

Acknowledgment

The content is solely the responsibility of the author and does not necessarily represent the official views of the National Institutes of Health (NIH), or the United States Government.

References

1. Appelhans BM, Pagoto SL, Peters EN, Spring BJ. HPA axis response to stress predicts short-term snack intake in obese women. *Appetite*. 2010;54:217–220.
2. Austin AW, Patterson SM, von Känel R. Hemoconcentration and hemostasis during acute stress: interacting and independent effects. *Ann Behav Med*. 2011;42:153–173.
3. Azuma T, Seki N, Tanabe N, et al. Prolonged effects of participation in disaster relief operations after the Mid-Niigata earthquake on increased cardiovascular risk among local governmental staff. *J Hypertens*. 2010;28:695–702.
4. Bachen E, Muldoon MF, Matthews KA, Manuck SB. Effects of hemoconcentration and sympathetic activation on serum lipid responses to brief mental stress. *Psychosom Med*. 2002;64:587–594.
5. Dimsdale JE, Herd JA, Hartley LH. Epinephrine mediated increases in plasma cholesterol. *Psychosom Med*. 1983;45:227–232.
6. Epel E, Jimenez S, Brownell K, Stroud L, Stoney C, Niaura R. Are stress eaters at risk for the metabolic syndrome? *Ann N Y Acad Sci*. 2004;1032:208–210.
7. Finney ML, Stoney CM, Engebretson TO. Hostility and anger expression in African American and European American men is associated with cardiovascular and lipid reactivity. *Psychophysiology*. 2002;39:340–349.
8. Fonseca VA, Smith H, Kuhadiya N, et al. Impact of a natural disaster on diabetes: exacerbation of disparities and long-term consequences. *Diabetes Care*. 2009;32:1632–1638.
9. Grant N, Hamer M, Steptoe A. Social isolation and stress-related cardiovascular, lipid, and cortisol responses. *Ann Behav Med*. 2009;37:29–37.
10. Greenlund KJ, Kiefe CI, Giles WH, Liu K. Associations of job strain and occupation with subclinical atherosclerosis: the CARDIA Study. *Ann Epidemiol*. 2010;20:323–331.
11. Itoh T, Nakajima S, Tanaka F, et al. Impact of the Japan earthquake disaster with massive Tsunami on emergency coronary intervention and in-hospital mortality in patients with acute ST-elevation myocardial infarction. *Eur Heart J Acute Cardiovasc Care*. 2014;3:195–203.
12. Jordan HT, Stellman SD, Morabia A, et al. Cardiovascular disease hospitalizations in relation to exposure to the September 11, 2001 World Trade Center disaster and posttraumatic stress disorder. *J Am Heart Assoc*. 2013;2:e000431.
13. Jovanović J, Stefanović V, Stanković DN, et al. Serum lipids and glucose disturbances at professional drivers exposed to occupational stressors. *Cent Eur J Public Health*. 2008;16:54–58.
14. Kang MG, Koh SB, Cha BS, Park JK, Baik SK, Chang SJ. Job stress and cardiovascular risk factors in male workers. *Prev Med*. 2005;40:583–588.
15. Kivim"aki M, Lawlor DA, Singh-Manoux A, et al. Common mental disorder and obesity: insight from four repeat measures over 19 years: prospective Whitehall II cohort study. *Br Med J*. 2009;339:b3765.
16. Kloner RA, Leor J, Poole WK, Perritt R. Population-based analysis of the effect of the Northridge earthquake on cardiac death in Los Angeles county, California. *J Am Coll Cardiol*. 1997;30:1174–1180.
17. Kuebler U, Trachsel M, von Känel R, Abbruzzese E, Ehlert U, Wirtz PH. Attributional styles and stress-related atherogenic plasma lipid reactivity in essential hypertension. *J Psychosom Res*. 2014;77:51–56.
18. Maduka IC, Neboh EE, Ufelle SA. The relationship between serum cortisol, adrenaline, blood glucose and lipid profile of undergraduate students under examination stress. *Afr Health Sci*. 2015;15:131–136.
19. Masson LF, McNeill G, Avenell A. Genetic variation and the lipid response to dietary intervention: a systematic review. *Am J Clin Nutr*. 2003;77:1098–1111.
20. Moura FA, Dutra-Rodrigues MS, Cassol AS, et al. Impact of seasonality on the prevalence of dyslipidemia: a large population study. *Chronobiol Int*. 2013;30:1011–1015.
21. Schwaberger G. Heart rate, metabolic and hormonal responses to maximal psycho-emotional and physical stress in motor car racing drivers. *Int Arch Occup Environ Health*. 1987;59:579–604.
22. Smith SJ, Cooper GR, Myers GL, Sampson EJ. Biological variability in concentrations of serum lipids: sources of variation among results from published studies and composite predicted values. *Clin Chem*. 1993;39:1012–1022.
23. Steptoe A, Brydon L. Associations between acute lipid stress responses and fasting lipid levels 3 years later. *Health Psychol*. 2005;24:601–607.
24. Stone NJ, Robinson JG, Lichtenstein AH, et al. 2013 ACC/AHA guideline on the treatment of blood cholesterol to reduce atherosclerotic cardiovascular risk in adults: a report of the American College of Cardiology/American Heart Association task force on practice guidelines. *Circulation*. 2014;129(Suppl 2):S1–S45.
25. Stoney CM, West SG, Hughes JW, et al. Acute psychological stress reduces plasma triglyceride clearance. *Psychophysiology*. 2002;39:80–85.
26. Stoney CM, Bausserman L, Niaura R, Marcus B, Flynn M. Lipid reactivity to stress: II. Biological and behavioral influences. *Psychophysiology*. 1999;18:251–261.
27. Stoney CM, Hughes JW. Lipid reactivity among men with a parental history of myocardial infarction. *Psychophysiology*. 1999;36:484–490.
28. Stoney CM, Matthews KA, McDonald R,H, Johnson CA. Sex differences in lipid, lipoprotein, cardiovascular, and neuroendocrine responses to acute stress. *Psychophysiology*. 1988;25:645–656.
29. Stoney CM, Niaura R, Bausserman L, Matacin M. Lipid reactivity to stress: I. Comparison of chronic and acute stress responses in middle-aged airline pilots. *Health Psychol*. 1999;18:241–250.
30. Strauss-Blasche G, Ekmekcioglu C, Marktl W. Serum lipids responses to a respite from occupational and domestic demands in subjects with varying levels of stress. *J Psychosom Res*. 2003;55:521–524.
31. Suarez EC, Harralson TL. Hostility-related differences in the associations between stress-induced physiological reactivity and lipid concentrations in young healthy women. *Int J Behav Med*. 1999;6:190–203.
32. Szczygielska A, Widomska S, Jaraszkiewicz M, Knera P, Muc K. Blood lipids profile in obese or overweight patients. *Ann Univ Mariae Curie-Skłodowska*. 2003;58:343–349.
33. Taggart P, Carruthers M. Hyperlipidaemia induced by the stress of racing driving. *Lancet*. 1971;1:363–366.
34. Thethi TK, Yau CL, Shi L, et al. Time to recovery in diabetes and comorbidities following Hurricane Katrina. *Disaster Med Public Health Prep*. 2010;4(Suppl 1):S33–S38.
35. Trasande L, Fiorino EK, Attina T, et al. Associations of World Trade Center exposures with pulmonary and cardiometabolic outcomes among children seeking care for health concerns. *Sci Total Environ*. 2013;444:320–326.
36. Trevisan M, Jossa F, Farinaro E, et al. Earthquake and coronary heart disease risk factors: a longitudinal study. *Am J Epidemiol*. 1992;135:632–637.
37. Verbeek R, Hovingh GK, Boekholdt SM. Non-high-density lipoprotein cholesterol: current status as cardiovascular marker. *Curr Opin Lipidol*. 2015;26:502–510.

38. Walvekar SS, Ambekar JG, Devaranavadagi BB. Study on serum cortisol and perceived stress scale in the police constables. *J Clin Diagn Res*. 2015;9:BC10–14.

39. Winning A, Glymour MM, McCormick MC, Gilsanz P, Kubzansky LD. Psychological distress across the life course and cardiometabolic risk. *J Am Coll Cardiol*. 2015;66:1577–1586.

40. Wirtz PH, Ehlert U, Bärtschi C, Redwine LS, von Känel R. Changes in plasma lipids with psychosocial stress are related to hypertension status and the norepinephrine stress response. *Metabolism*. 2009;58:30–37.

41. Wolf S, McCabe WR, Yamamoto J, Adsett CA, Schottstaedt WW. Changes in serum lipids in relation to emotional stress during rigid control of diet and exercise. *Circulation*. 1962;26:379–387.

42. Yamamoto H, Zheng KC, Ariizumi M. Influence of heat exposure on serum lipid and lipoprotein cholesterol in young male subjects. *Ind Health*. 2003;41:1–7.

43. Zhang CX, Tse LA, Deng XQ, Jiang ZQ. Cardiovascular risk factors in overweight and obese Chinese children: a comparison of weight-for-height index and BMI as the screening criterion. *Eur J Nutr*. 2008;47:244–250.

29

Corticosteroid Receptors

M.J.M. Schaaf[1], O.C. Meijer[2]

[1]Leiden University, Leiden, The Netherlands; [2]Leiden University Medical Center, Leiden, The Netherlands

Abstract

Corticosteroid effects are mediated via two types of "nuclear receptors": the mineralocorticoid receptor (MR) and the glucocorticoid receptor (GR). MRs acts as a high-affinity receptor for aldosterone and endogenous glucocorticoids; the GR is a somewhat lower affinity receptor for glucocorticoids. Both receptors have highly similar mechanisms of action that involve rapid nongenomic signaling and transcriptional regulation. Here, we focus on the latter. There is a substantial diversity in receptor isoforms, based on alternative splicing and use of alternative translation initiation sites. Post-translational modifications add to cell type and state-dependent activity of the receptors. Upon hormone binding the receptors migrate to the cell nucleus where they act in conjunction with transcription factors and coregulator proteins to regulate gene expression. These pathways allow extensive cross talk with other signaling pathways and may form the basis for more directed experimental and clinical interventions using signaling pathway selective modulators of receptor activity.

INTRODUCTION

In humans, the effects of corticosteroids are mediated by two intracellular receptors: the mineralocorticoid receptor (MR) and the glucocorticoid receptor (GR). Both receptors are members of the steroid receptor family, which in turn belong to the superfamily of nuclear receptors. The MR and GR are closely related. They have evolved from a single ancestral corticosteroid receptor that was present in jawless fish >470 million years ago through a process of gene duplication and subsequent mutation.[8] The ancestral corticosteroid receptor was able to bind both cortisol and aldosterone, even though this hormone had not appeared in nature yet. The MR currently present in tetrapods and bony fish has retained the binding characteristics of the ancestral receptor, enabling it to bind both cortisol and aldosterone, with the actual ligand depending on pre-receptor metabolism of cortisol.[58] The GR has reduced its sensitivity for cortisol and lost its capacity to bind aldosterone.

The MR and GR can exert rapid nongenomic effects as well as genomic effects mediated at or near the membrane,[19,21] or in the cytoplasm[17], as discussed elsewhere in this volume (Chapter 33). The vast majority of mechanistic MR/GR studies has been performed in the context of classic nuclear signaling, which involves nuclear localization and subsequent interactions with the DNA and other transcriptionally active proteins such as transcription factors and nuclear receptor coregulators. Here we focus our description on the transcriptional effects.

OVERALL STRUCTURE

As a result of their common ancestry, the human MR (hMR) and GR (hGR) genes are very similar in structure. The hMR gene (NR3C2) is located on chromosome 4 (spanning >450 kb), whereas the hGR gene (NR3C1) resides on chromosome 5 (spanning >80 kb). They both consist of nine exons and contain multiple alternative (noncoding) exon 1s. Two exons 1 have been identified for hMR (1α and 1β[59]) and seven (1A–1H[51]) for hGR. The usage of alternative exons 1 is suggested to be associated with activity of promoter regions located immediately downstream of an exon 1, which may play an important role in tissue-specific control of receptor gene transcription.[51] DNA methylation of the promoter regions may epigenetically program the expression of the receptors in a long-term fashion.[50]

Like all members of the nuclear receptor superfamily, the hMR and hGR consist of three major domains: a large N-terminal domain (NTD), a small DNA-binding domain (DBD), and a C-terminal ligand-binding domain (LBD). The NTD is poorly conserved between the two receptors but for each receptor highly conserved between species (Fig. 29.1). The hMR NTD consists of 602 aa and contains two domains that are

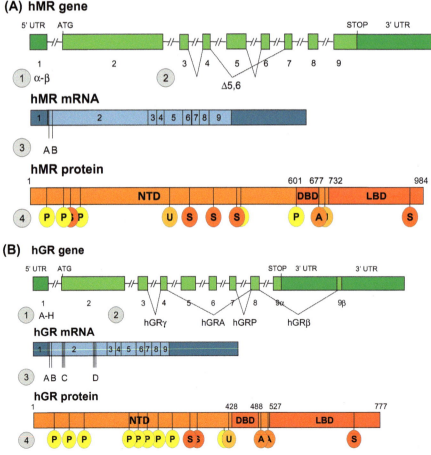

FIGURE 29.1 Structure of hMR (A) and hGR (B) gene, mRNA, and protein. Four mechanisms are indicated that result in different mRNA and/or protein isoforms: alternative promoter/exon 1 usage (1), alternative RNA splicing (2), alternative translation initiation (3), and post-translational modification (4).

important for the gene regulatory function of the receptor, the activation function-1a (AF-1a, aa 1–167) and the AF-1b domain (aa 445–602). The region between these AF-1 domains has been characterized as an inhibitory domain.[37] The NTD of the hGR also contains an AF-1 domain (aa 77–262), with a core region (aa 199–240) that is crucial for transactivation.[10] Although the hGR AF-1 has been shown to bind coregulatory proteins like p160 coactivators and DRIP/TRAP, specific binding motifs in this region have not been identified.[24] The DBDs of hMR (aa 601–677) and hGR (aa 421–486) are 94% identical and contain two zinc fingers that are formed by four cysteine residues that coordinate a zinc atom. The first zinc finger contains amino acid residues involved in sequence-specific interactions with the major groove of DNA, and the second zinc finger contains a domain that is important for receptor dimerization.[30] The LBDs of hMR and hGR show a lower level of sequence similarity but are structurally very similar.[6,28] Both LBDs are folded into a three-layered sandwich of 12 α-helices. The middle layer of helices is clustered at the top half of the domain but is absent from the bottom half, thus creating a hydrophobic pocket for ligand binding. Upon binding of an agonist, the most C-terminal helix, which contains part of the AF-2 domain, is packed tightly against the main domain of the LBD, in a conformation that allows binding of specific coactivators containing an LXXLL motif, an interaction that is essential for

transcriptional activity of the receptor. The LBD contains a dimerization domain that is mainly formed by beta sheets in the structure of the LBD.

RECEPTOR VARIANTS

Both MR and GR are expressed as several protein isoforms as a result of alternative translation initiation, alternative splicing, and post-translational modification, and the combination of these processes leads to an enormous number of possible receptor isoforms. First, alternative translation initiation from a single hGR mRNA produces several N-terminally truncated hGR variants. Eight alternative start codons can be used as translational start sites, resulting in eight receptor subtypes with progressively shorter NTDs: GR-A, GR-B, GR-C1-3, and GRD1-3. Interestingly, the different receptor isoforms each regulate a distinct cluster of target genes, indicating gene-specific differences in transcriptional activity between these isoforms.[29] In the human brain, the expression of these isoforms varies across postnatal developmental stages.[46] Two variants resulting from alternative translation initiation have been found in vitro for hMR, called MR-A and MR-B, and they appeared to have different transcriptional activity[37](Fig. 29.2).

Second, alternative RNA splicing of the results in four more possible hGR isoforms. The best-studied splice

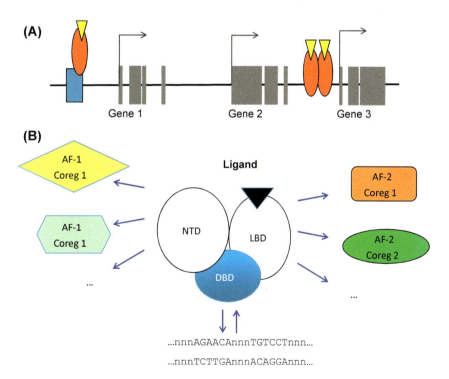

FIGURE 29.2 Nuclear signaling modes of MR and GR. (A) The receptors may act by binding directly to the DNA (often as dimers), or via "tethering" to other, nonreceptor, transcription factors. (B) Activation function-1 (AF-1) in the N-terminal domain and activation function-2 (AF-2) in the ligand-binding domain are the outputs of the receptor once bound to the DNA. They each interact with downstream complexes of coregulators, in a gene, cell-type, and state-dependent manner.

variant is hGRβ, which is a result of alternative splicing of the primary transcript in exon 9. The hGRβ isoform has a C-terminal sequence that differs from the C-terminus of the canonical hGR, also called hGRα. The hGRβ does not bind glucocorticoid agonists and does not activate gene transcription like hGRα. However, it has been shown to act as a dominant-negative inhibitor of the transcriptional activity of hGRα. In line with this dominant-negative activity, a correlation has been found between the expression of hGRβ in patients receiving antiinflammatory glucocorticoid therapy and resistance to this type of treatment.[44] Rodents lack GRβ in this form, but alternative splicing events lead to similar C-terminal truncated variants in other species.[45] Another splice variant of hGR is hGRγ which has an additional arginine residue in the DBD between the two zinc fingers as a result of usage of an alternative splice donor site between exon 3 and 4.[42] This receptor isoform has impaired transcriptional activity and its expression has been shown to be associated with glucocorticoid resistance.[40] Two other splice variants, hGR-A and hGR-P, are associated with glucocorticoid resistance in several types of cancer cells. Both isoforms lack a part of the LBD and are therefore not capable of binding ligands.[34]

Three alternative splice variants of hMR have been found. One is similar to hGRγ, resulting from usage of an alternative splice donor site between exon 3 and 4, leading to an insertion of four amino acids between the two zinc fingers.[7] Another hMR splice variant results from alternative usage of the splice donor site between exon 5 and 6 which leads to a 10 bp deletion in the mRNA and a truncated receptor not capable of ligand binding.[61] This variant appears not to affect the activity of the canonical hMR when coexpressed. Skipping exons 5 and 6 results in another receptor isoform that lacks a functional LBD, but this isoform has been shown to modulate the activity of the canonical hMR and hGR.[60]

Post-translational modification of hGR and hMR further adds to the repertoire of receptor subtypes. Both receptors are subject to phosphorylation at several sites and most of these phosphorylation events are dependent on ligand binding. The hGR is phosphorylated on at least one threonine and eight serine residues,[53] whereas five phosphorylation sites (four tyrosines and one serine) have been demonstrated for the hMR.[54] All these sites are located in the NTD, and many reside within the AF-1 domain. Kinases shown to be involved in hGR phosphorylation are MAPKs, cyclin-dependent kinases, casein kinase II, and glycogen synthase kinase 3β,[53] and protein kinase Cα has been shown to phosphorylate hMR.[54] Phosphorylation appears to modulate the transcriptional activity of the receptors, and this can be positive or negative depending on the phosphorylation site. This modulation results from changes in nuclear–cytoplasmic shuttling, cofactor recruitment, and receptor

degradation.[56] Sumoylation of hGR and hMR has been shown on specific lysine residues (three in hGR,[36] five in hMR[37]). This modification alters the transcriptional activity in a target gene-specific way, specifically regulating antiproliferative gene transcription for hGR. Ubiquitination has been reported for both hGR (on lysine 419) and hMR (on lysines 367 and 715). This modification targets the receptors for proteasomal degradation, which has an enhancing effect on the transcriptional activity of hGR[55] but a suppressive effect on hMR's activity.[49] Two lysine residues of hGR can be acetylated, which has been shown to have an inhibitory effect on the repression of NF-κB signaling by hGR.[18] Acetylation of hMR has been suggested, but its effect on hMR function has not been established yet.

The result of splicing, alternative translation, and post-translational modification is combinatorial. Other than in the activated immune system, the canonical full-length splice variants seem predominant.[14] Posttranslation modifications constitute an important level of regulation in cross talk with other factors, as for BDNF and the GR in hypothalamic neurons.[26]

FINDING NUCLEAR TARGETS

In the absence of ligand, both hMR and hGR are predominantly located in the cytoplasm, as part of a large multiprotein complex. A small fraction may be localized at or near the membrane to allow rapid nongenomic effects. For hGR, it is known that this complex includes chaperone proteins (hsp90, hsp70, hsp56, p23) and immunophilins of the FK506 family (FKBP51 and FKBP52), and these complexes are generally considered to be similar for all steroid receptors.[39] These proteins keep the receptor in a conformation that is unable to bind DNA but enables high-affinity ligand binding. The hMR binds cortisol with a very high affinity (the k_d is in the nanomolar range), approximately 10-fold higher than the affinity of the hGR for cortisol.[13] Ligand binding induces a conformational change of the receptor, which initiates transport of the receptor along the microtubules toward the nuclear pore complex.[15] Nuclear import through this complex is dependent on nuclear localization signals (NLSs) present in the receptor sequence . Three functional NLSs have been described For hMR,[56] and two have been identified for hGR.[53] In both cases, the most C-terminal NLS resides in the LBD and its function appears to be ligand dependent. These sequences are recognized by importins which bind to the NLSs of the receptors. The receptor–importin complex translocates to the nucleus where the importin dissociates from the receptor by the binding to RanGTP.[53] Nuclear export is mediated by a nuclear export signal (NES) which is located between the two zinc fingers of the DBD.[5] Thus, the subcellular localization of hMR and hGR is dependent on the rate of nuclear import and export.

Upon translocation, hMR and hGR have to find their target sequences in the vast amount of DNA inside the nucleus. Recent work in our laboratory on the intranuclear dynamics of hMR and hGR has shed some light on the mechanism behind their searching strategy. Using a combination of fluorescence recovery after photobleaching (FRAP) and single-molecule microscopy we have been able to perform a detailed analysis of the dynamics of hGR and hMR in the nucleus. For both receptors, three subpopulations of molecules were found. The first subpopulation seems to be freely diffusing through the nucleus, but since the diffusion rate of this fraction appears to be dependent on the ability of the receptor to bind DNA, we assume that this subpopulation undergoes brief (<1 ms) interactions with the DNA. Most likely, these interactions are nonspecific interactions with DNA reflecting the receptor scanning the DNA in search of its target sequence. Furthermore, two subpopulations were identified that showed longer immobilization times, ~0.5 and ~2.5 s. The subpopulation with the longest immobilization time is suggested to reflect DNA-bound transcriptionally active receptors, whereas the shorter immobilization time may be associated with a subpopulation that is bound to DNA indirectly or in a transient loosely DNA-bound state.[16]

Two main modes of transcriptional signaling are available to the MR and GR (Fig. 29.2A). First, they may bind to response elements in the DNA. These glucocorticoid response elements (GREs) are used both by MR and GR, even if the receptors have different target genes. Experiments in the rat suggest that, at least upon treatment with exogenous glucocorticoids under resting conditions, this is the dominant mode of signaling in the rat hippocampus for the GR.[38] The other mode is via protein–protein interactions with other—nonreceptor—transcription factors. This is an important mode of action for the GR in the activated immune system, where GR acts to dampen the activity of NF-kB and AP-1.[4] As AP-1 is a complex that consists of immediate early genes like c-fos, this mode of action may also be highly relevant for feedback type of actions in neuroendocrine processes.[23] The GR$^{dim/dim}$ mouse that has impaired binding of GR to GREs has substantial defects in gene regulation, but showed for example only mildly increased plasma ACTH levels.[41] Interpretation of these data is hampered by the fact that intact GR function in this mouse may reflect either NF-kB/AP-1 interaction of GR, or fully normal GRE binding at a subset of genes.[1]

The fact that there are different modes of binding to the DNA has major consequences for the "cistrome" of the receptors: the total of genomic loci where the receptors bind and may prime of induce changes in gene transcription. Cell-type specific chromatin organization is a major determinant, with overlap of genomic binding sites being as low as 5–10%.[20] However, also the cellular state (activation of other transcription factors) may lead to pronounced differences. Thus, the predominance of GRE-dependent GR signaling in the rat hippocampus may change when tested in activated conditions. This may also be reflected in recent data for the MR. The MR has always been considered to be much weaker at indirect DNA binding via other proteins,[32] but genome-wide binding profiles (from kidney[27]) suggest that GREs are absent from a substantial number of MR binding loci and that MR also may bind via other transcription factors.

TRANSCRIPTIONAL REGULATION

Coregulators

It is only after binding of MRs and GRs to the DNA that the actual signal transduction of the receptor takes place: the intermediate molecular steps that link the activated receptor to the effector mechanism, i.e., the transcription machinery. The proteins that bridge from the DNA-bound receptors to the transcription machinery are called nuclear receptor coregulators. In general, coregulators either modulate local chromatin structure (to allow or disallow binding of other factors) or they are part of a physical link to RNA polymerase II to affect initiation of elongation of RNA transcripts.[48] Bridging of the receptor to transcription start sites of target genes may involve very long distance interactions, as MR/GR-binding sites on the DNA may be many kilobases away from their putative target genes (even if only on very few occasions such GREs have unequivocally been linked to actual target genes[47]).

Coregulators have been proposed to act as signaling "hubs" and integrate information carried by steroids and other (membrane-bound) signaling pathways.[48] There are over 300 proteins that qualify as such, based on the fact that they do not have intrinsic transcriptional activity, but potentiate or attenuate the transcriptional effects of one or more nuclear receptor types. An estimated guess, based on in vitro interactions between GR and sets of such coregulators, is that about half of such proteins may interact with GR or MR.[3] Of note, even if GR may affect chromatin accessibility via its coregulators, in the majority of cases chromatin has to be made accessible by other transcription factors.[20] This suggests that steroids in most (but not all!) cases enhance or repress the activity of active genes, and on relatively few loci initiate transcription de novo. In addition, because the GRE can act as an allosteric modulator of the receptor, the

DNA also affects the recruitment of coregulators by the receptors.[33]

MR and GR share many coregulators, in particular those that act via the AF-2, such as steroid receptor coactivator-1 (SRC-1). AF-1 interacting coregulators have been more difficult to identify. Coregulators act in a receptor, cell-type, and gene-specific manner (Fig. 29.2B). This was illustrated by SRC-1, which is necessary for regulation of pituitary Pomc and brain Crh regulation via glucocorticoids, while regulation of other genes in the same areas—or in fact most of the HPA axis functionality—remain largely unaffected in absence of SRC-1.[25,57] For the MR, coregulators that interact dependent on cortisol or aldosterone binding have also been reported.[54] For effects of corticosteroids on the brain, it will be of interest to colocalize the expression of MR and GR with particular coregulators, in order to determine regions that respond in a parallel manner to steroid treatment.[31]

Target Genes

The actual target genes and noncoding RNAs[22] that are the final cellular mediators of MR- and GR-dependent effects differ markedly between tissues and cellular states. Reorganization of chromatin structure or presence, different sets of interacting proteins, and/or post-translational changes in MR/GR proteins can lead to a marked difference in the actual transcriptional response. A period of chronic stress led to 50% of initial corticosterone target genes becoming unresponsive, while an equal number of previously unresponsive genes emerged as reacting to the hormone.[11] Genes like *FKBP5*, *GILZ*, and *PER1* have been used as ubiquitously regulated markers for GR sensitivity.[43] *CRH* and *POMC* genes are well studied, but mechanisms and actual the physiological relevance of their regulation has remained in part unclear.[2] The relative importance of genomic steroid targets will have to be determined in specific experimental and clinical settings.

Ligands

As is also discussed in the chapter in ligands in this volume, the different signaling modes that have been described here form the basis for promising new ligands of MR and GR. The main drive has been historically to find GR ligands with antiinflammatory efficacy that lack side effects on metabolism and bone[4] but other approaches are conceivable, such as partial GR antagonism in brain or metabolic tissues, without increasing susceptibility for infection.[52] MR antagonist for cardiovascular disease that lack efficacy on potassium secretion is another variation on this theme.

GR selective receptor modulators may distinguish between processes that depend on direct DNA binding and those that involve protein–protein interaction.[12] Other synthetic ligands are based on differential recruitment of coregulator proteins.[3,9] These modulators in fact bare two types of promises. First, some may have real clinical benefit in particular settings. Second, by linking the effects of the selective modulators to their signaling profiles, they may be excellent tools to link MR/GR signaling pathways to physiological function (Fig. 29.2B).

CONCLUSION

Signaling via MR and GR is a complex cascade that involves ligand binds, nuclear translocation, DNA binding, and interactions with a substantial number of coregulators or other transcription factors. Receptor variants and post-translational modifications add to the regulatory repertoire of cells in different tissues. Of note, normally dose–response relationships are mathematically first order, pointing to a single rate-limiting step given a particular hormone concentration, which may differ per gene.[35] The mechanisms of action as described here are necessary to understand the insights in the processes described in this chapter will increase our insights in the crucial contextual nature of corticosteroid signaling, that includes cell specificity, developmental dependence, and cross talk with other signaling pathways.

References

1. Adams M, Meijer OC, Wang J, Bhargava A, Pearce D. Homodimerization of the glucocorticoid receptor is not essential for response element binding: activation of the phenylethanolamine N-methyltransferase gene by dimerization-defective mutants. *Mol Endocrinol*. 2003;17(12):2583–2592. Available from: http://doi.org/10.1210/me.2002-0305.
2. Aguilera G, Liu Y. The molecular physiology of CRH neurons. *Front Neuroendocrinol*. 2011. Available from: http://doi.org/10.1016/j.yfrne.2011.08.002.
3. Atucha E, Zalachoras I, van den Heuvel JK, et al. A mixed glucocorticoid/mineralocorticoid selective modulator with dominant antagonism in the male rat brain. *Endocrinology*. 2015;156(11):4105–4114. Available from: http://doi.org/10.1210/en.2015-1390.
4. Beck IME, Vanden Berghe W, Vermeulen L, Yamamoto KR, Haegeman G, De Bosscher K. Crosstalk in inflammation: the interplay of glucocorticoid receptor-based mechanisms and kinases and phosphatases. *Endocr Rev*. 2009;30(7):830–882. Available from: http://doi.org/10.1210/er.2009-0013.
5. Black BE, Holaska JM, Rastinejad F, Paschal BM. DNA binding domains in diverse nuclear receptors function as nuclear export signals. *Curr Biol*. 2001;11(22):1749–1758.
6. Bledsoe RK, Montana VG, Stanley TB, et al. Crystal structure of the glucocorticoid receptor ligand binding domain reveals a novel mode of receptor dimerization and coactivator recognition. *Cell*. 2002;110(1):93–105.

7. Bloem LJ, Guo C, Pratt JH. Identification of a splice variant of the rat and human mineralocorticoid receptor genes. *J Steroid Biochem Mol Biol.* 1995;55(2):159–162.

8. Bridgham JT, Carroll SM, Thornton JW. Evolution of hormone-receptor complexity by molecular exploitation. *Science.* 2006;312(5770):97–101. Available from: http://doi.org/10.1126/science.1123348.

9. Coghlan MJ, Jacobson PB, Lane B, et al. A novel antiinflammatory maintains glucocorticoid efficacy with reduced side effects. *Mol Endocrinol.* 2003;17(5):860–869. Available from: http://doi.org/10.1210/me.2002-0355.

10. Dahlman-Wright K, Almlöf T, McEwan IJ, Gustafsson JA, Wright AP. Delineation of a small region within the major transactivation domain of the human glucocorticoid receptor that mediates transactivation of gene expression. *Proc Natl Acad Sci USA.* 1994;91(5):1619–1623.

11. Datson NA, van den Oever JME, Korobko OB, Magarinos AM, De Kloet E, Mcewen B. Prior history of chronic stress changes the transcriptional response to glucocorticoid challenge in the dentate gyrus region of the male rat hippocampus. *Endocrinology.* 2013. Available from: http://doi.org/10.1210/en.2012-2233.

12. De Bosscher K, Vanden Berghe W, Beck IME, et al. A fully dissociated compound of plant origin for inflammatory gene repression. *Proc Natl Acad Sci USA.* 2005;102(44):15827–15832. Available from: http://doi.org/10.1073/pnas.0505554102.

13. De Kloet E. From receptor balance to rational glucocorticoid therapy. *Endocrinology.* 2014;155(8):2754–2769. Available from: http://doi.org/10.1210/en.2014-1048.

14. Derijk RH, Schaaf M, Stam FJ, et al. Very low levels of the glucocorticoid receptor beta isoform in the human hippocampus as shown by Taqman RT-PCR and immunocytochemistry. *Brain Res Mol Brain Res.* 2003;116(1–2):17–26.

15. Fitzsimons CP, Ahmed S, Wittevrongel CFW, et al. The microtubule-associated protein doublecortin-like regulates the transport of the glucocorticoid receptor in neuronal progenitor cells. *Mol Endocrinol.* 2008;22(2):248–262. Available from: http://doi.org/10.1210/me.2007-0233.

16. Groeneweg FL, van Royen ME, Fenz S, et al. Quantitation of glucocorticoid receptor DNA-binding dynamics by single-molecule microscopy and FRAP. *PLoS ONE.* 2014;9(3):e90532. Available from: http://doi.org/10.1371/journal.pone.0090532.

17. Gutièrrez-Mecinas M, Trollope AF, Collins A, et al. Long-lasting behavioral responses to stress involve a direct interaction of glucocorticoid receptors with ERK1/2-MSK1-Elk-1 signaling. *Proc Natl Acad Sci USA.* 2011;108(33):13806–13811. Available from: http://doi.org/10.1073/pnas.1104383108.

18. Ito K, Yamamura S, Essilfie-Quaye S, et al. Histone deacetylase 2-mediated deacetylation of the glucocorticoid receptor enables NF-kappaB suppression. *J Exp Med.* 2006;203(1):7–13. Available from: http://doi.org/10.1084/jem.20050466.

19. Jiang C-L, Liu L, Tasker JG. Why do we need nongenomic glucocorticoid mechanisms?. *Front Neuroendocrinol.* 2014;35(1):72–75. Available from: http://doi.org/10.1016/j.yfrne.2013.09.005.

20. John S, Sabo PJ, Thurman RE, et al. Chromatin accessibility predetermines glucocorticoid receptor binding patterns. *Nat Genet.* 2011;43(3):264–268. Available from: http://doi.org/10.1038/ng.759.

21. Karst H, Berger S, Erdmann G, Schütz G, Joëls M. Metaplasticity of amygdalar responses to the stress hormone corticosterone. *Proc Natl Acad Sci USA.* 2010;107(32):14449–14454. Available from: http://doi.org/10.1073/pnas.0914381107.

22. Kong X, Yu J, Bi J, et al. Glucocorticoids transcriptionally regulate miR-27b expression promoting body fat accumulation via suppressing the browning of white adipose tissue. *Diabetes.* 2015;64(2):393–404. Available from: http://doi.org/10.2337/db14-0395.

23. Kovács KJ, Földes A, Sawchenko PE. Glucocorticoid negative feedback selectively targets vasopressin transcription in parvocellular neurosecretory neurons. *J Neurosci.* 2000;20(10):3843–3852.

24. Kumar R, Thompson EB. Folding of the glucocorticoid receptor N-terminal transactivation function: dynamics and regulation. *Mol Cell Endocrinol.* 2012;348(2):450–456. Available from: http://doi.org/10.1016/j.mce.2011.03.024.

25. Lachize S, Apostolakis EM, van der Laan S, et al. Steroid receptor coactivator-1 is necessary for regulation of corticotropin-releasing hormone by chronic stress and glucocorticoids. *Proc Natl Acad Sci USA.* 2009;106(19):8038–8042. Available from: http://doi.org/10.1073/pnas.0812062106.

26. Lambert WM, Xu C-F, Neubert TA, Chao MV, Garabedian MJ, Jeanneteau FD. Brain-derived neurotrophic factor signaling rewrites the glucocorticoid transcriptome via glucocorticoid receptor phosphorylation. *Mol Cell Biol.* 2013;33(18):3700–3714. Available from: http://doi.org/10.1128/MCB.00150-13.

27. Le Billan F, Khan JA, Lamribet K, et al. Cistrome of the aldosterone-activated mineralocorticoid receptor in human renal cells. *FASEB J.* 2015;29(9):3977–3989. Available from: http://doi.org/10.1096/fj.15-274266.

28. Li Y, Suino K, Daugherty J, Xu HE. Structural and biochemical mechanisms for the specificity of hormone binding and coactivator assembly by mineralocorticoid receptor. *Mol Cell.* 2005;19(3):367–380. Available from: http://doi.org/10.1016/j.molcel.2005.06.026.

29. Lu NZ, Cidlowski JA. Translational regulatory mechanisms generate N-terminal glucocorticoid receptor isoforms with unique transcriptional target genes. *Mol Cell.* 2005;18(3):331–342. Available from: http://doi.org/10.1016/j.molcel.2005.03.025.

30. Luisi BF, Xu WX, Otwinowski Z, Freedman LP, Yamamoto KR, Sigler PB. Crystallographic analysis of the interaction of the glucocorticoid receptor with DNA. *Nature.* 1991;352(6335):497–505. Available from: http://doi.org/10.1038/352497a0.

31. Mahfouz A, Lelieveldt BPF, Grefhorst A, et al. Genome-wide co-expression of steroid receptors in the mouse brain: identifying signaling pathways and functionally coordinated region. *Proc Natl Acad Sci USA.* 2016;113(10):2738–2743.

32. Meijer OC, Williamson A, Dallman MF, Pearce D. Transcriptional repression of the 5-HT1A receptor promoter by corticosterone via mineralocorticoid receptors depends on the cellular context. *J Neuroendocrinol.* 2000;12(3):245–254.

33. Meijsing SH, Pufall MA, So AY-L, Bates DL, Chen L, Yamamoto KR. DNA binding site sequence directs glucocorticoid receptor structure and activity. *Science.* 2009;324(5925):407–410. Available from: http://doi.org/10.1126/science.1164265.

34. Moalli PA, Pillay S, Krett NL, Rosen ST. Alternatively spliced glucocorticoid receptor messenger RNAs in glucocorticoid-resistant human multiple myeloma cells. *Cancer Res.* 1993;53(17):3877–3879.

35. Ong KM, Blackford JA, Kagan BL, Simons SS, Chow CC. A theoretical framework for gene induction and experimental comparisons. *Proc Natl Acad Sci USA.* 2010;107(15):7107–7112. Available from: http://doi.org/10.1073/pnas.0911095107.

36. Paakinaho V, Kaikkonen S, Makkonen H, Benes V, Palvimo JJ. SUMOylation regulates the chromatin occupancy and anti-proliferative gene programs of glucocorticoid receptor. *Nucleic Acids Res.* 2014;42(3):1575–1592. Available from: http://doi.org/10.1093/nar/gkt1033.

37. Pascual-Le Tallec L, Lombès M. The mineralocorticoid receptor: a journey exploring its diversity and specificity of action. *Mol Endocrinol.* 2005;19(9):2211–2221. Available from: http://doi.org/10.1210/me.2005-0089.

38. Polman JAE, De Kloet E, Datson NA. Two populations of glucocorticoid receptor-binding sites in the male rat hippocampal genome. *Endocrinology.* 2013;154(5):1832–1844. Available from: http://doi.org/10.1210/en.2012-2187.

39. Pratt WB, Toft DO. Steroid receptor interactions with heat shock protein and immunophilin chaperones. *Endocr Rev*. 1997;18(3):306–360. Available from: http://doi.org/10.1210/edrv.18.3.0303.

40. Ray DW, Davis JR, White A, Clark AJ. Glucocorticoid receptor structure and function in glucocorticoid-resistant small cell lung carcinoma cells. *Cancer Res*. 1996;56(14):3276–3280.

41. Reichardt HM, Kaestner KH, Tuckermann JP, et al. DNA binding of the glucocorticoid receptor is not essential for survival. *Cell*. 1998;93(4):531–541.

42. Rivers C, Levy A, Hancock J, Lightman S, Norman M. Insertion of an amino acid in the DNA-binding domain of the glucocorticoid receptor as a result of alternative splicing. *J Clin Endocrinol Metab*. 1999;84(11):4283–4286. Available from: http://doi.org/10.1210/jcem.84.11.6235.

43. Sarabdjitsingh RA, Isenia S, Polman A, et al. Disrupted corticosterone pulsatile patterns attenuate responsiveness to glucocorticoid signaling in rat brain. *Endocrinology*. 2010;151(3):1177–1186. Available from: http://doi.org/10.1210/en.2009-1119.

44. Schaaf MJM, Cidlowski JA. Molecular mechanisms of glucocorticoid action and resistance. *J Steroid Biochem Mol Biol*. 2002;83(1–5):37–48.

45. Schaaf MJM, Champagne D, van Laanen IHC, et al. Discovery of a functional glucocorticoid receptor beta-isoform in zebrafish. *Endocrinology*. 2008;149(4):1591–1599. Available from: http://doi.org/10.1210/en.2007-1364.

46. Sinclair D, Webster MJ, Wong J, Weickert CS. Dynamic molecular and anatomical changes in the glucocorticoid receptor in human cortical development. *Mol Psychiatry*. 2011;16(5):504–515. Available from: http://doi.org/10.1038/mp.2010.28.

47. So AY-L, Bernal TU, Pillsbury ML, Yamamoto KR, Feldman BJ. Glucocorticoid regulation of the circadian clock modulates glucose homeostasis. *Proc Natl Acad Sci USA*. 2009;106(41):17582–17587. Available from: http://doi.org/10.1073/pnas.0909733106.

48. Stanisić V, Lonard DM, O'Malley BW. Modulation of steroid hormone receptor activity. *Prog Brain Res*. 2010;181:153–176. Available from: http://doi.org/10.1016/S0079-6123(08)81009-6.

49. Tirard M, Almeida OFX, Hutzler P, Melchior F, Michaelidis TM. Sumoylation and proteasomal activity determine the transactivation properties of the mineralocorticoid receptor. *Mol Cell Endocrinol*. 2007;268(1–2):20–29. Available from: http://doi.org/10.1016/j.mce.2007.01.010.

50. Turecki G, Meaney MJ. Effects of the social environment and stress on glucocorticoid receptor gene methylation: a systematic review. *Biol Psychiatry*. 2016;79(2):87–96. Available from: http://doi.org/10.1016/j.biopsych.2014.11.022.

51. Turner JD, Muller CP. Structure of the glucocorticoid receptor (NR3C1) gene 5' untranslated region: identification, and tissue distribution of multiple new human exon 1. *J Mol Endocrinol*. 2005;35(2):283–292. Available from: http://doi.org/10.1677/jme.1.01822.

52. Van den Heuvel JK, Boon MR, van Hengel I, et al. Identification of a selective glucocorticoid receptor modulator that prevents both diet-induced obesity and inflammation. *Br J Pharmacol*. 2016;173(11):1793–1804.

53. Vandevyver S, Dejager L, Libert C. Comprehensive overview of the structure and regulation of the glucocorticoid receptor. *Endocr Rev*. 2014;35(4):671–693. Available from: http://doi.org/10.1210/er.2014-1010.

54. Viengchareun S, Le Menuet D, Martinerie L, Munier M, Pascual-Le Tallec L, Lombès M. The mineralocorticoid receptor: insights into its molecular and (patho)physiological biology. *Nucl Recept Signal*. 2007;5:e012. Available from: http://doi.org/10.1621/nrs.05012.

55. Wallace AD, Cidlowski JA. Proteasome-mediated glucocorticoid receptor degradation restricts transcriptional signaling by glucocorticoids. *J Biol Chem*. 2001;276(46):42714–42721. Available from: http://doi.org/10.1074/jbc.M106033200.

56. Walther RF, Atlas E, Carrigan A, et al. A serine/threonine-rich motif is one of three nuclear localization signals that determine unidirectional transport of the mineralocorticoid receptor to the nucleus. *J Biol Chem*. 2005;280(17):17549–17561. Available from: http://doi.org/10.1074/jbc.M501548200.

57. Winnay JN, Xu J, O'Malley BW, Hammer GD. Steroid receptor coactivator-1-deficient mice exhibit altered hypothalamic-pituitary-adrenal axis function. *Endocrinology*. 2006;147(3):1322–1332. Available from: http://doi.org/10.1210/en.2005-0751.

58. Wyrwoll CS, Holmes MC, Seckl JR. 11β-hydroxysteroid dehydrogenases and the brain: from zero to hero, a decade of progress. *Front Neuroendocrinol*. 2011;32(3):265–286. Available from: http://doi.org/10.1016/j.yfrne.2010.12.001.

59. Zennaro MC, Keightley MC, Kotelevtsev Y, Conway GS, Soubrier F, Fuller PJ. Human mineralocorticoid receptor genomic structure and identification of expressed isoforms. *J Biol Chem*. 1995;270(36):21016–21020.

60. Zennaro MC, Souque A, Viengchareun S, Poisson E, Lombès M. A new human MR splice variant is a ligand-independent transactivator modulating corticosteroid action. *Mol Endocrinol*. 2001;15(9):1586–1598.

61. Zhou MY, Gomez-Sanchez CE, Gómez-Sánchez EP. An alternatively spliced rat mineralocorticoid receptor mRNA causing truncation of the steroid binding domain. *Mol Cell Endocrinol*. 2000;159(1–2):125–131.

30

Glucocorticoid Receptor: Genetics and Epigenetics in Veterans With PTSD

J.D. Flory, R. Yehuda

Icahn School of Medicine at Mount Sinai, New York, NY, United States

Abstract

Posttraumatic stress disorder (PTSD) is conditional on exposure to trauma. Despite this, there is strong evidence from twin studies and widespread agreement that genetic factors also contribute to risk for PTSD. Large-scale genomic studies of PTSD are planned. DNA can be modified by epigenetic processes via environmental inputs, thereby influencing how genes are expressed, making this an equally important area of research for PTSD. Results from a succession of studies of PTSD suggested that the negative-feedback system of the HPA axis is overly sensitive, implicating the glucocorticoid receptor (GR) in the pathophysiology of PTSD. The GR gene and the related gene, FKBP5, have emerged as significant actors in epigenetic research of PTSD. DNA methylation of GR is associated with the PTSD diagnosis and also predicts treatment response to psychotherapy in veterans. FKBP5 methylation interacts with early adversity to confer risk for PTSD. Finally, FKBP5 methylation is associated with symptom change following psychotherapy in veterans with PTSD.

The diagnosis of posttraumatic stress disorder (PTSD) was first established in 1980 (DSM-III)[3] in part, due to the fact that veterans of the Vietnam war were reporting persistent and distressing symptoms related to combat experiences that had occurred more than a decade earlier. PTSD was characterized as a disorder of re-experiencing the stressor (e.g., nightmares and intrusive recollections), numbing of emotional responsivity, and other symptoms thought to reflect the intensity and severity of the exposure (e.g., avoidance of cues and hypervigilance). The diagnosis was intended to reflect "universal effects" of extreme stress exposure, including combat, that persist after the stressor is no longer a threat. In addition, the symptoms were described as responses that would occur in "most people."

Epidemiological studies of PTSD began to appear in the research literature toward the end of the 1990s and challenged this assumption. These studies demonstrated that trauma exposure was widespread, with more than half of the US population reporting exposure to traumatic events; however, the prevalence of PTSD was relatively low <10%.[8,9,13] Studies of military personnel and war veterans suggested that the prevalence of PTSD was higher following combat trauma exposure relative to some civilian traumas (e.g., motor vehicle accidents). However, large numbers of veterans exposed to combat trauma do not report long-lasting effects as the prevalence appears to reach a maximal level of 30%.[10,17,20]

A second early assumption or expectation of the PTSD diagnosis was that it would be associated with activation of the hypothalamic–pituitary–adrenal (HPA) and sympathomedullary (SAM) pathways, reflecting the fight-or-flight or acute stress response. Yet the expected pattern of greater basal cortisol and catecholamines was not observed in that people with PTSD had higher catecholamines but lower basal cortisol compared to people without the PTSD diagnosis.[32] These observations lead to examination of individual differences and vulnerability factors that might explain the pathophysiology of PTSD with a focus on the HPA axis.[29] Briefly, when exposed to a stressor, the parvocellular neurons of the hypothalamus are stimulated to secrete corticotrophin-releasing hormone (CRH) and vasopressin (AVP) into the portal vessels system, followed by the synthesis and release of adrenocorticotropic hormone (ACTH) from the anterior pituitary gland. ACTH stimulates the adrenal cortex to synthesize and release glucocorticoids (GCs) into the periphery, including cortisol. Ultimately, the glucocorticoid receptor (GR) was implicated in the pathophysiology of PTSD, as inferred from results of the dexamethasone suppression test (DST), which showed that people with PTSD had an exaggerated cortisol suppression response to dexamethasone (DEX).[29] Similar results have been observed with in vitro measure of glucocorticoid sensitivity (IC50-$_{DEX}$).[29,31] In this assay, 50% inhibition of lysozyme activity is determined by incubating cultured peripheral blood mononuclear cells (PBMCs) with varying doses of DEX. In sum, results from a succession of studies (many in veterans) suggested that in PTSD, the negative-feedback system of the HPA axis is overly sensitive. This work provided a scientific rationale for the observations of lower circulating cortisol in people with PTSD and a context for examining the biological correlates of individual differences in response to trauma.

KEY POINTS

- The hypothalamic–pituitary–adrenal (HPA) axis is dysregulated in posttraumatic stress disorder (PTSD), reflecting hypersensitivity of the glucocorticoid receptor (GR).

- PTSD is conditional on exposure to a traumatic event, yet there is strong evidence that it is a heritable disorder. The search for genetic association with PTSD is ongoing.

- DNA function can also be modified by traumatic (and nontraumatic) exposure to environmental factors via epigenetic mechanisms such as DNA methylation, resulting in a change in gene expression.

- The GR and FKBP5 genes are implicated in pathophysiology of PTSD via epigenetic mechanisms.

PTSD AS A GENETIC DISORDER

Although the PTSD diagnosis is conditional on environmental exposure to trauma, compelling evidence from twin studies shows that genetic factors also contribute to the risk for PTSD as first demonstrated in a twin cohort of combat veterans.[11,24,25] One way to identify genes associated with PTSD risk or pathophysiology is to examine putative candidate genes that are related to known biological correlates of PTSD. However, it is often difficult to identify one or even a few genes that might be related to a complex behavior, neural circuit, or functional system. Some studies support this approach for genes that are related to the GR and HPA axis, including variation in the GR and CRH genes and in FK506-binding protein 5 (FKBP5),[5,7] a gene that regulates cortisol-binding affinity and nuclear translocation of GRs. Like many candidate gene association studies, the results have not been widely replicated for PTSD; for a review, see Ref. 1. In contrast, the genome-wide association study (GWAS) approach is a hypothesis generating method that allows for identification of a broad set of relevant biological processes that might otherwise not come into consideration. This comprehensive scan of the genome—usually single-nucleotide polymorphisms or SNPs—provides the opportunity to identify novel susceptibility factors for the development and persistence of PTSD and allows for identification of novel putative molecular mechanisms.

To date, only six GWAS studies of PTSD have been published,[2,4,12,19,22,28] including three samples of military personnel or veterans. No common loci or biological mechanism has been identified across the studies and the functional significance of the GWAS-identified loci has not been formulated. Moreover, the GWAS approach has not identified obvious GR-related genes. However, because of the large number of loci that are examined in a genome-wide analysis, large sample sizes are required and spurious findings can occur based on individual sampling characteristics (e.g., race, gender, type of trauma exposure). In 2013, the Psychiatric Genetics Consortium-PTSD workgroup was formed to bring together international investigators for the purpose of adequately powered GWAS studies of PTSD.[18] This collaboration is likely to yield new insights into the neurobiology of PTSD.

Candidate gene studies of PTSD have also demonstrated that genes can interact with the environmental factors to produce a certain response (i.e., a gene × environment interaction). For example, SNPs in the FKBP5 gene confer greater risk for PTSD in the presence of higher levels of childhood adversity.[7,27] However, what has emerged as a potentially more relevant concept for an environmental disorder such as PTSD is the idea that the environment modifies the way that particular genes function. Interestingly, trauma survivors often speak of being "transformed" by significant life events. The study of how the environments alter or influence gene expression may therefore provide a biological correlate of this clinical phenomenon.

PTSD AS AN EPIGENETIC DISORDER

Epigenetics refers to a transgenerationally transmissible functional change in the genome that can be due to environmental events and does not involve an alteration of sequence.[14] Epigenetic mechanisms include DNA methylation, hydroxymethylation, histone modification, and noncoding RNAs. DNA methylation—the covalent modification of DNA in which methyl groups are coupled to cytosine at CpG sites—is perhaps the best-studied epigenetic mechanism, due in part to its tractability to study.[16] Greater DNA methylation in specific gene regions (e.g., promoter) can be associated with lower transcriptional activity and, therefore, lower gene expression. Evidence from an animal model of early life events demonstrated that variation in maternal care regulates the methylation state of the GR exon 17 promoter in hippocampus, which in turn, regulates GR expression, the capacity for glucocorticoid negative feedback, and HPA axis responses to stress.[26]

As with association studies, investigations of epigenetic processes can be conducted by sampling DNA methylation (or other molecular mechanisms) across the genome or in selected candidate genes that are putatively related to the biology of PTSD (e.g., inflammatory processes, HPA axis). A recent review has summarized the epigenetic research literature with respect to PTSD and also includes a discussion of challenges regarding interpretation of epigenetic results. For example, epigenetic modifications are tissue-specific and results observed in whole blood or saliva samples may not replicate in the brain.[33]

With respect to candidate gene epigenetic research, the most commonly studied epigenetic process associated with HPA axis activity is cytosine methylation of the GR gene (NR3C1). GR gene expression is regulated by multiple promoter regions in NR3C1 with abundant glucocorticoid response elements. Results from a study of postmortem human brain tissue from victims of suicide showed that early life adversity was associated with DNA hypermethylation in the NR3C1 exon 1_F promoter region,[21] which is consistent with animal data cited above. We recently reported that male combat veterans with PTSD had significantly lower NR3C1 exon 1_F promoter methylation in PBMCs, compared to combat-exposed veterans who had never developed PTSD.[31] Moreover, NR3C1 exon 1_F promoter methylation was also significantly associated with three functional measures

of GC regulation that are associated with PTSD in combat veterans, including results from the DST, the in vitro measure of GR sensitivity described above ($IC_{50\text{-DEX}}$), and 24-h urinary cortisol excretion.

Additional mechanisms for the influence of childhood adversity on GR signaling and HPA axis function have been proposed. Notably, the FKBP5 protein regulates intracellular GR signaling by decreasing ligand binding and restricting GR translocation to the nucleus.[23] GR activation induces FKBP5 gene transcription, thus establishing an intracellular feedback loop that moderates GR sensitivity.[6] As noted above, genetic variants in FKBP5 interact with childhood adversity to predict the risk for developing PTSD.[7,27] Moreover, the DNA methylation state of selected CpGs across the FKBP5 gene is determined by an interaction between a sequence polymorphism and childhood adversity and also modulates sensitivity of FKBP5 to GR regulation.[15]

Although epigenetic modifications were initially believed to be permanent, subsequent work has demonstrated that DNA methylation is dynamic and potentially reversible. Research now demonstrates that PTSD also is not a permanent disability as symptoms can fluctuate, mediated by environmental circumstances, including treatment. This raises the possibility that some epigenetic processes are amenable to change over time and can be studied in the context of treatment. A proof-of-concept study was conducted in a small sample of veterans ($n = 16$) who participated in a psychotherapy clinical trial for PTSD to evaluate this possibility.[30] Although the treatment study was not designed to examine associations between epigenetic factors and symptom change, a subset of veterans consented to participate in this pilot arm of the study. Blood was sampled prior to the initiation of treatment, after 12 treatment sessions, and three months after the end of treatment for the purpose of examining NR3C1 exon 1_F and FKBP5 DNA methylation in PBMCs. Results showed that pretreatment NR3C1 exon 1_F promoter methylation was associated with treatment response (defined as no longer meeting diagnostic criteria for PTSD), but NR3C1 exon 1_F methylation did not differ between responders and nonresponders at the end of treatment or after the three-month follow-up period. In contrast, pretreatment DNA methylation of the FKBP5 promoter did not predict treatment response but declined in association with symptom recovery in veterans. These findings distinguish two epigenetic markers that may associate, respectively, with prognosis (GR gene methylation) and symptom severity (FKBP5 gene methylation).

A model for understanding the relationships observed in GR and FKBP5 methylation and their potential interactions in PTSD is presented in Ref. 30. In this model, early life adversity influences both GR and FKBP5 DNA methylation. In the context of PTSD, GR sensitivity is increased, likely resulting from reduced GR promoter methylation, which would ultimately result in lowered circulating cortisol levels and, therefore, low glucocorticoid signaling. The low cortisol levels would serve to further decrease FKBP5 gene expression through an intracellular loop mediated by glucocorticoid response elements in the FKBP5 gene. Lower FKBP5 gene expression could serve to sustain an increased GR sensitivity. A decline in FKBP5 promoter methylation, such as occurred in treatment responders in our proof-of-concept study, might allow for an increase in FKBP5 gene expression, which would, in turn, ultimately decrease GR sensitivity. Thus, we found that treatment responders showed decreased FKBP5 promoter methylation, suggestive of increased FKBP5 gene expression. Similarly, higher levels of GR promoter methylation, suggestive of lower GR expression, were also associated with a positive response to treatment. The mechanisms by which dynamic changes in GR sensitivity associate with changes in PTSD symptoms remains to be fully elucidated; however, these results suggest that the epigenetic mechanisms that regulate glucocorticoid signaling also associate with treatment outcome. A PTSD treatment study in veterans with a priori hypotheses regarding the association of treatment response to GR and FKBP5 DNA methylation (and gene expression) is in progress.

CONCLUSION

Genetic background contributes to whether one is exposed to trauma and develops PTSD and also appears to interact with environmental exposures. However, DNA function can also be modified by traumatic (and nontraumatic) exposure to environmental factors via epigenetic mechanisms such as DNA methylation, resulting in a change in gene expression. Much of the work that has demonstrated these effects has focused on the GR and FKBP5 genes and has occurred in the context of study of PTSD in veterans. There are still many unanswered questions regarding the extent to which genes and environmental contexts influence one another but the availability of new molecular techniques ensures a bright future for such work.

References

1. Almli LM, Fani N, Smith AK, Ressler KJ. Genetic approaches to understanding post-traumatic stress disorder. *Int J Neuropsychopharmacol*. 2014;17(2):355–370. http://dx.doi.org/10.1017/s1461145713001090.
2. Almli LM, Stevens JS, Smith AK, et al. A genome-wide identified risk variant for PTSD is a methylation quantitative trait locus and confers decreased cortical activation to fearful faces. *Am J Med Genet B Neuropsychiatr Genet*. 2015;168b(5):327–336. http://dx.doi.org/10.1002/ajmg.b.32315.
3. APA. *Diagnostic and Statistical Manual of Mental Disorders*. 3rd ed. Washington DC: American Psychiatric Association; 1980.
4. Ashley-Koch AE, Garrett ME, Gibson J, Liu Y, Dennis MF, Kimbrel NA, et al. Genome-wide association study of posttraumatic stress disorder in a cohort of Iraq–Afghanistan era veterans. *J Affect Disord*. 2015;184:225–234. http://dx.doi.org/10.1016/j.jad.2015.03.049.
5. Bachmann AW, Sedgley TL, Jackson RV, Gibson JN, Young RM, Torpy DJ. Glucocorticoid receptor polymorphisms and post-traumatic stress disorder. *Psychoneuroendocrinology*. 2005;30(3):297–306. http://dx.doi.org/10.1016/j.psyneuen.2004.08.006.
6. Binder EB. The role of FKBP5, a co-chaperone of the glucocorticoid receptor in the pathogenesis and therapy of affective and anxiety disorders. *Psychoneuroendocrinology*. 2009;34(Suppl 1):S186–S195. http://dx.doi.org/10.1016/j.psyneuen.2009.05.021.
7. Binder EB, Bradley RG, Liu W, Epstein MP, Deveau TC, Mercer KB, et al. Association of FKBP5 polymorphisms and childhood abuse with risk of posttraumatic stress disorder symptoms in adults. *JAMA*. 2008;299(11):1291–1305. http://dx.doi.org/10.1001/jama.299.11.1291.
8. Breslau N, Davis GC, Andreski P, Peterson E. Traumatic events and posttraumatic stress disorder in an urban population of young adults. *Arch Gen Psychiatry*. 1991;48(3):216–222.
9. Breslau N, Kessler RC, Chilcoat HD, Schultz LR, Davis GC, Andreski P. Trauma and posttraumatic stress disorder in the community: the 1996 Detroit area Survey of trauma. *Arch Gen Psychiatry*. 1998;55(7):626–632.
10. Dohrenwend BP, Turner JB, Turse NA, Adams BG, Koenen KC, Marshall R. The psychological risks of Vietnam for US veterans: a revisit with new data and methods. *Science*. 2006;313(5789):979–982. http://dx.doi.org/10.1126/science.1128944.
11. Goldberg J, True WR, Eisen SA, Henderson WG. A twin study of the effects of the Vietnam War on posttraumatic stress disorder. *JAMA*. 1990;263(9):1227–1232.
12. Guffanti G, Galea S, Yan L, Roberts AL, Solovieff N, Aiello AE, et al. Genome-wide association study implicates a novel RNA gene, the lincRNA AC068718.1, as a risk factor for post-traumatic stress disorder in women. *Psychoneuroendocrinology*. 2013;38(12):3029–3038. http://dx.doi.org/10.1016/j.psyneuen.2013.08.014.
13. Kessler RC, Sonnega A, Bromet E, Hughes M, Nelson CB. Posttraumatic stress disorder in the national comorbidity survey. *Arch Gen Psychiatry*. 1995;52(12):1048–1060.
14. Klengel T, Binder EB. Epigenetics of stress-related psychiatric disorders and gene × environment interactions. *Neuron*. 2015;86(6):1343–1357. http://dx.doi.org/10.1016/j.neuron.2015.05.036.
15. Klengel T, Mehta D, Anacker C, Rex-Haffner M, Pruessner JC, Pariante CM, et al. Allele-specific FKBP5 DNA demethylation mediates gene-childhood trauma interactions. *Nat Neurosci*. 2013;16(1):33–41. http://dx.doi.org/10.1038/nn.3275.
16. Klengel T, Pape J, Binder EB, Mehta D. The role of DNA methylation in stress-related psychiatric disorders. *Neuropharmacology*. 2014;80:115–132. http://dx.doi.org/10.1016/j.neuropharm.2014.01.013.
17. Kok BC, Herrell RK, Thomas JL, Hoge CW. Posttraumatic stress disorder associated with combat service in Iraq or Afghanistan: reconciling prevalence differences between studies. *J Nerv Ment Dis*. 2012;200(5):444–450. http://dx.doi.org/10.1097/NMD.0b013e3182532312.
18. Logue MW, Amstadter AB, Baker DG, Duncan L, Koenen KC, Liberzon I. The psychiatric genomics consortium posttraumatic stress disorder workgroup: posttraumatic stress disorder enters the age of large-scale genomic collaboration. *Neuropsychopharmacology*. 2015;40(10):2287–2297. http://dx.doi.org/10.1038/npp.2015.118.
19. Logue MW, Baldwin C, Guffanti G, Melista E, Wolf EJ, Reardon AF, et al. A genome-wide association study of post-traumatic stress disorder identifies the retinoid-related orphan receptor alpha (RORA) gene as a significant risk locus. *Mol Psychiatry*. 2013;18(8):937–942. http://dx.doi.org/10.1038/mp.2012.113.

20. Marmar CR, Schlenger W, Henn-Haase C, Qian M, Purchia E, Li M, et al. Course of posttraumatic stress disorder 40 years after the Vietnam war: findings from the national Vietnam veterans longitudinal study. *JAMA Psychiatry*. 2015;72(9):875–881. http://dx.doi.org/10.1001/jamapsychiatry.2015.0803.

21. McGowan PO, Sasaki A, D'Alessio AC, Dymov S, Labonte B, Szyf M, et al. Epigenetic regulation of the glucocorticoid receptor in human brain associates with childhood abuse. *Nat Neurosci*. 2009;12(3):342–348. http://dx.doi.org/10.1038/nn.2270.

22. Nievergelt CM, Maihofer AX, Mustapic M, Yurgil KA, Schork NJ, Miller MW, et al. Genomic predictors of combat stress vulnerability and resilience in US Marines: a genome-wide association study across multiple ancestries implicates PRTFDC1 as a potential PTSD gene. *Psychoneuroendocrinology*. 2015;51:459–471. http://dx.doi.org/10.1016/j.psyneuen.2014.10.017.

23. Stechschulte LA, Sanchez ER. FKBP51-a selective modulator of glucocorticoid and androgen sensitivity. *Curr Opin Pharmacol*. 2011;11(4):332–337. http://dx.doi.org/10.1016/j.coph.2011.04.012.

24. Stein MB, Jang KL, Taylor S, Vernon PA, Livesley WJ. Genetic and environmental influences on trauma exposure and posttraumatic stress disorder symptoms: a twin study. *Am J Psychiatry*. 2002;159(10):1675–1681. http://dx.doi.org/10.1176/appi.ajp.159.10.1675.

25. True WR, Rice J, Eisen SA, et al. A twin study of genetic and environmental contributions to liability for posttraumatic stress symptoms. *Arch Gen Psychiatry*. 1993;50(4):257–264.

26. Turecki G, Meaney MJ. Effects of the social environment and stress on glucocorticoid receptor gene methylation: a systematic review. *Biol Psychiatry*. 2016;79(2):87–96. http://dx.doi.org/10.1016/j.biopsych.2014.11.022.

27. Xie P, Kranzler HR, Poling J, et al. Interaction of FKBP5 with childhood adversity on risk for post-traumatic stress disorder. *Neuropsychopharmacology*. 2010;35(8):1684–1692. http://dx.doi.org/10.1038/npp.2010.37.

28. Xie P, Kranzler HR, Yang C, Zhao H, Farrer LA, Gelernter J. Genome-wide association study identifies new susceptibility loci for posttraumatic stress disorder. *Biol Psychiatry*. 2013;74(9):656–663. http://dx.doi.org/10.1016/j.biopsych.2013.04.013.

29. Yehuda R. Post-traumatic stress disorder. *N Engl J Med*. 2002;346(2):108–114. http://dx.doi.org/10.1056/NEJMra012941.

30. Yehuda R, Daskalakis NP, Desarnaud F, Makotkine I, Lehrner AL, Koch E, et al. Epigenetic biomarkers as predictors and correlates of symptom improvement following psychotherapy in combat veterans with PTSD. *Front Psychiatry*. 2013;4:118. http://dx.doi.org/10.3389/fpsyt.2013.00118.

31. Yehuda R, Flory JD, Bierer LM, Henn-Haase C, Lehrner A, Desarnaud F, et al. Lower methylation of glucocorticoid receptor gene promoter 1F in peripheral blood of veterans with posttraumatic stress disorder. *Biol Psychiatry*. 2015;77(4):356–364. http://dx.doi.org/10.1016/j.biopsych.2014.02.006.

32. Yehuda R, McFarlane AC. Conflict between current knowledge about posttraumatic stress disorder and its original conceptual basis. *Am J Psychiatry*. 1995;152(12):1705–1713. http://dx.doi.org/10.1176/ajp.152.12.1705.

33. Zannas AS, Provencal N, Binder EB. Epigenetics of posttraumatic stress disorder: current evidence, challenges, and future directions. *Biol Psychiatry*. 2015;78(5):327–335. http://dx.doi.org/10.1016/j.biopsych.2015.04.003.

31

Stress Effects on Learning and Memory in Humans

O.T. Wolf

Ruhr University Bochum, Bochum, Germany

Abstract

Stress leads to increased activity of the sympathetic nervous system (SNS) and of the hypothalamic-pituitary-adrenal (HPA) axis. Stimulation of the latter induces the release of glucocorticoids (GCs) from the adrenal cortex. The multiple effects of GCs in the brain are the focus of the current chapter. Acute stress effects comprise the enhancement of long-term memory consolidation and the simultaneous impairment of memory retrieval. Moreover, a qualitative shift away from a cognitive response style toward a more stimulus-driven habitual response style occurs. The acute response to GCs is altered in major depressive disorder and posttraumatic stress disorder. Chronic stress causes structural alterations in the hippocampus and the medial prefrontal cortex in the form of dendritic atrophy. However, remaining plasticity and thus the potential to reverse these changes appears to be more common than previously thought. The advanced understanding of the central nervous system effects of GCs will ultimately lead to progress in the treatment of mental and systemic diseases characterized by HPA axis dysregulation.

INTRODUCTION

Stress has a bad reputation in present-day societies. It is typically associated with physical and mental health problems. However, research over the past decades has illustrated that the impact of stress on brain functions such as learning and memory are far more complex than initially assumed. Stress may enhance or impair memory depending on several key modulators and mediators. Both quantitative and qualitative shifts take place.

Everyone has experienced episodes in which stress influenced their memory. There are situations in which we are unable to retrieve previously well-learned information, an example of how stress might interfere with our memory. In contrast, we tend to remember specific embarrassing, shameful, or frightening events from the past very well. This is an example of how stress can enhance our memory. Finally, there are conditions such as chronic stress or stress-associated mental disorders which are characterized by specific memory dysfunctions. The goal of the present chapter is to provide a brief and focused overview of current knowledge on the role of glucocorticoids (GCs) in mediating these stress effects. Advances in the field of psychoneuroendocrinology within the past decades have contributed to a more differentiated and balanced view of the effects of stress hormones on memory in animals and humans.

KEY POINTS

- Acute stress enhances long-term memory consolidation but impairs its retrieval
- Stress causes a shift to a more habitual—stimulus-driven—response style
- The impact of glucocorticoids on memory is altered in mental disorders
- Early life stress programs a vulnerable phenotype via epigenetic mechanisms
- Chronic stress induces dendritic atrophy in the hippocampus and the medial PFC
- In contrast, chronic stress leads to hypertrophy in the amygdala and the striatum
- Reinstating appropriate HPA signaling appears to be a promising therapeutic target in mental disorders associated with altered HPA axis activity.

DEFINITION OF STRESS

A common definition is that stress occurs when a person perceives a challenge to their internal or external balance (homeostasis[5]). Thus, a discrepancy between what "should be" and "what is" induces stress. A stressor can be physical (e.g., heat, thirst) or psychological (e.g., work overload, unemployment, mobbing, marital problems), as well as acute or chronic.[19] The subjective evaluation of the stressor and of available coping resources determines its impact on the individual.[16] Something perceived as a threat by one person might be perceived as an exciting opportunity by another. There is substantial interindividual variability in the response to stress. For humans as social animals, a threat to the social self (social evaluative threat), in combination with uncontrollability of the situation, is especially potent in triggering a stress response.[10]

THE TWO STRESS SYSTEMS: HPA AND SNS

Stress leads to neuroendocrine responses aimed at facilitating adaptation. In this context, the hypothalamic-pituitary-adrenal (HPA) axis and the sympathetic nervous system (SNS) play important roles. SNS activity leads to the rapid release of (nor) epinephrine from the adrenal medulla. This constitutes the first rapid response wave. Increased activity of the HPA axis leads to the release of GCs (cortisol in humans, corticosterone in most laboratory rodents) from the adrenal cortex. This response is slower and constitutes the second response wave.[5]

GCs are lipophilic hormones that can enter the brain, where they influence regions involved in cognitive functions (e.g., amygdala, hippocampus, prefrontal cortex (PFC), striatum). These effects are mediated by the two receptors for the hormone: the mineralocorticoid receptor (MR) and the glucocorticoid receptor (GR). They differ in their affinity for GCs and in their localization. While MR activation leads to enhanced neuronal excitability, GR activation causes a delayed suppression or normalization of the neuronal network.[14] In addition, GCs can exert rapid nongenomic effects which, in part, are mediated by recently described membrane-bound MRs[14] and GRs.[26] Thus GCs have time-dependent effects comprising of early rapid nongenomic effects and later occurring slower genomic effects.

After acute stress, the HPA axis' negative feedback system leads to GC levels returning to baseline values within hours.[5,10] In periods of chronic stress, persistent alterations of the HPA axis can occur, leading to continuously elevated cortisol concentrations. However, elevated cortisol levels as typically observed in major depressive disorder (MDD) are not always the consequence of chronic stress.[34] For example, reduced cortisol levels occur in several stress-associated somatoform disorders[11] as well as in posttraumatic stress disorder (PTSD).[34,40]

STRESS AND COGNITION: ACUTE EFFECTS

Stress affects the central processing of incoming information at multiple levels. Early influences on perception, attention, and working memory have been documented, as well as later effects on long-term memory. Initially, stress causes an increase in vigilance which goes along with a reduction in top-down cognitive control processes.[13] Working memory and executive functions are typically impaired during and shortly after stress exposure.[1] The present chapter will focus on the influence of stress on learning and memory. This topic has received a lot of attention during the past years, and our knowledge concerning it has consequently improved substantially.

Long-term memory can be subdivided into declarative or explicit and nondeclarative or procedural (implicit) memory. Based on its content, declarative memory can be further subdivided into episodic memory (recall of a

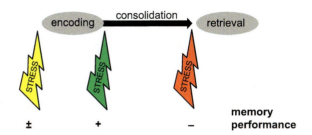

FIGURE 31.1   Phase-dependent effects of stress on long-term memory. Immediate postacquisition stress enhances memory consolidation, as indicated by the *green arrow* and the *plus sign*. This causes enhanced memory retrieval hours, days, or weeks later. In contrast, stress shortly before memory retrieval impairs long-term memory by temporarily blocking the accessibility of the memory trace. This is indicated by the *red arrow* and the *minus sign*. Preencoding stress has variable effects on long-term memory (thus *yellow* and +/−) which are mediated by the exact timing of the stressor, the emotionality of the learning material and the delay between encoding and retrieval.

specific event which can be located in space and time) and semantic memory (our knowledge of the world). The medial temporal lobe is critical for declarative memory, with the hippocampus being especially important for episodic memory.[21]

Long-term memory can also be subdivided into different memory phases, namely acquisition (or initial encoding), consolidation (or storage), and retrieval (or recall). During the 1990s, the literature regarding the effects of stress on episodic memory in humans was somewhat confusing, with groups reporting both enhancing as well as impairing effects of GCs on long-term memory. However, it has become evident that this is largely due to the fact that the different memory phases outlined above are modulated by GCs in an opposing fashion.[27,35]

GCs enhance memory consolidation; this process representing the adaptive and beneficial side of the action of GCs in the central nervous system (see Fig. 31.1). It has been conceptualized as the beneficial effects of "stress within the learning context" or "intrinsic stress." This wording emphasizes the fact that a stressful episode is remembered better than a nonstressful one, an effect which is mediated by the action of stress-released GCs on the hippocampal formation and which has been very well documented in rodents. Studies have shown that an adrenergic activation in the basolateral amygdala (BLA) is a prerequisite for the modulating effects of GCs on other brain regions (e.g., the hippocampus). Lesions in the BLA as well as beta blockade abolish the enhancing effects of postencoding GC administration.[27]

Comparable effects have been observed in humans: immediate postacquisition stress has repeatedly been linked to boosted memory consolidation. Converging evidence comes from pharmacological experiments. Cortisol administration shortly before memory encoding caused enhanced long-term memories especially for emotional learning material. Moreover, neuroimaging studies have characterized a stress-induced modulation

of the amygdala and hippocampal activity as the neural correlate of these effects.[35]

Preencoding stress or cortisol studies have led to a less consistent picture. Reports of enhancing as well as impairing effects can be found. In this case, the exact timing of the stressor, the emotionality of the learning material,[24] the relation of the learning material to the stressor, and the delay between encoding and retrieval[36] appear to be important modulatory factors.[35]

While an enhanced memory consolidation is adaptive and beneficial, this process appears to occur at the cost of impaired memory retrieval (see Fig. 31.1). Using a 24-h delay interval, researchers were able to show that stress or GC treatment shortly before retrieval testing impairs memory retrieval in rats in a water maze.[8] Further studies have revealed that, once again, an intact BLA as well as its adrenergic activation appear to be necessary for the occurrence of this negative GC effect.[27] Roozendaal has summarized these findings as indicative of stress putting the brain into a consolidation mode accompanied by impaired retrieval. Such a reduction in retrieval might facilitate consolidation by reducing interference. In humans, multiple studies have been able to demonstrate a stress-induced retrieval impairment using different stressors and different memory paradigms. Similar impairments occur after pharmacological administrations of GCs.[35] These behavioral effects are accompanied by reduced neural activity in the hippocampus, as demonstrated using functional magnetic resonance imaging.[23]

Of note, the beneficial effects on consolidation as well as the impairing effects on retrieval are more pronounced for emotionally arousing learning material. This observation fits the mentioned observation in rodents that GCs can only exert effects on memory in the presence of adrenergic activity in the amygdala. This arousal can apparently result from specifics of the learning material and/or specifics of the testing conditions.[15,27] In sum, studies in animals and humans converge on the idea that GCs acutely enhance long-term memory consolidation while impairing memory retrieval (see Fig. 31.1).

More recently, evidence has accumulated that stress may also influence extinction retrieval in classical (fear) conditioning paradigms.[18] The fear memory trace initially created during acquisition is not erased during extinction. Rather, extinction leads to a second inhibitory memory trace which is dependent on prefrontal brain regions. This inhibitory trace is considered to be context dependent and state dependent, as illustrated by several recovery phenomena (e.g., renewal, reinstatement, spontaneous recovery; see Ref. 25). Stress, via its impact on the amygdala and PFC, might impair extinction retrieval,[12] thus leading to the return of fear.[18] This hypothesis needs further evaluation but has substantial clinical relevance as it could explain the clinical observation that stress in patients is often associated with the reoccurrence of symptoms.

Cortisol as an Adjuvant Treatment for Psychotherapy?

De Quervain and colleagues have tested the hypothesis that the impairing impact of cortisol on emotional memory retrieval in combination with the enhancing effects of cortisol on memory consolidation might be beneficial in the context of the psychotherapeutic treatment of anxiety disorders. When patients are confronted with the fear-inducing stimulus during therapy (e.g., with a spider or with heights), cortisol may block the retrieval of previous fearful memories. Moreover, cortisol could enhance the storage of the new extinction memory trace.[7] Support for this model comes from several studies with different patient populations. Repeatedly, they were able to demonstrate that cortisol administration before exposure therapy sessions enhanced the effectiveness of the intervention compared to a placebo control condition.[6] These findings open up a new avenue of research aimed at supporting or boosting the success of psychotherapeutic treatments by using endogenous hormones.

Altered Acute Effects of GCs in Mental Disorders?

Several mental disorders, including MDD, PTSD, and borderline personality disorder (BPD), are characterized by dysregulation of the HPA axis and memory dysfunction. These pathological hallmarks appear to be associated with each other. However, as of today, only a few studies have investigated the acute effects of cortisol (or stress) in these disorders.

Patients with MDD often show HPA axis hyperactivity with elevated corticotropin-releasing hormone (CRH) levels, elevated cortisol levels, and impaired negative HPA axis feedback. An enhanced central CRH drive and/or impaired GR functioning might contribute to these neuroendocrine abnormalities.[34] In a series of studies, Wingenfeld and colleagues could reveal that MDD patients, in contrast to healthy controls, do not exhibit an impairing effect of cortisol on memory retrieval (see Fig. 31.2). This might be behavioral evidence for a reduced functioning of central GRs.[33]

Patients with PTSD, on the other hand, are characterized by different HPA-related findings. Here, enhanced negative feedback and reduced basal cortisol levels are typically found,[40] even though the empirical picture is somewhat heterogeneous. This might, in part, be due to different comorbidities and different vulnerability patterns. Strikingly, PTSD patients showed an enhanced rather than impaired memory retrieval after cortisol administration in several studies[33] (see Fig. 31.2). Similar findings have been obtained in patients with BPD. The interpretation of these findings is more challenging. One might suggest that the beneficial effects

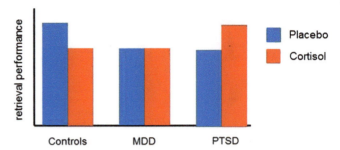

FIGURE 31.2 Effects of cortisol on memory retrieval. In healthy participants (left), cortisol impairs memory retrieval. In patients with MDD (middle), cortisol has no effect on memory retrieval. In patients with PTSD (right), cortisol enhances memory retrieval.

of acute cortisol elevations on memory processes mediated by the hippocampus could be due to enhanced GR functioning. However, findings on GR sensitivity in PTSD are inconclusive, and comparable studies on BPD patients are missing completely. Furthermore, the role of the MR as well as the MR/GR balance in this context is not well understood. Alternatively, memory improvement after cortisol administration could be interpreted in the context of inhibition of central corticotropin-releasing factor (CRF) release through cortisol administration.[33]

The question arises as to how this research is to transfer into treatment strategies. In PTSD patients, for example, there is initial evidence that GC administration might reduce the involuntary retrieval of aversive memories (flashbacks).[7] Moreover, cortisol administration in an emergency unit setting is apparently able to prevent the development of PTSD.[28] While these findings are promising, more research is needed to draw conclusions about the effectiveness, the underlying mechanisms, and long-term safety.

Stress and Multiple Memory Systems

In addition to its effect on memory quantity, stress also influences the quality of the memories formed.[31] For example, spatial memory tasks can often be solved using either a cognitive map strategy dependent on the hippocampus or using a stimulus–response strategy dependent on the striatum. Stress influences the participation of multiple memory systems in the solution of a given memory task by reducing the contribution of hippocampus-based declarative memories. At the same time, striatum-based, implicit stimulus–response learning is unaffected or potentially even boosted.[31] Interestingly, this shift appears to be mediated by the MR and is often able to rescue performance in the face of stress. A similar shift has been characterized during instrumental learning. Here, stress causes an increase in habitual behavior at the expense of PFC-mediated goal-directed behavior. As such, these shifts are typically adaptive since they may rescue performance at times when cognitive resources are compromised.[31] However, in vulnerable individuals, these stress-induced shifts might promote the development of mental disorders such as addiction and obsessive compulsive disorder.[30]

STRESS AND COGNITION: CHRONIC EFFECTS

The following paragraphs will focus on the impact of chronic stress on cognition. First, the long-term consequences of early life stress will be summarized. These changes have an impact throughout the lifespan leading up to old age. Next, the impact of chronic stress on memory in adulthood is reviewed.

Early Life Stress and Its Long-Term Consequences

There is mounting evidence to support the notion that early stress exposure is associated with accelerated neurodegenerative processes and early onset of memory decline in the course of aging.[17] Changes in stress susceptibility programmed early on in life might partially account for such deficits.[29]

Prenatal and postnatal stress exposure appears to be associated with a chronically increased reactivity of the HPA axis mediated by a reduced expression of central GRs.[20] Animal models show increased corticosterone concentrations and lower GR density in the hippocampus in the offspring of stressed mothers. Also, postnatal maternal separation and poorer maternal care have been linked to reduced GR gene expression in the hippocampus, which, in turn, is associated with reduced feedback sensitivity of the HPA axis. Recently, an epigenetic mechanism has been discovered in rodents that explains how environmental stimuli can impact gene expression. Variation in maternal care during the first week of life was associated with differences in GR gene promoter DNA methylation, leading to stable changes in GR gene expression.[41] Accumulating evidence suggests that the human GR gene is also influenced by early life programming.[32] Apparently, individuals with an increased stress susceptibility (reflecting genetic susceptibilities and/or early adversity) are especially vulnerable to stress-induced cognitive impairments later on in their adult life.[17]

Cognitive Effects of Chronic Stress During Adulthood

Animal research has provided important insights into the structural alterations caused by chronic stress in the brain. One main finding is that the integrity of the hippocampus and the medial PFC is compromised, while, in

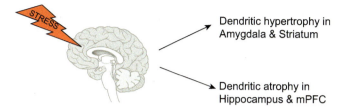

FIGURE 31.3 Effects of chronic stress on dendritic morphology. Three weeks of chronic stress causes dendritic atrophy in the hippocampus and medial PFC in rodents. In contrast, the amygdala and parts of the striatum become hypertrophic.

parallel, the amygdala (the "fear centre" of the brain) and parts of the striatum (the "habit centre" of the brain) become hyperactive.[27] In the hippocampus, chronic stress leads to a retraction of dendrites (dendritic atrophy), and similar effects occur in the medial PFC.[17] This atrophy appears to be reversible after stress termination, illustrating preserved room for neuroplasticity. In addition, stress causes reduced neurogenesis in the dentate gyrus and the mPFC. Even though the function of these new-born neurons is discussed controversially, a contribution to memory formation appears likely.[39] At the behavioral level, impaired performance in hippocampus-dependent memory tasks and PFC-dependent tasks (e.g., working memory, goal-directed actions, set-shifting capabilities) can be observed.[27]

In contrast to the hippocampus and the PFC, the amygdala becomes hypertrophic in conditions of chronic stress. Increases in dendritic arborization and spine density take place.[27] Similar effects (dendritic hypertrophy) have been observed in the striatum.[9] These alterations are summarized in Fig. 31.3. Moreover, CRF system activity in the amygdala, which is involved in anxiety, is enhanced. Chronically stressed animals show enhanced fear conditioning and are characterized by a more habitual and less goal-directed response style.[9] Thus, the balance between brain regions involved in cognition is altered by chronic stress.[17] While "analytic" cognitive functions mediated by the hippocampus and PFC are impaired, "affective" fear-related amygdala functioning and habit-related striatal functioning are enhanced.[34]

In humans, exposure to chronic stress (e.g., shift workers, airplane personnel, soldiers) is associated with deficits in cognition such as working memory and declarative memory.[17,34] These deficits can, in part, be explained by GC overexposure resulting from chronic stress. For example, several studies observed cognitive impairments after the administration of GCs over a period of several days. Further evidence comes from studies with patients receiving GC therapy for the treatment of autoimmune diseases. Whether the negative effects on memory reflect acute or chronic effects is sometimes hard to disentangle, and at least one study has shown a rapid reversal of the deficits after discontinuation of the GC treatment.[2] Data from patients with Cushing disease point in the same direction, with cognitive impairments and hippocampal volume reductions reported.[34] Hippocampal atrophy might be reversible once successful treatment has occurred. This would be in line with the remaining plasticity of this structure observed in rodent studies.[34]

INTERVENTION STRATEGIES

In laboratory animals, stress-induced dendritic atrophy in the hippocampus and PFC as well as reduced neurogenesis in the hippocampus can be prevented with antidepressants and anticonvulsants. Also, treatment with a GR antagonist is effective in preventing such stress-induced changes in neurophysiology. Similarly, memory impairments can be prevented with some of these drugs.[34]

In humans, chronic stress without an associated psychopathology could be alleviated by psychological stress intervention strategies. Possible examples are stress inoculation training or mindfulness-based stress reduction training. Initial evidence suggests that these psychological interventions can influence stress responsivity and may even lead to structural alterations in the human brain.[3]

Pharmacological treatment with beta blockers can prevent the effects of acute GC elevations on memory retrieval. It remains to be shown whether similar approaches are effective in conditions of chronic stress. In addition, GR antagonists and/or CRF antagonists might be candidate drugs. Moreover, drugs that influence the local GC metabolism in the brain could also be effective. For example, a pharmacological reduction of active GC concentrations in the hippocampus (inhibition of 11beta-HSD-1 synthesis) is efficient in preventing memory impairments in aging mice. In humans, a pilot study demonstrated that the 11beta-HSD-1 inhibitor carbenoxolone improved memory in older men and in patients with type 2 diabetes.[38]

In MDD, pharmacological agents that normalize HPA axis activity are being tested, with CRF antagonists and GR antagonists being of particularly great interest. In addition to pharmacological approaches, there is initial evidence that HPA axis dysfunction in MDD patients can be altered with the help of psychotherapy. This is of particular importance for patients who are known to respond less to pharmacotherapy, i.e., MDD patients with a history of early trauma.[22]

In sum, reinstating appropriate HPA signaling appears to be a promising approach both in chronically stressed animals and in human patients suffering from stress-related mental disorders.[4]

OUTLOOK

Research over the past decade has substantially helped to better understand the effects of GCs on memory. A more differentiated picture of stress effects on memory has evolved: oversimplifications in the form of statements such as "stress impairs memory" are no longer supported by the existing literature. An acute stress-induced GC increase enhances memory consolidation but impairs memory retrieval. These impairing effects of stress-induced GC secretion might prevent us from performing well during an exam and could also influence eyewitness testimonies. In addition, stress alters the quality of the memories formed.

Substantial sex differences have been observed in several animal studies, but evidence in humans is still sparse.[37] The direction of the effect appears to be task specific. More knowledge about sex differences should, in the long run, help understand sex differences in stress-associated disorders. Future studies need to investigate whether these sex differences are related to differences in the neuroendocrine stress response or to differences in the brain's response to endocrine stress messengers.

Chronic stress has mostly negative effects on both the brain and the body.[19] These observations are relevant to mental disorders as well as to the aging process. However, it is encouraging that research has repeatedly observed evidence for preserved plasticity and structural remodeling once the stress has ceased or GCs are back to normal levels. Therefore, successful interventions will not only be able to stop the aggravation of symptoms but also should often be able to reverse the underlying pathological alterations.

References

1. Arnsten AF. Stress signalling pathways that impair prefrontal cortex structure and function. *Nat Rev Neurosci.* 2009;10:410–422.
2. Coluccia D, Wolf OT, Kollias S, Roozendaal B, Forster A, de Quervain DJ. Glucocorticoid therapy-induced memory deficits: acute versus chronic effects. *J Neurosci.* 2008;28:3474–3478.
3. Davidson RJ, McEwen BS. Social influences on neuroplasticity: stress and interventions to promote well-being. *Nat Neurosci.* 2012;15:689–695.
4. De Kloet ER, Derijk RH, Meijer OC. Therapy Insight: is there an imbalanced response of mineralocorticoid and glucocorticoid receptors in depression? *Nat Clin Pract Endocrinol Metab.* 2007;3:168–179.
5. De Kloet ER, Joels M, Holsboer F. Stress and the brain: from adaptation to disease. *Nat Rev Neurosci.* 2005;6:463–475.
6. de Quervain DJ, Bentz D, Michael T, et al. Glucocorticoids enhance extinction-based psychotherapy. *Proc Natl Acad Sci USA.* 2011;108:6621–6625.
7. de Quervain DJ, Margraf J. Glucocorticoids for the treatment of post-traumatic stress disorder and phobias: a novel therapeutic approach. *Eur J Pharmacol.* 2008;583:365–371.

8. de Quervain DJ, Roozendaal B, McGaugh JL. Stress and gluco-corticoids impair retrieval of long-term spatial memory. *Nature*. 1998;394:787–790.

9. Dias-Ferreira E, Sousa JC, Melo I, et al. Chronic stress causes fron-tostriatal reorganization and affects decision-making. *Science*. 2009;325:621–625.

10. Dickerson SS, Kemeny ME. Acute stressors and cortisol responses: a theoretical integration and synthesis of laboratory research. *Psychol Bull*. 2004;130:355–391.

11. Fries E, Hesse J, Hellhammer J, Hellhammer DH. A new view on hypocortisolism. *Psychoneuroendocrinology*. 2005;30:1010–1016.

12. Hamacher-Dang TC, Uengoer M, Wolf OT. Stress impairs retrieval of extinguished and unextinguished associations in a predictive learning task. *Neurobiol Learn Mem*. 2013;104:1–8.

13. Hermans EJ, Henckens MJ, Joels M, Fernandez G. Dynamic adap-tation of large-scale brain networks in response to acute stressors. *Trends Neurosci*. 2014;37:304–314.

14. Joels M, Karst H, DeRijk R, De Kloet ER. The coming out of the brain mineralocorticoid receptor. *Trends Neurosci*. 2008;31:1–7.

15. Kuhlmann S, Wolf OT. A non-arousing test situation abolishes the impairing effects of cortisol on delayed memory retrieval in healthy women. *Neurosci Lett*. 2006;399:268–272.

16. Lazarus RS. Coping theory and research: past, present, and future. *Psychosom Med*. 1993;55:234–247.

17. Lupien SJ, McEwen BS, Gunnar MR, Heim C. Effects of stress throughout the lifespan on the brain, behaviour and cognition. *Nat Rev Neurosci*. 2009;10:434–445.

18. Maren S, Holmes A. Stress and fear extinction. *Neuropsychopharma-cology*. 2016;41:58–79.

19. McEwen BS. Protective and damaging effects of stress mediators. *N. Engl J Med*. 1998;338:171–179.

20. Meaney MJ. Maternal care, gene expression, and the transmission of individual differences in stress reactivity across generations. *Annu Rev Neurosci*. 2001;24:1161–1192.

21. Nadel L, Moscovitch M. Memory consolidation, retrograde amnesia and the hippocampal complex. *Curr Opin Neurobiol*. 1997;7:217–227.

22. Nemeroff CB, Heim CM, Thase ME, et al. Differential responses to psychotherapy versus pharmacotherapy in patients with chronic forms of major depression and childhood trauma. *Proc Natl Acad Sci USA*. 2003;100:14293–14296.

23. Oei NY, Elzinga BM, Wolf OT, et al. Glucocorticoids decrease hip-pocampal and prefrontal activation during declarative memory retrieval in young men. *Brain Imaging Behav*. 2007;1:31–41.

24. Payne JD, Jackson ED, Hoscheidt S, Ryan L, Jacobs WJ, Nadel L. Stress administered prior to encoding impairs neutral but enhances emotional long-term episodic memories. *Learn Mem*. 2007;14:861–868.

25. Quirk GJ, Mueller D. Neural mechanisms of extinction learning and retrieval. *Neuropsychopharmacology*. 2008;33:56–72.

26. Roozendaal B, Hernandez A, Cabrera SM, et al. Membrane-associated glucocorticoid activity is necessary for modulation of long-term memory via chromatin modification. *J Neurosci*. 2010;30:5037–5046.

27. Roozendaal B, McEwen BS, Chattarji S. Stress, memory and the amygdala. *Nat Rev Neurosci*. 2009;10:423–433.

28. Schelling G, Roozendaal B, de Quervain DJ. Can posttraumatic stress disorder be prevented with glucocorticoids? *Ann N Y Acad Sci*. 2004;1032:158–166.

29. Schlotz W, Phillips DI. Fetal origins of mental health: evidence and mechanisms. *Brain Behav Immun*. 2009;23:905–916.

30. Schwabe L, Dickinson A, Wolf OT. Stress, habits, and drug addiction: a psychoneuroendocrinological perspective. *Exp Clin Psychopharmacol*. 2011;19:53–63.

31. Schwabe L, Wolf OT. Stress and multiple memory systems: from 'thinking' to 'doing'. *Trends Cogn Sci*. 2013;17:60–68.

32. Turecki G, Meaney MJ. Effects of the social environment and stress on glucocorticoid receptor gene methylation: a systematic review. *Biol Psychiatry*. 2016;79:87–96.

33. Wingenfeld K, Wolf OT. Effects of cortisol on cognition in major depressive disorder, posttraumatic stress disorder and border-line personality disorder—2014 Curt Richter Award Winner. *Psychoneuroendocrinology*. 2015;51:282–295.

34. Wolf OT. The influence of stress hormones on emotional mem-ory: relevance for psychopathology. *Acta Psychol (Amst)*. 2008;127: 513–531.

35. Wolf OT. Stress and memory in humans: twelve years of progress? *Brain Res*. 2009;1293:142–154.

36. Wolf OT. Immediate recall influences the effects of pre-encoding stress on emotional episodic long-term memory consolidation in healthy young men. *Stress*. 2012;15:272–280.

37. Wolf OT. Effects of stress on learning and memory: evidence for sex differences in humans. In: Conrad CD, ed. *The Handbook of Stress: Neuropsychological Effects on the Brain*. Chichester, West Sussex, United Kingdom: Wiley-Blackwell; 2013:545–559.

38. Wyrwoll CS, Holmes MC, Seckl JR. 11β-Hydroxysteroid dehydro-genases and the brain: from zero to hero, a decade of progress. *Front Neuroendocrinol*. 2010;32:265–286.

39. Yau SY, Li A, So KF. Involvement of adult hippocampal neurogen-esis in learning and forgetting. *Neural Plast*. 2015;2015:717958.

40. Yehuda R. Post-traumatic stress disorder. *N Engl J Med*. 2002;346:108–114.

41. Zhang TY, Meaney MJ. Epigenetics and the environmental regulation of the genome and its function. *Annu Rev Psychol*. 2010;61:439–466.

CHAPTER

32

Sex Differences in Chronic Stress: Role of Estradiol in Cognitive Resilience

V. Luine[1], J. Gomez[2], K. Beck[3,4], R. Bowman[5]

[1]Hunter College of CUNY, New York, NY, United States; [2]National Institute on Drug Abuse, Baltimore, MD, United States; [3]VA NJ Health Care System, East Orange, NJ, United States; [4]Rutgers Medical School, Newark, NJ, United States; [5]Sacred Heart University, Fairfield, CT, United States

Abstract

Stress is associated with psychiatric disorders which show sex differences in expression. Application of chronic stressors to rodents has been used as a model for understanding the bases of these sex differences and for developing new treatments. Female rats are more resilient than males to chronic restraint stress in the cognitive domain, and estradiol, of both ovarian and neural origin, contributes to this resilience. For anxiety, chronic stress appears anxiogenic in both sexes, and limited information suggests that estradiol does not alter anxiety responses to stress. Depression in males is enhanced following different chronic stressors, but a lack of studies in females limits determining whether females are more resilient or susceptible to stress-dependent depression. These findings are presented and discussed within the framework of making the animal models more consistent with the behavioral phenotype of human females and for improving future research.

INTRODUCTION

Information is accumulating that stress, either acute or chronic, does not elicit identical neural and behavioral responses in the sexes.[1] These sex differences range from alterations in basic cellular responses to changes in higher order neural responses such as learning and memory abilities, and possibly anxiety and mood. While the vast majority of stress studies, as in most basic research, investigate only males, more recent studies have begun to include females. The importance of including females in stress research should have been apparent sooner because of well-documented sex differences in humans for coping with chronic stress and in the incidence of stress-related diseases. For example, females have a higher incidence of anxiety disorders, posttraumatic stress disorder (PTSD), and major depression than males while males show greater levels of alcohol and drug abuse than females.[2] A better understanding of how these sex differences emerge is critical because such information may inform the development of novel and more effective therapies for these disorders which are often precipitated by or related to stress.

> ## KEY POINTS
>
> - Learning and memory is impaired in male rats following chronic stress
> - Learning and memory is enhanced in female rats following chronic stress
> - Estradiol of both neural and ovarian origin contributes to cognitive stress resilience in females
> - Chronic stress increases anxiety in both sexes
> - Chronic stress is associated with sex differences in neuronal function
> - Sex differences in neural and behavioral responses to stress may contribute to known differences in the expression of neuropsychiatric disorders in humans

In this review, we highlight research from our own and other laboratories which show different neural and behavior responses to chronic stress in male and female rodents. At the neural level, a number of sex-dependent stress effects are emerging but their links to behaviors are not well described. In the cognitive realm, chronically stressed males are impaired in many learning and memory tasks involving the hippocampus and prefrontal cortex (PFC) while females are either not affected or show improved performance on such tasks. Chronic stress is usually anxiogenic in both sexes, but there are too few studies in rodent females to make definitive statements. Finally, sex differences in depressive behaviors have not been well investigated, and therefore whether sex differences following stress are present is currently unclear.

It has been postulated that the presence of higher estradiol levels in females underlies sex differences in responses to stress.[3] However, this hypothesis is complicated to assess because of the demonstration of neurally synthesized estrogens, which are especially abundant in hippocampal regions and which are also found in both sexes. Here, we consider how gonadally and neurally derived estradiol interacting within specific brain areas may influence stress responses. We hypothesize that estradiol may provide resilience against stress effects on cognition within prefrontal–hippocampal circuits.

LEARNING AND MEMORY IS IMPAIRED IN MALES FOLLOWING CHRONIC STRESS

In male rodents, 1–3 weeks of stress elicited by daily restraint or different daily stressors (unpredictable chronic stress, UCS) generally results in impaired learning and memory. Impairments in male learning have been shown using spatial learning and memory tasks like the eight-arm radial task (RAM) and the Morris water maze (MWM). These tasks are dependent on an intact hippocampus and also require interaction with the PFC.[4] Our laboratory was the first to report that 21 days of restraint stress for 6 h/day impairs the performance of male rats on the RAM.[5] Stressed males made significantly more errors, fewer correct choices, and earlier mistakes in completing the eight-arm choices than unstressed males (See Fig. 32.1A). These impairments are reversible because if stressed males are trained and tested beginning at 18 days poststress, no impairments are present.[5] In the MWM, a similar pattern of impaired learning is shown after chronic stressors[6,7] (Fig. 32.1B), and 21 days of UCS impaired males on the RAM.[8]

In tests assessing spatial memory (not learning), chronic stress also generally impairs male performance (see Ref. 9 for review). Spatial memory in both the object placement[10–13] (Fig. 32.1C) and the Y-maze task is impaired following 1–3 weeks of daily restraint[14–16] (Fig. 32.1D). Control males spend more time exploring the object in the new location rather than the old location suggesting they remember the old location, whereas stressed males spend the same amount of time exploring at both locations suggesting poor memory. Likewise, in the Y-maze task, stressed males do not explore the novel arm more than the other arms, whereas unstressed males do. Recognition memory, a nonspatial task, is also impaired following chronic stress to males. Stressed male rats are unable to significantly discriminate between known and new objects, an impairment of object recognition memory (Fig. 32.2A).[10,13,16] Temporal order recognition memory (TORM) is also impaired by 2 h of daily restraint for 5–7 days in male rats[17] (Fig. 32.2B). TORM is similar to object recognition but uses two sampling periods in order to assess explicit memory processes which are dependent on the PFC.

Thus, numerous studies show that chronic stress impairs both learning and memory in male rodents using a variety of cognitive tasks. An extensive discussion of these effects can be found in Conrad[9] where results are reviewed in relation to acquisition (learning), memory, and the details and nature of the tasks utilized. Conrad's conclusions are similar: the vast majority of studies show that chronic stress impairs cognition in male rodents, but effects can be moderated depending primarily on the type of tasks employed and type and duration of stressor.

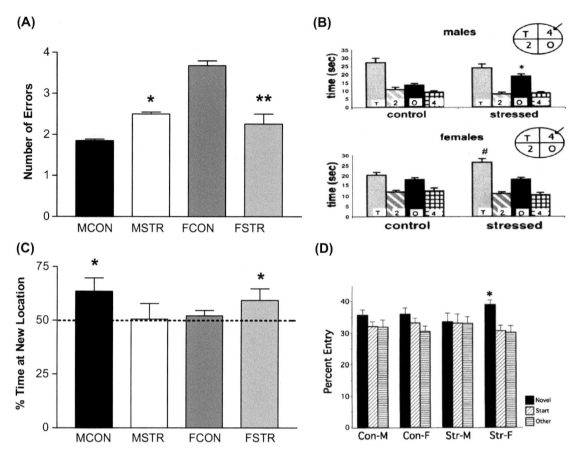

FIGURE 32.1 **Sex difference in chronic stress effects on spatial memory tasks. Adult male and female rats received 21 days of daily (6 h) restraint stress.** (A) Radial-arm maze. Experiments were conducted separately in the sexes, and data is pooled for males[15] and females.[18] The number of errors to complete the task significantly increased in stressed males and decreased in stressed females. (B) Morris water maze. Data are the mean time (s) spent in each quadrant (T = target; 2; O = opposite; 4) during the 60 s probe trial for both male and female groups. Stressed males were impaired by chronic stress and spent more time in the opposite quadrant than control males. Stress females were enhanced and spent more time in the target quadrant than control females (*Reprinted with permission from Kitraki E, Kremmyda O, Youlatos D, Alexis MN, Kittas C. Gender-dependent alterations in corticosteroid receptor status and spatial performance following 21 days of restraint stress. Neuroscience. 2004;125:47–55.*). (C) Object placement. The percentage of time exploring objects in the new location is shown. Control males and stressed females could significantly discriminate the old from the new location but stressed males and control females could not.[13] (D) Y maze. Data represent the percentage of entries made into the novel and other arms for 2–5 min. Stressed females entered the novel arm more than the start and other arms (*Data reprinted with permission from Conrad CD, Grote KA, Hobbs RJ, Ferayorni A. Sex differences in spatial and non-spatial Y-maze performance after chronic stress. Neurobiol Learn Mem. 2003;79:23–40.*).

LEARNING AND MEMORY IS ENHANCED IN FEMALES FOLLOWING CHRONIC STRESS

Because estrogens exhibit neurotrophic, antioxidant, and antiapoptotic effects and also promote some aspects of cognitive function,[3] we hypothesized that female rats might be less sensitive than males to the cognitive-impairing effects of chronic stress. Thus, adult females were treated to the same 21 days of daily 6-h restraint and cognitive testing as males.[13,14,18–20] Surprisingly, females were not only less sensitive to stress impairments of spatial memory, but chronic restraint stress enhanced performance on the RAM. As shown in Fig. 32.1A, stressed females made fewer errors in completing the task than nonstressed females, and they also more

rapidly reached learning criteria.[18] Similarly, 21 days of restraint stress enhanced female rodent performance in the MWM[6] (Fig. 32.1B) and UCS did not affect female performance on the radial arm water maze.[8] For spatial memory using the object placement test, control females did not significantly discriminate between the object in the new versus the old location, whereas stressed females did discriminate indicating an improvement in memory (Fig. 32.1C).[13,19] It should be noted that in the object placement task, like RAM, control females do not perform as well as control males. Males make two errors in completing the task while females make approximately twice as many errors, four (Fig. 32.1A). Male superiority in spatial memory tasks has been widely reported and is also present in humans.[4] Chronic stress eliminates this sex difference by impairing male and enhancing

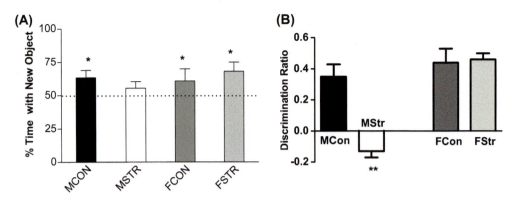

FIGURE 32.2 **Sex difference in chronic stress effects on nonspatial memory tasks. Male and female rats received daily (6 h) restraint stress.** (A) Object recognition. Subjects received 21 days of daily restraint, and experiments were conducted separately in the sexes. Entries are the mean of the percentage time spent exploring the new object. Dashed line at 50% indicates the chance performance of the task (same amount of time exploring the old and new object). All groups except the stressed male spend significantly more time exploring at the new object. *(Data pooled from separate experiments in male rats (From Beck KD, Luine VN. Food deprivation modulates chronic stress effects on object recognition in male rats: role of monoamines and amino acids. Brain Res. 1999;830, 56–71.) and female rats (Beck and Luine, unpublished); published data (From Bisagno V, Grillo CA, Piroli GG, Giraldo P, McEwen B, Luine VN. Chronic stress alters amphetamine effects on behavior and synaptophysin levels in female rats. Pharmac Biochem Behav. 2004;78, 541–550.)).* (B) Temporal order recognition memory. Male and female mice received 1 week of daily (6 h) restraint. Entries are the discrimination ratio (percentage of time exploring the novel, less recent, object). Males were impaired by stress but females were unaffected *(Modified data from Wei J, Yuen EY, Liu W, et al. Estrogen protects against the detrimental effects of repeated stress on glutamatergic transmission and cognition. Mol Psychiatry. 2014;19:588–598.).*

female performance. In the Y maze, memory is either not affected or enhanced by 7–21 days of restraint stress in females (Fig. 32.1D).[14,19,20]

For object recognition memory, stress does not affect females while it impairs males (Fig. 32.2A).[10,11,13,19,20] Like object recognition, TORM is not affected by stress in females but is impaired in males (Fig. 32.2B).[17] Most of these sex differences in response to stress, impairments in males and enhancements in females, are also maintained at shorter stress intervals.[7,11,16,20] In addition, longer periods of stress fail to impair female rats: no impairments in RAM, object recognition, or object placement were found after 28[18] or 35[21] days of stress in females.

Thus, female rodents show resilience to chronic stress, at least in terms of learning and memory, because they generally show enhanced cognitive function after stress or are unaffected, depending on the specific task. However, it should be noted that far fewer experiments have been conducted in female rodents than in males and that most experiments in females utilized restraint stress. Thus, chronic stress effects on female cognition need further investigation and confirmation.

ROLE OF ESTROGENS IN COGNITIVE RESILIENCE TO STRESS

Several lines of evidence support the idea that estradiol, of both ovarian and neural origin, may be responsible for female rodent's cognitive resilience to chronic stress. Because females have high circulating estradiol levels as compared to males, we tested the simple

prediction that ovariectomized (OVX) females would not show stress resilience, i.e., be impaired on RAM following chronic stress. As shown in Fig. 32.3A, OVX + stressed rats' performance was neither enhanced nor impaired which partially supported the hypothesis that estradiol confers stress resilience.[22] In addition, McLaughlin et al.[23] showed that OVX + chronically stressed rats still showed improvements on the Y maze as compared to nonstressed subjects. Consistent with our hypothesis, when estradiol was given to OVX-stressed females, RAM performance was enhanced (Fig. 32.3A).[22] At the time of these experiments, we hypothesized that resilience to stress in females must be due to both the activating effects of estradiol at adulthood and enduring, organizing effects of estradiol which occur during development in females. However, it is now known that significant concentrations of estradiol are synthesized locally in discrete regions of the brain by the enzyme aromatase from androgen precursors such as testosterone and/or directly from cholesterol, and, as such, levels in the hippocampus, hypothalamus, and PFC are higher than are present in the circulation. For example, Kato et al.[24] found that hippocampal 17β-estradiol levels are 4 nM in female rats during proestrus while circulating concentrations are only 0.1 nM and that hippocampal levels change over the estrus cycle. In addition, estradiol is still present in the hippocampus following ovariectomy.[24] Using hippocampal slices, Fester and Rune reported that hippocampal neurons synthesize estradiol, which maintains LTP and synapses in females but not males.[25] In females, inhibition of estradiol synthesis results in impairment of LTP and synapse loss, but these

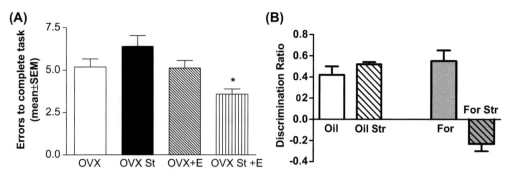

FIGURE 32.3 Effects of altering estradiol levels on stress effects in females. (A) Radial arm maze. Rats were ovariectomized and served as control (OVX), were stressed 6h daily for 21 days (OVX+St), received estradiol in silastic capsules (OVX+E), or were stressed daily and received estradiol (OVX St+E). ANOVA showed main effects of estradiol treatment (p<0.003) and an estradiol–stress interaction (p<0.002) indicating that estradiol-treated groups showed better performance and the estradiol-treated stressed females outperformed all other groups (p<0.05). The pattern of changes suggests that peripheral estradiol can provide resilience to stress (*Modified data from Bowman RE, Ferguson D, Luine VN. Effects of chronic restraint stress and estradiol on open field activity, spatial memory, and monoaminergic neurotransmitters in ovariectomized rats. Neuroscience. 2002;113:401–410.*). (B) Temporal order recognition task. Female mice received s.c. oil, oil and 7days of 6h daily restraint (oil Str), the aromatase inhibitor, Formastane (For), and Formestane 1h before daily stress (For Str). Data are discrimination ratio (see Fig. 32.2B). All groups except the For Str significantly discriminated suggesting that both peripheral and neural estradiol provide resilience to stress (*Modified data from Wei J, Yuen EY, Liu W, et al. Estrogen protects against the detrimental effects of repeated stress on glutamatergic transmission and cognition. Mol Psychiatry. 2014;19:588–598.*).

effects are not seen in males. Thus, the in situ production of estradiol and its sexually dimorphic effects indicates that neural estradiol can contribute to cognitive stress resilience in females such that estradiol is available to hippocampal and other sites even following ovariectomy. Estradiol may act as a neurosteroid as well as a gonadal steroid, and some evidence shows that it may also be a neuromodulator.[3] Thus, intrahippocampal estradiol synthesis is well poised to contribute to cognitive stress resilience in females. The basis for the apparent lack of effects by neural estradiol in males remains to be investigated.

Further evidence that female rats' cognitive resilience to stress depends on estradiol was provided by Zhen and colleagues who explored sex differences to stress in the TORM task.[17] As shown in Fig. 32.2B, chronic restraint stress impairs males, but not females, in the TORM task. Estrogen-dependent resilience in this sex difference is implicated because stress impairs TORM in females when estrogen receptors are inhibited or knocked out in the PFC. Interestingly, impairments of performance in stressed males were prevented by estradiol administration suggesting that peripheral (gonadally derived?) estradiol may confer resilience to stress, because as discussed above, neural estradiol does not appear to be active in the hippocampus of males. To further test estrogen's involvement, females were injected s.c. with the aromatase inhibitor, Formestane, which blocks both peripheral and neural synthesis of estradiol.[17] Under this condition, stress impaired TORM in females (Fig. 32.3B). When compared to experiments with RAM and Y maze in females, where only peripheral estrogen was removed by OVX (described above), these results provide some evidence that both peripherally and neurally derived estradiol provide resilience

against the detrimental effects of chronic stress on memory in females.

Thus, some evidence suggests that estradiol of both gonadal and neural origin may provide resilience against stress-induced cognitive impairments in female, but not male, rodents. In the next section, we explore stress effects on other behaviors.

CHRONIC STRESS INCREASES ANXIETY IN BOTH SEXES

As a prologue to this section, it needs to be pointed out that most studies seeking to understand how stress contributes to anxiety and depression usually begin with the statement that these conditions are up to two times more prevalent in human females than males. Then, the studies are usually conducted in male subjects, and few, if any, include both sexes. Table 32.1 provides a summary of the few studies where restraint stress was utilized to test effects on anxiety, and both sexes were usually analyzed. For inducing stress, subchronic and chronic variable stress (CVS, use of more than one type of stressor) is gaining popularity, but very few studies have analyzed both sexes and most were subchronic paradigms. The resident intruder stress model is also a popular paradigm; however, because females cannot be evaluated in this paradigm, sex differences cannot be investigated.

The current "gold standard" for anxiety measurement is the elevated plus maze (EPM) where low entries into or time spent on open arms indicate heightened anxiety. For the open field (OF), anxiety is measured by entrance into inner quadrants (high counts indicate less anxiety). Studies in Table 32.1 are arranged by increasing days of daily 6h restraint stress. Overall, stress generally

TABLE 32.1   Effects of Chronic Daily Restraint Stress on Anxiety-Like Behaviors on the Elevated Plus Maze (EPM) and the Open Field (OF) in Rats

Duration	Task	Males	Females	References
7 days	EPM	↑	=	Bowman et al.[11]
7 days	EPM	↑	=	Gomez et al.[16], Gomez and Luine[20]
21 days	OF, entry latency	Not done	↑	Bisagno et al.[19]
21 days	OF, center visits, dark phase	↑	↑	Beck and Luine[13]
21 days	OF, center visits, dark phase	↑	=	Huynh et al.[27]
21 days	EPM, dark phase	=	↑	Huynh et al.[27]
21 days	OF, center visits, light phase	=	=	Huynh et al.[27]
21 days	EPM, light phase	=	=	Huynh et al.[27]
35 days	OF, center visits	Not done	↑	Bowman and Kelly[21]
50 days*	EPM	↑	=	Noschang et al.[26]

Adult rats received 6 h/day of restraint for number of days shown except * which indicates restraint was for 1 h/day for 5 out of 7 days. Anxiety was measured on the elevated plus maze (EPM) and open field (OF) where upward arrows indicate an increase in anxiety measure and = indicates no change.

TABLE 32.2   Sex Differences in Chronic Stress Effects on Monoamine Metabolites in Brain Areas of Male and Female Rats

Area	MHPG		HVA		5-HIAA	
	Male	Female	Male	Female	Male	Female
CA1	↑ 153%	↑ 122%	↓ 63%	↑ 18%	↓ 45%	↑ 30%
CA3	↓ 41%	↑ 105%	↑ 46%	↓ 11%	↓ 22%	↑ 24%
Medial PFC	↓ 18%	↑ 21%	=	=	=	=
Basolateral amygdala	↓ 18%	↑ 21%	↓ 24%	↓ 4%	↓ 25%	=

Entries show significant increases (↑) or decreases (↓) in metabolites for norepinephrine (MHPG, 3-methoxy-4-hydroxyphenylglycol), dopamine (HVA, homovanillic acid), and serotonin (5-HIAA, 5-hydroxyindole acetic acid). An arrow of the same color but a lighter hue indicates a significant difference. Significant differences range from $p < 0.05$ to 0.001. Adult male and female rats served as controls or received 7 days of 6 h/day restraint stress.
Data from Bowman RE, Micik R, Gautreaux C, Fernandez L, Luine VN. Sex-dependent changes in anxiety, memory, and monoamines following one week of stress. Physiol Behav. 2009;97:21–29.

increases anxiety in both sexes on either the EPM or OF. It appears that males show stress-dependent increases in anxiety before females (7 days), but by 21 days females also show enhanced anxiety, though not in all studies.[26] Huynh et al. showed no effects on anxiety in either sex when evaluations were done during the light phase of the cycle, but, in the dark phase, females show enhanced anxiety on the EPM and males showed enhanced anxiety on the OF following stress.[27] Thus, the time during the day/night cycle when anxiety is measured may be an important variable.

Clearly, these animal studies do not model the human condition where women have a higher incidence of anxiety disorders than men. It should be noted that all rodent behavior tasks, both cognitive and anxiety-related, were developed in male, not female subjects. Thus, the tasks may tap into and accurately reflect male, but not female, attributes, and thus, female anxiety may not be accurately assessed (see Shansky for further discussion of this topic).[28] In addition, the task utilized may be critical in the outcome (Table 32.1), and the tasks may not accurately model anxiety as experienced by humans. Also not reflected in Table 32.1 is that stressor type may also influence outcomes. Since both sexes have not been directly compared using other stressors, this issue cannot be discussed. In future studies, the use of stressors that more closely resemble precipitators of human anxiety or PTSD might be valuable. Sexual aggression might better model women's stress experiences and exposure to predators might better model men's experiences in combat. Overall, future experiments need to at least include both sexes in order to determine the extent to which sex differences contribute to effects of chronic stress on anxiety and whether the increased vulnerability shown by human females is also present in rodents.

Whether estradiol imparts resilience or susceptibility to anxiety following stress is unknown. Estradiol

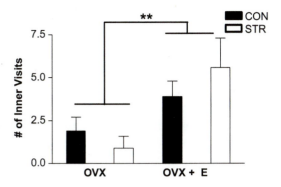

FIGURE 32.4 Effect of estradiol on anxiety on the open field in control or chronically stressed, ovariectomized rats. Rats were ovariectomized and received cholesterol or estradiol containing silastic capsules. Daily restraint or handling (control) was given for 21 days. Open field was tested 1 day after completion of stress and number of inner visits is plotted. There was a significant estradiol, but not stress, effect (p < 0.01); estradiol-treated subjects made more inner visits suggesting that estradiol does not alter stress effects on anxiety but decreases anxiety. *Modified data from Bowman RE, Ferguson D, Luine VN. Effects of chronic restraint stress and estradiol on open field activity, spatial memory, and monoaminergic neurotransmitters in ovariectomized rats. Neuroscience. 2002;113:401–410.*

treatment to OVX rats did not alter inner visits on the OF in stressed as compared to unstressed rats[22]; however, estradiol did increase inner visits in both unstressed and stressed subjects (Fig. 32.4). Anxiolytic effects of estradiol have been previously shown and may be mediated by estrogen binding to estrogen receptor beta.[29] Overall, at this time, it appears that sex differences in stress-induced anxiety are not strong, and estrogens do not appear to influence responses to stress in females. However, this conclusion is based on very limited evidence.

CHRONIC STRESS INCREASES DEPRESSION IN MALE RATS

Depression is assessed in rodents by preference for sucrose containing vs. plain water wherein less consumption of sucrose indicates anhedonia (loss of pleasure) and therefore depression. The forced swim test (FST), where time spent swimming or immobile is measured and increased immobility time indicates depression, is also utilized. The FST test was developed primarily for assessing potential antidepressant drugs and has been applied in only a few stress studies. In males, exposure to a variety of stressors (1–4 weeks) including predator stress, social defeat coupled with chronic isolation, and variable stressors is associated with increased male depression in the sucrose preference test.[30–33] Of these studies, only one examined females and found decreased sucrose preference during week 1 and 4 of stress while males

exhibited decreased preference from 1 to 4 weeks.[30] Few studies have utilized restraint stress to examine depression, but we examined sucrose preference in rats at day 7, 14, and 21 of daily restraint and found that stress significantly decreased sucrose preference in both sexes (Buenaventura, Khandaker, and Luine, unpublished) while Huynh et al. reported that this regimen did not alter sucrose preference or FST in either sex.[27] A shorter interval of restraint, 7 days, also did not affect either sex on the FST.[16,20] In contrast, 7 days of unpredictable stress decreased sucrose preference in female, but not male, mice.[34] Thus, studies suggest that chronic, but not subchronic stressors may cause depression in males, but effects in females have not been investigated sufficiently. Hence, a role for estradiol in moderating stress effects on depression cannot be discussed.

SEX DIFFERENCES IN STRESS EFFECTS ON NEURONAL FUNCTION

Chronic stress is associated with sex-specific alterations in areas important for learning and memory like the PFC and hippocampus and for mood and anxiety areas like the amygdala and PFC. However, the vast majority of neural studies, like the behavioral studies, fail to include females. A brief synopsis of important changes follows, and the pattern of results suggests that stress often downregulates neural activity or expression in males but either does not alter or upregulates females. Chronic stress differentially affects neuronal survival; in males short-term survival of new dentate gyrus neurons is decreased, whereas survival is increased in females.[35] Retraction/pruning of dendrites following stress has been well described in male subjects. Galea et al. confirmed retraction of CA3 apical dendrites following 21 days of restraint stress in males and found that females do not show changes in these dendrites but that basal CA3 dendrites are retracted.[36] McLaughlin et al. more recently showed that estradiol treatment to OVX rats prevents CA3 dendritic retractions and also increases CA1 spine density providing some morphological evidence for the idea that estradiol may confer cognitive resilience to stress.[23]

Several neurotransmitters show sexually dimorphic responses to stress. Glutamate activity is critical in learning and memory, and 21 days of restraint stress decreased glutamatergic neurotransmission and surface expression of glutamate receptors in the PFC of male, but not female, rats.[17] Chronic unpredictable mild stress for 21 days caused a 50% decrease in the endocannabinoid receptor, CB1, in the dorsal hippocampus of male rats, whereas an approximately 150% increase was found in females.[37]

Monoaminergic systems are among rapid and important stress responders/mediators, and following 21 days of restraint decreased levels of norepinephrine, dopamine, and serotonin (5-hydroxytryptamine, or 5-HT) levels are found in the hippocampus of males,[13,38] but opposite changes in these monoamines are found in females.[13] Recently, we examined effects of restraint for a shorter period, 1 week, and like the behavioral responses, monoaminergic systems showed robust sex differences (Table 32.2).[11] Changes in metabolites of the major amines, 3-methoxy-4-hydroxyphenylglycol (MHPG) for norepinephrine, homovanillic aid (HVA) for dopamine, and 5-hydroxyindole acetic acid (5-HIAA) for serotonin are shown by arrows in CA1, CA3, medial PFC, and basolateral amygdala in Table 32.2. Metabolites provide a measure of activity, but other assessments like in vivo release of transmitters also need to be made. In the majority of cases, metabolites decreased in males and increased in females following stress. The cortex showed the fewest changes, with only MHPG affected: an 18% decrease in males and a 21% increase in females. The largest changes were found in CA1 where both sexes showed an increase in the NE metabolite, MHPG, but males showed a significantly larger increase, 153%, than females, 122%, and both HVA and 5-HIAA decreased 45–63% in males and increased 18–30% in females. However, the largest difference between the sexes occurred in CA3 where MHPG decreased 41% in males and increased 105% in females. These remarkably different patterns in monoaminergic activity in the sexes following stress may be critically important in mediating memory and mood changes. Further experiments are, however, necessary to link the changes in activity to behavior.

CONCLUSIONS AND POSSIBLE CONTRIBUTIONS OF SEX DIFFERENCES IN THE STRESS RESPONSE AND ESTRADIOL TO MENTAL HEALTH

In rodent models, evidence shows that females are more resilient than males to chronic restraint stress in the cognitive domain and that estrogens of both ovarian and neural origin contribute to the resilience. For anxiety, chronic stress appears to be anxiogenic in both sexes and limited information suggests that estradiol does not alter anxiety responses to stress though estradiol can be anxiolytic when given alone. Depression in male rodents is enhanced following a number of different chronic stressors, and it is unclear whether females are more resilient or more susceptible than males since very few studies have examined females. The sex-dependent pattern of stress responses in rodents does not appear consistent with the general behavioral phenotype of human females who show more depression, anxiety, and PTSD than males.[2]

Several possibilities for this discrepancy have been put forward including that rodents may not successfully model humans, but a recent study showing depressive behavior in female, but not male, mice after subchronic CVS suggests that this species may be more appropriate for studies.[34] Another factor is that currently available behavioral tests in rodents were developed in males, and thus, results in females may not accurately tap into their attributes. On the other hand, it can be envisioned that heightened cognition in females following stress might facilitate remembrance of stressful events better than males and combined with the enhanced anxiety may, at a longer time interval poststress, culminate in depression in females. Thus, future experiments should assess stress-dependent effects at a longer intervals following cessation of stress.

Another consideration for future research is that stress is often coincident with other behaviors such as increased alcohol and drug consumption. We recently showed that combined exposure to stress and alcohol reverses the individual, deleterious effects of stress or alcohol alone on impaired spatial and nonspatial memory and increased anxiety in male rats.[16] In contrast, in female rats stress alone did not increase depression on the FST task but when combined with alcohol, depression was enhanced.[20] Likewise, the combination of stress and alcohol impaired female object recognition.[20] Chronic stress and amphetamine also interact to alter behavioral and neural responses in females,[19] but interactive effects in males have not been directly compared to females. Thus, studies that utilize stress with drugs, alcohol, or other stressors might also provide more consistency with humans.

In conclusion, while sex differences in cognitive responses to stress have been demonstrated in rats, the relationship of these effects to sex differences in incidences of anxiety and depression in humans remain largely unknown. However, more information about sex differences in neural function and behavior may be forthcoming since the National Institutes of Health (NIH) has recently unveiled policies to ensure that preclinical research funded by NIH includes both female and male subjects.[39] As indicated in this review, most current research does not include both sexes, and we have only scratched the surface on how stress is processed in female versus male brains. Only by using both sexes will it be possible to forge meaningful links between stress and mental functioning.

Acknowledgments

Experimental work discussed in this review was supported by the City University of New York, PSC-CUNY, NIH grants GM60654 (VL), GM60665 (VL), and RR03037 (HC). The authors thank the many undergraduate and graduate students who participated in the research.

References

1. Luine VN, Gomez JL. Sex differences in rodent cognitive processing and responses to chronic stress. In: Shansky R, ed. *Sex Differences in the Central Nervous System*. Elsevier; 2015:365–404.
2. Bangasser DA, Valentino RJ. Sex differences in stress-related psychiatric disorders: neurobiological perspectives. *Front Neuroendocrinol*. 2014;35:303–319.
3. Luine VN. Estradiol: mediator of memories, spine density and cognitive resilience to stress in female rodents, *J Steroid Biochem Mol Biol*. 2016;160:189–195.
4. Luine V. Recognition memory tasks in neuroendocrine research, *Behav Brain Res*. 2015;285:158–164.
5. Luine V, Villegas M, Martinez C, McEwen BS. Repeated stress causes reversible impairments of spatial memory performance. *Brain Res*. 1994;639:167–170.
6. Kitraki E, Kremmyda O, Youlatos D, Alexis MN, Kittas C. Gender-dependent alterations in corticosteroid receptor status and spatial performance following 21 days of restraint stress. *Neuroscience*. 2004;125:47–55.
7. McFadden LM, Paris JJ, Mitzelfelt MS, McDonough S, Frye CA, Matuszewich L. Sex-dependent effects of chronic unpredictable stress in the water maze. *Physiol Behav*. 2011;102:266–275.
8. Ortiz JB, Taylor SB, Hoffman AN, Campbell AN, Lucas LR, Conrad CD. Sex-specific impairment and recovery of spatial learning following the end of chronic unpredictable restraint stress: potential relevance of limbic GAD. *Behav Brain Res*. 2015;282:176–184.
9. Conrad CD. A critical review of chronic stress effects on spatial learning and memory. *Prog Neuropsychopharmacol Biol Psychiatry*. 2010;34:742–755.
10. Beck KD, Luine VN. Food deprivation modulates chronic stress effects on object recognition in male rats: role of monoamines and amino acids. *Brain Res*. 1999;830:56–71.
11. Bowman RE, Micik R, Gautreaux C, Fernandez L, Luine VN. Sex-dependent changes in anxiety, memory, and monoamines following one week of stress. *Physiol Behav*. 2009;97:21–29.
12. Gomez JL, Lewis M, Luine V. Alcohol access alleviates stress induced spatial memory impairments in male rats. *Alcohol*. 2012;46:499–504.
13. Beck KD, Luine VN. Sex differences in behavioral and neurochemical profiles after chronic stress: role of housing conditions. *Physiol Behav*. 2002;75:661–673.
14. Conrad CD, Grote KA, Hobbs RJ, Ferayorni A. Sex differences in spatial and non-spatial Y-maze performance after chronic stress. *Neurobiol Learn Mem*. 2003;79:23–40.
15. Wright RL, Conrad CD. Chronic stress leaves novelty-seeking behavior intact while impairing spatial recognition memory in the Y-maze. *Stress*. 2005;8:151–154.
16. Gomez JL, Lewis MJ, Sebastian V, Serrano P, Luine V. Alcohol administration blocks stress-induced impairments in memory and anxiety and alters hippocampal neurotransmitter receptor expression in male rats. *Hormones Behav*. 2013;63:659–661.
17. Wei J, Yuen EY, Liu W, et al. Estrogen protects against the detrimental effects of repeated stress on glutamatergic transmission and cognition. *Mol Psychiatry*. 2014;19:588–598.
18. Bowman RE, Zrull MC, Luine VN. Chronic restraint stress enhances radial arm maze performance in female rats. *Brain Res*. 2001;904:279–289.
19. Bisagno V, Grillo CA, Piroli GG, Giraldo P, McEwen B, Luine VN. Chronic stress alters amphetamine effects on behavior and synaptophysin levels in female rats. *Pharmac Biochem Behav*. 2004;78:541–550.
20. Gomez JL, Luine V. Female rats exposed to stress and alcohol show impaired memory and increased depressive-like behaviors. *Physiol Behav*. 2014;123:47–54.
21. Bowman RE, Kelly R. Chronically stressed female rats show increased anxiety but no behavioral alterations in object recognition or placement memory: a preliminary examination. *Stress*. 2012;15:524–532.
22. Bowman RE, Ferguson D, Luine VN. Effects of chronic restraint stress and estradiol on open field activity, spatial memory, and monoaminergic neurotransmitters in ovariectomized rats. *Neuroscience*. 2002;113:401–410.
23. McLaughlin KJ, Wilson JO, Harman J, et al. Chronic 17β-estradiol or cholesterol prevents stress-induced hippocampal CA3 dendritic retraction in ovariectomized female rats: possible correspondence between CA1 spine properties and spatial acquisition. *Hippocampus*. 2009;20:768–786.
24. Kato A, Hojo Y, Higo S, et al. Female hippocampal estrogens have a significant correlation with cyclic fluctuation of hippocampal spines. Oct 18 *Front Neural Circuits*. 2013;7:149. http://dx.doi.org/10.3389/fncir.2013.00149. [eCollection].
25. Fester L, Rune GM. Sexual neurosteroids and synaptic plasticity in the hippocampus. *Brain Res*. 2015;1621:162–169.
26. Noschang CG, Pettenuzzo LF, von Pozzer Toigo E, et al. Sex-specific differences on caffeine consumption and chronic stress-induced anxiety-like behavior and DNA breaks in the hippocampus. *Pharmacol Biochem Behav*. 2009;94:63–69.
27. Huynh TN, Krigbaum AM, Hanna JJ, Conrad CD. Sex differences and phase of light cycle modify chronic stress effects on anxiety and depressive-like behavior. *Behav Brain Res*. 2011;222:212–228.
28. Shansky RM. Sex differences in PTSD resilience and susceptibility: challenges for animal models of fear learning. *Neurobiol Stress*. 2015;1:60–65.
29. Oyola MG, Portillo W, Reyna A, et al. Anxiolytic effects and neuroanatomical targets of estrogen receptor-β (ERβ) activation by a selective ERβ agonist in female mice. *Endocrinology*. 2012;153:837–846.
30. Dalla C, Pitychoutis M, Kokras N, Papadopoulou-Daifoti Z. Sex differences in animal models of depression and antidepressant response. *Basic Clin Pharmacol Toxicol*. 2010;106:226–233.
31. Carnevali L, Mastorci F, Graiani G, et al. Social defeat and isolation induce clear signs of a depression-like state, but modest cardiac alterations in wild-type rats. *Physiol Behav*. 2012;106:142–150.
32. Gronli J, Murison R, Fiske E, et al. Effects of chronic mild stress on sexual behavior, locomotor activity and consumption of sucrose and saccharine solutions. *Physiol Behav*. 2005;84:571–577.
33. Burgado L, Harrell CS, Eacret D, et al. Two weeks of predatory stress induces anxiety-like behavior with co-morbid depressive-like behavior in adult male mice. *Behav Brain Res*. 2014;275:120–125.
34. Hodes GE, Pfau ML, Purushothaman I, et al. Sex differences in nucleus accumbens transcriptome profiles associated with susceptibility versus resilience to subchronic variable stress. *J Neurosci*. 2015;35:16362–16376.
35. Westenbroek C, Den Boer JA, Veenhuis M, Ter Horst GJ. Chronic stress and social housing differentially affect neurogenesis in male and female rats. *Brain Res Bull*. 2004;64:303–308.
36. Galea LA, McEwen BS, Tanapat P, Deak T, Spencer RL, Dhabhar FS. Sex differences in dendritic atrophy of CA3 pyramidal neurons in response to chronic restraint stress. *Neuroscience*. 1997;81:689–697.
37. Reich CG, Taylor ME, McCarthy MM. Differential effects of chronic unpredictable stress on hippocampal CB1 receptors in male and female rats. *Behav Brain Res*. 2009;203:264–269.
38. Sunanda R, Rao BS, Raju TR. Restraint stress-induced alterations in the levels of biogenic amines, amino acids and AchE activity in the hippocampus. *Neurochem Res*. 2000;25:1547–1552.
39. Clayton JA, Collins FS. Policy: NIH to balance sex in cell and animal studies. *Nature*. 2014;509:282–283.

33

Rapid and Slow Effects of Corticosteroid Hormones on Hippocampal Activity

M. Joëls, R.A. Sarabdjitsingh, F.S. den Boon, H. Karst

University Medical Center Utrecht, Utrecht, The Netherlands

Abstract

Hippocampal cells are continuously exposed to varying concentrations of corticosteroid hormones. These hormones can alter hippocampal activity by binding to mineralocorticoid receptor (MR) or glucocorticoid receptor (GR). Via nongenomic pathways, MRs generally enhance hippocampal excitability. The rapid nongenomic MR actions are important shortly after stress exposure and may help organisms selecting an appropriate response strategy. At the same time, slower genomic actions are started which several hours later, via GRs (among other factors), increase surface expression of AMPA receptors in hippocampal cells and strengthen glutamatergic signaling through pathways partly overlapping with long-term potentiation. This raises the threshold for subsequent induction of synaptic potentiation. Simultaneously, steady transfer of excitatory information is dampened. Synapses activated during stress are thus presumably strengthened and protected against excitatory inputs reaching the cells at a later time-point. This might facilitate consolidation of stress-related information. Thus, coordinated MR-mediated and GR-mediated actions on hippocampal neurons can promote behavioral adaptation.

CORTICOSTEROIDS AND THE BRAIN

The Stress Response

Potential threats from external or internal sources (stressors), be it physical or psychological in nature, are perceived by individuals and generally start a well-orchestrated response involving activation of the autonomic nervous system and the hypothalamus-pituitary-adrenal (HPA) axis. Actual or anticipated situations of threat are subjectively experienced as "stress."

Directly after stress, an immediate response is started via the sympathetic nervous system, of which the release of adrenaline from the adrenal medulla is the most prominent result. Slightly later, an extensive network of brain regions becomes activated which eventually leads to activation of parvocellular neurons in the paraventricular nucleus of the hypothalamus, causing the release of corticotropin-releasing hormone (CRH) into the portal

Stress: Neuroendocrinology and Neurobiology
http://dx.doi.org/10.1016/B978-0-12-802175-0.00033-4

vessels. CRH reaches the anterior pituitary where it leads to secretion of adrenocorticotropin hormone into the circulation which subsequently gives rise to the synthesis and release of hormones from the inner adrenal cortex. In humans this involves primarily release of cortisol; in rodents, corticosterone. Together this system is known as the HPA axis.

Exposure to stressful situations thus involves many mediators, including catecholamines, CRH, and corticosteroid hormones, which—in concert—help the organism to adapt to the changing environment. This chapter will focus on only one element in this complex response, i.e., the effect of corticosteroid hormones on electrical activity of neurons in the hippocampus.

Hormone Levels

In mammals, corticosteroids circulate in a circadian pattern, with hormone levels peaking at the end of the resting phase.[132] These circadian fluctuations actually represent the peaks of ultradian pulses with an interpulse interval of approximately 60 min. The pulsatile pattern is maintained across the blood–brain barrier and persists in corticosteroid target regions such as the hippocampus.[22] Corticosterone pulsatility is necessary to maintain normal GR signaling and HPA axis responsiveness to stress.[61,100]

KEY POINTS

- Corticosteroids binding to mineralocorticoid receptors in hippocampal cells rapidly increase glutamatergic transmission.

- Via glucocorticoid receptors, corticosterone slowly increases glutamatergic transmission in specific hippocampal synapses while attenuating overall information transfer. This increases the signal-to-noise ratio.

- In this slow time-domain the induction of synaptic plasticity is impaired, which may protect earlier encoded information.

- The rapid and slow corticosteroid effects on hippocampal excitability are thought to promote the selection of an appropriate behavioral strategy and retention of stress-related information after stress, respectively.

While corticosteroid hormones are secreted in ultradian or stress-induced surges, many other factors determine the eventual hormone concentration "seen" by hippocampal neurons. First, corticosterone-binding globulin (CBG), a low-capacity high-affinity plasma protein, serves as the principal carrier of circulating corticosteroids, together with albumin. Only a small fraction

of corticosteroids is thus available for binding to receptors. Shortly after stress, the concentration of CBG rises, which temporarily buffers the free corticosterone level.[90] Second, natural corticosteroids can pass the blood–brain barrier by diffusion due to their lipophilic nature. Yet, cortisol and many synthetic steroids like dexamethasone penetrate the brain poorly due to multidrug resistance (mdr) 1a P-glycoprotein activity.[40,71,79] Third, intracellular levels of corticosteroids are modulated by 11β-hydroxysteroid dehydrogenase (HSD) enzymes, which exist in two isoforms.[122] Type 2 (11βHSD-2) catalyses inactivation of glucocorticoids but is expressed only in a limited number of brain regions, not including the hippocampus. Yet, type 1 is ubiquitously present in the brain—also in the hippocampus—and promotes the reduction of inactive 11-keto derivatives like dehydrocorticosterone and cortisone into active corticosterone and cortisol, respectively.[105]

All in all, corticosterone is the main corticosteroid receptor ligand in the rodent brain and its levels in the hippocampus quite accurately follow fluctuations seen peripherally, though liable to modulatory mechanisms.

Corticosteroid Receptors: Genomic and Nongenomic Signaling

Corticosteroid receptors belong to the superfamily of nuclear receptors which act as transcriptional regulators. Two types of receptors exist in brain tissue,[93] i.e., the mineralocorticoid receptor (MR) and glucocorticoid receptor (GR). Although there are only single genes encoding for each of these receptors, multiple isoforms have been described (for reviews see Refs 18,94). GRs are quite ubiquitous in their distribution[93] and expressed in both neurons and glial cells. Brain regions with very high expression levels include the paraventricular nucleus of the hypothalamus, the CA1 hippocampal area and dentate gyrus, the central and cortical amygdala nuclei, lateral septum, and nucleus tractus solitarii. MRs display a much more restricted distribution, with high expression levels for instance in neurons of all hippocampal subfields and the lateral septum.

The MR has a high affinity for the endogenous hormones aldosterone, corticosterone, and cortisol, with a Kd of approximately 0.5 nM.[93] The affinity of the GR for corticosterone and cortisol is approximately 10-fold lower and manifold lower for aldosterone. Intracellular MRs are almost always substantially occupied by corticosterone (or cortisol), even with hormone concentrations during the trough of the circadian rhythm and ultradian pulses. By contrast, GRs are only partially occupied when corticosteroid levels are low due to the relatively low affinity of GRs for the natural ligands and gradually become occupied when hormone levels rise, e.g., at the peak of ultradian pulses or after stress.[61]

Activated MRs and GRs bind as homodimers to consensus sequences (glucocorticoid response elements, GREs) in the promoter of 1–2% of the genes and change transcription of these genes; they can also interact with other transcription factors, altering their efficacy.[14] GRs bind to two distinct populations of GREs in the hippocampus[87]: one population that shows GR binding over a wide range of corticosterone concentrations and another population that reveals GR binding only with very high corticosterone concentrations. Less is known about DNA binding of MRs. Coactivator and corepressor proteins may discriminate between MR and GR activation and thereby render site-specificity and context-specificity to the receptor-mediated signal.[133] The transcriptional patterns of hippocampal tissue following acute or chronic corticosterone elevations have been described in detail.[72] Interestingly, the pattern after acute corticosterone administration depends on the earlier stress history of the animal.[15,28]

In addition to this genomic mechanism of action, it is already known for decades that corticosteroids can change neuronal activity in a rapid manner that clearly must be nongenomic (see for reviews Refs 7,23). Whether or not this involves a different type of receptor is still a matter of debate. Some of the rapid corticosteroid actions clearly require the presence of the MR gene.[46,47] Other actions depend on the GR gene.[47,19,34] The receptor-mediating rapid effects appear to be accessible from the outside of the membrane.[19,46] Yet, electron microscopic proof for location close to or even within the membrane is still scarce[88] and biochemical isolation of a "membrane-receptor," unlike earlier success in amphibian brain,[78] has not been convincingly demonstrated to date in mammalian brain tissue. The pharmacological profile of these presumed membrane-located receptors is different from that described for intracellular MR and GR. For example, inhibitory effects thought to be mediated by GRs are not always blocked by the GR-antagonist RU38486[19] and the dose of corticosterone required to induce rapid effects on hippocampal cells is 10-fold higher than expected for an MR-dependent phenomenon.[46] These deviations in pharmacological profile are not unprecedented[78] (see for review Ref. 64). They may point to the existence of a different receptor molecule but could also be explained by the constraints related to the membrane localization and, e.g., the inability of chaperones to associate with the receptor.

In this chapter we will highlight the rapid and slow effects of corticosteroid hormones on hippocampal cells regarding: (1) active and passive membrane properties such as action potentials or resting membrane potential, respectively; (2) signaling through the main excitatory transmitter, glutamate, and the main inhibitory transmitter GABA; and (3) on the phenomenon of synaptic plasticity—by which synaptic transmission is increased

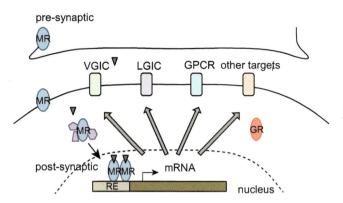

FIGURE 33.1 Corticosteroids (triangles) are lipophilic and easily enter the plasma membrane where they bind to intracellular receptors, that is, mineralocorticoid receptor (MR) and glucocorticoid receptor (GR). In the unbound form, these receptors are associated with other molecules such as heat shock proteins (angular shapes). Upon binding, the receptor complex dimerizes and translocates to the nucleus where the dimer binds response elements (REs) in responsive genes. Alternatively, MR (or GR) interacts with other transcription factors (not shown). Through both pathways, gene transcription is altered for a prolonged period of time. Corticosteroids transcriptionally regulate many molecules involved in neurotransmission, including voltage-gated ion channels (VGICs), ligand-gated ion channels (LGICs), G protein-coupled receptors (GPCRs, not discussed in this chapter) and, e.g., receptors for growth factors or ion pumps (here indicated as "other targets"; not discussed in this chapter). In addition to transcriptional regulation, there are nongenomic pathways through which corticosteroid receptors can affect information transfer. This involves receptors that are associated with the plasma membrane, either postsynaptically or presynaptically. *Reproduced with permission from Vogel S, Fernández G, Joëls M, Schwabe L. Cognitive adaptation under stress: a case for the mineralocorticoid receptor. Trends Cogn Sci. 2016;20:192–203.*

(long-term potentiation, LTP) or decreased (long-term depression, LTD) for a long period of time in an activity-dependent manner (see Fig. 33.1).

RAPID EFFECTS ON NEURONAL ACTIVITY

Active and Passive Membrane Properties

Passive and active membrane properties of hippocampal cells appear to be little affected by corticosterone[37] (see Table 33.1). With regard to voltage-gated ion channels, an early study reported rapid inhibition of L-type and N-type calcium currents in dissociated CA1 pyramidal neurons, but this was only seen with very high concentrations of cortisol and the natural ligand corticosterone was not very potent.[24] More recently, the voltage dependency of activation of a transient K-current (IA) in CA1 pyramidal cells was found to be shifted to the right by corticosterone via an MR-dependent postsynaptic mechanism, indicating that this channel is less activated during small depolarizations.[76] This will result in a higher likelihood to induce action potentials with excitatory input.

TABLE 33.1   Rapid Effects of Corticosteroid Hormones on Neural Activity in the Mammalian Brain

Area	In Vitro/In Vivo	Treatment	MR/GR	Effect	Basal Activity	Plasticity	References
CA1 area (guinea pigs)	In vitro	Cortisol, CORT		Very high cortisol concentrations (and to lesser extent CORT) inhibit and slow down activation of N-type and L-type calcium currents, via G protein and PKC-dependent process.			24
CA1 area (rats)	In vitro	ADX	MR?	In slices from ADX rats: low CORT doses increased pop. spike amplitude, higher doses less clear effect.	Enhanced		92
CA1 area (mice)	In vitro	CORT	MR	Almost immediate increase in mEPSC frequency, no change in mIPSC frequency. Via presynaptic MR and ERK pathway. Depends on Lsamp. Rightward shift in I-A activation curve, via postsynaptic MR, involves G proteins.	Enhanced		46,76,91
CA1 area (rats)	In vitro	ADX + CORT or ALDO	MR and GR	No change in passive or active membrane properties. High dose of CORT (MR + GR) reduces the amplitude of both the EPSP and slow IPSP, and the firing probability of synaptically driven action potentials, within 20 min, compared to low dose of ALDO (MR).	Stable via MR, reduced via GR		37
CA1 (rats)	In vitro	CORT		Reduction of pop spike amplitude with very high CORT dose. Starts after ~15 min, plateaus at 20–40 min.	Reduced		114
CA1 area (rats)	In vitro	Stress		Footshock just before slice preparation: mEPSC amplitude and frequency reduced. PPF reduced, but input-output unaffected. Very young rats (2–4 weeks)	Reduced		135,27
CA1 area (rats)	In vitro	CORT		Very high CORT concentration reduces (sharp electrode) or enhances (whole cell) IPSP amplitude. CORT effect depends on postsynaptic intracellular component?	Enhanced (?)		134,111
CA1 area (rats)	In vitro	DEX or restraint	GR (?)	Increased sIPSC frequency and amplitude, no change in mIPSC. Mimicked by DEX-BSA, requires G proteins, unaffected by MR or GR blockers. Via NO. Stress increases sIPSC frequency.	Reduced		34
Dorsal hipp. (Rats)	In vivo	CORT		Reduced firing rate appr. 20 min after peripheral CORT injection.	Reduced		86
CA1 area (rats)	In vivo	Stress		Stress impairs the stability of or reduces firing rates of place cells. No effect on location.	Decreased firing rate		56,81
CA1 area (mice)	In vitro	CORT	Not MR (?)	Enhanced LTP when CORT is given during (but not after) HFS. Depends on Lsamp.		Enhanced LTP	120,91
CA1 area (rats)	In vitro	CORT	GR	20–30 min after CORT onset: NMDA-fEPSP increased, AMPA-fEPSP unaffected. NMDA/AMPA EPSC ratio increased, no change in PPR; blocked by RU486, mimicked by DEX. No change in trafficking; More LTP and LTD.	Enhanced	Enhanced LTP, enhanced LTD	112

Region		Stimulus	Receptor	Effect			Refs
CA1 area (rats and mice)	In vivo	Stress	GR	LTP reduced or depotentiated and LTD facilitated shortly after novelty. Not linked to rise in CORT, but blocked by RU486 and requiring protein synthesis in one study.		LTD enhanced	124,125,69,127,33
DG (mice)	In vitro	CORT	MR	Quick increase in mEPSC frequency, no change in amplitude. Via MR.	Enhanced		80
DG (rats, ADX)	In vivo	CORT		Decreased fEPSP and LTP up to 30min after CORT.	Decreased	Decreased LTP	25
DG (rats)	In vitro	CORT		Increased LTP (early phase) only when GABA transmission is blocked	No change	Enhanced (early) LTP	89
Neonatal HIPPO culture	In vitro	CORT (BSA)	Not MR or GR	NMDA-evoked currents reduced by very high CORT concentrations. Not blocked by MR or GR antagonists. Also with CORT-BSA and intracellular CORT. Via cAMP/IP3? icv CORT reduces LTP <30min.	Reduced		63,136

ALDO, aldosterone; CORT, corticosterone; DEX, dexamethasone; eCB, endocannabinoid; HFS, high-frequency stimulation; nAA, nucleus accumbens; NO, nitric oxide; PPF, paired pulse facilitation; PPR, paired pulse response.
This table is based on Joëls M, Sarabdjitsingh RA, Karst H. Unraveling the time domains of corticosteroid hormone influences on brain activity: rapid, slow, and chronic modes. Pharmacol Rev. 2012;64:901–938.

Amino Acid Transmission

In principal hippocampal neurons, glutamate receptor-mediated responses are evoked by the arrival of action potentials and calcium release in the presynaptic terminal. However, to a limited extent glutamate transmission also occurs spontaneously, even in the absence of action potentials. This spontaneous background activity is apparent from the postsynaptic response to a spontaneously released synaptic vesicle containing glutamate, i.e., a so-called miniature excitatory postsynaptic current (mEPSC). Inhibitory transmission is mostly carried by GABA. Spontaneous release of GABA-containing vesicles is postsynaptically recorded as a spontaneous inhibitory postsynaptic current (sIPSC) or, in case of action potential blockade, a miniature (m)IPSC.

CA1 hippocampal pyramidal cells respond to corticosterone within minutes with enhanced mEPSC frequency[46] (Fig. 33.2). Other properties of the mEPSCs, such as amplitude, rise time or decay, are not consistently changed by the hormone. Follow-up experiments indicated that the hormone probably increases the release probability of glutamate-containing vesicles. Upon washout of corticosterone, the mEPSC frequency quickly returns to the pretreatment level. The rapid onset and offset of the response and the fact that corticosterone exerted very similar effects in the presence of a protein synthesis inhibitor support that the increased mEPSC frequency is mediated through a nongenomic pathway. Corticosterone conjugated to (membrane-impermeable) bovine serum albumin induced very similar effects on mEPSC frequency, while intracellular administration

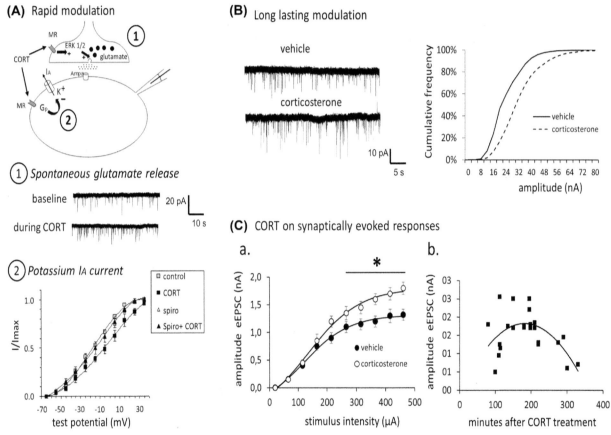

FIGURE 33.2 Rapid and delayed effects of corticosterone on the activity of CA1 hippocampal pyramidal cells. (A) In the hippocampus, corticosterone binds to presynaptic mineralocorticoid receptors (MRs) which via ERK1/2 signaling cause a rapid increase in the release probability of glutamate. This is reflected in enhanced frequency of the miniature excitatory postsynaptic currents (mEPSCs; middle). In addition, the hormone can bind to postsynaptically located MRs which through a G protein are coupled to transient potassium channels (IA). The hormone causes a rightward shift in the activation curve of this current (right), increasing the likelihood for postsynaptic action potential generation (*Based on Karst H, Berger S, Turiault M, Tronche F, Schutz G, Joels M. Mineralocorticoid receptors are indispensable for nongenomic modulation of hippocampal glutamate transmission by corticosterone. Proc Natl Acad Sci USA. 2005;102:19204–19207; Olijslagers JE, de Kloet ER, Elgersma Y, van Woerden GM, Joels M, Karst H. Rapid changes in hippocampal CA1 pyramidal cell function via pre- as well as postsynaptic membrane mineralocorticoid receptors. Eur J Neurosci. 2008;27:2542–2550.*). (B) Several (1–4) hours after exposure of hippocampal CA1 neurons to a brief pulse of corticosterone, the amplitude of mEPSCs is enhanced. This is also reflected in the cumulative frequency distribution of mEPSC amplitudes. (C) Synaptically evoked EPSCs were also enhanced in amplitude (a), but only in a restricted time-window after corticosterone exposure (b), appr. 3–4 h after treatment (*Based on Karst H, Joels M. Corticosterone slowly enhances miniature excitatory postsynaptic current amplitude in mice CA1 hippocampal cells. J Neurophysiol. 2005;94:3479–3486.*).

was ineffective.[46,76] This suggests that corticosterone binds to a molecule that is accessible from the outside of the cell, although actual proof for localization of the receptor in the membrane is still limited (see above). Pharmacological and genetic studies supported that the rapid effect depends on MR rather than GR[46] and requires expression of limbic system-associated membrane protein, Lsamp.[91] MRs mediating the rapid effect are probably localized on the presynaptic membrane and linked to the ERK1/2 signaling pathway.[76] A highly comparable MR-dependent raise in mEPSC frequency was observed in granule cells of the dentate gyrus.[80]

Corticosterone changes glutamate transmission in the CA1 hippocampal area not only by increasing the release probability, a presynaptic property, but also acts on the postsynaptic membrane: shortly after corticosterone administration, GluA2 subunits of the AMPA receptor exhibit increased lateral movement and a higher dwell-time in the postsynaptic density.[29] This effect too was found to be mediated by MRs, was mimicked by corticosterone-BSA, and persisted in the presence of a protein synthesis inhibitor. Both actions on glutamate transmission are expected to increase the (spontaneous) activity of hippocampal CA1 neurons. This would fit with the enhanced population spike amplitude observed extracellularly when treating slices from adrenalectomized rats (devoid of endogenous corticosteroids) with low doses of corticosterone, presumably activating MR.[92]

One study[112] reported that CA1 cells also respond more strongly to excitatory input at a slightly later time-point (i.e., 20–30 min after the start of corticosterone administration), although the characteristics were somewhat different: When excitatory postsynaptic currents were synaptically evoked (eEPSC), the ratio of responses via the NMDA versus AMPA receptors for glutamate was increased. Extracellularly, an increase in the field excitatory postsynaptic potential (fEPSP) evoked via NMDA receptor but not AMPA receptor was observed. The effect on NMDA and AMPA currents was mimicked by dexamethasone and blocked by the GR antagonist RU38486, suggesting GR involvement. The fact that RU38486 was effective raises the question whether these effects are truly nongenomic, because this drug is generally ineffective in blocking membrane receptor-mediated events.[19]

However, in this time-domain (10–60 min after corticosterone application), most studies reported reduced responses to synaptic input. For instance, spontaneous firing rate of hippocampal cells was reduced by corticosterone peripherally injected 20 min earlier.[86] This fits with data showing that various types of stress impair the stability or reduce the firing rate of hippocampal place cells in this time-domain.[56,81] In vitro administered corticosterone (at a very high dose) was reported to reduce the population spike amplitude in the CA1 area.[114]

Similarly, 20 min after corticosterone application the ability to evoke an action potential with synaptic stimulation and the amplitude of excitatory or inhibitory synaptic currents in CA1 neurons gradually declined.[37]

The overall activity of CA1 pyramidal cells of course depends on the balance between excitatory and inhibitory transmission. The rapid effects of corticosterone on GABAergic transmission, however, are somewhat difficult to interpret. GABAergic inhibitory responses were reduced when recording with sharp electrodes but slightly enhanced with whole cell recording, indicating that an intracellular molecule—which is dialyzed when recording in the whole cell mode—is important in mediating the reduced inhibition.[134,111] Hu et al.[34] reported that restraint stress and dexamethasone conjugated to albumin increase the frequency and amplitude of sIPSCs, but mIPSCs were unaffected. This effect involved postsynaptic G proteins, retrograde transport by nitric oxide, and was not blocked by either MR antagonist or GR antagonist.

Overall, corticosterone appears to rapidly (<10 min) enhance spontaneous excitatory activity, although synaptically evoked field potentials are not markedly altered by corticosterone administration.[120,89] Notably, this realm of the first hour after stress is also the time-domain in which other stress mediators are active and determine the excitability. For instance, CRH is known to quickly potentiate the population spike in the Schaffer projection to the CA1 hippocampal area.[6] Directly after these very rapid effects (i.e., 10–60 min after a rise in corticosterone level), hippocampal cell activity generally seems to be suppressed.

The rapid changes in hippocampal activity—especially those developing in minutes—may be relevant shortly after stress (see Concluding Remarks section) but also during the peaks of ultradian pulses, as was, e.g., shown for cortical cells too.[62] In this respect, it is of interest that a second pulse of corticosterone 1 h after the first evokes a highly comparable response in CA1 neurons, although some attenuation was observed during the third pulse.[47,102] Pulsatile exposure was also reported to be important to maintain the stability of hippocampal neuronal activity between the pulses.[101] Overall, it can be concluded that hippocampal cells respond strongly to a first pulse of corticosterone but that the influence of subsequent pulses on network activity slowly subsides.

Rapid Effects on Synaptic Plasticity

Rapid enhancement of LTP by corticosterone was observed in the CA1 and DG (summarized in Ref. 39; see Table 33.1), dependent on the presence of limbic system-associated membrane protein Lsamp.[91] Stress (as opposed to just corticosterone) appears to affect LTP differently. A series of papers showed that particularly exposure to a novel environment prevents induction of LTP, de-potentiates earlier induced LTP, and facilitates the induction of LTD (see for details Table 33.1). Remarkably, some of these effects could be blocked by RU38486 and depend on protein synthesis,[125] although their rapid onset (<10 min) and the fact that they occur independent of rises in corticosteroid level almost precludes a GR-dependent genomic effect.

SLOW EFFECTS ON NEURONAL ACTIVITY

Many studies have revealed hormone actions on neural activity in a much slower time-domain (with a delay of >1h), in line with the gene-mediated signaling pathway of GRs. We will only briefly summarize these studies, since they have been reviewed manifold over the past decades.[39]

Active and Passive Membrane Properties

Corticosteroids generally do not change passive or active membrane properties of CA1 pyramidal neurons in the dorsal hippocampus, such as resting membrane potential, input resistance, or characteristics of the action potential, at least not in tissue from ADX animals.[35,49] In the ventral-most part (20%) of the hippocampus, corticosterone lowers the threshold for action potential generation, resulting in higher excitability.[67]

By contrast, voltage-dependent calcium currents form a major target for glucocorticoids. In particular the amplitude of sustained high-voltage activated calcium currents was found to be enhanced by glucocorticoid treatment or stress, compared to the situation in which predominantly MRs are activated,[38,44,50] a process requiring protein synthesis and DNA binding of GR homodimers.[50,45] L-type rather than N-type calcium currents form a target for glucocorticoids.[11] In contrast to calcium currents, sodium and potassium currents are not much affected by corticosterone (see Table 33.2).

Circulating corticosteroid levels throughout the day also correlate with the calcium current amplitude in CA3 neurons.[57] By contrast, dentate granule cells do not show an enhanced calcium current amplitude after corticosterone exposure.[113]

Amino Acid Transmission

Slow corticosteroid effects on amino acid-mediated input to hippocampal cells have been studied in multiple ways: (1) by exposing cells to a depolarizing step, to mimic a steady excitatory input; (2) by examining

specific glutamatergic pathways; and (3) by studying GABAergic inputs.

Upon depolarization, CA1 neurons fire action potentials, but the frequency gradually accommodates due to activation of a slow calcium-dependent potassium current (IsAHP). After the depolarization, the IsAHP is slowly deactivated, resulting—in the current clamp mode—in a so-called after-hyperpolarization (AHP). High (compared to low) levels of corticosterone were found to enhance the amplitude of the IsAHP and AHP in dorsal hippocampal CA1 neurons, resulting in fewer spikes upon depolarization.[35,36,49] Similar effects were also seen in the CA3 area.[57] CA1 neurons in the ventral-most part of the hippocampus showed reduced firing frequency accommodation and more spikes upon depolarization after corticosterone administration.[67]

In CA1 and cultured hippocampal neurons, a pulse of corticosteroids enhanced the amplitude (but, different from the rapid effects, not the frequency) of mEPSCs recorded several hours after corticosteroid exposure, via GRs.[42,70] This concurs with a slow GR-dependent increase in surface expression of GluA2 subunits, which requires protein synthesis and occludes chemically induced LTP.[29,70] Interestingly, inhibitory properties, i.e., the sIPSC amplitude, are also enhanced in the dorsal hippocampus, via GRs[67]; in the ventral hippocampus, an MR-dependent reduction in sIPSC frequency was reported. The effects on sIPSCs, however, are seen in a somewhat faster time-domain than the effects on excitatory transmission.

Synaptically evoked responses recorded extracellularly in the various hippocampal areas are usually not found to be affected by corticosterone or stress, but enhancements or reduced activity were reported in a few studies (see Table 33.2 for details). Perhaps these corticosteroid actions are restricted to a limited number of synapses and thus not seen at a more general level.

In conclusion, at the level of single synapses corticosteroids slowly change glutamatergic transmission in a manner sharing characteristics of LTP. This does not imply that all excitatory information reaching hippocampal cells some hours after stress is facilitated. For instance, the larger spike frequency accommodation and AHP amplitude could serve as a brake to steady excitatory inputs, particularly when unrelated to the stressful event. Collectively this may enhance the signal-to-noise ratio of hippocampal synaptic transmission.

Slow Modulation of Synaptic Plasticity

Contrary to the rapid effects of corticosterone, slow actions hamper the induction of LTP in the CA1 hippocampal area in vitro, in vivo, or after exposure to stress, especially stress of an uncontrollable nature (see for

TABLE 33.2 Delayed Effects of Corticosteroid Hormones on Neural Activity in the Mammalian Brain

Area	In Vitro/In Vivo	Treatment	MR/GR	Effect	Basal Activity	Plasticity	References
CA1 area (rats and mice)	In vitro	Stress or CORT in vitro	MR and GR	Small Ca current amplitude with MR, small AHP, and more spikes during depolarization. With GR activation, slow enhancement in amplitude of L-type Ca current, bigger AHP, and fewer spikes with depolarization.	Increased with MR and decreased with GR		35,36,38,49,50,44,45, 31,11,113
CA1 area (rats)	In vitro	ADX + low CORT replacement	MR (?)	ADX animals, replacement with low dose of CORT: Higher membrane time constant, decreased slow AHP, and more action potentials during depolarization.	Increased (with MR?)		5
CA1 area (rats and mice)	In vitro	CORT	MR and GR	Small inward rectifying K-current (Ih?) with MR activation, large with GR activation. No effect on A current. Small effect of ADX on sodium currents, but no difference between MR and GR.			43,119
CA1 area (mice)	In vitro	CORT in vitro		Enhanced pop spike with low dose of CORT, decreased pop spike with high dose.	Enhanced (with MR?), decreased (with GR?)		95
CA1 area, dorsal (mice)	In vitro	CORT in vitro	GR	Slow enhancement in mEPSC amplitude, not frequency, via GR.	Enhanced (LTP-like)		42
CA1 area, dorsal and ventral (juvenile rats)	In vitro	CORT in vitro	MR and GR	Ventral hipp: lower AP threshold. Reduced sIPSC frequency, via MR; increased IPSC amplitude via GR. In dorsal hipp: lower input resistance, increased (m)IPSC amplitude via GR. Reduction in GABA-induced noise. No change in reversal potential. Onset: >25 min, peaks at 55 min.	Reduced in dorsal via GR, enhanced in ventral hipp via MR		67
CA1 area (rats and mice)	In vivo and in vitro	Stress or CORT in vitro, sometimes with MR and GR (ant) agonists	GR	LTP reduced by CORT in vitro and after uncontrollable stressors. Involvement of NMDA receptor. Requires intact BLA (but not CeA) in vivo. In vitro rescued by estradiol.	Generally no change; variable effects on PPR	Reduced LTP via GR	26,108,107,21,20, 53,54,137,123,48, 60,130,32,33,98, 77,10
CA1 area (ADX rats)	In vitro	MR and GR agonists	MR and GR	LTP enhanced by aldosterone, reduced by GR agonist.	No change	Enhanced LTP via MR, reduced via GR	85
CA1 area (rats and mice)	In vivo and in vitro	GR blockers in vivo	GR	LTD enhanced in slices made >30 min after stress, blocked by RU486. Decrease in Glu uptake and GluN2B crucial. Stress also enhances group I mGluR-dependent LTD via GR; effects of stress reversed by lithium.		Enhanced LTD via GR	123,127–129,55, 121,12,27,73
CA1 area (rats)	In vitro	MR and GR blockers in vivo, CORT in vitro	MR and GR	Swim stress (15 min), then slices: LTP in dorsal hipp reduced, LTD enhanced, GR dependent. LTP in ventral hipp (normally small) enhanced by stress, LTD converted to LTP, MR dependent. GR agonist in vitro (>80 min): more LTD in both areas; MR agonist (>80 min): From LTD to LTP in both areas.		Reduced LTP, enhanced LTD in dorsal hipp via GR; enhanced LTP, from LTD to LTP in ventral hipp via MR	66,68

Continued

II. ENDOCRINE SYSTEMS AND MECHANISMS IN STRESS CONTROL

TABLE 33.2 Delayed Effects of Corticosteroid Hormones on Neural Activity in the Mammalian Brain—cont'd

Area	In Vitro/In Vivo	Treatment	MR/GR	Effect	Basal Activity	Plasticity	References
CA3 area (rats)	In vitro	Circadian variations		Large calcium current and strong frequency accommodation during circadian peak.	Reduced (via GR?)		57
CA3 area (rats)	In vitro	ADX + CORT replacement		ADX associated with more nonbursting cells than in ADX + low CORT replacement (via MR?)			75
CA3 area (mice)	In vitro	Stress	ADX or GR blocker	Mossy fiber—CA3: Impaired LTP by stress, prevented by ADX or RU38486. Involving conversion of cAMP to adenosine which impairs transmitter release.	No change	Reduced LTP via GR	13
DG (rats)	In vitro	CORT		Restraining calcium influx via MR, no effect on calcium-current amplitude via GR			41,113
DG (rats)	In vivo	MR and GR agonists	MR and GR	Aldosterone enhances late LTP and prolongs LTP. GR agonists reveal LTD. In ADX rats reduced, restored by CORT; CORT to intact rats reduces LTP.	no change	Enhanced LTP via MR, reduced LTP (even LTD) by GR	82,83,84,109,126
DG (rats)	In vivo	Stress (prior to HFS)	MR and GR	Enhanced baseline responses and LTP. Stress + GR blocker gives strong LTP, stress + MR blocker gives LTD	Enhanced	Enhanced LTP via MR, reduced LTP (even LTD) via GR	48,4,118,131
DG (mice)	In vitro	Brief stress, MR and GR blockers pretreated	MR and GR	10min neck stress, slices prepared immediately afterward. LTP/LTD 3.5h later. LTP enhanced, LTD decreased. Both MR and GR antagonists block effect. No change in input-output curves, nor PPR.	No change	Enhanced LTP, reduced LTD	110
DG (rats)	In vivo	Stress (prior to HFS)		Reduced LTP after various types of stress or new environment (but not cold stress), very similar to effect of earlier LTP in nonstressed animals—stress induces LTP-like state. Depends on noradrenaline and CORT in BLA.	No change	Reduced LTP	106,8,2,3
DG (rats)	In vivo	Stress (prior to HFS) + CORT		Fear conditioning paradigm, CORT directly after training. If memory enhanced, then LTP induction also facilitated.		Enhanced LTP	1
DG (rats)	In vivo	Stress (after HFS)	MR and GR	Weak LTP impaired by handling (low stress, via GR) and prolonged by swim stress (high stress, via MR) 15min after HFS. Depends on processing of novel/contextual information. Requires intact BLA.		Prolonged LTP via MR, reduced via GR	58,59
Neonatal hipp culture	In vitro	CORT		Slow enhancement in mEPSC amplitude, frequency unaffected. Blocked by RU38486.	Enhanced (LTP-like) via GR		70

AHP, after-hyperpolarization; AP, action potential; BLA, basolateral amygdala; Ca, calcium; CeA, central amygdala; CORT, corticosterone; EC, entorhinal cortex; eCB, endocannabinoid; HFS, high-frequency stimulation; hipp, hippocampus; K-current, potassium current; NPY, neuropeptide Y; POMC, proopiomelanocortin; PPD, paired pulse depression; PPR, paired pulse response.
This table is reproduced with permission from Joëls M, Sarabdjitsingh RA, Karst H. Unraveling the time domains of corticosteroid hormone influences on brain activity: rapid, slow, and chronic modes. Pharmacol Rev. 2012;64:901–938.

review Refs 52,39; details in Table 33.2). This effect is a GR-dependent phenomenon[85] and depends on NMDA receptor-mediated transmission.[53] Corticosterone not only hampers LTP induction but also slowly promotes LTD (see Table 33.2). Enhanced LTP was reported for selective MR activation,[85] particularly in the ventralmost part of the CA1 hippocampal area.[66,68]

Reduced LTP after stress via a slow GR-mediated action was also seen in the hippocampal CA3 region, requiring the conversion of cAMP into adenosine.[13] Results in the dentate gyrus are more ambiguous. Some studies report enhanced LTP via MR activation or reduced LTP after GR activation (details in Table 33.2). Abrari et al.[1] reported facilitated induction of LTP in association with improved memory when rats were treated with a moderately high dose of corticosterone directly after training in a fear conditioning paradigm. However, other studies could not link stress-induced changes in LTP/LTD to known corticosteroid receptor types[110] or did not find changes after inescapable or cold stress at all.[106,117] Low stress conditions such as handling of the animal after LTP induction were found to impair weak LTP, whereas exposure to a high stress condition (forced swim) prolonged the late phase of LTP, via an MR-dependent process,[58] requiring an intact BLA.[59] The role of the basolateral amygdala (BLA) for LTP in the dentate was also apparent from other studies.[3]

In conclusion, in most cases stress or high corticosterone concentration was reported to impair the ability to induce synaptic strengthening. This is compatible with the view that first rapid effects of corticosterone lead to more glutamate release and associated postsynaptic changes, which subsequently through a slower GR-dependent mechanism promotes synaptic localization of GluA2 subunits. Those synapses activated during the stressful event and exposed to elevated levels of stress hormones will be strengthened for a considerable period of time, raising the threshold for subsequent synaptic strengthening. This could protect strengthened synapses from retrograde interference, so that the ongoing encoding process can occur relatively undisturbed.

CONCLUDING REMARKS

The above-described changes in neuronal activity of the hippocampus are likely to play a role in behavioral functions that critically depend on the hippocampus, such as contextual or spatial memory formation. This has been extensively studied, both in rodents and humans. We here address this issue, with special attention to the fact that corticosteroids may exert their effect on hippocampal function in two time-domains, i.e., a fast domain which primarily involves MR activation and a domain >1h later which depends on GR-mediated signaling.

Already in the 1990s, rodent studies showed that MRs are important for the appraisal of novel situations and selection of response strategies, thus promoting acquisition of important information.[16] It has been proposed that rapid actions via MR favor a shift toward simple spatial learning strategies, which involve striatal circuits rather than the hippocampus.[104,103] Recently, fMRI studies in humans indeed support that such a shift toward amygdala–striatal pathway—at the cost of hippocampal activity—takes place briefly after stress and that this shift depends on MRs.[104,115] These simple learning strategies directly after stress provide a good behavioral approach at the short term, yet may come at the price of decreased flexibility and increased vulnerability in the long term.[116] MRs are also important for control of emotional arousal and adaptive behaviors, since this is lost in the absence of forebrain MRs, so that anxiety-related responses remain augmented.[9] In line with this supposed adaptive role of MR, administration of an MR-antagonist prior to training in a contextual fear conditioning paradigm interfered with memory formation.[138]

Rapid effects not only play a role during acquisition of a task but also during retrieval. Stress or corticosteroid application prior to the retention of earlier learned information impairs retrieval of this information via a nongenomic pathway.[17,99] This effect involves the MR,[51] although interaction with other stress hormones is necessary to accomplish the full effect. Possibly, the rapid de-potentiation of LTP and shift toward LTD reported when animals are exposed to a (stressful) novel environment plays a role in this phenomenon,[124,69] although this rapid effect seems to be GR dependent rather than MR dependent.[125]

After these immediate effects mostly via MR (<10min after stress), corticosteroids affect hippocampal activity in a slightly later time-window (10–60min). In this intermediate time-domain too, hippocampus-dependent function at the circuit or behavioral level is altered by stress and glucocorticoids, probably through a process involving protein–protein interactions. For example, in the dentate gyrus acute stress starts a cascade in which ERK1/2 is phosphorylated, leading within 15min to activation of the nuclear kinases MSK1 and Elk-1.[30] This in turn results in histone acetylation and induction of the immediate early genes c-Fos and Egr-1. Histone acetylation via nongenomic actions of GR on phospho-CREB in the insular cortex was found to be important for object recognition, while such a process in the hippocampus plays a role in object location memory.[97]

The slow effects of corticosteroid hormones (>1h) on hippocampal function have been the subject of research for decades (see Ref. 65). There is abundant evidence that stress/glucocorticoids either directly[74] or conditionally[96] promote consolidation of spatial information. This optimally prepares the individual when confronted with a comparable stressor in future times.

Overall, stress affects hippocampal cellular and behavioral function in multiple ways, depending—among other things—on when exactly (relative to the stressor) hippocampal function is probed. We argue that all phases are important for a full-fledged cognitive response to stress, promoting the adaptive capacity of the individual.

References

1. Abrari K, Rashidy-Pour A, Semnanian S, Fathollahi Y, Jadid M. Post-training administration of corticosterone enhances consolidation of contextual fear memory and hippocampal long-term potentiation in rats. *Neurobiol Learn Mem.* 2009;91:260–265.
2. Akirav I, Richter-Levin G. Biphasic modulation of hippocampal plasticity by behavioral stress and basolateral amygdala stimulation in the rat. *J Neurosci.* 1999;19:10530–10535.
3. Akirav I, Richter-Levin G. Mechanisms of amygdala modulation of hippocampal plasticity. *J Neurosci.* 2002;22:9912–9921.
4. Avital A, Segal M, Richter-Levin G. Contrasting roles of corticosteroid receptors in hippocampal plasticity. *J Neurosci.* 2006;26:9130–9134.
5. Beck SG, List TJ, Choi KC. Long- and short-term administration of corticosterone alters CA1 hippocampal neuronal properties. *Neuroendocrinology.* 1994;60:261–272.
6. Blank T, Nijholt I, Eckart K, Spiess J. Priming of long-term potentiation in mouse hippocampus by corticotropin-releasing factor and acute stress: implications for hippocampus-dependent learning. *J Neurosci.* 2002;22:3788–3794.
7. Borski RJ. Nongenomic membrane actions of glucocorticoids in vertebrates. *Trends Endocrinol Metab.* 2000;11:427–436.
8. Bramham CR, Southard T, Ahlers ST, Sarvey JM. Acute cold stress leading to elevated corticosterone neither enhances synaptic efficacy nor impairs LTP in the dentate gyrus of freely moving rats. *Brain Res.* 1998;789:245–255.
9. Brinks V, Berger S, Gass P, de Kloet ER, Oitzl MS. Mineralocorticoid receptors in control of emotional arousal and fear memory. *Horm Behav.* 2009;56:232–238.
10. Cazakoff BN, Howland JG. Acute stress disrupts paired pulse facilitation and long-term potentiation in rat dorsal hippocampus through activation of glucocorticoid receptors. *Hippocampus.* 2010;20:1327–1331.
11. Chameau P, Qin Y, Spijker S, Smit AB, Joëls M. Glucocorticoids specifically enhance L-type calcium current amplitude and affect calcium channel subunit expression in the mouse hippocampus. *J Neurophysiol.* 2007;97:5–14.
12. Chaouloff F, Hemar A, Manzoni O. Acute stress facilitates hippocampal CA1 metabotropic glutamate receptor-dependent long-term depression. *J Neurosci.* 2007;27:7130–7135.
13. Chen CC, Yang CH, Huang CC, Hsu KS. Acute stress impairs hippocampal mossy fiber-CA3 long-term potentiation by enhancing cAMP-specific phosphodiesterase 4 activity. *Neuropsychopharmacology.* 2010;35:1605–1617.
14. Datson NA, Morsink MC, Meijer OC, de Kloet ER. Central corticosteroid actions: search for gene targets. *Eur J Pharmacol.* 2008;583:272–289.
15. Datson NA, Speksnijder N, Mayer JL, et al. The transcriptional response to chronic stress and glucocorticoid receptor blockade in the hippocampal dentate gyrus. *Hippocampus.* 2012;22:359–371.
16. de Kloet ER, Oitzl MS, Joels M. Stress and cognition: are corticosteroids good or bad guys? *Trends Neurosci.* 1999;22:422–426.
17. de Quervain DJ, Roozendaal B, McGaugh JL. Stress and glucocorticoids impair retrieval of long-term spatial memory. *Nature.* 1998;394:787–790.
18. Derijk RH, de Kloet ER. Corticosteroid receptor polymorphisms: determinants of vulnerability and resilience. *Eur J Pharmacol.* 2008;583:303–311.
19. Di S, Malcher-Lopes R, Halmos KC, Tasker JG. Nongenomic glucocorticoid inhibition via endocannabinoid release in the hypothalamus: a fast feedback mechanism. *J Neurosci.* 2003;23:4850–4857.
20. Diamond DM, Rose GM. Stress impairs LTP and hippocampal-dependent memory. *Ann NY Acad Sci.* 1994;746:411–414.
21. Diamond DM, Bennett MC, Fleshner M, Rose GM. Inverted-U relationship between the level of peripheral corticosterone and the magnitude of hippocampal primed burst potentiation. *Hippocampus.* 1992;2:421–430.
22. Droste SK, de Groote L, Atkinson HC, Lightman SL, Reul JM, Linthorst AC. Corticosterone levels in the brain show a distinct ultradian rhythm but a delayed response to forced swim stress. *Endocrinology.* 2008;149:3244–3253.
23. Evanson NK, Herman JP, Sakai RR, Krause EG. Nongenomic actions of adrenal steroids in the central nervous system. *J Neuroendocrinol.* 2010;22:846–861.
24. ffrench-Mullen JM. Cortisol inhibition of calcium currents in Guinea pig hippocampal CA1 neurons via G-protein-coupled activation of protein kinase C. *J Neurosci.* 1995;15:903–911.
25. Filipini D, Gijsbers K, Birmingham MK, Dubrovsky B. Effects of adrenal steroids and their reduced metabolites on hippocampal long-term potentiation. *J Steroid Biochem Mol Biol.* 1991;40:87–92.
26. Foy MR, Stanton ME, Levine S, Thompson RF. Behavioral stress impairs long-term potentiation in rodent hippocampus. *Behav Neural Biol.* 1987;48:138–149.
27. Gao Y, Han H, Xu R, Cao J, Luo J, Xu L. Effects of prolonged exposure to context following contextual fear conditioning on synaptic properties in rat hippocampal slices. *Neurosci Res.* 2008;61:385–389.
28. Gray JD, Rubin TG, Hunter RG, McEwen BS. Hippocampal gene expression changes underlying stress sensitization and recovery. *Mol Psychiatry.* 2014;19:1171–1178.
29. Groc L, Choquet D, Chaouloff F. The stress hormone corticosterone conditions AMPAR surface trafficking and synaptic potentiation. *Nat Neurosci.* 2008;11:868–870.
30. Gutièrrez-Mecinas M, Trollope AF, Collins A, et al. Long-lasting behavioral responses to stress involve a direct interaction of glucocorticoid receptors with ERK1/2-MSK1-Elk-1 signaling. *Proc Natl Acad Sci USA.* 2011;108:13806–13811.
31. Hesen W, Karst H, Meijer O, et al. Hippocampal cell responses in mice with a targeted glucocorticoid receptor gene disruption. *J Neurosci.* 1996;16:6766–6774.
32. Hirata R, Togashi H, Matsumoto M, Yamaguchi T, Izumi T, Yoshioka M. Characterization of stress-induced suppression of long-term potentiation in the hippocampal CA1 field of freely moving rats. *Brain Res.* 2008;1226:27–32.
33. Hirata R, Matsumoto M, Judo C, et al. Possible relationship between the stress-induced synaptic response and metaplasticity in the hippocampal CA1 field of freely moving rats. *Synapse.* 2009;63:549–556.
34. Hu W, Zhang M, Czeh B, Flugge G, Zhang W. Stress impairs GABAergic network function in the hippocampus by activating nongenomic glucocorticoid receptors and affecting the integrity of the parvalbumin-expressing neuronal network. *Neuropsychopharmacology.* 2010;35:1693–1707.

35. Joëls M, de Kloet ER. Effects of glucocorticoids and norepinephrine on the excitability in the hippocampus. *Science.* 1989;245:1502–1505.

36. Joëls M, de Kloet ER. Mineralocorticoid receptor-mediated changes in membrane properties of rat CA1 pyramidal neurons in vitro. *Proc Natl Acad Sci USA.* 1990;87:4495–4498.

37. Joëls M, de Kloet ER. Corticosteroid actions on amino acid-mediated transmission in rat CA1 hippocampal cells. *J Neurosci.* 1993;13:4082–4090.

38. Joëls M, Velzing E, Nair S, Verkuyl JM, Karst H. Acute stress increases calcium current amplitude in rat hippocampus: temporal changes in physiology and gene expression. *Eur J Neurosci.* 2003;18:1315–1324.

39. Joëls M, Sarabdjitsingh RA, Karst H. Unraveling the time domains of corticosteroid hormone influences on brain activity: rapid, slow, and chronic modes. *Pharmacol Rev.* 2012;64:901–938.

40. Karssen AM, Meijer OC, van der Sandt IC, et al. Multidrug resistance P-glycoprotein hampers the access of cortisol but not of corticosterone to mouse and human brain. *Endocrinology.* 2001;142:2686–2694.

41. Karst H, Joels M. Calcium currents in rat dentate granule cells are altered after adrenalectomy. *Eur J Neurosci.* 2001;14:503–512.

42. Karst H, Joels M. Corticosterone slowly enhances miniature excitatory postsynaptic current amplitude in mice CA1 hippocampal cells. *J Neurophysiol.* 2005;94:3479–3486.

43. Karst H, Wadman WJ, Joels M. Long-term control by corticosteroids of the inward rectifier in rat CA1 pyramidal neurons, in vitro. *Brain Res.* 1993;612:172–179.

44. Karst H, Wadman WJ, Joëls M. Corticosteroid receptor-dependent modulation of calcium currents in rat hippocampal CA1 neurons. *Brain Res.* 1994;649:234–242.

45. Karst H, Karten YJ, Reichardt HM, de Kloet ER, Schutz G, Joels M. Corticosteroid actions in hippocampus require DNA binding of glucocorticoid receptor homodimers. *Nat Neurosci.* 2000;3:977–978.

46. Karst H, Berger S, Turiault M, Tronche F, Schutz G, Joels M. Mineralocorticoid receptors are indispensable for nongenomic modulation of hippocampal glutamate transmission by corticosterone. *Proc Natl Acad Sci USA.* 2005;102:19204–19207.

47. Karst H, Berger S, Erdmann G, Schutz G, Joels M. Metaplasticity of amygdalar responses to the stress hormone corticosterone. *Proc Natl Acad Sci USA.* 2010;107:14449–14454.

48. Kavushansky A, Vouimba RM, Cohen H, Richter-Levin G. Activity and plasticity in the CA1, the dentate gyrus, and the amygdala following controllable vs. uncontrollable water stress. *Hippocampus.* 2006;16:35–42.

49. Kerr DS, Campbell LW, Hao SY, Landfield PW. Corticosteroid modulation of hippocampal potentials: increased effect with aging. *Science.* 1989;245:1505–1509.

50. Kerr DS, Campbell LW, Thibault O, Landfield PW. Hippocampal glucocorticoid receptor activation enhances voltage-dependent Ca2+ conductances: relevance to brain aging. *Proc Natl Acad Sci USA.* 1992;89:8527–8531.

51. Khaksari M, Rashidy-Pour A, Vafaei AA. Central mineralocorticoid receptors are indispensable for corticosterone-induced impairment of memory retrieval in rats. *Neuroscience.* 2007;149:729–738.

52. Kim JJ, Diamond DM. The stressed hippocampus, synaptic plasticity and lost memories. *Nat Rev Neurosci.* 2002;3:453–462.

53. Kim JJ, Foy MR, Thompson RF. Behavioral stress modifies hippocampal plasticity through N-methyl-D-aspartate receptor activation. *Proc Natl Acad Sci USA.* 1996;93:4750–4753.

54. Kim JJ, Lee HJ, Han JS, Packard MG. Amygdala is critical for stress-induced modulation of hippocampal long-term potentiation and learning. *J Neurosci.* 2001;21:5222–5228.

55. Kim JJ, Song EY, Kosten TA. Stress effects in the hippocampus: synaptic plasticity and memory. *Stress.* 2006;9:1–11.

56. Kim JJ, Lee HJ, Welday AC, et al. Stress-induced alterations in hippocampal plasticity, place cells, and spatial memory. *Proc Natl Acad Sci USA.* 2007;104:18297–18302.

57. Kole MH, Koolhaas JM, Luiten PG, Fuchs E. High-voltage-activated Ca2+ currents and the excitability of pyramidal neurons in the hippocampal CA3 subfield in rats depend on corticosterone and time of day. *Neurosci Lett.* 2001;307:53–56.

58. Korz V, Frey JU. Stress-related modulation of hippocampal long-term potentiation in rats: involvement of adrenal steroid receptors. *J Neurosci.* 2003;23:7281–7287.

59. Korz V, Frey JU. Bidirectional modulation of hippocampal long-term potentiation under stress and no-stress conditions in basolateral amygdala-lesioned and intact rats. *J Neurosci.* 2005;25:7393–7400.

60. Li HB, Mao RR, Zhang JC, Yang Y, Cao J, Xu L. Antistress effect of TRPV1 channel on synaptic plasticity and spatial memory. *Biol Psychiatry.* 2008;64:286–292.

61. Lightman SL, Conway-Campbell BL. The crucial role of pulsatile activity of the HPA axis for continuous dynamic equilibration. *Nat Rev Neurosci.* 2010;11:710–718.

62. Liston C, Cichon JM, Jeanneteau F, Jia Z, Chao MV, Gan WB. Circadian glucocorticoid oscillations promote learning-dependent synapse formation and maintenance. *Nat Neurosci.* 2013;16:698–705.

63. Liu L, Wang C, Ni X, Sun J. A rapid inhibition of NMDA receptor current by corticosterone in cultured hippocampal neurons. *Neurosci Lett.* 2007;420:245–250.

64. Lösel RM, Wehling M. Classic versus non-classic receptors for nongenomic mineralocorticoid responses: emerging evidence. *Front Neuroendocrinol.* 2008;29:258–267.

65. Lupien SJ, McEwen BS, Gunnar MR, Heim C. Effects of stress throughout the lifespan on the brain, behaviour and cognition. *Nat Rev Neurosci.* 2009;10:434–445.

66. Maggio N, Segal M. Striking variations in corticosteroid modulation of long-term potentiation along the septotemporal axis of the hippocampus. *J Neurosci.* 2007;27:5757–5765.

67. Maggio N, Segal M. Differential corticosteroid modulation of inhibitory synaptic currents in the dorsal and ventral hippocampus. *J Neurosci.* 2009;29:2857–2866.

68. Maggio N, Segal M. Differential modulation of long-term depression by acute stress in the rat dorsal and ventral hippocampus. *J Neurosci.* 2009;29:8633–8638.

69. Manahan-Vaughan D, Braunewell KH. Novelty acquisition is associated with induction of hippocampal long-term depression. *Proc Natl Acad Sci USA.* 1999;96:8739–8744.

70. Martin S, Henley JM, Holman D, et al. Corticosterone alters AMPAR mobility and facilitates bidirectional synaptic plasticity. *PLoS One.* 2009;4:e4714.

71. Meijer OC, de Lange EC, Breimer DD, de Boer AG, Workel JO, de Kloet ER. Penetration of dexamethasone into brain glucocorticoid targets is enhanced in mdr1A P-glycoprotein knockout mice. *Endocrinology.* 1998;139:1789–1793.

72. Morsink MC, Steenbergen PJ, Vos JB, et al. Acute activation of hippocampal glucocorticoid receptors results in different waves of gene expression throughout time. *J Neuroendocrinol.* 2006;18:239–252.

73. Niehusmann P, Seifert G, Clark K, et al. Coincidence detection and stress modulation of spike time-dependent long-term depression in the hippocampus. *J Neurosci.* 2010;30:6225–6235.

74. Oitzl MS, Reichardt HM, Joëls M, de Kloet ER. Point mutation in the mouse glucocorticoid receptor preventing DNA binding impairs spatial memory. *Proc Natl Acad Sci USA.* 2001;98:12790–12795.

75. Okuhara DY, Beck SG. Corticosteroids influence the action potential firing pattern of hippocampal subfield CA3 pyramidal cells. *Neuroendocrinology.* 1998;67:58–66.

76. Olijslagers JE, de Kloet ER, Elgersma Y, van Woerden GM, Joels M, Karst H. Rapid changes in hippocampal CA1 pyramidal cell function via pre- as well as postsynaptic membrane mineralocorticoid receptors. *Eur J Neurosci.* 2008;27:2542–2550.

77. Ooishi Y, Mukai H, Hojo Y, et al. Estradiol rapidly rescues synaptic transmission from corticosterone-induced suppression via synaptic/extranuclear steroid receptors in the hippocampus. *Cereb Cortex.* 2012;22(4):926–936.

78. Orchinik M, Murray TF, Moore FL. A corticosteroid receptor in neuronal membranes. *Science.* 1991;252:1848–1851.

79. Pariante CM. The role of multi-drug resistance p-glycoprotein in glucocorticoid function: studies in animals and relevance in humans. *Eur J Pharmacol.* 2008;583:263–271.

80. Pasricha N, Joels M, Karst H. Rapid effects of corticosterone in the mouse dentate gyrus via a nongenomic pathway. *J Neuroendocrinol.* 2011;23:143–147.

81. Passecker J, Hok V, Della-Chiesa A, Chah E, O'Mara SM. Dissociation of dorsal hippocampal regional activation under the influence of stress in freely behaving rats. *Front Behav Neurosci.* 2011;5:66.

82. Pavlides C, Watanabe Y, McEwen BS. Effects of glucocorticoids on hippocampal long-term potentiation. *Hippocampus.* 1993;3:183–192.

83. Pavlides C, Kimura A, Magarinos AM, McEwen BS. Type I adrenal steroid receptors prolong hippocampal long-term potentiation. *Neuroreport.* 1994;5:2673–2677.

84. Pavlides C, Kimura A, Magarinos AM, McEwen BS. Hippocampal homosynaptic long-term depression/depotentiation induced by adrenal steroids. *Neuroscience.* 1995;68:379–385.

85. Pavlides C, Ogawa S, Kimura A, McEwen BS. Role of adrenal steroid mineralocorticoid and glucocorticoid receptors in long-term potentiation in the CA1 field of hippocampal slices. *Brain Res.* 1996;738:229–235.

86. Pfaff DW, Silva MT, Weiss JM. Telemetered recording of hormone effects on hippocampal neurons. *Science.* 1971;172:394–395.

87. Polman JA, de Kloet ER, Datson NA. Two populations of glucocorticoid receptor-binding sites in the male rat hippocampal genome. *Endocrinology.* 2013;154:1832–1844.

88. Prager EM, Brielmaier J, Bergstrom HC, McGuire J, Johnson LR. Localization of mineralocorticoid receptors at mammalian synapses. *PLoS One.* 2010;5:e14344.

89. Pu Z, Krugers HJ, Joels M. Corticosterone time-dependently modulates beta-adrenergic effects on long-term potentiation in the hippocampal dentate gyrus. *Learn Mem.* 2007;14:359–367.

90. Qian X, Droste SK, Gutièrrez-Mecinas M, et al. A rapid release of corticosteroid-binding globulin from the liver restrains the glucocorticoid hormone response to acute stress. *Endocrinology.* 2011;152:3738–3748.

91. Qiu S, Champagne DL, Peters M, et al. Loss of limbic system-associated membrane protein leads to reduced hippocampal mineralocorticoid receptor expression, impaired synaptic plasticity, and spatial memory deficit. *Biol Psychiatry.* 2010;68:197–204.

92. Reiheld CT, Teyler TJ, Vardaris RM. Effects of corticosterone on the electrophysiology of hippocampal CA1 pyramidal cells in vitro. *Brain Res Bull.* 1984;12:349–353.

93. Reul JM, de Kloet ER. Two receptor systems for corticosterone in rat brain: microdistribution and differential occupation. *Endocrinology.* 1985;117:2505–2511.

94. Revollo JR, Cidlowski JA. Mechanisms generating diversity in glucocorticoid receptor signaling. *Ann NY Acad Sci.* 2009;1179:167–178.

95. Rey M, Carlier E, Soumireu-Mourat B. Effects of corticosterone on hippocampal slice electrophysiology in normal and adrenalectomized BALB/c mice. *Neuroendocrinology.* 1987;46:424–429.

96. Roozendaal B, de Quervain DJ, Schelling G, McGaugh JL. A systemically administered beta-adrenoceptor antagonist blocks corticosterone-induced impairment of contextual memory retrieval in rats. *Neurobiol Learn Mem.* 2004;81:150–154.

97. Roozendaal B, Hernandez A, Cabrera SM, et al. Membrane-associated glucocorticoid activity is necessary for modulation of long-term memory via chromatin modification. *J Neurosci.* 2010;30:5037–5046.

98. Ryan BK, Vollmayr B, Klyubin I, Gass P, Rowan MJ. Persistent inhibition of hippocampal long-term potentiation in vivo by learned helplessness stress. *Hippocampus.* 2010;20:758–767.

99. Sajadi AA, Samaei SA, Rashidy-Pour A. Intra-hippocampal microinjections of anisomycin did not block glucocorticoid-induced impairment of memory retrieval in rats: an evidence for non-genomic effects of glucocorticoids. *Behav Brain Res.* 2006;173:158–162.

100. Sarabdjitsingh RA, Isenia S, Polman A, et al. Disrupted corticosterone pulsatile patterns attenuate responsiveness to glucocorticoid signaling in rat brain. *Endocrinology.* 2010;151:1177–1186.

101. Sarabdjitsingh RA, Jezequel J, Pasricha N, et al. Ultradian corticosterone pulses balance glutamatergic transmission and synaptic plasticity. *Proc Natl Acad Sci USA.* 2014;111:14265–14270.

102. Sarabdjitsingh RA, Pasricha N, Smeets JA, et al. Hippocampal fast glutamatergic transmission is transiently regulated by corticosterone pulsatility. *PLoS One.* January 7, 2016;11(1):e0145858.

103. Schwabe L, Schachinger H, de Kloet ER, Oitzl MS. Corticosteroids operate as a switch between memory systems. *J Cogn Neurosci.* 2010;22:1362–1372.

104. Schwabe L, Tegenthoff M, Höffken O, Wolf OT. Mineralocorticoid receptor blockade prevents stress-induced modulation of multiple memory systems in the human brain. *Biol Psychiatry.* 2013;74:801–808.

105. Seckl JR, Walker BR. 11beta-hydroxysteroid dehydrogenase type 1 as a modulator of glucocorticoid action: from metabolism to memory. *Trends Endocrinol Metab.* 2004;15:418–424.

106. Shors TJ, Dryver E. Effect of stress and long-term potentiation (LTP) on subsequent LTP and the theta burst response in the dentate gyrus. *Brain Res.* 1994;666:232–238.

107. Shors TJ, Thompson RF. Acute stress impairs (or induces) synaptic long-term potentiation (LTP) but does not affect paired-pulse facilitation in the stratum radiatum of rat hippocampus. *Synapse.* 1992;11:262–265.

108. Shors TJ, Seib TB, Levine S, Thompson RF. Inescapable versus escapable shock modulates long-term potentiation in the rat hippocampus. *Science.* 1989;244:224–226.

109. Smriga M, Saito H, Nishiyama N. Hippocampal long- and short-term potentiation is modulated by adrenalectomy and corticosterone. *Neuroendocrinology.* 1996;64:35–41.

110. Spyrka J, Danielewicz J, Hess G. Brief neck restraint stress enhances long-term potentiation and suppresses long-term depression in the dentate gyrus of the mouse. *Brain Res Bull.* 2011;85:363–367.

111. Teschemacher A, Zeise ML, Zieglgansberger W. Corticosterone-induced decrease of inhibitory postsynaptic potentials in rat hippocampal pyramidal neurons in vitro depends on cytosolic factors. *Neurosci Lett.* 1996;215:83–86.

112. Tse YC, Bagot RC, Hutter JA, Wong AS, Wong TP. Modulation of synaptic plasticity by stress hormone associates with plastic alteration of synaptic NMDA receptor in the adult hippocampus. *PLoS One.* 2011;6:e27215.

113. van Gemert NG, Carvalho DM, Karst H, et al. Dissociation between rat hippocampal CA1 and dentate gyrus cells in their response to corticosterone: effects on calcium channel protein and current. *Endocrinology.* 2009;150:4615–4624.

114. Vidal C, Jordan W, Zieglgansberger W. Corticosterone reduces the excitability of hippocampal pyramidal cells in vitro. *Brain Res.* 1986;383:54–59.

115. Vogel S, Klumpers F, Krugers HJ, et al. Blocking the mineralocorticoid receptor in humans prevents the stress-induced enhancement of centromedial amygdala connectivity with the dorsal striatum. *Neuropsychopharmacology.* 2015;40:947–956.

116. Vogel S, Fernández G, Joëls M, Schwabe L. Cognitive adaptation under stress: a case for the mineralocorticoid receptor. *Trends Cogn Sci.* 2016;20:192–203.

117. Vouimba RM, Yaniv D, Diamond D, Richter-Levin G. Effects of inescapable stress on LTP in the amygdala versus the dentate gyrus of freely behaving rats. *Eur J Neurosci.* 2004;19:1887–1894.

118. Vouimba RM, Yaniv D, Richter-Levin G. Glucocorticoid receptors and beta-adrenoceptors in basolateral amygdala modulate synaptic plasticity in hippocampal dentate gyrus, but not in area CA1. *Neuropharmacology.* 2007;52:244–252.

119. Werkman TR, Van der Linden S, Joels M. Corticosteroid effects on sodium and calcium currents in acutely dissociated rat CA1 hippocampal neurons. *Neuroscience.* 1997;78:663–672.

120. Wiegert O, Joels M, Krugers H. Timing is essential for rapid effects of corticosterone on synaptic potentiation in the mouse hippocampus. *Learn Mem.* 2006;13:110–113.

121. Wong TP, Howland JG, Robillard JM, et al. Hippocampal long-term depression mediates acute stress-induced spatial memory retrieval impairment. *Proc Natl Acad Sci USA.* 2007;104:11471–11476.

122. Wyrwoll CS, Holmes MC, Seckl JR. 11beta-hydroxysteroid dehydrogenases and the brain: from zero to hero, a decade of progress. *Front Neuroendocrinol.* 2011;32:265–286.

123. Xiong W, Wei H, Xiang X, et al. The effect of acute stress on LTP and LTD induction in the hippocampal CA1 region of anesthetized rats at three different ages. *Brain Res.* 2004;1005:187–192.

124. Xu L, Anwyl R, Rowan MJ. Behavioural stress facilitates the induction of long-term depression in the hippocampus. *Nature.* 1997;387:497–500.

125. Xu L, Holscher C, Anwyl R, Rowan MJ. Glucocorticoid receptor and protein/RNA synthesis-dependent mechanisms underlie the control of synaptic plasticity by stress. *Proc Natl Acad Sci USA.* 1998;95:3204–3208.

126. Yamada K, McEwen BS, Pavlides C. Site and time dependent effects of acute stress on hippocampal long-term potentiation in freely behaving rats. *Exp Brain Res.* 2003;152:52–59.

127. Yang CH, Huang CC, Hsu KS. Behavioral stress modifies hippocampal synaptic plasticity through corticosterone-induced sustained extracellular signal-regulated kinase/mitogen-activated protein kinase activation. *J Neurosci.* 2004;24:11029–11034.

128. Yang CH, Huang CC, Hsu KS. Behavioral stress enhances hippocampal CA1 long-term depression through the blockade of the glutamate uptake. *J Neurosci.* 2005;25:4288–4293.

129. Yang J, Han H, Cao J, Li L, Xu L. Prenatal stress modifies hippocampal synaptic plasticity and spatial learning in young rat offspring. *Hippocampus.* 2006;16:431–436.

130. Yang CH, Huang CC, Hsu KS. Differential roles of basolateral and central amygdala on the effects of uncontrollable stress on hippocampal synaptic plasticity. *Hippocampus.* 2008;18:548–563.

131. Yarom O, Maroun M, Richter-Levin G. Exposure to forced swim stress alters local circuit activity and plasticity in the dentate gyrus of the hippocampus. *Neural Plast.* 2008;2008:194097.

132. Young EA, Abelson J, Lightman SL. Cortisol pulsatility and its role in stress regulation and health. *Front Neuroendocrinol.* 2004;25:69–76.

133. Zalachoras I, Houtman R, Meijer OC. Understanding stress-effects in the brain via transcriptional signal transduction pathways. *Neuroscience.* 2013;242:97–109.

134. Zeise ML, Teschemacher A, Arriagada J, Zieglgansberger W. Corticosterone reduces synaptic inhibition in rat hippocampal and neocortical neurons in vitro. *J Neuroendocrinol.* 1992;4:107–112.

135. Zhang J, Yang Y, Li H, Cao J, Xu L. Amplitude/frequency of spontaneous mEPSC correlates to the degree of long-term depression in the CA1 region of the hippocampal slice. *Brain Res.* 2005;1050:110–117.

136. Zhang Y, Sheng H, Qi J, et al. Glucocorticoid acts on a putative G protein-coupled receptor to rapidly regulate the activity of NMDA receptors in hippocampal neurons. *Am J Physiol Endocrinol Metab.* 2012;302(7):E747–E758.

137. Zhou J, Zhang F, Zhang Y. Corticosterone inhibits generation of long-term potentiation in rat hippocampal slice: involvement of brain-derived neurotrophic factor. *Brain Res.* 2000;885:182–191.

138. Zhou M, Bakker EH, Velzing EH, et al. Both mineralocorticoid and glucocorticoid receptors regulate emotional memory in mice. *Neurobiol Learn Mem.* 2010;94:530–537.

34

11β-Hydroxysteroid Dehydrogenases

J.R. Seckl

University of Edinburgh, Edinburgh, United Kingdom

Abstract

The action of glucocorticoids on target cells is dependent not only on circulating hormone levels but also on intracellular prereceptor metabolism by 11β-hydroxysteroid dehydrogenase (11β-HSD). 11β-HSD catalyzes the interconversion of active 11-hydroxy-glucocorticoids (cortisol, corticosterone) and inert 11-keto forms (cortisone, 11-dehydrocorticosterone). 11β-HSD type 2 potently inactivates glucocorticoids. In the adult CNS, high expression is largely confined to a few nuclei in the brain stem where 11β-HSD2 determines aldosterone-selective effects via mineralocorticoid receptors (MRs) on blood pressure and salt appetite. In the fetal brain 11β-HSD2 is widely expressed. It appears to protect nuclear glucocorticoid receptors (GRs) from premature exposure to glucocorticoids, thus determining the timing of cellular maturation. In contrast, 11β-HSD type 1 is absent during brain development but widespread in the adult brain. 11β-HSD1 catalyzes the reverse reaction, regenerating active glucocorticoids from circulating inert forms, thus locally amplifying glucocorticoid action. Levels of 11β-HSD1 rise in hippocampus and cortex with ageing and correlate with cognitive decline. Selective inhibition or knock out of 11β-HSD1 protects against age-related cognitive impairments in rodents, including in models of Alzheimer disease. 11β-HSD1 inhibition is being explored in the treatment of age-related cognitive and stress-related disorders.

DEFINITION: 11β-HYDROXYSTEROID DEHYDROGENASE

Elevated plasma cortisol or glucocorticoid pharmacotherapy has myriad adverse effects ranging from obesity, diabetes, and hypertension to depression and memory impairments. Until recently, glucocorticoid action upon target cells was believed to be determined by the product of circulating hormone levels and the density of glucocorticoid receptors (GRs) and mineralocorticoid receptors (MRs) in target cells. However, recent discoveries suggest a key role for local enzyme-mediated tissue metabolism of glucocorticoids in determining intracellular receptor activation. The main enzymes involved are the 11β-hydroxysteroid dehydrogenases (11β-HSDs).[1] These catalyze the interconversion of active cortisol and corticosterone (the former predominates in humans and most mammals, the latter is the sole form in rats and mice) and their inert 11-keto forms (cortisone, 11-dehydrocorticosterone). The role of these enzymes in determining glucocorticoid actions in CNS and periphery has become increasingly clear in health and in pathology.

- 11β-hydroxysteroid dehydrogenases (11β-HSDs) catalyze the interconversion of active 11-hydroxy-glucocorticoids (cortisol, corticosterone) and their inert 11-keto forms (cortisone, 11-dehydrocorticosterone).

- There are two isoforms, the products of separate, distantly related genes.

- 11β-HSD type 2 was the second isozyme characterized, but the first to be understood. It catalyzes the rapid inactivation (oxidation) of glucocorticoids. 11β-HSD2 generates aldosterone specificity for MRs in the kidney and in the brain stem, thus underlying the aldosterone-specific central effects on salt appetite and blood pressure.

- 11β-HSD2 is most highly expressed in the fetal CNS where it excludes glucocorticoids from intracellular receptors in developing cells. Bypass or inhibition of fetal brain 11β-HSD2 contributes to "fetal programming" of affective, cognitive, and attentional disorders in later life.

- In contrast 11β-HSD type 1 catalyzes predominantly the reverse ketoreductase reaction, which regenerates active glucocorticoids inside cells. 11β-HSD1 is widespread in the adult CNS.

- Levels of 11β-HSD1 are elevated in the ageing brain, notably in hippocampus and cortex. This correlates with and causes cognitive deficits with age. Selective inhibition of 11β-HSD1 improves cognitive function in aged rodents and, perhaps, in humans. Inhibitors are being explored in early clinical trials.

TWO ISOZYMES AND A SOLUTION TO THE MINERALOCORTICOID RECEPTOR PARADOX

Purified MR bind the mineralocorticoid aldosterone and the glucocorticoids cortisol and corticosterone with similar high affinity in vitro, but only aldosterone binds to MR in the kidney in vivo, though there is 100–1000× more glucocorticoid in the circulation. In resolution of this paradox, in 1988 groups in Edinburgh and Melbourne reported that 11β-HSD is a prereceptor enzymic mechanism that bestows aldosterone selectivity on intrinsically nonselective MRs in vivo.[2,3] The 11β-HSD activity associated with MR in the distal nephron acts as a dehydrogenase, rapidly inactivating physiological cortisol and corticosterone. This 11β-HSD activity forms a functional intracellular "barrier" to glucocorticoids, preventing their occupation of MR, allowing only the nonsubstrate aldosterone through. When the enzyme is congenitally deficient, the syndrome of apparent mineralocorticoid excess, or is inhibited by liquorice, glucocorticoids gain de novo access MR causing sodium retention, hypertension, and hypokalemia.

Although an NADP(H)-associated 11β-HSD activity (now called 11β-HSD1) and its encoding cDNA had been isolated from rat liver in the mid 1980s,[4] this was shown not to be responsible for renal MR selectivity; it is not located in the distal nephron with aldosterone-selective MR and its encoding gene, HSD11B1, is intact in patients with the syndrome of apparent mineralocorticoid excess.[5] Subsequently, a second 11β-HSD isozyme (11β-HSD2) was characterized, isolated, and its cDNA cloned from kidney and placenta in a variety of species.[6,7] This isozyme is a high-affinity 11β-dehydrogenase, which uses NAD as cofactor and is highly expressed in the distal nephron and placenta. Patients with the syndrome of apparent mineralocorticoid excess are homozygous or compound heterozygous for deleterious mutations in the HSD11B2 gene. 11β-HSD2 knockout mice faithfully recapitulated the syndrome.[8] This work underlined a more general principle that hormone access to nuclear receptors in vivo is often regulated by prereceptor metabolism. Other examples include 5-monodeiodinases (thyroid hormone receptors), 5α-reductase (androgen receptor), and aromatase (estrogen receptors).

11β-HSD2 IN THE ADULT CNS

11β-HSD2 is little expressed in the human or rodent forebrain.[9] Indeed, the vast majority of MR in the CNS are nonselective and bind glucocorticoids in vivo,[10] implying the absence of 11β-HSD2. However, there are some CNS actions selectively mediated by aldosterone implying MR "gated" by local 11β-HSD2, specifically blood pressure and salt appetite control. Thus, aldosterone-infused intracerebroventricularly (icv) elevates blood pressure in the rat, whilst the same dose given peripherally is ineffective, demonstrating a specific central action. Corticosterone icv does not reproduce this action and indeed antagonizes aldosterone hypertension. Similarly, aldosterone exerts central actions on salt appetite, which cannot be reproduced by corticosterone. Icv infusion of the 11β-HSD inhibitor carbenoxolone increases blood pressure in the rat, an effect reversed by an MR antagonist, suggesting central 11β-HSD2 prevents access of glucocorticoids to central MR.[11,12]

Although high 11β-HSD2 mRNA is only found in a few discrete sites in the adult rodent brain,[9] these loci also express MR and are believed to underlie the selective central effects of aldosterone to increase blood pressure (nucleus tractus solitarius) and modulate salt appetite (subcommissural organ, and less certainly the ventromedial

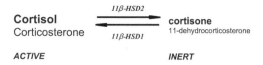

FIGURE 34.1 **11β-Hydroxysteroid dehydrogenases.** Type 1 Acts as a ketoreductase, regenerating active cortisol (corticosterone), whereas type 2 acts as a dehydrogenase, rapidly inactivating these physiological glucocorticoids.

hypothalamus and central nucleus of the amygdala). Importantly, selective knock-out of 11β-HSD2 in the adult mouse brain causes three times greater voluntary sodium consumption and hypertension via an MR-dependent mechanism.[13] Whether or not 11β-HSD2 expression is changed in CNS pathogenesis or in hypertension remains unexplored (Fig. 34.1).

11β-HSD2 IN THE DEVELOPING CNS

Glucocorticoids act as terminal differentiation signals in many fetal tissues, hence the efficacious use of maternally administered synthetic glucocorticoids (which are poor 11β-HSD2 dehydrogenase substrates) in premature labor to accelerate the maturation of fetal lungs and other organs. However, such early maturation comes with consequences; prenatal glucocorticoid administration reduces brain weight at birth. Given such widespread effects, it is unsurprising that GR and MR are highly expressed in the developing brain. Whether or not these receptors are occupied by endogenous glucocorticoids until late gestation is uncertain because there is also plentiful 11β-HSD2 in the CNS at midgestation in rodents and humans. This presumably functions to "protect" vulnerable developing cells from premature endogenous glucocorticoid action. Strikingly, 11β-HSD2 expression is dramatically switched-off in a CNS at the end of midgestation in a complex cell-specific pattern coinciding with the terminal stage of brain nucleus development.[14] This appears to be an exquisitely timed system of cellular protection from circulating glucocorticoids. Inhibition or knock-out of 11β-HSD2 in the fetal brain changes the timing of terminal cell maturation and the structure of the postnatal CNS.[15]

11β-HSD2 AND FETAL PROGRAMMING

Human epidemiological studies show that lower birth weight at term associates with an increased later occurrence of cardiometabolic (hypertension, type 2 diabetes, ischemic heart disease) and neuropsychiatric disorders (cognitive deficits, depression, anxiety, schizophrenia) in adult life. This phenomenon is called "fetal programming."[16]

In rodents, guinea pigs, sheep, nonhuman primates, and humans, excessive exposure to glucocorticoids during gestation, either bypass of plentiful fetoplacental 11β-HSD2 with nonsubstrate glucocorticoids such as dexamethasone, or use of 11β-HSD inhibitors (glycyrrhizic acid, the active component of liquorice, or its pharmaceutical derivative carbenoxolone), reduces birth weight. Though the weight deficit is regained after birth, the adult offspring shows increased blood pressure and serum insulin and glucose levels, as well as anxiety, depressive-like behaviors, and cognitive and attention deficits.[17] These effects are underpinned by, amongst other changes, a permanent increase in tissue sensitivity to glucocorticoids via increased GR in key metabolic organs, including liver and visceral adipose tissue, and altered GR and MR in various CNS regions.

FETAL PROGRAMMING OF THE HYPOTHALAMIC-PITUITARY-ADRENAL AXIS AND BEHAVIOR

Prenatal 11β-HSD2 inhibition or dexamethasone also "programs" the hypothalamic-pituitary-adrenal (HPA) axis, with permanently increased basal plasma glucocorticoid levels in the offspring in a variety of species including primates.[18] Intriguingly, low-birthweight humans also exhibit HPA hyperactivity in adult life.[19] The density of GR and MR in the hippocampus are usually reduced in such models[20] and expected to reduce the sensitivity of HPA axis negative feedback. Maternal undernutrition also lowers placental 11β-HSD2. This is anticipated to increase fetal exposure to maternal glucocorticoids, thus perhaps explaining the observed programming of adult offspring metabolic, cardiovascular, and HPA axis function. Similarly, maternal stress or anxiety reduces placental 11β-HSD2 and increases fetal glucocorticoid levels, including in humans.[21] However, the mechanisms are complex; for instance maternal undernutrition in mice also activates the fetal HPA axis per se, with reduced placental 11β-HSD2 contributing to but not fully explaining the programming effect.[22]

Overexposure to glucocorticoids in utero also leads to alterations in adult behavior to produce a phenotype reminiscent of anxiety/depression.[23] Additionally, cognitive functions are constrained. Prenatal glucocorticoid exposure also affects the developing dopaminergic system with clear implications for proposed developmental contributions to schizoaffective, attention-deficit hyperactivity, and extrapyramidal disorders.

Fetoplacental 11β-HSD2 plays an important role in protecting the brain and other organs from maternal (and fetal) glucocorticoid overexposure. Genetic deficiency of 11β-HSD2 in placenta and fetus reduces birth weight and programs brain and behavior in offspring compared with genetically intact littermates.[24] Both placental and fetal brain 11β-HSD2 play specific roles since knockout of 11β-HSD2 solely in the fetal brain programs reduced cognition and depressive-like behaviors, but not the HPA axis, cardiometabolic function, or anxiety-related behaviors.[25] Clearly the interplay between glucocorticoid actions on the mother, placenta, and fetal tissues is complex.

In humans, voluntary maternal liquorice consumption during pregnancy is associated with reductions in offspring cognition, increased cortisol levels, accelerated pubertal timing, and a substantially increased prevalence of attentional and other behavioral disorders in the offspring in childhood.[26,27] Preclinical science indicates a clear mechanistic basis for these findings. It might therefore be wise to consider regulating liquorice-based foodstuffs in pregnancy.

Mechanistically, prenatal dexamethasone or 11β-HSD inhibition increases corticotrophin-releasing hormone mRNA levels specifically in the central nucleus of the amygdala,[23] a key locus for the effects of the neuropeptide on the expression of fear and anxiety. Fetoplacental 11β-HSD inhibition also permanently increases GR and MR in the offspring amygdala. A direct relationship between brain GR levels and anxiety behaviors is supported by transgenic mice with selective loss of GR in the forebrain, which exhibit reduced fearfulness.

In considering how a transient exposure to glucocorticoids might lead to persisting changes in adult biology, alterations in cell number and changes in cellular gene expression have both been repeatedly reported. The latter is most plausibly ascribed to epigenetic changes (covalent modification of cytosine residues) in the promoters of target genes. Of these, much evidence has accumulated for stress/glucocorticoid programming of GR itself, with specific alternative gene promoters targeted by early life events. Notably, the perinatal environment, via a cascade of serotonergic neurotransmission and specific transcription factors, impact upon the brain-enriched exon 1_7 in rodents (exon 1F in humans) altering persistently its methylation state and hence accessibility to transcriptional control.[28,29] Other genes are similarly epigenetically impacted by the early environment including placental 11β-HSD2 itself.[30,31] Whether or not these correlations are causal of glucocorticoid and other aspects of fetal programming remains a key question for exploration.

11β-HSD1 IN THE BRAIN

11β-HSD1 is highly expressed in liver, adipose tissue, and the brain.[1] Studies in intact cells from liver, fat, lung, and other sites have shown that 11β-HSD1, though bidirectional in tissue homogenates, is usually a predominant 11β-ketoreductase in intact cells, a reaction direction driven by the excess levels of NADP(H) in the endoplasmic lumen where the enzyme is located. This reaction regenerates active glucocorticoids from inert 11-keto forms. A similar directional predominance is seen in the intact liver ex vivo and in rodents and humans in vivo.

In the brain, including in humans, 11β-HSD1-like bioactivity, immunoreactivity, and mRNA is widespread.[32,33] Intact hippocampal cells in culture show 11β–reductase activity. This reaction direction increases intracellular glucocorticoid levels. In searching for functions it was noted that liquorice-derived 11β-HSD inhibitors given to rats activate a series of regions of the CNS, including the hippocampus, a key glucocorticoid target with high GR and MR density where glucocorticoids act to suppress HPA activity and to modulate cognitive functions.

11β-HSD1 AND THE HPA AXIS

11β-HSD1 is expressed in the hippocampus, cortex, paraventricular nucleus of the hypothalamus (PVN), and pituitary, key sites of glucocorticoid negative feedback. 11β-HSD1 knockout mice (11β-HSD1$^{-/-}$) show adrenocortical hypertrophy and elevated basal ACTH levels at the nadir of the circadian rhythm. These effects are compatible with the requirement for increased glucocorticoid production from the adrenal cortex secondary to increased metabolic clearance of corticosterone because of the lack of glucocorticoid regeneration by 11β-HSD1, notably in the liver. Although original reports showed that basal plasma corticosterone levels are also elevated 11β-HSD1$^{-/-}$ mice, this effect is probably confined to mice of the 129 strain background and is not seen in other strains,[34] other species, or in humans treated with selective 11β-HSD1 inhibitors.[35] The difference appears to reflect variations in plasticity of GR, with most strains lacking 11β-HSD1 showing a compensatory rise in GR expression in the hippocampus and PVN whilst the opposite occurs in 129 strain mice. Overall the data suggest that 11β-HSD1 deficiency in the brain has little overt impact on plasma glucocorticoid levels per se unless other systems are also impaired.

Such changes may be important in pregnancy when the HPA axis becomes refractory to stressful stimulation. Interestingly, 11β-HSD1 activity is selectively increased in the PVN in pregnancy.[36] Such locally increased 11β-HSD1 may amplify glucocorticoid feedback and hence contribute to the attenuated HPA activity observed.

Other hypothalamic effects of 11β-HSD1 are mooted. 11β-HSD1 knockout mice have an increased appetite for high fat diet. 11β-HSD1 is expressed in the arcuate nucleus, key to appetite control.[37] Arcuate 11β-HSD1 is acutely induced by high fat feeding. 11β-HSD1-deficient mice show reduced mRNA expression encoding anorexigenic cocaine- and amphetamine-regulated transcript and melanocortin-4 receptor and increased orexigenic agouti-related peptide, suggesting increased appetitive drive. This regulation appears mediated through mu-opioid receptors, key controllers of reward. Hence arcuate 11β-HSD1 may contribute to the central adaption to a palatable energy dense challenge.

11β-HSD1 AND COGNITIVE AGING

A subgroup of aged rodents and humans shows cognitive decline associated with chronic elevation of plasma corticosterone.[38] If glucocorticoids are maintained at low levels in rats, by adrenalectomy and low-dose corticosterone replacement or perinatal programming of increased GR and MR in the hippocampus, the subsequent incidence of age-related cognitive impairments is reduced. The hippocampus requires glucocorticoids for neuronal function and survival but is especially vulnerable to the adverse effects of chronic glucocorticoid excess, which causes atrophy of dendrites and cognitive dysfunction. Thus chronically elevated glucocorticoids levels in Cushing syndrome associate with cognitive impairments, and shrinkage of the hippocampus. Therefore, chronic glucocorticoid overexposure has been implicated in the pathogenesis of age-related decline in hippocampus-associated cognitive functions.

In primary cultures of hippocampal cells, 11β-HSD1 potentiates kainic acid-induced neurotoxicity not only by corticosterone, but equally by intrinsically inert 11-dehydrocorticosterone, an effect prevented by the inhibitor carbenoxolone. 11-Dehydrocorticosterone in vivo increases kainic acid hippocampal toxicity in vivo, an effect also blocked by carbenoxolone. In vivo, aged 11β-HSD1$^{-/-}$ mice learn as well as young controls in hippocampus-associated spatial memory and recall tasks and avoid the cognitive decline seen in the majority of aged wild-type mice.[39] This cognitive protection in aged 11β-HSD1$^{-/-}$ mice associates with reduced intrahippocampal corticosterone levels, but unaltered plasma levels, indicating the potency of intracellular metabolism by 11β-HSDs in determining effective glucocorticoid action upon target receptors.

Intriguingly, 11β-HSD1 may play a role in pathogenesis, as rare (~1:200) haplotypes of the HSD11B1 gene associate with a risk of Alzheimer disease. Although common variants of HSD11B1 do not associate with cognition, total body 11β-HSD1 activity "predicts" subsequent brain volume loss and cognitive decline over the next 6 years.[40] Though the site of this effect is undefined, elevated 11β-HSD1 may contribute to cognitive decline with aging in humans. Indeed, in mice, aging associates with increased 11β-HSD1 mRNA in hippocampus and cortex, which correlates negatively with cognitive function.[41] Modest transgenic overexpression of 11β-HSD1 in the forebrain causes cognitive impairments with aging indicating a causal role.

Post-mortem human brain has high expression of 11β-HSD1 mRNA in the hippocampus, prefrontal cortex, and the cerebellum mirroring rodent findings. In two small, randomized, double-blind, placebo-controlled, crossover studies, administration of the 11β-HSD inhibitor carbenoxolone improved verbal fluency after four weeks in 10 healthy elderly men and improved verbal memory after 6 weeks in 12 elderly patients with type 2 diabetes.[33] Plasma cortisol was unaltered.

Moreover, short-term or long-term administration of a selective 11β-HSD1 inhibitor to aged mice reverses cognitive impairments. Similar effects are seen in the Tg2576 murine model of Alzheimer disease with some suggestion of slowing of the accumulation of amyloid Aβ pathological material.[42,43] The beneficial cognitive effects of 11β-HSD1 inhibition are mediated via reduced loading of GR (which mediates glucocorticoid-associated cognitive deficits).[44]

Whilst an initial trial of a selective 11β-HSD1 inhibitor in patients with moderate Alzheimer disease revealed no cognitive benefits, there remain doubts as to the effectiveness of the tested compound in the brain (as opposed to the periphery) or whether patients with earlier cognitive impairments might be beneficial, as in rodents. This hypothesis is subject to ongoing exploration in humans.

CONCLUSIONS

11β-HSDs interconvert active glucocorticoids and inert 11-keto forms. This seemingly obscure reaction is important in peripheral tissues and the CNS, determining glucocorticoid access to nuclear receptors and hence biological effects in health and disease. 11β-HSD1 is an 11β-ketoreductase that amplifies intracellular glucocorticoid levels in the brain and metabolic organs. Deficiency of 11β-HSD1 in mice reduces intracerebral glucocorticoid levels and prevents the emergence of cognitive deficits with aging. The importance in humans is under investigation.

11β-HSD2 is a potent 11β-dehydrogenase, with highly restricted expression in the adult CNS where it underpins aldosterone-selective actions on salt appetite and blood pressure. 11β-HSD2 plays an emerging role in preventing premature glucocorticoid impacts on the developing nervous system. Bypass or inhibition of 11β-HSD2 in development programs the brain and body, leading to cognitive, affective, and neuropsychiatric sequelae throughout post-natal life. The maintenance of fetoplacental 11β-HSD2 and the inhibition of 11β-HSD1 have therapeutic potential.

Glossary

11β-hydroxysteroid dehydrogenase (11β-HSD) An enzyme that catalyzes the interconversion of active cortisol (corticosterone in rats and mice) and inert cortisone (11-dehydrocorticosterone).

11β-HSD type 1 (11β-HSD1) An isozyme that predominantly catalyzes the reduction of inert cortisone to active cortisol in intact cells and organs.

11β-HSD type 2 (11β-HSD2) An isozyme that catalyzes the rapid dehydrogenation of active cortisol to inert cortisone.

Aromatase The enzyme catalyzing the conversion of androgens to estrogens.

Carbenoxolone A nonselective 11β-HSD inhibitor derived from the active component of liquorice.

Cushing syndrome Generic description of the phenotype observed with circulating glucocorticoid excess, due either to endogenous overproduction of cortisol or ACTH or to exogenous pharmacotherapy.

Glucocorticoid receptor (GR) Intracellular receptor for active cortisol.

Intracrine Process whereby intracellular enzymic transformations of steroids alter their access to nuclear receptors.

Mineralocorticoid receptor (MR) Intracellular receptor for active cortisol, selectivity for aldosterone in vivo generated by 11β-HSD2.

Syndrome of apparent mineralocorticoid excess Congenital syndrome of hypertension, hypernatremia, and hypokalemia due to inappropriate MR activation by glucocorticoids because of deleterious mutations of the 11β-HSD type 2 gene and hence lack of the enzyme in the kidney.

References

1. Chapman K, Holmes M, Seckl J. 11β-Hydroxysteroid dehydrogenases: intracellular gate-keepers of tissue glucocorticoid action. *Physiol Rev.* 2013;93(3):1139–1206.
2. Edwards CR, et al. Localisation of 11 beta-hydroxysteroid dehydrogenase–tissue specific protector of the mineralocorticoid receptor. *Lancet.* 1988;2(8618):986–989.
3. Funder JW, et al. Mineralocorticoid action: target tissue specificity is enzyme, not receptor, mediated. *Science.* 1988;242(4878):583–585.
4. Lakshmi V, Monder C. Purification and characterization of the corticosteroid 11 beta-dehydrogenase component of the rat liver 11 beta-hydroxysteroid dehydrogenase complex. *Endocrinology.* 1988;123(5):2390–2398.
5. Seckl JR. 11β-Hydroxysteroid dehydrogenase isoforms and their implications for blood pressure regulation. *Eur J Clin Invest.* 1993;23:589–601.
6. Albiston AL, et al. Cloning and tissue distribution of the human 11 beta-hydroxysteroid dehydrogenase type 2 enzyme. *Mol Cell Endocrinol.* 1994;105(2):R11–R17.
7. Brown RW, et al. Cloning and production of antisera to human placental 11β-hydroxysteroid dehydrogenase type 2. *Biochem J.* 1996;313:1007–1017.
8. Kotelevtsev Y, et al. Hypertension in mice lacking 11 beta-hydroxysteroid dehydrogenase type 2. *J Clin Invest.* 1999;103(5):683–689.
9. Robson AC, et al. 11 beta-hydroxysteroid dehydrogenase type 2 in the postnatal and adult rat brain. *Brain Res Mol Brain Res.* 1998;61(1–2):1–10.
10. Reul JMHM, de Kloet ER. Two receptor systems for corticosterone in rat brain: microdissection and differential occupation. *Endocrinology.* 1985;117:2505–2511.
11. Gomez-Sanchez EP. Intracerebroventricular infusion of aldosterone induces hypertension in rats. *Endocrinology.* 1986;118(2):819–823.
12. Gomez-Sanchez EP, Fort C, Thwaites D. Central mineralocorticoid receptor antagonism blocks hypertension in Dahl S/JR rats. *Am J Physiol.* 1992;262(1 Pt 1):E96–E99.
13. Evans LC, et al. Conditional deletion of Hsd11b2 in the brain causes salt appetite and hypertension. *Circulation.* 2016;133:1360–1370.
14. Diaz R, Brown RW, Seckl JR. Distinct ontogeny of glucocorticoid and mineralocorticoid receptor and 11beta-hydroxysteroid dehydrogenase types I and II mRNAs in the fetal rat brain suggest a complex control of glucocorticoid actions. *J Neurosci.* 1998;18(7):2570–2580.
15. Holmes MC, et al. 11 Beta-hydroxysteroid dehydrogenase type 2 protects the neonatal cerebellum from deleterious effects of glucocorticoids. *Neuroscience.* 2006;137(3):865–873.
16. Seckl JR. Glucocorticoids, developmental 'programming' and the risk of affective dysfunction. *Prog Brain Res.* 2008;167:17–34.
17. Harris A, Seckl J. Glucocorticoids, prenatal stress and the programming of disease. *Horm Behav.* 2011;59:p279–289.
18. de Vries A, et al. Prenatal dexamethasone exposure induces changes in nonhuman primate offspring cardiometabolic and hypothalamic-pituitary-adrenal axis function. *J Clin Invest.* 2007;117(4):1058–1067.
19. Phillips DIW, et al. Low birth weight predicts elevated plasma cortisol concentrations in adults from 3 populations. *Hypertension.* 2000;35(6):1301–1306.
20. Levitt NS, et al. Dexamethasone in the last week of pregnancy attenuates hippocampal glucocorticoid receptor gene expression and elevates blood pressure in the adult offspring in the rat. *Neuroendocrinology.* 1996;64(6):412–418.
21. O'Donnell KJ, et al. Maternal prenatal anxiety and down-regulation of placental 11β-HSD2. *Psychoneuroendocrinology.* 2012;37(6):818–826.
22. Cottrell EC, et al. Reconciling the nutritional and glucocorticoid hypotheses of fetal programming. *FASEB J.* 2012;26(5):1866–1874.
23. Welberg LA, Seckl JR, Holmes MC. Inhibition of 11 beta-hydroxysteroid dehydrogenase, the foeto-placental barrier to maternal glucocorticoids, permanently programs amygdala GR mRNA expression and anxiety-like behaviour in the offspring. *Eur J Neurosci.* 2000;12(3):1047–1054.
24. Holmes MC, et al. The mother or the fetus? 11 beta-hydroxysteroid dehydrogenase type 2 null mice provide evidence for direct fetal programming of behavior by endogenous glucocorticoids. *J Neurosci.* 2006;26(14):3840–3844.
25. Wyrwoll C, et al. Fetal brain 11 beta-hydroxysteroid dehydrogenase type 2 selectively determines programming of adult depressive-like behaviors and cognitive function, but not anxiety behaviors in male mice. *Psychoneuroendocrinology.* 2015;59:59–70.
26. Raikkonen K, et al. Maternal licorice consumption and detrimental cognitive and psychiatric outcomes in children. *Am J Epidemiol.* 2009;170(9):1137–1146.
27. Raikkonen K, et al. Maternal prenatal licorice consumption alters hypothalamic-pituitary-adrenocortical axis function in children. *Psychoneuroendocrinology.* 2010;35(10):1587–1593.
28. Weaver I, et al. Epigenetic programming by maternal behavior. *Nat Neurosci.* 2004;7:847–854.
29. McGowan PO, et al. Epigenetic regulation of the glucocorticoid receptor in human brain associates with childhood abuse. *Nat Neurosci.* 2009;12(3):342–348.
30. Green BB, et al. The role of placental 11-beta hydroxysteroid dehydrogenase type 1 and type 2 methylation on gene expression and infant birth weight. *Biol Reprod.* 2015;92(6):149.
31. Pena CJ, Monk C, Champagne FA. Epigenetic effects of prenatal stress on 11β-hydroxysteroid dehydrogenase-2 in the placenta and fetal brain. *PLoS One.* 2012;7(6):e39791.
32. Moisan MP, Seckl JR, Edwards CR. 11 beta-hydroxysteroid dehydrogenase bioactivity and messenger RNA expression in rat forebrain: localization in hypothalamus, hippocampus, and cortex. *Endocrinology.* 1990;127(3):1450–1455.
33. Sandeep TC, et al. 11 Beta-hydroxysteroid dehydrogenase inhibition improves cognitive function in healthy elderly men and type 2 diabetics. *Proc Natl Acad Sci USA.* 2004;101(17):6734–6739.

34. Carter RN, et al. Hypothalamic-pituitary-adrenal axis abnormalities in response to deletion of 11beta-HSD1 is strain-dependent. *J Neuroendocrinol*. 2009;21(11):879–887.

35. Rosenstock J, et al. The 11-beta-hydroxysteroid dehydrogenase type 1 inhibitor INCB13739 improves hyperglycemia in patients with type 2 diabetes inadequately controlled by metformin monotherapy. *Diabetes Care*. 2010;33(7):1516–1522.

36. Johnstone HA, et al. Attenuation of hypothalamic-pituitary-adrenal axis stress responses in late pregnancy: changes in feedforward and feedback mechanisms. *J Neuroendocrinol*. 2000;12(8):811–822.

37. Densmore VS, et al. 11β-Hydroxysteroid dehydrogenase type 1 induction in the arcuate nucleus by high-fat feeding: a novel constraint to hyperphagia? *Endocrinology*. 2006;147(9):4486–4495.

38. McEwen BS, et al. Corticosteroids, the aging brain and cognition. *Trends Endocrinol Metab*. 1999;10(3):92–96.

39. Yau JL, et al. Enhanced hippocampal long-term potentiation and spatial learning in aged 11beta-hydroxysteroid dehydrogenase type 1 knock-out mice. *J Neurosci*. 2007;27(39):10487–10496.

40. MacLullich AM, et al. 11β-hydroxysteroid dehydrogenase type 1, brain atrophy and cognitive decline. *Neurobiol Aging*. 2012;33(1):207. e1–207.e8.

41. Holmes MC, et al. 11 Beta-hydroxysteroid dehydrogenase type 1 expression is increased in the aged mouse hippocampus and parietal cortex and causes memory impairments. *J Neurosci*. 2010;30(20):6916–6920.

42. Sooy K, et al. Cognitive and disease-modifying effects of 11 beta-hydroxysteroid dehydrogenase type 1 inhibition in male Tg2576 mice, a model of Alzheimer's disease. *Endocrinology*. 2015;156(12):4592–4603.

43. Sooy K, et al. Partial deficiency or short-term inhibition of 11 beta-hydroxysteroid dehydrogenase type 1 improves cognitive function in aging mice. *J Neurosci*. 2010;30(41):13867–13872.

44. Yau JL, Noble J, Seckl JR. 11 Beta-hydroxysteroid dehydrogenase type 1 deficiency prevents memory deficits with aging by switching from glucocorticoid receptor to mineralocorticoid receptor-mediated cognitive control. *J Neurosci*. 2011;31(11):4188–4193.

35

Stress, Insulin Resistance, and Type 2 Diabetes

I. Kyrou[1,2,3,4], *H.S. Randeva*[1,2,3], *C. Tsigos*[4]

[1]Aston University, Birmingham, United Kingdom; [2]University Hospital Coventry and Warwickshire NHS Trust WISDEM, Coventry, United Kingdom; [3]University of Warwick, Coventry, United Kingdom; [4]Harokopio University, Athens, Greece

Abstract

Stress can be defined as a state of threatened body homeostasis which mobilizes a spectrum of adaptive physiologic and behavioral responses via the hypothalamic-pituitary-adrenal (HPA) axis and the sympathetic nervous system (SNS), i.e., the two main effector pathways of the stress system. These adaptive responses aim to reestablish the challenged body equilibrium (allostasis/cacostasis) and are programmed to promptly subside once this has been achieved. Growing evidence suggests that chronic stress significantly affects glucose homeostasis and is implicated in the dysregulation of metabolism over time. Indeed, various stress indices have been shown to exhibit a positive correlation with the increasing prevalence rates of both obesity and type 2 diabetes, particularly in Western societies. Recent data further indicate that chronic stress, associated with hypercortisolemia and prolonged SNS activation, promotes visceral accumulation of adipose tissue and insulin resistance, hence, contributing to the clinical presentation of central obesity, type 2 diabetes, and cardiovascular disease. On the other hand, obesity induces a systemic low-grade inflammation, mediated by proinflammatory adipokines which activate the acute phase reaction and act as an additional chronic stimulus to the stress system activation. Accordingly, a vicious cycle is formed, whereby chronic activation of the stress system contributes to obesity-related inflammation and insulin resistance, and vice versa.

INTRODUCTION

The survival of the individual relies on promptly composing a successful adaptive response to imposed stressors. However, prolonging such adaptive responses, as in chronic stress, may have detrimental effects on glucose and metabolic homeostasis over time.[1-6] Glucocorticoids and catecholamines, the principal hormonal effectors of the stress system, mediate the majority of these effects and can progressively lead to various metabolic syndrome manifestations, including obesity-related diabetes. Accumulating evidence suggests that a significant positive association exists between increased cortisol levels and secretion of proinflammatory factors (i.e., adipokines and cytokines) by adipose tissue depots following weight gain. Hence, chronic hypercortisolemia, even mild, appears to contribute to increasing visceral adiposity and insulin resistance, while circulating adipokines trigger the acute phase reaction, promote a generalized proinflammatory state, and act as a chronic stimulus to HPA axis, thus, forming a deleterious vicious cycle.[4-6] This chapter reviews the main components of the stress system and potential mechanisms implicated in stress-related metabolic dysregulation and presents available clinical data linking chronic stress to the development of insulin resistance and type 2 diabetes.

KEY POINTS

- Stress constitutes a state of threatened body homeostasis, mobilizing an adaptive central and peripheral response which is typically self-contained once the threat from the acute stressor(s) is effectively addressed.
- The stress system response is mediated primarily by activation of the hypothalamic-pituitary-adrenal (HPA) axis and the sympathetic nervous system (SNS).
- Chronic stress-induced hypercortisolism, even mild and/or functional, and prolonged SNS activation promote increased visceral adiposity, insulin resistance, and several behavioral changes, contributing to the development of type 2 diabetes and metabolic syndrome.
- Obesity, particularly central/visceral, induces a chronic, low-grade, proinflammatory state which is mediated by a proinflammatory circulating adipokine profile and acts as a chronic stressor inducing prolonged/chronic activation of the stress system.
- Chronic stress exhibits a positive correlation with the increasing prevalence rates of both obesity and type 2 diabetes, particularly in Western societies.

THE STRESS RESPONSE

Stress can be defined as a state in which the established normal/healthy body homeostasis (eustasis) is threatened or perceived to be threatened following exposure to potentially adverse forces (stressors).[1–3] To defend homeostasis against such threats and reestablish the disturbed equilibrium, a repertoire of physiologic and behavioral responses is rapidly mobilized, constituting the "adaptive stress response." Whether exposed to extrinsic or intrinsic stressors, this adaptive response, referred to by Hans Selye as "the general adaptation syndrome," attains a relatively stereotypic, nonspecific nature aiming to mobilize available body resources against the stressor(s). As such, during acute stress, attention is enhanced and most of the central nervous system (CNS) functions focus mainly on the perceived threat(s). In parallel, catabolism and inhibition of growth and reproduction are promoted in the periphery, while the cardiac output, respiration, and cardiovascular tone are increased and blood flow is redirected to temporarily provide higher blood/oxygen perfusion to the threatened body site(s) and to the aroused CNS, heart, and muscles.

The stress system as a whole relies on a highly complex and interconnected neuroendocrine, cellular, and molecular infrastructure, which is located in both the CNS and vital organs of the periphery. The central control stations of this system are strategically positioned in the hypothalamus and the brain stem, including primarily the parvocellular corticotropin-releasing hormone (CRH) and arginine vasopressin (AVP) neurons of the paraventricular nuclei (PVN) of the hypothalamus, as well as the locus coeruleus/norepinephrine system (central sympathetic system) (Fig. 35.1).[1–4] Moreover, the hypothalamic-pituitary-adrenal (HPA) axis and the efferent sympathetic/adrenomedullary system constitute the two main effector limbs via which the CNS triggers and coordinates the adaptive responses of the peripheral organs and tissues during exposure to stressors.

The Hypothalamic-Pituitary-Adrenal Axis

The integrity and normal activity of the HPA axis are essential for composing an effective stress response. Within this axis, the hypothalamus regulates the secretion of the adrenocorticotropic hormone (ACTH) from the anterior pituitary primarily via CRH, while AVP constitutes a potent synergistic factor in stimulating ACTH secretion (Fig. 35.1).[7,8] Both CRH and AVP are secreted into the hypothalamo-hypophyseal portal system in a synchronized manner which is characterized by a highly precise circadian rhythm. Under normal resting conditions, the amplitude of the CRH and AVP hypothalamic pulses increase early in the morning, hence, resulting in increased ACTH secretion and finally in the cortisol secretory bursts noted in the systemic circulation.[9] This diurnal rhythm is regulated by changes in lighting, physical activity, and feeding schedules and can be markedly disrupted by imposed stressors.[10]

Further down the HPA axis, circulating ACTH constitutes the key regulator of glucocorticoid secretion by the cortex of the adrenals. In turn, glucocorticoids, mainly cortisol in humans, are the final mediators of the HPA axis and play a vital role in inducing the adaptive response to stress and regulating the whole body homeostasis.[1–4] Of note, glucocorticoids, via their widespread glucocorticoid receptors, act on positive and negative glucocorticoid-responsive elements (GREs) to activate or repress, respectively, multiple genes, many of which are directly or indirectly implicated in crucial metabolic pathways. Accordingly, cortisol appears to upregulate a number of genes encoding lipoprotein receptors and enzymes of glucose, lipid, and amino acid metabolism, while also

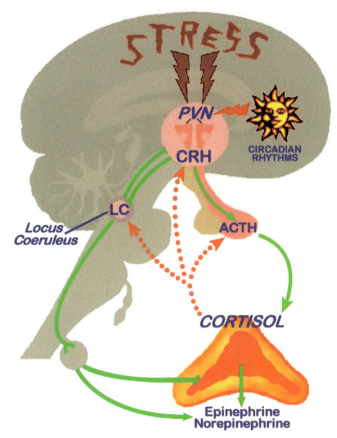

TABLE 35.1 Selected Genes Which Are Involved in Key Endocrine/Immune/Metabolic Pathways/Functions and Are Regulated by Glucocorticoids

Downregulated Genes (Affected Pathway/Function)	Upregulated Genes (Affected Pathway/Function)
CRH (HPA axis regulation)	GLUT4 (glucose transport)
POMC (HPA axis and appetite control)	Hepatic PEPCK (gluconeogenesis)
IL-6 (proinflammatory response)	Glucose-6-phosphatase (gluconeogenesis)
TNF-α (proinflammatory response)	Leptin (energy homeostasis)
IL-8 (proinflammatory response)	Hormone-sensitive lipase (lipolysis—lipid metabolism)
Adiponectin (antiinflammatory response)	Lipoprotein lipase (lipolysis—lipid metabolism)
Osteocalcin (bone metabolism)	VLDL receptor (lipoprotein metabolism)
Prolactin (reproduction)	Tyrosine aminotransferase (amino acid catabolism)
	Tryptophan oxygenase (amino acid catabolism)

CRH, corticotropin-releasing hormone; *GLUT4*, glucose transporter 4; *HPA*, hypothalamic-pituitary-adrenal; *IL-6*, interleukin-6; *IL-8*, interleukin-8; *PEPCK*, phosphoenolpyruvate carboxykinase; *POMC*, proopiomelanocortin; *TNF-α*, tumor necrosis factor-α; *VLDL*, very low-density lipoprotein.

FIGURE 35.1 Schematic representation of the main components of the stress system, highlighting the major central stations and peripheral effector pathways mediating the adaptive stress response. *ACTH*, adrenocorticotropic hormone; *CRH*, corticotropin-releasing hormone; *LC*, locus coeruleus/norepinephrine-sympathetic system; *PVN*, paraventricular nucleus. Stimulation is represented by *solid green lines* and inhibition by *dashed red lines*.

suppresses other genes encoding CRH, proopiomelanocortin, and various adipokines/cytokines (e.g., adiponectin and interleukin-6 (IL-6)) (Table 35.1).[11–15] It must be highlighted that glucocorticoids control the basal HPA axis activity and are essential for the termination of the stress response, by providing the required negative feedback at the hypothalamic and pituitary level, as well as at higher brain centers (Fig. 35.1).[1–4]

The Sympathetic Nervous System

The autonomic nervous system equips the body with a rapidly responding mechanism which controls all vital organs and functions (e.g., rapid regulation of cardiovascular, respiratory, gastrointestinal, renal, and endocrine activity) via either the sympathetic nervous system (SNS) or the parasympathetic nervous system or the combined activity of both.[1–4] Sympathetic innervation of peripheral organs is derived from the efferent preganglionic fibers, whose cell bodies lie in the intermediolateral column of the spinal cord. These nerves synapse in the bilateral

chains of sympathetic ganglia with postganglionic sympathetic neurons which richly innervate the smooth muscle of the vasculature, as well as multiple other organs and tissues, including the heart, skeletal muscles, kidneys, gut, and adipose tissue. Notably, the SNS also offers an additional humoral component to the stress response by providing most of the circulating epinephrine and part of the norepinephrine from the adrenal medulla (Fig. 35.1).

Metabolic Dysregulation Related to Chronic Stress

In the context of the adaptive response to stress, the activation of the HPA axis and the SNS, on one hand, is meant to be proportional to the force of the imposed stressor(s) in order to be sufficient to restore the disturbed homeostasis, while, on the other hand, it is also programmed to be of limited duration.[1–4] From an evolutionary perspective, this ability to self-contain the adaptive stress response in a timely fashion is considered equally essential for survival, since it ensures that the adaptive catabolic, antireproductive, antigrowth, and immunosuppressive stress-related effects are only transient. Chronic stress, expressed through prolonged activation of the HPA axis and the SNS, eventually renders these adaptive effects detrimental by extending their duration.

Chronic hypercortisolemia is considered to play a pivotal role in the deleterious consequences of chronic stress. Indeed, in order to mobilize every available body energy resource against the imposed stressor(s), glucocorticoids exert a broad spectrum of primarily catabolic effects as part of the adaptive stress response.[4–6] As such, hepatic gluconeogenesis is enhanced and the circulating glucose levels are increased, while lipolysis is also induced (although hypercortisolemia favors abdominal and dorsocervical fat accumulation). Furthermore, protein degradation is promoted at multiple organs/tissues, including skeletal muscles and bones, in order to provide amino acids which can serve as additional substrates for oxidative pathways. In addition to these direct catabolic actions, glucocorticoids also exert indirect effects in the same direction by antagonizing the anabolic effects of sex steroids, insulin, and growth and thyroid hormones on their target organs/tissues (Fig. 35.2).[4–6] Under normal conditions, this stress-induced shift of metabolism toward a generalized catabolic state typically subsides in a timely manner once the threat from the imposed stressor(s) is contained. However, prolonged stress can progressively cause, via chronic hypercortisolemia, visceral adipose tissue accumulation, decreased lean body mass (i.e., muscle and bone mass), and insulin resistance.[4–6,16–18] Interestingly, the phenotype of Cushing's syndrome, characterized by abdominal and trunk fat accumulation, decreased lean body mass and various other metabolic syndrome manifestations (e.g., dyslipidemia, hypertension, insulin resistance, and type 2 diabetes), is present in a variety of pathophysiologic conditions which are described as pseudo-Cushing's states. Under conditions of chronic stress, the development of this phenotype can be attributed to stress-induced mild hypercortisolism which may be further combined with increased peripheral tissue sensitivity to glucocorticoids. The latter appears to also play a pivotal role in the detrimental consequences of chronic hypercortisolemia and depends on a number of factors, including (1) the tissue-specific concentration and activity of glucocorticoid receptors, (2) the circulating and tissue levels of cortisol-binding globulin (CBG), and (3) the local expression of enzymes such as 11β-hydroxysteroid dehydrogenase type 1 (11β-HSD1; an enzyme that locally catalyzes the conversion of the inactive cortisone to cortisol and appears to be increased in adipose tissue depots of obese individuals, potentially contributing to the development of central obesity and metabolic syndrome).[18–20] Of note, it has been shown that the basic transcription factor CLOCK (i.e., a histone acetyltransferase which is a central component of the self-oscillating transcription factor loop that generates circadian rhythms) represses the nuclear glucocorticoid receptor transcriptional activity by acetylating lysine residues within the lysine cluster located in the hinge

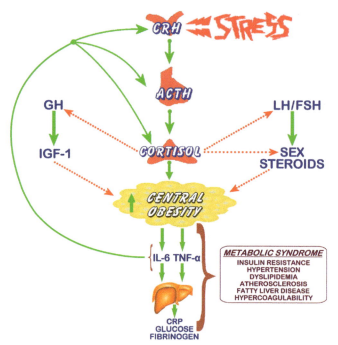

FIGURE 35.2 Schematic representation of interactions between chronic activation of the hypothalamic-pituitary-adrenal (HPA) axis and central/visceral obesity which may lead to metabolic syndrome manifestations. Chronic HPA activation and hypercortisolism, even mild and/or functional, promotes adipose tissue accumulation and particularly visceral obesity, while it suppresses both the gonadal and the growth hormone (GH) axis. In obesity, the secretion of pro-inflammatory adipokines/cytokines is proportional to the adipose tissue mass, especially from central adipose tissue depots, creating an unremitting inflammatory load which induces the production of acute-phase reactants in the liver and is linked to the development of the metabolic syndrome. In turn, proinflammatory adipokines/cytokines, such interleukin-6 (IL-6) which is the main circulating/endocrine cytokine, act as an additional chronic stimulus on the HPA axis, hence, forming a vicious cycle with detrimental effects on metabolism over time. *ACTH*, adrenocorticotropic hormone; *CRH*, corticotropin-releasing hormone; *CRP*, C-reactive protein; *FSH*, follicle-stimulating hormone; *LH*, luteinizing hormone; *IGF-1*, insulin-like growth factor-1; *TNF-α*, tumor necrosis factor-α. Stimulation is represented by *solid green lines* and inhibition by *dashed red lines*.

region of the glucocorticoid receptors.[20] This CLOCK-mediated repression oscillates during the day in inverse phase to the normal diurnal rhythm of the HPA axis, thus, progressively increasing the glucocorticoid sensitivity of target tissues after the morning hours (peak glucocorticoid sensitivity in the evening hours) and acting as a target tissue counterregulatory mechanism to the diurnal fluctuation of circulating glucocorticoids. As such, even mild elevations of circulating cortisol levels in the evening hours, as frequently noted in chronic stress, may result in functional hypercortisolism with disproportionately more potent glucocorticoid effects due to the increased underlying glucocorticoid sensitivity of target tissues. This stress-related functional hypercortisolism further promotes the clinical manifestations of the metabolic syndrome.[20,21]

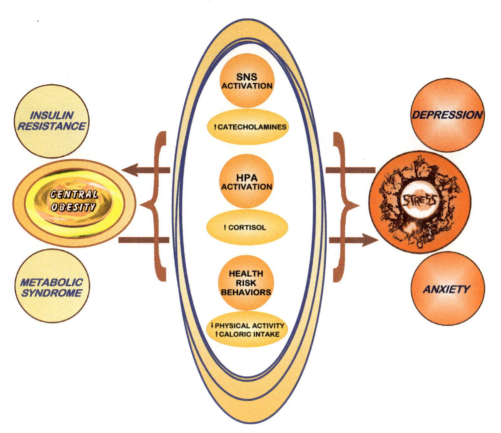

FIGURE 35.3 Schematic representation of reciprocal links between chronic stress activation and derangement of metabolic homeostasis. Chronic stress, manifested with symptoms of depression and/or anxiety, induces prolonged activation of the hypothalamic-pituitary-adrenal (HPA) axis and the sympathetic nervous system (SNS) which together with detrimental changes in certain health behaviors (e.g., sedentary lifestyle, comfort/binge eating) may progressively lead to central/visceral obesity, insulin resistance, and metabolic syndrome manifestations over time, and vice versa.

Increased sympathoadrenal system activity is also an important pathophysiologic component of chronic stress which appears to contribute to the development of impaired glucose tolerance, while particularly increasing the risk for acute cardiovascular events (e.g., myocardial infraction and stroke).[22,23] Finally, a number of behavioral changes which impact on both physical activity (e.g., increased hours of sleep and a sedentary lifestyle) and dietary habits (e.g., increased portion size, comfort/binge eating, and alcohol abuse) exhibit a strong positive correlation with chronic stress disorders and may lead to progressive weight gain and dysregulation of glucose and lipid metabolism (Fig. 35.3).[4–6]

CHRONIC STRESS, DEPRESSION AND TYPE 2 DIABETES

A growing body of epidemiological data suggests that a bidirectional association exists between obesity-related diabetes and chronic stress disorders, such as depression. Melancholic depression represents the prototypic example of chronic stress system hyperactivation which may

result in multiple detrimental somatic effects.[1–4] Indeed, patients with depression have been shown to develop metabolic syndrome manifestations (e.g., central obesity, hypertension, dyslipidemia, and impaired glucose tolerance), atherosclerosis, and osteopenia, as well as certain types of infections and cancers.[4–6] Importantly, without appropriate treatment, the life expectancy of these patients is significantly reduced by 15–20 years, even after excluding suicides.[1,2]

Available evidence also shows that being depressed at baseline increases the risk of subsequently developing type 2 diabetes, while this association is not fully explained by other known diabetes risk factors, including obesity, family history of type 2 diabetes, physical activity levels, smoking, and alcohol consumption.[24–27] Of note, based on meta-analysis data which included prospective studies of depression predicting incident type 2 diabetes, depression is suggested to increase the type 2 diabetes risk by 60%.[28] Moreover, major stressful life events appear to be associated with the onset of type 2 diabetes. As documented in the Hoorn Study, a high number of relatively common major life events during a preceding 5-year period were related to both increased visceral adiposity and higher prevalence of previously unknown type 2 diabetes.[29] Indeed, results from the Hoorn Study also showed that, in a middle-aged and elderly population-based cohort, individuals reporting more stressful life events at baseline exhibited a significantly increased risk for developing metabolic syndrome during 6.5 years of follow-up.[30] In addition, the results of a case–control retrospective study in premenopausal women revealed that women with rapid onset of weight gain following exposure to a major stressful event were characterized by the development of obesity and overactivated adrenocortical function.[31] A number of additional studies have also indicated that chronic work-related stress predicted both generalized and central obesity during midlife, an association which was largely independent from other covariates.[32,33] Interestingly, a cohort of UK female students who exhibited weight gain over their first year at the university were reported to exhibit significantly higher levels of perceived stress.[34] In accord with that finding, data from a cohort of 5473 junior and high school students in China also showed that depressive and anxiety symptoms and life stress were significantly associated with unhealthy eating behaviors, even after adjusting for age, gender, body mass index, and various other socioeconomic factors.[35] Thus, it is becoming increasingly evident that prolonged and/or major stress and depression may, at least in certain cases, precede the development of central obesity and the onset of type 2 diabetes, indicating a potential role in the underlying pathophysiologic mechanisms.

On the other hand, patients with type 2 diabetes are reported to exhibit depression rates up to twofold higher compared to individuals without diabetes. Indeed, meta-analysis data showed that presence of diabetes almost doubles the odds of comorbid depression, a stable finding which was not affected by factors such as gender, assessment method, subject source, or type of diabetes.[36]

Finally, it must be highlighted that effective depression treatment in diabetic patients is shown to significantly lower hyperglycemia, independently of the differential effects of such treatment on body weight.[37] Interestingly, controlled trials of cognitive-behavioral therapy demonstrated that improvements in depression scores resulted in significantly better glycemic control.[38] Moreover, successful behavioral stress management and/or use of anxiolytic medications in type 2 diabetic patients have been reported to contribute to sustained improvements in glycemic control.[39]

OBESITY-RELATED INFLAMMATION AS A CHRONIC STRESS STATE

Over the last decade, it has been established that obesity induces a low-grade inflammatory state which initially starts inside the expanding adipose tissue depots and eventually becomes systemic/generalized. Notably, this proinflammatory state is heightened with weight gain, particularly with increased central adiposity, and persists for as long as the excessive body weight is maintained. Compelling evidence links this unremitting obesity-related inflammatory stress to the pathogenesis of insulin resistance, type 2 diabetes, and atherosclerosis.[40–43]

It has been also known for several decades that inflammatory stress is associated with concurrent activation of the HPA axis. Indeed, cytokines and other humoral mediators of inflammation are potent activators of the central stress response, thus, creating a feedback loop through which the immune/inflammatory system communicates with the brain.[44] The three main inflammatory cytokines, i.e., tumor necrosis factor-α (TNF-α), interleukin-1β (IL-1β), and IL-6 can stimulate the HPA axis alone or in synergy with each other. Among these, IL-6 is considered the main circulating/endocrine cytokine and appears to play a key role in the immune stimulation of the HPA axis (Fig. 35.2).[45,46] Interestingly, some of the activating effects of cytokines on the HPA axis may be exerted indirectly by stimulation of the central catecholaminergic pathways, while activation of peripheral nociceptive, somatosensory, and visceral afferent fibers may lead to stimulation of both the catecholaminergic and CRH neuronal systems via ascending spinal pathways.

As noted with other adaptive stress responses, a typical inflammatory response is programmed to subside once the initial threat (immune stressor) is contained. As such, at conventional inflammation sites (e.g., at sites of injury or local infection) cytokines are secreted by immune cells transiently over a certain period in order to prevent a prolonged local activation of macrophages which would have detrimental effects.[47] However, obesity-related inflammation is persistent and unremitting due to a vicious cycle between adipocytes and macrophages inside each adipose tissue depot. In this context, the accumulation of adipose tissue in obesity is followed by a continuous secretion of adipokines and chemokines from adipocytes into the systemic circulation. In turn, this leads to chemoattraction and recruitment of mononuclear cells from the bloodstream into the expanding adipose tissue depots, hence, creating a growing population of resident macrophages.[48,49] These resident macrophages also release locally cytokines (e.g., TNF-α, IL-6), which further stimulate the secretion of proinflammatory adipokines, suppress the expression of adipokines with antiinflammatory properties (e.g., adiponectin), and promote insulin resistance in adipocytes.[48–50] Over time, this process leads to a chronic, low-grade, systemic inflammation characterized by increased release of proinflammatory adipokines/cytokines into the circulation which exert adverse metabolic effects on various tissues and organs, including the endothelium, liver, and skeletal muscles.[40–43] Thus, obese patients typically exhibit increased circulating levels of proinflammatory adipokines (e.g., leptin, IL-6, and TNF-α) which correlate directly to the manifestations of the metabolic syndrome.[43,51,52] Indeed, high circulating levels of both CRP and IL-6 levels have been associated to increased risk of type 2 diabetes.[53] In addition, other common markers of inflammation (e.g., fibrinogen and haptoglobin) have also been associated prospectively with the development of diabetes in adults, while chronic subclinical inflammation appears to be independently associated with insulin resistance even in nondiabetic subjects. Conversely, circulating levels of adiponectin have been shown to hold an inverse correlation to cardiometabolic complications.[54] Overall, these findings support the notion that the described obesity-related proinflammatory cascade constitutes a chronic stressor which is closely linked to a spectrum of cardiometabolic complications.

CONCLUSION

Chronic stress appears to contribute via various neuroendocrine mechanisms to the development of central/visceral obesity and consequently to obesity-related insulin resistance and type 2 diabetes. Conversely, excess accumulation of visceral adipose tissue contributes to the development of a chronic stress state, primarily of proinflammatory nature. The exact underlying mechanisms which facilitate this vicious cycle and lead to the dysregulation of the metabolic homeostasis over time are yet to be fully clarified. Better understanding of these links is expected to provide significant insight into the pathogenesis of both chronic stress and obesity-related cardiometabolic disease and help the development of novel and more targeted therapeutic interventions.

References

1. Chrousos GP. Stress and disorders of the stress system. *Nat Rev Endocrinol*. 2009;5(7):374–381.
2. Tsigos C, Chrousos GP. Hypothalamic-pituitary-adrenal axis, neuroendocrine factors and stress. *J Psychosom Res*. 2002;53(4): 865–871.
3. Chrousos GP, Gold PW. The concepts of stress and stress system disorders. Overview of physical and behavioral homeostasis. *JAMA*. March 4, 1992;267(9):1244–1252.
4. Charmandari E, Tsigos C, Chrousos G. Endocrinology of the stress response. *Annu Rev Physiol*. 2005;67:259–284.
5. Kyrou I, Tsigos C. Stress hormones: physiological stress and regulation of metabolism. *Curr Opin Pharmacol*. December 2009;9(6):787–793.
6. Kyrou I, Chrousos GP, Tsigos C. Stress, visceral obesity, and metabolic complications. *Ann NY Acad Sci*. 2006;1083:77–110.
7. Vale W, Spiess J, Rivier C, Rivier J. Characterization of a 41-residue ovine hypothalamic peptide that stimulates secretion of corticotropin and beta-endorphin. *Science*. September 18, 1981;213(4514):1394–1397.
8. Lamberts SW, Verleun T, Oosterom R, de Jong F, Hackeng WH. Corticotropin-releasing factor (ovine) and vasopressin exert a synergistic effect on adrenocorticotropin release in man. *J Clin Endocrinol Metab*. 1984;58(2):298–303.
9. Veldhuis JD, Iranmanesh A, Johnson ML, Lizarralde G. Amplitude, but not frequency, modulation of adrenocorticotropin secretory bursts gives rise to the nyctohemeral rhythm of the corticotropic axis in man. *J Clin Endocrinol Metab*. 1990;71(2):452–463.
10. Chrousos GP. Ultradian, circadian, and stress-related hypothalamic-pituitary-adrenal axis activity-a dynamic digital-to-analog modulation. *Endocrinology*. 1998;139(2):437–440.
11. Meijer OC. Understanding stress through the genome. *Stress*. June 2006;9(2):61–67.
12. Bamberger CM, Schulte HM, Chrousos GP. Molecular determinants of glucocorticoid receptor function and tissue sensitivity to glucocorticoids. *Endocr Rev*. 1996;17(3):245–261.
13. Dostert A, Heinzel T. Negative glucocorticoid receptor response elements and their role in glucocorticoid action. *Curr Pharm Des*. 2004;10(23):2807–2816.
14. Wang M. The role of glucocorticoid action in the pathophysiology of the metabolic syndrome. *Nutr Metab (Lond)*. February 2, 2005;2(1):3.
15. Degawa-Yamauchi M, Moss KA, Bovenkerk JE, et al. Regulation of adiponectin expression in human adipocytes: effects of adiposity, glucocorticoids, and tumor necrosis factor alpha. *Obes Res*. 2005;13(4):662–669.
16. Björntorp P. Do stress reactions cause abdominal obesity and comorbidities? *Obes Rev*. 2001;2(2):73–86.
17. Björntorp P, Rosmond R. The metabolic syndrome-a neuroendocrine disorder? *Br J Nutr*. 2000;83(suppl 1):S49–S57.
18. Putignano P, Pecori Giraldi F, Cavagnini F. Tissue-specific dysregulation of 11beta-hydroxysteroid dehydrogenase type 1 and pathogenesis of the metabolic syndrome. *J Endocrinol Invest*. 2004;27(10):969–974.
19. Anagnostis P, Athyros VG, Tziomalos K, Karagiannis A, Mikhailidis DP. Clinical review: the pathogenetic role of cortisol in the metabolic syndrome: a hypothesis. *J Clin Endocrinol Metab*. August 2009;94(8):2692–2701.
20. Kino T, Chrousos GP. Circadian CLOCK-mediated regulation of target-tissue sensitivity to glucocorticoids: implications for cardiometabolic diseases. *Endocr Dev*. 2011;20:116–126.
21. Chrousos GP. The role of stress and the hypothalamic-pituitary-adrenal axis in the pathogenesis of the metabolic syndrome: neuroendocrine and target tissue-related causes. *Int J Obes Relat Metab Disord*. 2000;24(suppl 2):S50–S55.
22. Dimsdale JE. Psychological stress and cardiovascular disease. *J Am Coll Cardiol*. April 1, 2008;51(13):1237–1246.
23. Grippo AJ, Johnson AK. Stress, depression and cardiovascular dysregulation: a review of neurobiological mechanisms and the integration of research from preclinical disease models. *Stress*. 2009;12(1):1–21.
24. Everson-Rose SA, Meyer PM, Powell LH, et al. Depressive symptoms, insulin resistance, and risk of diabetes in women at midlife. *Diabetes Care*. 2004;27(12):2856–2862.
25. Golden SH, Williams JE, Ford DE, et al. Depressive symptoms and the risk of type 2 diabetes: the Atherosclerosis Risk in Communities study. *Diabetes Care*. 2004;27(2):429–435.
26. Anderson SE, Cohen P, Naumova EN, Must A. Association of depression and anxiety disorders with weight change in a prospective community-based study of children followed up into adulthood. *Arch Pediatr Adolesc Med*. 2006;160(3):285–291.
27. Carnethon MR, Biggs ML, Barzilay JI, et al. Longitudinal association between depressive symptoms and incident type 2 diabetes mellitus in older adults: the cardiovascular health study. *Arch Intern Med*. April 23, 2007;167(8):802–807.
28. Mezuk B, Eaton WW, Albrecht S, Golden SH. Depression and type 2 diabetes over the lifespan: a meta-analysis. *Diabetes Care*. 2008;31(12):2383–2390.

29. Mooy JM, de Vries H, Grootenhuis PA, Bouter LM, Heine RJ. Major stressful life events in relation to prevalence of undetected type 2 diabetes: the Hoorn Study. *Diabetes Care*. 2000;23(2):197–201.

30. Rutters F, Pilz S, Koopman AD, et al. Stressful life events and incident metabolic syndrome: the Hoorn study. *Stress*. 2015;18(5):507–513.

31. Vicennati V, Pasqui F, Cavazza C, Pagotto U, Pasquali R. Stress-related development of obesity and cortisol in women. *Obesity (Silver Spring)*. September 2009;17(9):1678–1683.

32. Chandola T, Brunner E, Marmot M. Chronic stress at work and the metabolic syndrome: prospective study. *BMJ*. March 4, 2006; 332(7540):521–525.

33. Brunner EJ, Chandola T, Marmot MG. Prospective effect of job strain on general and central obesity in the Whitehall II Study. *Am J Epidemiol*. April 1, 2007;165(7):828–837.

34. Serlachius A, Hamer M, Wardle J. Stress and weight change in university students in the United Kingdom. *Physiol Behav*. November 23, 2007;92(4):548–553.

35. Hou F, Xu S, Zhao Y, et al. Effects of emotional symptoms and life stress on eating behaviors among adolescents. *Appetite*. September 2013;68:63–68.

36. Anderson RJ, Freedland KE, Clouse RE, Lustman PJ. The prevalence of comorbid depression in adults with diabetes: a meta-analysis. *Diabetes Care*. 2001;24(6):1069–1078.

37. Surwit RS, van Tilburg MA, Zucker N, et al. Stress management improves long-term glycemic control in type 2 diabetes. *Diabetes Care*. 2002;25(1):30–34.

38. Lustman PJ, Griffith LS, Freedland KE, Kissel SS, Clouse RE. Cognitive behavior therapy for depression in type 2 diabetes mellitus. A randomized, controlled trial. *Ann Intern Med*. October 15, 1998;129(8):613–621.

39. Brieler JA, Lustman PJ, Scherrer JF, Salas J, Schneider FD. Antidepressant medication use and glycaemic control in comorbid type 2 diabetes and depression. *Fam Pract*. February 2016;33(1):30–36.

40. Gregor MF, Hotamisligil GS. Inflammatory mechanisms in obesity. *Annu Rev Immunol*. 2011;29:415–445.

41. Hotamisligil GS. Inflammation and metabolic disorders. *Nature*. 2006;444:860–867.

42. Makki K, Froguel P, Wolowczuk I. Adipose tissue in obesity-related inflammation and insulin resistance: cells, cytokines, and chemokines. *ISRN Inflamm*. 2013;2013:139239.

43. Maury E, Brichard SM. Adipokine dysregulation, adipose tissue inflammation and metabolic syndrome. *Mol Cell Endocrinol*. 2010;314:1–16.

44. Turnbull AV, Rivier CL. Regulation of the hypothalamic-pituitary-adrenal axis by cytokines: actions and mechanisms of action. *Physiol Rev*. 1999;79(1):1–71.

45. Tsigos C, Papanicolaou DA, Kyrou I, Defensor R, Mitsiadis CS, Chrousos GP. Dose-dependent effects of recombinant human interleukin-6 on glucose regulation. *J Clin Endocrinol Metab*. 1997;82(12):4167–4170.

46. Tsigos C, Papanicolaou DA, Defensor R, Mitsiadis CS, Kyrou I, Chrousos GP. Dose effects of recombinant human interleukin-6 on pituitary hormone secretion and energy expenditure. *Neuroendocrinology*. 1997;66(1):54–62.

47. Chrousos GP. The hypothalamic-pituitary-adrenal axis and immune-mediated inflammation. *N Engl J Med*. May 18, 1995;332(20):1351–1362.

48. Lee J. Adipose tissue macrophages in the development of obesity-induced inflammation, insulin resistance and type 2 diabetes. *Arch Pharm Res*. 2013;36:208–222.

49. Weisberg SP, McCann D, Desai M, Rosenbaum M, Leibel RL, Ferrante Jr AW. Obesity is associated with macrophage accumulation in adipose tissue. *J Clin Invest*. 2003;112(12):1796–1808.

50. Wellen KE, Hotamisligil GS. Obesity-induced inflammatory changes in adipose tissue. *J Clin Invest*. 2003;112(12):1785–1788.

51. Tilg H, Moschen AR. Adipocytokines: mediators linking adipose tissue, inflammation and immunity. *Nat Rev Immunol*. 2006;6:772–783.

52. Matsuzawa Y. The metabolic syndrome and adipocytokines. *FEBS Lett*. May 22, 2006;580(12):2917–2921.

53. Pradhan AD, Manson JE, Rifai N, Buring JE, Ridker PM. C-reactive protein, interleukin 6, and risk of developing type 2 diabetes mellitus. *JAMA*. July 18, 2001;286(3):327–334.

54. Li S, Shin HJ, Ding EL, van Dam RM. Adiponectin levels and risk of type 2 diabetes: a systematic review and meta-analysis. *JAMA*. July 8, 2009;302(2):179–188.

36

Steroid Hydroxylases

J. Hofland, F.H. de Jong

Erasmus MC, Rotterdam, The Netherlands

Abstract

Steroid hormones play an important role in the survival of the individual and of species. This is realized by the regulation of salt and sugar homeostasis by mineralocorticosteroids and glucocorticosteroids and of reproductive processes by sex hormones. This chapter discusses the roles of members of the cytochrome P450 (CYP) enzyme family in the biosynthesis of the hormonally active steroids: CYP11A1 for all steroids, CYP21A2 and CYP11B2 for the mineralocorticoid aldosterone, CYP17A1, CYP21A2, and CYP11B1 for the glucocorticoid cortisol, CYP17A1 for androgens and CYP19A1 for estrogens. Furthermore, the importance of the cholesterol transporter steroidogenic acute regulatory protein and the electron donors P450 oxidoreductase and ferredoxin is indicated. The localization and regulation of the expression of the CYPs and the effects of inactivating mutations in the genes encoding these enzymes are described. Finally, the role of steroid-catabolizing CYPs is discussed.

INTRODUCTION

The cytochrome P450 (*CYP*) genes form a large group, encoding a class of heme-thiolate enzymes active in the hydroxylation of endogenous and exogenous substances. These enzymes are important in detoxification and catabolism of pharmacological substances[35] and also in the biosynthesis of steroids.[22] Homology between the various genes is relatively high: *CYPs* in the same family by definition share more than 40% sequence identity

and members of the same subfamily share more than 55% identity.[25] In the human genome, 18 families of *CYP* genes, encoding 57 genes have been identified.[26]

In steroidogenesis, CYP proteins are instrumental in conveying molecular oxygen to designated positions in the steroid skeleton, in this way introducing hydroxyl groups. Subsequently, adjacent carbon–carbon bonds can be broken. By these mechanisms, specific CYPs determine which final hormonal steroids will be produced. The specific roles of these enzymes in the pathway of steroidogenesis are indicated in Fig. 36.1. The electrons necessary to drive this process are supplied by NADPH and are passed on to mitochondrial CYPs by way of the flavoprotein ferredoxin reductase to the iron-sulfur protein ferredoxin (Fig. 36.2A).[27] For the CYPs in the microsomal cell fraction, the role of ferredoxin is taken over by P450 oxidoreductase (POR).[27] In addition, cytochrome b5 may play an additional role to support 17,20-lyase activity (Fig. 36.2B).[18]

CHOLESTEROL SIDE-CHAIN CLEAVAGE ENZYME: CYP11A1

The enzyme CYP11A1 performs the conversion of cholesterol to pregnenolone, by first hydroxylating cholesterol at the 20α and 22 carbon atoms and then lysing the bond between these carbon atoms. The enzyme, of which only one form is known, has been found in the classical steroidogenic organs: adrenal glands, the gonads, and the placenta,[22] but also in heart[7] and brain.[20] Intracellularly, it is localized in the inner mitochondrial membrane. Apparently, this localization is extremely important for the activity of the enzyme: transfection studies with constructs

that direct the protein into the endoplasmic reticulum failed to show activity.[5]

Regulation of the expression of *CYP11A1* differs among tissues. In organs where stimulation of *CYP11A1* transcription place under the influence of adrenocorticotropic hormone (ACTH) or gonadotropins (zonae fasciculata and reticularis in the adrenal glands, testes, and ovaries) cyclic AMP (cAMP)-responsive elements

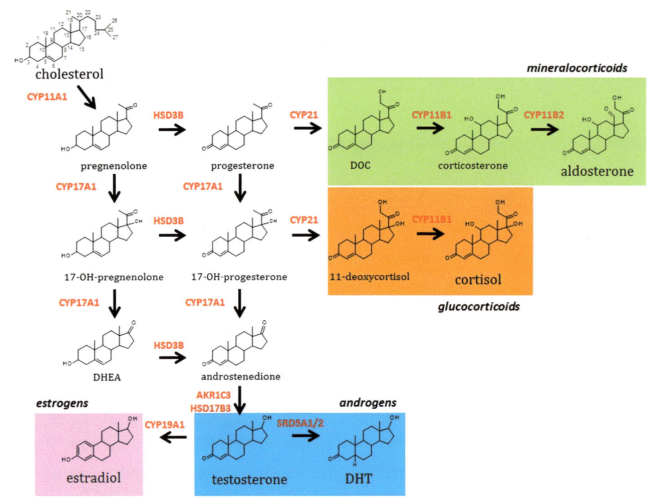

FIGURE 36.1 **Human steroidogenic pathway.** Steroid hormones are produced through several enzymatic reactions, catalyzed by cytochrome P450 (CYP) and hydroxysteroid dehydrogenase (HSD) enzymes, from the common precursor cholesterol. Nomenclature of the steroidogenic reactions is derived from the number of the affected carbon atom. Depending on the activity of the steroidogenic enzymes different subclasses of steroid hormones can be produced, that have specific nuclear receptor-binding capabilities.

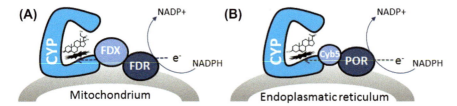

FIGURE 36.2 **Electron transfer for cytochrome P450 (CYP)-mediated enzymatic activity.** (A) Mitochondrial CYP enzymes (CYP11A1, CYP11B1/2) utilize ferredoxin and ferredoxin reductase to transfer electrons from nicotinamide adenine dinucleotide phosphate (NADPH) in order to catalyze the hydroxylation of the steroid substrate. (B) CYP enzymes localized in the endoplasmic reticulum (CYP17A1, CYP21A2, CYP19A1) employ the coenzyme P450 oxidoreductase (POR) for electron transfer, whereas cytochrome b5 (Cyb5) also confers a specific, but essential allosteric function for the 17,20-lyase activity of CYP17A1.

induce expression of *CYP11A1*,[21] whereas in the adrenal zona glomerulosa the calcium/protein kinase C (PKC) pathway plays a role.[2] Steroid production catalyzed by adrenocortical CYP11A1 thus forms a key element of the renin-angiotensin-aldosterone system (RAAS) and hypothalamus-pituitary-adrenal (HPA) axis (Fig. 36.3). Furthermore, a number of transcription factors, such as steroidogenic factor-1 (SF-1) and specificity protein-1 (Sp-1), play a role in the stimulation of the transcription of *CYP11A1*.[16] In contrast, placental expression of *CYP11A1* is constitutive.[14]

The transcriptional regulation of the amount of enzyme in the glands takes a relatively long time (2–4 h), whereas the stimulation of steroidogenesis is much faster in the adrenal glands and the gonads. These fast effects of increased cAMP levels are mediated through effects on the transport of cholesterol from the outer to the inner mitochondrial membrane by the steroidogenic acute regulatory protein (StAR) after its phosphorylation.[40] StAR is expressed in steroidogenic cells in adrenals, testes, and ovaries and in the brain,[40] but not in the human placenta[31]; the placental syncytiotrophoblasts apparently use an alternative way to transport cholesterol to the inner mitochondrial membrane.[30] The inability to lyse the C20—C22 bond of cholesterol, as present in congenital lipoid adrenal hyperplasia, may be due to inactivating mutations in *CYP11A1*, but can also be caused by inactivating mutations of StAR, indicating the crucial importance of this protein in steroidogenesis.[23]

17-HYDROXYLASE: CYP17A1

After the conversion of cholesterol to pregnenolone in the mitochondria, further conversion of pregnenolone takes place in the endoplasmic reticulum by the conversion to 17-hydroxy-pregnenolone (17-OH pregnenolone) by CYP17A1. Transfer of electrons from NADPH in this reaction is made possible by the coenzyme POR (Fig. 36.2B). Alternatively, pregnenolone can be converted to progesterone by 3β-hydroxysteroid dehydrogenase (HSD3B); progesterone can also be hydroxylated at the C17 position by CYP17A1. In the presence of the coenzyme cytochrome b5, 17-OH pregnenolone can be converted to dehydroepiandrosterone (DHEA) by the same CYP17A1, which also catalyzes the 17,20-lyase reaction. For the human enzyme 17-OH pregnenolone rather than 17-OH progesterone is the preferred substrate for the lyase reaction.[19]

CYP17A1 is at the bifurcations of the steroidogenic pathways to mineralocorticoids, glucocorticosteroids, and the sex steroids in the adrenal glands: in the absence of CYP17A1 it is not possible to synthesize cortisol and

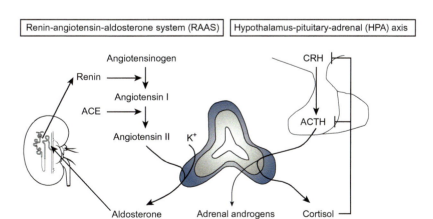

FIGURE 36.3 **The adrenal cortex as part of the RAAS and the HPA axis.** The production of aldosterone in the zona glomerulosa is controlled by renin and angiotensin II. Aldosterone stimulates a.o. sodium reabsorption in the kidney, leading to an increase in blood pressure. Adrenocortical production of cortisol and adrenal androgens in the zonae fasciculata and reticularis is controlled by pituitary adrenocorticotropic hormone (ACTH), which in turn is stimulated by hypothalamic corticotropin-releasing hormone (CRH) secretion. Besides the widespread effects of cortisol, the glucocorticoid also exerts negative feedback on the production of CRH and ACTH.

androgens, leading to the production of aldosterone in the zona glomerulosa of the adrenal gland (Fig. 36.1). In the zonae fasciculata and reticularis, CYP17A1 is present, making it possible to synthesize cortisol after 17-hydroxylation. CYP17A1 is expressed together with cytochrome b5 in the zona reticularis, enabling the lyase reaction and therefore the production of DHEA by removing the C20—C21 side chain of 17-OH pregnenolone. CYP17A1 is also expressed in combination with cytochrome b5 in the Leydig cells in the testes and in ovarian theca cells. Finally, CYP17A1 can be expressed in neural tissue, enabling the local production of DHEA, which might be important as a neurosteroid through its interactions with γ-aminobutyric acid (GABA) and N-methyl-D-aspartate (NMDA) receptors.[39] Most of these studies have been performed in rodents, but Schonemann et al.[37] detected CYP17A1 RNA and protein, together with POR, in human fetal brain tissue.

The expression of the CYP17A1 gene is mainly stimulated by cAMP[45] and can be inhibited by angiotensin II-induced production of activin, a member of the transforming growth factor-β (TGF-β) ligand superfamily, in the zona glomerulosa of the adrenal gland.[13] Bone morphogenetic protein-4, another member of the TGF-β family, which is mainly expressed in the zona glomerulosa of the adrenal gland has a similar effect on the expression of CYP17A1.[34] The absence of CYP17A1 expression in the zona glomerulosa prevents the production of cortisol and leads to the production of the mineralocorticoids.

The separate regulation of 17-hydroxylase and 17,20-lyase activities of CYP17A1 has been resolved on the basis of the finding of mutated genes, causing the production of 17-hydroxylated steroids without removal of the C20—C21 side chain. These mutations are localized in the region of the protein, which interacts with cytochrome b5. Cytochrome b5 enhances electron transfer by POR[12]; lyase activity appears to be more vulnerable to the disruption of interaction with this protein than the 17-hydroxylase reaction. Similarly, the changing ratios between the production of glucocorticoids and adrenal androgens, mainly DHEA-sulfate, during human development are likely to be due to a changed expression of cytochrome b5.[33]

The pivotal role of CYP17A1 in steroidogenesis is emphasized by the effects of inactivating mutations in this enzyme: a complete block of 17-hydroxylation leads to the inability to synthesize cortisol, whereas 11-deoxycorticosterone (DOC) will be produced in abundance, giving rise to hypertension. Furthermore, sex steroids cannot be produced, causing underviril-ization in boys and absence of puberty in both male[43] and female[41] subjects. As expected, similar phenotypes were observed in patients with mutations of the coenzyme cytochrome b5,[18] whereas POR deficiency yields

a different phenotype, because of the combined involvement of this factor in the actions of CYP17A1, CYP21A2, and CYP19A1.[1]

21-HYDROXYLASE: CYP21A2

CYP21A2 catalyzes the hydroxylation of carbon atom 21 to form DOC from progesterone and 11-deoxycortisol from 17-OH progesterone (Fig. 36.1). 21-hydroxylase is essential for the production of mineralocorticosteroids and glucocorticoids and diverts steroidogenesis away from sex steroids. This enzyme is localized in the endoplasmic reticulum and, like CYP17A1, is also dependent on the cofactor POR for electron transfer from NADPH.[22] Adrenal expression of CYP21A2 is primarily controlled by the ACTH/cAMP/protein kinase A (PKA) pathway, although angiotensin II acting through a PKC-dependent pathway also stimulates CYP21A2 expression in cells of the zona glomerulosa.

Extra-adrenal 21-hydroxylase activity has been described, particularly in the liver, but this enzyme activity could be catalyzed by other CYP enzymes as CYP21A2 expression appears to be highly specific for the adrenal cortex. Using sensitive RT-PCR techniques, low levels of CYP21A2 mRNA have been detected in several regions of the central nervous system, including amygdala, caudate nucleus, corpus callosum, thalamus, hippocampus, cerebellum, and spinal cord. Brain 21-hydroxylase activity, e.g., progesterone to DOC conversion, has, however, been linked to CYP2D6, not CYP21A2.[17]

The CYP21A2 gene is located on chromosome 6p in close proximity to the pseudo-gene CYP21A1P, which is an untranslated ancient gene duplication. Genetic recombination between the genes leads to frequent exchanges of bases and thus to a high prevalence of mutations in the CYP21A2 gene, causing one of the most prevalent autosomal-recessive inherited diseases. Absent or reduced CYP21A2 activity impairs the production of aldosterone and cortisol combined with adrenal androgen overproduction due to the absence of glucocorticoid feedback at the level of the pituitary. 21-Hydroxylase deficiency causes over 90% of congenital adrenal hyperplasia (CAH), also known as adrenogenital syndrome, occurring at an incidence of 1:15,000–20,000 births. Depending on the residual level of enzyme activity or expression, the clinical phenotype can range from salt-wasting to simple-virilizing and nonclassical CAH.[46] In the former, complete knockout of CYP21A2 function leads to virilization in utero, giving rise to ambiguous genitalia at birth and postnatal renal salt loss with hyperkalemia, acidosis, and risk of hypotension, coma, and death due to undetectable glucocorticoid and mineralocorticoid levels. These children should be diagnosed soon after birth through measurement of serum 17-OH progesterone levels and put on lifelong glucocorticoid and mineralocorticoid replacement. Phenotype is less severe in simple-virilizing disease as aldosterone synthesis is maintained, but the cortisol deficiency causes elevated androgen levels. Nonclassical CAH is characterized by moderate 21-hydroxylase deficiency and predominantly causes hyperandrogenism and menstrual irregularities in female patients. It has been suggested that the mild hypocortisolism in CAH patients triggers compensatory changes in corticotropin-releasing hormone increasing their vulnerability to stressors, both physical and emotional.[9]

11β-HYDROXYLASES: CYP11B1 AND CYP11B2

Hydroxylation of C11 in the C ring of the 21-hydroxylated steroids DOC and 11-deoxycortisol can be catalyzed by the two isoenzymes CYP11B2 (aldosterone synthase) and CYP11B1 (11β-hydroxylase), respectively. The human enzymes show a 93% amino acid homology and their tandem location at chromosome 8q is thought to derive from an early duplication. Multiple CYP11B enzymes are common to mammals and there is a certain species diversity regarding the function of the individual enzymes.[36]

Whereas both human enzymes possess 11-hydroxylase and 18-hydroxylase activities, only CYP11B2 catalyzes 18-oxidase, the final step of aldosterone synthesis from 18OH-corticosterone. CYP11B1 expression is restricted to the two inner adrenocortical zones but is most important in cortisol synthesis in the zona fasciculata.[32] Additionally, it is likely to play a role in the production of 11-androstenedione from androstenedione in the zona reticularis.[6] The recent creation of CYP11B subtype-specific antibodies showed detailed protein localization of these enzymes within the normal adrenal cortex and adrenal hyperplasia and tumors for the first time.[28] CYP11B2-positive cells are present in small confined clusters in the zona glomerulosa, localized among adrenal stem, progenitor, and stromal cells. As adrenocortical cells divide near the capsule and migrate centripetally, cells first express CYP11B2 followed by a switch to CYP11B1 and CYP17A1 coexpression. The factors causing this phenotypic switch remain elusive. Like CYP11A1, the 11β-hydroxylases are located within the inner mitochondrial membrane. Therefore, NADPH electron transfer is facilitated by ferredoxin and ferredoxin reductase.[22]

Regulation of CYP11B1 and CYP11B2 levels is dependent on the HPA axis and RAAS system, respectively (Fig. 36.3). CYP11B1 is the most potently ACTH-stimulated gene in human adrenocortical cells and is elevated in cortisol-producing adrenal adenomas, which can be caused by constitutively active PKA catalytic subunits. CYP11B2, on the other hand, is primarily regulated by angiotensin II and potassium levels through PKC and stimulation of intracellular calcium concentrations and activation of calmodulin. Expression of CYP11B2 is augmented in bilateral adrenal hyperplasia, characterized by an excess of aldosterone-secreting cell clusters, and in Conn adenomas.[24] CYP11B2 immunohistochemistry has been proposed as a marker for aldosterone-producing adenomas in order to differentiate from a nonfunctional adenoma with a background of bilateral hyperplasia. The recently discovered gene mutations in potassium (KCNJ5) and calcium channels (CACNA1D) or ATPases (ATP1A1, ATP2B3) underlying approximately 50% of aldosterone-producing adenomas all increase intracellular calcium concentration which in turn stimulates transcription of CYP11B2.[4] Primary hyperaldosteronism can also be caused by a hybrid CYP11B1/CYP11B2 gene having an ACTH-responsive promoter and 18-oxidase activity. This genetic crossover results in glucocorticoid-remediable aldosteronism (GRA); patients can be treated with dexamethasone.[10] Effects of posttranslational modifications and cofactor levels on CYP11B1 or CYP11B2 activity are largely unknown but could play a significant role. This is emphasized in adrenocortical carcinomas that are characterized by poor 11β-hydroxylase activity in the presence of CYP11B1 expression.

Similar to CYP21A2, mutations in CYP11B1 can cause CAH.[36] Accounting for less than 10% of CAH patients, CYP11B1 mutations can lead to hyperandrogenism and adrenal insufficiency and are marked by elevated levels of 11-deoxycortisol. These patients require lifelong glucocorticoid replacement. Since aldosterone synthase catalyzes several reactions from corticosterone to aldosterone, mutations in CYP11B2 can abrogate either 18-hydroxylase activity or both 18-hydroxylase and 18-oxidase activity. This distinction between type I and type II aldosterone synthase deficiency can be made through measurement of 18OH-corticosterone but has little clinical consequence as both patient groups require mineralocorticoid supplementation.

Given the pivotal role of the RAAS system in controlling blood pressure, many studies have looked at the role of CYP11B2 in hypertension and congestive heart failure. Several single nucleotide polymorphisms in CYP11B2 have been linked to the occurrence of hypertension, but larger follow-up studies have failed to replicate these findings.[44] Local presence of aldosterone production in cardiac tissue was postulated to contribute to the positive effects of mineralocorticoid receptor antagonists on top of ACE inhibition, but intracardiac expression of CYP11B2 could not be detected.[7]

Expression of both CYP11B enzymes is limited but detectable in several compartments of the central nervous system, with higher levels and regional variation for CYP11B1.[47] Given the widespread distribution and effects of glucocorticoid and mineralocorticoid receptors in brain tissue, local de novo aldosterone and cortisol production could cause significant physiological effects. Although local production of corticosterone and aldosterone has been shown in rat cerebellum and hippocampus, 11β-hydroxylase activities in human brain have not been detected. Intra- or paracrine activation of glucocorticoids in the brain is more likely to occur as a consequence of the cortisol–cortisone shuttle by the 11β-hydroxysteroid dehydrogenases.[8]

The vital and highly specific role of these enzymes in aldosterone and cortisol production has led to the development of 11β-hydroxylase inhibitors for the treatment of hypertension or Cushing's syndrome. Well-known drugs that inhibit CYP11B1 include the fungicide ketoconazole, the anesthetic agent etomidate, and metyrapone. All three can be used for the treatment of hypercortisolism, whereas metyrapone is also frequently used to assess overall HPA axis activity through measurement of 11-deoxycortisol levels after 24h of dosing.[11] A novel CYP11B1 and CYP11B2 inhibitor, osilodrostat, was developed for the treatment of primary hyperaldosteronism and hypertension, but has been recently also successfully been implemented in Cushing's syndrome.[3]

AROMATASE: CYP19A1

The final steroid-hydroxylating enzyme, which plays a role in the biosynthesis of steroids, is CYP19A1. It is responsible for the hydroxylation of carbon atom C19 of testosterone and androstenedione, the subsequent cleavage of the bond between C19 and C10, and subsequent redistribution of the electrons over the A ring to form the aromatic steroids estradiol and estrone, respectively. Only one copy of the gene has been found. It has a complex structure at the 5′ end, allowing start of transcription at a number of sites, under the control of various promoter sequences.[38] This means that in various tissues, different mechanisms can be used for the stimulation of the transcription of the gene: in the gonads and adrenals, a cAMP-mediated mechanism is responsible for the stimulation of *CYP19A1* by gonadotropins and ACTH, respectively, whereas in fat cells, cortisol can stimulate the production of estrogens. *CYP19A1* is also expressed in a large number of other tissues, such as uterus, muscle, bone, and brain. In the central nervous system, estrogens formed by this enzyme may play a role in sexual differentiation.

Defects of *CYP19A1* are very rare. However, the clinical picture in patients with mutations in this gene shows profound changes of parameters of gonadal function, bone metabolism, and aspects of the metabolic syndrome, indicating the role of estrogens in the regulation of numerous aspects of homeostasis.[15] These effects are mimicked by the results of aromatase inhibitor treatment of patients with estrogen receptor-positive breast cancer as reviewed by To et al.[42]

STEROID CATABOLISM BY CYTOCHROME P450s

Apart from their role in the biosynthesis of steroid hormones, enzymes of the CYP family also play important roles in the breakdown of steroids.[29] The most important enzymes in this respect are those of the CYP1 and CYP4 subfamilies. Members of the CYP1 subfamily are important in the hydroxylation of mainly estrogens, where both C2 and C4 may be attacked. CYP4 family members hydroxylate corticosteroids and androgens at the 6 position and can also attach a hydroxyl group at C16. It is important to be aware of the fact that expression of all of these enzymes can be induced by xenobiotics, leading to aberrant metabolism of the steroid hormones and, for example, changes in the ratios between the serum levels of steroids produced in the adrenal gland, which might lead to erroneous conclusions of disturbed steroidogenesis.

References

1. Arlt W, Walker EA, Draper N, et al. Congenital adrenal hyperplasia caused by mutant P450 oxidoreductase and human androgen synthesis: analytical study. *Lancet*. 2004;363(9427):2128–2135.
2. Barrett PQ, Bollag WB, Isales CM, McCarthy RT, Rasmussen H. Role of calcium in angiotensin II-mediated aldosterone secretion. *Endocr Rev*. 1989;10(4):496–518.
3. Bertagna X, Pivonello R, Fleseriu M, et al. LCI699, a potent 11beta-hydroxylase inhibitor, normalizes urinary cortisol in patients with Cushing's disease: results from a multicenter, proof-of-concept study. *J Clin Endocrinol Metab*. 2014;99(4):1375–1383.
4. Beuschlein F. Regulation of aldosterone secretion: from physiology to disease. *Eur J Endocrinol*. 2013;168(6):R85–R93.
5. Black SM, Harikrishna JA, Szklarz GD, Miller WL. The mitochondrial environment is required for activity of the cholesterol side-chain cleavage enzyme, cytochrome P450scc. *Proc Natl Acad Sci USA*. 1994;91(15):7247–7251.
6. Bloem LM, Storbeck KH, Schloms L, Swart AC. 11beta-hydroxyandrostenedione returns to the steroid arena: biosynthesis, metabolism and function. *Molecules*. 2013;18(11):13228–13244.
7. Chai W, Hofland J, Jansen PM, et al. Steroidogenesis vs. steroid uptake in the heart: do corticosteroids mediate effects via cardiac mineralocorticoid receptors? *J Hypertens*. 2010;28(5):1044–1053.
8. Chapman K, Holmes M, Seckl J. 11beta-hydroxysteroid dehydrogenases: intracellular gate-keepers of tissue glucocorticoid action. *Physiol Rev*. 2013;93(3):1139–1206.
9. Charmandari E, Merke DP, Negro PJ, et al. Endocrinologic and psychologic evaluation of 21-hydroxylase deficiency carriers and matched normal subjects: evidence for physical and/or psychologic vulnerability to stress. *J Clin Endocrinol Metab*. 2004;89(5):2228–2236.
10. Dluhy RG, Lifton RP. Glucocorticoid-remediable aldosteronism. *J Clin Endocrinol Metab*. 1999;84(12):4341–4344.

11. Feelders RA, Hofland LJ, de Herder WW. Medical treatment of Cushing's syndrome: adrenal-blocking drugs and ketoconazole. *Neuroendocrinology*. 2010;92(suppl 1):111–115.

12. Geller DH, Auchus RJ, Miller WL. P450c17 mutations R347H and R358Q selectively disrupt 17,20-lyase activity by disrupting interactions with P450 oxidoreductase and cytochrome b5. *Mol Endocrinol*. 1999;13(1):167–175.

13. Hofland J, Steenbergen J, Hofland LJ, et al. Protein kinase C-induced activin A switches adrenocortical steroidogenesis to aldosterone by suppressing CYP17A1 expression. *Am J Physiol Endocrinol Metab*. 2013;305(6):E736–E744.

14. Huang N, Miller WL. LBP proteins modulate SF1-independent expression of P450scc in human placental JEG-3 cells. *Mol Endocrinol*. 2005;19(2):409–420.

15. Jones ME, Boon WC, Proietto J, Simpson ER. Of mice and men: the evolving phenotype of aromatase deficiency. *Trends Endocrinol Metab*. 2006;17(2):55–64.

16. Kawabe S, Yazawa T, Kanno M, et al. A novel isoform of liver receptor homolog-1 is regulated by steroidogenic factor-1 and the specificity protein family in ovarian granulosa cells. *Endocrinology*. 2013;154(4):1648–1660.

17. Kishimoto W, Hiroi T, Shiraishi M, et al. Cytochrome P450 2D catalyze steroid 21-hydroxylation in the brain. *Endocrinology*. 2004;145(2):699–705.

18. Kok RC, Timmerman MA, Wolffenbuttel KP, Drop SL, de Jong FH. Isolated 17,20-lyase deficiency due to the cytochrome b5 mutation W27X. *J Clin Endocrinol Metab*. 2010;95(3):994–999.

19. Lee-Robichaud P, Akhtar ME, Akhtar M. Control of androgen biosynthesis in the human through the interaction of Arg347 and Arg358 of CYP17 with cytochrome b5. *Biochem J*. 1998;332(Pt 2):293–296.

20. Mellon SH, Griffin LD, Compagnone NA. Biosynthesis and action of neurosteroids. *Brain Res Brain Res Rev*. 2001;37(1–3):3–12.

21. Mellon SH, Vaisse C. cAMP regulates P450scc gene expression by a cycloheximide-insensitive mechanism in cultured mouse Leydig MA-10 cells. *Proc Natl Acad Sci USA*. 1989;86(20):7775–7779.

22. Miller WL, Auchus RJ. The molecular biology, biochemistry, and physiology of human steroidogenesis and its disorders. *Endocr Rev*. 2011;32(1):81–151.

23. Miller WL, Bose HS. Early steps in steroidogenesis: intracellular cholesterol trafficking. *J Lipid Res*. 2011;52(12):2111–2135.

24. Nanba K, Tsuiki M, Sawai K, et al. Histopathological diagnosis of primary aldosteronism using CYP11B2 immunohistochemistry. *J Clin Endocrinol Metab*. 2013;98(4):1567–1574.

25. Nebert DW, Adesnik M, Coon MJ, et al. The P450 gene superfamily: recommended nomenclature. *DNA*. 1987;6(1):1–11.

26. Nebert DW, Russell DW. Clinical importance of the cytochromes P450. *Lancet*. 2002;360(9340):1155–1162.

27. Nebert DW, Wikvall K, Miller WL. Human cytochromes P450 in health and disease. *Philos Trans R Soc Lond B Biol Sci*. 2013;368(1612):20120431.

28. Nishimoto K, Nakagawa K, Li D, et al. Adrenocortical zonation in humans under normal and pathological conditions. *J Clin Endocrinol Metab*. 2010;95(5):2296–2305.

29. Niwa T, Murayama N, Imagawa Y, Yamazaki H. Regioselective hydroxylation of steroid hormones by human cytochromes P450. *Drug Metab Rev*. 2015;47(2):89–110.

30. Olvera-Sanchez S, Espinosa-Garcia MT, Monreal J, Flores-Herrera O, Martinez F. Mitochondrial heat shock protein participates in placental steroidogenesis. *Placenta*. 2011;32(3):222–229.

31. Pollack SE, Furth EE, Kallen CB, et al. Localization of the steroidogenic acute regulatory protein in human tissues. *J Clin Endocrinol Metab*. 1997;82(12):4243–4251.

32. Rainey WE. Adrenal zonation: clues from 11beta-hydroxylase and aldosterone synthase. *Mol Cell Endocrinol*. 1999;151(1–2):151–160.

33. Rainey WE, Carr BR, Sasano H, Suzuki T, Mason JI. Dissecting human adrenal androgen production. *Trends Endocrinol Metab*. 2002;13(6):234–239.

34. Rege J, Nishimoto HK, Nishimoto K, Rodgers RJ, Auchus RJ, Rainey WE. Bone morphogenetic Protein-4 (BMP4): a paracrine regulator of human adrenal C19 steroid synthesis. *Endocrinology*. 2015;156(7):2530–2540.

35. Rendic S, Guengerich FP. Survey of human oxidoreductases and cytochrome P450 enzymes involved in the metabolism of xenobiotic and natural chemicals. *Chem Res Toxicol*. 2015;28(1):38–42.

36. Schiffer L, Anderko S, Hannemann F, Eiden-Plach A, Bernhardt R. The CYP11B subfamily. *J Steroid Biochem Mol Biol*. 2015;151:38–51.

37. Schonemann MD, Muench MO, Tee MK, Miller WL, Mellon SH. Expression of P450c17 in the human fetal nervous system. *Endocrinology*. 2012;153(5):2494–2505.

38. Simpson ER. Sources of estrogen and their importance. *J Steroid Biochem Mol Biol*. 2003;86(3–5):225–230.

39. Starka L, Duskova M, Hill M. Dehydroepiandrosterone: a neuroactive steroid. *J Steroid Biochem Mol Biol*. 2015;145:254–260.

40. Stocco DM. StAR protein and the regulation of steroid hormone biosynthesis. *Annu Rev Physiol*. 2001;63:193–213.

41. ten Kate-Booij MJ, Cobbaert C, Koper JW, de Jong FH. Deficiency of 17,20-lyase causing giant ovarian cysts in a girl and a female phenotype in her 46,XY sister: case report. *Hum Reprod*. 2004;19(2):456–459.

42. To SQ, Knower KC, Cheung V, Simpson ER, Clyne CD. Transcriptional control of local estrogen formation by aromatase in the breast. *J Steroid Biochem Mol Biol*. 2015;145:179–186.

43. Van Den Akker EL, Koper JW, Boehmer AL, Themmen AP, Verhoef-Post M, Timmerman MA, et al. Differential inhibition of 17alpha-hydroxylase and 17,20-lyase activities by three novel missense CYP17 mutations identified in patients with P450c17 deficiency. *J Clin Endocrinol Metab*. 2002;87(12):5714–5721.

44. Verwoert GC, Hofland J, Amin N, et al. Expression and gene variation studies deny association of human HSD3B1 gene with aldosterone production or blood pressure. *Am J Hypertens*. 2015;28(1):113–120.

45. Waterman MR. Biochemical diversity of cAMP-dependent transcription of steroid hydroxylase genes in the adrenal cortex. *J Biol Chem*. 1994;269(45):27783–27786.

46. White PC, Speiser PW. Congenital adrenal hyperplasia due to 21-hydroxylase deficiency. *Endocr Rev*. 2000;21(3):245–291.

47. Yu L, Romero DG, Gomez-Sanchez CE, Gomez-Sanchez EP. Steroidogenic enzyme gene expression in the human brain. *Mol Cell Endocrinol*. 2002;190(1–2):9–17.

Manipulating the Brain Corticosteroid Receptor Balance: Focus on Ligands and Modulators

E.R. de Kloet[1], N.V. Ortiz Zacarias[2], O.C. Meijer[1]

[1]Leiden University Medical Center, Leiden, The Netherlands; [2]Leiden Academic Center for Drug Research, Leiden, The Netherlands

Abstract

Part I, Chapter 3 of the Handbook was dedicated to the complementary function of mineralocorticoid receptor (MR) and glucocorticoid receptor (GR) signaling pathways that need to operate in balance for effective stress coping, adaptation, and circadian organization. This chapter focuses on the therapeutic application of MR:GR ligands and modulators for (1) substitution therapy of adrenal-deficient patients which requires to mimic episodic secretions of the endogenous glucocorticoids, (2) pharmacotherapy to treat symptoms of, for example, inflammatory and autoimmune disorders, with synthetic glucocorticoids that may have severe psychiatric side effects, and (3) diagnostic purpose as well as drug target for treatment of neuropsychiatric disorders. The chapter is concluded with perspectives toward novel modulators that target coregulators of MR and GR with greater selectivity, enhanced efficacy, and reduced side effects.

INTRODUCTION

Glucocorticoids secreted by the adrenals as end products of the hypothalamus-pituitary-adrenal (HPA) axis are crucial for the coordination of brain and body functions during stress coping, adaptation, and circadian events. These actions exerted by the adrenocortical hormone occur in concert with the other mediators of the HPA axis, which include hypothalamic corticotropin-releasing hormone (CRH) and vasopressin as well as pituitary ACTH and endorphins derived from the proopiomelanocortin (POMC) precursor.[1] The principal glucocorticoids in man are secreted in a ratio of cortisol:corticosterone 20:1[2] and collectively abbreviated here as CORT. Corticosterone is the principal glucocorticoid hormone in rats and mice.

KEY POINTS

- Cortisol and corticosterone (CORT) bind to mineralocorticoid receptors (MR) and glucocorticoid receptors (GR) colocalized in abundance in limbic brain.

- MR and GR operate in complementary fashion to coordinate circadian events and responsiveness of the stress response system during ultradian/circadian CORT secretions.

- There is an important unmet medical need in limiting the serious adverse effects of common glucocorticoid therapy on cognition, mood, and metabolism.

- Novel receptor ligands and modulators are becoming available that target MR and GR with greater precision and selectivity.

- MR and GR and their genetic variants are used for diagnosis and treatment of stress-related brain disorders.

CORT action during the circadian cycle is engaged in coordination of sleep stages and management of resources that range from appetite and food intake to storage of energy substrates and from mobilization of energy to allocation in demanding cells, circuits, and tissues.[3] Throughout the day, the HPA axis operates in an anticipatory mode. Hence, surges of CORT are highest in the morning for man (evening in rodents) during the circadian peak labeled as cortisol awakening response (CAR) in anticipation of an upcoming busy day or other challenges. CORT secretion oscillates hourly in order to stay tuned and resilient for stressors. These ultradian variations subside toward nighttime to allow sleep, recovery, and processing of past experiences.[4]

Besides its episodic variation CORT is also secreted in response to stressors and subsequently feeds back on brain circuits involved in processing of stressful information that initially led to its own secretion via the HPA axis. By feeding back, CORT promotes coping and adaptation aimed to restore homeostasis and allostasis (see Box 37.1). This action of CORT is mediated by the glucocorticoid receptor (NR3C1, GR) and the mineralocorticoid receptor (NR3C2, MR).[5,6]

The CORT receptors have the following characteristics (see Chapters 29 and 33). First, MR and GR are unevenly distributed and coexpressed in abundance in limbic brain particularly in the hippocampus. MR has a 10-fold higher affinity than GR for CORT and displays

BOX 37.1

Cellular homeostasis primarily is engaged in maintaining a stable equilibrium which is a conditio sine qua non for physiological parameters such as Na/K balance, pH, and body temperature. A situation (stressor) that is not predicted or controlled and exceeds the adaptive capacity of the organism triggers a stress response that promotes physiological and behavioral adaptations. This psychological process of check and balance with a strong anticipatory component proceeds in the brain and is aimed to maintain a labile equilibrium, termed allostasis. Thus, CORT action is engaged in facilitating "adaptation to change" very much the same way as a juggler tries to balance a disk on a stick; CORT provides the energy to maintain a labile equilibrium, a process that is termed allostatic load[68] (McEwen, Volume 1 Chapter 5).

an enormous diversity in function that depends on cell type and context.[5] Highest density of MR is in the limbic structures, notably the hippocampus. GR is widely distributed in neurons and glial cells with the highest expression in the typical stress centers: PVN, limbic-frontocortical regions, and ascending bioamine neurons.

Second, MR is a promiscuous receptor and binds with high-affinity CORT, aldosterone, and also deoxycorticosterone/cortisol and progesterone. However, even under low basal resting conditions, the concentration of CORT in the cell nucleus of hippocampus neurons is at least 20-fold higher than aldosterone. A discreetly localized aldosterone-selective MR is involved in the regulation of salt appetite and Na/K balance.[7] In the limbic brain a low-affinity membrane variant of MR mediates rapid nongenomic actions of CORT on the presynaptic release of glutamate.[8,9]

Third, the low-affinity GR responds to stress and circadian peak levels of CORT by three modes of operation: a rapid inhibitory membrane action involving endocannabinoids and slow genomic actions by either transrepression or transactivation of stress-induced signaling pathways (Fig. 37.1).[10]

The action of CORT mediated by these different receptor populations has three complementary modes of operation in managing stressful information: the standby mode, the onset, and the adaptation phase of the stress response. During the standby mode, the prestress basal ultradian and circadian oscillations of CORT maintain responsivity and resilience to upcoming stressful experiences.[11] Genomic MR maintains the threshold or sensitivity of the system, while the transient ultradian GR signals are energizing. During the onset phase, MR mediates rapid CORT action on the appraisal of salient

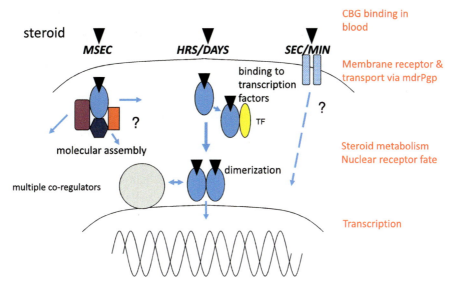

FIGURE 37.1 Factors affecting corticosteroid action. Glucocorticoids and mineralocorticoids are steroids that bind to intracellular receptors which mediate the effect of the steroids on gene transcription. In brain, the receptors also mediate rapid steroid effects on neurotransmission. The various levels the steroid action is modulated include (1) binding of cortisol, corticosterone, and also prednisone to CBG in blood and tissues (pituitary), (2) passage of the blood–brain barrier and export of all steroids except aldosterone and corticosterone by multidrug resistance P-glycoprotein (mdr-Pgp), (3) intracellular conversion by 11β-hydroxysteroid dehydrogenases type 1 or type 2 (11-HSD1 or 11-HSD2), (4) binding to the multimeric proteins, including FKBP5, containing mineralocorticoid receptors (MR) and glucocorticoid receptors (GR), (5) dissociation of multimeric protein and receptor, (6) GR and MR dimerization or association of receptor monomers with other transcription factors, and (7) interaction of MR and GR with multiple coregulators at the genomic level and regulation of gene transcription.

BOX 37.2

The corticosteroid receptor balance hypothesis (CoRe Balance Hypothesis) refers to "the central action of cortisol and corticosterone on stress coping and adaptation, which is mediated by mineralocorticoid receptors (MR) and glucocorticoid receptors (GR). Upon imbalance of MR:GR-regulated limbic–cortical signaling, the initiation and/or management of the neuroendocrine stress response is compromised. At a certain threshold this may lead to a condition of HPA axis dysregulation and impaired behavioral adaptation, which can enhance susceptibility to stress-related neurodegeneration and mental disorders (de Kloet, Volume 1, Chapter 3)."[14,56]

---

information, the selection of an appropriate coping style, and management of emotional arousal and reactivity.[9,11] The termination or adaptation phase requires additional GR activation by stress levels of CORT to prevent the initial stress reaction from overshooting and to promote behavioral adaptation and memory storage, in the preparation of future encounters.[12] This knowledge forms the basis of the MR:GR balance hypothesis (Box 37.2).[5,13,14]

The knowledge of the properties and mode of action of MR and GR is of importance since clinically, substitution therapy of adrenal deficient patients with receptor ligands is aimed toward maintenance of the balanced activation of both receptor types. Hence it is critical to substitute with naturally occurring cortisol, corticosterone, and aldosterone—or their analogs—in a manner that mimics the secretory patterns and bioavailability of these hormones during stress adaptation and circadian/ultradian variations.[15]

In contrast, pharmacotherapy is mostly symptomatic, thus treating the symptoms rather than the cause. For this purpose, synthetic steroids were developed with a much greater potency than parent CORT. These analogs were directed to treat specific pathologies to correct MR-mediated and/or GR-mediated processes that are either excessive or inadequate. Thus, while synthetic GR ligands were effective in the treatment of inflammatory disorders and autoimmunity, these steroids produced after prolonged use steroid diabetes, cognitive impairment, psychosis, and mood disorders as severe side effects.[16–18] Likewise, potent MR agonists were developed that act on ion homeostasis, and also on the nonselective MR in brain heart and fat. Hence, there is an urgent unmet medical need for a more targeted glucocorticoid and mineralocorticoid treatment, devoid of side effects, and that achieves a balanced activation of MR and GR. Here the receptor modulator comes in (Box 37.3).

The efficacy of glucocorticoids and mineralocorticoids depends on interaction with corticosteroid-binding globulin (CBG), passage of the blood–brain barrier, intracellular conversion by 11β-hydroxysteroid dehydrogenases type 1 or type 2 (11-HSD1 or 11-HSD2), and

Selective Steroid Receptor Modulators are synthetic ligands that induce a conformation that differs from (full or partial) agonists or antagonists. As a consequence, only a part of the interactions of the receptor with downstream signaling factors can take place. This results in agonism on some receptor-mediated processes, but antagonism on others. An example is the Selective Estrogen Receptor Modulator tamoxifen—an antagonist in breast tumors but agonist in bone. The promise of modulators is separation of wanted and unwanted effects mediated by GR and/or MR[16,19,34] (Chapter 29).

interaction of MR:GR with molecular machinery at the membrane and/or genomic level (Fig. 37.1; Chapter 29). In the next sections the focus is on the action of GR and MR ligands and modulators, respectively. Since these compounds modify complex feedback loops and poorly penetrate the blood–brain barrier, the neuropsychopharmacology of the MR:GR balance is often complex. Therefore the functional aspects of MR and GR ligands in human studies are considered collectively in section Human Studies With MR:GR Ligands and Modulators. Section Perspectives concerns the perspectives in application of selective receptor agonists, antagonists, and modulators.

GLUCOCORTICOID RECEPTOR LIGANDS AND MODULATORS

Hench, Kendall, and Reichstein were awarded the Nobel Prize in 1950, "for their discoveries relating to the hormones of the adrenal cortex, their structure and biological effects." The first synthetic glucocorticoids were already developed in the 1950s. As of today, synthetic glucocorticoids (e.g., dexamethasone and prednisone) are commonly used for their effective anti-inflammatory and immunosuppressive spectrum of applications.

As the late professor Marius Tausk stated already in 1951: "Cortisone treatment is appropriate where the defensive reactions of the organism cause more damage than the agent against which they defend" and used a metaphor that corticosteroids "protect against the water damage caused by the fire brigade."[5] For this purpose synthetic glucocorticoids were developed that had much more potent inflammatory and immunosuppressive actions. Since CORT has via MR and GR a pleiotropic action, this enhanced potency in suppressing proinflammatory genes also can affect numerous other processes

such as suppression of the HPA axis causing consequently adrenal atrophy, cognitive impairment, steroid diabetes, immunosuppression, and disturbed daily and sleep-related activities. Moreover, the residual mineralocorticoid action on salt retention/sodium homeostasis was also a concern in the development of novel glucocorticoid analogs, especially for long-term treatment. For applications, side effects, and future directions, see Barnes.[16]

Much effort was made to develop more selective-acting glucocorticoids.[19] The strategies involved were based on target selectivity in tissues: only directed to inflammation, but not to adrenal atrophy, steroid diabetes, or other side effects. Michael Karin in 1994 discovered the distinction between direct DNA-binding-dependent transactivation and the transrepression mechanism that depends on protein–protein interaction of GR with, for example, proinflammatory transcription factors. This implied the development of a class of Selective Glucocorticoid Receptor Modulators often referred to as "dissociated ligands."[20]

The distinct phenotype of the GR (dim/dim) mouse mutant—in which dimerization of the GR is prevented—supported this distinction. These mutants are compromised in energy metabolism,[21] cognitive performance, and calcium conductance,[22,23] while the GR-regulated immune response is maintained. However, more recent studies have shown that often DNA binding of GR also can contribute to anti-inflammatory efficacy making the transactivation/transrepression distinction somewhat blurred.[24] Fortunately, novel types of selective GR modulators have been developed that differentiate between wanted and unwanted effects in other ways.[25]

Chemical Aspects

All synthetic glucocorticoids are structurally related to hydrocortisone (cortisol) and corticosterone. Most of them are 21-carbon polycyclic compounds with a steroid structure, a fused 17-carbon atom ring system with an aliphatic side chain at carbon-17, a keto group at carbon-3, 1 or 2 double bonds in the A-ring, and an oxygen at carbon-11 (Fig. 37.2). Different side groups determine the differences in glucocorticoid and mineralocorticoid properties. Reducing the salt-retaining property was the main effort in the early development of new glucocorticoids. For example, the fluorination of dexamethasone (Dex) highly increases its anti-inflammatory potency, but unfortunately also its salt-retaining properties. The inserted 16α-methyl group, however, markedly reduces the mineralocorticoid properties, negating the effect of the 9α-fluoro atom. Prednisolone is a dehydrogenated derivative of cortisol with an enhanced glucocorticoid and declined mineralocorticoid activity. The most selective and

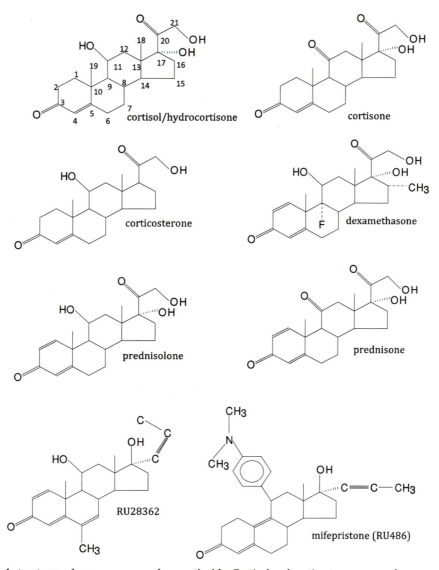

FIGURE 37.2 **Chemical structures of some common glucocorticoids.** Cortisol and corticosterone are endogenous glucocorticoids. Cortisol is hydrocortisone. Cortisone is an inactive endogenous derivative of cortisol. Prednisolone, prednisone, dexamethasone (Dex), and RU28362 are synthetic glucocorticoids. Mifepristone is a glucocorticoid antagonist.

Cortisol	11,17,21-trihydroxy-,(11beta)-pregn-4-een-3,20-dion
Corticosterone	(11β)-11,21-dihydroxypregn-4-ene-3,20-dione
Dexamethasone	(9α-fluoro-11β,17α,21-trihydroxy-16α-methylpregna-1,4-diene-3,20-dione)
Prednisolone	(11β,17α,21-trihydroxypregna-1,4-diene-3,20-dione)
RU28362	(11β,17ß-dihydroxy-6-methyl-17α(1-propynyl)-androsta-1,4,6-triene-3-one)
Mifepristone-RU486	[17β-hydroxy-11β-(4-dimethylaminophenyl)-17α-(1-propynyl))-estra-4,9-dien-3-one]

"pure" ligand RU28362 has only been used in experimental research up to now.

Selective GR modulators have also been based on steroid backbones, such as the hemisuccinate of 21-hydroxy-6,19-epoxyprogesterone.[26] The early dissociated ligand compound A was derived from plants and has a much more simple chemistry. Many subsequent ligands are based on phenylpyrazoles and arylpyrazoles,[27] and often display high selectivity in binding in combination with selective modulator activity.

An effective receptor antagonist of glucocorticoids is mifepristone (RU486). This compound was originally developed as a progesterone antagonist and used for abortions up to 63 days of pregnancy. Mifepristone also expresses high affinity for GR, but binding generally does not result in stimulation of transcription. RU486 antagonizes glucocorticoid effects in vitro and inhibits the effect of Dex on HPA activity in vivo. RU486 has weak agonistic properties in some in vitro settings, and possibly in in vivo settings, but only in the absence of glucocorticoid agonists. Mifepristone is rapidly metabolized in rodents, but very slow in man because of binding to a glycoprotein. Since mifepristone is recognized by P-glycoprotein (Pgp), large doses (>100 mg per kg in the mouse) need to be administered for central GR blockade. Mifepristone is a candidate drug for treatment of psychotic depression and currently used for treating symptoms of Cushing disease.[28]

Pharmacokinetics

Glucocorticoids are small, highly lipophilic molecules and therefore hardly soluble in water, but easily soluble in ethanol, polyethylene glycol (PEG), or dimethyl sulfoxide (DMSO). Because of their strong lipophilicity, the orally administered glucocorticoids are readily absorbed. Prednisone and cortisone are biologically inactive. They are quickly converted to the active compounds prednisolone and cortisol/hydrocortisone, respectively, by

dehydrogenation at the carbon-11 hydroxyl group in the liver. In the blood, synthetic glucocorticoids are mainly bound to albumin, a common plasma protein with low affinity. Unlike the endogenous glucocorticoids, most synthetic glucocorticoids are not bound to CBG. This plasma protein with a high affinity for naturally occurring corticosteroids, but low presence in plasma, binds prednisolone but not Dex.[29]

Plasma half-life is increased in the presence of a double bond (prednisolone) or a fluorine atom (Dex), because of decreased rate of metabolism. The plasma half-lives are between 2 and 4h (Table 37.1). Because of the modifying action on gene transcription, the duration of the biological effects is much longer than the plasma half-life. Thus, it is not the presence of the drug itself in the circulation but the long-lasting interaction of the ligand receptor complex and subsequent activation of protein synthesis, which accounts for slow onset and long-lasting biological actions.

Synthetic glucocorticoids are metabolized in the same way as the endogenous corticosteroids. The main metabolizing tissue is the liver, where steroids are conjugated with glucuronic acid or sulfuric acid. The unchanged parent drug and its metabolites are secreted in the urine, for over 60% (in the case of Dex) up to 90% (in the case

TABLE 37.1   Characteristics of Several Common Glucocorticoids and Mineralocorticoids

Compound	Relative Affinity of Glucocorticoid Receptors	Relative Potency		Half-life (h)		Comments
		Glucocorticoid	Mineralocorticoid	Plasma	Biological	
Hydrocortisone (cortisol)	1	1	1	1.5	8–12	Endogenous glucocorticoids (gc) of man
						Drug of choice for replacement
Cortisone	0.01	0.8	0.8	0.5	8–12	Inactive until converted to hydrocortisone
Corticosterone	0.85	0.3–0.5	5–15		8–12	Endogenous gc of rodents
						Not used in man
Dexamethasone	7.1	25–30	0.5	3–4	36–72	Drug of choice for suppression of ACTH production
Prednisolone	2.2	3.5–4	(0.6–)0.8	2–3	12–36	
Prednisone	0.05	4	(0.6–)0.8	3–3.5	12–36	Inactive until converted to prednisolone
Methylprednisolone	11.9	5	0.5	3	12–36	
RU 28362	9		0			Pure GR agonist; not used in man
Betamethasone	5.4	25–30	0.5	6.6	36–54	
Triamcinolone	1.9	5	0.1	1.5	12–36	
Mifepristone (RU486)	7	–	–	18		GR antagonist
Aldosterone		0.07	18,5	15′		Endogenous mineralocorticoids (mc)
Fludrocortisone		6.3	185	3,5		Potent mc agonist. Moderate gc agonist

of prednisolone) of the administered dose within 24 h. Changes in plasma half-life of glucocorticoids may result from changed hepatic metabolism due to liver or renal malfunction. Partial excretion via the kidney allows estimations of 24 h production from urine.

The passage of synthetic glucocorticoids through the blood–brain barrier is poor because of the presence of multidrug resistance 1a (mdr1a) P-glycoprotein (Fig. 37.3). It was shown that this drug efflux pump, which is expressed at the apical membranes of endothelial cells of the blood–brain barrier, hampers the access of moderate amounts of Dex and prednisolone to the brain. Ablation of the gene in the mdr1a Pgp knockout (−/−) mice resulted in enhanced uptake of tracer amounts of Dex, prednisone, and also of cortisol. While this was expected for the rodent, cortisol was hampered in brain penetration in human also. The ratio of corticosterone/cortisol rose from 0.05 in blood to 0.3 in brain.[2,30] Accordingly, the Pgp barrier does not lead to exclusion of cortisol and synthetic glucocorticoids, but to an underexposure of the brain relative to other tissues.

Pharmacodynamics

GR is expressed in brain areas involved in processing of stressful information and in control of HPA axis. In adaptive behavior, the role of GR has been extensively studied in contextual fear conditioning, the Morris water maze, and in retention of acquired immobility during exposure to the forced swim stressor. If glucocorticoids are administered immediately after the learning experience the consolidation of the acquired response is promoted, when behavior is measured 24 h later.[31]

The retention of acquired immobility during exposure to a forced swim stressor was promoted by Dex that fully restored the deficient behavior of ADX rats in a dose of 10 µg Dex sc/100 g body weight sc. The cognitive effect of this dose of Dex mimics that exerted by stress levels of endogenous CORT, if administered in a 50-fold higher dose of 5 mg CORT/100 g body weight. Central action of dexamethasone in promoting retention of the acquired immobility response is blocked by systemic pretreatment of 1.0–10 mg mifepristone/100 g body weight. Further studies showed that more than a million times lower dose of mifepristone (1 ng) was effective in blocking memory storage in adrenal-intact rats when administered locally in the dentate gyrus.[31]

The main target of Dex in blocking stress-induced HPA axis activation is in the pituitary corticotrophs, where ^{3}H-Dex accumulates when administered to ADX animals. Because of the presence of mdr1a P-glycoprotein efflux transporter at the blood–brain barrier, tracer amounts of the synthetic glucocorticoid are hampered in uptake and retention in cell nuclei of hippocampus neurons. The pattern is reversed for ^{3}H-corticosterone that is retained very well in hippocampus, but not at all in the pituitary. Hence the pituitary appears to be not a likely feedback site for genomic effects of endogenous CORT.[30]

Schmidt et al.[32] demonstrated that in mice carrying a conditional knockout of the GR gene specifically targeted at the pituitary corticotrophs, the circulating corticosterone levels were not different from their intact controls. Corticosterone is not retained in pituitary because of the presence of intracellular CBG-like molecules. Dex does not bind to CBG and can bind to pituitary GR to exert potent inhibition of HPA axis activity. In the mouse PVN parvocellular neurons, the deletion of GR-exon 3 in the mouse PVN almost abolished GR expression. In contrast to the pituitary GR(−/−), these PVN GR(−/−) were at adulthood resistant to Dex suppression and showed

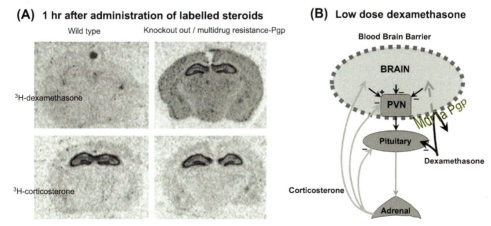

(A) 1 hr after administration of labelled steroids

Wild type   Knockout out / multidrug resistance-Pgp

[3]H-dexamethasone

[3]H-corticosterone

(B) Low dose dexamethasone

Blood Brain Barrier

BRAIN

PVN

Mdr1a Pgp

Pituitary

Dexamethasone

Corticosterone

Adrenal

FIGURE 37.3 Hypothalamus-pituitary-adrenal (HPA) axis and the role of multidrug resistance P-glycoprotein. (A) Autoradiogram of ^{3}H-corticosterone and ^{3}H-cortisol uptake 1 h after administration of a tracer dose (0.7 µg) to an adrenalectomized mouse. Left, wild-type mouse; Right multidrug resistance P-glycoprotein (−/−) mutant. ^{3}H-dexamethasone shows a pattern similar to ^{3}H-cortisol in contrast to tracer corticosterone which is not affected by Pgp.[2,69] (B) Implications of hampered penetration of dexamethasone in brain. Dexamethasone targets pituitary in suppression of stress-induced HPA axis activity and can reach PVN-GR sites only in higher doses.[5,30]

elevated ACTH and CORT levels during the circadian peak and in response to a restraint stressor.[33]

Selective modulators of GR may target distinct signaling pathways in the brain. The compound C108297 was shown to differentiate between GR-mediated CRH upregulation in the amygdala (where it had no efficacy) and downregulation in the PVN (where it was efficacious).[25] Such compounds suppress via the PVN the HPA axis and at the same time block the GR-dependent amygdala fear pathway, a dissociation that may be beneficial in clinical settings. It is, however, as yet still impossible to predict the effects that different selective modulators exert on the brain on the basis of their molecular interaction profiles.[34]

MINERALOCORTICOID RECEPTOR LIGANDS AND MODULATORS

The naturally occurring mineralocorticoid is aldosterone which is secreted as end product of the renin–angiotensin system. Change in blood volume, Na^+ depletion, and sympathetic nervous activity cause release of renin from the kidney. This enzyme cleaves angiotensinogen released from the liver in a substrate for angiotensin-converting enzyme from the lung. Then angiotensin II is generated which is—together with K^+ and ACTH—the trigger for aldosterone secretion from the zona glomerulosa of the adrenal cortex.[35]

The potent synthetic agonist MR agonist fludrocortisone is important for the management of adrenal-deficient patients (Table 37.2). Recently fludrocortisone was found effective as add-on in the treatment of the depressed patient. Increased CORT stimulation of the kidney MR may occur because of impaired 11-HSD2 function or blockade of 11-HSD2 by glycyrhetinic acid (carbenoxolone) present in licorice.

MR antagonists spironolactone, eplerenone, and in future perhaps nonsteroidal finerenone are used for treatment of mild renal dysfunction and appeared since the RALES trials in the late 1990s life-saving for patients with heart failure and left ventricular systolic dysfunction postmyocardial infarction. Spironolactone blocks the action of excess aldosterone, as occurs in primary and secondary hyperaldosteronism. Currently, the prevalence of hyperaldosteronism may account for as much as 10% of hypertension, and this may even be an underestimate. Funder[36] therefore advocates that low-dose MR antagonists should be mandatory as a first-line treatment of all hypertensives. Eplerenone has similar indications to treat heart failure as spironolactone, as appeared from the Ephesus trial, and more recently, may be used for new indications to treat metabolic disease, renal disease, and stroke. Finerenone was tested in Phase II clinical trials for chronic heart failure (ARTS-HF) with positive results. It is also developed to ameliorate renal dysfunction in type 2 diabetic mellitus patients.[37] Drospirenone, the progesterone derivative in contraceptives and postmenopausal replacement therapy (Yasmin), has high affinity for the MR.

Chemical Aspects

Fludrocortisone is cortisol with a fluorine atom at the 9α-position (Fig. 37.4). The insertion of fluorine at this position enhances both the binding affinity and potency of this synthetic derivative. Fludrocortisone showed a more than 10-fold increase in glucocorticoid anti-inflammatory activity when compared to cortisone acetate in ADX rats. Mineralocorticoid activity was 4.5–9 times more active than deoxycorticosterone in maintaining the life of ADX dogs. The increase in glucocorticoid and mineralocorticoid activity of the 9α-halogenation has been proposed to be related to a change in the molecular structure of this steroid, in which the fluorine atom in 9α-position results in a change of the A-ring conformation, which is bent underneath the plane of the molecule to a greater extent than in cortisol, which could influence the way it interacts with the receptor.

Spironolactone contains at the C17 position a γ-lactone (Aldactone) and the closely related γ-hydroxy derivative which is a metabolite (Soldactone). Spironolactone is a progesterone derivative with reduced affinity for the progesterone receptor, but still causing progesterone-like effects (EC 740 nM). Spironolactone is a weak inverse MR agonist in the absence of CORT or aldosterone. It prevents MR to reach its active conformation preventing the binding to coactivators.[38] In adrenal-intact individuals, it is the most common and widely used mineralocorticoid antagonist (IC 24 nM) and a potassium-sparing diuretic. It is an antiandrogen (IC 77 nM), but except at high doses has no (anti) glucocorticoid activity (IC 2410 nM) and no estrogen activity.

TABLE 37.2   Characteristics of Mineralocorticoid Antagonists

Steroid	Selectivity	Potency	Half-life	Distribution	Specificity
Spironolactone	Low	High	Long	6× kidney	No
Eplerenone	High	Low	Short 4 h	3× kidney	No
Finerenone	High	High	?	Equal	CV>kidney

From Jaisser F, Farman N. Emerging roles of the mineralocorticoid receptor in pathology: toward new paradigms in clinical pharmacology. Pharmacol Rev. 2016;68:49–75.[70]

Eplerenone is a 9,11-epoxy derivative of spironolactone. The drug was developed as a second-generation more specific MR antagonist than spironolactone as a K⁺-sparing diuretic, but has a 20-fold lower affinity than spironolactone, but does not bind to plasma proteins. Accordingly, in vivo eplerenone has about 60% of the biological activity of spironolactone. Eplerenone blocks the interaction of MR with coactivators. The drug has no affinity for the GR, androgen receptors, or progesterone receptors. Eplerenone requires twice a day dosing and is much more expensive.

Finerenone is a third-generation dihydropyridine-derived nonsteroidal MR antagonist. The compound has higher affinity to MR than spironolactone (EC 50 = 18 nM)

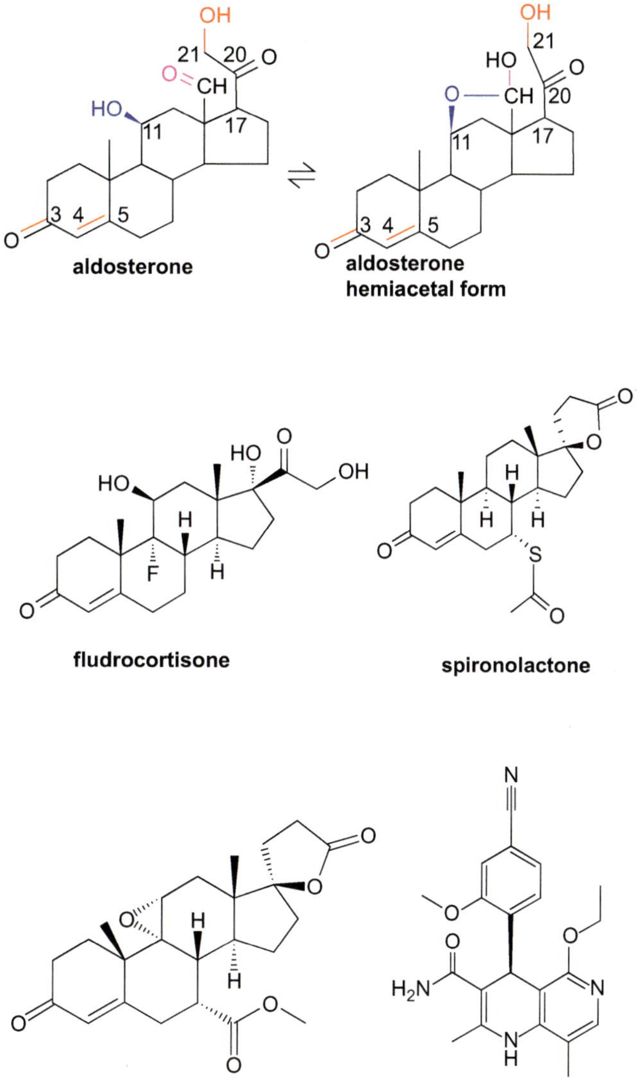

FIGURE 37.4 **Chemical structures of some common mineralocorticoids.** Aldosterone is the endogenous mineralocorticoid, fludrocortisone (or 9α-fluorocortisol or 9α-fluorohydrocortisone) is a mineralocorticoid agonist with moderate glucocorticoid agonist activity. Spironolactone, eplerenone, and finerenone are 1st, 2nd, and 3rd generation mineralocorticoid antagonists.

Aldosterone	11β,21-Dihydroxy-3,20-dioxopregn-4-en-18-al
Fludrocortisone	9α-fluoro-11β,17α,21-trihydroxy-4-pregnene-3,20-dione
Spironolactone	7α-acetylthio-3-oxo-17α-pregn-4-ene-21,17-carbolactone
Eplerenone	(Pregn-4-ene-7,21-dicarboxylic acid, 9,11-epoxy-17-hydroxy-3-oxo, γ-lactone, methyl ester (7α, 11α, 17α))
Finerenone	(BAY 94-8664) (4S)-4-(4-cyano-2-methoxyphenyl)-5-ethoxy-2,8-dimethyl-1,4-dihydro-1,6-naphthyridine-3-carboxamide

and lacks affinity for the other steroid receptors. Binding of finerenone to the MR generates a highly unstable complex, unable to bind to coregulators. It has a somewhat less K-sparing effect than spironolactone.[36]

Pharmacokinetics

The selectivity of the MR depends on access of either aldosterone or CORT to the receptor. Cortisol plasma concentration in man and that of corticosterone in rodents is 1–40 μg/100 mL, while circulating aldosterone levels are a 1000-fold lower, in the range of 5–20 ng/100 mL plasma. Of these steroids CORT is bound for 90% to CBG, leaving still the challenge that for the aldosterone-selective MR, aldosterone has to fight a 100-fold excess of CORT. How is this possible? .

11β-Hydroxysteroid Dehydrogenase Type 1 and 2 (11-HSD1 and 11-HSD2)

Aldosterone access is regulated by the intracellular enzyme 11β-hydroxysteroid dehydrogenase type 2 (11-HSD2), which oxidizes the 11-hydroxy group of CORT to the bio-inactive 11-keto group. Aldosterone forms a hemiacetal bridge that protects the 11β-hydroxyl group from being dehydrogenated and inactivated by this enzyme. However, as argued by Funder et al., the efficiency of 11-HSD2 will be insufficient to exclude CORT from binding to the aldosterone-selective MR. As an additional specificity-conferring mechanism, Funder[39] proposed that the dehydrogenase generates excess NADPH which brings about a redox state thought unfavorable for activation of the CORT–MR complex[40] (see Fig. 37.5).

In limbic brain structures, MR occupancy achieved in cell nuclei of hippocampus of adrenalectomized animals injected with tracer doses of ^{3}H-aldosterone or ^{3}H-corticosterone can be mutually blocked by excess of either one of these hormones, but not by dexamethasone. Moreover, in these nonepithelial cells, the 11-HSD1 isoform operates as a hydrogenase to regenerate bioactive CORT creating a condition favorable for CORT–MR activation. Thus, for example, in the limbic structures it is estimated that CORT constitutes 90% of the various ligands bound to MR and this ratio was found in cell nuclei. During hyperaldosteronemia or with rising levels of progesterone around ovulation and pregnancy, the relative occupancy of MR by CORT may decrease.

Fludrocortisone

Studies have revealed that, unlike cortisol, fludrocortisone is not inactivated by 11-HSD2 oxidation in the kidney, since conversion to the inactive metabolite 9α-fluorocortisone is greatly impaired, leading to a free access of fludrocortisone to the renal MR. More recently it has been confirmed that 9α-fluorination diminishes 11-HSD2 oxidation, enhancing the active form of the steroid. In this way 11β-HSD seem to be important pharmacokinetic determinants of the activity of MR ligands.[41]

Clinically, fludrocortisone is used orally in its acetate form (fludrocortisone acetate; Florinef). After oral administration, it is rapidly absorbed and peak plasma levels are reached within 120 min; it presents two different phases of elimination between 2 and 6 h after administration, with two different elimination half-lives at 1.6 and 4.1 h. More important is its biological half-life of 18–36 h, explaining its long-lasting effects in the body. Since the maximum effect of this drug is reached around 3–6 h after oral administration,[42,43] it would be useful to administer it in the early morning just before the cortisol reaches its peak in the circadian rhythm.

Antimineralocorticoids

Spironolactone has a short half-life of 1.5 h. It is rapidly metabolized in the liver for more than 80% to 7α-thiomethylspironolactone, which carries most of

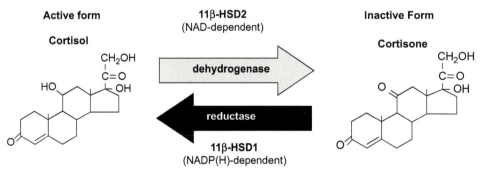

FIGURE 37.5   **11β-hydroxysteroid dehydrogenase type 1 and 2.** 11-HSD2 is a dehydrogenase that inactivates glucocorticoids by converting cortisol to cortisone using NAD$^+$ as cofactor. In brain, 11-HSD2 is expressed in discrete regions, e.g., n. tractus solitarii and periventricular areas. 11-HSD1 requiring NADPH regenerates the active 11-hydroxy forms of glucocorticoids from the inactive 11-keto in order to amplify their local concentration. 11-HSD1 is widely distributed with the highest expression in cerebellum, hippocampus, and cortex. *From Chapman K, Holmes M, Seckl J. 11β-hydroxysteroid dehydrogenases: intracellular gate-keepers of tissue glucocorticoid action.* Physiol Rev. *2013;93:1139–1206.*

the bioactivity of the parent steroid. This compound is further converted to bioactive canrenone which is the major circulating metabolite and has a much longer half-life of 16.5 h. Canrenone is more potent than spironolactone and constitutes the bioactive component of soluble K canrenoate. Canrenone inhibits steroidogenic enzymes such as 11β-hydroxylase, cholesterol side-chain cleavage enzyme, 17α-hydroxylase, and 21-hydroxylase. In vitro, canrenone binds to and blocks the androgen receptor (AR). The antiandrogen effects of spironolactone are considered to be largely due to its unchanged form, rather than due to one of its metabolites.[44]

Pharmacodynamics

Grossmann et al.[45] evaluated the mineralocorticoid potencies of several steroids using a transactivation assay in CV-1 cells and compared these with their glucocorticoid potency transfected with human GR or human MR (Fig. 37.6). In the transactivation assays, fludrocortisone was found to be 6.3 times more potent than cortisol in GR-mediated action, while regarding MR, it exhibited 10 times more potency than aldosterone and 185 times more potency than cortisol, presenting more potent overall transactivation via the MR.

Regarding the binding affinity, fludrocortisone binds with higher affinity to the MR than to the GR in the hippocampal cytosol and in binding competition assays at the human MR, fludrocortisone binds to the MR with even higher affinity than the endogenous ligands cortisol and aldosterone.[44]

The genomic MR is concerned with a tonic function regulating the threshold and/or sensitivity of the stress response system. Central blockade of MR with an intracerebroventricular (icv) or intrahippocampal infusion of MR antagonist increases the basal activity of the HPA axis and enhances the stress-induced release of HPA axis hormones within an hour of administration. MR antagonists attenuate the autonomic outflow as appears from the blockade of a stress-induced pressor response that could also be eliminated by renal denervation. MR antagonists display anxiolytic and antiaggressive activity.[13]

Rapid nongenomic effects of CORT on acquisition and retrieval of fear-motivated behavior are mediated by MR. This effect was reversed dose-dependently by MR antagonists administered icv prior to CORT.[11] MR antagonist affects the selection of an appropriate coping strategy to deal with a real or anticipated challenge. Using the Morris water maze, it was found that MR antagonists given 15 min before the retrieval test could alter the search strategy because in this free swim trial

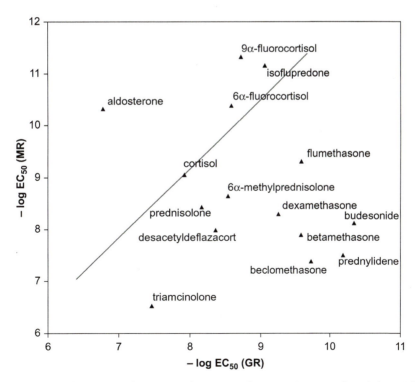

FIGURE 37.6 **Mineralocorticoid vs. glucocorticoid potency.** Glucocorticoid potency increases from left to right and the mineralocorticoid potency increases from bottom to top. The diagonal line separates typical glucocorticoids from mineralocorticoids. Selectivity increases with the perpendicular distance from the diagonal line: to the bottom right for glucocorticoids, to the top left for mineralocorticoids. *Reprinted with permission from Grossmann C, Scholz T, Rochel M, et al. Transactivation via the human glucocorticoid and mineralocorticoid receptor by therapeutically used steroids in CV-1 cells: a comparison of their glucocorticoid and mineralocorticoid properties.* Eur J Endocrinol. *2004;151:397–406.*

the mice did not discriminate anymore between the quadrant where the platform originally was located and the opposite quadrant. Thus the MR antagonist can alter the search strategy.[46]

HUMAN STUDIES WITH MR:GR LIGANDS AND MODULATORS

Section Human Studies With MR:GR Ligands and Modulators is concerned with indications for treatment of brain pathology either (section Substitution Therapy Benefits From Concomitant MR and GR Stimulation) as a consequence of lack of adrenal steroids requiring substitution therapy or (section Pharmacotherapy With Synthetic Glucocorticoids) as collateral damage during and after glucocorticoid pharmacotherapy. Section Concomitant MR Stimulation During Glucocorticoid Therapy is directed to approaches that may help to ameliorate the brain deficits accompanying the current practice in substitution and pharmacotherapy. In section Brain MR and GR: Targets for Therapy either brain MR and/or GR is considered a target for therapy of depression and anxiety disorders. Finally, section Diagnostic Use describes a reflection on the exploitation of MR and/or GR as diagnostic tool of brain pathology.

We choose to collapse MR-based and GR-based therapeutical and diagnostic strategies because CORT targets both receptors. Therefore substitution as well as pharmacotherapy needs to factor in these mutually dependent signaling pathways that are activated by CORT and that need to be in balance for optimal performance to ensure effective stress coping and adaptation.

Substitution Therapy Benefits From Concomitant MR and GR Stimulation

The principle of substitution therapy is to mimic as good as possible the episodic variation in circulating CORT and to ensure that at least the organization of daily activities and sleep stages are maintained. For this purpose, delivery methods are being developed that should release the steroids according to a circadian or even an ultradian pattern. This would also apply to adrenal-deficient patients such as Addison disease, adrenogenital syndrome, or other deficiencies in the synthesis of endogenous CORT.[15] Fludrocortisone and cortisol are used in substitution therapy for activation of GR and MR. In spite of this replacement PAI patients report more psychological morbidity and Quality of Life (QoL) impairments in comparison with controls. In addition, maladaptive personality traits, depression vulnerability, and impaired QoL were more severe with higher doses of cortisol.[47] Verbal memory improved significantly

with fludrocortisone in comparison with the group that received only cortisol replacement.[48]

Pharmacotherapy With Synthetic Glucocorticoids

Currently about 1% of the Western world population receives synthetic glucocorticoids.[17] There is no doubt that glucocorticoids are still the treatment of choice in inflammatory disorders and autoimmunity, and are lifesaving in case of brain edema (Box 37.4).

Prolonged glucocorticoid therapy with high doses cause hyperglycemia, immunosuppression, osteoporosis, myopathy, and adrenal atrophy. In 15.7 of 100 patient years of all courses of glucocorticoid therapy, severe psychiatric side effects occur. These include cognitive impairment, emotional and mood disorders, a twofold increased risk of depression, a fourfold increased risk of mania, delirium, confusion, or disorientation and nearly a sevenfold increased risk of suicide. Moreover, after cessation of excess glucocorticoid exposure patients may have enduring psychiatric complaints.[17,18]

The mechanism underlying glucocorticoid-induced brain pathology proceeds along three mutually dependent pathways. First, even though synthetic glucocorticoids are hampered in brain penetration, high doses very well may excessively activate GR-dependent aspects of mood and cognition. Second, the suppression of ultradian and circadian rhythmicity implies altered resilience to acute stressors. This deficit may persist, because of lasting adrenal atrophy after discontinuation of glucocorticoid therapy. Third, the profound suppression of endogenous CORT by synthetic glucocorticoids and its consequent reduced bioavailability to activate the brain MR and thus the function of this receptor in processing of salient information.[5]

Concomitant MR Stimulation During Glucocorticoid Therapy

As noted in section Substitution therapy benefits from concomitant MR and GR stimulation, concomitant

---

BOX 37.4

A short distance below the South Summit, as the climbers descended into thick clouds and falling snow, Pittman collapsed again and asked Fox to give her an injection of a powerful steroid called dexamethasone. "Dex," as it is known, can temporarily negate the deleterious effects of altitude (from Jon Krakauer—*Into Thin Air*).

---

replacement of adrenal-deficient patients with the MR agonist fludrocortisone in addition to CORT improved mood and cognition. Several lines of evidence support this notion. It was found that short-term exposure to Dex had an "energizing" effect by increasing activation, arousal, and concentration. This positive effect of acute Dex mediated by GR switched to emotional arousability and negative feelings (anger, sadness) if the treatment with the synthetic glucocorticoid was prolonged. Coadministration with CORT counteracted this dysphoric Dex effect. CORT enhanced a measure for "high spirits" and promoted a euphoric mood, likely via activation of MR. In support, in genetic studies an MR gene variant (haplotype 2) was identified which encodes a more bioactive hippocampal MR and appeared associated with enhanced optimism, decreased rumination, and protection to depression.[49]

Born et al. found that low amounts of CORT enhanced and the antimineralocorticoid canrenoate decreased the percentage of slow wave sleep (SWS) occurring in the beginning of the night. The antimineralocorticoid disinhibited also HPA axis activity during early sleep. Together these findings suggest that the suppression of HPA axis activity and the increase in SWS both involve hippocampal MR. A very interesting phenomenon is furthermore that activation of hippocampal MR by fludrocortisone can promote during SWS declarative memory consolidation of previously encoded information. This effect mediated by MR is counteracted by GR activation, which reduces the percentage of SWS and REM sleep.[50]

Liston et al.[51] showed recently in Dex-treated rats that circadian oscillations of CORT are a prerequisite for learning-dependent synaptic plasticity. Life-imaging with transcranial two-photon microscopy revealed that in the Dex-treated rats, additional intermittent administration of CORT increased dendritic spine formation and pruning in vivo in the cerebral cortex which could explain enhanced cognitive performance.

Brain MR and GR: Targets for Therapy

Several lines of evidence point to the modulation of the MR as a novel approach for the treatment of depression and possibly also for anxiety disorders. First, a decrease of MR expression in limbic structures was found in patients with major depression.[52] Klok et al.[52] compared brain tissue from patients suffering from major depression and nondepressed subjects, finding significantly lower MR transcript levels (30–50% lower) in the hippocampus, inferior frontal gyrus, and cingulate gyrus, while no changes were observed in the amygdala or nucleus accumbens. Moreover, differences in MR expression along the anteroposterior axis have been found in patients with depression; with a significant decrease of MR mRNA levels and MR/GR ratio in the anterior

hippocampus, but no significant changes in the posterior hippocampus.[53] Further studies suggest that changes in MR expression in patients with depression are region dependent; and the HPA hyperactivity observed in these patients might result from the decreased MR expression.

A second line of evidence emerges from the finding that treatment with antidepressants increases the expression of MR in rat hippocampus. Seckl and Fink[54] assessed the changes in MR and GR mRNA expression in rat hippocampus after treatment with three different antidepressants: amitriptyline, desipramine, and citalopram. They found that, after 14 days of treatment, the three drugs significantly increased the MR mRNA levels by 87%, 60%, and 17% respectively. Expression of GR was also found to increase with amitriptyline and desipramine, although to a lesser extent. This increase of MR expression seems to be induced throughout the hippocampal regions (CA1, CA2, CA3, and dentate gyrus) as observed after chronic treatment with amitriptyline. Long-term or chronic treatment has been found to induce a significant increase of hippocampal MR after treatment with imipramine, fluoxetine, or amitriptyline. Together, these results suggest that increased expression of hippocampal MR may lead to normalization of HPA axis or MR/GR balance.[55]

In support of a role for MR in treatment of depression, preliminary results from a clinical trial studying the effects of antimineralocorticoids in combination with amitriptyline in depressed patients, suggested that the use of the MR-antagonist spironolactone for the first 10 days diminished the antidepressive effects of amitriptyline.[56] Otte et al. (2009)[57] performed a proof-of-concept study in patients with major depression, in order to test whether the addition of the MR agonist fludrocortisone or the MR antagonist spironolactone, daily during 3 weeks, accelerates the onset of action and improves efficacy of escitalopram in a 5-week treatment. In their study, addition of fludrocortisone was found to decrease plasma cortisol levels, although lower cortisol levels were observed in the group of responders compared with nonresponders. Furthermore, this agonist significantly accelerated the treatment response by 6 days in the group of patients who responded to antidepressant treatment. Spironolactone did not accelerate the onset of action and was associated with an increase in plasma cortisol levels.

All these previous studies have suggested the stimulation of MR function, via MR agonists such as fludrocortisone, to be a promising new approach to improve the treatment of depression due to normalization of the HPA axis or recovery of MR/GR balance. Previous studies found that fludrocortisone is able to inhibit the HPA axis in healthy humans with and without first depleting the MRs from endogenous cortisol with metyrapone, which inhibits cortisol synthesis. Furthermore, fludrocortisone differentially inhibits the HPA axis in psychotic major depression (PMD), finding diminished inhibition in this group of patients compared to healthy controls, suggesting that the treatment response will also depend on the subtype of depression.[58] Moreover, the modulation of MR has been also investigated in patients with treatment-resistant depression, using spironolactone and the MR/GR agonist prednisolone. In these patients, hypercortisolism and possible down-regulation of MR was found, which suggests that the use of MR agonists, such as fludrocortisone, could be beneficial in their therapy.[59]

Fludrocortisone seems to act as an accelerator of the antidepressive effects of SSRIs or other antidepressants, but it does not seem to improve the psychopathology on its own. Otte and colleagues reported that in young depressed patients and in healthy younger and older individuals, fludrocortisone had beneficial effects. However, in older depressed patients MR stimulation by fludrocortisone impaired verbal learning and visuospatial memory performance. In another study, 18 healthy women and 20 patients with bipolar disorder and underwent two tests of social cognition, the Multifaceted Empathy Test (MET) and the Movie for the Assessment of Social Cognition (MASC). Fludrocortisone enhanced emotional empathy across groups. Stimulation of MR enhanced emotional empathy in healthy women and in BPD patients measuring cognitive and emotional facets of empathy.[60]

Given that fludrocortisone is not a selective MR agonist, it cannot be excluded that some of the observed effects might be mediated by GR activation. This effect would be compatible with previous studies in which GR agonists have demonstrated antidepressant effects; for example, both dexamethasone and hydrocortisone were found to improve the Hamilton depression scale scores in depressed patients treated with these agonists, possibly by blocking the HPA axis via negative feedback and depleting the brain from endogenous cortisol. Consistent with these findings, fludrocortisone might also mediate HPA inhibition via GR agonism and cause depletion of cortisol from GR and MR in the brain; then, fludrocortisone could refill and activate mainly the MR, due to its greater transactivation potency in the MR compared to the GR.[45] As Otte et al.[57] have suggested, the use of a higher dose in a shorter time frame could lead to a larger GR agonism.

Since fludrocortisone is not a brain-MR selective agonist, it also activates the renal MR resulting in marked sodium retention and several fluids and electrolyte disturbances; therefore, careful monitoring of patients receiving this drug should be considered.

Finally, the MR:GR balance can also be favorably affected by mifepristone (RU486). The blockade of GR by mifepristone would leave more MR available for stimulation by endogenous CORT. In 2001 mifepristone was reported to rapidly relieve severely depressed patients from psychotic symptoms. However, in the 2016 published analysis of the Cochrane database it was stated: "Good evidence is insufficient to conclude whether antiglucocorticoid drugs provide effective treatment for psychosis. Some global state findings suggest a favourable effect for mifepristone, and a few overall adverse effect findings favour placebo. Additional large randomised controlled trials are needed to justify findings." Furthermore, the FDA approved Korlym (mifepristone) in the United States as a once-daily oral medication for treatment of hyperglycemia secondary to hypercortisolism in adult patients with endogenous Cushing syndrome.

Diagnostic Use

For diagnostic purpose the Dex suppression test (DST) developed by Carroll in the 1970s is widely used to test HPA axis function.[61] A low dose of Dex results in suppression of the ACTH and CORT release. Failure to suppress implies hypersecretion of these hormones due to, for example, Cushing syndrome. In psychiatry the DST was used to test glucocorticoid resistance in patients with enhanced HPA axis activity as is the case in major depressive disorder. For this purpose the a.m. CORT escape from suppression of Dex administered the night before was the criterion. Alternatively, also glucocorticoid supersensitivity was measured in this way in individuals that had suffered from early life trauma, as may occur in PTSD patients. Since there are many conditions with altered HPA axis activity beyond depression and PTSD, interest waned for the test as routine support for the diagnosis of psychiatric disorders. Currently, the cortisol awakening response (CAR) and measurement of hair cortisol are being standardized as measure for HPA axis dysregulation.[62,63]

The so-called combined Dex-CRH test developed by Florian Holsboer[64] has an impressive predictive validity for relapse and remission of patients diagnosed for major depressive disorders with a sensitivity of about 80%. Disruption of HPA axis regulation assessed preceded relapse, while prior to remission the HPA axis as tested with the Dex-CRH was normalized. In the Dex-CRH test the synthetic glucocorticoids is administered at 23:00 h and in the p.m. phase the next day the response to CRH is tested. In depressed patients the escape from Dex suppression will be amplified by CRH displaying an exaggerated CORT and ACTH response.[64] Actually, the Dex-CRH test is for testing pituitary function which is in line with the preferred pituitary site of action of Dex in the suppression of stress-induced HPA axis activity. In their book "Endocrine Psychiatry, Solving the riddle of Melancholia," Shorter and Fink give an in-depth account of the rise and fall of the dexamethasone suppression test.[61]

PERSPECTIVES

Progress in the pharmacotherapy will benefit from further insight into the action mechanism of glucocorticoids, both on the genomic and membrane level. Using this knowledge, combinatorial approaches may assist in the development of a precision medicine targeting the defunct MR-mediated and/or GR-mediated action of CORT. This may include anti-inflammatory ligands for the GR and/or its coregulators devoid of, for example, metabolic and neuroendocrine side effects. Alternatively, novel compounds that target specifically brain MR and/or GR activity to affect anxiety-driving circuitry or circuits involved in mood, cognition, or reward are urgently needed.

There is an unmet medical need to improve the QoL of patients suffering from adrenal insufficiency, because the current replacement therapies still need to be optimized to mimic the endogenous ultradian and circadian rhythms of CORT. Likewise, patients on glucocorticoid therapy may benefit from CORT as add-on to reactivate the MR which is essential for coping with stress. The few studies that use the CORT add-on reported so far are promising with respect to improved mood, sleep hygiene, and cognitive performance.[49–51,70a] One may consider even corticosterone rather than cortisol because of its better brain penetration and metabolic profile.[2,30]

MR and GR can also be used to stratify depressed patients. Common functional haplotype of the MR based on rs5522 and rs2070951 SNPs (the gain of function haplotype 2 (CA), frequency ≈0.41) is associated with dispositional optimism, decreased rumination, and reduced thoughts of hopelessness as well as protection to depression[65] and moderation of childhood maltreatment.[66] Stratification can be further achieved with HPA axis, biological, psychological, and symptom measures of a particular endophenotype characteristic of a subtype of depression.[67]

Acknowledgments

The support to E.R. de Kloet by the Royal Netherlands Academy of Arts and Sciences, COST Action ADMIRE BM1301 and STW Take-off 14095 is gratefully acknowledged. The authors are grateful to Dr A.M. Karssen for his contribution to the current topic in the Encyclopedia of Stress (2nd Edition pp. 704–708, 2007).

Declaration of Interest

E.R. de Kloet is on the scientific advisory Board of Dynacorts Therapeutics and Pharmaseed Ltd., and owns stock of Corcept Therapeutics.

References

1. Uchoa ET, Aguilera G, Herman JP, Fiedler JL, Deak T, de Sousa MB. Novel aspects of glucocorticoid actions. *J Neuroendocrinol*. 2014;26:557–572.
2. Karssen AM, Meijer OC, van der Sandt IC, et al. Multidrug resistance P-glycoprotein hampers the access of cortisol but not of corticosterone to mouse and human brain. *Endocrinology*. 2001;142:2686–2694.
3. Pecoraro N, Dallman MF, Warne JP, et al. From Malthus to motive: how the HPA axis engineers the phenotype, yoking needs to wants. *Prog Neurobiol*. 2006;79:247–340.
4. Russell GM, Lightman SL. Can side effects of steroid treatments be minimized by the temporal aspects of delivery method? *Expert Opin Drug Saf*. 2014;13:1501–1513.
5. De Kloet ER. From receptor balance to rational glucocorticoid therapy. *Endocrinology*. 2014;155:2754–2769.
6. McEwen BS, Gray JD, Nasca C. 60 years of neuroendocrinology: redefining neuroendocrinology: stress, sex and cognitive and emotional regulation. *J Endocrinol*. 2015;226:T67–T83.
7. Geerling JC, Loewy AD. Aldosterone in the brain. *Am J Physiol Renal Physiol*. 2009;297:F559–F576.
8. Karst H, Berger S, Turiault M, Tronche F, Schütz G, Joëls M. Mineralocorticoid receptors are indispensable for nongenomic modulation of hippocampal glutamate transmission by corticosterone. *Proc Natl Acad Sci USA*. 2005;102:19204–19207.
9. Joëls M, Karst H, DeRijk R, de Kloet ER. The coming out of the brain mineralocorticoid receptor. *Trends Neurosci*. 2008;31:1–7.
10. Hill MN, Tasker JG. Endocannabinoid signaling, glucocorticoid-mediated negative feedback, and regulation of the hypothalamic-pituitary-adrenal axis. *Neuroscience*. 2012;204:5–16.
11. Joëls M, Sarabdjitsingh RA, Karst H. Unraveling the time domains of corticosteroid hormone influences on brain activity: rapid, slow, and chronic modes. *Pharmacol Rev*. 2012;64:901–938.
12. Sapolsky RM, Romero LM, Munck AU. How do glucocorticoids influence stress responses? Integrating permissive, suppressive, stimulatory, and preparative actions. *Endocr Rev*. 2000;21:55–89.
13. De Kloet ER, Vreugdenhil E, Oitzl MS, Joëls M. Brain corticosteroid receptor balance in health and disease. *Endocr Rev*. 1998;19:269–301.
14. De Kloet ER, Joëls M, Holsboer F. Stress and the brain: from adaptation to disease. *Nat Rev Neurosci*. 2005;6:463–475.
15. Johannsson G, Falorni A, Skrtic S, et al. Adrenal insufficiency: review of clinical outcomes with current glucocorticoid replacement therapy. *Clin Endocrinol (Oxf)*. 2015;82:2–11.
16. Barnes PJ. Glucocorticosteroids: current and future directions. *Br J Pharmacol*. 2011;163:29–43.
17. Fardet L, Petersen I, Nazareth I. Suicidal behavior and severe neuropsychiatric disorders following glucocorticoid therapy in primary care. *Am J Psychiatry*. 2012;169:491–4977.
18. Judd LL, Schettler PJ, Brown ES, et al. Adverse consequences of glucocorticoid medication: psychological, cognitive, and behavioral effects. *Am J Psychiatry*. 2014;171:1045–1051.
19. Coghlan MJ, Elmore SW, Kym PR, Kort ME. The pursuit of differentiated ligands for the glucocorticoid receptor. *Curr Top Med Chem*. 2003;3:1617–1635.
20. De Bosscher K, Van Craenenbroeck K, Meijer OC, Haegeman G. Selective transrepression versus transactivation mechanisms by glucocorticoid receptor modulators in stress and immune systems. *Eur J Pharmacol*. 2008;583:290–302.
21. Reichardt HM, Kaestner KH, Tuckermann J, et al. DNA binding of the glucocorticoid receptor is not essential for survival. *Cell*. 1998;93:531–541.
22. Karst H, Karten YJ, Reichardt HM, de Kloet ER, Schütz G, Joëls M. Corticosteroid actions in hippocampus require DNA binding of glucocorticoid receptor homodimers. *Nat Neurosci*. 2000;3:977–978.

23. Oitzl MS, Reichardt HM, Joëls M, de Kloet ER. Point mutation in the mouse glucocorticoid receptor preventing DNA binding impairs spatial memory. *Proc Natl Acad Sci USA*. 2001;98:12790–12795.

24. Vandevyver S, Dejager L, Tuckermann J, Libert C. New insights into the anti-inflammatory mechanisms of glucocorticoids: an emerging role for glucocorticoid-receptor-mediated transactivation. *Endocrinology*. 2013;154:993–1007.

25. Zalachoras I, Houtman R, Atucha E, et al. Differential targeting of brain stress circuits with a selective glucocorticoid receptor modulator. *Proc Natl Acad Sci USA*. 2013;110:7910–7915.

26. Alvarez LD, Martí MA, Veleiro AS, et al. Hemisuccinate of 21-hydroxy-6,19-epoxyprogesterone: a tissue-specific modulator of the glucocorticoid receptor. *Chem Med Chem*. 2008;3:1869–1877.

27. Shah N, Scanlan TS. Design and evaluation of novel nonsteroidal dissociating glucocorticoid receptor ligands. *Bioorg Med Chem Lett*. 2004;14:5199–5203.

28. Garner B, Phillips LJ, Bendall S, Hetrick SE. Antiglucocorticoid and related treatments for psychosis. *Cochrane Database Syst Rev*. 2016;1:CD006995.

29. *Williams textbook of endocrinology. Chapter 15. The adrenal cortex*. Elsevier; 2012.

30. Karssen AM, Meijer OC, Berry A, Sanjuan Piñol R, de Kloet ER. Low doses of dexamethasone can produce a hypocorticosteroid state in the brain. *Endocrinology*. 2005;146:5587–5595.

31. De Kloet ER, Molendijk ML. Coping with the forced swim stressor: towards understanding an adaptive mechanism. *Neural Plast*. 2016:850719.

32. Schmidt MV, Sterlemann V, Wagner K, et al. Postnatal glucocorticoid excess due to pituitary glucocorticoid receptor deficiency: differential short- and long-term consequences. *Endocrinology*. 2009;150:2709–2716.

33. Laryea G, Schütz G, Muglia LJ. Disrupting hypothalamic glucocorticoid receptors causes HPA axis hyperactivity and excess adiposity. *Mol Endocrinol*. 2013;27:1655–1665.

34. Atucha E, Zalachoras I, van den Heuvel JK, et al. A mixed glucocorticoid/mineralocorticoid selective modulator with dominant antagonism in the male rat brain. *Endocrinology*. 2015;156:4105–4114.

35. Gomez-Sanchez EP. Brain mineralocorticoid receptors in cognition and cardiovascular homeostasis. *Steroids*. 2014;91:20–31.

36. Funder JW. Mineralocorticoid receptor antagonists: emerging roles in cardiovascular medicine. *Integr Blood Press Control*. 2013;6:129–138.

37. Bauersachs J, Jaisser F, Toto R. Mineralocorticoid receptor activation and mineralocorticoid receptor antagonist treatment in cardiac and renal diseases. *Hypertension*. 2015;65:257–263.

38. Bledsoe RK, Madauss KP, Holt JA, et al. A ligand-mediated hydrogen bond network required for the activation of the mineralocorticoid receptor. *J Biol Chem*. 2005;280:31283–31293.

39. Funder JW. Mineralocorticoid receptors: distribution and activation. *Heart Fail Rev*. 2005;10:15–22.

40. Chapman K, Holmes M, Seckl J. 11β-hydroxysteroid dehydrogenases: intracellular gate-keepers of tissue glucocorticoid action. *Physiol Rev*. 2013;93:1139–1206.

41. Diederich S, Scholz T, Eigendorff E, et al. Pharmacodynamics and pharmacokinetics of synthetic mineralocorticoids and glucocorticoids: receptor transactivation and prereceptor metabolism by 11beta-hydroxysteroid-dehydrogenases. *Horm Metab Res*. 2004;36:423–429.

42. Otte C, Jahn H, Yassouridis A, et al. The mineralocorticoid receptor agonist, fludrocortisone, inhibits pituitary-adrenal activity in humans after pre-treatment with metyrapone. *Life Sci*. 2003;73:1835–1845.

43. Banda J, Lakshmanan R, Vvs SP, Gudla SP, Prudhivi R. A highly sensitive method for the quantification of fludrocortisone in human plasma using ultra-high-performance liquid chromatography tandem mass spectrometry and its pharmacokinetic application. *Biomed Chromatogr*. 2015;29:1213–1219.

44. Sutanto W, de Kloet ER. Mineralocorticoid receptor ligands: biochemical, pharmacological, and clinical aspects. *Med Res Rev*. 1991;11:617–639.

45. Grossmann C, Scholz T, Rochel M, et al. Transactivation via the human glucocorticoid and mineralocorticoid receptor by therapeutically used steroids in CV-1 cells: a comparison of their glucocorticoid and mineralocorticoid properties. *Eur J Endocrinol*. 2004;151:397–406.

46. Oitzl MS, de Kloet ER. Selective corticosteroid antagonists modulate specific aspects of spatial orientation learning. *Behav Neurosci*. 1992;106:62–71.

47. Tiemensma J, Andela CD, Biermasz NR, Romijn JA, Pereira AM. Mild cognitive deficits in patients with primary adrenal insufficiency. *Psychoneuroendocrinology*. 2016;63:170–177.

48. Schultebraucks K1, Wingenfeld K, Otte C, Quinkler M. The role of fludrocortisone in cognition and mood in patients with primary adrenal insufficiency (Addison's disease). *Neuroendocrinology*. 2016;103:315–320.

49. Plihal W1, Krug R, Pietrowsky R, Fehm HL, Born J. Corticosteroid receptor mediated effects on mood in humans. *Psychoneuroendocrinology*. 1996;21:515–523.

50. Groch S, Wilhelm I, Lange T, Born J. Differential contribution of mineralocorticoid and glucocorticoid receptors to memory formation during sleep. *Psychoneuroendocrinology*. 2013;38:2962–2972.

51. Liston C, Cichon JM, Jeanneteau F, Jia Z, Chao MV, Gan WB. Circadian glucocorticoid oscillations promote learning-dependent synapse formation and maintenance. *Nat Neurosci*. 2013;16:698–705.

52. Klok MD, Alt SR, Irurzun Lafitte AJ, et al. Decreased expression of mineralocorticoid receptor mRNA and its splice variants in postmortem brain regions of patients with major depressive disorder. *J Psychiatr Res*. 2011a;45:871–878.

53. Medina A, Seasholtz AF, Sharma V, et al. Glucocorticoid and mineralocorticoid receptor expression in the human hippocampus in major depressive disorder. *J Psychiatr Res*. 2013;47:307–314.

54. Seckl JR, Fink G. Antidepressants increase glucocorticoid and mineralocorticoid receptor mRNA expression in rat hippocampus in vivo. *Neuroendocrinology*. 1992;55:621–626.

55. Reul JM, Stec I, Söder M, Holsboer F. Chronic treatment of rats with the antidepressant amitriptyline attenuates the activity of the hypothalamic-pituitary-adrenocortical system. *Endocrinology*. 1993;133:312–320.

56. Holsboer F. The corticosteroid receptor hypothesis of depression. *Neuropsychopharmacology*. 2000;23:477–501.

57. Otte C, Hinkelmann K, Moritz S, et al. Modulation of the mineralocorticoid receptor as add-on treatment in depression: a randomized, double-blind, placebo-controlled proof-of-concept study. *J Psychiatr Res*. 2010;44:339–346.

58. Lembke A, Gomez R, Tenakoon L, et al. The mineralocorticoid receptor agonist, fludrocortisone, differentially inhibits pituitary-adrenal activity in humans with psychotic major depression. *Psychoneuroendocrinology*. 2013;38:115–121.

59. Juruena MF, Pariante CM, Papadopoulos AS, Poon L, Lightman S, Cleare AJ. The role of mineralocorticoid receptor function in treatment-resistant depression. *J Psychopharmacol*. 2013;27:1169–1179.

60. Wingenfeld K, Wolf OT. Effects of cortisol on cognition in major depressive disorder, posttraumatic stress disorder and borderline personality disorder – 2014 Curt Richter Award Winner. *Psychoneuroendocrinology*. 2015;51:282–295.

61. Shorter E, Fink M. *Endocrine psychiatry: solving the riddle of Melancholia*. Oxford University Press; 2010.

62. Stalder T, Kirschbaum C, Kudielka BM, et al. Assessment of the cortisol awakening response: expert consensus guidelines. *Psychoneuroendocrinology*. 2016;63:414–432.

63. Wester VL, van Rossum EF. Clinical applications of cortisol measurements in hair. *Eur J Endocrinol*. 2015;173:M1–M10.

64. Ising M, Horstmann S, Kloiber S, et al. Combined dexamethasone/corticotropin releasing hormone test predicts treatment response in major depression - a potential biomarker? *Biol Psychiatry*. 2007;62:47–54.

65. Klok MD, Giltay EJ, Van der Does AJ, et al. A common and functional mineralocorticoid receptor haplotype enhances optimism and protects against depression in females. *Transl Psychiatry*. 2011b;1:e62.

66. Vinkers CH, Joëls M, Milaneschi Y, et al. Mineralocorticoid receptor haplotypes sex-dependently moderate depression susceptibility following childhood maltreatment. *Psychoneuroendocrinology*. 2015;54:90–102.

67. Hellhammer D, Hero T, Gerhards F, Hellhammer J. Neuropattern: a new translational tool to detect and treat stress pathology I. Strategical consideration. *Stress*. 2012;15:479–487.

68. McEwen BS, Wingfield JC. What is in a name? Integrating homeostasis, allostasis and stress. *Horm Behav*. 2000;57:105–111.

69. Meijer OC, de Lange EC, Breimer DD, de Boer AG, Workel JO, de Kloet ER. Penetration of dexamethasone into brain glucocorticoid targets is enhanced in mdr1A P-glycoprotein knockout mice. *Endocrinology*. 1998;139:1789–1793.

70. Jaisser F., Farman N.. Emerging roles of the mineralocorticoid receptor in pathology: toward new paradigms in clinical pharmacology. Pharmacol Rev. 2016;68:49–75.

70a. Warris LT, van den Heuvel-Eibrink MM, Aarsen FK, et al. Hydrocortisone as an intervention for dexamethasone-induced adverse effects in pediatric patients with acute lymphoblastic leukemia: results of a double-blind, randomized controlled trial. *J Clin Oncol*. 2016;34:2287–2293.

38

Stress and the Central Circadian Clock

M.S. Bartlang[1], G.B. Lundkvist[2]

[1]University of Würzburg, Würzburg, Germany; [2]Max Planck Institute for Biology of Ageing,
Cologne, Germany

Abstract

Circadian clocks are internal molecular time-keeping mechanisms that enable organisms to adjust their behavior and physiology to the 24-h environment. In addition to the circadian system, the stress system effectively restores the internal dynamic equilibrium of living organisms, called homeostasis, in light of any threatening stimulus (i.e., stressor). Any dysregulation in either system or disturbance of their molecular interrelation might lead to severe health effects. Importantly, both systems communicate with and feedback on each other at various physiological and neuronal levels. Thus, any disturbance or uncoupling possibly contributes to the development of several somatic and affective disorders. In this chapter, we discuss the biological function of the circadian and the stress system, their interactions, and the clinical implications of their uncoupling or dysregulation. For the sake of clarity and focus, we will only address the mammalian central circadian clock in the brain.

INTRODUCTION

All inhabitants of our planet are exposed to recurrent environmental changes generated by the rotation of the earth in the solar system. In order to anticipate these predictable fluctuations, organisms have evolved an evolutionary conserved internal time-keeping system, i.e., the circadian clock. In mammals, the circadian system is organized in a hierarchical fashion: a central pacemaker is located in the bilateral suprachiasmatic nucleus (SCN) of the hypothalamus, whereas subsidiary peripheral clocks exist in virtually all tissues and organs (see Fig. 38.1) (for review see Ref. 7).

Importantly, most species are also exposed to unpredictable changes in the environment, such as decreased food resources, increased predator numbers, or sudden

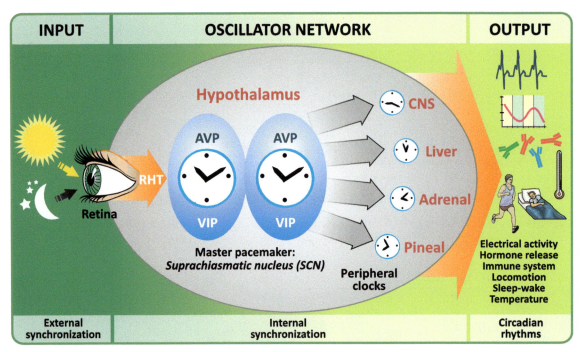

FIGURE 38.1 **Organization of the mammalian circadian system.** The circadian system consists of three major components: the inputs (so-called *Zeitgebers*), the rhythm generator in the oscillator network, and the outputs. Furthermore, it is organized in a hierarchical fashion with the bilateral suprachiasmatic nucleus (SCN) being the master pacemaker and subsidiary peripheral clocks functioning as slave clocks. The SCN is synchronized by the external 24 h cycle or other nonphotic *Zeitgebers* and produces sustained and synchronous cellular rhythmicity to coordinate rhythms in both central and peripheral tissues. This, in turn, results in rhythmic behavioral and physiological outputs, such as the sleep-wake cycle. The multioscillator network is synchronized through several lines of communication. While light is the primary input for the SCN (via the RHT), peripheral oscillators are synchronized by timing cues from the SCN and in some cases other timing cues such as food availability. *AVP*, arginine-vasopressin; *CNS*, central nervous system; *RHT*, retinohypothalamic tract; *VIP*, vasoactive intestinal peptide.

changes in social states. To adapt to these sudden alterations, the acute activation of the stress response system comprising the hypothalamic-pituitary-adrenal (HPA) axis and the sympathetic nervous system (SNS) represents a fundamental survival mechanism. In contrast, chronic activation of the stress system might result in severe somatic and affective disorders (for review see Ref. 5).

At first glance, the circadian and the stress system seem to represent two separate bodily control systems that are involved in adaptation to predictable and unpredictable stimuli, respectively. However, both systems are fundamental for survival and, thus, communicate with each other at various physiological and neuronal levels.

THE CENTRAL MOLECULAR CLOCK

The mammalian SCN is formed by a heterogeneous bilateral network of 10,000–50,000 single neurons[14] and serves as a master pacemaker in the control of a wide array of behavioral and physiological rhythms (e.g., sleep-wake, locomotion, cardiovascular function, endocrine processes) (see Box 38.1 and Fig. 38.1) (for review see Ref. 7). According to peptide expression, projection patterns, and neuronal morphology, each unilateral SCN can be divided into two subregions: a dorsomedial shell and a ventrolateral core

region. Shell neurons produce primarily arginine-vasopressin (AVP) and are mainly innervated by limbic areas, the hypothalamus, and the core region. In contrast, vasoactive intestinal peptide (VIP) is mainly synthesized in the core region that receives most of the input from the retina and brain regions that receive photic input.[32]

The Molecular Clockwork

In order to adjust to the environmental changes, cellular rhythms of roughly 24 h are generated in every single cell by transcriptional–translational autoregulatory loops that drive recurrent molecular oscillations in mRNA and protein levels of the so-called clock genes (see Fig. 38.2). In the SCN, the single neuronal oscillators are coupled to each other by paracrine and synaptic interaction to produce a coherent rhythmic output and to prevent the single oscillators from desynchronizing. In this way, the clock machinery, which also includes ion fluxes such as calcium, electrical activity, and cAMP, drives the biological activities in the body, such as daily hormonal fluctuations, the sleep-wake cycle, and the immune system activity (for review see Ref. 15). The transcription factors circadian locomotor output cycles kaput (CLOCK) and brain and muscle aryl hydrocarbon receptor nuclear translocator-like protein 1

BOX 38.1

HOW DO WE KNOW THAT THE SCN CONTROLS THE DAILY GC RHYTHM?

A series of electrical lesion studies in rodents in the 1970s provided clear evidence that the SCN possesses a primary role in the generation of mammalian circadian rhythms. Following selective disruption of the master SCN clock, locomotor activity, drinking behavior, body temperature, and GC secretion was completely absent.[31,44] Despite the lack of neuronal connections between the grafted SCN and the host brain, transplantation of donor SCN tissue into hosts with lesioned SCN remarkably restored these rhythms, suggesting that a diffusible secreted factor might be responsible for transmitting the circadian signal from the SCN.[40] However, while circadian rhythms of locomotor activity, drinking behavior, and body temperature could be restored by SCN transplants, circadian GC secretion was not, implicating that, in addition to secreted factors, neuronal efferents are required for generation of certain circadian rhythms.[30]

(BMAL1) constitute the positive limb of the feedback loop. Independent of an organism's activity phase, the CLOCK:BMAL1 heterodimer binds, via PAS domains, to the E-box promoter elements of clock genes and clock-controlled genes at the beginning of a circadian day, thus activating their transcription. The clock genes *Period* (*Per*; isoforms *Per1* and *Per2*) and *Cryptochrome* (*Cry*; isoforms *Cry1* and *Cry2*), in turn, constitute the negative limb: their mRNA is translated into proteins in the cytoplasm of the cell over the course of the day. Upon reaching a certain threshold, the protein products form heterodimers and homodimers that feedback to the nucleus by binding to the CLOCK:BMAL1 protein complex to autorepress the expression of their own genes at the beginning of the circadian night. Thereafter, constitutive degradation decreases the PER and CRY protein levels and as soon as these levels fall below a defined threshold required for sufficient autorepression, a new transcriptional cycle can be initiated. Several posttranslational modifications like histone acetylation, phosphorylation, ubiquitination, and methylation seem to be required for the delay between transcriptional activation and repression which is essential for a precise and functional circadian rhythm. In addition to this central molecular loop, multiple accessory regulatory loops contribute to a proper clock function. One of those loops is composed of nuclear receptors from the reverse erythroblastosis virus (REV-ERB) and retinoic acid receptor-related orphan receptor (ROR), families that are transcriptionally regulated by the positive limb and, in turn, activate (ROR) or repress (REV-ERB) transcription of Bmal1. The PER2 protein fine-tunes this process by interacting with REV-ERB to synchronize the negative and positive limbs of the transcriptional–translational feedback loops. In general, all these clock-related transcription factors impact a broad spectrum of physiological functions like sleep-wake behavior, thermoregulation, or nutrition by regulating the transcription rate of several clock-responsive genes (for review see Ref. 37).

KEY POINTS

- The circadian system is our control system for predictable environmental changes (e.g., day/night, seasons). It consists of the central clock in the hypothalamic suprachiasmatic nucleus (SCN) and subsidiary peripheral clocks
- The stress system is our control system for unforeseen changes in the environment (e.g., predators). It comprises the hypothalamic-pituitary-adrenal (HPA) axis and the sympathetic nervous system (SNS)
- Both systems are fundamental for survival and communicate with each other at various levels; disruptions in either system can lead to somatic and affective disorders
- The central circadian clock activates the HPA axis, thereby controlling the daily glucocorticoid release from the adrenal cortex
- Acute and chronic stress can affect core clock components within the SCN; disruptions of the molecular circadian clockwork is often related to depressive-like behavior

The internal clock machinery works autonomously and relatively precisely in each cell. However, in order to stay tuned to geophysical time the internal "clock time" needs to be adjusted and reset daily by environmental signals, a process called entrainment. Changes in illumination (occur for instance at dawn and dusk) and ambient temperature are the most important environmental synchronizers, the so-called *Zeitgebers*. In addition, nutrition, social factors, or even stress are further entrainment factors and might dominate the solar cycle in certain situations (for review see Ref. 7).

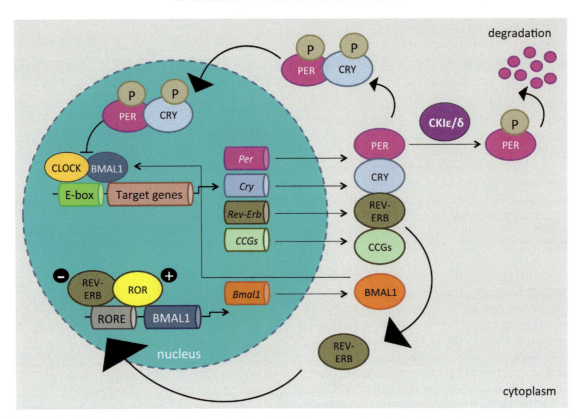

FIGURE 38.2 A simplified view of the molecular circadian clock. The mammalian clock relies on two interlocked transcriptional and translational feedback loops. The positive limb is formed by the transcription factors CLOCK and BMAL1 that bind to E-box sequences in the promoters of clock genes like *Per*, *Cry*, and *Rev-Erb*, thus activating their transcription at the beginning of a circadian day. As a result, PER proteins accumulate in the cytoplasm and become phosphorylated by kinases like casein kinases ε and δ and glycogen synthase kinase 3β. The phosphorylated forms of PER proteins are unstable and are degraded. Late in the subjective day, CRY accumulates in the cytoplasm and promotes the formation of stable CKI/PER/CRY complexes that enter the nucleus at the beginning of a circadian night and repress CLOCK:BMAL1-mediated transcription. To initiate a new circadian cycle, PER and CRY proteins get degraded and the CLOCK:BMAL1-mediated transcription starts again. The interacting positive and negative feedback loops of circadian genes warrant low levels of PER and CRY and, concomitantly, high levels of BMAL1 at the beginning of a new circadian day. Additional regulation of Bmal1 further tunes the loop. The CLOCK:BMAL1 heterodimer inhibits *Bmal1* transcription, and the REV-ERB protein enters the nucleus to suppress the transcription of the *Bmal1* gene during the day. At night, REV-ERB protein levels are low, allowing *Bmal1* transcription to take place. ROR, another nuclear receptor, competes with the REV-ERB for the RORE binding site and activates the transcription of *Bmal1*. BMAL1, brain and muscle aryl hydrocarbon receptor nuclear translocator-like protein; CCGs, clock-controlled genes; CKI, casein kinase; CLOCK, circadian locomotor output cycles kaput; *Cry*, cryptochrome; E-box, enhancer box; P, phosphorylated; *Per*, Period; REV-ERB, reverse erythroblastosis virus; ROR, retinoic acid receptor-related orphan receptor; RORE, retinoic acid receptor-related orphan receptor response element. *Italic notation* for genes, capital notation for respective proteins.

STRESS

In order to retain a constant internal environment in response to external environmental changes, organisms maintain a complex dynamic equilibrium, also called homeostasis. If this equilibrium state is disrupted by a threatening stimulus, i.e., stressor, a stress response is generated to regain homeostasis. This includes the activation of the SNS and the HPA axis.[39] Upon activation of the SNS, large quantities of the catecholamines norepinephrine and epinephrine are secreted into nearby capillaries within seconds. Once released into the blood, these hormones amplify the "fight or flight" reaction in order to provide sufficient energy (for review see Ref. 10). While the neuronal innervation of end organs through the SNS provides an immediate response to stressor exposure, activation of the HPA axis is slower (within minutes) and more persistent in its actions. In response to stressful stimuli, parvocellular neurons of the paraventricular nucleus of the hypothalamus (PVN) secrete releasing hormones, such as corticotropin-releasing hormone (CRH) and AVP into the portal circulation at the median eminence. These releasing hormones then act synergistically on the anterior pituitary by binding to their respective receptors (CRH-R1 and AVP-R1b), thereby triggering the secretion of adrenocorticotropic hormone (ACTH) from preformed granules into the peripheral circulation. Upon reaching the adrenal cortex, ACTH activates the synthesis and secretion of glucocorticoids (GC) (see Fig. 38.3). Once released, circulating GC modulate the expression of approximately 10% of our genes and exert widespread actions in the body which are essential for the maintenance of homeostasis and enable an organism to prepare for, respond to, and cope with physical and emotional stress (for review see Ref. 26). GC exert

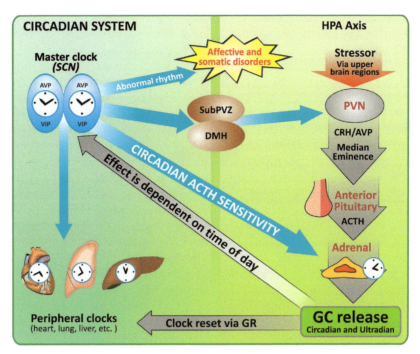

FIGURE 38.3   Interactions between the circadian system and the HPA axis. The hypothalamic PVN receives homeostatic/stress inputs from the brain stem and limbic areas. Upon activation, parvocellular CRH/AVP-secreting neurons project to the median eminence where they terminate in close proximity to a capillary plexus. CRH and AVP are directly secreted into these vessels and transported via the hypophyseal portal system to corticotropes in the anterior pituitary gland. As a result, ACTH is released into the venous circulation. When ACTH reaches the adrenal cortex, it activates the synthesis and secretion of GC that, in turn, act on different levels of the HPA axis via negative feedback. The circadian system and the HPA axis communicate with each other at various levels. The central mammalian clock in the SCN controls the daily activity of the HPA axis, thus leading to a circadian and ultradian GC release from the adrenal cortex. The SCN also controls the diurnal sensitivity of the adrenal gland to the incoming ACTH message. The local adrenal clock contributes to the rhythmic GC secretion. Secreted GC resets peripheral clocks via GRs. Acute/chronic stressor exposure might affect core clock components in the SCN. This process, however, appears to be dependent on the time of day when the stressor occurs. The stress effects on the SCN might lead to increased susceptibility for affective and somatic disorders. *ACTH*, adrenocorticotropic hormone; *AVP*, arginine-vasopressin; *CRH*, corticotropin-releasing hormone; *DMH*, dorsomedial hypothalamic nucleus; *GC*, glucocorticoid(s); *GR*, glucocorticoid receptor; *HPA axis*, hypothalamic-pituitary-adrenal axis; *PVN*, paraventricular nucleus of the hypothalamus; *SCN*, suprachiasmatic nucleus; *subPVZ*, subparaventricular zone; *VIP*, vasoactive intestinal peptide.

their actions via intracellular receptors, namely the glucocorticoid receptor (GR) and the mineralocorticoid receptor (MR). When activated, the cytoplasmic GR/MR, previously in an inactive complex state, undergoes a conformational change promoting the translocation from the cytoplasm into the nucleus. Within the nucleus, the activated receptors stimulate transcription of GC target genes by binding to glucocorticoid response elements (GRE). In order to reset the activated HPA axis system and restore homeostasis, secreted GC exert negative feedback inhibition at different levels of the HPA axis by binding to the GR/MR.[36]

SITES OF INTERACTION BETWEEN THE CIRCADIAN CLOCK SYSTEM AND THE STRESS SYSTEM

The Circadian and Ultradian Release of Glucocorticoids

In addition to the aforementioned stress-related secretion of GCs, these cholesterol-derived molecules are further released in a circadian manner under nonstressed conditions with an increased concentration prior to the active period of the day (early morning in humans, early evening in rats and mice).[4,47] Furthermore, the circadian pattern is overlaid by an ultradian rhythm, i.e., a rhythm with a period significantly shorter than 24h, with a pulse frequency (the rate of hormonal release) averaging between 60 and 90 min.[47,48]

The daily GC rhythm is controlled by the central circadian clock in a multimodal fashion: (1) the SCN controls the activity of the HPA axis by conveying excitatory and inhibitory information through synaptic contacts to the medioparvocellular PVN, where the CRH- and AVP-expressing neurons are located[18]; (2) the clock control over GC secretion operates via the sympathetic input to the adrenal gland, thereby modulating the sensitivity of the target organ to the incoming ACTH message[35,45]; and (3) the peripheral autonomous adrenal clock itself confers rhythmic expression of several clock genes as well as encoding molecules involved in the steroidogenic pathway and the ACTH signal transduction cascade in the *zona glomerulosa* and *zona fasciculata* of the adrenal gland. Thereby, mechanisms (2) and (3) are the main routes of SCN control over GC secretion. However, each of these paths might dominate under different conditions. The adrenal clockwork, for instance, appears to be essential for circadian GC production in mice in constant darkness, while under normal light–dark conditions light information is capable of regulating daily changes in GC production even in the absence of a functional local adrenal clock.[43]

The ultradian pulses that underlie the circadian GC rhythm occur with a relatively constant, roughly hourly, frequency, whereas the pulse amplitude (the amount of hormonal release) is variable. Besides the circadian input, homeostatic as well as stress-related signals impact the amplitude of these secretory episodes. The rising GC levels at the beginning of the active phase result from increases in the amplitude of the pulses, reaching its maximum just before awakening and declining thereafter to reach a trough early in the sleep phase (for review see Ref. 19). It has been demonstrated that the pulsatility of GC secretion is crucial for the appropriate stress response. The time of stressor application, for instance, determines the physiological stress response depending on the phase of an endogenous basal pulse. Rats that were exposed to white noise for 10 min responded with additional GC secretion if endogenous basal GC levels increased just prior to stressor exposure. In contrast, no or neglected GC responses were noticeable when basal endogenous GC levels were falling at stressor initiation.[48] These findings suggest that the basal GC pulsatility dynamically interacts with the ability of an organism to mount a stress response. Termination of the rising phase of the GC pulse by rapid feedback inhibition via the GR and MR may be essential in the generation of such pulses. While the GR mediates the acute effects of GC, the MR has an approximately 10-fold higher affinity than the GR and is involved in permissive or long-term activation during the peak of circadian GC concentration (for review see Ref. 21).

Circadian Responsiveness of the Hypothalamic-Pituitary-Adrenal Axis to Stress

Besides the ultradian pulsatility of the GC rhythm, the stress response might also be influenced by the fluctuating sensitivity of the HPA axis over the day. In addition, the type of stressor seems to play a crucial role. While psychological stressors, such as novel environment or restraint, elicit the largest HPA axis response during the inactive phase of rats (i.e., early day), physical stressors like hypoglycemic shock provoke greater stress responses at the onset of activity (i.e., early evening). One likely explanation for these differences might be the fact that physical stress is relayed to the PVN mainly via the brain stem, while processing psychological stress information requires the interpretation from higher brain centers involving the limbic system. The central clock in

the SCN might differentially interfere with these signals from different brain areas by enhancing input from psychological stressors during day and inhibiting physical stress input at the same time (for review see Ref. 8).

Besides the commonly accepted interaction of the stress response and the time of day of stressor exposure in relation to psychological and physical stressors, recent studies suggest that this also applies for psychosocial stressors such as social defeat. Exposure of mice to 19 days of social defeat at the beginning of the active phase at Zeitgeber time (ZT)13–15 (ZT0 is defined as the time when lights are turned on) results in a more negative outcome (decreased social preference, reduced home–cage activity during the dark phase, more severe chemically induced colitis effects) as compared to stressor exposure at ZT1–3.[2] In contrast, stress responses to intruder/resident confrontations of golden hamsters during the rest period (ZT2) as measured by heart rate, core body temperature, and general activity were significantly stronger compared to stressor exposure during the activity time (ZT14).[12]

Although we and others began to study the interactions between the type and duration of stressors and the time of the day and their effects on parameters like HPA axis activity, physiological rhythms, or the immune status, the exact mechanisms underlying the diurnal differences in the stress response remain to be elucidated.

Can Stress Directly Affect the Central Clock?

Several studies have shown that circadian rhythms at the output level are strongly affected by acute and chronic stressor exposure. For instance, altered rhythms in sleep-wake behavior,[33] body temperature,[46] locomotor activity,[13] and hormone secretion[9] have been reported after exposure to chronic mild, shaker, and restraint stress, respectively. An imbalance between normally precisely orchestrated physiological and behavioral rhythms might be either attributable to alterations in SCN activity or might arise from stress-induced changes in peripheral suboscillators. While various studies have examined stress effects on peripheral clocks (see Box 38.2), investigations on the central pacemaker are scarce. This might be ascribed to the fact that, in contrast to peripheral clocks, GR expression could not be detected in the SCN.[1]

Acute Stress and the Suprachiasmatic Nucleus

With respect to the acute stressful situation, it is very reasonable that the SCN is devoid of GRs since elsewise the SCN rhythm could be perturbed any time an organism is stressed. In support, early studies indicated that acute social defeat does not perturb

BOX 38.2

GRs—A CHECKPOINT FOR THE INTERACTION BETWEEN GC AND THE CIRCADIAN SYSTEM AT THE PERIPHERAL LEVEL

In order to properly adjust to acute stressful situations, peripheral oscillators need to be reset in such situations. In contrast to the SCN, peripheral nonbrain tissues mostly contain GRs and MRs. Clock genes, such as *Per1*, and possibly *Per2*, possess a GRE in their promoter region.[1,41] Following binding to the GRE, ligand-activated GRs/MRs can phase-shift the expression rhythm of several clock genes and clock-related genes, leading to the resetting of circadian rhythms, for instance in the liver, heart, and kidney (for review see Refs. 34,50) Direct interactions between clock factors and the GR at peripheral target tissues may also occur. CLOCK/BMAL1 can acetylate lysine residues in the hinge region of the GR. This posttranslational modification renders GRs unable to fully interact with GRE, thus modifying the transcription rate of GC-responsive genes. Moreover, CRY has also been shown to prevent GR function via direct binding, resulting in decreased transactivation potential on a GRE-controlled luciferase reporter gene.[23]

---

the central oscillator in the SCN as measured by body temperature and activity output,[29] and it has been assumed that the central SCN clock cannot be affected by acute stress. However, there are strong indications that the SCN brain clock is responsive to stress under certain circumstances and not completely insensitive as initially believed. In early studies it was shown that exogenous GC enhances AVP and VIP mRNA expression in the SCN,[24] and AVP release within the master pacemaker was increased following 10 min of forced swim and water immersion,[11] strongly suggesting a direct and rapid effect on SCN function. Furthermore, it has been shown that acute exposure to predator scent stress leads to an upregulation of PER1 and PER2 protein expression in the SCN of male rats.[22] Although predator scent stress is sensed by the olfactory bulb that contains a self-sustained, SCN-independent clock and the signaling from the olfactory bulb to the SCN might be indirect, this finding clearly shows that a certain type of acute stressor can indeed impact clock gene expression in the SCN.

Chronic Stress and the Suprachiasmatic Nucleus

Also with regard to chronic stressful situations, there is conflicting evidence of stress effects on the central clock. Chronic stressor exposure including forced swim, restraint, and social stress did not affect the central clock although it impinged the amplitude of the activity rhythm in mice.[42] In contrast, a reduction in PER2 oscillation amplitude has been reported in the rat SCN following 4 weeks of chronic unpredictable stress.[16] The same stress paradigm also led to a decreased CLOCK and BMAL1 protein expression in the central clock.[17] Likewise, seven days of repeated restraint decreased the PER2 protein expression in the mouse SCN.[20]

Stress Effects on the Suprachiasmatic Nucleus Might be Dependent on the Time of Day

What could be the reason for these contradictory results regarding stress effects on the SCN clock? One explanation might be that the SCN sensitivity to stress follows a circadian rhythm; i.e., mammals may be differentially sensitive/responsive to stress depending on what time of day they are exposed to the stressor. In support, our groups have recently shown that 19 days of repeated social defeat in mice has completely different effects on the SCN clock depending on whether the animals are exposed in the beginning of the active (night) or the nonactive (day) phase. Social defeat at ZT13–15 increased the PER2 rhythm amplitude in the SCN and PER2 protein expression in the posterior part of the central clock, whereas social defeat at ZT1–3 did not have any significant effect on PER2 rhythm and expression.[3] This truly indicates that a GC signal evoked by stressor exposure can be sensed by and perturb core clock components in the SCN, but only at certain times of the day. The time-dependent stress signal is likely perceived via intermediary GR-containing brain areas like the PVN, dorsomedial hypothalamic nucleus, or the raphe nuclei.[27]

Stress and Depressive-Like Behavior Might be Linked to the Circadian Clock

Regardless of the predictability level of the stressor and the effect on protein expression and oscillation amplitude of core clock components, alterations of the molecular clockwork within the SCN were often correlated with depressive-like behavior. The more pronounced effects of stressor exposure at ZT13–15 on the SCN molecular clock were also correlated with more severe effects on behavior, such as reduced activity at the beginning of the active phase and lack of social preference, indicating depressive-like behavior. Moreover, effects of immune system functions were significantly more severe following stressor exposure at early night (ZT13–15) compared to early day (ZT1–3).[2]

A link between circadian rhythm disturbances and mood disorders is beyond doubt. Virtually all of the successful treatments for mood disorders seem to affect circadian rhythms and it is assumed that stabilization and/or resetting of these rhythms by treatments are crucial for therapeutic efficacy (for review see Ref. 28). For instance, the effects of seasonal affective disorder, a mood disorder that correlates with the extremely shortened daily light period during the winter season, can be alleviated by bright light therapy. Patients suffering from seasonal affective disorder have abnormal levels of the dark phase hormone melatonin and exhibit a delayed chronobiological cycle.[49] Since clock genes, especially *Per1* and *Per2*, are inducible by light in the SCN, early morning light can phase-advance behavioral and endocrine rhythms, thus alleviating depressive-like symptoms.[25]

SUMMARY AND FUTURE DIRECTIONS

Findings presented in this chapter highlight the complexity of the multilevel interaction of the circadian and the stress system, as summarized in Fig. 38.3. The circadian clock controls the activity of the HPA axis through multisynaptic contacts between the SCN and the PVN which provides the basis for the activity-related circadian GC release. In addition, the central clock alters the sensitivity of the adrenal gland to the incoming ACTH message, thereby influencing the secretion of GC.

Vice versa, stress affects core clock components such as PER2 and CLOCK/BMAL1 within the central oscillator, which is often correlated with depressive-like behavior. In addition, rotating shift-work and frequent jet lag are provoked by our modern-life habits. Individuals who are exposed to these lifestyles have a greater risk of developing both somatic and affective disorders. Human epidemiological studies have shown that rotating shift nurses exhibit a higher risk to suffer from breast cancer compared to day shift nurses.[38] Likewise, shift-work experience is associated with a higher incidence of major depressive disorder.[6]

Given the interrelated influence of these bodily control systems, a proper functioning of the systems is critical for a healthy body and mind. It becomes clear that the balance between the two systems is crucial for the maintenance of proper circadian rhythms such as the GC rhythms that exert widespread function within our bodies. Further studies are required to understand the exact molecular interactions between these two systems and their interplay in the development of human pathology to resolve the outstanding issues.

References

1. Balsalobre A, Brown SA, Marcacci L, et al. Resetting of circadian time in peripheral tissues by glucocorticoid signaling. *Science*. 2000;289:2344–2347.
2. Bartlang MS, Neumann ID, Slattery DA, et al. Time matters: pathological effects of repeated psychosocial stress during the active, but not inactive, phase of male mice. *J Endocrinol*. 2012;215:425–437.
3. Bartlang MS, Savelyev SA, Johansson AS, Reber SO, Helfrich-Forster C, Lundkvist GB. Repeated psychosocial stress at night, but not day, affects the central molecular clock. *Chronobiol Int*. 2014;31:996–1007.
4. Cheifetz PN. The daily rhythm of the secretion of corticotrophin and corticosterone in rats and mice. *J Endocrinol*. 1971;49:xi–xii.
5. Chrousos GP. Stress and disorders of the stress system. *Nat Rev Endocrinol*. 2009;5:374–381.
6. Cole RJ, Loving RT, Kripke DF. Psychiatric aspects of shiftwork. *Occup Med*. 1990;5:301–314.
7. Dibner C, Schibler U, Albrecht U. The mammalian circadian timing system: organization and coordination of central and peripheral clocks. *Annu Rev Physiol*. 2010;72:517–549.
8. Dickmeis T. Glucocorticoids and the circadian clock. *J Endocrinol*. 2009;200:3–22.
9. Dubovicky M, Mach M, Key M, Morris M, Paton S, Lucot JB. Diurnal behavioral and endocrine effects of chronic shaker stress in mice. *Neuro Endocrinol Lett*. 2007;28:846–853.
10. Elenkov IJ, Wilder RL, Chrousos GP, Vizi ES. The sympathetic nerve – an integrative interface between two supersystems: the brain and the immune system. *Pharmacol Rev*. 2000;52:595–638.
11. Engelmann M, Ebner K, Landgraf R, Wotjak CT. Swim stress triggers the release of vasopressin within the suprachiasmatic nucleus of male rats. *Brain Res*. 1998;792:343–347.
12. Gattermann R, Weinandy R. Time of day and stress response to different stressors in experimental animals. Part I: golden hamster (Mesocricetus auratus Waterhouse, 1839). *J Exp Anim Sci*. 1996;38:66–76.
13. Gorka Z, Moryl E, Papp M. Effect of chronic mild stress on circadian rhythms in the locomotor activity in rats. *Pharmacol Biochem Behav*. 1996;54:229–234.
14. Guldner FH. Numbers of neurons and astroglial cells in the suprachiasmatic nucleus of male and female rats. *Exp Brain Res*. 1983;50:373–376.
15. Hastings MH, Brancaccio M, Maywood ES. Circadian pacemaking in cells and circuits of the suprachiasmatic nucleus. *J Neuroendocrinol*. 2014;26:2–10.
16. Jiang WG, Li SX, Zhou SJ, Sun Y, Shi J, Lu L. Chronic unpredictable stress induces a reversible change of PER2 rhythm in the suprachiasmatic nucleus. *Brain Res*. 2011;1399:25–32.
17. Jiang WG, Li SX, Liu JF, et al. Hippocampal CLOCK protein participates in the persistence of depressive-like behavior induced by chronic unpredictable stress. *Psychopharmacology*. 2013;227:79–92.
18. Kalsbeek A, Buijs RM. Output pathways of the mammalian suprachiasmatic nucleus: coding circadian time by transmitter selection and specific targeting. *Cell Tissue Res*. 2002;309:109–118.
19. Kalsbeek A, van der Spek R, Lei J, Endert E, Buijs RM, Fliers E. Circadian rhythms in the hypothalamo-pituitary-adrenal (HPA) axis. *Mol Cell Endocrinol*. 2012;349:20–29.
20. Kinoshita C, Miyazaki K, Ishida N. Chronic stress affects PERIOD2 expression through glycogen synthase kinase-3beta phosphorylation in the central clock. *Neuroreport*. 2012;23:98–102.
21. Kolbe I, Dumbell R, Oster H. Circadian clocks and the interaction between stress axis and adipose function. *Int J Endocrinol*. 2015;2015:693204.
22. Koresh O, Kozlovsky N, Kaplan Z, Zohar J, Matar MA, Cohen H. The long-term abnormalities in circadian expression of Period 1 and Period 2 genes in response to stress is normalized by agomelatine administered immediately after exposure. *Eur Neuropsychopharmacol*. 2012;22:205–221.

23. Lamia KA, Papp SJ, Yu RT, et al. Cryptochromes mediate rhythmic repression of the glucocorticoid receptor. *Nature.* 2011;480:552–556.

24. Larsen PJ, Vrang N, Moller M, et al. The diurnal expression of genes encoding vasopressin and vasoactive intestinal peptide within the rat suprachiasmatic nucleus is influenced by circulating glucocorticoids. *Brain Res Mol Brain Res.* 1994;27:342–346.

25. Lewy AJ, Sack RL, Miller LS, Hoban TM. Antidepressant and circadian phase-shifting effects of light. *Science.* 1987;235:352–354.

26. Lightman SL. The neuroendocrinology of stress: a never ending story. *J Neuroendocrinol.* 2008;20:880–884.

27. Malek ZS, Sage D, Pevet P, Raison S. Daily rhythm of tryptophan hydroxylase-2 messenger ribonucleic acid within raphe neurons is induced by corticoid daily surge and modulated by enhanced locomotor activity. *Endocrinology.* 2007;148:5165–5172.

28. McClung CA. Circadian genes, rhythms and the biology of mood disorders. *Pharmacol Ther.* 2007;114:222–232.

29. Meerlo P, van den Hoofdakker RH, Koolhaas JM, Daan S. Stress-induced changes in circadian rhythms of body temperature and activity in rats are not caused by pacemaker changes. *J Biol Rhythms.* 1997;12:80–92.

30. Meyer-Bernstein EL, Jetton AE, Matsumoto SI, Markuns JF, Lehman MN, Bittman EL. Effects of suprachiasmatic transplants on circadian rhythms of neuroendocrine function in golden hamsters. *Endocrinology.* 1999;140:207–218.

31. Moore RY, Eichler VB. Loss of a circadian adrenal corticosterone rhythm following suprachiasmatic lesions in the rat. *Brain Res.* 1972;42:201–206.

32. Moore RY. Entrainment pathways and the functional organization of the circadian system. *Prog Brain Res.* 1996;111:103–119.

33. Moreau JL, Scherschlicht R, Jenck F, Martin JR. Chronic mild stress-induced anhedonia model of depression; sleep abnormalities and curative effects of electroshock treatment. *Behav Pharmacol.* 1995;6:682–687.

34. Nader N, Chrousos GP, Kino T. Interactions of the circadian CLOCK system and the HPA axis. *Trends Endocrinol Metab.* 2010;21:277–286.

35. Oster H, Damerow S, Kiessling S, et al. The circadian rhythm of glucocorticoids is regulated by a gating mechanism residing in the adrenal cortical clock. *Cell Metab.* 2006;4:163–173.

36. Ratman D, Vanden Berghe W, Dejager L, et al. How glucocorticoid receptors modulate the activity of other transcription factors: a scope beyond tethering. *Mol Cell Endocrinol.* 2013;380:41–54.

37. Reppert SM, Weaver DR. Coordination of circadian timing in mammals. *Nature.* 2002;418:935–941.

38. Schernhammer ES, Laden F, Speizer FE, et al. Rotating night shifts and risk of breast cancer in women participating in the nurses' health study. *J Natl Cancer Inst.* 2001;93:1563–1568.

39. Selye H. Stress and the general adaptation syndrome. *Br Med J.* 1950;1:1383–1392.

40. Silver R, LeSauter J, Tresco PA, Lehman MN. A diffusible coupling signal from the transplanted suprachiasmatic nucleus controlling circadian locomotor rhythms. *Nature.* 1996;382:810–813.

41. So AY, Bernal TU, Pillsbury ML, Yamamoto KR, Feldman BJ. Glucocorticoid regulation of the circadian clock modulates glucose homeostasis. *Proc Natl Acad Sci USA.* 2009;106:17582–17587.

42. Solberg LC, Horton TH, Turek FW. Circadian rhythms and depression: effects of exercise in an animal model. *Am J Physiol.* 1999;276:R152–R161.

43. Son GH, Chung S, Choe HK, et al. Adrenal peripheral clock controls the autonomous circadian rhythm of glucocorticoid by causing rhythmic steroid production. *Proc Natl Acad Sci USA.* 2008;105:20970–20975.

44. Stephan FK, Zucker I. Circadian rhythms in drinking behavior and locomotor activity of rats are eliminated by hypothalamic lesions. *Proc Natl Acad Sci USA.* 1972;69:1583–1586.

45. Ulrich-Lai YM, Arnhold MM, Engeland WC. Adrenal splanchnic innervation contributes to the diurnal rhythm of plasma corticosterone in rats by modulating adrenal sensitivity to ACTH. *Am J Physiol.* 2006;290:R1128–R1135.

46. Ushijima K, Morikawa T, To H, Higuchi S, Ohdo S. Chronobiological disturbances with hyperthermia and hypercortisolism induced by chronic mild stress in rats. *Behav Brain Res.* 2006;173:326–330.

47. Weitzman ED, Fukushima D, Nogeire C, Roffwarg H, Gallagher TF, Hellman L. Twenty-four hour pattern of the episodic secretion of cortisol in normal subjects. *J Clin Endocrinol Metab.* 1971;33:14–22.

48. Windle RJ, Wood SA, Shanks N, Lightman SL, Ingram CD. Ultradian rhythm of basal corticosterone release in the female rat: dynamic interaction with the response to acute stress. *Endocrinology.* 1998;139:443–450.

49. Winkler D, Pjrek E, Praschak-Rieder N, et al. Actigraphy in patients with seasonal affective disorder and healthy control subjects treated with light therapy. *Biol Psychiatry.* 2005;58:331–336.

50. Yamamoto T, Nakahata Y, Tanaka M, et al. Acute physical stress elevates mouse period1 mRNA expression in mouse peripheral tissues via a glucocorticoid-responsive element. *J Biol Chem.* 2005;280:42036–42043.

39

Nongenomic Effects of Glucocorticoids: Translation From Physiology to Clinic

H. Gong, L. Liu, C.-L. Jiang

Second Military Medical University, Shanghai, P.R. China

Abstract

Glucocorticoids (GCs) are a class of steroid hormones that have been known to be widely used clinically as antiinflammatory, immunosuppressive, antishock drugs. Unfortunately, they can also produce numerous and potentially serious adverse effects that limit their usage. Thus, it is necessary to search for novel GCs with a better benefit-risk ratio compared to conventional GCs. GCs are believed traditionally to take effects mainly via the so-called genomic mechanisms, which are also largely responsible for GCs' side effects. However, an ever-growing body of evidence indicates that some effects of GCs can be mediated by the nongenomic mechanism. Theoretically, the discovery of nongenomic mechanisms of GCs provides novel approaches for the development of GCs to treat various diseases safely. The new GC drugs will take clinical effects mainly through the nongenomic mechanisms instead of classical genomic mechanisms to reduce side effects.

INTRODUCTION

Glucocorticoids (GCs) are steroid hormones that are released by the adrenal cortex in response to stress and affect multiple regulatory processes, such as metabolism, development, cognitive function, and other aspects of physiology. Cortisone (17-hydroxy-11-dehydrocorticosterone) was first used successfully to treat rheumatoid arthritis in 1948.[11] Since then, due to their potent antiinflammatory, immunosuppressive, antishock, and antiallergic effects, synthetic GCs have been extremely widely used in the clinical treatment of a large number of patients suffering from various inflammatory and neoplastic diseases as well as in organ transplantation.

However, their clinical use is limited by numerous, unpredictable and potentially serious side effects especially with high dosage and prolonged usage. For this reason, great efforts have been made to improve synthetic GC drugs with fewer side effects for several decades. Although both the fluorinated GCs (e.g., dexamethasone and betamethasone) and other novel GC drugs (e.g., prednisolone and methylprednisolone) have been synthesized to reinforce the antiinflammatory and reduce adverse side effects to some extent, people still hope to search for more ideal methods of the development of new GC drugs.

Since the genomic mechanism, especially transactivation, of GCs is thought to be responsible for numerous undesirable side effects,[25] trying to use nongenomic mechanisms of GCs more intensively, therefore, may represent a novel strategy for the development of GCs with low side effect profile. The new GC drugs will take clinical effects mainly through nongenomic mechanisms and do not execute the classical genomic mechanism to diminish unwanted side effects.[12]

THE CLASSICAL GENOMIC MECHANISM OF GC ACTIONS AND THE ADVERSE EFFECTS OF GCs

For many years, it has been believed that GCs take effects mainly through the delayed and prolonged genomic pathway. GCs can regulate gene expression both positively and negatively. Both of these effects are mainly mediated by the GC receptor (GR), which is part of nuclear hormone receptor superfamily and exists as multiprotein complex in the cytoplasm. The lipophilic GCs can easily cross the plasma membrane, enter the cell, bind to GR, and form the hormone–receptor complex. Then, the complex translocates into the nucleus, where it can exert its genomic effects in three primary ways as follows: (i) transcriptional repression or activation by binding to specific sequences, glucocorticoid-response elements (GREs), in the promoter region of target genes, (ii) transcriptional repression or activation through protein–protein interactions with other transcription factors, such as nuclear factor-κB (NF-κB), activator protein-1 (AP-1), and several signal transducers and activators of transcription (STATs), (iii) posttranscriptional modification of transcription factors via transcription activating

the expression of antiinflammatory molecules such as GC-induced leucine zipper (GILZ), MAPK phosphatase-1 (MKP-1), and tristetraprolin (TTP).[1,13]

Besides the therapeutic effects, endogenous or exogenous GC excess is also associated with numerous and sometimes irreversible side effects (Table 39.1). GCs affect the metabolism and lead to redistribution of body fat, diabetes, obesity, insulin resistance, and impaired glucose tolerance. In muscle tissues, GCs can elicit the atrophy of muscle by increasing the rate of protein degradation by the ubiquitin–proteasome system and autophagy-lysosome system.[3] GCs also have a wide variety of adverse effects on bone (osteoporosis, fractures, etc.), skin (skin thinning, epidermal atrophy, etc.), eyes (ocular hypertension, glaucoma, cataract, etc.), cardiovascular system (hypertension, atherosclerosis, etc.), and central nervous system (depression, mood swings, etc.). In addition, other notable side effects include preterm, avascular necrosis, etc.[5,25]

THE NONGENOMIC EFFECTS OF GC ACTIONS RELATED TO CLINICAL USAGES

Besides the above-mentioned genomic effects, which are believed to be responsible for most therapeutic effects, mounting evidence suggests that GCs also affect various functions via rapid, nongenomic mechanisms. The nongenomic GC mechanisms have been exploited in clinical therapy, where it has become increasingly evident that nongenomic GC activity may be relatively more important in mediating the therapeutic effects of intermediate-to-high doses of GCs, especially in high-dose pulsed GC administration.[4,16]

Recent research on the development of GCs with specific nongenomic mechanisms with fewer side effects

TABLE 39.1   Side Effects of GCs in Different Organ Systems

Muscle Diseases	Bone Impairment	Skin Diseases	Oculopathy	Cardiovascular Diseases	Central Nervous System Disorders	Metabolic Disorders	Other Side Effects
Muscle atrophy	Osteoporosis	Skin thinning	Cataract	Atherosclerosis	Mood swings	Diabetes	Preterm
	Fractures	Epidermal atrophy	Glaucoma	Hypertension	Euphoria	Redistribution of body fat	Avascular necrosis
	Avascular necrosis	Striae rubrae distensae	Ocular hypertension	Hyperlipidemia	Depression	Insulin resistance	Infection risk
		Perioral dermatitis	Exophthalmos	Vasculitis	Cerebral atrophy	Obesity	Water–electrolyte imbalance
		Hypertrichosis		Thrombosis	Steroid dependence	Impaired glucose tolerance	Peptic ulcer

will provide promising clinical applications, including different organ systems and suppression of inflammation.[6,24] The findings of beneficial or adverse effects of GCs based on nongenomic pathways will broaden our knowledge regarding their physiological/pathological roles. Since GCs are widely used clinically as mentioned above, whether nongenomic mechanisms play a part in these clinical applications is attractive for new drug development. It seems that rapid nongenomic GC effects play an important role, because clinical effects can be rapidly observed following GC administration especially with high-dose application.

The Nongenomic Effects of GCs in Antiinflammatory and Immunosuppressive Actions

Although nongenomic steroid effects have been widely recognized recently, relatively few studies have reported the nongenomic effect of GCs on antiinflammation and immunosuppression. GCs exert rapid effects on a variety of immunocytes. GCs can directly regulate leukocyte adhesion and locomotion though a nongenomic mechanism that is independent of modulation of gene expression.[23] GCs regulate thymocyte apoptosis through a nongenomic GC signaling pathway. In T cells, GCs rapidly inhibit the signal transmission pathway mediated by T-cell receptors (TCR) using a nongenomic mechanism that requires the binding of GCs to membrane receptors and not nuclear receptors and GCs can also modulate T cells' cytoskeletal architecture by nongenomic mechanisms.[20,21]

Dexamethasone is a synthetic member of the GCs class of hormones, which is commonly used to treat chronic inflammatory disorders, severe allergies, and other disease states. A single injection of high-dose dexamethasone could activate endothelial nitric oxide synthase (eNOS) and exhibited antiinflammatory effects to suppress systemic inflammation through the nongenomic signaling pathway.[22] Dexamethasone also can induce Src kinase Lck and downstream kinase activation, augment chemokine signaling, and function in resting human T cells.[8]

Our previous studies have indicated the nongenomic mechanisms of GC actions in immune cells, including macrophages, neutrophils, etc. GCs can exert their antiinflammatory and immunosuppressive effects in these cells via nongenomic mechanisms. Macrophages are multifunctional cells that play important roles in inflammation and the immune response. Their most important function is phagocytosis, which is accompanied by the generation of reactive oxygen species (ROS), such as superoxide anion. It was shown that GCs inhibited phagocytosis by macrophages following a long-duration (a few hours to days) pretreatment.[9] GCs were found to rapidly inhibit both the uptake of neutral red

and superoxide anion production by macrophages in less than 30 min. These effects were insensitive to the competitive GR antagonist RU486 (mifepristone, which potently inhibits the binding of GCs to GR) and the translation inhibitor actidione. GC coupled to bovine serum albumin (GC–BSA) was able to mimic the rapid inhibitory effects of GC. The results indicated that GC could rapidly inhibit phagocytosis and superoxide anion production by mouse peritoneal macrophages by a rapid nongenomic mechanism.[18]

Few cells play as prominent a role as the neutrophil in the inflammatory response and, therefore, the effect of GCs on neutrophils from human and other species has been an area of great interest. Human neutrophils contain three main lysosomal granules, azurophil granules, specific granules, gelatinase granules, and secretory vesicles. Specific proteolytic and digestive enzymes capable of destroying the extracellular matrix and bacterial debris are stored inside these granules, which, therefore, are involved in immune and inflammatory processes as well as in a variety of diseases and tissue injuries.[2] Both high doses of 6a-methylprednisolone and hydrocortisone showed rapid inhibitory effects on human neutrophil degranulation activated by N-formyl-methionyl-leucylphenylalanine (fMLP), which is a synthetic analog of a chemotactic peptide derived from a variety of bacteria. Neither RU486 nor actidione altered the inhibitory effects of GCs. The results demonstrate that high doses of GCs exert rapid inhibitory effects on human neutrophil degranulation at the cellular level via a new mechanism that is independent of GR occupation or protein synthesis.

The Nongenomic Effects of GCs in Antishock Action

It is well established that GCs possess good therapeutic effects during endotoxic shock and septic shock. Moreover, adrenal steroid hormones and catecholamines play central roles in both maintaining survival in times of stress and in regulating normal physiologic responsiveness, especially in regulating the cardiovascular system. GCs play an important role in the control of vascular smooth muscle tone and blood pressure through their permissive effects that potentiate vasoactive responses to catecholamines.[29] Clinically, the use of GCs as adjunctive therapy for severe sepsis and septic shock involves a rapid permissive action. The nongenomic effects of GCs manifested in the immune or other systems may also be present in the cardiovascular system. However, despite their recognized importance on the cardiovascular system, few studies have addressed their regulatory action through nongenomic mechanism and more extensive work is required in this field.[15]

Our previous work has found that GCs exert their permissive actions to norepinephrine in blood pressure through nongenomic mechanisms. Dexamethasone potentiated the responses to norepinephrine of adrenalectomized septic rats in 10 min. The rapid potentiating effect of GCs was observed not only in vivo, but also in isolated cells. The GCs rapidly promoted NE-induced phosphorylation of MLC20 in vascular smooth muscle cells by a nongenomic mechanism involving an increase in the activation of the RhoA/ROCK, ERK, and p38 signaling pathways.[30] GCs could also potentiate the norepinephrine-induced shrinkage and actin cytoskeleton rearrangement of single cells from resistance mesenteric arteries, which was similar to the finding reported by Koukouritaki and colleagues.[14]

The Nongenomic Effects of GCs in Antiallergic Action

GCs are the most potent antiinflammatory agents available for allergic diseases, including allergic rhinitis and asthma, and are routinely believed to require several hours to take effect through regulation of gene expression. GCs are effective in the treatment of allergic rhinitis. Nasal itching, a pathophysiologically complex sensation, was markedly reduced following either of the two GCs, betamethasone and methylprednisolone, within 10 min after administration of the study drug, therefore presumably via nongenomic mechanisms.[28]

We reported that GCs could inhibit allergic asthma within 10 min using changes of lung resistance and dynamic lung compliance of guinea pigs sensitized with ovalbumin and challenged with the same antigen given by aerosol.[31] The rapid inhibitory effect of GCs on allergic asthma is mediated by inhibiting airway mast cell degranulation within 10 min in the guinea pig allergic asthma model. Histamine is thought to be one of the major mediators in the allergic reaction, and IgE-mediated histamine releases from mast cells and plays a pivotal role in allergic diseases. GCs rapidly inhibited IgE-mediated exocytosis and histamine release of mast cells, which could be mimicked by membrane-impermeable BSA-conjugated corticosterone, and neither RU486 nor actidione blocked the rapid GC action. We further demonstrated that the GC nongenomic effect was not mediated by direct actions on the secretory machinery, but by a reduction in the intracellular $[Ca^{2+}]_i$ response.[17,33] GCs also rapidly inhibited the histamine-induced contractions of airway smooth muscle in a process mediated by nongenomic mechanisms.[27]

Taken together, these findings indicate that there exist nongenomic mechanisms of GCs during their clinical application as antiinflammatory, immunosuppressive, antishock, and antiallergic drugs.

THE NONGENOMIC MECHANISM OF GC ACTIONS AND DEVELOPMENT OF NEW DRUGS

As mentioned above, genomic mechanisms, including transrepression and transactivation, are responsible for most of the therapeutic effects of GCs. It is also widely believed that most antiinflammatory effects of GCs are mediated by gene transrepression, whereas the adverse effects of GCs are mediated by gene transactivation. In other words, both the clinical effects and the side effects of GCs can be exerted through classical genomic mechanisms.

Nevertheless, the discovery of nongenomic mechanisms of GCs is a major breakthrough in stress research, and further insights into these mechanisms may open novel approaches for the therapy of various diseases.[10] Many studies have indicated that the therapeutic aspects could be mediated through nongenomic mechanisms.

Membrane-bound GRs are present on different immune cell types and might be potential candidate targets for GC therapy. Drugs that specifically target membrane-bound steroid receptors may therefore be of therapeutic value.[19]

Thus, theoretically, the specific activation of therapeutically relevant nongenomic mechanisms might produce lesser side effects, and/or improve therapy. A new approach of optimizing GC therapy could develop drugs selectively affecting nongenomic mechanisms.[10,24] Therefore, new GCs, which exhibit their clinical pharmacological effects through nongenomic mechanism instead of genomic mechanism, may reduce the side effect profile.

Hydrocortisone can be conjugated with glycine to achieve a larger molecular structure that cannot pass through the cell membrane due to increased hydrophilicity, and then take effects via nongenomic mechanisms. Evaluation of antiinflammatory efficacy showed that hydrocortisone-conjugated glycine (HG) could inhibit neutrophil degranulation within 15 min. A luciferase reporter assay demonstrated that HG did not activate the GREs within 30 min, which verified that the rapid effects were independent of activation of the genomic pathway. Furthermore, HG could not only inhibit the IgE-mediated histamine release from mast cells within 30 min that could not be blocked by RU486 and actidione, but also rapidly alleviated the allergic reaction in the asthma model via a nongenomic pathway.[32] Although HG did not activate the GREs within 30 min, the hydrocortisone analog activated the GREs after 2 h because of dissociation of glycine to hydrocortisone according to our results. It is strongly needed to synthesize more compounds that are difficult to dissociate from the conjugation. Therefore, nongenomic mechanism of GCs provides us a novel strategy for the development of GCs.

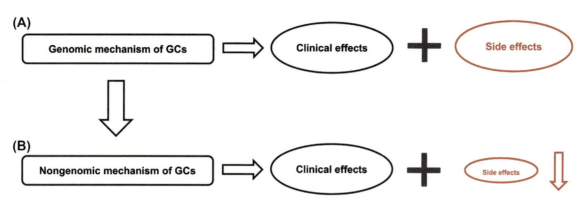

FIGURE 39.1 The new strategy of development of novel GCs. (A) Classical genomic mechanisms of GCs mediate both clinical effects and side effects. (B) The antiinflammatory, immunosuppressive, antiallergic, and antishock effects of GCs can be mediated through nongenomic mechanisms. Thus, nongenomic mechanisms of GCs may provide us a new strategy for development of novel drugs, which take clinical pharmacological effects mainly via nongenomic mechanisms, and produce lesser side effects because of not executing the classical genomic mechanism.

CONCLUSION

With the increasing therapeutic use of GCs in clinical practice, great efforts have been made to diminish GC-induced adverse effects, thus improving the benefit/risk ratio of the drugs. Besides approaches to optimize the use of conventional GC drugs that are already available and develop new innovative GR ligands,[26] new GCs selective to nongenomic mechanisms should also be exploited.

Screening distinctive GCs by using the ligands that specifically affect nongenomic mechanisms would probably lead to a decrease in side effects and an increase in therapy. A new type of GCs, which were synthesized through traditional GC molecules conjugating macromolecules, could not go through the cell membrane to exclude the classical genomic effect, and takes effect mainly via nongenomic mechanisms. However, even though the genomic effects are eliminated, the nongenomic effects are still very wide. Apart from their beneficial therapeutic nongenomic effects, GCs also exert nongenomic effects in the central nervous system at the levels of behavior, neural system activity, individual neuron activity, and subcellular signaling activity, such as rapidly increase novelty-related locomotor activity and aggression, rapidly inhibit the acoustic startle response, and retrieval of long-term memory in rats.[7] Therefore, these findings provide a new strategy of development of novel GCs (Fig. 39.1).

There is the novel strategy for dissociating nongenomic from genomic GC ligands and developing new GCs with a lower side effect profile via targeting nongenomic mechanisms. What also deserves expectation is that this new idea may contribute to improving the current situation of GC usages, which will be of considerable clinical significance. However, detailed knowledge on the underlying molecular mechanisms is still lacking. It will take many years to understand the molecular mechanisms. Furthermore, future study should also focus on the possibility of the cross-talk between genomic and nongenomic pathways that might still induce subsequent side effects.

Acknowledgment

This work is supported by the National Natural Science Foundation of China (31371200 and 81571169).

References

1. Beato M, Chavez S, Truss M. Transcriptional regulation by steroid hormones. *Steroids*. 1996;61(4):240–251.
2. Borregaard N, Cowland JB. Granules of the human neutrophilic polymorphonuclear leukocyte. *Blood*. 1997;89(10):3503–3521.
3. Braun TP, Marks DL. The regulation of muscle mass by endogenous glucocorticoids. *Front Physiol*. 2015;6:12. http://dx.doi.org/10.3389/fphys.2015.00012.
4. Buttgereit F, Wehling M, Burmester GR. A new hypothesis of modular glucocorticoid actions: steroid treatment of rheumatic diseases revisited. *Arthritis Rheum*. 1998;41(5):761–767.
5. Cooper MS, Seibel MJ, Zhou H. Glucocorticoids, bone and energy metabolism. *Bone*. 2015. http://dx.doi.org/10.1016/j.bone.2015.05.038.
6. De Bosscher K, Beck IM, Haegeman G. Classic glucocorticoids versus non-steroidal glucocorticoid receptor modulators: survival of the fittest regulator of the immune system? *Brain Behav Immun*. 2010;24(7):1035–1042.
7. Evanson NK, Herman JP, Sakai RR, Krause EG. Nongenomic actions of adrenal steroids in the central nervous system. *J Neuroendocrinol*. 2010;22(8):846–861.
8. Ghosh MC, Baatar D, Collins G, et al. Dexamethasone augments CXCR4-mediated signaling in resting human T cells via the activation of the Src kinase Lck. *Blood*. 2009;113(3):575–584.
9. Grasso RJ, Klein TW, Benjamin WR. Inhibition of yeast phagocytosis and cell spreading by glucocorticoids in cultures of resident murine peritoneal macrophages. *J Immunopharmacol*. 1981;3(2):171–192.
10. Haller J, Mikics E, Makara GB. The effects of non-genomic glucocorticoid mechanisms on bodily functions and the central neural system. A critical evaluation of findings. *Front Neuroendocrinol*. 2008;29(2):273–291.

11. Hench PS, Kendall EC, Slocumb CH, Polley HF. The effect of a hormone of the adrenal cortex (17-hydroxy-11-dehydrocorticosterone; compound E) and of pituitary adrenocorticotropic hormone on rheumatoid arthritis. *Proc Staff Meet Mayo Clin.* 1949;24(8):181–197.

12. Jiang CL, Liu L, Li Z, Buttgereit F. The novel strategy of glucocorticoid drug development via targeting nongenomic mechanisms. *Steroids.* 2015;102:27–31.

13. Kadmiel M, Cidlowski JA. Glucocorticoid receptor signaling in health and disease. *Trends Pharmacol Sci.* 2013;34(9):518–530.

14. Koukouritaki SB, Margioris AN, Gravanis A, Hartig R, Stournaras C. Dexamethasone induces rapid actin assembly in human endometrial cells without affecting its synthesis. *J Cell Biochem.* 1997;65(4):492–500.

15. Lee SR, Kim HK, Youm JB, et al. Non-genomic effect of glucocorticoids on cardiovascular system. *Pflugers Arch.* 2012;464(6):549–559.

16. Lipworth BJ. Therapeutic implications of non-genomic glucocorticoid activity. *Lancet.* 2000;356(9224):87–89.

17. Liu C, Zhou J, Zhang LD, et al. Rapid inhibitory effect of corticosterone on histamine release from rat peritoneal mast cells. *Horm Metab Res.* 2007;39(4):273–277.

18. Long F, Wang YX, Liu L, Zhou J, Cui RY, Jiang CL. Rapid nongenomic inhibitory effects of glucocorticoids on phagocytosis and superoxide anion production by macrophages. *Steroids.* 2005;70(1):55–61.

19. Lowenberg M, Stahn C, Hommes DW, Buttgereit F. Novel insights into mechanisms of glucocorticoid action and the development of new glucocorticoid receptor ligands. *Steroids.* 2008;73(9–10):1025–1029.

20. Lowenberg M, Verhaar AP, Bilderbeek J, et al. Glucocorticoids cause rapid dissociation of a T-cell-receptor-associated protein complex containing LCK and FYN. *EMBO Rep.* 2006;7(10):1023–1029.

21. Muller N, Fischer HJ, Tischner D, van den Brandt J, Reichardt HM. Glucocorticoids induce effector T cell depolarization via ERM proteins, thereby impeding migration and APC conjugation. *J Immunol.* 2013;190(8):4360–4370.

22. Murata I, Ooi K, Shoji S, et al. Acute lethal crush-injured rats can be successfully rescued by a single injection of high-dose dexamethasone through a pathway involving PI3K-Akt-eNOS signaling. *J Trauma Acute Care Surg.* 2013;75(2):241–249.

23. Pitzalis C, Pipitone N, Perretti M. Regulation of leukocyte-endothelial interactions by glucocorticoids. *Ann N Y Acad Sci.* 2002;966:108–118.

24. Song IH, Buttgereit F. Non-genomic glucocorticoid effects to provide the basis for new drug developments. *Mol Cell Endocrinol.* 2006;246(1–2):142–146.

25. Stahn C, Buttgereit F. Genomic and nongenomic effects of glucocorticoids. *Nat Clin Pract Rheumatol.* 2008;4(10):525–533.

26. Strehl C, Buttgereit F. Optimized glucocorticoid therapy: teaching old drugs new tricks. *Mol Cell Endocrinol.* 2013;380(1–2):32–40.

27. Sun HW, Miao CY, Liu L, et al. Rapid inhibitory effect of glucocorticoids on airway smooth muscle contractions in Guinea pigs. *Steroids.* 2006;71(2):154–159.

28. Tillmann HC, Stuck BA, Feuring M, et al. Delayed genomic and acute nongenomic action of glucocorticosteroids in seasonal allergic rhinitis. *Eur J Clin Invest.* 2004;34(1):67–73.

29. Yang S, Zhang L. Glucocorticoids and vascular reactivity. *Curr Vasc Pharmacol.* 2004;2(1):1–12.

30. Zhang T, Shi WL, Tasker JG, et al. Dexamethasone induces rapid promotion of norepinephrine-mediated vascular smooth muscle cell contraction. *Mol Med Rep.* 2013;7(2):549–554.

31. Zhou J, Kang ZM, Xie QM, et al. Rapid nongenomic effects of glucocorticoids on allergic asthma reaction in the Guinea pig. *J Endocrinol.* 2003;177(1):R1–R4.

32. Zhou J, Li M, Sheng CQ, et al. A novel strategy for development of glucocorticoids through non-genomic mechanism. *Cell Mol Life Sci.* 2011;68(8):1405–1414.

33. Zhou J, Liu DF, Liu C, et al. Glucocorticoids inhibit degranulation of mast cells in allergic asthma via nongenomic mechanism. *Allergy.* 2008;63(9):1177–1185.

PART III

DIURNAL, SEASONAL, AND ULTRADIAN SYSTEMS

Circadian Rhythm Effects on Cardiovascular and Other Stress-Related Events

R. Manfredini[1], B. Boari[1], R. Tiseo[1], R. Salmi[2], F. Manfredini[1]

[1]University of Ferrara, Ferrara, Italy; [2]General Hospital of Ferrara, Ferrara, Italy

Abstract

Circadian rhythmicity is ensured by the circadian timing system based on input signals (environmental cues), an intrinsic rhythm generator and output rhythms. This system prepares organisms for changes in their physical environments, and to respond to environmental factors in a temporally appropriate matter. The hypothalamic-pituitary-adrenal (HPA) axis; the autonomic sympathetic nervous system; and the cardiovascular, metabolic, and immune systems contributing to maintain biological homeostasis during environmental or physiological challenges are characterized by the presence of a daily rhythm or by significant daytime variations. In combination to the individual differences in terms of diurnal preferences (chronotype), conditions evoking desynchronization of human circadian rhythms (shift–work, jet–lag exposure) or different stressors may induce significant changes in sympathetic–parasympathetic balance and tone of HPA axis. This condition in humans might negatively affect the cardiovascular system both chronically (by accelerating atherosclerosis) and acutely, via multiple mechanisms, triggering the onset of acute life-threatening events particularly in the morning.

CHRONOBIOLOGY AND CIRCADIAN RHYTHMS

Circadian rhythms represent ubiquitous biological oscillation of approximately 24-h periods, highly conserved in rather all living organisms. The daily timekeeping system is called "circadian" from the Latin "circa diem," which means "approximately a day," deriving from duration of a cycle of earth rotation. At the same time, this system is both autonomous and self-sustainable, and also continuously entrained by external time cues (called "synchronizers").[1] Circadian clocks are cell autonomous, transcriptionally based, molecular mechanisms that confer the selective advantage of anticipation, enabling organisms to prepare for changes in their physical environments and respond to environmental factors in a temporally appropriate manner. The mammalian circadian timing system consists of three basic components: (1) input signals (environmental cues), (2) a circadian oscillator as an intrinsic rhythm generator, and (3) output rhythms. The hypothalamic suprachiasmatic nucleus (SCN) is considered the central circadian oscillator both anatomically and functionally. It receives photic information from the eyes via the retinohypothalamic tract and synchronizes the circadian timing system with environmental time. Such self-sustaining nature requires the presence of a genetic mechanism known as the molecular circadian clockwork, and clock genes are required for generating and maintaining the circadian rhythm both at the organism and even cellular level.[2] The circadian system is strictly and hierarchically organized. At the top of the mammalian circadian timing system, the SCN is composed by neurons with self-sustaining rhythmic capacity, but also most tissues and peripheral organs express their own clock genes, and even cultured cells in vitro retain their rhythmicity.[3]

KEY POINTS

- The endogenous circadian system confers to living organisms the selective advantage of anticipation, enabling to prepare and respond to environmental factors in a temporally appropriate manner.

- Many physiological processes are affected by circadian rhythms, and individual differences can be summarized under the concept of "chronotype": morningness/eveningness (M/E). Chronotypes differ with regards to many aspects of behavioral, personality, and lifestyle.

- The occurrence of cardiovascular events is unevenly distributed during the 24h, but shows a much greater incidence than expected in the morning hours. A temporal coincidence with peaks of some unfavorable triggers and circadian misalignment may play a role.

- A novel recently described cardiac syndrome occurring in post-menopausal women, "Takotsubo cardiomyopathy" or "stress cardiomyopathy," whose pathophysiology is not well understood, might be triggered by multifactorial mechanisms including rapid elevation of circulating catecholamine central triggered by emotional and/or physical stress. Similarly to myocardial infarction, a morning preference has been reported also for TTC.

- Stress, via multiple mechanisms, can exert major effects upon the circulatory system, and be responsible for life-threatening events particularly in the morning hours.

STRESS AND CIRCADIAN RHYTHMS

Living organisms try to maintain biological homeostasis during environmental or physiological challenges and protect from internal or external stress by using several mechanisms, such as the hypothalamic-pituitary-adrenal (HPA) axis; the sympathetic adrenomedullary system (SAM); and the cardiovascular (CV), metabolic, and immune systems. The main systems responsible for activation of the stress response reside in the hypothalamus and brainstem. They include corticotropin-releasing hormone (CRH) and arginine–vasopressin neurons in the paraventricular nucleus of the hypothalamus, and the locus coeruleus–norepinephrine systems in the pons and medulla.[4] The HPA axis is activated by CRH from the paraventricular nucleus of the hypothalamus, which prompts the release of corticotropin from the pituitary, stimulating the production of glucocorticoids (mainly cortisol) from the adrenal cortex and, to a lesser extent, mineralocorticoids and androgens.[5] Cortisol is an adrenal steroid hormone that controls a variety of physiological processes, i.e., metabolism, immune response, CV activity, and brain function. The HPA axis is a major neuroendocrine circuit of the stress response system, and adrenal cortisol synthesis and secretion are tightly regulated by upstream hormones secreted from the hypothalamus and the pituitary. In addition to the response to stress, cortisol is characterized by the existence of an evident daily rhythm. In fact, circulating cortisol levels are higher during the activity period (day for diurnal species and night for nocturnal species) and peak levels are linked to the beginning of the activity period. Such daily variation, although still not fully understood in its molecular bases, is generated by multimodal forms of regulation, including SCN, autonomic nervous system, and intrinsic mechanisms. However, cortisol rhythm depends on the rhythmic release of ACTH, and the daily rhythm in nonstress levels of plasma cortisol usually displays a 5- to 10-fold higher amplitude from the trough to peak levels in rodents, whereas the plasma ACTH rhythm is relatively lower or even not significantly different throughout the day.[6] The two main systems, HPA and SAM, have different vascular and metabolic effects, even potentially harmful, since they can increase BP, decrease insulin sensitivity, and activate hemostasis. Thus, stress may induce significant changes in sympathetic-parasympathetic balance and tone of HPA, which might negatively affect the CV system both acutely (by triggering the onset of acute events) and chronically (by accelerating atherosclerosis).[7] In addition to the just described diurnal variation in cortisol levels, circulating catecholamine concentrations also show large variation, characterized by diurnal increase and night reduction. Cortisol and catecholamines play a role in the complex series of factors underlying the common finding of an

TABLE 40.1 Relevant Papers on Cardiovascular Events and Circadian Variation

Event	Author	Journal	Year	Main Finding: Peak
Myocardial infarction	Muller et al.	N Engl J Med	1985	Morning (6 a.m.–noon)
	Willich et al.	Circulation	1989	No morning peak with prior β-adrenergic blockade
	Willich et al.	Circulation	1991	Association of wake time and onset
	Peters et al.	J Am Coll Cardiol	1993	Secondary evening peak
	Manfredini et al.	Am J Emerg Med	2004	Worst outcome for morning MIs
Cardiac death & arrest	Muller et al.	Circulation	1987	SCD: Morning peak (7 a.m.–11 a.m.)
	Willich et al.	Am J Cardiol	1992	First 3 h after awakening
	Levine et al.	JAMA	1992	6 a.m.–noon
Stroke	Marler et al.	Stroke	1989	10 a.m.–noon
	Kelly-Hayes et al.	Stroke	1995	8 a.m.–noon
	Casetta et al.	Arch Neurol	2002	Morning, independent of patients' features
	Casetta et al.	JAMA	2002	Morning (hemorrhagic)
	Manfredini et al.	Chronobiol Int	2005	Ischemic and hemorrhagic stroke: Pathophysiology
Aortic aneurysm rupture or dissection	Gallerani et al.	J Thorac Cardiovasc Surg	1997	Morning (thoracic aorta)
	Manfredini et al.	Lancet	1999	Morning (abdominal aorta)
	Mehta et al.	Circulation	2002	Morning (thoracic aorta)
	Manfredini et al.	J Vasc Surg	2004	Thoracic and abdominal aneurysm: Pathophysiology
Pulmonary embolism	Gallerani et al.	Eur heart J	1992	Morning
	Manfredini et al.	Eur J Med	1993	Morning
	Sharma et al.	Am J Cardiol	2001	Morning

early morning increase of ischemic heart events and cerebrovascular events (Table 40.1). Importantly, dysregulation in the clock system and the HPA axis may cause similar pathologic manifestations, including obesity, metabolic syndrome, and CV disease, by uncoupling circulating cortisol concentrations from tissue sensitivity to cortisol.[8] Moreover, disruption of sleep–wake cycle may interfere with autonomic balance and function of the HPA as well,[9] and sleep deprivation is associated with several metabolic disorders, including obesity and insulin resistance.

STRESS AND INDIVIDUAL CHRONOTYPE

In humans and other mammals, many physiological processes are affected by circadian rhythms. Individual differences in these chronobiological rhythms can be summarized under the concept of morningness/eveningness (M/E). Depending on the diurnal preference, the phase position, and period of the circadian rhythms, people may either be morning- or evening-orientated (M- or E-type, often also called "lark" or "owl," respectively)

or, in most cases, be neutral chronotype. Chronotypes differ with regards to their sleeping behavior, personality, mental health, smoking and dietary habits, school achievements, and so on. The Morningness–Eveningness Questionnaire is the most widely used tool to determine individual chronotype.[10] Free cortisol daytime levels in relation to morningness, performed in healthy middle-aged adults, are higher in M-relative to E-types,[11] and after a stressor-test, E-types had lower salivary cortisol levels and flattened diurnal curve in comparison with M-types.[12] Furthermore, the enhanced physiological arousal in E-types might contribute to increased vulnerability to psychological distress.[13] Several studies have shown that eveningness is associated with negative psychological outcomes, including depressive and anxiety symptoms, as well as impaired subjective sleep quality, increased circadian misalignment, and daytime sleepiness. Interestingly, short sleep duration and decreased sleep quality are emerging risk factors for metabolic diseases and obesity, and E-types tend to have unhealthy eating habits more frequently than M-types. In obese individuals with less than 6.5 h sleep per night, eveningness was associated with eating later in the day, increase in body mass index, and food portion size.[14] Shift-work represents, together with jet-lag, the classical example of desynchronization of human circadian rhythms. Daily cycles of sleep/wake, hormones, and physiological processes are often misaligned with behavioral patterns during shift work. Chronotype modulates the effects of working times, and M-types show shortened sleep duration during night shifts, high social jet lag, and higher levels of sleep disturbance, as well as E-types during early shifts.[15] An experimental attempt of abolition of most strenuous shifts for extreme chronotypes (i.e., mornings for E-types, nights for M-types) led to a significant increase of self-reported sleep duration and quality and increased well-being ratings on workdays.[16]

STRESS AND CIRCADIAN VARIATION OF CARDIOVASCULAR EVENTS

A growing body of evidence showed that occurrence of CV events is unevenly distributed during the 24h.[17] Myocardial infarction (MI) and sudden cardiac death (SCD), acute aortic rupture or dissection, stroke, and pulmonary embolism are much greater in incidence than expected in the morning (Table 40.1). For aortic disease, a recent meta-analysis estimated, in the morning hours, an absolute increased risk of 58% with respect to the remaining hours and 139% with respect to the night hours.[18] Such temporal patterns result from corresponding temporal variation in pathophysiologic mechanisms and cyclic environmental triggers that elicit the onset of clinical events.[19] Twenty-four–hour rhythmic organization of CV functions is such that defense mechanisms against acute events are incapable of providing the same degree of protection during the day and night. Instead, temporal gates of excessive susceptibility exist, particularly in the morning and to a lesser extent evening (in diurnally active persons), to aggressive mechanisms through which overt clinical manifestations may be triggered. When peak levels of critical physiologic variables are aligned together at the same circadian time, the risk of acute events becomes significantly elevated such that even relatively minor, and usually harmless, physical and mental stress can precipitate dramatic life-threatening clinical events.[20]

CIRCADIAN VARIATION OF TRIGGERING FACTORS

In fact, MI, SCD, and stroke exhibit a circadian variation with morning peak after waking up. In addition, physical exertion and mental stress are common precipitants, since they can influence BP, HR, plasma epinephrine levels, coronary blood flow, platelet aggregability, and endothelial function. Ischemia occurs when myocardial oxygen demand exceeds oxygen supply, and upregulation of sympathetic output and catecholamines increase myocardial oxygen demand and decrease myocardial oxygen supply, and favor thrombosis.[21] Increases in BP and ventricular contractility enhance intravascular shear stress and cause rupture of vulnerable atherosclerotic plaques. The BP circadian rhythm has a clear genetic basis and is driven by many neurohumoral factors characterized by intrinsic circadian rhythmicity. Moreover, the fact that the BP can lose or reverse its nocturnal fall in a variety of pathologic conditions in which sleep and activity show only minor change gives further evidence for an endogenous, although not predominant, component.[22] In addition to BP, many clinical triggers of MI have been identified, including blood and coagulation, vascular system, and external or internal stress factors. Table 40.2 summarizes some most relevant morning triggering factors. Moreover, circadian misalignment has been implicated in the development of metabolic and CV disease. Molecular- and genetic-based studies suggest that the cardiomyocyte circadian clock influences multiple myocardial processes, including transcription, signaling, growth, metabolism, and contractile function.[23] In animal models, the cardiomyocyte circadian clock has recently been linked to the pathogenesis of heart disease in response to adverse stresses, such as ischemia/reperfusion,[24] and circadian variation in BP and HR is disrupted in mice in which core clock genes are deleted or mutated.[25]

A NOVEL STRESS-RELATED MODEL OF CARDIAC DISEASE WITH POSSIBLE CIRCADIAN VARIATION: TAKOTSUBO CARDIOMYOPATHY

A novel cardiac syndrome, characterized by transient left ventricular and dysfunction, has been described in Japanese patients. It has been named "Takotsubo cardiomyopathy" (TTC), from the Japanese terms indicating the particular shape of the end-systolic left ventricle in ventriculography resembling that of the round-bottom and narrow-neck pot used for trapping octopuses.[26] Other definitions of this syndrome are "apical ballooning," "acute stress cardiomyopathy," or "broken heart." The diagnostic criteria include (1) transient hypokinesis, akinesis, or dyskinesis in the left ventricular midsegments with or without apical involvement; regional wall motion abnormalities that extend beyond a single epicardial vascular distribution; and, frequently but not always, a stressful trigger; (2) absence of obstructive coronary disease or angiographic evidence of acute plaque rupture; (3) new ECG abnormalities (ST-segment elevation and/or T-wave inversion) or modest elevation in cardiac troponin; and (4) absence of myocarditis or pheochromocytoma.[26] Current estimates of TTC prevalence is

TABLE 40.2 Relevant Papers on Cardiovascular Triggering Factors and Circadian Variation

Factor	Author	Journal	Year	Main Findings: Morning Hours
Heart & blood pressure	Gordon et al.	J Clin Invest	1966	↑Plasma renin activity
	Turton et al.	Clin Chim Acta	1974	↑Plasma catecholamines
	Millar-Craig et al.	Lancet	1978	↑BP
	Kamath et al.	Am J Cardiol	1991	↑HR
	Bursztyn et al.	Am J Hypertens	1994	↑BP after afternoon siesta
	Kario et al.	Am J Hypertens	2004	Morning BP surge
Blood & coagulation	Weizman et al.	J Clin Endocrinol Metab	1971	↑Plasma cortisol
	Ehrly et al.	Biorheology	1973	↑Blood viscosity
	Tofler et al.	N Engl J Med	1987	↑PLT aggregation
	Brezinski et al.	Circulation	1988	↑PLT aggregation, upright posture
	Andreotti et al.	Am J Cardiol	1988	↓Endogenous fibrinolysis
Vascular system	Panza et al.	N Engl J Med	1991	↑α-Sympathetic activity (↑vascular tone)
	Quyyumi et al.	Circulation	1992	↓Ischemic threshold
	Somers et al.	N Engl J Med	1993	Sympathetic nerve activity during sleep REM phase
	El-Tamini et al.	Circulation	1995	↑Vasomotor activity in dysfunctional coronary endothelium
	Otto et al.	Circulation	2004	↓Endothelial function
Other, external	Behar et al.	Am J Med	1993	↑External triggers
	Krantz et al.	Circulation	1996	↑Daily activities
Other, internal	Willich et al.	Circulation	1991	↑Stress or emotional upset
	Krantz et al.	JAMA	2000	↑Mental stress
	Virag et al.	Front Physiol	2014	Circadian disturbances
	Takeda et al.	Cell Mol life Sci	2015	Desynchronization of clock genes

BP, Blood pressure; *HR*, heart rate; *PLT*, platelet.

approximately 1–3% of patients (up to 6–9% if only women are considered) with suspicion of acute coronary syndrome (ACS). Usually it occurs preferably in post-menopausal women (90%), with mean age ranging 60–75 years.[27] The most common presenting clinical symptoms of TTC are chest pain and dyspnea, although syncope, palpitations, hypotension, cardiogenic shock, nausea and vomiting, ventricular arrhythmias, and/or cardiac arrest may also be included.[27] The clinical onset is usually preceded by an emotional and/or physical stress in approximately two-thirds of patients.[28] Physical stress is more frequent in men, while emotional stress or no identifiable trigger is more prevalent in women. However, in up to 30% of patients no preceding emotional or physical stressful events can be identified (Fig 40.1).

The pathophysiology of TTC is not well understood, and multifactorial mechanisms are involved. A rapid elevation of circulating catecholamine central triggered by emotional and/or physical stress may play a key pathogenic role, and TTC is associated with higher circulating catecholamine concentrations than during MI.[29] On one hand, catecholamines have a series of direct effects upon the myocardium, i.e., cellular damage, contraction band necrosis, defects in perfusion, and altered cellular metabolism may represent the major determinants of sympathetically mediated myocardial reversible dysfunction in susceptibility patients with TTC.[30] Again, high levels of circulating epinephrine may trigger a switch in intracellular signal trafficking in ventricular cardiomyocytes, from G(s) protein to G(i) protein signaling via the β2-adrenoceptor. Although this switch to β2-adrenoceptor-G(i) protein signaling protects against the proapoptotic effects of intense activation of β1-adrenoceptors, it is also negatively inotropic, and especially at the apical myocardium, in which β-adrenoceptor density is greatest.[31] The high catecholamine levels in the acute phase of TTC and common emotional triggers suggest a dysregulated stress response system. In patients with TTC and healthy controls, tested to verify responses to mental stress, significantly higher norepinephrine and dopamine levels during both mental

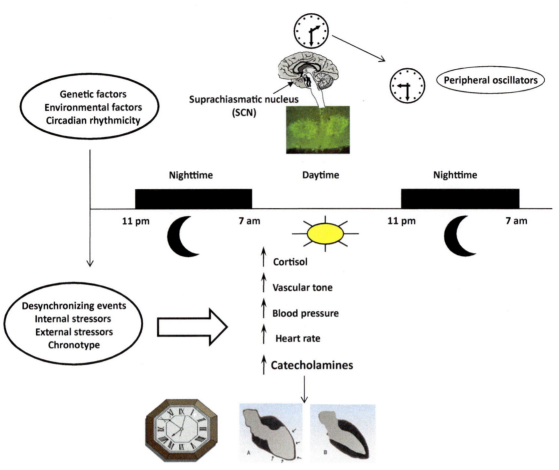

FIGURE 40.1   Morning preference for Takotsubo cardiomyopathy: hypothesis for a circadian model.

stress were found in TTC compared with healthy subjects. No evidence was found for a dysregulated HPA axis, and TTC patents showed blunted emotional arousal to mental stress, suggesting that catecholamine hyperreactivity and not emotional hyperreactivity to stress may play a role in myocardial vulnerability.[32]

The striking clinical similarity with MI includes temporal findings. In fact, a morning preference has been reported also for TTC.[33,34] Once again, a series of morning triggering actors shared with MI may play a role. Moreover, catecholamines show a circadian periodicity, with a urinary peak in late morning,[35] and in healthy women under routine life-style, the excretion of norepinephrine was higher during the working hours (9a.m.–3p.m.).[36] As previously reported, the stress hormone cortisol, in healthy individuals, also exhibits a diurnal pattern with an early morning peak and a nocturnal nadir, even if an age-related elevation in the morning has been observed in women, but not in men.[37] Acute stressful events, independent of traditional risk factors, may play a triggering role on the occurrence of ACS, and chronic stress may play a role as well.[38] Alterations in cortisol awakening and stress responses are sensitive markers for the basal activity and responsiveness of the HPA axis in psychopathological conditions. The circadian system and the stress response system play together a crucial role in the adaptation of the organism to environmental challenges. The central master clock utilizes the light as its primary synchronizer, whereas peripheral clocks are influenced by neurohumoral factors. Interestingly, norepinephrine is utilized as a signal within the cardiomyocyte.[38] Acute exposure of cardiomyocytes to norepinephrine, mimicking a burst of sympathetic activity, have been shown to induce oscillations in three circadian clock components (*bmall1*, *rev-erbaα*, and *per2*), as well as the circadian clock-regulated gene *dbp*. A tight link between catecholamines and molecular clock exists: (1) the molecular clock influences sympathoadrenal function, and norepinephrine and epinephrine exhibit a diurnal variation, with a higher levels during the active phase; (2) genes relevant to CAs synthesis and disposition are under the control of the molecular clock; (3) BMAL1 and CLOCK are indispensable for the circadian rhythm of BP; and (4) the circadian clock modulates selective stress response.[39] In conclusion, stress is capable of exerting major effects upon the circulatory system, particularly in the morning, via multiple mechanisms, and be responsible for life-threatening events in humans.[40]

References

1. Dunlap JC. Molecular bases for circadian clocks. *Cell.* 1999;96:271–290.

2. Ko CH, Takahashi JS. Molecular components of the mammalian circadian clock. *Hum Mol Gen.* 2008;2:R271–R277.

3. Nagoshi E, Saini C, Bauer C, Laroche T, Nef F, Schibler U. Circadian gene expression in individual fibroblasts: cell-autonomous and self-sustained oscillators pass time to daughter cells. *Cell.* 2004;119:693–705.

4. Tsigos C, Chrousos GP. Hypothalamic-pituitary-adrenal axis, neuroendocrine factors and stress. *J Psychosom Res.* 2002;53:865–871.

5. Mathe G. The need of a physiologic and pathophysiologic definition of stress. *J Psychosom Res.* 2000;54:119–121.

6. Watts AG, Tanimura S, Sanchez-Watts G. Corticotropin-releasing hormone and arginine vasopressin gene transcription in the hypothalamic paraventricular nucleus of unstressed rats: daily rhythms and their interactions with corticosterone. *Endocrinology.* 2004;145:529–540.

7. Brotman DJ, Golden SH, Wittstein IS. The cardiovascular toll of stress. *Lancet.* 2007;370:1089–1100.

8. Kino T. Circadian rhythms of glucocorticoid hormone actions in target tissues: potential clinical implications. *Sci Signal.* 2012;5:4.

9. Meerlo P, Sgoifo A, Suchecki D. Restricted and disrupted sleep: effects on autonomic function, neuroendocrine stress systems and stress responsivity. *Sleep Med Rev.* 2008;12:197–210.

10. Horne JA, Ostberg O. A self-assessment questionnaire to determine morningness-eveningness in human circadian rhythms. *Int J Chronobiol.* 1976;4:97–110.

11. Kudielka BM, Bellingrath S, Hellhammer DH. Further support for higher salivary cortisol levels in "morning" compared to "evening" persons. *J Psychosom Res.* 2007;62:595–596.

12. Oginska H, Fafrowicz M, Golonka K, Marek T, Mojsa-Kaja J, Tucholska K. Chronotype, sleep loss, and diurnal pattern of salivary cortisol in a simulated daylong driving. *Chronobiol Int.* 2010;27:959–974.

13. Roeser K, Obergfell F, Meule A, Vogele C, Schlarb AA, Kubler A. Of larks and hearts—morningness/eveningness, heart rate variability, and cardiovascular stress response at different times of day. *Physiol Behav.* 2012;106:151–157.

14. Lucassen EA, Zhao X, Rother KI, et al. Evening chronotype is associated with changes in eating behavior, more sleep apnea, and increased stress hormones in short sleeping obese individuals. *PLoS One.* 2013;8:e56519.

15. Juda M, Vetter C, Roenneberg T. Chronotype modulates sleep duration, sleep quality, and social jet lag in shift-workers. *J Biol Rhythms.* 2013;28:141–151.

16. Vetter C, Fischer D, Matera JL, Roenneberg T. Aligning work and circadian time in shift workers improves sleep and reduces circadian disruption. *Curr Biol.* 2015;25:907–911.

17. Smolensky MH, Portaluppi F, Manfredini R, et al. Diurnal and twenty-four hour patterning of human diseases: cardiac, vascular, and respiratory diseases, conditions, and syndromes. *Sleep Med Rev.* 2015;21:3–11.

18. Vitale J, Manfredini R, Gallerani M, et al. Chronobiology of acute aortic rupture or dissection: a systematic review and a meta-analysis of the literature. *Chronobiol Int.* 2015;32:385–394.

19. Manfredini R, Boari B, Salmi R, et al. Twenty-four-patterns in occurrence and pathophysiology of acute cardiovascular events and ischemic heart disease. *Chronobiol Int.* 2013;30:6–16.

20. Portaluppi F, Manfredini R, Fersini C. From a static to a dynamic concept of risk: the circadian epidemiology of cardiovascular events. *Chronobiol Int.* 1999;16:33–49.

21. Manfredini R, Gallerani M, Portaluppi F, Fersini C. Relationships of the circadian rhythms of thrombotic, ischemic, hemorrhagic, and arrhythmic events to blood pressure rhythms. *Ann N Y Acad Sci.* 1996;783:141–158.

22. Fabbian F, Smolensky MH, Tiseo R, Pala M, Manfredini R, Portaluppi F. Dipper and non-dipper blood pressure 24-hour patterns: circadian rhythm-dependent physiologic and pathophysiologic mechanisms. *Chronobiol Int.* 2013;30:17–30.

23. Durgan DJ, Young ME. The cardiomyocyte circadian clock: emerging roles in health and disease. *Circ Res.* 2010;106:647–658.

24. Takeda N, Maemura K. The role of clock genes and circadian rhythm in the development of cardiovascular diseases. *Cell Mol Life Sci.* 2015;72:3225–3234.

25. Curtis AM, Cheng Y, Kapoor S, Reilly D, Price TS, Fitzgerald GA. Circadian variation of blood pressure and the vascular response to asynchronous stress. *Proc Natl Acad Sci USA.* 2007;104:3450–3455.

26. Prasad A, Lerman A, Rihal CS. Apical ballooning syndrome (Tako-Tsubo or stress cardiomyopathy): a mimic of acute myocardial infarction. *Am Heart J.* 2008;155:408–417.

27. Bossone E, Savarese G, Ferrara F, et al. Takotsubo cardiomyopathy: overview. *Heart Fail Clin.* 2013;9:249–266.

28. Summers MR, Prasad A. Takotsubo cardiomyopathy: definition and clinical profile. *Heart Fail Clin.* 2013;9:111–122.

29. Wittstein IS, Thiemann DR, Lima JA, et al. Neurohumoral features of myocardial stunning due to sudden emotional stress. *N Engl J Med.* 2005;352:539–548.

30. Nef HM, Mollmann H, Kostin S, et al. Tako-Tsubo cardiomyopathy: intraindividual structural analysis in the acute phase and after functional recovery. *Eur Heart J.* 2007;28:2456–2464.

31. Lyon AR, Rees PS, Prasad S, Poole-Wilson PA, Harding SE. Stress (Takotsubo) cardiomyopathy—a novel pathophysiological hypothesis to explain catecholamine-induced acute myocardial stunning. *Nat Clin Pract Cardiovasc Med.* 2008;5:22–29.

32. Smeijers L, Szabó BM, van Dammen L, et al. Emotional, neurohormonal, and hemodynamic responses to mental stress in Tako-Tsubo cardiomyopathy. *Am J Cardiol.* 2015;115:1580–1586.

33. Bossone E, Citro R, Eagle KA, Manfredini R. Tako-tsubo cardiomyopathy: is there a preferred time of onset? *Intern Emerg Med.* 2011;6:221–226.

34. Manfredini R, Salmi R, Fabbian F, Manfredini F, Gallerani M, Bossone E. Breaking heart: chronobiologic insights into takotsubo cardiomyopathy. *Heart Fail Clin.* 2013;9:147–156.

35. Descovich GC, Montalbetti N, Kuhl JF, Rimondi S, Halberg F, Ceredi C. Age and catecholamine rhythms. *Chronobiologia.* 1974;1:163–171.

36. Hansen AM, Garde AH, Skovgaard LT, Christensen JM. Seasonal and biological variation of urinary epinephrine, norepinephrine, and cortisol in healthy women. *Clin Chim Acta.* 2001;309:25–35.

37. Van Cauter E, Leproult R, Kupfer DJ. Effects of gender and age on the levels and circadian rhythmicity of plasma cortisol. *J Clin Endocrinol Metab.* 1996;81:2468–2473.

38. Roohafza H, Talaei M, Sadeghi M, Mackie M, Sarafzadegan N. Association between acute and chronic life events on acute coronary syndrome: a case-control study. *J Cardiovasc Nurs.* 2010;25:E1–E7.

39. Durgan DJ, Hotze MA, Tomlin TM, et al. The intrinsic circadian clock within the cardiomyocyte. *Am J Physiol Heart Circ Physiol.* 2006;289:H1530–H1541.

40. Manfredini R, Boari B, Salmi R, Malagoni AM, Manfredini F. Circadian rhythm effects on cardiovascular and other stress-related events. In: Fink G, ed. *Encyclopedia of Stress, Second Edition.* Oxford: Academic Press; 2007:500–505.

CHAPTER

41

Seasonal Variation in Stress Responses*

J.C. Borniger[1], Y.M. Cisse[1], R.J. Nelson[1], L.B. Martin[2]

[1]The Ohio State University – Wexner Medical Center, Columbus, OH, United States; [2]University of South Florida, Tampa, FL, United States

Abstract

Winter is energetically demanding; thermogenic requirements are high and resources are scarce. This prolonged imbalance in energy intake and expenditure causes a stress response and glucocorticoid release. A chronic stress response typically results in pathology, but anticipation of seasonal stressors allows for the reorganization of endocrine, metabolic, immune, and neural processes to reduce energetic requirements. These trade-offs are mediated by differential investment in immune function and reproduction during times of low (fall and winter) and high (spring and summer) energy availability, respectively. Many nontropical animals monitor day length (i.e., photoperiod), in order to coordinate these seasonal adaptations. Much of the photoperiodic signal is transduced by pineal melatonin, as it is secreted inversely proportional to the absolute day length. This chapter will discuss seasonal changes in glucocorticoid release, binding globulin capacity, and receptor expression and their ultimate contributions to differential stress responses.

INTRODUCTION

Stressors are not constant, but rather, vary over time. In addition to daily changes in stress, seasonal changes in the quality and quantity of stress are common. Many energetic adaptations have evolved that attenuate the stress response during the winter when energy shortages may limit the ability of organisms to cope successfully with specific types of stressors. The temporal organization of such stress responses therefore must be considered to obtain a holistic understanding of the mechanisms underlying the stress response.

The main endocrine features of the stress response involve hormones that act to suppress energy storage and promote energy utilization from adipose and liver stores. Epinephrine and the glucocorticoids are the major hormonal components of the stress response. Animals require a relatively steady supply of energy to sustain biological functions. Although these energy demands are somewhat continual, both daily and seasonal fluctuations in energy requirements occur as physiological and behavioral activities wax and wane. Animals typically do not feed continuously to supply energy to cells; rather, adaptations have evolved to allow animals to store and conserve energy when they are engaged in nonfeeding activities. Eventually, however, the chronic need for energy depletes these stores and requires animals to obtain food. In most habitats, significant daily and seasonal fluctuations in food availability occur. For example, outside of the

---

*This article is a revision of the previous edition article by R.J. Nelson and L.B. Martin, volume 3, pp 427–431, © 2007, Elsevier Inc.

tropics, food availability is generally low during the winter when thermogenic energy demands are typically high. Consequently, energy intake and energy expenditures are never perfectly balanced because of fluctuations in both energy requirements and availability. Whenever a significant imbalance of energy is perceived, a stress response in the form of secreted glucocorticoids ensues.

Reproduction is sufficiently demanding that a stress response is often associated with breeding in terms of elevated glucocorticoid concentrations. That is, glucocorticoids are released to support the energetic needs of reproduction including territorial defense or courtship behaviors. However, when energy is insufficient to support reproduction and thermogenesis or other energetic demands, the stress response becomes prolonged. A chronic stress response can result in pathology. Most animals restrict breeding to specific times of the year when food is abundant and survival and reproductive success are most likely. The inhibition of winter breeding is one of the central components of a suite of energy-saving seasonal adaptations evolved among nontropical animals. In addition to the seasonal onset and cessation of reproductive function among nontropical vertebrates, it is also well established that endocrine, metabolic, growth, neural, immune, and thermogenic processes undergo seasonal changes to promote winter energy conservation. Mechanisms in virtually every physiological and behavioral system have evolved to cope with the winter energetic bottleneck resulting from increased thermogenic energy requirements during periods of low energy availability. Presumably, animals possessing these energy-conserving adaptations enhance their survival and ultimately increase their reproductive success. Generally, glucocorticoid concentrations are higher during the winter than the summer, and individuals display more elevated stress responses during the winter than the summer. However, because coping strategies are engaged in anticipation of the winter stressors, many individuals do not show exaggerated stress responses during the winter.

How do animals determine time of year? In most cases, the photoperiod (day length) acts as the principal external cue to determine time of year. Photoperiod measurement is important for organizing the seasonal collapse and regrowth of the reproductive system in mammals and birds from temperate and boreal latitudes. Indeed, changes in day length provide the most error-free signal for predicting favorable conditions. Animals can ascertain the time of year, and whether winter is approaching or receding by using just 2 bits of information: the absolute day length and whether days are getting longer or shorter. Day length is encoded by the hormone melatonin, which is secreted primarily at night from the pineal gland in both diurnal and nocturnal animals (Box 41.1).

SEASONAL ENERGETIC ADAPTATIONS

Many aspects of life are energetically challenging, and populations have evolved timing mechanisms to parse out energetically demanding activities at different times of the year. For example, birds engage in predictable cycles of migration, breeding, parenting, molting, and overwintering. These energetically expensive activities generally do not overlap, and, again, all these activities elicit classical stress responses (i.e., elevated activation of the hypothalamic-pituitary-adrenal (HPA) axis).

One 1936 study is particularly informative in understanding the basis of Hans Selye's thinking about stress responses.[25] Rats were housed in low-temperature conditions for 2 days, 2 weeks, or 2 months. After 2 days in these conditions, Selye noted the hypertrophy of the adrenal cortex; reduced eosinophil (a type of immune cell circulating in the blood) numbers; and atrophy of the thymus, spleen, and lymph nodes. In other words, the rats displayed a classic stress response. After 2 weeks in the low temperatures, it was discovered that rats no longer displayed these symptoms; essentially, they had become cold adapted and could tolerate low temperatures without exhibiting any further stress responses. After approximately 2 months of exposure to low temperatures, however, the rats lost this acquired resistance to low temperatures and died.

This process of coping with adversity was termed the general adaptation syndrome (GAS), and consists of three stages: (1) the alarm reaction, (2) resistance, and finally, (3) exhaustion. During the alarm reaction, stressors are detected. Coping with the stressor occurs during the resistance stage. The exhaustion stage is characterized by the termination of the stress response and the onset of stress pathology, which in extreme cases can lead to death. Although Selye believed that the exhaustion phase was due to the termination of the stress response, current ideas about stress indicate that the pathological effects of stress are due to the prolonged activation of the response. These three stages of the GAS correspond to the response of the rats to low temperatures after 2 days (alarm reaction), 2 weeks (resistance), and 2 months (exhaustion), respectively. If individuals have the opportunity to gradually move to low temperatures or are allowed to prepare for low-temperature conditions by first being housed for 8–10 weeks in short-day lengths, then typically stress responses do not occur.[35]

BOX 41.1

PHOTOPERIOD SIGNAL TRANSDUCTION IN MAMMALS

In mammals, day length is transduced into a physiological signal by pineal melatonin. Indeed, pinealectomy blocks all photoperiod-induced changes in phenotype in most mammalian species studied thus far,[39] and exogenous treatment of pinealectomized animals with melatonin causes them to develop a phenotype similar to untreated pineal-intact animals maintained in short-day conditions (e.g., Refs. 43,40).

Light information reaches the pineal gland through a polysynaptic pathway beginning in the retina. In contrast to birds and other nonmammalian vertebrates where light can act directly on photoreceptors in the mediobasal hypothalamus, septal regions of the telencephalon, and pineal gland,[38] mammals possess no extraocular photoreceptors[41] and all photic cues are processed through the retina. Intrinsically photosensitive retinal ganglion cells (ipRGCs) are primarily involved in circadian timekeeping and the pupillary light reflex.[37] They relay the light signal along the retinohypothalamic tract to the suprachiasmatic nuclei (SCN). From there, the SCN modulate ipRGC input and send projections to the paraventricular nucleus of the hypothalamus, to the intermediolateral cell column of the spinal cord, then to the superior cervical ganglion. Sympathetic post-ganglionic noradrenergic projections innervate the pineal gland to influence melatonin secretion.[42] SCN input to the pineal gland is inhibitory, that is, light-induced activation of SCN signaling inhibits pineal melatonin secretion. Lack of light input at night eliminates this inhibitory signaling and allows for the nightly rise in melatonin. Melatonin is released directly into circulation and cerebrospinal fluid allowing for endocrine signaling throughout the organism.

Additionally, in many summer breeding species, long-day signaling causes the stimulation of thyrotrophin (TSH) in the pars tuberalis (PT) of the anterior pituitary. Ependymal cells called tanycytes express type 2 deiodinase (DIO2), which converts the prohormone thyroxine (T4) into the bioactive triodothyronine (T3). T3 stimulates gonadotropin-releasing hormone (GnRH) release from the median eminence of the hypothalamus, leading to downstream stimulation of the reproductive system through luteinizing and follicle-stimulating hormone secretion.[38] Melatonin inhibits this pathway.

SEASONAL ADJUSTMENTS IN GLUCOCORTICOID SECRETION

Glucocorticoids are released in response to stressful stimuli and can affect reproduction, digestion, growth, immunity, and virtually any other physiological processes. Predictably, seasonal variation in glucocorticoid secretion is more often the rule than the exception, probably because of the pivotal role these hormones play in orchestrating the distribution of resources among competing physiological activities.

Most animals studied to date display some seasonal variation in stress responses. For instance, Indian barnacles (*Balanus balanoides*) exhibit elevated antioxidant enzyme activity (a component of this species' stress response) prior to the monsoon season compared with during it.[20] Further, almost all wild amphibians, reptiles, and birds studied to date show marked changes in glucocorticoid concentrations during the year, with maximal values observed during the breeding season. Individuals of most mammalian species exhibit similar basal glucocorticoid concentrations year-round in the wild. However, this apparent pattern may reflect the paucity of studies on wild rodents rather than a lack of seasonality of this parameter. Indeed, rodents, such as deer mice (*Peromyscus* sp.), bred and maintained in the lab exhibit high corticosterone concentrations when kept in long-day (breeding) conditions. Other species, however, such as Siberian hamsters (*Phodopus sungorus*), display elevated corticosterone concentrations when housed in short photoperiods.

Although it is clear from these studies that stress responses are temporally labile, two questions remain: (1) is seasonal variation omnipresent within and among species (e.g., over a species' range), and (2) are other aspects of the stress response as seasonally labile as corticosteroid hormone concentrations? In terms of spatial variability, the results are mixed. Presumably, if photoperiod has significant effects on stress responses, then animals living at higher latitudes would exhibit higher circulating corticosteroid concentrations during breeding than animals at low latitudes simply because of the large variation in annual photoperiod. Redpolls (*Carduelis flammea*) are birds that live in the challenging arctic conditions of Alaska. Both males and females showed the typical marked elevation in blood corticosterone concentrations when captured for approximately 1h, indicating that they responded similarly to many other vertebrates.[32] However, the amount of the elevation in post-capture corticosterone varied across the year;

corticosterone concentrations were generally blunted during January and maximal during June when birds were breeding. Generally, birds that had the highest fat stores secreted the lowest amounts of glucocorticoids post-capture. Similar patterns have been reported for other birds living at high latitudes including the white-crowned sparrows (Zonotrichia leucophrys gambelii),[22] snow buntings (Plectrophenax nivalis), and Lapland long-spurs (Calcarius lapponicus).[33] Across latitudes, however, results are mixed. Some species show little variability in circulating glucocorticoid concentrations (e.g., pied fly-catchers, Ficedula hypoleuca, and mountain chickadees, Poecile atricapilla), whereas others (e.g., house sparrows, Passer domesticus, and white-crowned sparrows, Z. leuco-phrys) exhibit spatial variation.[30] Rufous-collared sparrows (Zonotrichia capensis) are a tropical species that live where (presumably) seasonal differences would be independent of photoperiod; nonetheless, corticosterone levels also vary seasonally, with higher concentrations observed during breeding than during molt.[30] However, in contrast to Zonotrichia species inhabiting more temperate climates, they do not show sex differences in corticosterone secretion or binding globulin capacity. Additionally, a negative relationship exists between molt duration and suppression of the HPA-axis. Species with more flexible breeding seasons (e.g., zebra finch, rock dove, red crossbill) show less HPA suppression than do more seasonal breeders. This pattern is hypothesized to reflect a life-history "trade-off" where competition exists between molting and the stress response for a limiting resource (e.g., energy, protein, behavioral repertoire). The opportunistic breeders white-plumed honeyeaters (Lichenostomus penicillatus) show no seasonal variation in baseline or stress-induced corticosterone concentrations and are able to maintain "breeding levels" of corticosterone during molt.[5] This suggests that corticosterone suppression is not necessary for forming high-quality feathers. In common with spatial variability, the social system of animals can also affect seasonal variation in glucocorticoid responses. A study of greylag geese (Anser anser) revealed that social status and physiological stress are not related in the same way in all seasons of the year.[14]

Fewer studies on seasonal variation in glucocorticoid concentrations in wild mammals have been conducted; thus, generalities have yet to emerge. In wild male red-backed voles (Myodes rutilus), total corticosterone plasma concentrations vary seasonally between the reproductive and nonreproductive seasons.[9] In degus (Octodon degus), a diurnal and social mammal native to Chile, capture-induced elevations in glucocorticoids are higher during the breeding than the nonbreeding season.[21] However, another study on wild degus reported seasonal variation in stress-induced glucocorticoid concentrations, and a sexual dimorphism existed such that males displayed the highest response during the nonbreeding season and females during late gestation.[1] In tuco-tucos (Ctenomys talarum), cortisol, but not corticosterone, values seem to vary seasonally, and stressors also had mixed effects on these two glucocorticoids.[29]

In addition to hormones, other aspects of the gluco-corticoid stress response show significant spatial and temporal flexibility. In particular, the HPA axis appears to be substantially malleable among wild songbirds. Individuals of different species can apparently modu-late the levels of glucocorticoids they produce at dif-ferent times of the year by changing the sensitivity of adrenergic tissues to upstream regulatory hormones such as corticotropin-releasing hormone or adreno-corticotropic hormone (ACTH). For example, Japanese quail dampen their glucocorticoid response to ACTH in response to short days.[12] Alternatively, some species are capable of modulating steroid-binding globulin levels in their blood across the year instead, although it remains unclear whether these globulins promote or inhibit the actions of the glucocorticoids to which they bind. One recent perspective argues for limited emphasis on ste-roid binding globulin capacities, because other blood-borne factors bind them, binding capacity can vary with local enzymatic activity and tissue temperature, and free and total glucocorticoids operate over different time-frames.[23] Finally, glucocorticoid receptors also appear to be seasonally variable. In house sparrows, intracellular glucocorticoid and mineralocorticoid receptor expres-sion in the hippocampus and spleen (but not skin) varies seasonally.[15] Splenic and thymic glucocorticoid receptor expression also varies relative to melatonin concentra-tion in Northern palm squirrels (Funambulus pennanti).[11] Glucocorticoid receptor expression is also increased in the hippocampus,[31] but not amygdala or prefrontal cor-tex,[24] with no concurrent changes in cortisol concentra-tions of Siberian hamsters following short-day exposure.

WHY ARE STRESS RESPONSES SEASONALLY VARIABLE?

Presumably, stress responses help animals prepare for or recover from stressors. Why, then, are stress responses variable during the year when all individuals could benefit from optimal stress responses? Several hypoth-eses have been proposed. First, glucocorticoid stress responses are expensive and hence may not be afford-able at all times of the year. Indeed, pyruvate, a readily usable energy source, blocked the stress-evoked sup-pression of cell-mediated and humoral immune activ-ity in mice (Mus musculus).[18] Second, stress responses may be more important at some times of year than oth-ers, such as during breeding or other activities in which abundant self-antigens will be revealed to the immune system. Indeed, glucocorticoids have been argued to have evolved immunosuppressive effects to prevent misdirected attacks against self in times of adversity. Third, high corticosterone concentrations, particularly when maintained for extended periods (i.e., chronic), can be immunosuppressive; over the short term, however, glucocorticoids can be immunoenhancing. In general, winter survival in small animals is expected to require a positive balance between short-day enhancement of immune defenses and glucocorticoid-induced immunosuppression. Immunosuppression caused by high glucocorticoids may be induced by many winter-related factors, including overcrowding, high competition for scarce resources, low temperatures, low food availability, high predator pressure, and lack of shelter. Overall, the balance between enhanced immune function (i.e., to the point where autoimmune disease becomes a danger) and stress-induced immunosuppression (i.e., to the point where opportunistic pathogens and parasites overwhelm the host) must be met for animals to survive and become reproductively successful.

Several studies support this hypothetical construct. For instance, hamsters maintained in short photoperiods (and exposed to a restraint stressor) showed more rapid wound healing[13] and trafficking of leukocytes to the skin compared to unrestrained short-day animals (or restrained and unrestrained long-day hamsters).[2] In one passerine bird, cutaneous immune responses to a T-cell mitogen (phytohemagglutinin) were less dramatic in animals induced into breeding than in those in a nonbreeding condition.[16] Moreover, corticosterone appears to be immunosuppressive in the same species only in winter (not summer) conditions. Such immune–endocrine interrelationships are not always uniform between the sexes within a species. For example, acute stress enhances cell-mediated immune activity (delayed-type hypersensitivity to dinitrofluorobenzene) in female, but not male, Siberian hamsters.[3] Indeed, ample evidence indicates that immune activity can vary dramatically between sexes; thus, it is plausible that the glucocorticoid mediators of immune defenses might be labile as well.

As elevated glucocorticoids might be a reliable indicator of how organisms deal with environmental perturbations (i.e., allostatic load), the "cort-fitness hypothesis" was introduced. The hypothesis suggests that under identical environmental conditions, individuals of poor quality (i.e., low fitness) should display high baseline glucocorticoid concentrations. Further expanding on this hypothesis, Boonstra and colleagues coalesced much of previous research into three primary hypotheses for glucocorticoid variation across reproductive and nonreproductive seasons.[9] The first of these is the "enabling hypothesis," which predicts that glucocorticoids enable energetic investment into reproduction. In contrast, the "inhibitory hypothesis" focuses on potential detrimental effects of glucocorticoids on reproduction, where elevated concentrations could inhibit the HPA axis and suppress reproduction. A more nuanced class of hypotheses put seasonal regulation of glucocorticoids into the context of life-history theory, where investment in reproduction reduces survival. Termed the "cost of reproduction hypothesis," it posits that changes in glucocorticoids across reproductive and nonreproductive seasons may reflect a decline in the condition of the HPA axis resulting from reproductive investment. There is ample support for the "enabling hypothesis" in wild birds, reptiles, and amphibians as glucocorticoid concentrations are often higher during the reproductive season than during the nonreproductive season. However, in mammals, a generalizable trend is not apparent, with mixed results supporting all three of the aforementioned hypotheses. One recent study supports the inhibitory hypothesis in a long-lived avian species (*Rissa tridactyla*); females treated with subcutaneous corticosterone fledged fewer chicks, although this study did not account for potential negative feedback mechanisms influencing HPG function.[19] A generalizable pattern may not emerge unless we begin to synthesize data across multiple time-scales, from homeostatic (short) to developmental and predictive (long) adaptive responses. As Woods and Wilson write, "Fitness, not stability, is the most important currency for understanding the processes shaping physiological systems over evolutionary time."[34] Applying this perspective to seasonal variation in glucocorticoid responses will facilitate an understanding of how and why the HPA axis is seasonally regulated in wild organisms.

One other hormone that is integrally involved in vertebrate photoperiodism, melatonin, can affect the seasonal interplay between corticosterone and immune function. Generally, melatonin enhances immune activity,[6] whereas glucocorticoids (at chronically high levels) compromise immune activity.[26] Under certain circumstances, melatonin treatment may ameliorate the immunosuppressive effects of glucocorticoids. For instance, Siberian hamsters attenuated the duration of fever during simulated short winter day lengths, presumably to conserve energy.[10] Hamsters housed in long days were injected with saline or melatonin before lights off each day for 1 or 6 weeks; then fever was assessed following injections of bacterial lipopolysaccharide. The fever duration was attenuated (32%) only in hamsters that decreased body mass, increased glucocorticoid concentrations, and displayed gonadal regression in response to 6 weeks of melatonin.[4] These results suggest that long-term exposure to long-duration melatonin signals induce the physiological changes required for short-day immune responses. Indeed, melatonin and glucocorticoids have reciprocal effects on each other's receptor expression; melatonin downregulates glucocorticoid receptor expression while cortisol downregulates melatonin receptor 1 and 2 expression in the spleen and thymus of the Northern palm squirrel (*Funambulus pennanti*).[11] These physiological changes may also be achieved via interactions among hormones with secondary photoperiodic roles, such as cortisol, leptin, and thyroid hormones. Leptin, a physiologically salient cue of energy availability, reverses the effects of short-day–induced decreased food intake and serum anti-KLH IgG in Siberian hamsters.[8] Thyroid hormone (TH) signaling regulates photoperiodic changes in reproduction[36] and immunity[28] through seasonal alterations in expression of iodothyronine deiodinases, enzymes responsible for the conversion between active and inactive forms of TH[27] (Box 41.2).

In addition to their effects on immune function, glucocorticoids can dramatically affect behavior, particularly behaviors associated with reproduction. Generally, stress responses are up- and downregulated across seasons, but the way in which this happens depends on the social and physical environments in which species live. For instance, songbirds living at high latitudes are thought to mount weak corticosterone responses to stress because the abandonment of offspring that it might lead to would be maladaptive given that these species typically only have one opportunity per year to breed.[17] In the tropics, however, corticosterone–behavior relationships may be different because reproductive efforts are generally less constrained by time.

Considered together, both the incidence and responses to stressors vary on a seasonal basis. The environmental regulation of seasonal changes in many

---

BOX 41.2

PHOTOPERIODISM

Photoperiodism is the ability of plants and animals to measure environmental day length (photoperiod), typically by monitoring night length. This process underlies a biological calendar. Photoperiod is inversely correlated to the nightly duration of melatonin secretion in animals. Melatonin is secreted from the pineal gland exclusively during the dark of night, thus serving as an endogenous signal of night length. The biological ability to measure photoperiod permits organisms to ascertain the time of year and develop seasonally appropriate physiological and behavioral adaptations. Although the specific mechanisms that underlie the ability to measure day length differ among taxa, individuals that respond to day length can precisely, and reliably, ascertain the time of year with just 2 bits of data: (1) the length of the daily photoperiod, and (2) whether day lengths are increasing or decreasing. Because the same photoperiod occurs twice a year (i.e., 21 March and 21 September), animals must be able to discriminate between these two dates; many photoperiodic vertebrates have solved this problem by developing an annual alteration between two physiological states termed photoresponsiveness and photorefractoriness.[53] For example, photoperiodic rodents born in the spring will grow to adult size, undergo puberty, and become reproductive in 6–8 weeks, whereas a sibling born in the autumn will not grow or undergo puberty for 4–5 months. For individuals of many species, the annual change in day length provides a reliable environmental cue for time of year, allowing animals to anticipate predictable seasonal stressors such as low food availability, increased pathogen prevalence, and decreased temperature experienced in winter. Such

environmental changes are considered ultimate factors that directly affect an individual's survival.[46] The proximate factor of day length, although not directly affecting survival, serves as a reliable anticipatory cue for these seasonal changes in ultimate factors.

Photoperiodic responses have been identified in virtually all organisms that experience seasonal environmental changes: first in plants,[50] followed by birds,[51] and finally mammals.[45,47] Because seasonal breeding cycles are perhaps the most salient seasonal rhythm, the research emphasis was initially focused on seasonal changes in mammalian reproductive function, but other factors such as immune function, metabolism, and brain function vary seasonally and are now active areas of research.

Removal of the pineal gland eliminates photoperiodic responses but does not alter circadian entrainment. Administering melatonin to pinealectomized mammals for short (<6 h/day) durations is sufficient to elicit long-day physiological responses; stimulating reproduction in long-day breeders (Siberian hamsters[49]) and suppressing reproduction in short-day breeders (sheep[48]). Conversely, long (>8 h/day) duration melatonin infusions induce short-day physiological responses. Melatonin receptors are present in high density on a variety of brain regions (pars tuberalis, SCN, and thalamic nuclei[55]). Some seasonal gating mechanisms are exerted through direct binding to melatonin receptors[44] in a system and species-dependent manner. But melatonin may also act indirectly by seasonally altering release of other hormones, such as thyroid hormones that play a key role in regulating both immune and reproductive function in Siberian hamsters.[52,54]

traits, including reproduction, metabolism, and immunity, is primarily mediated by photoperiod. From a physiological perspective, the pineal hormone melatonin appears to coordinate seasonal adjustments in stress responses including changes in reproductive, metabolic, and immune function. To what extent the immune-enhancing effects of melatonin are unique to seasonally breeding rodents and whether it is a generalized feature of seasonal adjustments in stress responses among humans remain largely unspecified. Although physiological responses are important mediators of seasonal and photoperiodic changes in stress responses, behavioral alterations might also have an important role in reducing their energetic costs. The careful integration of the behavioral and physiological mechanisms underlying the seasonality of stress responsiveness at both ultimate and proximate levels should provide important and novel insights into the interaction among seasonal environmental factors, stressors, immune function, and the pattern of diseases and mortality. To date, most efforts have not advanced much beyond describing the seasonal changes that occur in responses to stressors. Indeed, a recent review concluded that a consensus does not exist for the endocrine profile of chronically stressed wild animals.[7] The next step is to identify the cellular and molecular mechanisms that mediate the effects of season and photoperiod on stress responsiveness and link variability in these mechanisms to fitness.

Glossary

Glucocorticoids A class of steroid hormones secreted from the adrenal cortices that serves to promote energy use. Corticosterone is the glucocorticoid usually secreted by rodents and birds, whereas cortisol is typically secreted by primates.

Melatonin An ancient and pleiotropic indoleamine hormone secreted at night from the pineal gland in both diurnal and nocturnal species. It is important in the regulation of daily and seasonal biological rhythms. Extrapineal melatonin is produced in the retina, skin, gut, and immune competent cells.

Photoperiod Day length or the amount of light per day.

Photoperiodism The response to the length of day and night, including the ability to determine day length (usually by measuring night length) in both plants and animals; first proposed by botanists Garner and Allard in 1922.

Pineal gland An endocrine gland that secretes melatonin, usually located between the telencephalon and diencephalon in vertebrates, also called the epiphysis.

References

1. Bauer CM, Hayes LD, Ebensperger LA, Romero LM. Seasonal variation in the degu (*Octodon degus*) endocrine stress response. *Gen Comp Endocrinol*. 2014;197:26–32.
2. Bilbo SD, Dhabhar FS, Viswanathan K, Saul A, Yellon SM, Nelson RJ. Short day lengths augment stress-induced leukocyte trafficking and stress-induced enhancement of skin immune function. *Proc Natl Acad Sci USA*. 2002;99:4067–4072.
3. Bilbo SD, Nelson RJ. Sex differences in photoperiodic and stress-induced enhancement of immune function in Siberian hamsters. *Brain Behav Immun*. 2003;17:462–472.
4. Bilbo SD, Nelson RJ. Melatonin regulates energy balance and attenuates fever in Siberian hamsters. *Endocrinology*. 2013;143:2527–2533.
5. Buttemer WA, Addison BA, Astheimer LB. Lack of seasonal and moult-related stress modulation in an opportunistically breeding bird: the white-plumed honeyeater (*Lichenostomus penicillatus*). *Hormones Behav*. 2015;76:34–40.
6. Carrillo-Vico A, Lardone PJ, Alvarez-Sánchez N, Rodríguez-Rodríguez A, Guerrero JM. Melatonin: buffering the immune system. *Int J Mol Sci*. 2013;14:8638–8683.
7. Dickens MJ, Romero LM. A consensus endocrine profile for chronically stressed wild animals does not exist. *Gen Comp Endocrinol*. 2013;191:177–189.
8. Drazen DL, Demas GE, Nelson RJ. Leptin effects on immune function and energy balance are photoperiod dependent in Siberian hamsters (*Phodopus sungorus*). *Endocrinology*. 2013;142:2768–2775.
9. Fletcher QE, Dantzer B, Boonstra R. The impact of reproduction on the stress axis of free-living male northern red backed voles (*Myodes rutilus*). *Gen Comp Endocrinol*. 2015;224:136–147.
10. Fonken LK, Bedrosian TA, Michaels HD, Weil ZM, Nelson RJ. Short photoperiods attenuate central responses to an inflammogen. *Brain Behav Immun*. 2012;26:617–622.
11. Gupta S, Haldar C. Physiological crosstalk between melatonin and glucocorticoid receptor modulates T-cell mediated immune responses in a wild tropical rodent, *Funambulus pennanti*. *J Steroid Biochem Mol Biol*. 2013;134:23–36.
12. Hazard D, Couty M, Faure JM, Guemene D. Daily and photoperiod variations of hypothalamic-pituitary-adrenal axis responsiveness in Japanese quail selected for short or long tonic immobility. *Poult Sci*. 2005;84:1920–1925.
13. Kinsey SG, Prendergast BJ, Nelson RJ. Photoperiod and stress affect wound healing in Siberian hamsters. *Physiol Behav*. 2003;78:205–211.
14. Kotrschal K, Hirschenhauser K, Mostl E. The relationship between social stress and dominance is seasonal in greylag geese. *Anim Behav*. 1998;55:171–176.
15. Lattin CR, Waldron-Francis K, Romero LM. Intracellular glucocorticoid receptors in spleen, but not skin, vary seasonally in wild house sparrows (*Passer domesticus*). *Proc Biol Sci*. 2013;280:20123033.
16. Martin Ii LB, Gilliam J, Han P, Lee K, Wikelski M. Corticosterone suppresses cutaneous immune function in temperate but not tropical House Sparrows, *Passer domesticus*. *Gen Comp Endocrinol*. 2005;140:126–135.
17. Meddle SL, Owen-Ashley NT, Richardson MI, Wingfield JC. Modulation of the hypothalamic-pituitary-adrenal axis of an Arctic-breeding polygynandrous songbird, the Smith's longspur, *Calcarius pictus*. *Proc Biol Sci*. 2003;270:1849–1856.
18. Neigh GN, Bowers SL, Pyter LM, Gatien ML, Nelson RJ. Pyruvate prevents restraint-induced immunosuppression via alterations in glucocorticoid responses. *Endocrinology*. 2013;145:4309–4319.
19. Nelson BF, Daunt F, Monaghan P, et al. Protracted treatment with corticosterone reduces breeding success in a long-lived bird. *Gen Comp Endocrinol*. 2015;210:38–45.
20. Niyogi S, Biswas S, Sarker S, Datta A. Seasonal variation of antioxidant and biotransformation enzymes in barnacle, *Balanus balanoides*, and their relation with polyaromatic hydrocarbons. *Mar Environ Res*. 2001;52:13–26.
21. Quispe R, Villavicencio CP, Addis E, Wingfield JC, Vasquez RA. Seasonal variations of basal cortisol and high stress response to captivity in *Octodon degus*, a mammalian model species. *Gen Comp Endocrinol*. 2014;197:65–72.
22. Romero LM, Wingfield JC. Seasonal changes in adrenal sensitivity alter corticosterone levels in Gambel's white-crowned sparrows (*Zonotrichia leucophrys gambelii*). *Comp Biochem and Physiol C Pharmacol Toxicol Endocrinol*. 1998;119:31–36.

23. Schoech SJ, Romero LM, Moore IT, Bonier F. Constraints, concerns and considerations about the necessity of estimating free glucocorticoid concentrations for field endocrine studies. *Funct Ecol.* 2013;27:1100–1106.

24. Scotti M-AL, Rendon NM, Greives TJ, Romeo RD, Demas GE. Short-day aggression is independent of changes in cortisol or glucocorticoid receptors in male Siberian hamsters (*Phodopus sungorus*). *J Exp Zool A Ecol Genet Physiol.* 2015;323:331–342.

25. Selye H. A syndrome produced by diverse nocuous agents. *Nature.* 1936;138:32.

26. Sorrells SF, Caso JR, Munhoz CD, Sapolsky RM. The stressed CNS: when glucocorticoids aggravate inflammation. *Neuron.* 2009;64:33–39.

27. St Germain DL. Iodothyronine deiodinase. *Trends Endocrinol Metab.* 1994;5:36–42.

28. Stevenson TJ, Onishi KG, Bradley SP, Prendergast BJ. Cell-autonomous iodothyronine deiodinase expression mediates seasonal plasticity in immune function. *Brain Behav Immun.* 2014;36:61–70.

29. Vera F, Antenucci CD, Zenuto RR. Cortisol and corticosterone exhibit different seasonal variation and responses to acute stress and captivity in tuco-tucos (*Ctenomys talarum*). *Gen Comp Endocrinol.* 2011;170:550–557.

30. Wada H, Moore IT, Breuner CW, Wingfield JC. Stress responses in tropical sparrows: comparing tropical and temperate *Zonotrichia.* *Physiol Biochem Zool.* 2006;79:784–792.

31. Walton JC, Grier AJ, Weil ZM, Nelson RJ. Photoperiod and stress regulation of corticosteroid receptor, brain-derived neurotrophic factor, and glucose transporter GLUT3 mRNA in the hippocampus of male Siberian hamsters (*Phodopus sungorus*). *Neuroscience.* 2012;213:106–111.

32. Wingfield JC, Deviche P, Sharbaugh S, et al. Seasonal changes of the adrenocortical responses to stress in redpolls, *Acanthis flammea*, in Alaska. *J Exp Zool.* 1994;270:372–380.

33. Wingfield JC, Suydam R, Hunt K. The adrenocortical responses to stress in snow buntings (*Plectrophenax nivalis*) and Lapland longspurs (*Calcarius lapponicus*) at Barrow, Alaska. *Comp Biochem Physiol C Pharmacol Toxicol Endocrinol.* 1994;108:299–306.

34. Woods AH, Wilson KJ. An elephant in the fog: unifying concepts of physiological stasis and change. In: *Integrative Organismal Biology.* John Wiley & Sons, Inc.; 2014:119–135.

35. Yellon SM, Kim K, Hadley AR, Tran LT. Time course and role of the pineal gland in photoperiod control of innate immune cell functions in male Siberian hamsters. *J Neuroimmunol.* 2005;161: 137–144.

36. Yoshimura T. Thyroid hormone and seasonal regulation of reproduction. *Front Neuroendocrinol.* 2013;34:157–166.

37. Berson DM. Strange vision: ganglion cells as circadian photoreceptors. *Trends Neurosci.* 2003;26:314–320.

38. García-Fernández JM, Cernuda-Cernuda R, Davies WI, et al. The hypothalamic photoreceptors regulating seasonal reproduction in birds: a prime role for VA opsin. *Front Neuroendocrinol.* 2015;37:13–28.

39. Hazlerigg DG, Wagner GC. Seasonal photoperiodism in vertebrates: from coincidence to amplitude. *Trends Endocrinol Metab.* 2006;17:83–91.

40. Hiebert SM, Green SA, Yellon SM. Daily timed melatonin feedings mimic effects of short days on testis regression and cortisol in circulation in Siberian hamsters. *Gen Comp Endocrinol.* 2006;146:211–216.

41. Nelson RJ, Zucker I. Absence of extraocular photoreception in diurnal and nocturnal rodents exposed to direct sunlight. *Comp Biochem Physiol A Physiol.* 1981;69:145–148.

42. Teclemariam-Mesbah R, Ter Horst GJ, Postema F, Wortel J, Buijs RM. Anatomical demonstration of the suprachiasmatic nucleus-pineal pathway. *J Comp Neurol.* 1999;406:171–182.

43. Walton JC, Chen Z, Travers JB, Nelson RJ. Exogenous melatonin reproduces the effects of short day lengths on hippocampal function in male white-footed mice, *Peromyscus leucopus. Neuroscience.* 2013;248:403–413.

44. Badura LL, Goldman BD. .Central sites mediating reproductive responses to melatonin in juvenile male Siberian hamsters. *Brain Res.* 1992;598:98–106.

45. Baker JR, Ranson RM. Factors affecting the breeding of the field mouse (*Microtus agrestis*). Part I. Light. *Proc R Soc Lond Ser B.* 1932;110:313–323.

46. Baker JR. The evolution of breeding seasons. In: DeBeer GB, ed. *Evolution: Essays on Aspects of Evolutionary Biology.* Oxford, UK: Clarendon Press; 1938:161–177.

47. Bissonnette TH. Modification of mammalian seasonal cycles: Reactions of ferrets (*Putorius vulgaris*) of both sexes to electric light added after dark in November and December. *Proc R Soc B Biol Sci.* 1932;110:322–336.

48. Bittman EL, Karsch FJ. Nightly duration of pineal melatonin secretion determines reproductive response to inhibitory day length in the ewe. *Biol Reprod.* 1984;40:118–126.

49. Carter DS, Goldman BD. Antigonadal effects of timed melatonin infusion in pinealectomized male Djungarian hamsters (*Phodopus sungorus sungorus*): duration is the critical parameter. *Endocrinology.* 1983;113:1261–1267.

50. Garner WW, Allard HA. Effect of relative length of day and night and other factors of the environment on growth and reproduction in plants. *J Agric Res.* 1920;18:553–606.

51. Rowan W. Relation of light to bird migration and developmental changes. *Nature.* 1925;115:494–495.

52. Stevenson TJ, Prendergast BJ. Photoperiodic time measurement and seasonal immunological plasticity. *Front Endocrinol.* 2015;37:76–88.

53. Prendergast BJ, Zucker I, Nelson RJ. Seasonal rhythms of mammalian behavioral neuroendocrinology. In: Pfaff D, Arnold A, Etgen A, Fahrbach S, Moss R, Rubin R, eds. *Hormones, Brain, and Behavior.* 2nd ed. San Diego: Academic Press; 2009:507–538.

54. Prendergast BJ, Pyter LM, Kampf-Lassin A, Patel PN, Stevenson TJ. Rapid induction of hypothalamic iodothyronine deiodinase expression by photoperiod and melatonin in juvenile Siberian hamsters (*Phodopus sungorus*). *Endocrinology.* 2012;154:831–841.

55. Weaver DR, Provencio I, Carlson LL, Reppert SM. Melatonin receptors and signal transduction in photorefractory Siberian hamsters (*Phodopus sungorus*). *Endocrinology.* 1991;138:1086–1092.

Further Reading

1. Breuner CW, Orchinik M. Downstream from corticosterone: seasonality of binding globulins, receptors, and behavior in the avian stress response. In: Dawson A, ed. *Avian Endocrinology.* New Delhi: Narosa Publishing; 2002:385–399.

2. Breuner CW, Orchinik M, Hahn TP, et al. Differential mechanisms for regulation of the stress response across latitudinal gradients. *Am J Physiol.* 2003;285:R594–R600.

3. Breuner CW, Delehanty B, Boonstra R. Evaluating stress in natural populations of vertebrates: total CORT is not good enough. *Funct Ecol.* 2013;27:24–36.

4. Cohen AA, Martin LB, McWilliams SR, Wingfield JC, Dunne JA. Physiological regulatory networks: ecological roles and evolutionary constraints. *Trends Ecol Evol.* 2012;27:428–435.

5. Liebl AL, Shimizu T, Martin LB. Covariation among glucocorticoid regulatory elements varies seasonally in house sparrows (Passer domesticus). *Gen Comp Endocrinol.* 2013;183:32–37.

6. Nelson RJ, Demas GE, Klein SL, Kriegsfeld LJ. *Seasonal Patterns of Stress, Immune Function, and Disease.* New York: Cambridge University Press; 2002.

7. Nelson RJ. Seasonal immune function and disease responses. *Trends Immunol.* 2004;24:187–192.

8. Romero LM, Platts SH, Schoech SJ, et al. Understanding stress in the healthy animal – potential paths for progress. *Stress.* 2015;18:491–497.

9. Romero LM. Seasonal changes in plasma glucocorticoid concentrations in free-living vertebrates. *Gen Comp Endocrinol.* 2002;128:1–24.

10. Romero LM. Physiological stress in ecology: lessons from biomedical research. *Trends Ecol Evol.* 2004;19:249–255.

11. Wingfield JC, Romero LM. *Tempests, Poxes, Predators, and People: Stress in Wild Animals and How They Cope.* New York: Oxford University Press; 2016.

Seasonal Rhythms

L.M. Romero, C.M. Bauer, R. de Bruijn, C.R. Lattin

Tufts University, Medford, MA, United States

Abstract

Many free-living species show seasonal rhythms in baseline and stress-induced glucocorticoid release. Glucocorticoid titers, as well as glucocorticoid-binding proteins (CBG) and intracellular glucocorticoid receptors, vary with different life history stages over the course of the year. Concentrations are often highest during breeding. Seasonal glucocorticoid rhythms appear to be regulated by a mixture of hypothalamic, pituitary, adrenal, and glucocorticoid feedback mechanisms. Three hypotheses have been proposed to explain why seasonal glucocorticoid rhythms exist: to mediate seasonally different energetic needs, to initiate seasonally appropriate stress-induced behaviors, and/or to prepare for subsequent stressors whose frequency varies seasonally. Future work distinguishing between these hypotheses will help us understand how seasonal glucocorticoid rhythms likely aid the short-term survival of free-living animals. Finally, biomedical studies that ignore seasonal variation in glucocorticoid responses miss the rich variation in these responses that likely affect humans as well.

EVIDENCE FOR SEASONAL GLUCOCORTICOID RHYTHMS

Glucocorticoid secretion in response to noxious stimuli is so common that it is often used as a working definition of what stimuli are stressors. In a laboratory setting, the magnitude of glucocorticoid secretion to stressors applied in similar circadian, developmental, and psychological contexts within the same species is remarkably consistent despite individual variation. In the majority of free-living species studied to date, however, glucocorticoid release is modulated seasonally.[23] In other words, the magnitudes of both basal and stress-induced glucocorticoid concentrations vary throughout the year. An example, for white-crowned sparrows (*Zonotrichia leucophrys*), is presented in Fig. 42.1.

In studies of free-living wild animals, glucocorticoids are measured by directly sampling from freshly caught individuals. Because glucocorticoids start to increase 2–3 min after initiation of a stressor, plasma samples collected within a few minutes of capture are assumed to reflect prestress, or baseline, glucocorticoid concentrations. These initial samples are usually referred to as "baseline" rather than "basal" because the immediate past history of a wild animal is rarely known, thereby making a true basal sample, as defined in laboratory

Stress: Neuroendocrinology and Neurobiology
http://dx.doi.org/10.1016/B978-0-12-802175-0.00042-5

studies, impossible. Stress-induced glucocorticoid concentrations then reflect the animal's response to capture, handling, captivity, etc. and are assumed to be a measure of the hypothalamic-pituitary-adrenal (HPA) axis's ability to respond to a stressor. This "capture-stress" protocol can then be used to compare glucocorticoid responses across individuals, species, and seasons.[33]

These studies provide clear evidence that a majority of reptilian, amphibian, avian, and mammalian species seasonally modulate glucocorticoid concentrations.[23] Males and females of the same species sometimes differ in whether they show seasonal rhythms, and occasionally a species will show a seasonal rhythm in baseline but not stress-induced glucocorticoid concentrations (or visa versa). However, approximately 75% of all species that have been studied have at least one sex with a seasonal rhythm of glucocorticoid release.

The evidence for an annual glucocorticoid rhythm is generally much stronger for baseline than stress-induced concentrations, primarily because more studies have focused on this time point. Furthermore, avian species often show a seasonal rhythm in baseline but not stress-induced glucocorticoid concentrations, suggesting that baseline and stress-induced glucocorticoid concentrations are regulated differently. Laboratory studies indicate that baseline and stress-induced glucocorticoids have different regulatory mechanisms and serve different physiological functions that are mediated by separate receptors (e.g., Ref. 6). It is not

surprising, therefore, that the HPA axis might be regulated differently under baseline and stress conditions and that these separate regulatory pathways can show different annual cycles.

Of those species that do show a seasonal rhythm, the seasonal peak is usually during the breeding period. There are three notable caveats. First, the most robust rhythm in birds is a nadir during the prebasic molt when they replace their feathers. The effect is often very pronounced, with many animals essentially failing to secrete glucocorticoids during molt (Fig. 42.1). This is thought to protect the bird from the protein catabolism effects of glucocorticoids during the protein mobilization of feather replacement.[7] The current hypothesis is that the downside for long-term survival of glucocorticoid-induced degradation of growing feathers outweighs any short-term survival benefits of secreting glucocorticoids in response to an acute stressor. Second, desert-breeding birds show a seasonal glucocorticoid peak concurrent to the rainy season, suggesting that seasonal rhythms can be timed to seasonal environmental cues as well as seasonal physiological cues (e.g., Refs. 8,35). Third, even though most mammal species show a seasonal rhythm, there is no consensus season when mammals tend to have elevated glucocorticoid concentrations.[23] This may result from the often-dissociated timing of mate selection, copulation, pregnancy, and lactation. A systematic study of the impact of these different life history periods on seasonal glucocorticoid release has not yet been attempted for wild free-living mammals.

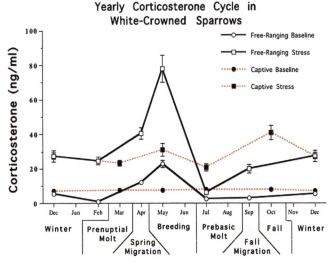

FIGURE 42.1 Seasonal rhythm of baseline and stress-induced corticosterone from free-living and captive white-crowned sparrows. *Reprinted with permission from Romero LM, Wingfield JC. Alterations in hypothalamic-pituitary-adrenal function associated with captivity in Gambel's white-crowned sparrows* (Zonotrichia leucophrys gambelii). Comp Biochem Physiol. *1999;122B:13–20.*

KEY POINTS

- There is increasing evidence that glucocorticoids, and their regulatory system the HPA axis, show annual rhythms in free-living species.

- Glucocorticoids are thought to modulate the response to environmental changes, and as such are expected to adapt to seasonal changes in energy needs, behavioral responses, and exposure to stressors.

- To further our understanding of how glucocorticoids help animals survive, there is a need for a cohesive hypothesis that integrates the different aspects of glucocorticoid physiology, and has explanatory power across all life history stages.

- Seasonal modulation of glucocorticoids often disappears in captive and domesticated animals, which is important to consider when working with laboratory models of stress.

Many wild species held in long-term captivity also show seasonal rhythms, but these rhythms do not usually match what is found in free-living animals (e.g., Ref. 9). This likely reflects that both domestication and captivity alter HPA axis function in ways that are different for each species, making it difficult, if not impossible, to ascertain natural annual variation from these populations.

EVIDENCE FOR SEASONAL RHYTHMS DOWNSTREAM OF GLUCOCORTICOID RELEASE

Glucocorticoid release is only one aspect of glucocorticoid physiology. After release, the steroids must be transported to the target tissues and bind to the appropriate receptors in those tissues. The impact of glucocorticoids on the animal is an integration of these three steps (release, transport, receptor binding), as well as others.[32] Seasonal changes in glucocorticoid-binding globulins (CBGs), mineralocorticoid receptors (MRs), and/or glucocorticoid receptors (GRs), could potentially augment or counteract any seasonal variation in glucocorticoid titers. Current evidence suggests that these aspects of glucocorticoid function do indeed vary seasonally, but the overall impact of these changes is not yet clear.

The capacity of CBG has been found to vary seasonally and to be regulated by gonadal testosterone.[4] Free steroids, the glucocorticoids unbound to CBG, are able to diffuse across cellular membranes and bind to intracellular receptors and thereby initiate physiological changes. This implies that free steroid concentrations may be more biologically relevant than total steroid concentrations.[2] CBG titers often increase or decrease in parallel to glucocorticoid titers, so that when total and free steroid concentrations are compared, the seasonal rhythm in total glucocorticoids often changes or disappears entirely.[2] On the other hand, CBG may act as a long-term reservoir that can extend glucocorticoid availability.[18,30] If CBG does primarily act as a reservoir, increases in CBG that parallel glucocorticoid increases would thereby augment glucocorticoid access to receptors. Current research is attempting to distinguish between these possibilities.

There are also seasonal changes in the density of GR and MR in different target tissues, although when the peak in receptor density occurs appears to be tissue-, species-, and even population-specific.[3,12,16] Receptor concentrations do not track glucocorticoid titers in a straightforward manner—that is, receptor density is not merely low when hormone titers are high, nor are receptor concentrations always high when plasma hormone levels are high. One set of studies examined seasonal patterns in GR and MR binding in 13 different tissues of wild-caught house sparrows and found only two overall patterns across different tissue types.[15–17] First, GR was more likely to show seasonal modulation than MR. Second, receptor concentrations in several tissues peaked immediately prior to breeding. Increased or decreased receptor density during certain times of the year may be due to animals' needs to augment or moderate glucocorticoid effects on specific tissues during different life history stages. Studies in a wider range of free-living animal species should help clarify how seasonal modulation of downstream components of the HPA axis alters the effects of seasonally varying plasma hormones.

MECHANISMS REGULATING ANNUAL RHYTHMS OF GLUCOCORTICOIDS

Progress is just beginning in understanding how HPA axis function is regulated seasonally. What induces these changes, be it photoperiod, temperature, food availability, etc., is currently unknown. Plasma glucocorticoid concentrations often are correlated with adrenal mass, and early studies showed increases in adrenal mass during breeding (e.g., Ref. 31). Consequently, the ability of adrenal tissue to respond to ACTH changes seasonally. A few studies in birds indicate that there are multiple regulatory points in the HPA axis. Some species seasonally regulate glucocorticoid release from the adrenal, others seasonally regulate ACTH release from the pituitary, and still others regulate CRH and arginine vasotocin (the avian congener of arginine vasopressin) release from the hypothalamus.[23] There is also some evidence that the efficacy of negative feedback changes seasonally.[14] These regulatory control points may be related to how sensitive each species is to adverse environmental conditions.[26]

Glucocorticoids also interact with the gonadal system. Although glucocorticoids inhibit gonadal hormone release, exogenous testosterone elevates glucocorticoids in at least one free-living bird species.[29] This suggests a complex interaction between the gonadal and adrenal systems, with glucocorticoids downregulating gonadal androgen release concurrent with gonadal androgens elevating glucocorticoid release. Gonadal androgens, therefore, could potentially be an important physiological regulator of seasonal glucocorticoid rhythms.

WHY DO SEASONAL GLUCOCORTICOID RHYTHMS EXIST?

Modulation of glucocorticoid release presents us with an interesting paradox: if glucocorticoid release is so important for survival, then how do animals survive stressors at certain times of the year when they essentially fail to release glucocorticoids (see Fig. 42.1)? Three major hypotheses have been proposed to explain seasonal glucocorticoid rhythms (Fig. 42.2).

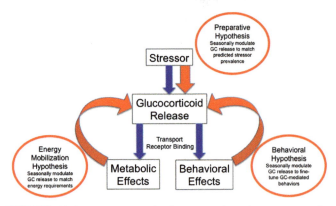

FIGURE 42.2 Summary of the three major hypotheses proposed to explain why glucocorticoids vary seasonally. The standard response to a stressor is depicted in blue and the proposed regulatory mechanisms are depicted in red. *GC*, glucocorticoids.

1. The Energy Mobilization Hypothesis focuses on the metabolic effects of glucocorticoids. Glucocorticoids play an important role in energy mobilization, especially during chronic stress,[28] so glucocorticoid concentrations should be highest during energetically costly times of the year. Under this hypothesis, high glucocorticoid concentrations during breeding result from breeding exerting the highest energetic demands during the year. Females of many species, for example, expend substantial energy during breeding, and breeding males of many species have increased energy costs associated with testosterone and territorial defense. Moreover, increased energy mobilization could explain seasonal peaks during nonbreeding periods if these periods are more energetically costly than breeding. Two examples, however, mammalian hibernation and avian molt, highlight the insufficiency of the Energy Mobilization Hypothesis in providing a universal explanation for seasonal glucocorticoid rhythms.[23] Hibernating mammals require dramatically increased fat stores to survive the winter, but it is not clear that these energy requirements for accumulating and depositing these fat stores are greater or less than lactation. Furthermore, the Energy Mobilization Hypothesis would predict that hibernators that accumulate fat would have higher glucocorticoid titers than hibernators that rely upon cached food, but data suggest the opposite may be true.[25] Second, feather replacement during molt in birds requires substantial energy expenditure,[20] yet this period is the nadir of the annual rhythm. The Energy Mobilization Hypothesis can be a powerful explanatory mechanism for certain species during some times of the year, but it clearly is insufficient to fully explain seasonal glucocorticoid rhythms.

2. The Behavior Hypothesis focuses on glucocorticoid's behavioral effects (Fig. 42.2). It posits that annual glucocorticoid rhythms result from different requirements for expressing (or not expressing) glucocorticoid-mediated behaviors at different times of the year. For example, an important glucocorticoid-mediated behavior in wild animals might be fleeing an area and relocating during storms. This may be an excellent strategy for most of the year, but could be catastrophic for an individual's overall fitness when relocation requires abandoning young. Consequently, the Behavior Hypothesis suggests that the need to seasonally regulate the expression, or nonexpression, of those behaviors drives glucocorticoid rhythms.

There can also be subtle differences between species. There is good evidence that the degree of parental care in birds negatively correlates with baseline glucocorticoid concentrations.[1] In addition, short-lived species with few potential breeding attempts tend to have lower glucocorticoid concentrations than long-lived species that have many potential breeding opportunities.[11] Alternatively, species with limited breeding opportunities appear to decouple glucocorticoid release with glucocorticoid impact on the reproductive system.[33] Extreme examples of this include species that have only one breeding attempt in their lifetimes, such as several salmon species, where glucocorticoid concentrations are extraordinarily high yet breeding progresses as normal, but there are many less extreme examples.[34] Studies on glucocorticoid effects during breeding led to the proposal of the Brood Value Hypothesis[1] that posits that individuals will modulate glucocorticoid responses depending upon how many broods they likely have remaining during their life. Individuals will upregulate glucocorticoid release, and thus bias personal survival, with many remaining reproductive attempts, whereas individuals will downregulate release, and thus bias offspring survival, with few remaining reproductive attempts.

The Behavior Hypothesis, however, does not fit a more general annual rhythm in glucocorticoid concentrations. Glucocorticoid titers are highest in most species during breeding (e.g., Fig. 42.1), exactly when glucocorticoid concentrations should be lowest if seasonal glucocorticoid rhythms were primarily to regulate the behavioral consequences of glucocorticoid release. Consequently, this hypothesis may be useful in explaining modulation of glucocorticoid concentrations within a single season, but not across seasons.

3. The Preparative Hypothesis focuses on the coordinating aspects of glucocorticoid physiology and posits that seasonal rhythms in glucocorticoid concentrations serve to modulate the priming of other stress pathways during periods with different potential exposure to stressors (Fig. 42.2). Glucocorticoids are only one part of the stress response. Other hormones (especially epinephrine and norepinephrine), neurotransmitters (e.g., 5-HT, CRH), opioid peptides, cytokines (e.g. IL-6), as well as other brain functions, exert their effects rapidly with the onset of stress without involving the pituitary or adrenals. However, glucocorticoids have a permissive effect on many of these systems and thereby facilitate better performance under stress. Higher baseline glucocorticoid concentrations may augment the priming effects on these systems in preparation for further stressors. This is the traditional explanation for timing the circadian peak for the beginning of the active period. These permissive effects may help prepare the organism for subsequent stressors.

If the Preparative Hypothesis is correct, higher glucocorticoid concentrations should coincide with seasons where animals can predict that stressors will be more likely to occur. The available data fit this prediction moderately well. The risk of exposure to many stressors in wild animals likely varies seasonally. During the breeding season, for example: predation risk may increase when caring for young; competition for mates, with the risk of frequent fights, increases during breeding; and disease may become more prevalent when breeding individuals congregate. Consequently, glucocorticoid concentrations may vary seasonally in order to mediate changes in preparedness in the nonglucocorticoid pathways of the stress response. Evidence for the Preparative Hypothesis is slowly building (e.g., Ref. 25).

The three explanatory hypotheses presented above are derived from different aspects of glucocorticoid physiology (Fig. 42.2). The Energy Mobilization Hypothesis emphasizes glucocorticoid's metabolic effects in compensating for increased energetic demands, or the allostatic load, experienced by the animal. The Behavior Hypothesis emphasizes the desired acute behavioral effects of glucocorticoids, specifically the importance of avoiding stress-induced behaviors that will interfere with breeding. The Preparative Hypothesis emphasizes the regulatory role glucocorticoids have in preparing other stress systems for times of the annual cycle when there is a high likelihood of experiencing stressors. Integrating these hypotheses into a cohesive physiology will provide the foundation for understanding how elevated glucocorticoid concentrations aid in survival.

AVIAN MIGRATION: TESTING SEASONAL GLUCOCORTICOID RHYTHM HYPOTHESES

Migratory birds undergo two migrations per year: a fall migration to wintering grounds and a spring migration to breeding grounds. While birds travel the same total distance during both migrations, spring migration is usually faster with fewer stops in between. Therefore, we can assume that spring migration is more energetically expensive than fall migration (Fig. 42.3). Because inclement weather and low food availability are more common during spring migration, we can also assume that migrants have an increased chance of encountering predictable stressors during spring versus fall migration (Fig. 42.3).

In general, migratory birds tend to have higher corticosterone (CORT) levels at stopover sites during spring migration compared to fall migration.[21] We ask which of the three seasonal glucocorticoid hypotheses best explains this pattern:

The Energy Mobilization Hypothesis predicts that baseline CORT is highest during energetically expensive times of year. Therefore, we would expect baseline CORT to be higher during the spring stopovers and that high baseline CORT during spring migration enhances hyperphagia and fat deposition. However, the majority of lab and field studies have not found a positive effect

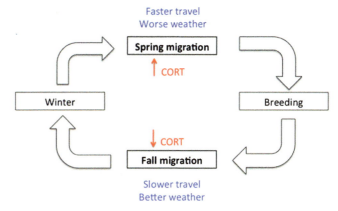

FIGURE 42.3 Seasonal schedule for migratory birds. Compared to fall migration, spring migration is often faster and more direct since early arrival to breeding grounds is advantageous. Because of this, spring migration can be considered more energetically expensive than fall migration. Inclement weather and low food availability are also more common during spring vesus fall migration. Therefore, migrants are more likely to encounter predictable stressors during spring migration than fall migration. The majority of studies have found that corticosterone (CORT) levels are higher in migrants during spring versus fall stopovers.

of baseline CORT on food intake and lipogenesis in migrating birds.[10,22]

The Behavior Hypothesis predicts that CORT is highest during the times of year when glucocorticoid-mediated behaviors increase survival, as glucocorticoids can have permissive effects on survival-promoting behaviors, such as fleeing severe storms. While high CORT during spring migration might help birds prioritize survival over reproduction before significant reproductive investment has occurred, increased CORT might also delay arrival to breeding grounds, which can result in reduced reproductive output. Therefore, the Behavior Hypothesis makes no clear a priori predictions about CORT levels during spring versus fall stopovers.

The Preparative Hypothesis predicts that CORT is highest during times of year when predictable stressors are more likely. Therefore, we would predict that CORT is higher during spring stopovers, as there is a higher chance of encountering inclement weather during spring versus fall migration. In conclusion, the Preparative Hypothesis best explains seasonal patterns in migratory CORT levels, but this has yet to be empirically tested and will need confirmation that migratory CORT levels are not intermediate points during life history stage transitions (see Fig. 42.1).

ADAPTIVE SIGNIFICANCE

Proper concentrations of glucocorticoids seem crucial for survival. Too little and even mild stressors will kill animals, but too much can lead to suppressed gonadal and immune function, and ultimately disease. Consequently, a balance must be struck between needing glucocorticoids to survive stressors and modulating glucocorticoid secretion to prevent deleterious exposures. What is becoming clear is that there is a seasonal rhythm in what is the "right amount" of glucocorticoids. Understanding why the "right amount" changes seasonally will help us understand why glucocorticoids are released, both basally and during stress, and how they help animals survive.

Seasonal variation in glucocorticoid concentrations, however, presents two challenges to our understanding of the role of glucocorticoids in the stress response. The first challenge is to the major assumption that release of glucocorticoids helps animals survive stressors.[28] For many species, there is a period during the year when glucocorticoids have a dramatically damped response to stressors. Whether or not this damped response threatens survival is currently unknown. The second challenge is in understanding why glucocorticoid titers are usually at their seasonal peak during the breeding season. Breeding often requires substantial energetic resources, which fits with newer concepts of allostasis, reactive

scope, and especially allostatic load, as explanations of glucocorticoid release.[19,24] However, there are other energetically costly periods during the annual cycle that are not coincident with elevated glucocorticoids. A better understanding of these seasonal changes is likely to lead to a better understanding of why glucocorticoids are secreted during stress.

IMPORTANCE FOR BIOMEDICAL RESEARCH

Species routinely adjust their physiology to cope with different life history stages (e.g., pregnancy, migration, hibernation) and it is perhaps not surprising that seasonal adjustments in glucocorticoid release and function occur as well. In fact, it appears to be a general phenomenon shared by many free-living species. However, much of this rich seasonal variation in responses disappears when animals are brought into captivity, likely masked by the chronic stress of captivity itself (see example in Fig. 42.1; Ref. 5). Furthermore, domestication can exert profound effects on HPA function, including glucocorticoid release (e.g., Ref. 13). This could have an impact on many common laboratory models of stress, including studies on laboratory rodents. Focusing solely on a 12:12 light:dark cycle, as is the default light cycle for housing most laboratory animals, will provide only a partial view of the role of glucocorticoid release during a stress response. Without an understanding of the natural seasonal variation in glucocorticoid release, it will be difficult to integrate glucocorticoid titers into a full picture of the physiology of a species.

References

1. Bókony V, Lendvai AZ, Liker A, Angelier F, Wingfield JC, Chastel O. Stress response and the value of reproduction: are birds prudent parents? *Am Nat.* 2009;173(5):589–598. http://dx.doi.org/10.1086/597610.
2. Breuner CW, Delehanty B, Boonstra R, Fox C. Evaluating stress in natural populations of vertebrates: total CORT is not good enough. *Funct Ecol.* 2013;27(1):24–36. http://dx.doi.org/10.1111/1365-2435.12016.
3. Breuner CW, Orchinik M. Seasonal regulation of membrane and intracellular corticosteroid receptors in the house sparrow brain. *J Neuroendocrinol.* 2001;13:412–420.
4. Breuner CW, Orchinik M. Beyond carrier proteins: plasma binding proteins as mediators of corticosteroid action in vertebrates. *J Endocrinol.* 2002;175:99–112.
5. Calisi RM, Bentley GE. Lab and field experiments: are they the same animal? *Horm Behav.* 2009;56(1):1–10. http://dx.doi.org/10.1016/j.yhbeh.2009.02.010.
6. de Kloet ER, Oitzl MS, Joels M. Functional implications of brain corticosteroid receptor diversity. *Cell Mol Neurobiol.* 1993;13(4):433–455.
7. DesRochers DW, Reed JM, Awerman J, et al. Exogenous and endogenous corticosterone alter feather quality. *Comp Biochem Physiol Part A Mol Integr Physiol.* 2009;152:46–52.

8. Deviche P, Beouche-Helias B, Davies S, Gao S, Lane S, Valle S. Regulation of plasma testosterone, corticosterone, and metabolites in response to stress, reproductive stage, and social challenges in a desert male songbird. *Gen Comp Endocrinol.* 2014;203:120–131. http://dx.doi.org/10.1016/j.ygcen.2014.01.010.

9. Dickens MJ, Bentley GE. Stress, captivity, and reproduction in a wild bird species. *Horm Behav.* 2014;66(4):685–693. http://dx.doi.org/10.1016/j.yhbeh.2014.09.011.

10. Eikenaar C, Bairlein F, Stowe M, Jenni-Eiermann S. Corticosterone, food intake and refueling in a long-distance migrant. *Horm Behav.* 2014;65(5):480–487. http://dx.doi.org/10.1016/j.yhbeh.2014.03.015.

11. Hau M, Ricklefs RE, Wikelski M, Lee KA, Brawn JD. Corticosterone, testosterone and life-history strategies of birds. *Proc R Soc B Biol Sci.* 2010;277(1697):3203–3212.

12. Krause JS, McGuigan MA, Bishop VR, Wingfield JC, Meddle SL. Decreases in mineralocorticoid but not glucocorticoid receptor mRNA expression during the short Arctic breeding season in free-living Gambel's white-crowned sparrow (*Zonotrichia leucophrys gambelii*). *J Neuroendocrinol.* 2015;27(1):66–75. http://dx.doi.org/10.1111/jne.12237.

13. Kunzl C, Sachser N. The behavioral endocrinology of domestication: a comparison between the domestic Guinea pig (*Cavia aperea* f. *porcellus*) and its wild ancestor, the cavy (*Cavia aperea*). *Horm Behav.* 1999;35:28–37.

14. Lattin CR, Bauer CM, de Bruijn R, Romero LM. Hypothalamus-pituitary-adrenal axis activity and the subsequent response to chronic stress differ depending upon life history stage. *Gen Comp Endocrinol.* 2012;178(3):494–501. http://dx.doi.org/10.1016/j.ygcen.2012.07.013.

15. Lattin CR, Romero LM. Seasonal variation in corticosterone receptor binding in brain, hippocampus, and gonads in house sparrows (*Passer domesticus*). *Auk.* 2013;130(4):591–598. http://dx.doi.org/10.1525/auk.2013.13043.

16. Lattin CR, Romero LM. Seasonal variation in glucocorticoid and mineralocorticoid receptors in metabolic tissues of the house sparrow (*Passer domesticus*). *Gen Comp Endocrinol.* 2015;214:95–102. http://dx.doi.org/10.1016/j.ygcen.2014.05.033.

17. Lattin CR, Waldron-Francis K, Romero LM. Intracellular glucocorticoid receptors in spleen, but not skin, vary seasonally in wild house sparrows (*Passer domesticus*). *Proc R Soc Biol Sci Ser B.* 2013;280:20123033.

18. Malisch JL, Breuner CW. Steroid-binding proteins and free steroids in birds. *Mol Cell Endocrinol.* 2010;316(1):42–52. http://dx.doi.org/10.1016/j.mce.2009.09.019.

19. McEwen BS, Wingfield JC. The concept of allostasis in biology and biomedicine. *Horm Behav.* 2003;43(1):2–15.

20. Murphy ME, King JR. Energy and nutrient use during molt by white-crowned sparrows Zonotrichia-leucophrys-gambelii. *Ornis Scand.* 1992;23(3):304–313.

21. Raja-aho S, Lehikoinen E, Suorsa P, et al. Corticosterone secretion patterns prior to spring and autumn migration differ in free-living barn swallows (*Hirundo rustica* L.). *Oecologia.* 2013;173(3):689–697. http://dx.doi.org/10.1007/s00442-013-2669-9.

22. Ramenofsky M. Hormones in migration and reproductive cycles of birds. In: Norris DO, Lopez KH, eds. *Hormones and Reproduction of Vertebrates.* Birds; vol. 4. San Diego: Elsevier Academic Press; 2011:205–237.

23. Romero LM. Seasonal changes in plasma glucocorticoid concentrations in free-living vertebrates. *Gen Comp Endocrinol.* 2002;128:1–24.

24. Romero LM, Dickens MJ, Cyr NE. The reactive scope model – a new model integrating homeostasis, allostasis, and stress. *Horm Behav.* 2009;55:375–389.

25. Romero LM, Meister CJ, Cyr NE, Kenagy GJ, Wingfield JC. Seasonal glucocorticoid responses to capture in wild free-living mammals. *Am J Physiol Regul Integr Comp Physiol.* 2008;294(2):R614–R622.

26. Romero LM, Reed JM, Wingfield JC. Effects of weather on corticosterone responses in wild free-living passerine birds. *Gen Comp Endocrinol.* 2000;118:113–122.

27. Romero LM, Wingfield JC. Alterations in hypothalamic-pituitary-adrenal function associated with captivity in Gambel's white-crowned sparrows (*Zonotrichia leucophrys gambelii*). *Comp Biochem Physiol.* 1999;122B:13–20.

28. Sapolsky RM, Romero LM, Munck AU. How do glucocorticoids influence stress-responses? Integrating permissive, suppressive, stimulatory, and preparative actions. *Endocr Rev.* 2000;21:55–89.

29. Schoech SJ, Ketterson ED, Nolan V. Exogenous testosterone and the adrenocortical response in dark-eyed juncos. *Auk.* 1999;116:64–72.

30. Schoech SJ, Romero LM, Moore IT, Bonier F. Constraints, concerns and considerations about the necessity of estimating free glucocorticoid concentrations for field endocrine studies. *Funct Ecol.* 2013;27(5):1100–1106. http://dx.doi.org/10.1111/1365-2435.12142.

31. Sheppard DH. Seasonal changes in body and adrenal weights of chipmunks (*Eutamias*). *J Mammal.* 1968;49:463–474.

32. Wingfield JC. Ecological processes and the ecology of stress: the impacts of abiotic environmental factors. *Funct Ecol.* 2013;27:37–44.

33. Wingfield JC, Romero LM. Adrenocortical responses to stress and their modulation in free-living vertebrates. In: McEwen BS, Goodman HM, eds. *Handbook of Physiology; Section 7: The Endocrine System.* Coping with the Environment: Neural and Endocrine Mechanisms; vol. IV. New York: Oxford Univ. Press; 2001:211–234.

34. Wingfield JC, Sapolsky RM. Reproduction and resistance to stress: when and how. *J Neuroendocrinol.* 2003;15(8):711.

35. Wingfield JC, Vleck CM, Moore MC. Seasonal changes of the adrenocortical response to stress in birds of the Sonoran Desert. *J Exp Zool.* 1992;264:419–428.

43

Ultradian Rhythms

F. Spiga, J. Pooley, G. Russell, S.L. Lightman

University of Bristol, Bristol, United Kingdom

Abstract

The hypothalamic-pituitary-adrenal (HPA) axis regulates circulating levels of glucocorticoid hormones to provide a rapid response and defense against stress. Under basal (i.e., unstressed) conditions, glucocorticoids are released with an ultradian pattern that results in rapid ultradian oscillations of hormone levels both in the blood and within target tissues, including the brain. In this review we discuss the origin and regulation of ultradian HPA rhythm, both at system levels and at the level of the adrenal cortex, how it affects the physiology of the organism, both at transcriptional and behavioral levels, and what its clinical relevance is.

INTRODUCTION

Glucocorticoid hormones (corticosterone in the rat, cortisol in humans) are the end products of the hypothalamic-pituitary-adrenal (HPA) axis and are essential for regulating the organism's homeostasis and its response to stress. Glucocorticoids are synthesized in the adrenal cortex in response to adrenocorticotropic hormone (ACTH) release from the anterior pituitary, which is, in turn regulated by the release of corticotropin-releasing hormone (CRH) and arginine vasopressin (AVP) from the paraventricular nucleus of the

hypothalamus (PVN). Upon release into the general circulation, glucocorticoids exert their effects by binding specific receptors, the glucocorticoid and mineralocorticoid receptors (GR and MR, respectively). In addition to the well-known metabolic, cardiovascular, immune-suppressive and antiinflammatory effects, glucocorticoids also regulate their own production through negative feedback mechanisms within the HPA axis including the inhibition of synthesis and release of ACTH from the anterior pituitary,[14] and inhibition of CRH release by direct modulation of neuronal activity both in the PVN as well as other brain structures.[42]

ULTRADIAN RHYTHM OF THE HPA AXIS

Under basal (i.e., unstressed) conditions glucocorticoids secretion is characterized by a circadian pattern with hormone levels peaking during the active phase of the animal. In addition to this, studies in several species, including the rat and human, have revealed that glucocorticoids are released dynamically from the adrenal gland[12,13] resulting in an ultradian pulsatile rhythm in the blood. In the rat, corticosterone pulses have a near-hourly frequency and changes in the amplitude of these pulses throughout the 24-h cycle determine the circadian variation of hormone secretion (Fig. 43.1).

The ultradian rhythm of glucocorticoids is an important factor in determining the neuroendocrine, behavioral, and genomic response to stressors. Furthermore, CORT pulsatility is crucial for physiological activation of GR and MR, and for optimal transcriptional responses of glucocorticoid-responsive genes. Studies in the rat have shown that variation in both the amplitude and the frequency of corticosterone pulses occurs in a number of physiological and pathological conditions, including aging and chronic inflammatory disease (reviewed in Ref. 37). Therefore, it is possible that these changes in the pattern of glucocorticoid release may be associated with the disrupted physiological functions observed in these conditions.

In this chapter we will review recent finding about the origin and regulation of glucocorticoid pulsatility, the importance of pulsatility for gene expression and behavior, and finally we will address the clinical relevance of pulsatility.

THE ORIGIN AND REGULATION OF GLUCOCORTICOID PULSATILITY

While it is known that the circadian rhythm of glucocorticoid secretion is under the control of the suprachiasmatic nucleus (SCN), the mechanism underlying the ultradian rhythm of glucocorticoids, and how this rhythm is maintained at different levels of the HPA axis, has been less investigated. Recent studies using rats with disrupted SCN activity have shown that the glucocorticoid ultradian pulse generator is independent of the SCN.[46] Studies in human and in rat have also shown that in addition to glucocorticoids, ACTH is also released in a pulsatile manner, with the ACTH pulse amplitude increasing throughout the day as a result of the circadian input from the SCN, controlling the circadian secretion of CRH.[49]

KEY POINTS

- Ultradian rhythm of glucocorticoids originates from dynamic interactions between the pituitary and the adrenal cortex.
- Pulsatile ACTH secretion is fundamental for maintaining normal ultradian rhythm of glucocorticoids.
- Ultradian rhythm of glucocorticoids is essential for normal glucocorticoid-responsive gene activity and for behavioral response to stress.
- Disruption of ultradian rhythm of glucocorticoids is associated with disease in humans.

Using a combination of mathematical modeling and in vivo experimental approaches, we have recently produced evidence suggesting that the pituitary-adrenal system in the rat possesses an endogenous oscillatory mechanism generating pulses of ACTH and

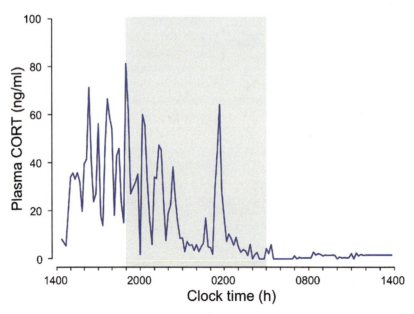

FIGURE 43.1   **Ultradian rhythm of corticosterone in the rat.** The data represent an example of 24-h corticosterone profile of plasma from adult male rat. Shaded region indicates the dark phase. Under basal (i.e., unstressed) conditions, corticosterone levels are characterized by an ultradian rhythm of near-hourly pulses of hormone secretion during the 24-h cycle. Shaded region indicates the dark phase. *Data adapted from Walker JJ, Spiga F, Waite E, et al. The origin of glucocorticoid hormone oscillations.* PLoS Biol. *2012;10(6):e1001341.* http://dx.doi.org/10.1371/journal.pbio.1001341.

corticosterone at a physiological ultradian frequency that is independent of pulsatile release of CRH and/or AVP from the PVN.[47,48] This hypothesis is based on the knowledge of an ACTH-driven positive feedforward pathway that leads to the synthesis and secretion of glucocorticoids from the adrenal cortex, and a glucocorticoid-mediated negative feedback regulates ACTH secretion at the level of the anterior pituitary, a mechanism presumably mediated by activated GR. Indeed we have shown that ultradian ACTH and corticosterone oscillations can be induced by constant infusion of physiological doses of CRH, whereas constant infusion of a higher dose of CRH, which would be expected in response to a severe stressor, resulted in an elevated and sustained level of CORT. Further, a recent study using cultured rat anterior pituitary cells perifused with constant levels of CRH has shown that CRH-induced ACTH secretion is indeed subjected to rapid inhibition after incubation with a pulse of corticosterone, resulting is pulsatile ACTH secretion. The same study also suggests that this rapid nongenomic mechanism involves the activation of a cell membrane-associated GR.[7]

We have also investigated the importance of pulsatile ACTH in generating pulses of glucocorticoids. Using a model of suppressed endogenous ACTH secretion, we found that while the adrenal gland responds rapidly to pulses of ACTH, with pulsatile activation of the steroidogenic pathway in vivo in the rat that parallels pulsatile secretion of CORT, this dynamic response is absent when an identical dose of ACTH is infused at a constant rate.[36] Moreover, the responsiveness of the adrenal gland to a pulse of ACTH equivalent to that seen during an acute stress response is also reduced in rats infused with constant ACTH.[34] These observations suggest that the adrenal is adapted to respond rapidly to individual pulses of ACTH and that optimal adrenal responsiveness depends on pulsatile ACTH.

ACTH released form the anterior pituitary induces the synthesis of glucocorticoids by binding and activating the melanocortin type-2 receptor (MC2R),[21] the ACTH-specific cell surface G protein-coupled receptor. This leads to both nongenomic and genomic processes that regulate both the rapid acute and long-term synthesis of glucocorticoids. Studies investigating the effects of a single ultradian pulse of ACTH on the dynamics of steroidogenic gene transcription, in relationship to the dynamics of CORT secretion in the rat, show that the adrenal steroidogenic pathway is highly dynamic in response to a pulse of ACTH, with rapid changes in the levels of StAR, CYP11A1, and MRAP transcription, measured as changes in levels of primary transcript (hnRNA),[18,35] that are associated with rapid and transient phosphorylation of CREB, and rapid dephosphorylation and nuclear translocation of the CREB coactivator, transducer of regulated CREB activity 2 (CRTC2).[40]

These findings show that pulsatile events leading to the dynamic transcription of steroidogenic genes in the adrenal gland parallel the pulsatile secretion of CORT. This evidence suggests that in basal (unstressed) conditions endogenous pulsatile ACTH results in pulsatile expression of steroidogenic proteins throughout the circadian cycle and that increase in both the amplitude of ACTH pulses and/or responsiveness of the adrenal to ACTH during the circadian peak will lead to increased amplitude in the transcription of these genes, ultimately resulting in the observed circadian variation in steroidogenic proteins levels.[23]

But what happens in stress and disease conditions? Several clinical studies have shown that ultradian rhythm of glucocorticoids become disrupted in pathological situations and recently we have shown that a hyperresponsiveness of the adrenal to ACTH occurs in critical illness that may be associated with an inflammatory response. Consistent with this, immunological stress in the rat can disrupt the normal dynamics of adrenal response to ACTH, as shown by sustained increase in both plasma and adrenal glucocorticoid levels in rats injected with LPS, despite normal levels of ACTH. Elevated glucocorticoid levels are associated with increased expression of StAR and MRAP, and a decrease in StAR transcriptional repressor DAX-1, suggesting that LPS can induce an increase in adrenal steroidogenic activity. What is remarkable is that the effect on both glucocorticoid biosynthesis and on steroidogenic gene expression is not observed following an injection of a high dose of ACTH that is able to produce plasma ACTH levels that are comparable to these observed after LPS administration.[34]

These findings are consistent with a dissociation occurring between the ACTH drive from the pituitary and the resultant glucocorticoid response from the adrenal often observed in disease, presumably due to factors other than ACTH affecting the responsiveness of the adrenal to ACTH, as it has been described for examples for a number of pathogens and cytokines (reviewed in Ref. 3).

IMPORTANCE OF PULSATILITY FOR GENE EXPRESSION AND BEHAVIOR

In vivo microdialysis has established ultradian glucocorticoid secretion produced by the adrenal gland is passed to the cellular microenvironment. Subcutaneous tissue and the brain have been shown to possess ultradian rhythms of free corticosterone, which has been demonstrated to be synchronized with blood corticosterone levels in rats.[24] This section reviews the functional significance of this rhythm at the molecular and behavioral levels.

Two nuclear hormone receptors respond to physiological glucocorticoids. The glucocorticoid receptor (GR) and mineralocorticoid receptor (MR) are cytoplasmic in the absence of ligand and translocate into the nucleus following activation by hormone binding. High-affinity MR and lower-affinity GR are ligand-activated transcription factors, binding short DNA palindromic sequences known as glucocorticoid response elements (GREs); a mode of action described as genomic.

Following injection of an intravenous dose of corticosterone designed to mimic an ultradian pulse MR remains elevated in the nucleus for at least the full 60 min interpulse interval.[5] Thus far, this appears to confirm predictions of a continuously activated receptor during the circadian peak that does not respond to the ultradian rhythm.[26] How DNA binding of the MR is influenced by the ultradian rhythm, and the effect this has on MR-mediated transcriptional activation has not yet been examined.

Conversely, the behavior of GR in response to the ultradian rhythm has been extensively studied. Studies have shown that with application of an intravenous pulse of corticosterone hippocampal GR rapidly translocates into the nucleus and it is reduced to almost preinjection levels through proteasome-dependent clearance by 60 min.[5] Studies using several cell lines, however,

have shown that GR remains in the nucleus following the first pulse and is recomplexed between pulses with intranuclear chaperones.[39] GR DNA-binding studies consistently shows association of GR with DNA at pulse peaks while binding is lost rapidly following hormone washout. Cyclical GR binding is associated with simultaneous cycles of p300 and CBP histone acetyltransferase recruitment, H3 and H4 acetylation, and dynamic chromatin remodeling/accessibility changes. Some changes in chromatin accessibility outlast the interpulse period suggesting chromatin carries a memory of the pulse with currently unknown functional relevance. RNA polymerase II recruitment occurs in phase with these cycles suggesting a transcriptional role for the ultradian rhythm.[4] There is also recent evidence that corticosterone pulses additionally reorganize long-range DNA interactions, which may contribute to gene regulation.[4–6,38,39] Cyclic GR DNA binding at an ultradian (hourly) frequency results in "gene pulsing." Transcription (nascent RNA production) of several *trans*-activated genes occurs in bursts coinciding with hormone pulse peaks (Fig. 43.2).[6,39] While gene pulsing occurs under constant hormone for related estrogen, progesterone, and androgen receptors (ER, PR, and AR, respectively), for GR gene pulsing appears to depend on secretory rhythmicity. In AtT-20 cells, GR cyclically associates with the

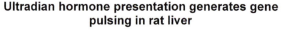

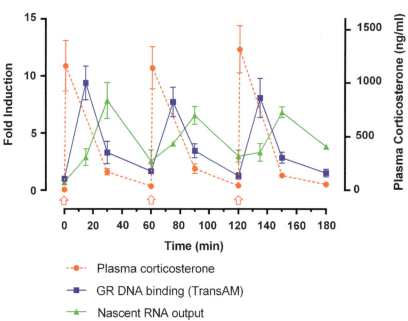

FIGURE 43.2 **Gene pulsing in response to ultradian corticosterone administration.** Pulses of corticosterone delivered (*arrows*) to adrenalectomized rats by intravenous injection produces cycles in plasma hormone concentration (*red*), liver GR DNA binding measured by TransAM DNA-binding ELISA (*blue*), and bursts of liver *Per1* nascent RNA transcription (*green*) associated with the pulse peaks. *Redrawn from data in Stavreva DA, Wiench M, John S, et al. Ultradian hormone stimulation induces glucocorticoid receptor-mediated pulses of gene transcription. Nat Cell Biol. 2009;11(9):1093–1102.* http://dx.doi.org/10.1038/ncb1922.

negative GRE in the POMC gene promoter[4] predicting "inverse pulsing" where transcription is associated with hormone troughs. These phenomena have proved difficult to study in cell lines since simply changing the media can induce transcriptional changes, but animal studies and cellular flow through apparatus should offer eventual clarification.

Gene pulsing is common, proposed as a general feature of gene expression with a range of genes shown by single cell analysis to express in bursts rather than continuously, even under constant stimulation.[17] This arrangement may confer a low signal-noise ratio as well as providing the ability to finely tune gene expression at a lower energy cost relative to constant transcription. Periodic stimulation coordinates cells in culture while weak responses arise from asynchrony under chronic stimulation.[1] Thus, the ultradian pattern may allow cells within an organ to coordinate their overall responses. Continuous activation can induce GR downregulation in a variety of tissue and cell types, precipitating the loss of negative feedback and desensitized gene responses (weak or absent response to second stimulation).

Over multiple pulses in cell lines and animal tissues, pulsatile corticosterone presentation at constant amplitude appears to maintain a steady-state mRNA level for regulated genes that is highly responsive to changing hormone levels.[6,39] On the whole animal, where pulse amplitude variation generates the circadian rhythm, the effect may be mirrored in circadian patterns of gene expression while leaving the system extremely reactive to stressors. Indeed, nuclear translocation and gene responses to acute corticosterone are disrupted when slow release corticosterone pellets interfere with the ultradian rhythm.[30] In HeLa cells, pulsatile application of cortisol produces differential regulation of target genes relative to constant stimulation, with some genes responding to pulsed or constant stimulation only.[20] Though yet to be revealed experimentally, mathematical computations predict that the same gene can be upregulated or downregulated by the pulse pattern depending on the amount of transcription factor available and the binding affinity of the specific site.[50]

MR and GR have additionally been reported to possess rapid responses to corticosteroids occurring through membrane-associated receptors that do not require gene transcription or the passage of the hormone through the cellular membrane.[41] Such responses are described as nongenomic. Rapid increases in miniature excitatory postsynaptic currents (mEPSC) frequency in CA1 of the hippocampus mediated by nongenomic actions of the membrane-bound MR may allow for moderate oscillations of neuronal excitability in phase with the ultradian rhythm.[15,16] Similarly, nongenomic GR-mediated actions rapidly increase GluA2-AMPA receptor content within postsynaptic densities of hippocampus CA1 following

one pulse of corticosterone and suppress LTP.[31] A second pulse given in the ultradian frequency range normalizes these outcomes. These observations lead to the possibility that ultradian glucocorticoid release may maintain the potential for frequency-dependent information coding within this region or produce oscillations in the activity that subserve memory coding.[33] These possibilities deserve further testing, including application of a third/fourth pulse to determine if glutamatergic transmission does in fact oscillate, presumably with a longer frequency than the ultradian signal.

Additional rapid glucocorticoid actions have been described in multiple tissue types including increases in locomotor activity under novelty, suppression of reproductive behaviors, enhanced long-term potentiation in CA1, and facilitation of learning following acute glucocorticoid, inhibition of GABAergic input to PVN magnocellular neurons, inhibition of stimulated insulin release from pancreatic beta cells, suppression of a histone deacetylase complex, inhibition of smooth muscle contractility in the trachea, and cardio-protective nitric oxide synthase stimulation.[41] Due to the time scale of these effects, it is possible that some may occur cyclically with glucocorticoid release, though these possibilities require formal testing with appropriately low doses of hormone.

In the coming years it is likely that ultradian dynamics will be examined in other signaling cascades mediated by glucocorticoid hormones. There has been little investigation of how ultradian GR activation influences functions dependent on negative GREs, tethering of GR to DNA by other proteins, or non-genomic GR actions.[41] The potential for ultradian dynamics in systems where GR inhibits other transcription factors via protein–protein interaction has not been investigated. Furthermore the role of the MR in the genomic response to the ultradian rhythm is yet to be elucidated.

Physiologically, the effect of pulsatile secretion in the maintenance of tissue reactivity and prevention of desensitization extends to the stress response. Blunted ACTH responses to noise stress are observed in adrenalectomized rats infused with constant relative to pulsatile corticosterone,[29] suggesting pulses maintain the HPA axis in a state of "readiness" to respond. As predicted by mathematical models,[19] animals infused with higher pulse amplitudes produced smaller responses.[29] Yet as the higher dose was delivered over the same time frame, it is difficult to determine if this effect is due to the higher amplitude or the faster rate of rise, the latter being linked to fast feedback sensitivity. This requires clarification given computations by others have predicted the opposite, i.e., that higher amplitude oscillations produce larger stress responses.[32]

Gene expression responses (GR, MR, c-fos) to stress are complex, specific to brain regions, dependent on ultradian pulse amplitude, and whether the stress was delivered in the rising or falling phase of the pulse.[29] Sensitivity to the phase of the ultradian pulse was apparent for c-fos induction both in the brain and the pituitary but particularly the amygdala. In basolateral amygdala (BLA) neurons, c-fos mRNA increases during the rising phase of a pulse while decreasing in the falling phase. As c-fos induction has been repeatedly associated with neuronal activation, it is tempting to speculate an ultradian activation profile of the BLA. As yet there is no direct evidence to link ultradian c-fos induction to BLA excitability, neurotransmission, or a related physiological outcome. Several observations are, however, worthy of note in this respect. High BLA c-fos in the pulse peaks parallels a higher magnitude stress response during the rising phase of the ultradian pulse, while low BLA c-fos in the falling phase mirrors smaller than average responses to stress.[29] This phase dependence appears to be a mathematical product of the system[25,32] so is also likely in humans. Behavioral correlates of stress (increased general activity, grooming, rearing, risk assessment) are additionally sensitive to the phase and amplitude of the ultradian rhythm with more exaggerated behaviors observed in rats receiving noise stress during the rising phase of a simulated 50 ng/ml pulse.[29] Male rats are also more aggressive toward an intruder when introduced during the rising phase versus the falling phase.[9]

Mathematical models make several predictions regarding the influence of stress on the ultradian rhythm. Acute stress produces a phase shift in the ultradian rhythm which has also been shown experimentally.[25] Acute stress applied at different stages of an ultradian pulse can influence subsequent pulse amplitudes. Increases or decreases are dependent on the phase angle at which the stressor was applied.[19] Chronic stress via increased CRH drive produces increases or decreases in pulse amplitude depending on the set point at the commencement of stress. The suspicion that stress history may alter pulse dynamics is worthy of additional interrogation, particularly in view of the known paradox that hypocortisolemia or hypercortisolemia can accompany mental health disorder.[19]

Pulsatile release may additionally be advantageous in a signaling system with profound cytotoxic effects. Extended elevations in glucocorticoid hormones deplete neurons, various cell types of the immune lineage, and photoreceptor supporting cells in the retina. Though further work is needed to expand this concept, pattern-dependent regulation of cell survival as a functional aspect of glucocorticoid signaling may additionally possess a cell-type specific dimension, as pulsatile cortisol application reduces proliferation and increases apoptosis in HeLa cells relative to constant stimulation.[20]

Ultradian activity thus appears to cycle transcriptional activity, neuroendocrine, and behavioral processes between states, which ultimately preserve reactivity of the system to challenge. Temporal separation of rapid nongenomic actions and slower genomic actions of glucocorticoids, which frequently have opposing effects, has been proposed to be involved in the generation of such systems, and the loss of this relationship may account for deleterious effects of prolonged glucocorticoid exposure.[27,33] The ultradian glucocorticoid rhythm is also sexually diergic (at least in the rodent) with regard to amplitude, is defined through both genetic and epigenetic factors, and adapts to changing physiological and pathophysiological needs. Greater understanding of the normal function of this rhythm will inevitably guide improved utilization of these steroids in the clinic and our understanding of the physiological effects of maladapted glucocorticoid signaling in disease states.

CLINICAL RELEVANCE OF PULSATILITY

In the previous sections we have reviewed what is known about the origin and regulation of glucocorticoids ultradian rhythms, and their importance for gene expression and behavioral responses to stress. But what is the clinical relevance of glucocorticoid pulsatility? Pulsatile patterns of secretion are known to be of crucial importance in other systems, pulsatile gonadotropin-releasing hormone is used to induce ovulation in fertility treatments while long-acting analogs are used to downregulate gonadal hormone secretion to prepare for in vitro fertilization. As already described, glucocorticoid pulsatility appears to be just as important in HPA regulatory control, and glucocorticoid pulsatility is highly conserved across species. As observed for corticosterone in rodent studies, cortisol, the predominant glucocorticoid in humans, in addition to its circadian rhythm is also characterized by a dynamic pulsatile/ultradian secretion. Mathematical modeling techniques have demonstrated that in human, cortisol is secreted roughly every 60–90 min with the highest amplitude and frequency just prior to awakening.[44] Cortisol is strongly bound to plasma proteins, in particular corticosteroid-binding globulin (CBG), which makes the pattern of cortisol presentation even more crucial. CBG is saturated at fairly low levels of cortisol (400–500 nmol/L); therefore at the circadian peak it is already saturated.[10] Consequently any pulse above this threshold will result in proportionately higher levels of free as opposed to bound hormone. As described in the animal model, pulses are seen not only in the blood stream, but synchronous pulses have also been confirmed via in vivo microdialysis in subcutaneous tissue in humans.[2]

This pulsatility was previously thought to be noise in the system or artifact due to the stress associated with the processes involved in blood sampling; however, with the advent of stress-free automated blood sampling and in vivo microdialysis, it is now possible to examine disease states under truly basal conditions.[2,11] Obstructive sleep apnea (OSA) is a prime example of this. OSA is associated with significant cardiovascular and metabolic morbidity. Patients with OSA have altered secretory dynamics, reflected through altered pulse mass, frequency, and degree of disorder, which post-CPAP (continuous positive airways) treatment returns toward normal parameters.[11] A similar pattern of alteration has also been documented in Cushing's, a disorder characterized by a pathological excess of cortisol production resulting in central obesity, insulin resistance, cardiovascular disease, depression, and osteoporosis.[43]

One should also examine conditions requiring glucocorticoid replacement; there is an ever-growing cohort of patients on long-term glucocorticoid replacement therapy. Patients with Addison's and hypopituitarism have an age-related mortality (predominately from cardiovascular, infectious disease, and cancer) that is twice that of the background population. In real terms, this is the same risk as smoking. Morbidity is equally impacted upon, with health-related quality of life especially early morning mental and physical fatigue being a particular issue for patients. The primary aim of glucocorticoid replacement therapy is to mimic as closely as possible normal physiological patterns of secretion. These patients receive oral replacement therapy in the form of oral hydrocortisone two to three times a day in an attempt to replicate the circadian pattern of cortisol secretion. As these patients are receiving what is considered to be optimal treatment, why does this disparity still exist? As described earlier, cortisol is an anticipatory hormone that rises just prior to wakening. With the current oral immediately absorbed glucocorticoids, it is impossible to mirror precisely this anticipatory rise. There is also some evidence that the pattern of replacement is important with patients reportedly feeling better on three/four times daily in comparison to less frequent dosing. Some patients even wake up at 3 a.m. to take a dose of hydrocortisone to produce a circadian rise.[28] Modified, dual release, and circadian subcutaneous infusions are currently being marketed.[22,45] These go a long way in more closely replicating the circadian pattern of cortisol secretion; indeed, there have been claims of total dose reduction and associated improved metabolic outcomes. However, impact on quality of life parameters has been variable. Open studies have shown an improvement,[22] while a blinded trial using continuous hydrocortisone replacement was unable to document any significant subjective benefit.[8] None of these treatments, however, address the underlying pulsatile pattern of secretion. This suggests that although there are

steps being made in the right direction, the remaining disparity could still be due to the pattern of presentation. A pulsatile pattern of replacement is currently being trialed in patients (EudraCT Number: 2012-001,104-37), the outcomes of which are as yet unknown.

Glucocorticoids are also widely used therapeutically in the treatment of autoimmune and inflammatory conditions. Despite their efficiency in treating the underlying condition, a staggering proportion of patients experience side effects even at comparatively low dosages. This is a significant rate-limiting factor as it can reduce their tolerability and pose challenges in deciphering what a symptom of disease activity is versus steroid side effect. In addition prolonged glucocorticoid treatment can induce glucocorticoid resistance in a subset of susceptible patients, adding a further degree of complexity.[28] Bearing in mind the phenomena of gene pulsing and differential gene activation patterns[39] that has been described in depth in this chapter, oscillating levels of glucocorticoids appear to allow the HPA axis to be held in a state of constant dynamic equilibrium. This prevents abnormal prolonged duration of activation or downregulation of glucocorticoid-regulated genes. This suggests that it could be the pattern of presentation used in glucocorticoid-based therapeutics in addition to the total dose and duration of treatment that contributes to the development of the glucocorticoid side effect profiles.

In summary, it appears that glucocorticoid chronobiology has considerable clinical importance and that understanding the impact pulsatility has on epigenetics may get us closer to teasing out the complex relationship between HPA dysfunction and disease states and therapeutics.

References

1. Ashall L, Horton CA, Nelson DE, et al. Pulsatile stimulation determines timing and specificity of NF-kappaB-dependent transcription. *Science*. 2009;324(5924):242–246. http://dx.doi.org/10.1126/science.1164860.
2. Bhake RC, Leendertz JA, Linthorst AC, Lightman SL. Automated 24-hours sampling of subcutaneous tissue free cortisol in humans. *J Med Eng Technol*. 2013;37(3):180–184. http://dx.doi.org/10.3109/03091902.2013.773096.
3. Bornstein SR, Engeland WC, Ehrhart-Bornstein M, Herman JP. Dissociation of ACTH and glucocorticoids. *Trends Endocrinol Metab*. 2008;19(5):175–180. http://dx.doi.org/10.1016/j.tem.2008.01.009.
4. Conway-Campbell BL, George CL, Pooley JR, et al. The HSP90 molecular chaperone cycle regulates cyclical transcriptional dynamics of the glucocorticoid receptor and its coregulatory molecules CBP/p300 during ultradian ligand treatment. *Mol Endocrinol*. 2011;25(6):944–954. http://dx.doi.org/10.1210/me.2010-0073.
5. Conway-Campbell BL, McKenna MA, Wiles CC, Atkinson HC, de Kloet ER, Lightman SL. Proteasome-dependent down-regulation of activated nuclear hippocampal glucocorticoid receptors determines dynamic responses to corticosterone. *Endocrinology*. 2007;148(11):5470–5477. http://dx.doi.org/10.1210/en.2007-0585.

6. Conway-Campbell BL, Sarabdjitsingh RA, McKenna MA, et al. Glucocorticoid ultradian rhythmicity directs cyclical gene pulsing of the clock gene period 1 in rat hippocampus. *J Neuroendocrinol*. 2010;22(10):1093–1100. http://dx.doi.org/10.1111/j.1365-2826.2010.02051.x.

7. Deng Q, Riquelme D, Trinh L, et al. Rapid glucocorticoid feedback inhibition of ACTH secretion involves ligand-dependent membrane association of glucocorticoid receptors. *Endocrinology*. 2015;156(9):3215–3227. http://dx.doi.org/10.1210/EN.2015-1265.

8. Gagliardi LN, Thynne M, Borch T, et al. Continuous subcutaneous hydrocortisone infusion therapy in addison's disease: a randomized, placebo-controlled clinical trial. *J Clin Endocrinol Metab*. 2014;99(11):4149–4157.

9. Haller J, Halasz J, Mikics E, Kruk MR, Makara GB. Ultradian corticosterone rhythm and the propensity to behave aggressively in male rats. *J Neuroendocrinol*. 2000;12(10):937–940.

10. Hammond GL, Smith CL, Underhill DA. Molecular studies of corticosteroid binding globulin structure, biosynthesis and function. *J Steroid Biochem Mol Biol*. 1991;40(4–6):755–762.

11. Henley DE, Russell GM, Douthwaite JA, et al. Hypothalamic-pituitary-adrenal axis activation in obstructive sleep apnea: the effect of continuous positive airway pressure therapy. *J Clin Endocrinol Metab*. 2009;94(11):4234–4242. http://dx.doi.org/10.1210/jc.2009-1174.

12. Jasper MS, Engeland WC. Synchronous ultradian rhythms in adrenocortical secretion detected by microdialysis in awake rats. *Am J Physiol*. 1991;261(5 Pt 2):R1257–R1268.

13. Jasper MS, Engeland WC. Splanchnic neural activity modulates ultradian and circadian rhythms in adrenocortical secretion in awake rats. *Neuroendocrinology*. 1994;59(2):97–109.

14. Jones MT, Hillhouse EW, Burden JL. Dynamics and mechanics of corticosteroid feedback at the hypothalamus and anterior pituitary gland. *J Endocrinol*. 1977;73(3):405–417.

15. Karst H, Berger S, Erdmann G, Schutz G, Joels M. Metaplasticity of amygdalar responses to the stress hormone corticosterone. *Proc Natl Acad Sci USA*. 2010;107(32):14449–14454. http://dx.doi.org/10.1073/pnas.0914381107.

16. Karst H, Berger S, Turiault M, Tronche F, Schutz G, Joels M. Mineralocorticoid receptors are indispensable for nongenomic modulation of hippocampal glutamate transmission by corticosterone. *Proc Natl Acad Sci USA*. 2005;102(52):19204–19207. http://dx.doi.org/10.1073/pnas.0507572102.

17. Levine JH, Lin Y, Elowitz MB. Functional roles of pulsing in genetic circuits. *Science*. 2013;342(6163):1193–1200. http://dx.doi.org/10.1126/science.1239999.

18. Liu Y, Smith LI, Huang V, et al. Transcriptional regulation of episodic glucocorticoid secretion. *Mol Cell Endocrinol*. 2013;371(1–2):62–70. http://dx.doi.org/10.1016/j.mce.2012.10.011.

19. Markovic VM, Cupic Z, Vukojevic V, Kolar-Anic L. Predictive modeling of the hypothalamic-pituitary-adrenal (HPA) axis response to acute and chronic stress. *Endocr J*. 2011;58(10):889–904.

20. McMaster A, Jangani M, Sommer P, et al. Ultradian cortisol pulsatility encodes a distinct, biologically important signal. *PLoS One*. 2011;6(1):e15766. http://dx.doi.org/10.1371/journal.pone.0015766.

21. Mountjoy KG, Mortrud MT, Low MJ, Simerly RB, Cone RD. Localization of the melanocortin-4 receptor (MC4-R) in neuroendocrine and autonomic control circuits in the brain. *Mol Endocrinol*. 1994;8(10):1298–1308. http://dx.doi.org/10.1210/mend.8.10.7854347.

22. Oksnes M, Bjornsdottir S, Isaksson M, et al. Continuous subcutaneous hydrocortisone infusion versus oral hydrocortisone replacement for treatment of Addison's disease: a randomized clinical trial. *J Clin Endocrinol Metab*. 2014;99(5):1665–1674. http://dx.doi.org/10.1210/jc.2013-4253.

23. Park SY, Walker JJ, Johnson NW, Zhao Z, Lightman SL, Spiga F. Constant light disrupts the circadian rhythm of steroidogenic proteins in the rat adrenal gland. *Mol Cell Endocrinol*. 2013;371(1–2):114–123. http://dx.doi.org/10.1016/j.mce.2012.11.010.

24. Qian X, Droste SK, Lightman SL, Reul JM, Linthorst AC. Circadian and ultradian rhythms of free glucocorticoid hormone are highly synchronized between the blood, the subcutaneous tissue, and the brain. *Endocrinology*. 2012;153(9):4346–4353. http://dx.doi.org/10.1210/en.2012-1484.

25. Rankin J, Walker JJ, Windle R, Lightman SL, Terry JR. Characterizing dynamic interactions between ultradian glucocorticoid rhythmicity and acute stress using the phase response curve. *PLoS One*. 2012;7(2):e30978. http://dx.doi.org/10.1371/journal.pone.0030978.

26. Reul JM, van den Bosch FR, de Kloet ER. Differential response of type I and type II corticosteroid receptors to changes in plasma steroid level and circadian rhythmicity. *Neuroendocrinology*. 1987;45(5):407–412.

27. Russell GM, Kalafatakis K, Lightman SL. The importance of biological oscillators for hypothalamic-pituitary-adrenal activity and tissue glucocorticoid response: coordinating stress and neurobehavioural adaptation. *J Neuroendocrinol*. 2015;27(6):378–388. http://dx.doi.org/10.1111/jne.12247.

28. Russell GM, Lightman SL. Can side effects of steroid treatments be minimized by the temporal aspects of delivery method? *Expert Opin Drug Saf*. 2014;13(11):1501–1513. http://dx.doi.org/10.1517/14740338.2014.965141.

29. Sarabdjitsingh RA, Conway-Campbell BL, Leggett JD, et al. Stress responsiveness varies over the ultradian glucocorticoid cycle in a brain-region-specific manner. *Endocrinology*. 2010;151(11):5369–5379. http://dx.doi.org/10.1210/en.2010-0832.

30. Sarabdjitsingh RA, Isenia S, Polman A, et al. Disrupted corticosterone pulsatile patterns attenuate responsiveness to glucocorticoid signaling in rat brain. *Endocrinology*. 2010b;151(3):1177–1186. http://dx.doi.org/10.1210/en.2009-1119.

31. Sarabdjitsingh RA, Jezequel J, Pasricha N, et al. Ultradian corticosterone pulses balance glutamatergic transmission and synaptic plasticity. *Proc Natl Acad Sci USA*. 2014;111(39):14265–14270. http://dx.doi.org/10.1073/pnas.1411216111.

32. Scheff JD, Calvano SE, Lowry SF, Androulakis IP. Transcriptional implications of ultradian glucocorticoid secretion in homeostasis and in the acute stress response. *Physiol Genomics*. 2012;44(2):121–129. http://dx.doi.org/10.1152/physiolgenomics.00128.2011.

33. Schwabe L, Joels M, Roozendaal B, Wolf OT, Oitzl MS. Stress effects on memory: an update and integration. *Neurosci Biobehav Rev*. 2012;36(7):1740–1749. http://dx.doi.org/10.1016/j.neubiorev.2011.07.002.

34. Spiga F, Lightman SL. Dynamics of adrenal glucocorticoid steroidogenesis in health and disease. *Mol Cell Endocrinol*. 2015. http://dx.doi.org/10.1016/j.mce.2015.02.005.

35. Spiga F, Liu Y, Aguilera G, Lightman SL. Temporal effect of adrenocorticotropic hormone on adrenal glucocorticoid steroidogenesis: involvement of the transducer of regulated cyclic AMP-response element-binding protein activity. *J Neuroendocrinol*. 2011;23(2):136–142. http://dx.doi.org/10.1111/j.1365-2826.2010.02096.x.

36. Spiga F, Waite EJ, Liu Y, Kershaw YM, Aguilera G, Lightman SL. ACTH-dependent ultradian rhythm of corticosterone secretion. *Endocrinology*. 2011b;152(4):1448–1457. http://dx.doi.org/10.1210/en.2010-1209.

37. Spiga F, Walker JJ, Terry JR, Lightman SL. HPA axis-rhythms. *Compr Physiol*. 2014;4(3):1273–1298. http://dx.doi.org/10.1002/cphy.c140003.

38. Stavreva DA, Coulon A, Baek S, et al. Dynamics of chromatin accessibility and long-range interactions in response to glucocorticoid pulsing. *Genome Res*. 2015;25(6):845–857. http://dx.doi.org/10.1101/gr.184168.114.

39. Stavreva DA, Wiench M, John S, et al. Ultradian hormone stimulation induces glucocorticoid receptor-mediated pulses of gene transcription. *Nat Cell Biol*. 2009;11(9):1093–1102. http://dx.doi.org/10.1038/ncb1922.

40. Takemori H, Kanematsu M, Kajimura J, et al. Dephosphorylation of TORC initiates expression of the StAR gene. *Mol Cell Endocrinol*. 2007;265-266:196–204. http://dx.doi.org/10.1016/j.mce.2006.12.020.

41. Tasker JG, Di S, Malcher-Lopes R. Minireview: rapid glucocorticoid signaling via membrane-associated receptors. *Endocrinology*. 2006;147(12):5549–5556. http://dx.doi.org/10.1210/en.2006-0981.

42. Ulrich-Lai YM, Herman JP. Neural regulation of endocrine and autonomic stress responses. *Nat Rev Neurosci*. 2009;10(6):397–409. http://dx.doi.org/10.1038/nrn2647.

43. van den Berg G, Pincus SM, Veldhuis JD, Frolich M, Roelfsema F. Greater disorderliness of ACTH and cortisol release accompanies pituitary-dependent Cushing's disease. *Eur J Endocrinol*. 1997;136(4):394–400.

44. Veldhuis JD, Iranmanesh A, Lizarralde G, Johnson ML. Amplitude modulation of a burstlike mode of cortisol secretion subserves the circadian glucocorticoid rhythm. *Am J Physiol*. 1989;257(1 Pt 1):E6–E14.

45. Verma S, Vanryzin C, Sinaii N, et al. A pharmacokinetic and pharmacodynamic study of delayed- and extended-release hydrocortisone (Chronocort) versus conventional hydrocortisone (Cortef) in the treatment of congenital adrenal hyperplasia. *Clin Endocrinol (Oxf)*. 2010;72(4):441–447. http://dx.doi.org/10.1111/j.1365-2265.2009.03636.x.

46. Waite EJ, McKenna M, Kershaw Y, et al. Ultradian corticosterone secretion is maintained in the absence of circadian cues. *Eur J Neurosci*. 2012;36(8):3142–3150. http://dx.doi.org/10.1111/j.1460-9568.2012.08213.x.

47. Walker JJ, Spiga F, Waite E, et al. The origin of glucocorticoid hormone oscillations. *PLoS Biol*. 2012;10(6):e1001341. http://dx.doi.org/10.1371/journal.pbio.1001341.

48. Walker JJ, Terry JR, Lightman SL. Origin of ultradian pulsatility in the hypothalamic-pituitary-adrenal axis. *Proc Biol Sci*. 2010;277(1688):1627–1633. http://dx.doi.org/10.1098/rspb.2009.2148.

49. Watts AG, Tanimura S, Sanchez-Watts G. Corticotropin-releasing hormone and arginine vasopressin gene transcription in the hypothalamic paraventricular nucleus of unstressed rats: daily rhythms and their interactions with corticosterone. *Endocrinology*. 2004;145(2):529–540. http://dx.doi.org/10.1210/en.2003-0394.

50. Wee KB, Yio WK, Surana U, Chiam KH. Transcription factor oscillations induce differential gene expressions. *Biophys J*. 2012;102(11):2413–2423. http://dx.doi.org/10.1016/j.bpj.2012.04.023.

Index